KT-379-294
SCIENCE

GENERAL EDITORS
DR MARK STEER, HAYLEY BIRCH
AND DR ANDREW IMPNEY

A CASSELL BOOK

An Hachette Livre UK Company

First published in the UK 2008 by
Cassell Illustrated, a division of
Octopus Publishing Group Ltd.
2-4 Heron Quays,
London E14 4JP

A CIP catalogue record for this book is available from the British Library.

ISBN-13: 978-1-84403589-2 (UK Edition)

Distributed in the United States and Canada by
Sterling Publishing Co., Inc
387 Park Avenue South, New York, NY 10016-8810

ISBN-13: 978-1-84403641-7 (U.S. Edition)

10 9 8 7 6 5 4 3 2 1

Project Editor: Jo Wilson
Copy-editor Ceri Perkins
Index: Pamela Ellis

Production: Caroline Alberti
Creative Director: Geoff Fennell
Designer: John Round

Commissioning Editor: Laura Price
Publisher: Mathew Clayton

Printed in China

CONTENTS

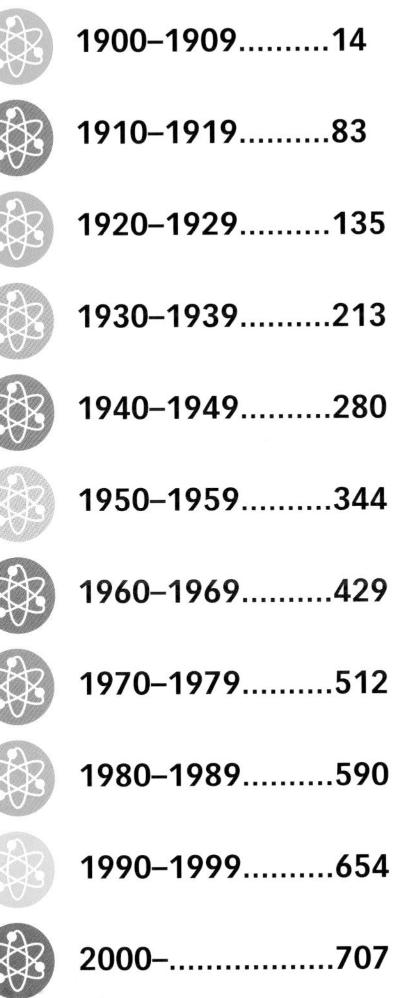

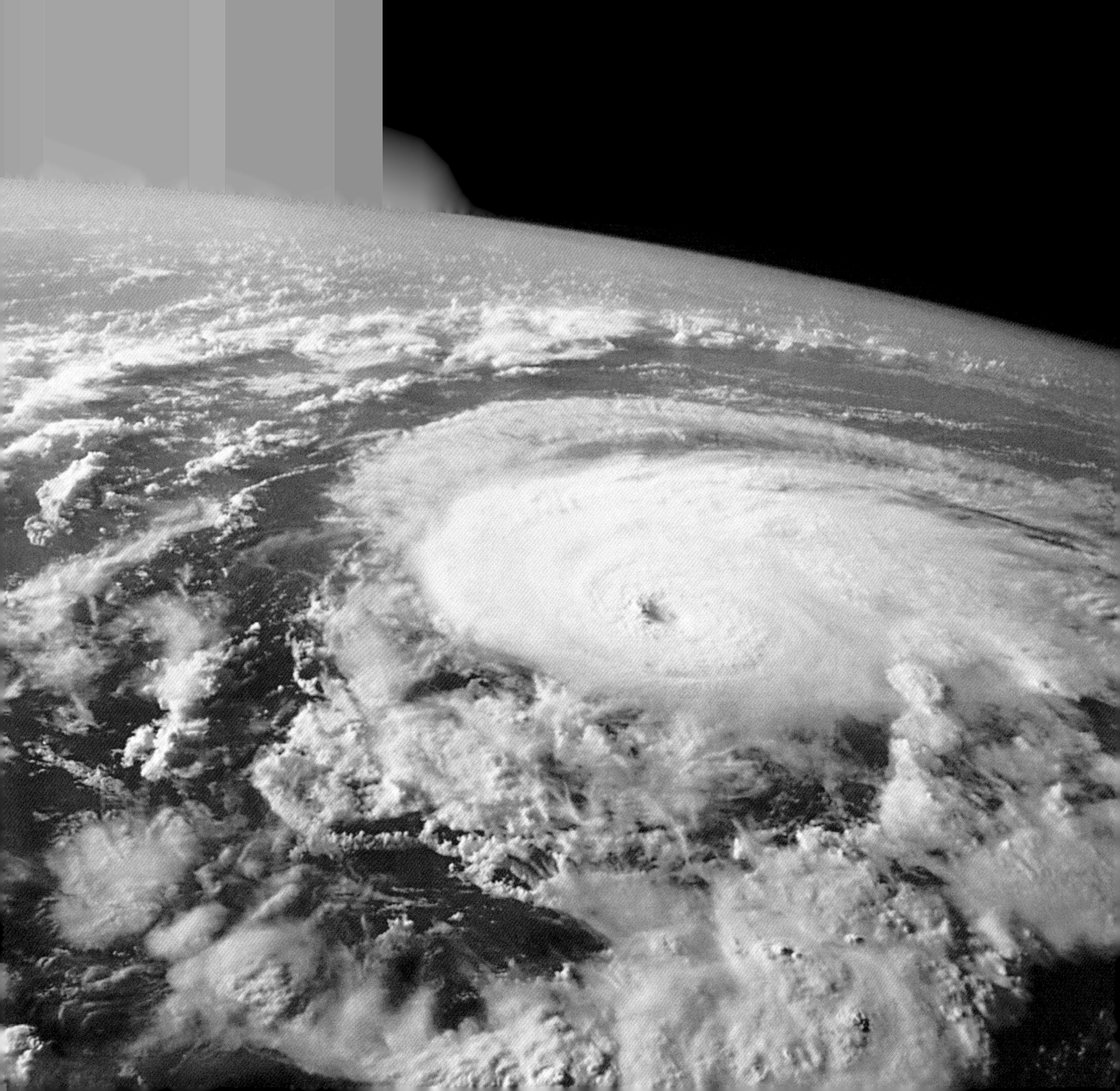

CONTRIBUTORS

Fabian Acker has written for *Book of Knowledge*, *New Scientist, The Times*, The *Sunday Times*, BBC's *Science Now*, and other scientific journals. He won the *Sunday Times'* Travel Writer of the Year Award in 1991, and BT's Technology Journalist of the Year in 1993.

William Addison studied chemistry for many years at the University of Bristol, using tightly focused laser beams to manipulate nanoparticles. He also knows how to dismantle a laptop using only a manicure set.

Neal Anthwal works as a post-doctoral researcher in London. After a fling in anthropology, he worked on evolution and development in systems from mouse jaws to trout muscle. He aims to combine a career in developmental biology with bass guitar.

Kat Arney gained a degree and PhD from the University of Cambridge. She works as a science communicator for Cancer Research UK, producing their monthly podcast. A freelance writer, she also co-presents BBC radio show *The Naked Scientists*.

Katherine Ball studied English and Environmental Science at the University of the West of England in Bristol. She has written extensively about many aspects of science and currently works in PR.

Jim Bell graduated in Zoology and Science Communication. He worked as a presenter at the National Space Centre and At-Bristol in the UK, and has a somewhat unhealthy obsession with fish.

Hayley Birch is senior editor of *Null Hypothesis the Journal of Unlikely Science*, but still finds time to write freelance for publications including *Nature*, BBC *Focus* magazine and the *Daily Telegraph*.

Riaz Bhunnoo has a strong interest in scientific penmanship. He studied for a Master's degree in chemistry at Southampton University and loved it so much he stayed for more. He went on to synthesize an anti-cancer compound as part of his PhD and is exploring the wonders of biological sciences.

Richard Bond has degrees in biology, history and philosophy of science from the University of London. He has written widely on science policy and on science and society, notably in *Wavelength: the Journal of Science, Society and the Media.* He provides the humorous "Other Lab" column for *Null Hypothesis.*

Christopher Booroff was tempted to study "The Universe", but opted for an MSci in physics with space science at University College London, where he began writing for *Null Hypothesis*. A fellow of the Royal Astronomical Society he regrets his decision; Master of the Universe sounds impressive.

Adrian Bowyer is senior lecturer in the Engineering Department at the University of Bath. His current research area is self-replicating machines and he is inventor and developer of the RepRap open-source replicating rapid prototyper.

Matt Brown is editor of Nature Network, a social networking site for scientists from the publishers of *Nature*. He has an academic background in chemistry and biochemistry and worked as an editor and writer for years.

Shamini Bundell has been a zoologist for many years and now has a certificate from Cambridge to prove this. When not babbling about how awesome science is she does creative artsy things to make up for the time spent on facts and figures.

Raychelle Burks discovered her love of science writing through a science writing course. Research and writing occupy her days as a chemistry Ph.D. candidate at the University of Nebraska-Lincoln, where she analyzes illegal and illicit substances of forensic interest.

Catherine Charter is involved in plant pathology research and writes for *Null Hypothesis*.

Julie Clayton is a freelance science writer who has worked as assistant producer for BBC television, and written for *Nature*, *New Scientist* and other journals.

Nicola Currie is a postgraduate student, freelance writer and poet from Cambridge. Interests include politics, religion and the arts she has written for newspapers and online.

Kellye Curtis is finishing her PhD in nanoelectronics at the University of Cambridge. She also sings and plays glockenspiel in Cambridge's premier/only cardi-core band, The Puncture Repair Kit.

Josh Davies has written a blurb about himself in 40 words. It was impossible to summarize his life in such a short space so he decided to spend his time talking about the word count. (This is exactly 40 words)

Simon Davies is a chemistry graduate of the University of Bristol, working as science technician in Norwich and as a writer.

Mike Davis is a freelance writer with a penchant for anything unusual, interesting or downright unlikely. With numerous articles, gags and commentaries, to his name, the translation of "English" into "English" for overseas clients is high on his list of achievements.

Nathan Dennison is a geneticist researching the paratransgenic control of malaria at Stockholm University. He writes, reads and researches all things DNA.

Arran Frood has worked in science media for 10 years, and contributes news and features for *New Scientist*, *Focus* and BBC Online. He joined *Nature* in 2005 and created the *History of the Journal Nature* website. A full-time freelance science writer and editor, he dabbles in fiction.

Matt Gibson has programmed computers since his parents bought him a Sinclair ZX81. A Warwick University graduate, his journalistic work includes everything from book reviews to celebrity interviews.

Christina Giles is a science writer and editor with a BSc in biochemistry and an MSc in science communication. She honed her skills in the heady world of motor caravan journalism.

Katie Giles studied pharmacology before she decided working with people was preferable to pipettes. She qualified as a doctor in 2005 and works in intensive care. When not saving lives (and putting in drips and catheters) she writes on a variety of scientific and medical themes.

Arthur Goldsmith is a Reader at the University of Bristol. For 30 years he has led a renowned research team studying learning behavior and development in birds.

Barney Grenfell is a freelance writer and science-communicator. He communicated science at the London Science Museum and At-Bristol science centre before taking the plunge into the world of freelance.

David Hall is a scientist with expertise in rodent biology and bird conservation. He has worked in Mauritius and has a PhD from Bristol University. He works for *Null Hypothesis* and enjoys travel and sport.

Gavin Hammon obtained a B.Sc. in astrophysics – which only cost him $40 and five minutes to reply to the email. He then undertook a Ph.D. in what he thought was nuclear physics, but turned out to be unclear physics. He is a Patent Attorney in Germany.

Maria Hampshire is a specialist medical writer and editor with a degree in human physiology, PhD research into eye diseases, and experience in healthcare publishing.

Eleanor Hullis is the *Null Hypothesis* Young Science Writer of the Year, 2008. Recently ranked 19th in the world in Saddleseat Equitation, she is about to kick-start an illustrious career by reading Spanish and Russian at the University of Cambridge.

Hannah Isom studied biomedical sciences at the University of Manchester and worked briefly in medical marketing. She studies print journalism and spends her free time writing.

Ceri Harrop is studying for a PhD in Biochemistry. She enjoys sharing science: teaching at museums and at science festivals and on the radio. She wants to combine careers in science and communication.

David Hawksett works in Government science policy, science media, and education. As the science and technology consultant for Guinness World Records he was official witness to the first privately funded manned spaceflight. He writes for Lucasfilm and works for Starchaser Industries.

Rebecca Hernandez studied biochemistry at the University of Washington and did an 8-year stint in Seattle's biotech sector. As well as bench science she now dabbles in freelance writing and science education, and travels, plays drums and dances flamenco.

Andrew Impey is cofounder and Managing Director of *Null Hypothesis, the Journal of Unlikely Science*. A doctor of ducks he regularly lectures on animal behaviour and aquatic ecology in a variety of universities and academic institutions.

Gareth Jones is Professor of Evolutionary Biology at the University of Bristol. He has studied bats and their echolocation in six continents over the past 23 years. He has published 150 scientific papers and discovered three new bat species.

opposite Human DNA (deoxyribonucleic acid) sequence as a series of colored bands.

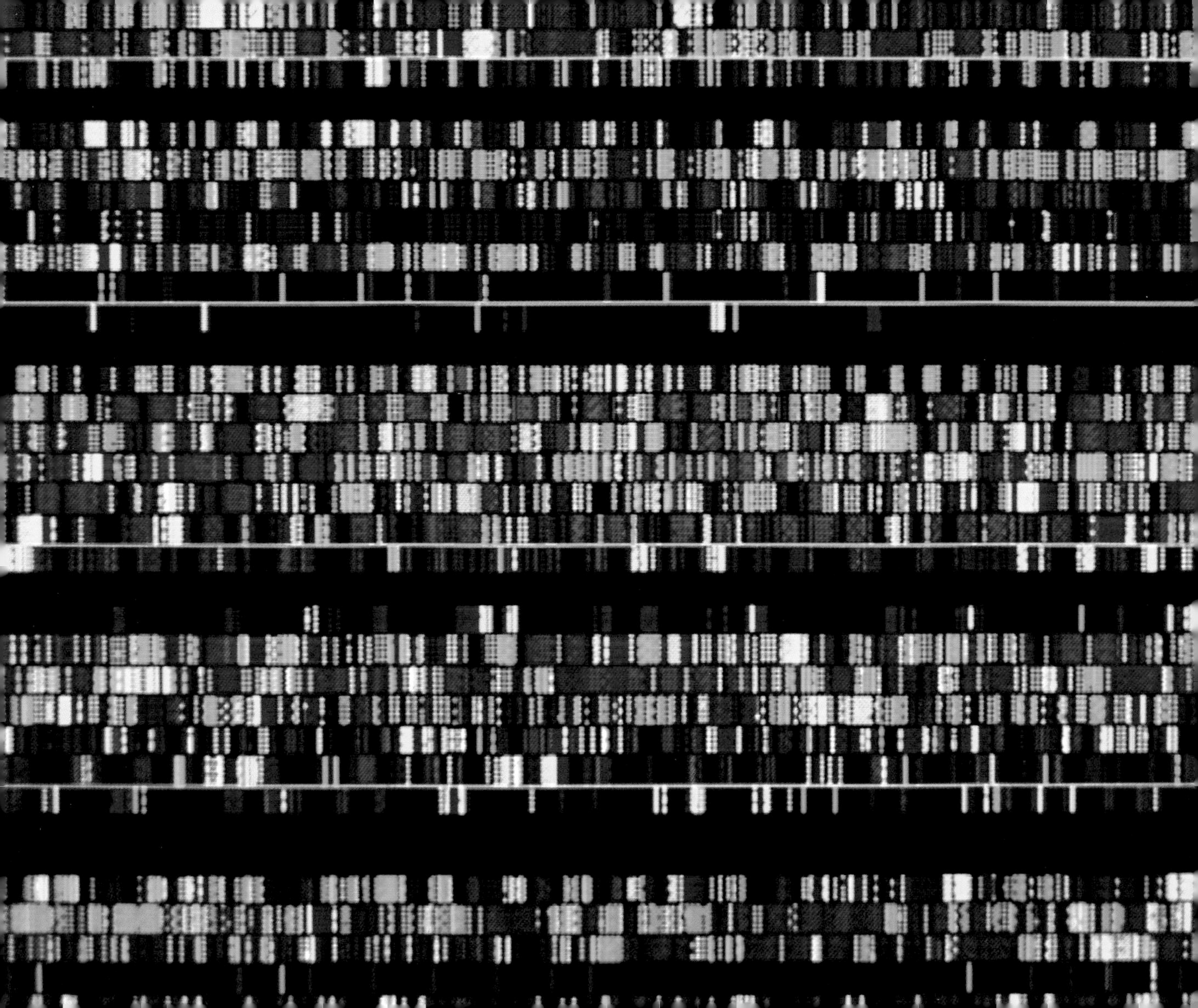

Fiona Kellagher studied English Literature at Warwick University. She taught English as a second language and is an editor.

Douglas Kitson studied Mathematics and Physics at the University of Warwick, and headed to the University of the West of England in Bristol to do an MSc in Science Communication. He has rather untidy hair.

Andrey Kobilnyk has been asking questions involving the word "why" since he learned the word. He lives in London and is editor of FirstScience.com

Umia Kukathasan is Irish born, Sri Lankan blooded and brought up in Essex. She escaped to London to play b-ball, and debate the impact of science with Maurice Wilkins. She's plotting to break into guerrilla-science: facilitating learning when least expected...

Jamie F. Lawson is an evolutionary psychologist with a BA in Anthroplogy and Archaeology, an MSc in Evolutionary Anthropology, and aPhD in Psychology at the University of St Andrews in 2008. He specializes in mating strategy and facial attractiveness in humans.

Chris Lochery studied Creative Writing at Leeds University. After graduating he got a job as a science-based children's entertainer, on the birthday party circuit and now writes.

Helen McBain graduated from UCL with a degree in chemistry and mathematics. She developed her writing skills at the Institute of Physics, and is a press officer for the British Government, working on science issues.

opposite Enrico Fermi

B. James McCallum has degrees in biology from Davidson College and medicine from the University Of South Carolina School Of Medicine. He is a Fellow of the American College of Physicians, and Assistant Professor of Clinical Internal Medicine in Columbia, S.C.

Emma Norman is studying Environmental Science and Environmental Water Management. After messing about in rivers, she is undertaking a PhD looking at rock falls and coastal erosion.

Kate Oliver is studying organic semiconductors at UCL before moving to CERN to embark on a career in science communication. She is the only pink-haired physicist she knows.

Anne Pawsey studies physics at the University of Bristol and L'Ecole Nationale Supérieure de Physique de Grenoble. In her free time she writes, plays bassoon and skis.

Becky Poole is a post-doctoral researcher at the University of Bristol investigating effects of the environment on gene expression in bread wheat. She ice-skates and volunteers for the British Association for the Advancement of Science.

Helen Potter is studying for a PhD in biosynthesis at Cambridge University. She has been writing freelance for over a year and still thinks science is cool.

Steve Robinson is studying for a journalism postgraduate diploma at Cardiff University having completed a Biological Sciences degree at Oxford, specializing in cell and developmental biology. A gadget-lover, and football fanatic, he writes a regular blog.

Leila Sattary escaped Glasgow for the island nation of St Andrews where she spent 5 years learning about complicated physics. She has worked for Glasgow and Oxford Universities and enjoys rock music and traveling on trains.

Eric Schulman is a PhD astronomer and the author of *A Briefer History of Time* (1999). He was the Armchair Astrophysics columnist for *Mercury* magazine and has been on radio and television in North America and Europe.

Faith Smith is a zoologist based at Bristol University. She has combined her love of animals and travel by taking part in conservation trips across the globe, and has spent the last six months studying ant behaviour in a lab in Bristol.

Stuart Smith spent his youth in the swamps of South Florida before studying biology and comparative physiology. He then became a physician in South Carolina, where he lives (in another swamp) with his family.

Mark Steer is cofounder of *Null Hypothesis*. He completed a PhD on the behavior of theoretical bees and lectures on conservation science at University of the West of England. He can play the piano: badly.

James Urquhart is a science writer with an MSc in Science Communication from Imperial College. He knows biology and the history and philosophy of science and medicine, climbs hills, and wants to write a book.

Richard Van Noorden has an MSc in natural sciences from Cambridge University. He's written news and features for *Nature* and *Chemistry World*, broadcasts the wonders of chemistry on the *Naked Scientists* radio show and the *Chemistry World* podcast.

Harriet Ward is a freelance writer and science educator specializing in child psychology and *Doctor Who*.

Sarah Watson works as a data manager for a cardiovascular disease research team at Cambridge University. She studied Mathematics at the University of Bath, graduating as a Master of Mathematics.

Hannah Welham is currently based at Bristol University, studying Zoology. She has been writing for Null Hypothesis and numerous other publications while studying.

Vicky West is a freelance science writer and presenter, with interest in conservation and environmental issues. Her projects include anything from sustainable agriculture in Brazil to fabrics made of beer bi-products.

Melissa Wilson is a traveling writer. Trained as a biologist, she has recently completed two years working as a journalist in Japan.

Jo Wimpenny is completing a DPhil at Oxford University. Her thesis examines tool-use and cognition in New Caledonian crows – trying to find ways to outwit them.

Learning is **Logan Wright's** oxygen: slightly poisonous and highly explosive, it's ultimately an essential survival fuel. He's often found reading books while playing guitar and pretending to be a human metronome.

Ed Yong is an award-winning science writer who works at Cancer Research UK. He won the *Daily Telegraph* Science Writer competition in 2007, has written for *Nature*, *New Scientist* and *the Economist*, and writes the blog Not Exactly Rocket Science at scienceblogs.com/notrocketscience.

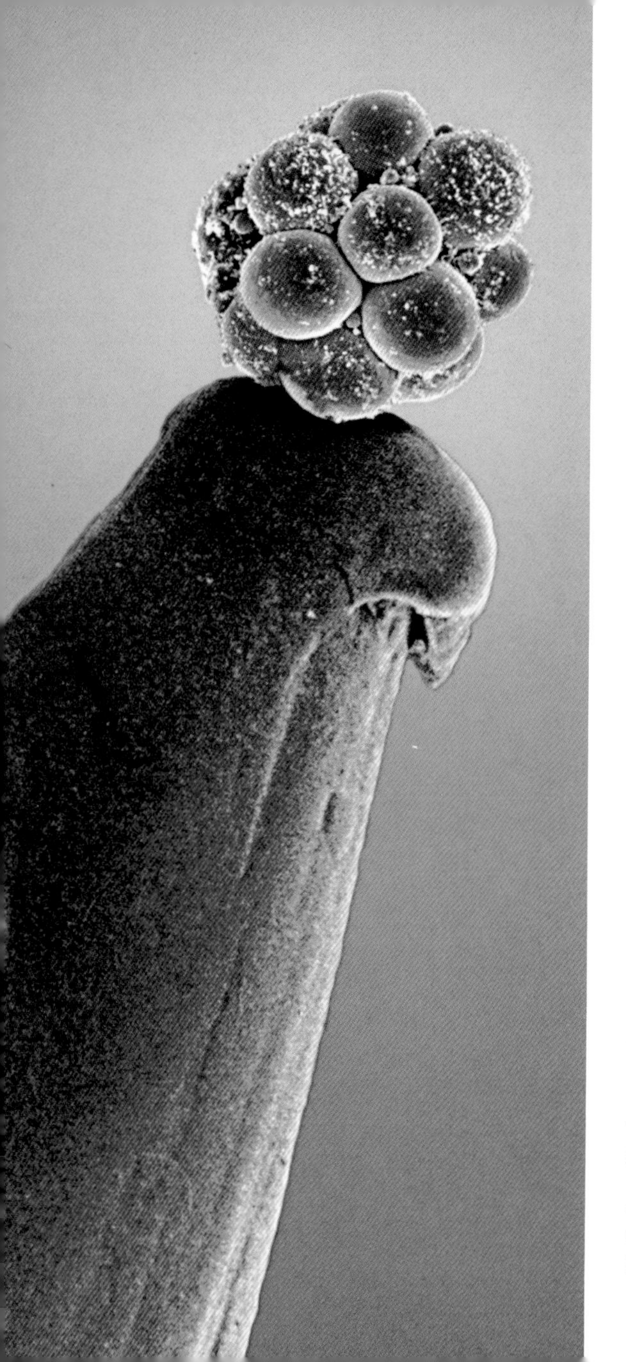

INTRODUCTION

The turn of the twentieth century signaled a time for change. Preceding decades had been colored by an increasing feeling that science, especially physics, was over. The passing of the century heard Lord Kelvin, one of the greatest scientists of the time, say, "There is nothing new to be discovered in physics now, all that remains is more and more precise measurement." That very same year, however, a new order would start to be established. Towards the end of 1900, German physicist Max Planck would suggest that energy was made up of little packets called quanta. Science was about to plunge from the macro-scale into a new quantum world.

Planck was most likely working by oil lamp when he penned his paper that would change the world. If he was lucky enough to have one of Edison's new electric light bulbs it would have been very inefficient and rather short-lived. Over the remaining 47 years of his life, Planck would witness electrical appliances coming of age, and by the time he died the first computers – the century's single most important invention – would be crunching through their first calculations.

The emergence of quantum physics, along with the theory of relativity, is arguably the most sensational advance of the last few centuries. These epoch-making scientific revolutions vie for attention in the coming pages alongside a slew of other scientific and technological achievements. This last century was when mankind entered the Earth's deepest realms, took to her skies and even walked on her moon. We made the connection between Darwin's theory of evolution and Mendel's discovery of genetic inheritance, identified DNA as life's blueprint and even worked out how to manipulate it to our own ends and even working out how to manipulate it to our own ends. Agricultural revolutions have allowed the world's population to skyrocket and medical breakthroughs have enabled us to extend our lifespan.

Given the innumerable changes over the last century, it was an extraordinary challenge for us to choose just one thousand moments to be included in this book and no doubt some will disagree about the relative importance of some of the entries. We have attempted to focus not just on the pure scientific advances, but the social events which shaped academic endeavour, technological change and the public's attitudes to science. The result is a book where, for instance, Richard Feynman rubs shoulders with Captain Kirk. Feynman was one of the most influential physicists since Einstein, while *Star Trek* defined society's vision of the future, and proved to be most prophetic when we look at many of the current gadgets and gizmos we have about us today.

Science relies on the slow gathering of data and ideas to which many people contribute and in many instances it is difficult to pin an advance down to a single event. However through highlighting a particular discovery, invention, person, publication or event in this way, we have been able to summarize the process whilst picking on one particular link in an ongoing chain. This book is designed to celebrate the very heights of human achievement over humankind's most innovative century. However, progress has not been without its cost. As you read through the decades different themes emerge. At the beginning of the century there was a prevailing view that man's activity simply could not have any discernable impact on nature. This was a period of vast industrial expansion and technological advance spurred on by two world wars. It wasn't until the 1950s and 60s that environmental concerns started to become widespread, helped in part by the success of Rachel Carson's revolutionary book *Silent Spring*. Indiscriminate pesticide use, leaded petrol fumes, acid rain and destruction of the ozone layer are all problems that arose largely during the last century and most have, at least in part, been dealt with. Other problems, however, persist.

In 1896, the ingenious Swedish scientist Svante Arrhenius forecast that carbon dioxide released from coal-fired power stations could, given a long enough time period, cause average global temperatures to increase. Arrhenius wasn't overly concerned about his calculations since, at the rates of fuel use in the late nineteenth century, it would have taken three thousand years for global temperatures to rise by 5–6°C. Current estimates, following 110 years during which world population has risen from around 1.6 billion to well over 6.5 billion, state that the same temperature increase could have occurred by the end of this century. The damage that would be caused by such a rapid increase is likely to be widespread and drastic. Finding a way to combat increasing global temperatures should be science's, and mankind's, main priority for the 21st century.

That said, our nascent century promises to be no less extraordinary than the last. It has been an interesting task for us as editors to try and spot which of the latest advances will be the most telling. What will power the cars of the future, how will genetic engineering alter our lives and, maybe most significantly, what are the bounds of nanotechnology? Currently one of the hottest research topics in science, nanotechnology could well revolutionize our lives over the next century as much as electricity revolutionized the last. This is invention and innovation on the tiniest of scales, set to change all areas of life from medical treatments and power generation to ultra-fast computers and even self-cleaning paint

During the twentieth century the world changed beyond all recognition; underpinning every change lay scientific discovery and technological endeavour. Society continues to change at a breakneck speed. Humankind will move beyond the borders of our Earth to colonize other planets, discover non-polluting ways to produce boundless energy and invent ever more impressive ways to cheat death. We may even find Einstein's ultimate prize: a grand theory of everything.

Dr Mark Steer, Hayley Birch and Dr Andrew Impney

Key Event
The Black Death hits the West Coast

In 1899, a ship from Hong Kong docked in San Francisco, bringing with it the deadly infectious disease, bubonic plague. The ship was quarantined, but two individuals escaped, and when their bodies were recovered, they showed signs of the plague. No new cases were found in San Francisco until nine months later, when an autopsy identified plague-like germs in the body of a dead Chinese laborer found in a Chinatown hotel. The source was unknown, but rats from the ship were probably responsible.

At the time, San Francisco was an important industrial center, and the business community was unsupportive of Chinatown being quarantined. The authorities were in denial and assured other cities all was well. However, the disease continued to spread, and only with the intervention of the surgeon general, and the arrival of a new governor in 1903, did clean-up operations begin. By 1904, 122 people had died.

In 1906, a massive earthquake hit San Francisco. The resulting damage meant many had to live in temporary camps, which also became home to massive populations of rats and fleas, the very animals that new research overseas had linked to the transmission of plague.

With the urgent need to control the spread of disease, a bounty of five cents per rat – soon increased to ten cents – was offered in 1907. A public health report stated 250,000 pieces of rat poison and 35,000 traps were used weekly. With these interventions, the epidemic eventually died out in 1909.

Katie Giles

Date 1900

Country USA

Why It's Key
Demonstrated how public health interventions can control lethal disease outbreaks.

opposite The rat flea *Xenopsylla cheopis*

Key Discovery
Gamma radiation

While studying the properties of beta radiation, the French chemist Paul Villard made an intriguing observation. He noticed that in experiments where a beam of beta rays was refracted (passed through a medium of different density), there were often traces of another, unrefracted beam in the results.

Villard set up another experiment, using the newly discovered element radium as a source. He focused a beam of radiation from the radium through a series of glass plates and a magnetic field, to be recorded finally on photographic film. The unrefracted beam appeared again. It did not seem to respond to any external magnetic or electric fields, and would even show up on the photographic film when it was placed behind 0.2 milimeters of lead.

Villard suggested that the radiation he had found was a new type of more penetrating X-ray. He concluded that the three distinct types of radium beams – easily absorbed rays, a dividable stream of charged electrons, and his new super-penetrating X-rays – were analogous to the three types of radiation emitted by cathode ray tubes. With this observation, Villard had correctly generalized radiation into the three types we now know as alpha, beta, and gamma. There was, however, very little interest in his discovery or theory, perhaps because it was outside the current scientific paradigm.

In 1903, Ernest Rutherford, having studied the penetrative power of the beams, named them gamma rays and this term soon fell into common usage. Villard, however, remains pretty much forgotten.

Kate Oliver

Date 1900

Scientist Paul Villard

Nationality French

Why It's Key A new type of very penetrating radiation, and the first suggestion that all radiation observed was one of three things: alpha particles, beta particles, or electromagnetic waves.

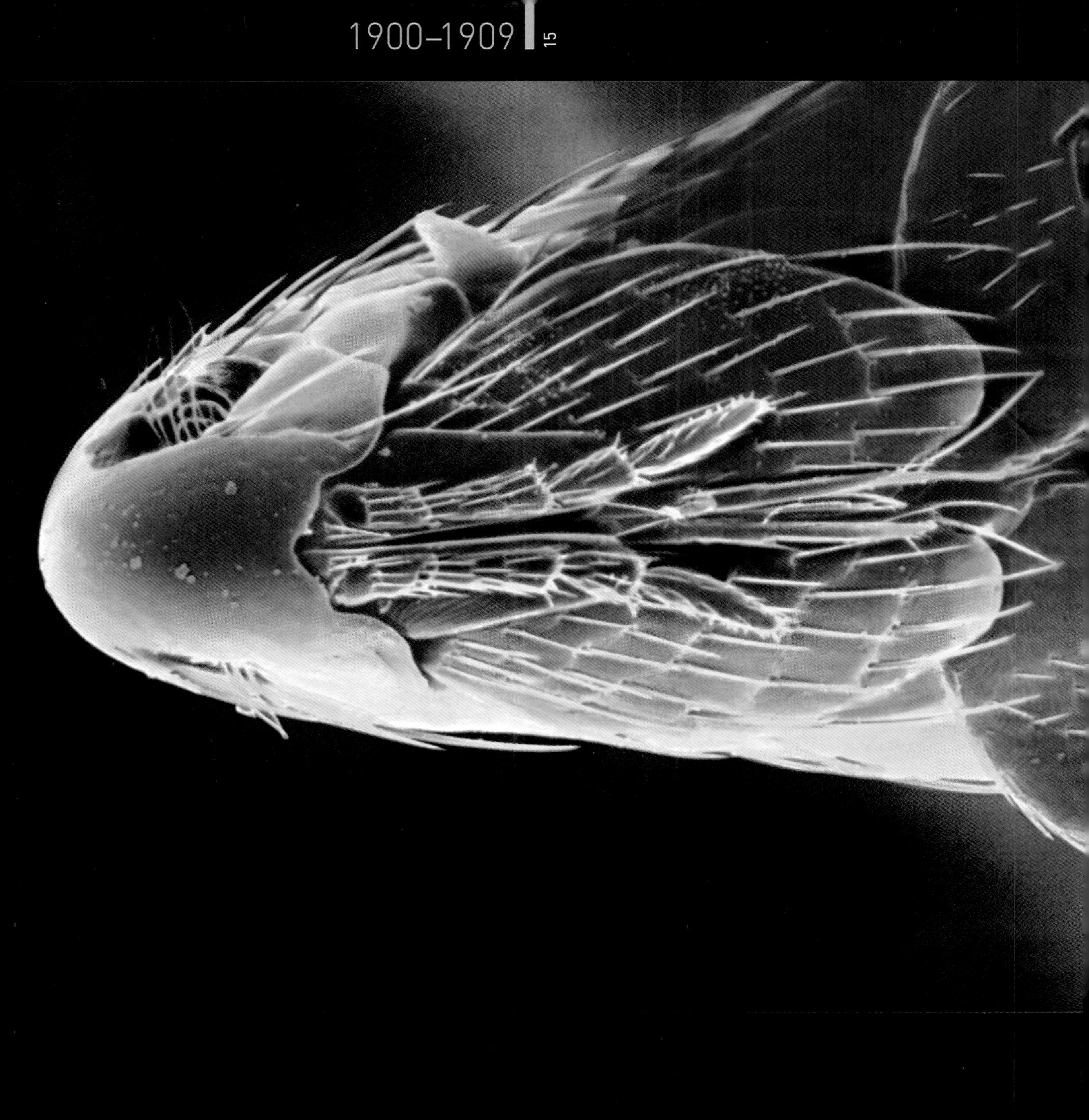

Key Invention The Zeppelin

Not everyone realizes that the giant behemoths of the sky know as Zeppelins are named for their most successful pioneer, the Prussian military genius Count Ferdinand von Zeppelin. Previous attempts at lifting a hydrogen-filled "airship" off the ground had met with partial success, but it was not until Zeppelin himself entered the arena that the technology began to prove itself. Having purchased the plans for a rigid dirigible from the widow of the Croatian inventor David Schwartz, Zeppelin set out to revolutionize air travel.

Using the extreme buoyancy of hydrogen gas in air, Zeppelin calculated that a cloth-covered airship, containing seventeen separate pockets of hydrogen gas and supported by a lightweight aluminum skeleton, would be able to stay aloft, even with the added mass of two internal combustion engines.

After just two years of development and refinement, the Count was ready for his first true test flight. On July 2, 1900, near Lake Constance in Germany, Zeppelin's prototype LZ-1 took to the skies on the very first un-tethered flight of a true airship. The vehicle was massive; measuring around 128 meters in length, and holding an internal volume of 11,298 cubic meters of hydrogen. With five intrepid passengers on board, the LZ-1 reached an altitude of 390 meters and traveled six kilometers in seventeen minutes.

It took another decade for Zeppelin's first commercial airship, the Deutschland, to enter service and, during World War I, his creations were used as bombers.

David Hawksett

Date 1900

Scientist Count Ferdinand von Zeppelin

Nationality Prussian

Why It's Key Revolutionized the concept of traveling "by balloon," and was one of the first practical uses of hydrogen gas.

Key Event **Subatomic bombshell** Max Planck announces quantum theory

During the first year of the new century, a German physicist named Max Planck introduced a concept that challenged the very foundations of physical science. His idea formed the basis of the most important physical theory of the twentieth century: Quantum Theory.

In the lead up to 1900, physicists were of the view that they had everything pretty much sewn up; the Universe was made up of "stuff" (particles), which was acted upon by forces (gravitation and charge). Problem solved; Universe conquered; on to the next problem.

This hubris stemmed from the fact that many of the predictions made by classical physics had been borne out through experimentation. Little did anyone suspect that science would be turned on its head by a middle-aged, conservative physicist.

It was a quiet revolution. In seeking a theoretically rigorous explanation of results obtained from experiments done on black-body radiation – radiation emitted by objects that absorb electromagnetic radiation – Planck had to start thinking about energy in a new way, as "quanta." These were discrete packets of energy rather than the waves that had previously been supposed. Planck published this idea in a paper presented to the German Physical Society on December 14, 1900. It formed the basis for the theory that explains nearly all chemistry and physics, and that has shaped our contemporary scientific view of the Universe in which we live.

Barney Grenfell

Date 1900

Country Germany

Why It's Key Quantum theory was the single most important scientific advance of the twentieth century. It's fitting that it was also one of the first.

opposite Max Planck (right) with Niels Bohr

Key Person **Sigmund Freud**
Father of psychoanalysis

An early pioneer in the study of the mind, and deemed "the father of psychoanalysis," Sigmund Freud (1856–1939) made the link between people's words, thoughts, and dreams, to their unconscious.

The son of German Jews, Freud originally wanted to be a lawyer, and later a doctor. But he wasn't so keen on the messy nature of the physical body, so he eventually decided to focus his attention on the mysteries of the human mind. He studied the difficulty of speaking and comprehending language, known as aphasia, and was also known to self-experiment with, and advocate, cocaine. However, it was his later work on hysteria for which he gained most kudos. He initially believed neurosis stemmed from childhood trauma, treatable with the "talking cure." But a year after the death of his father, Freud moved on from the trauma theory of neurosis. He adopted the notorious theories of infantile sexuality and the Oedipus complex – where a child competes with one parent for the affections of the other parent. In 1900, Freud's most famous book (though it didn't sell so well upon release) was published. *The Interpretation of Dreams* explained his theories on the symbolic significance and unconscious forces involved in dreaming. The book purported that every action or thought has a purpose, even if it may be unconscious.

His work is controversial, and many of his findings are now disregarded. But, in an era where the workings of the brain, consciousness, and psychology were largely unknown, Freud's psychoanalysis was a stepping stone for modern psychology.

Umia Kukathasan

Date 1900

Nationality Austrian

Why He's Key Freud's work may now be considered largely flawed with inconsistency and bad scientific method, but his probing imagination helped formulate modern psychology.

Key Invention **The Brownie**
Light in a box

The Kodak Box camera, first introduced in 1900, was only effective because of the advent of light sensitive material (film) flexible enough to be rolled inside lightproof paper. Earlier cameras had used heavy glass plates which required complicated processing to develop, making it difficult for the unskilled amateur to take photographs. When roll film became available at the end of the nineteenth century, the development of a light, portable camera became feasible, and exposed film could now be sent by post for processing.

Originally in the form of a 9.5 centimetre high box, the Brownie had a fixed lens at the front and a roll of film opposite the lens at the back. Aperture, focus, and timing controls were preset, and film was manufactured to meet these parameters. Any object, from distances of about 2.5 meters to infinity was in focus, which helped the inexperienced user to take photographs easily and successfully.

This simple construction enabled the camera to be mass-produced; the original version sold 245,000 units before being discontinued in favor of more advanced models. Changes were made mainly to the housing, which originally was tough cardboard. In the earliest models the object to be photographed was viewed through a small right-angled prism on the top of the camera, so that the user had to look down to view the image through the prism. Later models had inscribed guidelines on the prism to help the user ensure that the image seen was the image photographed.

Fabian Acker

Date 1900

Scientist Kodak

Nationality USA

Why It's Key Turned photography into a hobby for the masses, with small, affordable cameras.

Key Event **Dark clouds draw in**
Physics will never be the same again

By the end of the nineteenth century, physicists were beginning to feel rather smug and self-satisfied. During a speech in 1900, Lord Kelvin, the pre-eminent scientist of his day, claimed, "Physics is essentially complete: There are just two dark clouds on the horizon." While he expected these problems to be cleared up fairly easily, he had actually foretold the two biggest discoveries of the next century – discoveries that would turn physics on its head.

The first problem was the discovery that the speed of light was constant; no matter whether the Earth was moving toward a beam of light or away from it, light speed remained the same. Classical physics had no way of explaining this conundrum – it needed what was to become the most celebrated theory of modern times. Behind Kelvin's first cloud lay Einstein's Theory of Relativity; this says light is constant while time and distance are not.

The second problem was concerned with how objects absorb energy and radiate it back out again. Kelvin pointed out that the methods physicists were using to predict how much energy would be radiated gave numbers that were much too high; in some cases the predictions even approached infinity. Something clearly wasn't right. It was again down to Einstein to find the solution by positing that energy could only be emitted in precise packets, or quanta. And thus quantum physics appeared from behind that second cloud – the classical view of the world was well and truly shattered.

Kate Oliver

Date 1900

Country UK

Why It's Key A summary of the state of nineteenth-century physics provides a timely warning that no matter how sorted you think things are, they can always get more complicated.

Key Person **David Hilbert**
The man with a Hit List

"We must know – we shall know," announced David Hilbert (1862–1943) as he presented what he considered the twenty-three most important problems at the Second International Congress of Mathematicians in Paris in 1900.

A German mathematician, David Hilbert, and his students created the mathematical infrastructure which later became instrumental for much of quantum mechanics and general relativity. His most famous new concept was "Hilbert space," which, instead of describing the three-dimensions that we are used to operating in, described infinitely dimensional space.

Of Hilbert's twenty-three problems, ten have been solved and seven partially solved with controversial solutions. Two are still unresolved, and the remaining four have since been deemed too vague to ever be completely finished. His problems ranged across pure and applied mathematics and while some were quickly dealt with, others took or may very well take many decades. The sixth problem calls for the axiomization of physics – the description of everything by a set of well defined laws, which by any account, may take some time to figure out. His eighth problem, the infamous Riemann hypothesis, is now a Millennium Prize Problem and many suspect it will stay near the top of the list of important unsolved problems for years to come. Many aspects of the twenty-three problems are still of great interest and relevance and Hilbert's insight enthused and guided mathematicians throughout the twentieth century toward what he correctly highlighted as the most important problems.

Leila Sattary

Date 1900

Nationality German

Why He's Key Led to foundation mathematics for quantum mechanics and general relativity.

Key Discovery **Aurora Borealis** The puzzle of the northern lights

For thousands of years people have marveled at the amazing spectacle of the northern lights, or aurora borealis; a fantastic display of coruscating colors that illuminates the sky in spectacular fashion. And for as long as people have marveled, they have also suggested explanations as to what these amazing displays might be. But it was not until the 1970s that a theory for how auroras are created was actually confirmed – over seventy years after it was first suggested.

In 1900, Kristian Birkeland, an avid researcher and experimental scientist, suggested a theory that the lights observed in the sky were formed by charged particles, issued from sunspots, coming into contact with Earth's atmosphere and then being shaped by lines of geomagnetic force. He demonstrated this experimentally by placing a charged metal sphere in an evacuated chamber and observing that the colored discharge changed when he magnetized the sphere; the greater the magnetic field he produced, the more focused in an area the discharge became.

The problem with Birkeland's theory was that it could not be confirmed from the ground, only from a position above the Earth's ionosphere – one of the outermost layers of the atmosphere, which begins 80 kilometers above the surface. It would be almost seventy years before this was to happen.

The first complete map of the magnetic "Birkeland currents," as they came to be known, was made in 1974 by Alfred Zmuda and James Armstrong, some fifty-seven years after Birkeland's death.

Barney Grenfell

Date 1900

Scientist Kristian Birkeland

Nationality Norwegian

Why It's Key Birkeland's theory explains one of the most fantastic natural phenomena on earth. In recognition, the Norwegian 200 Krone note bears a picture of Birkeland in front of the aurora borealis.

opposite Kristian Birkeland with his apparatus for studying the cause of the aurora borealis

Key Discovery **Blood types**

In 1901, Austrian Karl Landsteiner found the solution to a puzzle that, for more than two hundred years, had perplexed doctors who attempted to save lives by performing blood transfusions. The problem was that patients often went into shock or became jaundiced in reaction to new blood. Closer examination had revealed that red blood cells were clumping together and blocking blood vessels. From 1901 to 1903 Landsteiner reported that mixing the blood of different people in the laboratory caused the red cells of some but not others – to clump upon mixing.

Further investigations, including some using his own blood and that of lab colleagues, enabled Landsteiner to explain this. In 1909, he grouped people into four main blood groups: A, B, AB, and O. This led doctors to perform blood group testing of donors and recipients before a transfusion and, from 1915 onwards, blood transfusion became a relatively safe and routine procedure – in time to save countless lives on the battlefields of World War I.

We now know that people of different ethnic origins vary in the frequency of different blood groups, with group O being the commonest in the UK, and group B being more prevalent in Asian communities. Group O is the oldest blood group, and was present in humans in the Stone Age. Groups A and B emerged later, in people migrating from one world region to another – such as from Africa to Europe – between 25,000 and 10,000 years BCE.

Julie Clayton

Date 1901

Scientist Karl Landsteiner

Nationality Austrian

Why It's Key The discovery of the ABO blood group antigens transformed medicine by enabling life-saving blood transfusions to become routine practice.

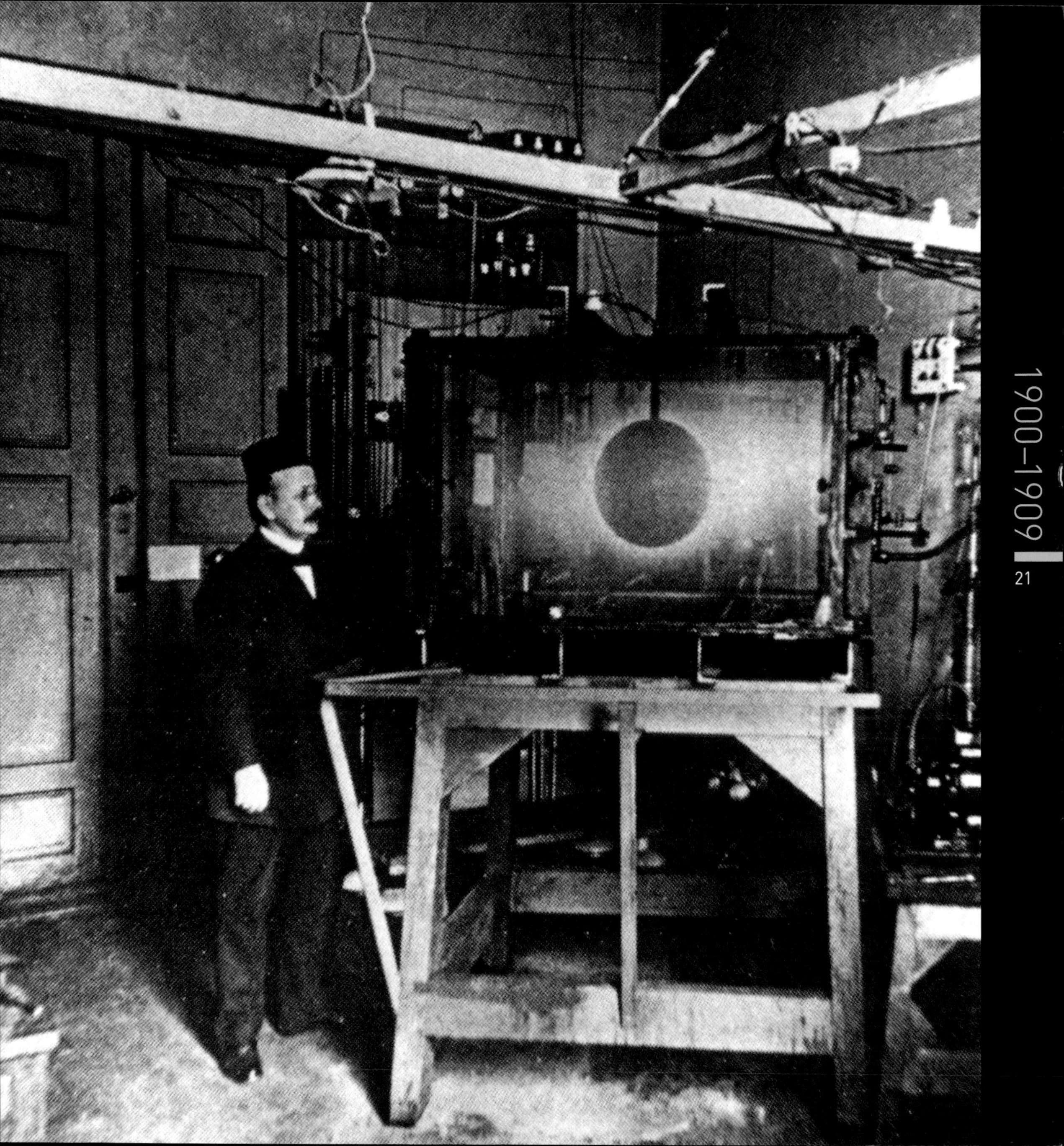

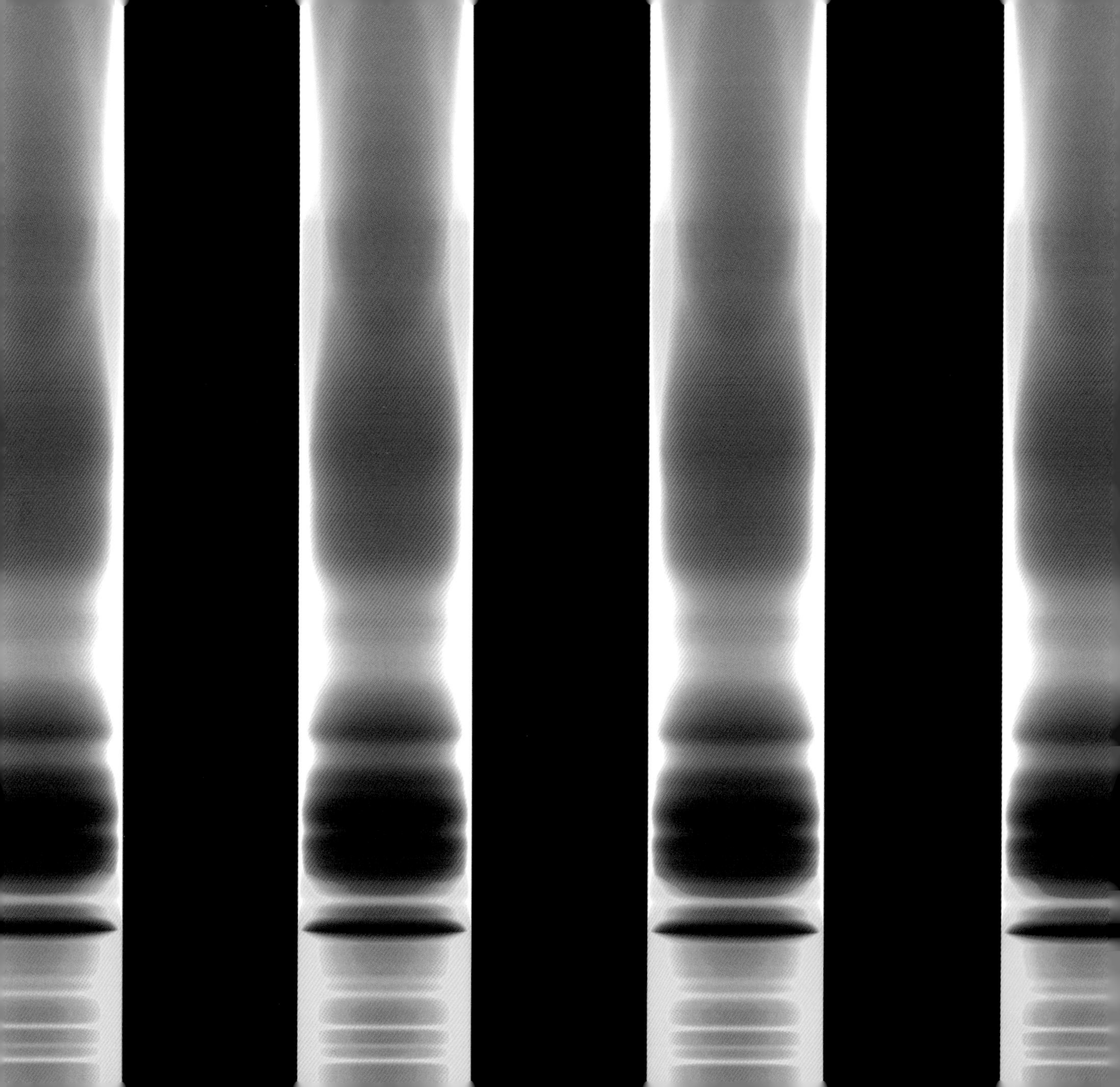

Key Discovery **Chromatography technique for organic compound separation**

For a long time, dye makers used chromatography as a way of checking the make-up of batches of dye. A white string dipped into a vat of dye will start to absorb the liquid, but the various colors each travel a different distance, leaving a range of colored bands on the thread.

It wasn't until 1901, however, that the Russian chemist Mikhail Tsvet – Tsvet being the Russian word for color – used a similar technique to separate plant pigments, and coined the term chromatography, meaning "color writing." Instead of a piece of string, he used a powder-filled column and dissolved the pigments he wanted to separate in an organic solvent. When he poured the liquid through the tube, the pigments were deposited as coloured bands. Chromatography is now one of the standard ways scientists use to separate the components of a mixture so that they can be analyzed. But unfortunately for Tsvet, the value of his work went unrecognized for many years. This was partly his own fault; he only published in Russian or in obscure German botanical journals. And his scientific integrity was questioned when the first team to try to repeat his experiment failed. It emerged later they had used a solvent that destroyed the plant pigments, so didn't get any results.

By the 1940s, chromatography was becoming an established scientific technique. But it was a pair of British scientists who picked up a Nobel Prize for developing it, leaving Tsvet largely unrecognized.

Anne-Claire Pawsey

Date 1901

Scientist Mikhail Tsvet

Nationality Russian

Why It's Key Tsvet's work on separating plant pigments went unnoticed for decades. He is now known as the father of chromatography.

opposite Computer artwork shows separation of different chemicals by chromatography.

Key Publication ***The Mutation Theory*** Imperfections lead to evolution

In the early twentieth century, the rediscovery of Gregor Mendel's work with pea plants created a hotbed of activity surrounding genes, evolution, and inherited traits. The theory of evolution was rapidly advancing, not least thanks to the contributions of a Dutch botanist named Hugo de Vries.

De Vries' book *The Mutation Theory* suggested that new species could arise spontaneously. He had observed that new characteristics could appear within the space of a generation and he pinned the blame on mutations. De Vries began developing his theory during a succession of plant breeding experiments with evening primrose. He noticed that new varieties of the plant would occasionally occur, as if from nowhere, and he believed that these were the results of changes within particles that could be passed on from one generation to the next – what we now would call genetic material. He came to expect these new varieties to prosper – if the changes were favorable in terms of the plant's survival and procreation – and to remain until further mutations occurred.

The Dutchman's work seemed to contradict that of Darwin's theory of evolution and gained him support from anti-Darwin campaigners. However, the new plant varieties he observed turned out to be caused not by mutations, but by the rearrangement of DNA within an individual plant's reproductive cells. It is now accepted that mutations do indeed contribute to evolution, but the reality is a gradual selection of these mutations over many generations.

Nathan Dennison

Publication *The Mutation Theory*

Date 1901

Author Hugo de Vries

Nationality Dutch

Why It's Key Although de Vries' original mutation theory soon lost support, he is credited with having rediscovered Mendel's work on inheritance.

Key Discovery
Alzheimer's Disease

It's the turn of the century and a prominent German-speaking physician has just described a new disorder of the mind in a woman dubbed Auguste D. This may sound familiar, but this time the doctor isn't Freud and he isn't concerned with dreams or mothers or phallic symbols; this is Alois Alzheimer, and he has just described the disease that bears his name today.

Although Alzheimer was far from the first person to describe dementia, and may not have even been the first to describe the characteristic protein tangles that are seen in the brain on autopsy, his name has become synonymous with this debilitating disease that causes memory loss, confusion, and, ultimately, death.

Alzheimer's disease is believed to affect up to half of all people over the age of eighty-five, and a significant proportion of those under that. Despite its prominence, there is still no definitive diagnostic test that can be performed on individuals to confirm the disease. The diagnosis is based on a collection of symptoms and can only truly be made on autopsy – just as in 1906, when Alzheimer described it following the death of poor old Auguste D, whom he had studied for the last five years of her life. By the same token, a cure for Alzheimer's remains elusive, though some treatments designed to slow its progression have been developed.

B. James McCallum

Date 1901

Scientist Alois Alzheimer

Nationality German

Why It's Key Alzheimer's disease is one of the most common causes of dementia. It still cannot be prevented or cured.

opposite Colored computer tomography scans show brain atrophy due to Alzheimer's causing enlarged ventricle cavities (white, at centre of brain)

Key Event
First Nobel Prizes awarded

"The merchant of death is dead." Those are the words of a French newspaper's obituary of Alfred Nobel in 1888. Unimpressed by the Swedish chemist's record of "finding more ways to kill more people faster than ever before," including the invention of dynamite and gelignite, and his role as a weapons manufacturer, the newspaper report was scathing. It must have made especially grim reading for Nobel himself, who wasn't actually dead. The newspaper had printed the obituary prematurely.

Many people believe this experience inspired Alfred Nobel to leave behind a more life-affirming legacy: the Nobel Prizes. When he died for real in 1896, he left the majority of his vast fortune – more than US$100 million in today's money – to create the awards, to be given every year for contributions to physics, chemistry, medicine or physiology, literature, and peace.

The first prizes were awarded in 1901. At least two of those first winners made contributions to the world which you'll still recognize today: Wilhelm Röntgen won the Physics prize for his work discovering an interesting new type of radiation – familiar to us now as X-rays; and Jean Henry Dunant, founder of the International Committee of the Red Cross and the Geneva Convention, shared the first Peace prize with Frèdèric Passy.

Matt Gibson

Date 1901

Country Sweden

Why It's Key The Nobel Prizes became one of the most respected accolades in science, and have helped raise the public's awareness of scientific achievements every year since they were started.

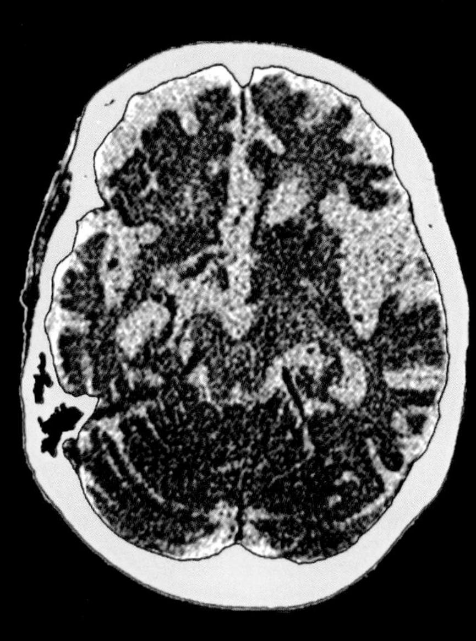
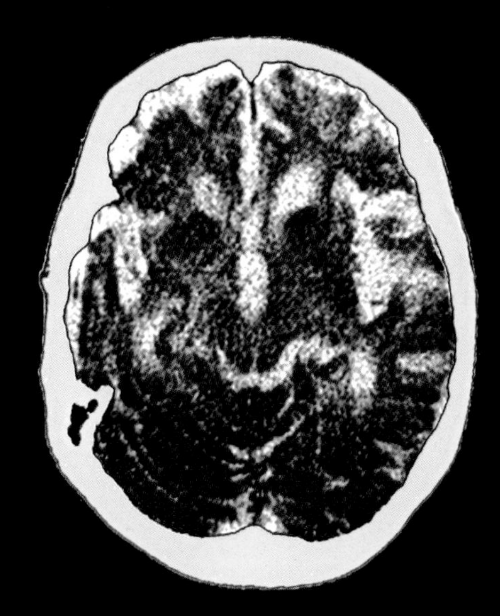
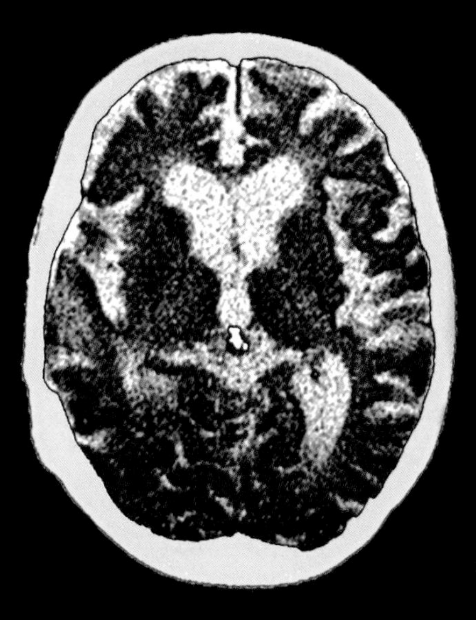
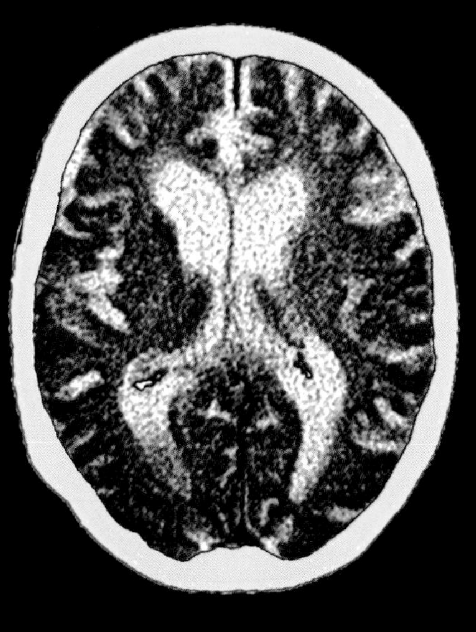
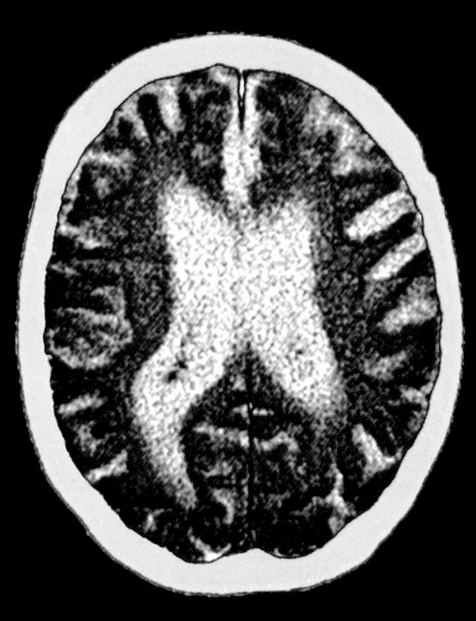
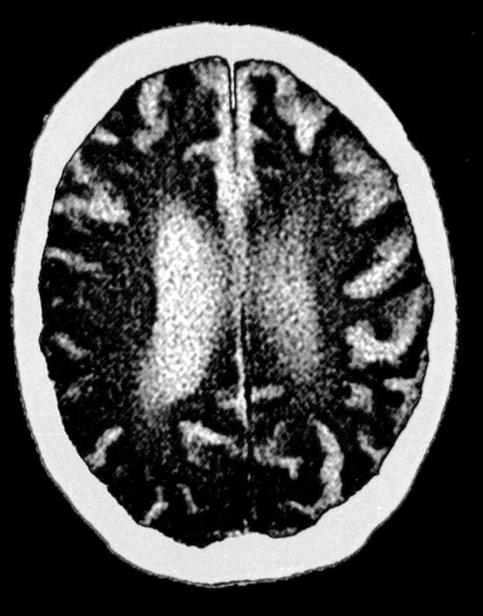

Key Event **Calling Canada**
Marconi receives the first transatlantic radio signal

On December 12, 1901, the Italian inventor Guglielmo Marconi waited with bated breath at an abandoned hospital on the cliffs of Newfoundland. He had set up his primitive receiving equipment in the hope of hearing a faint signal from his transmitter at Poldhu, Cornwall – more than 2,700 kilometers away across the Atlantic Ocean.

It was the culmination of seven years' work, inspired when Marconi read about the achievements of Henrich Hertz, the father of radio waves. Marconi had already successfully established radio links from the Isle of Wight to Bournemouth, as well as sending the first radio transmissions across the English Channel. These milestones had given him the confidence to invest £50,000 in his Atlantic experiment – at the time an incredible sum.

The high winds of Newfoundland made it difficult for Marconi to operate his aerial, which was held aloft by a kite, and several attempts at communication with his Cornish station had already failed. But on this day, at the appointed time, and with his ear pressed against the telephone headset, he finally heard the Morse code signal for the letter "s."

"The chief question was whether wireless waves would be stopped by the curvature of the earth. All along I had been convinced that this was not so. The first and final answer came at 12:30 when I heard ... dot ... dot ... dot," he later recalled.

Today thousands of amateur radio enthusiasts still make pilgrimages to Poldhu to honor "the father of wireless."

David Hawksett

Date 1901

Country UK

Why It's Key Began the modern era of instant global communications.

opposite Transmitter used in Marconi's broadcast. Wooden racks at center right hold banks of capacitors and transformers are at left. The spark gap that provided the signal is at far right.

Key Person **Thomas Edison**
Providing portable power for the planet

Thomas Alva Edison (1847–1931) was a business-man and inventor who had over a thousand registered patents in the USA. With a career spanning over sixty years, Edison is acknowledged as a truly remarkable man, whose inventions have changed the way people live the world over, and continue to do so.

For a man who didn't learn to talk until he was four years old and had only three months' official schooling, Edison's accomplishments were impressive. His earliest invention was the phonograph in 1877, the first machine to play recorded sound. He is also credited with significant improvements to electric lighting. But perhaps his greatest contribution to today's society is patent number 879612: the alkaline storage battery. Taking over ten years to fully develop, Edison intended to replace bulky lead-acid batteries with more slimline versions that would provide power for electric cars. After thousands of tests on various metals, he developed a potassium hydroxide electrolyte that would react with nickel electrodes to produce power. Poor early performances of the battery led to redesign and re-launch in 1910. Unfortunately this was too late for electric cars, as Edison's friend Henry Ford had just designed a reliable gasoline engine car.

Despite this setback, the battery became a real money spinner for Edison, finding employment in train lights and power lamps for mines. His battery was also the starting point for the lithium storage batteries used today, thus earning him a reputation as "The Father of the Electrical Age."

Nathan Dennison

Date 1901

Nationality American

Why He's Key Edison satisfied the energy-hungry twentieth century with his inventions. His inventing prowess was unquestionable, with invention after invention still in use to this day.

CAUTION
VERY DANGEROUS
STAND CLEAR.
Ht2

Key Invention
Vacuum cleaner

The first powered "vacuum cleaner," was conceived, named, built, and patented by English engineer Hubert Cecil Booth in 1901. It paved the way for a future of efficient cleaning and more hygienic living and working conditions. Before Booth's innovation, cleaning required much time and effort: dusting, carpet beating, furniture removal, sweeping, and more dusting. And even then, much of the dust would float around only to resettle.

The demonstration of a powered "blowing" cleaner in 1901 inspired Booth to make a better machine using suction. His first "experiment" involved him placing his mouth on the back of a restaurant seat and sucking. He almost choked. Booth's "Puffing Billies" were cumbersome beasts comprising a gasoline driven, 5-horse-power piston engine that could create sufficient suction using a small leather diaphragm pump. Far from being convenient, they were housed in horse-drawn carts with 30-meter long hoses that extended from the street into the building. As electricity supplies to buildings increased, and small electric motors and better designed fans became available, Booth and others were able to create the portable and compact cleaners we are familiar with today.

An American, James Murray Spangler, was the first to achieve this in 1907. However, he sold the rights because he lacked the capital to market the product. The buyer was William Henry Hoover, a shrewd businessman and the first manufacturer to successfully mass-produce vacuum cleaners.

James Urquhart

Date 1901

Scientist Hubert Booth

Nationality British

Why It's Key Made cleaning more convenient and efficient, and greatly improved sanitation by removing vast quantities of germ-laden dust from the home, work-place, and public places such as theaters, hotels, and trains.

Key Event
Entering the Mercedes era

The modern car is the brainchild of many individuals, but few inventors were as crucial in its development as Wilhelm Maybach and Emil Jellinek. Together, the two are credited with creating the first modern automobile engines at around the turn of the nineteenth century.

Wilhelm Maybach was an early employee of Gottlieb Daimler's company, DMG, where he proved to be an outstanding engineer and designer. He was instrumental in the design of such things as the spray carburetor, the tubular radiator fan, the honeycomb radiator, and the phoenix engine. Though his relationship with DMG was somewhat rocky – he quit at one point, and was forced out of his position as head engineer on multiple occasions – he still ended up as the technical expert who made the modern car.

Jellinek's role is a bit less concrete. An adventurer who made a fortune by trading North African native products, he was engrossed in the new sport of auto racing. He was so impressed by DMG's products that he pushed the company to make even faster and more powerful machines. On one early race car, he stenciled the name of his daughter, Mercedes – a name which stuck with the car and engine. It was partially at Jellinek's prompting that Maybach developed the early modern automobile; as a result, the two men are credited with its invention.

B. James McCallum

Date 1901

Country Germany

Why It's Key Maybach and Jellinek ushered in a new era for automobiles.

Key Discovery **The Antikythera Mechanism**
World's oldest analog computer, complete with cogs

In 1902, Greek archaeologist Valerios Stais was working on artefacts recovered from a shipwreck discovered off the coast of Antikythera Island. Two years previously, a diver had found the wrecked Roman cargo ship, and various objects, such as statues, had been retrieved. Among these items, Stais noticed what looked like a heavily corroded gear wheel.

What he had found was an ancient geared mechanism. A further eighty-one fragments were recovered, including thirty hand-cut bronze gears, the largest fragment of which had twenty-seven cogs. The device these came from would have been around 33 centimeters high, 17 centimeters wide, 9 centimeters thick, and mounted in a wooden frame. The first proper scientific investigation was by British historian Derek Price, who concluded the object was of Greek origin and its purpose was as an astronomical instrument. A modern study of the device published in 2006, revealed more of the secrets of the Antikythera Mechanism.

The Anglo-Greek team used advanced imaging techniques on the fragments, such as three-dimensional X-ray microfocus tomography. The device turned out to be far more complex than originally thought. It incorporated the Greek zodiac and an Egyptian calendar, and had two dials on the back displaying accurate timings of lunar cycles and eclipses, even compensating for velocity changes in the Moon's elliptical orbit. There is also evidence it incorporated planetary motions. Constructed around 100–150 BC, it predates other mechanical devices of this complexity by around 1000 years.

David Hawksett

Date 1901

Scientist Valerios Stais

Nationality Greek

Why It's Key At nearly two thousand years old, the Antikythera Mechanism has been referred to as the world's oldest analog computer, but it took some sophisticated modern technology to discover its function.

Key Event **The filing of fingerprints**
A whorled first

On July 1, 1901, the Assistant Commissioner of the Criminal Investigation Department, Edward Richard Henry, began setting up the first UK finger-print bureau.

Head back a few years to Calcutta: the old system used to identify people was the Bertillon system, a laborious process that required many measurements including height, left foot, right ear, and the width of the person's cheeks. Convinced of the potential of identification by fingerprints, Henry, who was the Inspector General of police in Bengal, sought a way of classifying them.

Azizul Haque and Hem Chandra Bose, working at the Calcutta Anthropometric Bureau, developed the Henry System of fingerprint classification. By assigning the fingers numbers and counting the unique whorl patterns on them, there were 1,024 possible groups, enabling them to be easily filed and searched.

In June 1897, British India began identifying criminals by fingerprints. The new system was so successful that a committee reviewed Scotland Yard's identification practices and abandoned the Bertillon system in favor of the Henry system. In 1901, Henry was transferred to Scotland Yard and began setting up the bureau. Started by just three people, it now employs around six hundred and has played a vital role in criminal investigations, including the Great Train Robbery of 1963.

Douglas Kitson

Date 1901

Country UK

Why It's Key Vital to modern policing, the collection of fingerprints streamlined the processes for catching criminals, identifying bodies, and helping match people with amnesia to their true identities.

Key Person **Wilhelm Conrad Röntgen**
X-ray discoverer

Having shown no particular excellence as a child, and been thrown out of high school after being wrongly accused of drawing a caricature of a teacher, Wilhelm Röntgen (1845–1923) took a non-traditional path through education. He eventually entered Zurich Polytechnic, graduating with a doctorate in mechanical engineering. Later he would always insist on working on his own and constructed many of his own pieces of equipment. It must have been rather satisfying then, after these initial difficulties, to be the first person ever to receive the Nobel Prize for Physics, in 1901.

While experimenting with one of the new-fangled cathode ray tubes, Röntgen noticed that even if the tube was shrouded in black paper and the room was dark, a photo-reactive plate would still start to glow when the tube was turned on. Similarly, photographic plates could still become exposed even if the cathode ray tube was covered, depending on the materials placed between the tube and the plate. Rays of something other than light must be shooting from the tube, he reasoned.

Since he had no idea what these rays were, Röntgen named them "X-rays," X being the general symbol for an unknown. But X-rays were not thought of in their familiar medical context until Röntgen's wife let him take a "Röntgenogram" of her hands. Her dense bones and metallic ring showed up in bright unexposed white, and the flesh in fainter grays. The benefits of being able to see the internal structure of the human body non-invasively were obvious.

Kate Oliver

Date 1901

Nationality German

Why He's Key Röntgen was the first physicist to win the Nobel Prize.

opposite First X-ray photograph of a human being, made in 1895. It shows the hand of Röntgen's wife, who is wearing a ring.

Key Invention
Handheld vibrators create a buzz

The various incarnations of the vibrator over the years – from steam powered to remote controlled – have provided plenty of scope for amusement, in more ways than one. However the growth of the female massage industry also acts as a measuring stick for social change.

As early as the 1700s, women had been visiting spas or "water-cure resorts" and were encouraged to do so by their doctors. What they sought was not simply peace and relaxation, but relief from what was termed "hysteria" or anorgasmia (clinical signs included irritability and "excessive bicycling"). At European spas, women were treated by pelvic douche; crudely, directing powerful jets of water at the genitals.

When the first electromechanical vibrators appeared on the market in the 1880s, many were purchased by these water-cure resorts. Doctors also employed vibrators to help alleviate the symptoms of hysteria. It wasn't until the early twentieth century, however, that the patent for the first handheld, electrical massage appliance was granted, paving the way for the incredible range of home pleasure devices that exist today.

Doctors had always belittled, and expressed a reluctance to provide, the physical therapies which they themselves recommended; genital massaging was a chore very often delegated to female physiotherapists and midwives. But as consumer goods companies started stocking the appliances, the demand for such therapies waned – women were doing it for themselves.

Hayley Birch

Date 1902

Scientist Oscar A. En Holm

Nationality American

Why It's Key Created as a medical cure, the vibrator has since given women a license to take their pleasure into their own hands.

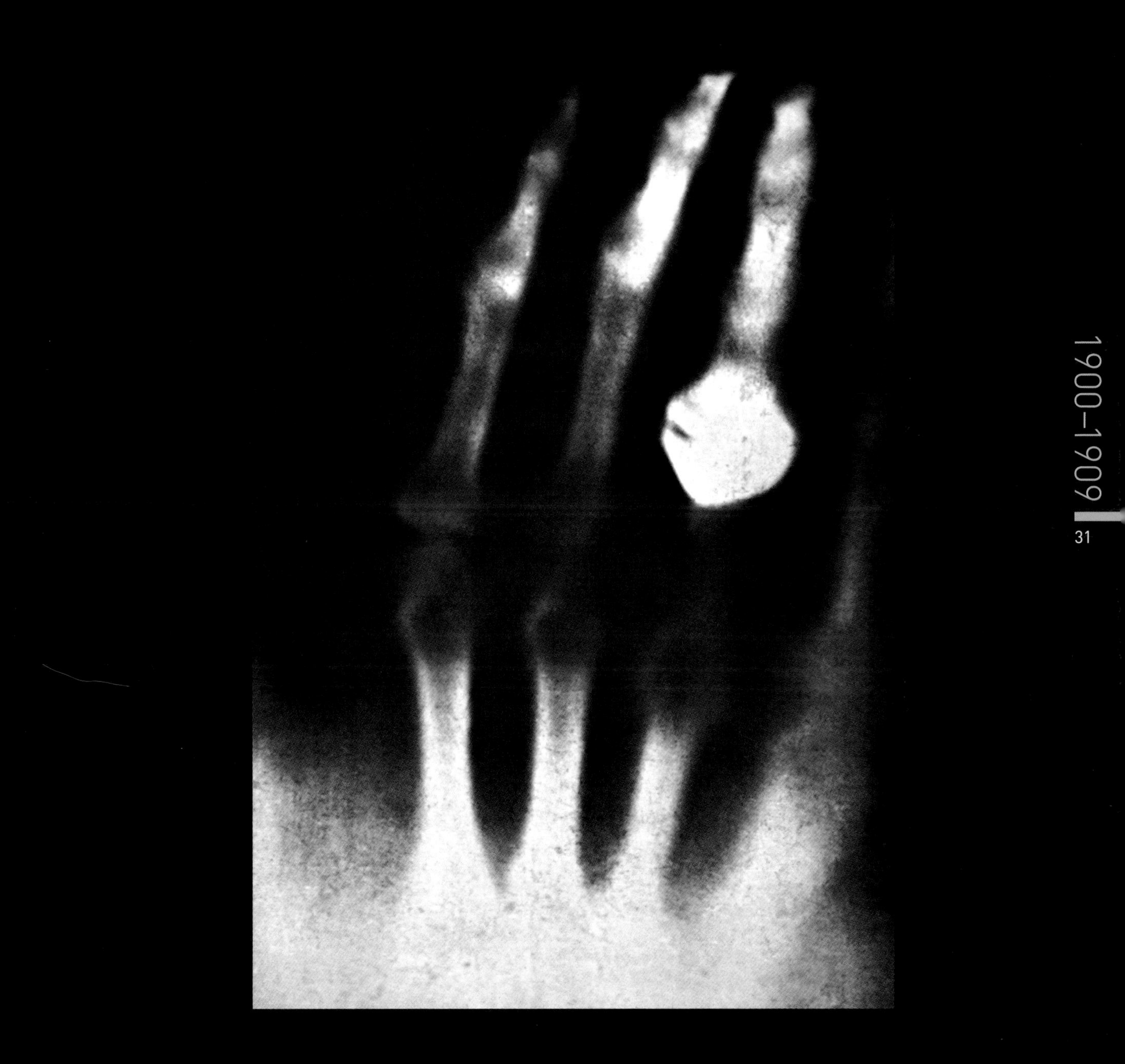

Key Discovery **Make mine a cold one**
Liquid oxygen and nitrogen obtained from air

German engineer and scientist, Carl von Linde is a man who should, perhaps, be the patron saint of lager drinkers. He is credited with designing, at the behest of his friend (and president of the German Brewers Union) Gabriel Sedlmayr, the world's first practical refrigerator, in 1876, thereby enabling year-round brewing of this temperature-dependent beer.

The success of Linde's refrigeration enterprise enabled him to concentrate on his experiments concerning the removal of heat from gases and liquids at low temperatures. In 1895, he produced a machine that used the Joule-Thomson effect (whereby a gas which expands will cool down), to produce commercial amounts of liquefied gases, which up until then could be made in only very small quantities. Linde's method sees air sucked into a machine, where it is compressed, pre-cooled, and then decompressed, causing it to cool further. In a countercurrent heat exchanger, air which has already been cooled is itself used to cool more compressed air, and so on. The process is repeated until the air is so cool that it becomes liquefied.

By 1902, Linde had developed a method for separating pure liquid oxygen and nitrogen from liquid air, again in commercially viable amounts. Linde's achievements changed the status of liquid oxygen from laboratory curiosity to international commodity, and also kick-started the medical and industrial gases industry.

Mike Davis

Date 1902

Scientist Carl von Linde

Nationality German

Why It's Key Linde's discoveries had vital commercial applications in virtually every sphere of science: food storage; medicine; manufacturing; and, of course, limitless supplies of cold beer.

Key Publication **"The Cause and Nature of Radioactivity"**

In 1901, young physicist Ernest Rutherford and his even younger colleague Frederick Soddy were investigating the radioactive properties of the element thorium at the University of Montreal. Thorium samples were known to be radioactive, but what radioactivity actually entailed was a mystery.

During one experiment, Soddy managed to separate thorium nitrate into two chemicals; one which wasn't at all radioactive, and one which was very much the opposite. He called the highly radioactive chemical thorium X. Excited by their new discovery, the pair wrote up their findings and then went home for a three week break.

When they returned to the lab, they discovered, much to their amazement, that the thorium X samples had ceased to be radioactive, and that the non-radioactive samples had returned to their initial level of radioactivity. The pair studied this phenomenon in detail and eventually determined that atoms of thorium in the non-radioactive sample were breaking down into atoms of the highly radioactive thorium X (which we now know as the element radium) and helium. The thorium X, in turn, breaks down into different elements, eventually leaving lead. Rutherford and Soddy borrowed a term from alchemy to describe how elements change into other elements: transmutation.

They published their findings in 1902 in a paper called "The Cause and Nature of Radioactivity." By describing transmutation for the first time, these two young scientists lifted the lid on the phenomenon of radioactive decay.

Anne-Claire Pawsey

Publication "The Cause and Nature of Radioactivity"

Date 1902

Authors Ernest Rutherford, Frederick Soddy

Nationality Canadian

Why It's Key For the first time, physicists understood the processes of radioactivity, and the methods by which some elements can turn into others.

Key Discovery ***T. rex***
A monster find

Barnum Brown was an eccentric paleontologist. Often to be found in his full length fur coat during digs, he is most famous for his role in the discovery of *Tyrannosaurus rex* fossils.

In 1900, Brown and his team unearthed a fossil in eastern Wyoming that consisted of teeth, jaw, and skull parts of the aforementioned beast. A colleague, Henry Osborn, who worked with Brown at the American Natural History Museum, named this find *Dynamosaurus imperiosus*. Two years later, Brown uncovered a more complete skeleton in the Hall Creek Formation in Montana. This earth-shattering discovery was of a two-legged carnivorous beast of truly terrifying proportions; Henry Osborn gave it the grand title of *Tyrannosaurus Rex*, or "tyrant lizard king." Osborn was in a hurry to publish his findings on both *T. rex* and *Dynamosaurus imperiosus*, due to competition with a fellow paleontologist, and in 1905, both beasts were described in the same journal. But after further analysis, he realized that the two species were likely to be the same creature, which was finally christened *T. rex*.

It is now thought that other individuals had found *T. rex* fossils, including teeth and vertebrae, as far back as 1874, but it is Brown who is generally credited for this monstrous discovery. *T. rex* still generates great excitement today and discoveries of specimens continue – Sue, a 12.8 meter-long *T. rex* emerged from a dig in South Dakota in 2000.

Katie Giles

Date 1902

Scientists Barnum Brown, Henry Osborn

Nationality American

Why It's Key A monstrous find in the world of paleontology, and still the kids' favorite dino.

Key Discovery **Hormones**
Getting on Pavlov's nerves

The English scientist Ernest Starling and his brother-in-law, William Bayliss, were two of the greatest physiologists of their generation. Starling is best remembered for his contributions to circulatory physiology; his understanding of the complex relationship between the heart and the smallest of the blood vessels is still the basis of contemporary circulatory theory. But the implications of one of the pair's lesser known discoveries are just as far-reaching.

Together, Starling and Bayliss discovered what causes the secretion of digestive compounds when food comes into contact with the small intestine. Their work not only identified what would later become known as the first "hormone," it also uncovered holes in Ivan Pavlov's Nobel Prize-winning research. Pavlov believed digestive secretions produced by the pancreas were controlled exclusively by the nervous system. But Starling and Bayliss isolated a chemical substance they called "secretin," which was able to trigger pancreatic secretion, even when all the surrounding nerves had been removed, and also when they injected it into an anesthetized dog. Although Pavlov had to concede he was wrong, he made no mention of "secretin" in his Nobel Prize-winning lecture in 1904.

Far from being a one trick pony, Bayliss went on to study the use of saline injections to counteract shock, a procedure said to have saved the lives of many wounded soldiers in World War I, and which earned him a knighthood in 1922.

Stuart M. Smith

Date 1902

Scientists Ernest Starling, William Bayliss

Nationality British

Why It's Key An early development in the study of endocrinology. Hormones, the body's chemical messengers, were now on the agenda.

Key Invention
Air conditioning

The world's first air conditioning unit was installed on the premises of a Boston publishing company as early as 1902. Its creator, engineering graduate, Willis Carrier, designed it to help ink dry in the hot, humid conditions of the printing house. Air conditioning systems soon began to show up in all sorts of glamorous locations, including Broadway and, in 1938, the White House.

Despite the endorsement of the rich and famous, it wasn't until the 1950s that the air conditioning business really took off. New domestic air conditioners encouraged families to stay indoors during the summer months, watching television and playing with other new toys and gadgets instead of enjoying the company of neighbors and friends in the humidity and sweltering temperatures outdoors.

By 2000 the Carrier Corporation had global sales of more than US$8 billion and employed some 45,000 people. Today, however, the specter of climate change is beginning to make major energy guzzlers such as air conditioning look less appealing.

Carrier's original system used ammonia, a poisonous gas and irritant, to cool air; he later exchanged ammonia for dialene. Whilst modern systems are much safer than the old ammonia-based units, however, they are by no means risk free. Bacteria living in air conditioning units can pose a real threat to human health if they are allowed to thrive. Contaminated air conditioners are in fact the primary cause of outbreaks of Legionnaires' disease, a severe and potentially fatal pneumonia-like disease.

Hayley Birch

Date 1902

Scientist Willis Carrier

Nationality American

Why It's Key Revolutionized the home and working conditions of millions of people worldwide.

opposite Willis Carrier and his air conditioner

Key Discovery **Paired chromosomes**
The carriers of heredity

Chromosomes – dense coils of DNA – were first observed in 1842 by Swiss botanist Karl von Nageli, although their genetic significance was initially overlooked. Similarly, Gregor Mendel's study into inheritance in 1866 had little impact at the time. These two concepts were finally united in 1902 by a biologist who suggested that paired chromosomes acted as units of inheritance, leading to a revolution in genetics.

Mendel had bred thousands of pea plants and studied the inheritance of their characteristics, proving that certain ratios of characters emerged on cross-breeding. This suggested that there was a discrete unit of inheritance rather than a "blending" of traits from the parent plants, as previously thought. Though his work was rediscovered in 1900, the physical unit of inheritance remained elusive.

In 1902, a young scientist called Walton Sutton was studying the cells of grasshoppers at his base at the University of Kansas in his hometown. He was examining a process called meiosis, which is involved in producing sperm and egg cells. The chromosomes appeared to occur in distinct pairs, which segregated into the different cells. The following year he concluded that paired chromosomes may be the physical unit of inheritance Mendel had suggested.

His theory remained controversial until 1915, when Thomas Morgan proved it with his work on the fruit fly, *Drosophila*. Sutton's chromosome theory of inheritance brought together cellular studies and genetics, allowing scientists to investigate the physical nature of how traits are passed on through generations.

Steve Robinson

Date 1902

Scientist Walton Sutton

Nationality American

Why It's Key Sutton's idea finally unified theoretical units of heredity with the physical cellular world.

Key Discovery
Dual-layer atmosphere

Our atmosphere is made up of a number of distinct layers, each with different temperature and pressure ranges. Yet just over a hundred years ago scientists still believed that the atmosphere was a single uniform layer. It was due to the diligent work of one Léon Philippe Teisserenc de Bort that a more accurate picture emerged.

He had been working in the Bureau Central Météorologique since shortly after its creation in 1878, but resigned from his position in 1892 in order to better carry out his experimental work.

Teisserenc de Bort pioneered many advanced techniques for gathering atmospheric data; constructing delicate, sophisticated instruments attached to large sounding balloons. These were launched from a specially constructed hanger, and with them he was able to gain altitudes which were previously unattainable.

Over an exhaustive two hundred launches, de Bort observed that temperature decreased steadily up to an altitude of eleven kilometers. Beyond that, however, the temperature stabilized. He concluded that there were at least two atmospheric layers, which he named the troposphere and stratosphere. A further three layers have since been added to the outer reaches of the atmosphere – the mesosphere, thermosphere, and exosphere – all out of reach of de Bort's balloons.

Barney Grenfell

Date 1902

Scientist Léon Philippe Teisserenc de Bort

Nationality French

Why It's Key Through de Bort's in-depth observation of the structure of the atmosphere we have gained a greater understanding of the ways weather systems develop.

opposite The lower three layers of the atmosphere (from bottom) are the troposphere (orange), the stratosphere, and mesosphere (both pale blue)

Key Discovery **The Verneuil process**
Manufacturing rubies

Auguste Victor Louis Verneuil was professor of applied chemistry at the Museum of Natural History in Paris. During his time there, he worked on his "flame fusion" process, later named the Verneuil Process. Although the work was completed by 1892, he didn't announce it until ten years later.

Rubies and sapphires are natural varieties of the mineral corundum, but very rare and expensive ones. The Verneuil Process produces corundum artificially. Since it is exceptionally hard, second only in strength to diamond, artificial corundum is used in many industrial applications as well as jewelry.

The process involves melting fine powders over a very hot oxyhydrogen flame; the temperatures needed are at least 2,000 degrees Celsius. Small amounts of oxides are added depending on what color you want your crystal to be – chromium oxide for red rubies, or ferric oxide for blue sapphires. As the powder melts, it falls onto a rod below where it re-crystallizes and forms the new gemstones.

Although modern methods are available, Verneuil's discovery is considered to be the founding step of modern industrial crystal growth technology and still remains widely used today. The gemstones produced are practically identical to the naturally occurring stones. However, minute gas bubbles that form on cooling, and curved growth lines (not the parallel lines in natural crystals) are the giveaway signs, but only experts can tell the difference.

David Hall

Date 1902

Scientist Auguste Verneuil

Nationality French

Why It's Key This new process allowed previously expensive rubies to be produced easily and cheaply for use in industry and jewelry.

Key Discovery **Anaphylaxis**
Once sensitized, twice shy

Today, millions of people are beset by allergies to things as diverse as milk, bee venom, and penicillin. But skip back a century ago and few people knew what allergies were or why they happened. In 1900, the man who would change all that, Charles Richet, was cruising aboard the yacht of Prince Albert I of Monaco. He had accepted an offer from the prince to study the venom of the Portuguese man-of-war jellyfish. Richet had seen Louis Pasteur vaccinate chickens against cholera with weakened bacteria and hoped to do the same with animal venom.

Back in Paris, Richet, together with Paul Portier, injected dogs with small doses of sea anemone venom, and re-injected them after they had recovered. To their horror, they found that, far from immunizing the dogs, they had made them hyper-sensitive to the poison – they died from a violent reaction out of all proportion to the small dose given.

Richet dubbed the reaction "anaphylaxis" and found that it could be triggered by a variety of poisons. The sensitivity needed some time to build up, but once it had, it was permanent and could be transferred between animals through injections.

He envisioned the build-up of a mystery substance that made the animals hyper-sensitive; today we know this as an antibody called immunoglobulin E (IgE). Richet's discovery has since saved many lives by leading the way to testing patients for potentially life-threatening allergies.

Ed Yong

Date 1902

Scientists Charles Richet, Paul Portier

Nationality French

Why It's Key Formed the basis for almost all later research on allergies, which has saved the lives of many hypersensitive people.

Key Person **Hermann Emil Fischer**
A man with natural chemistry

Hermann Emil Fischer (1852–1919) was a remarkable scientist who made numerous contributions to the world of chemistry over a century ago. Born in Germany, where his father, disappointed that he wasn't interested in joining the family lumber trade, sent him off to be a student of chemistry, believing his son too stupid to become a businessman. Fischer proved himself to be quite the opposite, however, by making major breakthroughs in many areas of chemistry.

One key achievement was his discovery of the chemical structures of different sugar molecules, and their interrelations. He also made important progress in the study of protein structure and function, discovering some new amino acids and studying the way that they join together chemically to make proteins. Even on holiday in the Black Forest, Fischer was thinking chemistry, analyzing the enzymes and other substances present in local lichens.

Fischer was a scientist of the highest calibre. With a natural understanding of scientific principles; he was conscientious and demanded thorough experimental proof of hypotheses. He ended his career as Chair of Chemistry at the University of Berlin, receiving many awards in his lifetime, including the 1902 Nobel Prize in Chemistry. Fischer's home life wasn't without its traumas though. His wife died after seven years together and of their three sons, two tragically died young. His surviving son, Hermann Otto Laurenz, continued the scientific tradition, becoming Professor of Biochemistry at Berkeley University, USA.

Katie Giles

Date 1902

Nationality German

Why He's Key Talented scientist who made great progress in many areas of organic chemistry.

Key Event ***Le Voyage dans la Lune***
The first science fiction film airs

Given a couple of seconds to think about it, most people will make the link between Edison and his long-lasting light bulb. That's what electrical genius, Thomas A. Edison, is best remembered for. Few would suspect him as the man who made a hit of the first science fiction movie. But in fact, it's one of Edison's lesser-known accomplishments.

Le Voyage dans la Lune – literal translation: *A Trip to the Moon* – was a silent movie shot in black and white by the French film maker, Georges Méliès. Edison didn't have anything to do with its production. But having financed some early forays into film projection at the end of the nineteenth century, he fancied himself as something of an industry guru. He had some technicians make copies of the movie – which were then distributed across the United States – before its French director had even set foot in America. The eight-minute wonder is widely considered to be the first film of the science fiction genre. *Le Voyage dans la Lune*'s visual effects were ahead of its time, but its representation of future space exploration will draw at least a twitch of a smile from today's viewer. While six intrepid astronauts prepare to propel themselves skywards using a giant cannon, a flock of hat-waving sailor girls flutter around in daring panty-sized shorts.

Despite the film's success, Edison's shrewd move left Méliès bankrupt – he never made a penny from any of the American showings.

Hayley Birch

Date 1902

Country France/USA

Why It's Key Spawned a genre that gave us some of our best-loved films – *Star Wars* and of course, *Flight of the Navigator.*

Key Publication **Chaos theory**
The answer's out there, but you can't predict it

Frenchman Henri Poincaré is often regarded as the "last universalist." He contributed in many areas of mathematics, theoretical physics, and philosophy of science, although he is most respected for his foundational ideas on chaos theory, which revolutionized the world of mathematics.

In a bizarre show of scientific enquiry, King Oscar II of Sweden offered a cash prize to the first person to show that the solar system was dynamically stable and obeyed Newtonian physics – the physics of regular, visible objects traveling at normal speeds. The challenge of this was to explain the motion of three bodies in the same system; the advance from two to three bodies had eluded scientists since the time of Newton himself. Poincaré entered and won the competition although he did not succeed in solving the problem. The work he published impressed the judges so much that they gave him the prize anyway. For the first time Poincaré found an instance of Newton's physics where bodies did not settle into patterns or find a definite endpoint.

He published three books between 1903 and 1908 showing how systems were very sensitive to initial conditions. Predicting how three objects will interact becomes impossible because small differences in the initial conditions produce big differences in the final motion. Concepts of chaos theory have been used in almost all areas of science and beyond including weather, the stock exchange, animal populations, planetary motion, and epileptic seizures. The three-body-problem was solved in 1912 by Karl Sundman.

Leila Sattary

Publication "Science and Method"

Date 1903

Author Henri Poincaré

Nationality French

Why It's Key A key theory describing how some systems are ultimately unpredictable since small changes can have a huge effect in the long term.

Key Invention
Electrocardiogram machine

The electrocardiogram, or ECG, is a machine that records the electrical impulses generated by the muscles of the heart with each contraction. In a healthy human heart, these impulses occur in a very orderly and precise fashion, so by measuring and mapping them, doctors can get a good idea of the overall functionality and health of the organ.

These days, ECG machines are as common in modern hospitals as beds, but this was not always the case. In fact, the first ECG machine was rather primitive, and may well have required the inhabitants of a nearby swamp to make it work. In 1838, the Italian Professor Carlo Matteucci proved for the first time that our every heartbeat is accompanied by an electrical impulse. His protoype ECG consisted of a frog's leg with some nerves still attached: the leg muscles contracted with every heart beat. However this apparatus had a limited shelf life and was soon replaced by an array of better alternatives. Problems still arose, however, because collecting the results was exceedingly time-consuming.

It wasn't until 1895, when Willem Einthoven invented the string galvanometer, which used a silver-coated glass filament to record pulses of electricity, that ECG became a useful clinical tool. Einthoven went on to develop the first practical electrocardiogram machine in 1903 and was subsequently awarded the Nobel Prize for Medicine for his invention in 1924. Early machines weighed around 270 kilograms and required five people to operate them; the same device today is the size of a laptop computer.

James McCallum

Date 1903

Scientist Willem Einthoven

Nationality Dutch

Why It's Key The electrocardiogram provides a quick, inexpensive, painless, and non-invasive way to evaluate the electrical function of the heart. It has become the backbone of modern cardiology.

opposite Willem Einthoven with his string galvanometer in his laboratory

Key Event **The Wright flight**
Planes take to the skies

Humans have always looked to the sky with wonder and dreamed of flight. Over the ages, innumerable inventors attempted to construct flying machines; all failed, until 1903. By the time Wilbur and Orville Wright began experimenting with flight, there was already a significant library of information available on what inventions would not fly. Carefully learning from the mistakes of others, the Wrights set out to make history.

Unlike other flight seekers, the Wrights tested components individually, trying to maximize their time and money to craft a flawless flyer. Even so, they had built several non-motorized planes before completing their historic 1903 model, the Wright Flyer I. By adding a small gasoline engine and a carefully constructed propeller to a previous model, they found success. On December 17, 1903, the brothers successfully completed four separate flights. The first, by Orville, traveled 37 meters in twelve seconds at a speed of just 11 kilometers per hour. By the fourth flight, however, the brothers were getting the hang of things and Wilbur flew for fifty-nine seconds, covering 260 meters. Shortly after this fourth and final flight of the day, a strong gust of wind caught the plane and flipped it over several times causing severe damage.It never flew again.

The following day, only four newspapers reported the Wright brothers' achievement, and the news was widely dismissed as exaggerated. In the following years, while the brothers did find acclaim, bitter legal disputes over their patents saw public opinion turn against them.

Logan Wright

Date 1903

Country America

Why It's Key Air travel and transport is commonplace today, bringing the world together like never before.

Key Person **Antoine Henri Becquerel**
Discovered radioactive rocks

Antoine Henri Becquerel (1852–1908) was born in Paris, to a renowned family of scientists. His grandfather had invented a method of extracting metal from ore using electrolysis, and his father was a professor of applied physics. By the age of forty, Henri had become the third of his family to be professor of physics at the Museum National d'Histoire Naturelle. He was fascinated by light and devoted himself to studying how it was absorbed by crystals, as well as effects such as polarization and phosphorescence.

In 1895, Wilhelm Röntgen discovered X-rays, and Becquerel decided to test for any links between this new form of light and natural phosphorescence. He decided to use uranium salts, as he had inherited some from his father, and exposed them to sunlight before putting them on unexposed photographic plates wrapped in opaque paper. Images of the uranium crystals were visible on the plates after he developed them and he realized that the uranium salts were emitting a radiation that could pass through opaque paper. His initial thoughts were that the salts were soaking up sunlight and re-emitting them as X-rays, fogging the plates.

Bequerel's next round of experiments were delayed by overcast skies, so he put his next set of salts back in the drawer on top of the photographic plate. When he developed these plates he was amazed to find they were fogged and that the salts did not need to absorb sunlight to radiate in this strange way. He had discovered natural radioactivity.

David Hawksett

Date 1903

Nationality French

Why He's Key The 1903 Nobel Prize winner in Physics, Becquerel was one of the main discoverers of radioactivity.

Key Event **Mount Wilson Observatory founded**
The birthplace of astrophysics

Toward the end of the nineteenth century, Harvard was promised US$50,000 to build the finest observatory in the world. A 101.6-centimeter telescope, the largest ever, was ordered and a team of scientists and a smaller, 33-centimeter telescope were installed on the summit of California's Mount Wilson during the winter of 1889. But all proceeded to be ruined; damaged by the harsh rain, snow, and tourists. The setback was too much for some and Harvard withdrew from Mount Wilson, abandoning the site and the planned telescope.

While Mount Wilson was an observatory without a telescope, George Ellery Hale was in possession of a 152.4-centimeter glass lens and no mounting. Hale applied for funding from the Carnegie Institute to build a new observatory, and – in spite of his misgivings about the site – set off for Mount Wilson in 1904, then only accessible by foot or cantankerous donkey.

From these seeds, the Mount Wilson Observatory was founded, equipped with living quarters, an ever-improving telescope, and an ever-growing reputation. In fact it was with the 254-centimeter Hooker telescope that Edwin Hubble made his groundbreaking observation that the Universe was expanding.

The observatory was a cradle for Hale's new science of astrophysics, developing the previous astronomy of simply observing the position of lights in the sky into a coherent model of the structure, motion, and interactions of the massive bodies in the Universe.

Kate Oliver

Date 1904

Country USA

Why It's Key A remote mountaintop, beset by tourists, donkeys, and difficulties with the weather, became the location for a totally new type of astronomy.

opposite The Hooker telescope at Mount Wilson observatory

Key Person **Heike Kamerlingh Onnes**
Cryogenically cool

Born in the Netherlands, Heike Kamerlingh Onnes (1853–1926) became one of the leading figures in experimental physics after founding a major cryogenics lab at the University of Leiden.

Originally set up as a center for the study of electrostatic Van der Waals forces between molecules, the cryogenics lab went on to make a number of revolutionary steps throughout the early 1900s. As one of the largest of its kind in the world, the laboratory also put Leiden on the map as a pivotal center in the field of low-temperature physics.

In 1908, Onnes became the first person to cool a gas – helium – to its liquid state. This is now regarded as having been the first step in the quest for absolute zero, the temperature at which a system has the lowest energy possible. Onnes was also a pioneer of the theory of superconductivity, in which a metal's resistance to current is now known to drop to zero at very low temperatures. Working on mercury at the Leiden laboratory, he provided a perfect demonstration of the technique, with research that was rewarded with a Nobel Prize in 1913.

Superconductivity is now used as a major component in the workings of magnetic resonance imaging (MRI) scanners which provide highly detailed images for medical diagnoses; mobile phone base stations; and nuclear magnetic resonance (NMR) machines. Onnes may have been based in Leiden, but the ground-breaking consequences of his life's work are now seen the world over.

Hannah Welham

Date 1904

Nationality Dutch

Why He's Key With the founding of his cryogenics lab in 1904, Onnes set out on a path to the supercool.

Key Person
Ivan Pavlov and his dribbling dogs

Ivan Petrovich Pavlov's (1849–1936) name has become almost synonymous with one particular animal. Although his experiments with salivating dogs are probably his best known, Pavlov's expertise spanned psychology, physiology, and neurology. Nonetheless, it was the dribbling dogs that won him the 1904 Nobel Prize in Medicine, the first time the prize had ever been awarded to a physiologist.

Pavlov's famous experiment involved collecting the saliva produced by dogs' salivary glands. When presented with food, a hungry dog produces large amounts of saliva, as most dog owners can testify to. Pavlov combined the presentation of food with an unrelated stimulus, like a flashing light or ringing bell, and found he could produce a reflex response that was conditional on previous experience; the dogs would begin to salivate simply at the sound of the bell, or flash of the light, even in the absence of food.

Although scientists were aware of unconditional reflexes such as instinctively removing your hand from a hot surface, or the contracting of your pupil in bright light, Pavlov showed a reflex could be developed based on experience – a conditional reflex response. The dogs were not making reasoned decisions to produce saliva because of an abstract event, like a ringing bell, they were responding unthinkingly. Scientists gained a new perspective on behavior, in animals and humans. We are aware that conditional responses can occur on a physiological level, as with the salivating dogs, but also an emotional level – an important concept for many areas of psychology, including phobia treatment.

Jim Bell

Date 1904

Nationality Russian

Why He's Key Bridged the gap between physiology and psychology, and influenced subsequent work in both subject areas.

opposite Ivan Pavlov and his staff demonstrate conditioned reflex phenomenon with a dog.

Key Person **William Ramsey**
A Nobel Prize for the noble gases

William Ramsey (1852–1916) was the first Briton to receive the Nobel Prize for Chemistry, just eight years after Alfred Nobel's death. It was well deserved, Ramsey's contribution to inorganic chemistry was the discovery of a whole new part of the Periodic Table, not merely one new element but, through his collaborations with William Strutt and Morris Travers, four new elements.

Ramsey was born in Glasgow in October 1852. He was schooled in his home city but moved to the University of Tubingen in Germany for his doctoral thesis. On returning to Scotland, he took up a place at Anderson's University for some years, before being appointed chemistry professor at University College Bristol. But it was during his next appointment, at University College London, that he was to perform his Nobel Prize-winning research on inert gases. Towards the end of the eighteenth century, Henry Cavendish had carried out experiments that caused him to suggest the existence of an unreactive, undetectable gas that made up a small part of the composition of air. Much later experiments performed by Lord Rayleigh confirmed this, by showing differences in the densities of nitrogen prepared from ammonia and air.

Ramsay was intrigued by these discoveries and, working together with Rayleigh, isolated and identified a new element, an inert gas, which was named argon – "the lazy one" in Greek. Ramsay went on to discover and name an additional three inert gases; helium, krypton, and neon.

Barney Grenfell

Date 1904

Nationality British

Why He's Key Ramsay's four new elements and their context in the periodic table represented a great advance in the field of inorganic chemistry.

Key People
The Wright brothers

Wilbur and Orville Wright are probably best known for their historic first flight in a motorized airplane, arguably the first ever to do so. But of no less importance, though less well known, is their contribution to aeronautics through their patented wing-design.

The Wrights started out as printers. Orville, with Wilbur's help, had designed and constructed a printing press, and it was into this that the young brothers initially invested much of their creative energy. From printing, they moved into the manufacture, sale, and repair of bicycles, opening the Wright Bicycle Exchange in the late 1800s. It wasn't until the very end of that century that the brothers began to show an interest in aeronautics. Despite being best known for their powered flight, the Wright brothers' early efforts were all with gliders. They performed countless experiments using non-powered craft, even constructing a wind tunnel to test their efforts. They believed that the key to mastering flight was the perfection of an effective control system; power could come later.

Their diligent experimentation, in the form of numerous manned and unmanned test flights, informed the development of their flying machines, and led them to the three axis-control system. This allowed the pilot to dictate lateral, up-and-down, and side-to-side movement, making manned flights much safer than they had previously been, and allowing for the development of more powerful motor driven craft.

Logan Wright

Date 1904

Nationality American

Why They're Key The three-axis control system, for which the Wright brothers were eventually granted patent number 821,393s, can still be found in use in airplane design today.

Key Invention **The telemobiloscope**
The radar nobody wanted

Traveling was pretty dangerous at the beginning of the twentieth century, particularly during fog or at night. Nowadays we rely on all manner of devices to protect ships and aircraft from collisions, but this was not always the case. Radar is one of the most significant of these devices, bouncing radio waves off objects to let you know they're there. When you think how important it is now, it's surprising that the first versions of it were ignored by industry and navies.

In April 1904, Christian Huelsmeyer applied for a patent in Germany for his telemobiloscope, which transmitted radio waves from a spark gap. These waves were beamed out of a reflector in the direction of the user's choice. If the waves hit a metallic object they would bounce back to the telemobiloscope's receiver, and the device would ring a bell until the object had moved out of the beam. There was, however, a potential difficulty; with a river full of ships beaming radio waves in every direction there was danger of jumbling up signals and receiving false warnings. This was addressed by time limiting the mechanism; the machine only responds to radio waves it receives within certain intervals.

The telemobiloscope could detect ships up to three kilometers away, but, despite the successful tests and good public opinion, the German navy and shipping lines were not interested. The result was a wait of thirty-one years, until the version of radar we use today was invented.

Douglas Kitson

Date 1904

Scientist Christian Huelsmeyer

Nationality German

Why It's Key A precursor to radar – a technology that prevents massive amounts of damage and danger.

Key Publication
Sunspot butterfly diagram

Sunspots are areas on the Sun's surface – or photosphere – that are cooler than their surroundings. In the early 1860s, English astronomer Richard Carrington published his "Observations of the Spots on the Sun from 1853 to 1861," reporting that sunspots migrate from the poles toward the equator as part of a cycle, a phenomenon called Spörer's Law, after German astronomer Gustav Spörer. In 1887, Spörer realized the number of sunspots between 1645 and 1715 was a thousand times smaller than normal. This period is now called the Maunder Minimum, after another astronomer Edward Maunder, who worked in the solar department of the Greenwich Royal Observatory beginning in 1873.

By 1890, Maunder had called attention to Spörer's discovery of "a very remarkable interruption of the ordinary course of the spot cycle." The Maunder Minimum coincided with the coldest part of the Little Ice Age, suggesting a connection with Earth's climate.

In 1904, Maunder and his wife Annie created a diagram showing the location of sunspots over the twenty-eight years between 1874 and 1902. As Maunder put it, "an examination of the diagram brings out Spörer's Law with remarkable clearness." We now know sunspots are caused by the Sun's magnetic field, but the physics behind sunspot migration is not fully understood. Meanwhile, solar physicists have been benefiting from the clarity of Maunder's butterfly diagram for over a century.

Eric Schulman

Publication "Observations of the Spots on the Sun from 1853 to 1861"

Date 1904

Author Edward Maunder

Nationality British

Why It's Key Illustrates the migration of sunspots over the 11-year sunspot cycle and highlights the importance of sunspots in dictating our climate.

Key Experiment

Every element has its own X-ray spectrum

Charles Glover Barkla was a British physicist who moved around almost as frequently as the particles in the gas he studied, working at Liverpool, Cambridge (under J.J. Thompson, discoverer of the electron), Liverpool again, King's College London, and finally Edinburgh.

At the time, the nature of X-rays was a mystery. Some thought they were a form of light, others believed they were a stream of particles, like the cathode ray. Barkla proved they behaved more like light, by polarizing X-rays scattered by gas. He also found that some gases produced a secondary, unpolarized type of radiation when hit with X-rays. The nature of these rays depended on the gas he used. Furthermore, he found that the penetrability of the secondary rays correlated with the gaseous element's position in the periodic table, and hence its atomic weight. We now know that the effect is more about the atom's structure than its weight. When X-rays pass through a gas, they collide with electrons, which are blasted out of the atom. As the remaining electrons settle down into a stable state, they shake off excess energy as X-rays of a longer wavelength than the original ray. Different elements have different electronic structures, so the penetrability of the emitted X-rays is characteristic of, and unique to, the gaseous element.

Thus, every element has its own X-ray spectrum. Today, the phenomenon is used to identify the constitution of unknown materials, and X-ray fluorescence spectrometers have even flown to Mars.

Matt Brown

Date 1904

Scientist Charles Glover Barkla

Nationality British

Why It's Key Barkla laid the groundwork for later scientists to pick apart the structure of the atom.

Key Invention

Vacuum tube

When is an invention not an invention? When it's a jar-shaped "vacuum tube" or "thermionic valve," of the type used by John Fleming to study practical uses of the "Edison effect" – the name of a phenomenon that Edison observed in 1883 whilst working on his new incandescent lamp.

In early electric lamps, carbon from the filament would become deposited on the inside of the glass, blocking the light. Edison added an extra electrode which he hoped would attract the carbon vapour and keep it off the glass. It didn't work, but by so doing he discovered that an electrical current can flow *from* the hot element to the extra (cold) electrode, but not the other way. Edison saw no specific commercial value in the effect, but, as was his wont, he patented it anyway. It was, in reality, the first vacuum tube in all but name.

This discovery excited Professor John Ambrose Fleming's interest. By 1890 he had determined that the "lamp" functioned as a rectifier – a means of converting alternating current (AC) to direct current (DC) – and had soon invented what we now call a diode, but Fleming called a valve because it forced electricity to flow exclusively in one direction.

In 1904 Fleming had a brainwave; he could use his valve to increase dramatically the sensitivity of radio receivers. Vacuum tubes finally had a use, and soon came of age with Lee de Forest's invention of the triode. Before long vacuum tubes were being used in everything from hearing aids to televisions.

Mike Davis

Date 1904

Scientist John Fleming

Nationality British

Why It's Key Rejuvenated a stalling radio/telegraphy industry and paved the way for numerous spin-off electronic components.

opposite Sir John Fleming with his thermionic valve, or diode

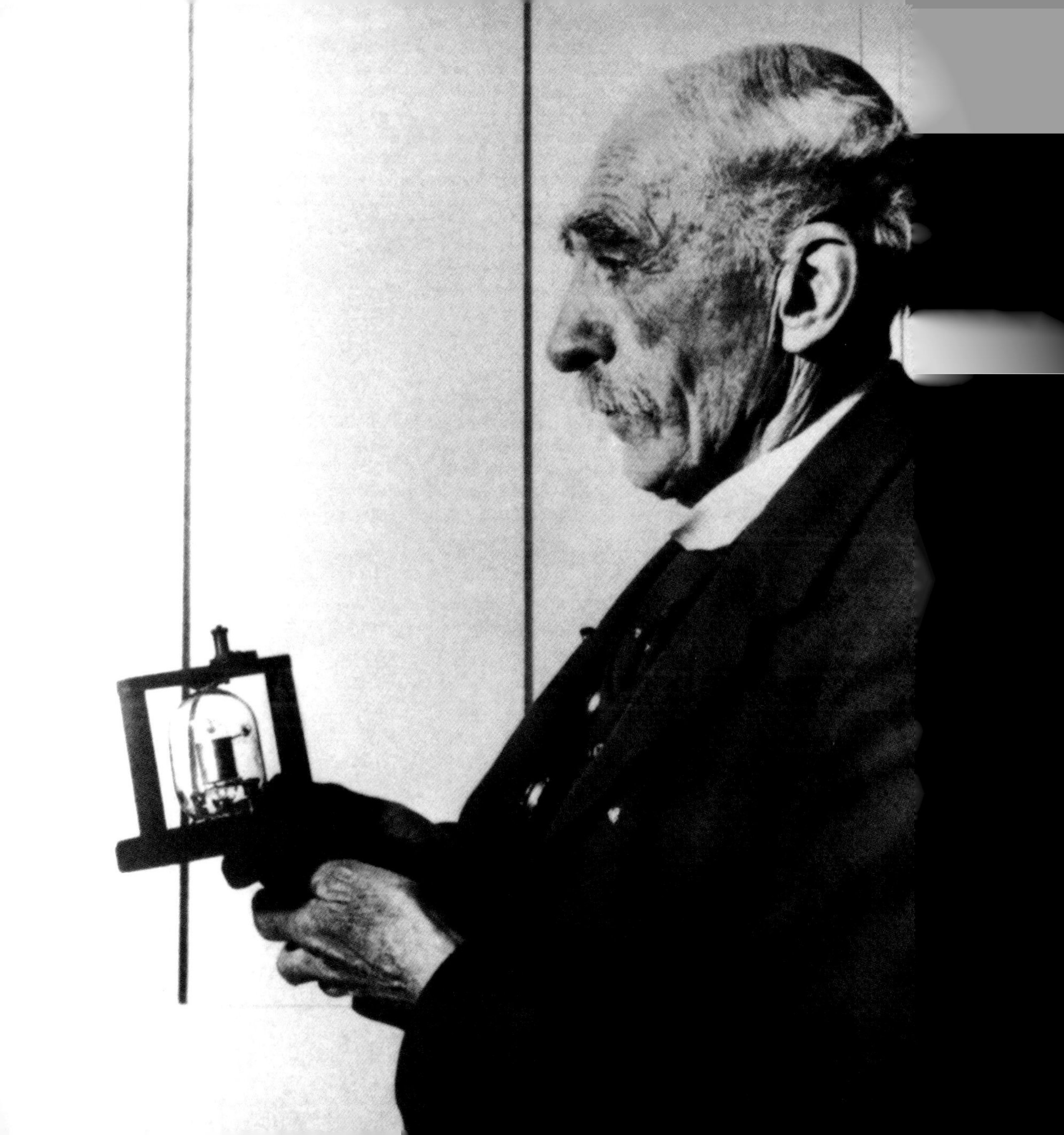

Key Discovery
Gender is determined by chromosomes

The early 1900s were exciting times for biology as scientists strived to understand the mechanisms of inheritance. By the turn of the century, the German biologist Theodor Boveri had proposed that chromosomes – discrete packages of DNA found in the nuclei of cells – were the agents of heredity. This idea was supported by Walter Sutton, who in 1902 demonstrated that chromosomes are found in pairs: one inherited from Mum and one from Dad. His discovery also fitted neatly with the model of inheritance proposed by Gregor Mendel.

Looking more closely at the chromosome pairs, it was apparent that one pair was different from the rest. Indeed, in some cases it wasn't a pair at all. In 1905 Edmund Beecher Wilson reported that, in two types of insect, females had one more chromosome (the X chromosome) than males. At the same time, biologist Nettie Stevens noticed that, for one pair of chromosomes, male mealworms had one small (Y) and one large (X) chromosome, while females had two X chromosomes. The two researchers had independently discovered that sex is determined by chromosomes. The system described by Stevens was later shown to operate in all mammals, further supporting the hypothesis that chromosomes are responsible for passing characteristics from parent to offspring.

These early experiments mark the emergence of the field of "genetics," a term coined in 1905 by the British biologist William Bateson to describe the study of heredity. The findings are still relevant today and form the foundations of modern genetic research.

Becky Poole

Date 1905

Scientists Edmund Beecher Wilson, Nettie Stevens

Nationality American

Why It's Key These early experiments form the basis of our current understandin of sex determination and th mechanisms of inheritance, giving rise to the field of genetics.

Key Event Einstein's miraculous year for physics
Annus Mirabilis

"I want to know how God created this world. I am not interested in this or that phenomenon, in the spectrum of this or that element. I want to know his thoughts; the rest are details." Albert Einstein was in his element in 1905. While still working at a patent office, he published five fundamental papers, all within a few months, which changed the face of physics forever. The two most famous papers led to the Theory of Special Relativity. In the first of these he stated that the speed of light was constant, and in the second he asserted the equivalence of energy and mass, leading to the most famous equation in science: E = mc2.

His other papers challenged the wave theory of light, stating that it should also be considered a stream of particles, or "quanta," marking an important step toward quantum mechanics. Einstein cited the photoelectric effect – the emission of electrons from matter which has absorbed electromagnetic radiation – as an example of a situation where light behaves like a particle and he won the 1921 Nobel Prize in Physics for this contribution. He also laid a mathematical basis for Brownian Motion – the random motion of tiny objects, such as pollen grains, suspended in gas – which supported atomic theory. To top it all off, Einstein showed how to calculate Avogadro's number (which allows chemists to calculate the number of molecules in a given mass of any substance) as well as the size of molecules. Not bad for a patent clerk, third class.

Although not acknowledged at the time, 1905 was so significant for science that 2005, its hundredth anniversary, was named International Year of Physics.

Leila Sattary

Date 1905

Country German-Swiss

Why It's Key A most important year for scientific discovery, and the making c Albert Einstein.

Key Invention **The birth of the IQ test**
Binet pioneers intelligence testing

The first bright spark to attempt to measure intelligence was a chap called Sir Francis Galton, who believed that a person's brain power was directly linked to attributes such as sight, hearing, and, strangely enough, breathing capacity. However, after a series of experiments in 1888 failed to reveal any correlation, his theory was swiftly abandoned as complete nonsense.

It would eventually take a self-taught French psychologist to establish the basis of a high IQ. In 1904, the French government called upon Alfred Binet to devise a test for school children that would identify those who were slow learners and in need of special education. He had been studying child psychology in his own home, setting tasks for his two young daughters and observing their intellectual function.

He used his valuable first-hand experience to design a thirty question test of increasing difficulty, assigning each one a rating based on the average abilities of children at different ages. With the help of his colleague, Théodore Simon, he tested a sample of average-ability children from local schools, using their scores to assign each one a mental age as a benchmark of their intelligence. The experiments led to the invention of the Binet-Simon intelligence scale in 1905 – the first modern intelligence test. This scale would later form the basis of the modern Intelligence Quotient (IQ) test.

Hannah Isom

Date 1905

Scientists Alfred Binet, Théodore Simon

Nationality French

Why It's Key The Binet-Simon intelligence scale became the first standard intelligence test for children, and provided the basis of the modern IQ test.

Key Publication
Wave-particle duality of light

Einstein was bothered by the interaction of light with atoms. Atoms were known to be tiny particles, and according to the most successful theory of the nineteenth century – Maxwell's theory of electromagnetism – light was definitely a wave. But how could a spread-out fuzzy patch of energy like light be created by a highly localized change: for example an atom's electron dropping down an energy level? Perhaps, Einstein suggested, light was made up of separate packets of energy.

Einstein's contemporaries thought he was trying to fix what wasn't broken. All previous measurements were consistent with a wave picture. But, said Einstein, those measurements are averages over a long period of time. Perhaps a time-average isn't appropriate for talking about things that happen in instants, like the emission of light from an atom. In those sorts of rapid-change circumstances, Einstein's model proved accurate. The idea was that packets or particles of light – photons – could knock electrons away from their atoms. This explained the photo-electric effect, where zinc under a certain frequency of light gains a positive charge; due to the loss of negatively charged electrons, the neutral zinc atom gains a positive charge. The fact that the light needed to be of a certain frequency to get this effect justified Einstein's statement that the energy of the photon corresponded to the frequency of the light – before a certain frequency, the photons didn't have the welly to kick out electrons.

Einstein's theory of the light particle was initially ignored, but it was for this he received the Nobel Prize.

Kate Oliver

Publication "On a Heuristic Viewpoint Concerning the Production and Transformation of Light"

Date 1905

Author Albert Einstein

Nationality German-Swiss

Why It's Key With this paper, Einstein unwittingly started the quantum revolution in physics, explained a few puzzling effects, and bagged himself a Nobel Prize.

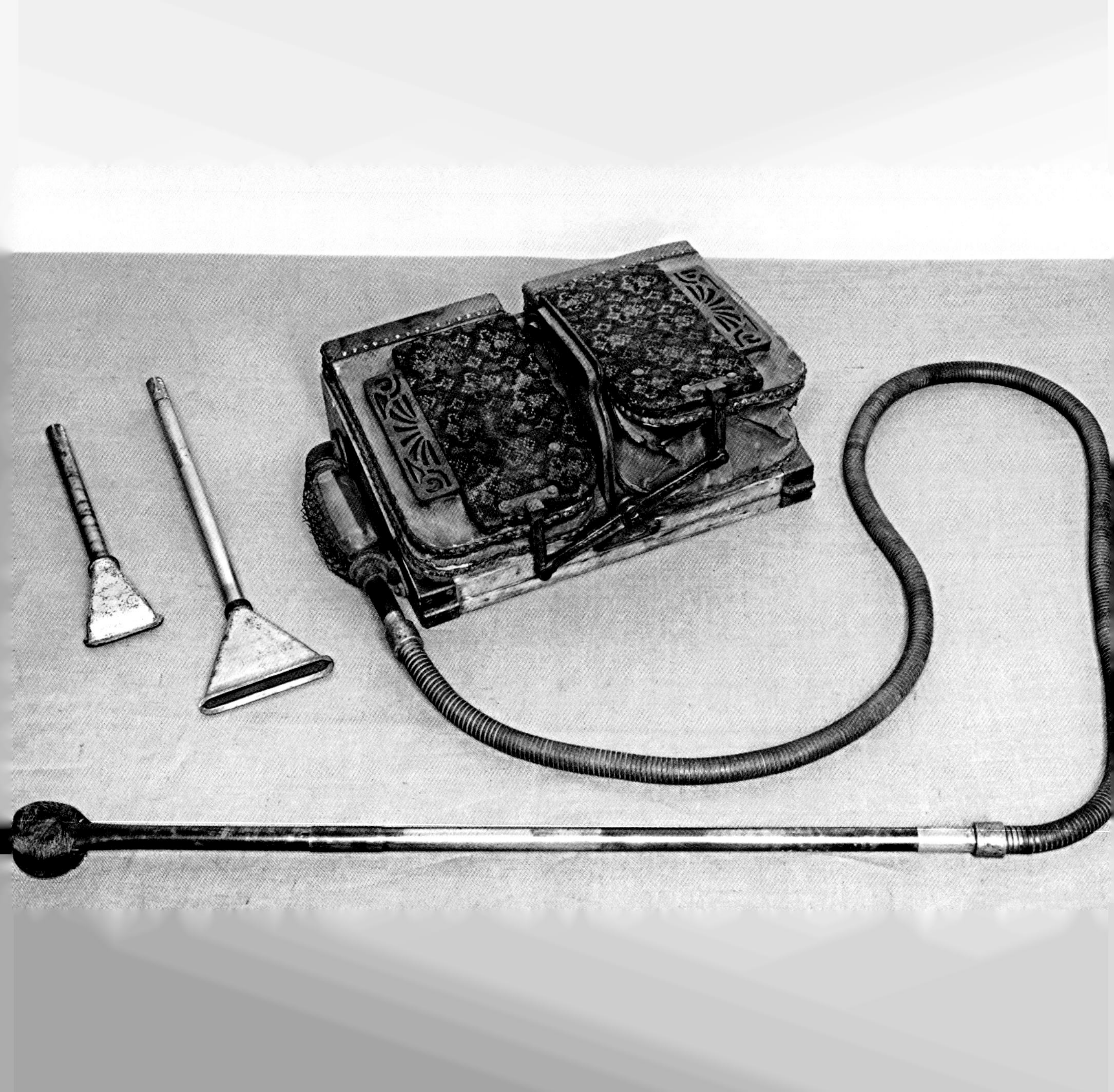

Key Invention **Pipe dream** The first home vacuum cleaner

During the 1800s, more and more homes began to install carpets as they became increasingly available and affordable. But with this new luxury came a problem: how to keep them clean.

Hubert Booth had invented "Puffing Billies" – petrol driven cleaners – in 1901, but it wasn't until 1905 that the first practical domestic vacuum cleaner starting blazing a dust-free trail across many a living room carpet.

Despite being manually powered, Walter Griffiths' new design, dubbed an "Improved Apparatus for Removing Dust from Carpets," laid down the foundations for modern-day versions of the vacuum cleaner. His prototype uniquely incorporated a pipe out of the body of the machine. This could be detached for cleaning and fitted with various head attachments to cater for different surfaces, as well as more hard-to-reach places. To suck up the dust, an operator would have to work something resembling a pair of bellows. Although this might seem a tiresome pursuit by today's standards, it made a vast improvement on previous models, some of which had required that cleaning machines be kept in the cellar, with pipes installed in each room. Crucially, Griffith's appliance was compact enough for it to be operated by just one person.

Unfortunately for Griffiths, however, his fortune was to be swept somewhat from under him by William Hoover, who went on to make millions from the vacuum industry.

Emma Norman

Date 1905

Scientist Walter Griffiths

Nationality British

Why It's Key One step on from the first powered model, Griffith's machine was the original pipe vacuum, and arguably the first mechanical cleaner of a practical size for consumers.

opposite Walter Griffiths' foot-operated vacuum cleaner

Key Invention **Norwegians nail nitrogen nutrients**

At the turn of the twentieth century, many countries, even in the Western world, had difficulty feeding their populations. Methods were needed to increase crop production, and mineral fertilizers were seen as the key. Like a vitamin pill for plants, adding nitrogen nutrients to the soil in the form of fertilizer is an obvious way to boost crop yields. But nitrates are hard to make since nitrogen is an unreactive gas, with its molecules held together by strong chemical bonds.

In 1905, Norwegian industrialist Kristian Birkeland and businessman Sam Eyde exploited a method to "fix" nitrogen from the atmosphere into a compound that could be used by plants. This first nitrogen fertilizer was produced in Nottoden, Norway, in 1905. This white, soluble salt was known as Norgessalpeter, the Norwegian name for calcium nitrate, $Ca(NO_3)_2$. In the Birkeland-Eyde process, nitrogen and oxygen from the air are fused into nitric oxide (NO) by passing a strong alternating current through a special type of furnace, known as an electric arc furnace. Oxidation and dissolution in water then produce the nitrate. The process requires significant amounts of electricity and cold water, and the pair had to commission a nearby hydroelectric power station for the task.

Once you can make nitrate, you can also make ammonium nitrate – a key component of ammunition. The technology was therefore in strong demand during World War I. Norwegian industry cornered the market for nitrogen compounds, and most of the world's calcium nitrate is still made in Porsgrunn in south-east Norway.

Matt Brown

Date 1905

Scientists Kristian Birkeland, Sam Eyde

Nationality Norwegian

Why It's Key Opened the door to modern agriculture, and eased food shortages in many parts of the world. Birkeland and Eyde's company, Norsk Hydro, is still producing nitrogen to this day.

Key Discovery
Friendly bacteria

Yoghurt, in its original form, could only be produced in Bulgaria and some isolated surrounding areas. It was a mystery as to why and unlocking the mystery opened new avenues of microbiological research.

Born in Bulgaria in 1878, Stamen Grigorov was well placed to solve this mystery. At the age of just twenty-seven, whilst working in an institute of microbiology in Geneva, he made his most important discovery – a certain strain of bacillus (rod-shaped) bacteria which we now know to be central to the production of natural yoghurt. It was a microbe which would only grow in the climatic conditions peculiar to the Balkan regions.

The bacterial strain, named *Lactobacillus bulgaricus*, produces the chemical acetaldehyde while it ferments, which helps to give yoghurt its characteristic smell. But it has a more important role than just supplying a pleasant niff. The bacterium feeds on the lactic acid found in milk. This not only helps to preserve the white stuff but also breaks down lactose into more simple sugars. People who suffer from lactose intolerance often find that such yoghurts can help their digestion.

Isolating the bacteria meant that, for the first time, yoghurt could be manufactured on a large scale, which was handy, since Grigorov had also found that yoghurt could help in the treatment of diseases such as tuberculosis, ulcers, and various gynaecological conditions. It was one of the first instances that scientists had realized bacteria has a beneficial role to play in maintaining health.

David Hall

Date 1905

Scientist Stamen Grigorov

Nationality Bulgarian

Why It's Key Bacteria aren't just dirty germs. And yoghurt can be made all over the world.

opposite Color-enhanced scanning electron micrograph (SEM) of *Lactobacillus bulgaricus* in yogurt

Key Publication Special Relativity
A new world order

Over two centuries before Albert Einstein was born, Sir Isaac Newton was formulating mathematical descriptions of the world around him. So-called Newtonian mechanics stood the test of time and were generally accepted until, in the early twentieth century, Einstein was unsatisfied with some of Newton's results. Einstein realized that while Newtonian mechanics could be applied to almost any system in the Universe, they failed when objects were traveling at speeds close to the speed of light. He set about deriving a theory which could describe objects traveling at all speeds.

"On the Electrodynamics of Moving Bodies" was the original name of Einstein's paper, published in September 1905. It was Einstein's colleague, Max Planck, who recommended using the term "relativity" since it described the principle of the paper; the principle being how the laws of physics change for observers moving relative to each other. Einstein later added the term "special" to his 1905 paper since the contents of the original document were just a special case of the broader general relativity.

Two months later, in November 1905, Einstein published a paper entitled "Does the inertia of a body depend upon its energy content?" In this work, Einstein suggested that as an object of a given mass (m) radiates some energy (E) and the mass of the object decreases by the amount E/c^2, where c^2 represents the speed of light squared. Following a simple rearrangement, Einstein reached his now world famous mass-energy equivalence formula, $E=mc^2$.

Gavin Hammond

Publication *"On the Electrodynamics of Moving Bodies"*

Date 1905

Scientist Albert Einstein

Nationality German-Swiss

Why It's Key Einstein's fundamental work is possibly the most important discovery of the twentieth century. $E=mc^2$ shows that mass can become energy, a feature that is exploited in the nuclear power industry.

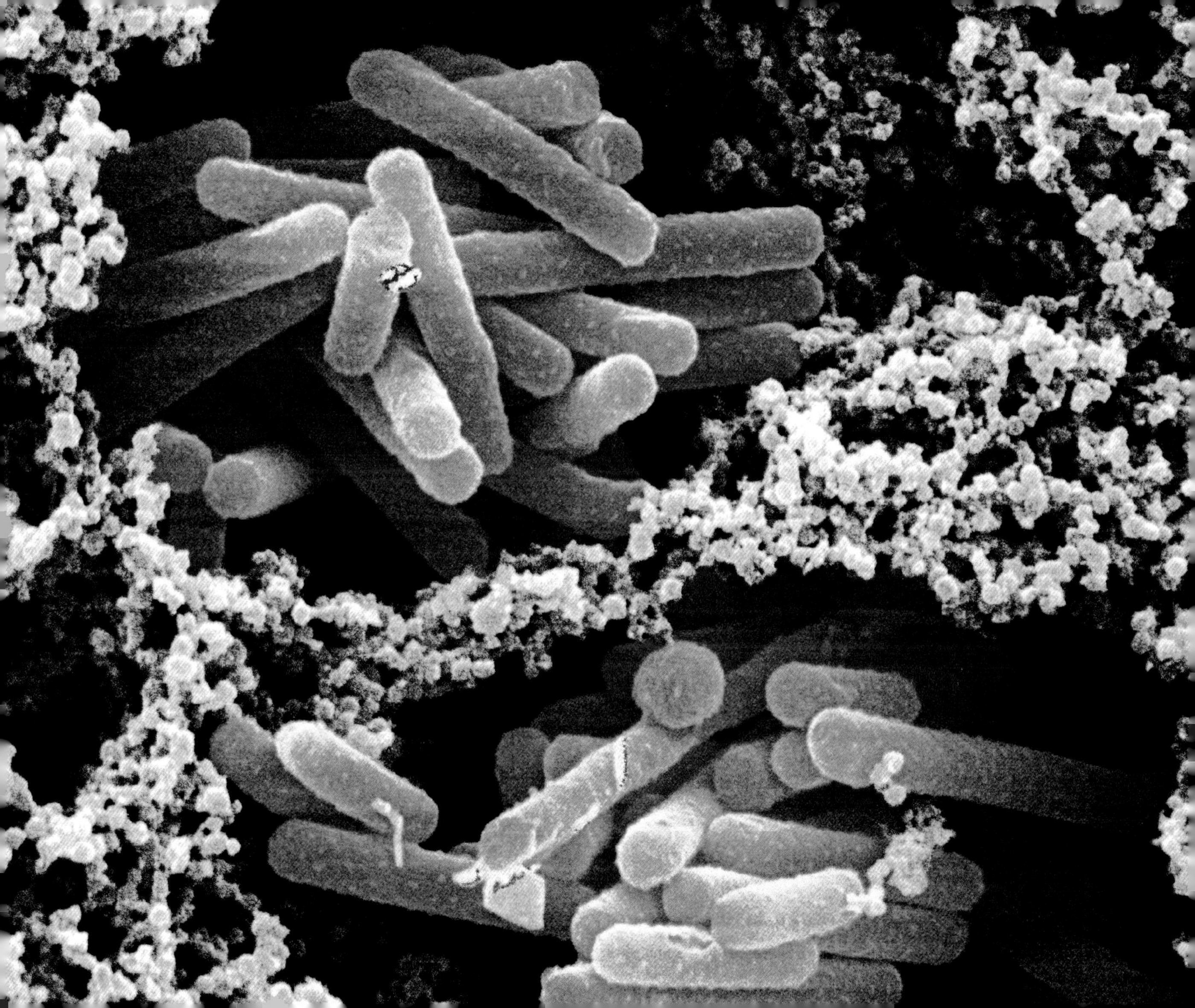

Key Publication **Atoms exist!**
Random movement and Brownian Motion

Brownian Motion, originally known as Brownian Movement, is the constant motion of particles in non-specific directions. This phenomenon was first noted by the British botanist Robert Brown, in 1827, as he observed the movement of pollen grains suspended in water. But the importance of his discovery was not truly recognized until further research confirmed that Brownian Motion was a clear demonstration of the existence of molecules in continuous motion.

In 1905, Albert Einstein proposed the mathematical laws governing the movement of particles. The theory of Brownian Motion attracted the attention of physicists, since it was thought to indirectly confirm the existence of atoms and molecules. With Brownian Motion now being treated as a practical mathematical model, the theory became more generally accepted, leading to more scientific theories and applications being developed.

Brownian Motion is used today in market analysis; the value of an item is often unrelated to past or future changes, a situation comparable to the movement of a particle. As Brown had observed, the direction of particle movement is unrelated to past or future directions, so the mathematical model of Brownian Motion can be used to develop a hypothesis on which to base decisions.

The distribution of aerosol particles can also be predicted using Einstein's mathematical model. Importantly this has helped to establish the possible environmental and physiological effects of airborne pollutants.

Sarah Watson

Publication "On the Motion of Small Particles Suspended in a Stationary Liquid"

Date 1905

Author Albert Einstein

Nationality German-Swiss

Why It's Key The first proof that atoms actually exist.

Key Event
Found: the biggest diamond in the world

Late one afternoon in 1905, Frederick Wells, the superintendent of the prolific Premier Mine in South Africa, was doing his daily rounds. His attention was captured by something shining very brightly in the corner of his eye. .

He was about four meters below ground, and had to sharply crane his neck to get a closer look at the sparkling object protruding from one of the walls. He scaled the wall and extracted what appeared to be a large piece of glass with his pocket knife. Tests later showed he had stumbled across the largest gem-quality diamond in the world. Weighing 3,106 carats it was named the Cullinan diamond after the owner of the mine. Wells was rewarded with US$10,000 and the stone was sold for US$800,000 to the Transvaal Government, who later presented it to British King Edward VII on his sixty-sixth birthday.

Edward sent the stone to the Asscher Brothers in Amsterdam for cutting; they had previously handled what was now the second largest diamond in the world, the Excelsior. They studied the Cullinan diamond for months before making the first cut. The diamond was split into three major parts and the largest of these was a pear-shaped stone weighing 530 carats. This was named Cullinan I and is the largest cut diamond in the world. Some claim that the Cullinan is only a fragment of the whole diamond and that the other part is still buried deep underground in South Africa.

Riaz Bhunnoo

Date 1905

Country South Africa

Why It's Key The Cullinan is the largest gem-quality diamond in the world. The largest fragment, the Cullinan I, now features in the head of the Royal Sceptre in the British Crown Jewels.

Key Discovery **Novocaine as an anesthetic**
Numbness without the hook

Imagine going to your dentist to get a tooth pulled and being asked to take some cocaine. It sounds like a crazy scenario but could still be a reality were it not for the work of German chemist, Alfred Einhorn.

In 1884, cocaine was the local anesthetic of choice and it proved to be much more effective than the previous option – a lump of ice. However, despite cocaine's ability to stop nerve cells from sending signals, its image was soon colored when scientists discovered how addictive it was. At the turn of the century, the race was on to find an alternative.

Scientists used the chemical structure of cocaine as a basis, and began tweaking it to try and design a related substance with all the benefits and none of the risks. The first attempts were too irritating to be useful, but in 1905 Alfred Einhorn finally hit upon a winner when he successfully synthesized procaine.

Even though its anesthetic effects were weaker than those of cocaine, procaine didn't cause euphoria, it wasn't addictive, and it acted quickly. Best of all, it was the first local anesthetic that could be injected. A surgeon named Heinrich Braun introduced it into medical use, where it was most commonly referred to by its trade name – Novocain. Today, the drug is mostly referred to as novocaine and is still used in dentistry and minor surgeries. It has formed the basis for even more effective anesthetics like lidocaine, which are less likely to produce allergic reactions.

Ed Yong

Date 1905

Scientist Alfred Einhorn

Nationality German

Why It's Key Allowed doctors to safely perform minor surgeries without the dangers of general anesthetic or the addictive properties of cocaine.

Key Person **Nettie Stevens**
Pioneer of X/Y genetics

Nettie Stevens (1861–1912) was one of the first female scientists to make a name for herself in the biological sciences. She was born in 1861 and raised in the small town of Cavendish, Vermont. Despite the grim educational opportunities for women at this time, Stevens excelled in mathematics and science. She attended a teachers' college and taught for a number of years before she decided to further her own education.

Stevens earned a Master's degree in biology in 1900, and moved to Bryn Mawr College in Philadelphia, where she studied under Thomas Hunt Morgan. After receiving her PhD, and a glowing recommendation from Morgan, in 1903, she moved to the Carnegie Institute, Pennsylvania, to perform original research on sex determination.

Initially working on mealworms, Stevens showed, in 1905, that when an egg was fertilized with a sperm containing an X chromosome, it became a female; with a Y chromosome sperm, it became a male. This simple explanation was the basis of sex determination.

Coincidentally, a former Bryn Mawr researcher, Edwin Wilson, published a very similar paper the same year, after Stevens, and was later awarded the Nobel Prize for his findings along with Stevens' former mentor, Thomas Hunt Morgan. Many felt that Stevens was robbed of the prize; however Wilson did continue to acknowledge her work throughout his career.

Stevens died from breast cancer in 1912, unable to occupy a professorship back at Bryn Mawr which had been created especially for her.

Rebecca Hernandez

Date 1905

Nationality American

Why She's Key Discovered the basis for one of the principal theories in biology.

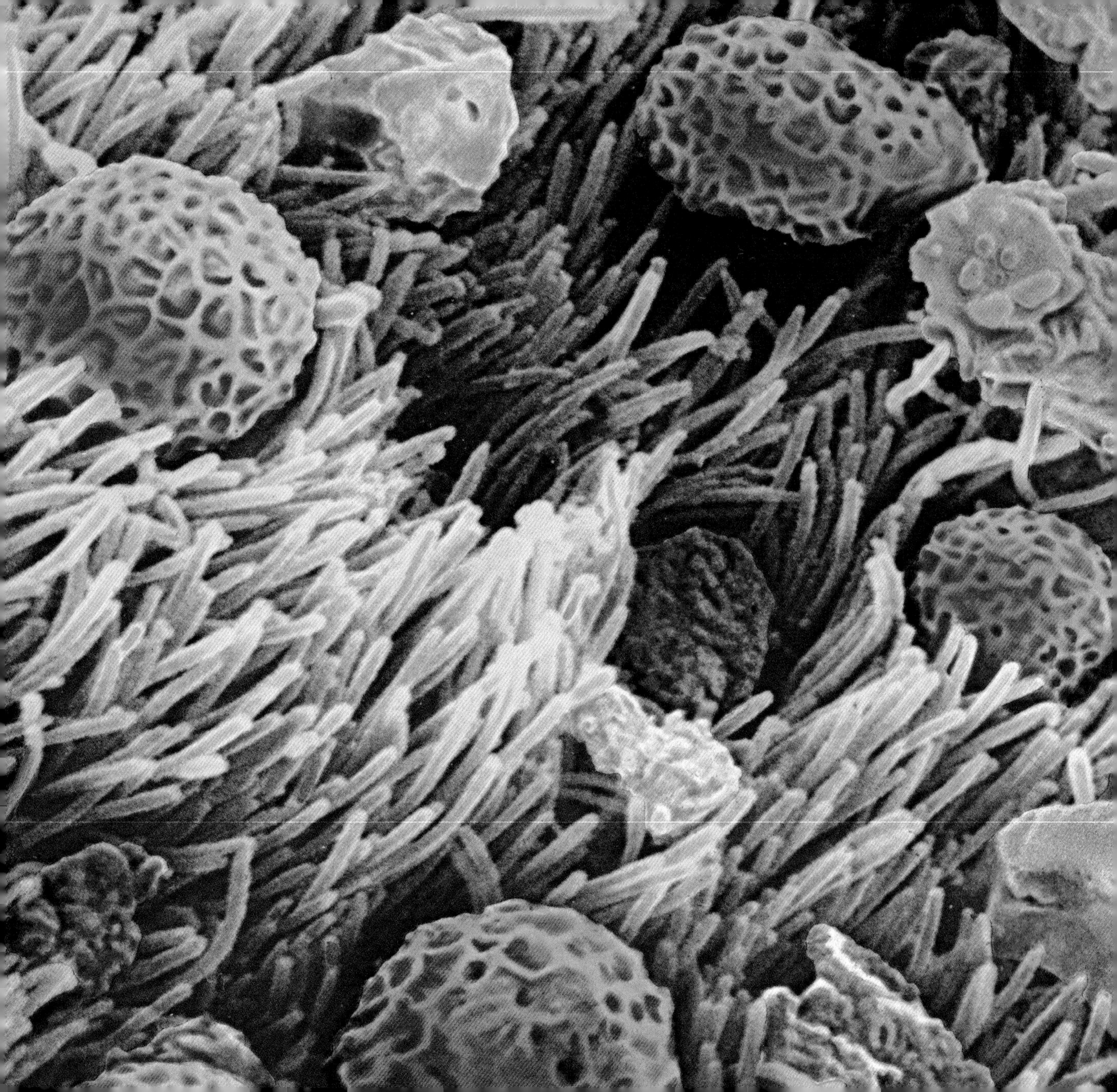

Key Invention
Electric washing machine

Before the invention of the electric washing machine, washing clothes was a chore to say the least. Victorian women stood over wooden wash tubs for hours on end, mashing up soap and washing and wringing their dirty laundry by hand and willpower alone. By the end of nineteenth century, some lucky Americans had tubs with hand-operated dolly sticks and gas-heated fires to provide hot water for the wash, but still the situation wasn't ideal.

Living at home in Chicago in the early 1900s, inventor Alva Fisher was set the task (by his mother) of making a useful appliance for the home. Targeting the family's ineffective washing machine, he started by attaching a simple electric motor to the washing tub. The old style wringer and dolly-sticks were then attached to the motor, and pumps fixed on to add and drain water. By hooking the whole device up to a hose, he found that a wash and drain could be completed using only electricity.

Initially, a few electrical hitches in the design led to some unexpected sparks and shocks in Fisher's kitchen, but by moving the water pumps to the side of the tub, away from the electric motor, he overcame this problem. Fisher's design was eventually sold to the Westinghouse Corporation, and the age of electrical domestic appliances was born.

Hannah Welham

Date 1906

Scientist Alva Fisher

Nationality American

Why It's Key It marked the advent of easy-use electrical appliances in the home.

Key Event
The term "allergy" is coined

At the turn of the twentieth century, scientists were only just starting to understand the complexities of the human immune system. In 1906, Clemens Peter Von Pirquet and Béla Schick came one step closer when they coined the term "allergy". The term was used to describe an exaggerated immune response, misdirected toward a harmless substance.

Pirquet and Schick came across their momentous discovery while studying serum sickness in children at their hospital in Vienna. The scientists found that if children were injected with serum – the clear portion of blood – from horses, they would become sick after around ten days. When injected a second time with the horse serum, the children would become sick much more rapidly, even within the same day. They named this phenomenon "allergy," from the Greek *allos* meaning changed or altered state, and *ergon* meaning work. In modern science, an allergy is caused by a hypersensitivity of the immune system to substances – or allergens – that should not cause disease under normal circumstances. Common allergens include food, drugs, pollen, pet dander, and insect stings.

The symptoms resulting from an allergic reaction can vary from mild swelling, itching, and rashes, to a potentially fatal reaction called anaphylaxis. Pirquet and Schick found that an allergic reaction couldn't occur unless the person had previously been exposed to the allergen. Upon this first exposure, B-cells (a type of white blood cell) produce antibodies, which remember the allergen and cause a more rapid response if it is encountered in the future.

Hannah Isom

Date 1906

Country Austria/USA

Why It's Key As well as being instrumental in advancing the field of general immunology, Pirquet's work eventually led to the development of the Mantoux test, used to this day to diagnose tuberculosis.

opposite SEM of the surface of the trachea with breathed in pollen and dust, airborne particles that may cause asthma or hay fever (allergic rhinitis).

Key Publication
The third law of thermodynamics

The laws of thermodynamics are not easy to understand. The big picture isn't too tough, but when you get down to how atoms behave with temperature changes, you need to start talking about entropy. Entropy tells you how chaotic the system is. At high entropy, everything is zooming around like a fish tank full of piranhas at feeding time; at low entropy the atoms are happy just where they are.

In 1906, Walther Nernst published a third law of thermodynamics to describe what happens to atoms around absolute zero, which is a chilly -273.15 degrees Celsius. As you pump heat out to cool something down, its entropy decreases, and the atoms stop whizzing around as much.

What Nernst realized is that as you get near absolute zero, the amount you can change the entropy by each time gets smaller as well. This means that at absolute zero your experimental item will have reached a minimum state of entropy – as low as it can go – but to get to zero, you will have to go through an infinite number of coolings, which you simply cannot do. Also, if you want to do anything with your coolest-of-the-cool item, like move it or poke it, you can't, because that would increase the entropy again.

In 1921, Nernst was given the Nobel Prize for progress in thermochemistry. His work has had profound effects on the worlds of electronics, magnets, and just about every experiment dealing with really low temperatures.

Douglas Kitson

Date 1906

Author Walther Nernst

Nationality German

Why It's Key Absolute zero is indeed very "cool." The third law of thermodynamics is highly important and fundamental to physics as we know it.

opposite Walter Nernst in his laboratory

Key Person
Santiago Ramón y Cajal and Camillo Golgi

At the turn of the twentieth century, the contents of the human brain were a mystery, despite the best efforts of phrenologists. Enter Santiago Ramón y Cajal (1852–1934) and Camillo Golgi (1843–1926), the forefathers of modern neuroscience.

Camillo Golgi was an Italian physician working in the pathology lab at the University of Pavia, when he became frustrated with the lack of a technique that would allow researchers to see nerve cells. After experimenting with a range of metals, Golgi found that silver could effectively stain individual nerve cells in a slice of brain tissue. This discovery meant he could start to map the fine structure of the brain, which he believed to be made up of a continuous web of interconnected cells. Ramón y Cajal also used Golgi's technique to investigate gray matter, but he came to a different conclusion. A fiery and pugnacious man – he apparently destroyed the gates of his home town with a cannon – Ramón y Cajal believed that the brain was made up of billions of separate nerve cells, and that they communicated through specific junctions, called synapses. Later research showed that Ramón y Cajal's idea was a more accurate reflection of reality than Golgi's. However, the recent discovery that electrical nerve impulses can actually jump between cells without using synapses suggests that there might have been some truth in Golgi's model.

Whatever their differences, both men were awarded the 1906 Nobel Prize in Physiology or Medicine, paving the way for modern neuroscience.

Kat Arney

Date 1906

Nationality Spanish, Italian

Why It's Key A silver stain revealed the structure of the brain and led to modern neuroscience.

742 667
11,8 23,7 45,3 56,7 71,9

Key Discovery
Evidence that the Earth has a fluid core

By the start of the twentieth century, the existence of a dense core inside the Earth had been deduced. The outer core of the Earth was discovered via seismological studies in 1906 by Irish-born geologist, Richard Dixon Oldham.

The destructive power of earthquakes has provided much of our knowledge about the Earth's internal structure. Earthquake waves traverse every part of the Earth's interior, and by studying how long these waves take to reach different points on the surface, it is possible to reconstruct their paths and speeds at specific points in the interior.

Oldham observed two distinct types of seismic wave; surface and body. Within the body group – which propagate deeply within the Earth – he discovered primary, or "P", waves, which create compressional movement; and secondary, or "S", waves, which produce a shearing motion in a direction perpendicular to the P wave. While P waves can pass through gaseous, liquid, or solid material, "S" waves can only penetrate solid matter.

During his survey of the 1897 Assam earthquake in India, Oldham observed abnormally long travel times of seismic waves. He correctly surmised that because they were being slowed down – or in the case of the "S" waves, absorbed completely – they must be encountering a central, apparently fluid core; the Earth's outer core is now commonly supposed to be made up of molten iron and nickel.

Mike Davis

Date 1906

Scientist Richard Dixon Oldham

Nationality Irish

Why It's Key Oldham's research helped other scientists map the Earth's various layers and build an accurate picture of our planet's construction.

Key Event
SOS adopted as international distress signal

Here's a question for you: What does SOS stand for? If you answered "save our souls," then, perhaps surprisingly, you would be wrong. It's not as easy a question as it first seems.

Most people can tell you that SOS is the international Morse code distress signal. To be completely accurate, however, the code is not really "SOS". Although . . . or "dot dot dot" signifies an S, and – – – or "dash dash dash" signifies an O, there is a gap between the letters when spelling a word in Morse code. The signal . . . – – – . . . – – – . . . – – – . . . , and so on, is actually a procedural code. The phrases "save our souls," "survivors on shore," and so on, are actually backronyms: that is, they became associated with SOS AFTER it became the signal indicating an emergency situation. Before the widespread use of radios and the introduction of "mayday, mayday, mayday" as the distress call when lives are in danger, ships communicated by Morse code. Commonplace from the 1890s until the late 1920s, although more recently it has been superseded by satellite technology.

Until 1906, there was no international consensus on emergency code, although the first distress call was announced by the Marconi Company a few years previously as "CQD." "CQ" was an existing land telegraph signal for a message of interest to all radios, and "D" stood for distress. Despite the international agreement in 1906, CQD was used for years afterwards. The *Titanic* is reported to have transmitted both SOS and CQD after it struck that fateful iceberg in 1912.

Jim Bell

Date 1906

Country International

Why It's Key SOS is still widely recognized today as an international code for distress. It can also be used in a variety of forms, both visual and acoustic.

Key Person **Alfred Wegener**
The father of continental drift

Alfred Wegener (1880–1930), famous for his once controversial theory of continental drift, was fascinated by weather. After completing his PhD in astronomy, Wegener began experimenting with wind and was soon swept away by his interest in meteorology. He set a world record in an international balloon contest in 1906 – by flying for fifty-two hours straight – and was inspired to find a way to use weather balloons to track air masses. His interest in the weather took him on numerous expeditions to Greenland. On one of his trips, in 1912, he studied polar air circulation, which eventually led him to make the longest ever crossing of the ice cap on foot.

It wasn't until 1910 that Wegener began formulating his ideas on continental drift. After stumbling upon some data in the University of Marberg library on matching fossils in widely separated continents, the idea that the continents used to be a single land mass never left him. Though he served in World War I and was wounded twice, he published his findings on continental drift in 1915.

Wegener was to meet an unfortunate end on his fourth expedition to Greenland in 1930. In a severe blizzard, his team got separated. His body was not found until six months later and the cause of death was registered as heart failure due to over-exertion. After Wegener's death, support for his theory of continental drift grew. In 1960, Hess proposed the mechanism of sea-floor spreading to explain the movement of continents, and Wegener's theory is now generally accepted as fact.

Faith Smith

Date 1906

Nationality German

Why He's Key His theories provided evidence for the notion that the continents were once all joined in one single land mass.

Key Event
First successful human blood transfusion

One of the great medical advancements of the last century was the discovery of blood types, and the subsequent development of transfusion medicine. While at the turn of the century it was already known that mixing human blood could result in the cells clumping together, and ultimately being destroyed, the mechanism behind this phenomenon remained a mystery.

In 1901, Karl Landsteiner demonstrated that there were certain molecules on the surface of red blood cells, called antigens, which caused an immune response if introduced in to the body of a person who did not have the antibodies to match. Further experimentation showed that there were two common antigens in this system – A and B – which yield four common blood types: Type A (A antibodies), Type B (B Antibodies), Type AB (A and B antibodies), and Type O (neither A nor B antibodies).

The first successful blood transfusion occurred soon after Landsteiner's discovery, and was actually, in effect, minor surgery. American surgeon George Washington Crile took patients to the operating room to expose blood vessels in both the recipient and the donor. A short tube was then used to link the two and allow blood to flow from the artery of the donor to the vein of the recipient. Subsequent advances have allowed for the procedure to be performed away from operating theatres, by simply using syringes and a bit of pipe.

James McCallum

Date 1907

Country USA

Why It's Key Blood transfusions save countless lives every day.

Key Invention **The fax machine**
Letters by phone

In the digital age of the early twenty-first century, sending text and images to the other side of the world is so easy that we take it for granted; it is simply a matter of writing an email, attaching an image and clicking "send."

It hasn't always been this easy. Just over a hundred years ago, another innovation in telecommunications was in development. It was overlooked for the majority of the twentieth century, but is still used today: the fax machine.

Fax (short for facsimile) has actually existed in a rudimentary form since the mid-1800s. Scottish inventor Alexander Bain patented a device that could scan the etched surface of a metal "document" and transmit the information over a telegraph wire. Further developments of Bain's original "contact" system were numerous throughout the second half of the nineteenth century.

In the early 1900s, inventors Arthur Korn and Édouard Belin independently demonstrated the first photoelectric fax systems. Korn discovered he could convert the various gray tones on a photograph into electric signals by using the light-sensitive element selenium. And by 1907, both men had sent inter-city faxes across Europe.

It took almost another seventy years of slow progress before the technology was made small and sleek enough to be commercially attractive. The 1980s, however, saw a massive increase in popularity as Korn and Belin's invention finally bore fruit.

Barney Grenfell

Date 1907

Scientists Arthur Korn, Édouard Belin

Nationality German, French

Why It's Key Expanded the boundaries of communication, allowing images and text to be copied and sent instantaneously across large distances.

opposite Édouard Belin with his telephotograph machine, an early form of fax machine.

Key Discovery **Amino acids**
Forming new bonds

Hermann Emil Fischer was born into a wealthy family and his father was keen for him to join the family's lumber business. Luckily for the science world , he followed his heart and instead studied chemistry, where he had a profound effect on the field and contributed hugely to our understanding of acid bases.

Fischer began to look at the structures of amino acids in the late 1800s, but at the start of the twentieth century he became increasingly interested in the bonds between the acids – peptide bonds. He worked on separating amino acids into groups and, in the process, discovered new types – the cyclic amino acids, proline and oxyproline. He also attempted to fuse the acids and in doing so discovered the peptide bonds which joined them together. We now know that each amino acid is book-ended on the one side by two oxygen molecules and a hydrogen molecule, and on the other, by two hydrogen molecules and a nitrogen molecule. As amino acids bond to form protein chains, these opposing sides meet, releasing hydrogen and oxygen in the form of water, hence the reason why this reaction is dubbed "condensation."

Fischer's work also revealed the existence of dipeptides, and later tripeptides and polypeptides. His studies had a profound effect on the world of biochemistry as they showed the huge complexity of amino acids and the immense amount of variation available in protein chains. As key components of proteins, amino acids are essential building blocks for all forms of life.

Katherine Ball

Date 1907

Scientist Hermann Emil Fischer

Nationality German

Why It's Key Showed how amino acids formed chains using peptide bonds, allowing the future synthesis of proteins.

Key Publication **What if gravity and acceleration are the same thing?**

We see gravity every day; we might even notice that light things and heavy things fall at the same rate. But in physics, we discover nothing to explain why this is the case. We can predict how strong a gravitational field will be using Newton's equations, but the nature of gravity remains mysterious.

Einstein suggested that the experience of seeing everything accelerate at an equal rate – as we do with gravity – can be explained in an alternative way: by saying that you are accelerating instead. All the laws of nature we have found in a world with gravity should hold true if we translate them into a world where there is just constant acceleration in place of it. Let's try it.

When a series of waves accelerates toward us, its peaks appear closer together; and when it's accelerating away, the peaks spread out. When this happens for a wave of light, its color shifts toward the blues when the wavelengths squash up (blue-shifting) and towards the reds when they stretch away (red-shifting). If gravitation and acceleration are equivalent, a light beam escaping from a gravitational field will also be being pulled backwards by it, moving against the acceleration and having its peaks stretched apart. So light that has escaped from a gravitational field should be red-shifted compared to the same light emitted in empty space.

This thought experiment produced a testable prediction for Einstein's theory and explained what is now recognized as gravitational redshift – an important astronomical measurement.

Kate Oliver

Publication "On the Relativity Principle and the Conclusions Drawn from it"

Date 1907

Nationality German-Swiss

Author Albert Einstein

Why It's Key A thought-experiment, taken to its limits, produces a testable prediction that would provide evidence for Einstein's theories. The idea that acceleration and gravity are ultimately the same explains gravitational redshift.

Key Discovery **Radiometric dating** Earth is 2.2 billion years old

"How old is the Earth?" was a question that had been bothering the scientific community for years. By 1907, all manner of ages had been suggested in answer, ranging from around 4,000 to a few million years. It was a Yale professor, Bertram Boltwood, who finally proposed a more realistic birth date for Earth.

Boltwood had a keen interest in the burgeoning science of radioactivity, having attended lectures on the subject by Ernest Rutherford. He carried out pioneering work identifying the shared properties of various chemical elements – work that would later give rise to the discovery of isotopes. The discovery for which Boltwood is best known, however, is scientifically dating the Earth.

It had already been established that radioactive elements break down into other elements; Boltwood showed that uranium is broken down finally into lead. His stroke of genius was in devising a method for dating rock samples by measuring the proportion of uranium to lead in them.

Using his new technique, Boltwood dated a number of samples, obtaining results that showed an age of between 535 million and 2.2 billion years old. His findings suggested the Earth was considerably older than anyone had previously imagined. More recent readings with improved equipment and older rock samples have dated the earth at around 4.5 billion years old, the figure commonly accepted today.

Barney Grenfell

Date 1907

Scientist Bertram Boltwood

Nationality American

Why It's Key For the first time an accurate method of scientifically measuring the Earth's age could be put forward.

Key Publication
Is Mars habitable?

It might never have crossed your mind, but there's a good reason why fictional little green men always seem to have come from Mars and not Jupiter, Saturn, or Venus. In 1877 the Italian astronomer Giovanni Schiaparelli reported, in Italian, that he had spotted a network of "canali" – channels – on the surface of Mars. One overenthusiastic mistranslation later, however, and Schiaparelli was reported to have claimed that there were canals on Mars.

One man who was particularly intrigued by this was Percival Lowell, who is also now remembered for correctly predicting the existence of Pluto by using the wrong data. Lowell set out to show that the presence of the canals proved intelligent life lived on the red planet. He published two books presenting an extraordinary amount of "knowledge" about the workings of Martian society. For many in the science world, this was too much. One man who took umbrage at Lowell's claims was Alfred Russel Wallace, the man (now largely forgotten) who also devised the theory of evolution by natural selection independently of Charles Darwin. In what is now increasingly recognized as one of the first truly scientific studies of the habitability of another planet – in essence kick-starting the field of exobiology – Wallace systematically demolished Lowell's claims in his book *Is Mars Habitable?*. The rarefied atmosphere, freezing temperatures, and aridity of Mars could not support intelligent life, he argued.

Whilst Lowell's theories received short shrift from the scientific fraternity, they were pounced on with glee by science fiction authors – Martians were here to stay.

Mark Steer

Publication *Is Mars Inhabitable?*

Date 1907

Author Alfred Russel Wallace

Nationality British

Why It's Key The Martian myth is born, as is the field of exobiology.

Key Experiment **Of sponges and sieves**
Sociable cells

The human body contains about 50 trillion cells, all merrily carrying out their own special functions and making sure that we can eat, sleep, reproduce, and watch TV whenever a suitable occasion arises. 1.5 billion years ago, however, the world was inhabited solely by single-celled organisms. The question of how multicellular organisms evolved was one of the first challenges for Darwinists. And the first answers came from an experiment involving a sponge and sieve.

Sponges are one of the simplest types of animal. They have no true tissues – no muscles, nerves or organs – and only eight different types of cell (humans have around 210). In 1907, biologist Henry Wilson separated all the cells in a sponge's body by shoving it through a fine silk sieve. Having popped the resulting sponge purée back into water Wilson watched through his microscope as each of the cells started to act independently, as though each was an individual organism in its own right. When the free-swimming cells encountered each other, however, they joined forces, forming a colony. Over the course of the experiment the cells all agglomerated to build an entirely new sponge.

Here was evidence, not only that cells were the building blocks of all living things, but of how multicellular organisms evolved: colonies of single-celled organisms gradually becoming more and more dependent upon each other. While a sponge's cells can live without each other for a while, our cells, for instance, find it a much harder task.

Mark Steer

Date 1907

Scientist Henry Van Peters Wilson

Nationality American

Why it's Key Showed that cells are the fundamental building blocks of bodies and hinted at how multicellular organisms might have evolved.

Key Event **A haze of color**
The first color photography process

Marketed by the Lumière brothers in 1907, the autochrome lumière was the first ever practical color photography process. As a result, images could easily be captured in all their vibrant glory without the artificial aid of an artist. The brothers used the method of additive color synthesis – creating color by adding proportions of the primary colors of light – to create photographic images.

The first autochromes were glass plates covered in a tacky substance, and dusted with a mixture of dyed red, green, and blue starch grains. They were treated with carbon black (soot) to fill in the gaps between irregular shaped grains, and coated in a silver halide emulsion. Once exposed, the images were viewed with a special device called a diascope. This would allow light to pass through the autochrome, and the image was projected onto a mirror. The resulting images had a wide tonal gradation, but were characteristically hazy with stray colors.

Autochromes dominated the color photography market until the 1930s, when film based methods were introduced. A few years later they were almost completely replaced with the hi-tech, multi-layer, subtractive color film from Kodak.

Despite this, love for the dream-like effect of autochrome photography lives on. In 2006, the film, *The Illusionist*, which sought to recreate the visual qualities of the technique, was nominated for an Oscar.

Faith Smith

Date 1907

Country France

Why It's Key The earliest color photography process, bringing pictures to life for the first time.

opposite Auguste and Louis Jean Lumière in their laboratory at Lyon, France

Key Invention **Triode thermionic amplifiers**
Turning up the volume

American inventor and scientist Lee De Forest received more than 180 US patents in his lifetime, but by far his most important was for the triode.

Diodes had previously been invented by British engineer John Ambrose Fleming. In 1906, however, De Forest made his own version, the Audion – a diode tube which maintained an electromagnetic current between two electrodes; a positive anode and a negative cathode. He soon improved upon this and added a third electrode – a grid – in between the cathode filament and anode plate.

This three element tube – which contained a partial vacuum – outperformed the simpler diode. Although Fleming's diode device was able to convert radio waves to an electrical signal, De Forest's Audion was also an amplifier. This new function was perhaps the most significant development in radio since the experimental work of Marconi and Tesla, and, in 1907, he received a patent for the Audion as a detector of radio signals, an audio amplifier, and an oscillator for transmitting.

Like many inventors at the time, De Forest spent years and a fortune in court trying to protect his inventions. But his Audion triode amplifier opened up the possibilities of wireless telephony, and allowed the clear broadcast of voice, music, and any other audio signal for the first time. Triodes remained a major part of all electronics until the invention of the transistor in 1947.

David Hawksett

Date 1907

Scientist Lee De Forest

Nationality American

Why It's Key One of the most important advances in electronics in the twentieth century.

Key Person
Albert Michelson

He carried out an experiment that would change the way that we looked at the Universe and provided the springboard for Einstein to develop the theory of special relativity, so it's a shame then that Albert Michelson (1852–1931) always felt his single most important contribution to science was a failure.

Having grown up in the rough mining towns of the Southwest United States, Michelson entered the U.S. Naval Academy at the age of seventeen and soon showed a great propensity for physics, if not for seamanship. At that time, scientific wisdom held that the Earth was floating in a mysterious substance called "aether" as it orbited the Sun. Michelson set out to prove this was true. In a series of experiments, which culminated in his collaboration with chemist Edward Morley, he hoped to confirm aether's existence by measuring how much the speed of light was affected by traveling through it. Unfortunately for him, all that the now famous Michelson-Morley experiment showed in 1887 was that aether didn't exist and that the speed of light was a constant. These were the results Einstein would need to produce the theory of special relativity; Michelson, however, would remain miffed about the non-existence of aether.

But it wasn't just for carrying out a "failed" experiment that Michelson became the first American to win a Nobel Prize in 1907. His enduring interest in measuring the speed of light bagged him the honor, and by his death in 1933 he had refined his estimates to within 2 kilometers per hour of the real total value, 299,792 kilometers per hour.

Mark Steer

Date 1907

Nationality American

Why He's Key A master of optics, Michelson carried out one of the most important experiments in the history of physics, made extraordinarily precise measurements of the speed of light, and was the first man to measure a star.

Key Invention **Detergent**
Good science comes out in the wash

Before the advent of modern washing powders, to get your laundry "whiter than white" you'd have to scrub it with soap before leaving it outdoors to bleach in the sun. But, with the invention of Persil, clothes got the double-whammy of detergent and bleach in one product.

The breakthrough came in 1907, when the company Henkel & Cie, of Düsseldorf, Germany, announced the first "self-acting" washing powder. The revolutionary product mixed a soap detergent with the bleaching agent sodium perborate and a sodium silicate. Abbreviating the names of these latter ingredients gives "Persil," first advertised on June 6, of that year.

Molecules of perborate contain a special attachment of oxygen, known as a peroxygen bond. This makes the compound stable when in a powder form. When added to water, however, the perborate breaks down to a peroxide – a powerful oxidizing bleach, strong enough to turn towels and tea cloths a whiter white than even the most diligent housewife could ever have achieved. Coupled with silicates to soften the water, and detergent powder, Persil had the power to dominate the market for several years.

The formulation may have changed, and modern fabrics are made from materials undreamt of a hundred years ago, but Persil remains a leading brand of washing powder to this day. Coincidentally, the world's first electric washing machine was also released in the same year.

Matt Brown

Date 1907

Scientists Henkel & Cie

Nationality German

Why It's Key Whiter whites for less hard graft.

Key Discovery **The Haber process** Salvation from starvation

The prediction of worldwide food shortages thanks to population growth caused German chemist Fritz Haber to return to his notebooks in the early part of the century. He knew that a new source of nitrogen for fertilizers was needed to improve crop yields. At the time, the only source of nitrogen was from Chilean saltpeter (naturally occurring deposits of sodium nitrate), and this wouldn't last forever.

Haber had already conducted experiments in which he had produced small quantities of ammonia by reacting nitrogen and hydrogen with each other. Nitrogen gas makes up 78 per cent of the atmosphere and hydrogen can be easily produced from water, so there was plenty of raw material. Ammonia could easily be turned into nitrates to use in fertilizers and explosives; this reaction, thought Haber, was the key.

The reaction is based on an equilibrium being reached between the two gases; Haber set to work discovering which conditions yielded the highest quantity of ammonia. He tried multiple different temperatures, pressures, and catalysts. In 1908, he came to the conclusion that the best conditions are found at 450 degrees Celsius, 250 atmospheres of pressure, and with an osmium or uranium catalyst.

The following year, the Haber process was already being used on an industrial scale by Carl Bosch. Nitrates could now be manufactured in abundance. They formed the basis of the inorganic fertilizers that would enrich the land, helping to feed the world's population, but also the production of explosives which would devastate populations during World War I.

Simon Davies

Date 1908

Scientist Fritz Haber

Nationality German

Why It's Key The production of fertilizers on a large scale prevented the mass starvation of a growing world population, and the capacity for Germany to produce explosives extended World War I by at least two years.

Key Invention **Cellophane**

Switzerland has a tradition of science and invention. As well as devising unusual clocks and knives, the ever inventive Swiss have also come up with milk chocolate and life insurance, and boast more Nobel Prize winners per capita than any other country. Their contribution to food does not begin and end with chocolate, however. October 19, 1872, was a momentous day in the history of food packaging technology. On that day, in Zurich, a young boy named Jacques Brandenburger was born. He would grow up to change the world – in a small way, at least.

Little Jacques was something of an early developer. At the tender age of twenty-two he became the youngest holder of a doctorate in Switzerland. Great things were clearly expected of him and luckily the world did not have to wait too long for Jacques to make his mark. In 1908 he invented a product that would earn him the affectionate moniker "Mister Cellophane".

Cellophane might not have been the solution to any great conflict; it might not have answered any of life's great questions; but it is a very useful substance. Made from plant cellulose, it is flexible, permeable to water vapor (unless treated), and resistant to fat, grease, and oil. Plus, it is completely biodegradable.

Although the popularity of Cellophane waned slightly from the 1970s onwards, recent interest in biodegradable and sustainable materials has seen a move away from polypropylene films, and something of a renaissance for Cellophane. Perhaps it could save the world after all.

Jim Bell

Date 1908

Scientist Jacques Brandenburger

Nationality Swiss

Why It's Key Cellophane could prevent world hunger by stopping people's sandwiches from going off, and – through its biodegradability – help combat climate change. Maybe.

Key Invention **Great Geiger**
The first radioactivity meter

You may be familiar with the characteristic crackles and pops of a Geiger counter, which indicate the levels of potentially dangerous and invisible radiation. Discovered in 1896, radioactivity was to revolutionize physics but, as none of the human senses were capable of detecting radiation, the race was on to develop instruments that could measure it.

Hans Geiger's first version of his famous radiation counter was constructed in 1908, while working at the University of Manchester under the directorship of his collaborator, Ernest Rutherford, the "father of nuclear physics." It was a sealed metal tube containing a gas, with a window at one end and a wire running down the middle. The wire was electrically charged in such a way that any radioactive particles passing through the device would collide with the gas causing a spark. As each spark was identical, it was then fairly simple to count the number of sparks per second.

His first counter was only capable of detecting alpha particles, which consist of two protons and two neutrons – the same as a nucleus of a helium atom. In 1928 Geiger, along with colleague Walther Müller, was able to create a souped-up version that could detect other types of radioactive particles. The basic design of this version, also known as the Geiger-Müller counter, is still one of the main methods of detecting radiation used today.

David Hawksett

Date 1908

Scientist Hans Geiger

Nationality German

Why It's Key Provided, for the first time, an instant method of measuring the hazard level of ionizing radioactivity.

opposite Rutherford (right) and Geiger with his radiation counting instrument, 1908

Key Event **The great Siberian explosion**
Tunguska

On a chilly morning in June 1908, passengers aboard the trans-Siberian express were quietly enjoying the spectacular scenery, when an extremely bright fireball suddenly ripped across the sky. An enormous explosion occurred shortly afterwards some 563 kilometers away, which shook the train so much that it was forced to screech to a halt.

An explosion had occurred, in the region of Tunguska, that was so intense that tremors of earthquake proportions registered strongly in Irkutsk, 885 kilometers away, and weakly in Washington DC, on the other side of the world. 64 kilometers from the blast centre, people were thrown into the air by the shockwave and ceilings collapsed. The force of the blast dwarfed the atomic bombs that would be dropped in World War II, and was similar in intensity to a hydrogen bomb. Astonishingly, nobody died because the blast point was so isolated.

Scientists examined the site of the explosion and were astounded to find forty miles of flattened forest. They expected to find a crater where an object had struck, but there was only a patch of trees with their branches ripped off, suggesting that the explosion had occurred in the atmosphere. The atomic bomb dropped on Hiroshima exploded in mid-air and left a similar imprint. In addition, as the blast settled down, black rain fell as had happened in Hiroshima. The most widely accepted explanation for the explosion is that a meteorite disintegrated upon entering the earth's atmosphere and exploded six kilometers above the face of the Earth.

Riaz Bhunnoo

Date 1908

Country Russia

Why It's Key This represented the biggest explosion in the world before the hydrogen bomb. It led to serious consideration of defense strategies, in case a meteorite should one day decide to head to Earth.

Key Person **Paul Eugen Bleuler**
Coined the terms "schizophrenia" and "autism"

Eugen Bleuler (1857–1939) was one of the most important psychiatrists of the twentieth century, known for his contributions to our understanding of mental illness. He challenged the view that psychiatric illness was usually the result of brain damage, proposing instead that there were often underlying psychological causes.

As director of the Burghölzli Asylum in Zürich, Switzerland, he coined the term "schizophrenia" in 1908, in a study of more than six hundred of his own patients. Bleuler was unusual in spending long hours talking to his patients, often well into the night. Contrary to what was believed at the time, Bleuler argued that the disorder then known as dementia praecox was not, in fact, a specific disease. Nor was it necessarily incurable or inevitably leading to full blown dementia, as had previously been supposed. Instead, he defined schizophrenia as a group of diseases, symptoms of which included random trains of thought and split personalities. Bleuler also introduced two important concepts linked to schizophrenia. "Autism" described the way some patients seemed to lose contact with reality, often resulting in bizarre fantasies. "Ambivalence" described the way some schizophrenics believed contradictory things at the same time, unable to see the contradiction.

The psychiatrist Carl Jung worked with Bleuler in the early 1900s. Together they became followers of Sigmund Freud, who, more than anyone, showed that understanding the human mind is not simply a case of simply studying the human brain.

Richard Bond

Date 1908

Nationality Swiss

Why He's Key Led to a better understanding of mental illness and prompted more effective and more sympathetic treatments.

Key Event **America on Wheels**
The first Ford T

He might not have invented the car or the assembly line, but the way Henry Ford used them both transformed America. It was revolutionary to use a business model that lowered product price, and the company's profit margins, in order to increase the number of sales, and it worked. Cars stopped being the handmade symbols of luxury that they once were, and became a necessary part of daily life. Ford also paid his workers enough money so that they could afford to buy the car, giving the product an in-house market.

The first Ford Model T came off the production line in Detroit on October 1, 1908, with a twenty horsepower engine and a top speed of 72 kilometers per hour. By modern standards it was pretty complicated to drive, despite having only two gears and reverse.

There were three pedals on the floor; one to choose first or second gear (as long as the lever was forward), one for reverse, and one to brake. The accelerator wasn't on the floor at all, but on the steering column instead. Sounds complicated? Under pressure, a confused driver was liable to stamp on all three floor pedals, lock up the drive train, and bring the car to a skidding halt.

Henry Ford is famously quoted saying, "You can have any color, as long as it's black," but, there's no evidence he ever really said it. In fact, when the Model T was first made, it was available in green, red, blue, and gray. But not black.

Douglas Kitson

Date 1908

Country USA

Why It's Key By making a small profit, but selling many – rather than a big profit and selling few – Henry Ford's Model T made having a car essential to American life.

opposite Workers constructing a Model T engine on an assembly line in a Ford Motor Company factory

Key Discovery **The Burgess Shale**
Walcott discovers a window to our past

High in the Canadian Rockies lies a window to our distant past. There, in 1909, American paleontologist Charles Walcott discovered the famed Burgess Shale, a treasure trove of perfectly preserved fossils over 500 million years old.

Walcott was the secretary of the government-funded Smithsonian Institute in Washington. In 1907, he traveled to British Columbia to investigate stories of "stone bugs" found by railway workers. He returned to the site two years later and exposed the source of the fossils, a site he dubbed the Burgess Shale.

The Burgess Shale was formed when a huge underwater mudflow buried the animals at the bottom of a reef. Their bodies quickly filled with particles of mud, encasing and so protecting them from the bacteria that would otherwise have degraded them. Even the fragile, soft-bodied organisms had been fossilized intact.

Over the next decade, Walcott collected more than 65,000 remarkable fossil specimens, many of which were unknown to science. It wasn't until the 1980s that scientists realized that these were not just new species, as Walcott had assumed, but quite possibly belonged to entirely new categories of animal life.

The Burgess Shale revolutionized our ideas about the diversity of organisms on Earth. We now appreciate that life was far more varied and diverse in the Middle Cambrian than we previously thought; some of the more bizarre forms of life found at the Burgess Shale remain very difficult to classify.

Steve Robinson

Date 1909

Scientist Charles Walcott

Nationality Canada

Why It's Key The rare discovery of perfectly preserved fossils provided an unprecedented insight into life half a billion years ago.

Key Discovery **Mohorovičić's Moho**
Boundary between Earth's crust and mantle identified

In 1909, Croatian seismologist Andrija Mohorovičić made a startling discovery. While researching seismic activity along the Earth's surface, he noticed that seismic waves always seemed to speed up at the same deep point below the Earth's surface. Considering that seismic waves should travel at uniform speed in constant media, Mohorovičić concluded that the waves must be passing through two separate environments. Could the Earth's surface be made up of more than one layer?

Due to the apparent acceleration of seismic waves at this constant depth below the surface, Mohorovičić was able to say that the waves must be passing from a high density layer to a low density layer. These two layers are now known to be the Earth's mantle and crust, with the Mohorovičić (or Moho) discontinuity as the boundary between them. This boundary is described as discontinuous because it marks a point of change in the speed of waves passing through it.

Mohorovičić discovery shed light on a topic that had confounded geologists for almost a century. The thought of the Earth existing as a single layer is now one of the past, and today we can construct a much more accurate image of the Earth's inner structure. Perhaps more significantly, the effect of the Mohorovičić discontinuity on the behavior of seismic waves has proved critical to our understanding of potentially devastating earthquakes.

Hannah Welham

Date 1909

Scientist Andrija Mohorovičić's

Nationality Croatian

Why It's Key Mohorovicic discovered that we live on a thin solid strip atop a churning mass of molten rock.

Key Invention **Bakelite**
Born of fire and mystery

When you next turn on your TV, use the telephone or run a comb through your hair, you might want to mutter a quiet "thank you" to Leo Hendrik Baekeland. You see, in 1909, he invented Bakelite by combining phenol, or carbolic acid, and formaldehyde and quite literally changed the world, lifting the curtain on the age of plastics and altering for ever the way countless things were made.

Baekeland's phenolic resin initially found general use as an easily moldable electrical insulator. Its cheapness and ability to be molded into virtually any shape meant it was quickly adopted by designers and manufacturers of everything from radios to radiator caps. Steering wheels, doorknobs and even Chanel jewelry appeared in a Bakelite form. The material would become an iconic material in Art Deco design.

Bakelite (or polyoxybenzylmethylenglycolanhydride, if you prefer) was the first truly synthetic plastic; "...a composition, born of fire and mystery, having the rigor and brilliance of glass, the luster of amber from the Isles..." was how *Time* magazine described it in its September 22, 1924 issue.

In spite of the best efforts of Baekeland's lawyers, the Bakelite trademark became so abused and widespread in its use that it became a generic term for all similar materials. It would even appear, for example, with a lower case "b" in dictionaries.

Mike Davis

Date 1909

Scientist Leo Hendrik Baekeland

Nationality Belgian

Why It's Key Heralded the dawn of the age of plastics and revolutionized manufacturing techniques.

Key Invention **Ductile tungsten**
The original light bulb moment

Since 1809, a revolution in the production of light had been brewing – the electric light bulb. Humphrey Davey and Thomas Edison, among others, had contributed to the development of this important product, but by the beginning of the twentieth century, there was still an important problem: the filaments burned out too quickly.

Carbon filaments were the best available, but they didn't last long at the temperatures of 2,100 degrees Celsius required to give enough light. But the General Electric Company in the United States were about to change all that. They hired the brilliant scientist William Coolidge to tackle the problem.

Several other scientists had begun to use tungsten in filaments instead of carbon because, as well as having a high melting point (3,410 degrees Celsius, not much less than carbon), it evaporates at a much slower rate, prolonging its life. But tungsten presented one little problem: the method used to produce the filaments caused it to be so brittle that it could not be bent without breaking. There was no way of shaping it into a filament.

Coolidge's great breakthrough came in 1909 with the development of a new technique by which bendable or "ductile" tungsten wire could be produced. Since this discovery, the vast majority of electric light bulbs have used a ductile tungsten filament. It is only in the last few years that fluorescent "low energy" bulbs, which use gas-filled tubes, have started to supersede incandescent bulbs.

Simon Davies

Date 1909

Scientist William D. Coolidge

Nationality American

Why It's Key Without this discovery, the mass production of electric light bulbs would have been impossible.

Key Discovery
Cure for syphilis

Syphilis, perhaps the most infamous of sexually transmitted diseases, has been around for hundreds of years and has historically attracted much stigma. Infection begins with a single sore called a chancre at the point of contact, followed by a rash, fevers, and headaches. As the disease progresses into its late stages, it attacks the body's organs causing paralysis, insanity, and even death.

In 1909, German biologist Paul Ehrlich, discovered a "magic bullet" to cure syphilis, earning him the accolade as one of the fathers of chemotherapy. As with many great discoveries, Ehrlich happened upon the cure for syphilis almost by chance. He was working to find a cure for another disease, "sleeping sickness," by testing a series of arsenic-derived compounds on mice. After testing more than nine hundred compounds, it was his assistant, Sahachiro Hata, who decided to revisit compound number 606. Hata had noticed that this compound, arsphenamine, was effective against the newly discovered syphilis-causing bacterium, *Treponema pallidum*. The pair began testing 606 on mice, guinea pigs, and rabbits, and were able to cure them of syphilis completely. The drug was released in 1910 under the name Salvarsan and became widely used. However, it soon became apparent that this bullet was not as magical as first thought. Soon, patients started suffering a relapse of their illness and, furthermore, the side-effects caused by the toxic arsenic were very unpleasant. In 1928, another accidental discovery – penicillin – took over from Salvarsan as treatment of choice for syphilis.

Hannah Isom

Date 1909

Scientist Paul Ehrlich

Nationality German

Why It's Key The birth of chemotherapy treatment in medicine.

opposite Colored transmission electron micrograph (TEM) showing a Treponema pallidum bacterium (orange) in penile skin (blue)

Key Event **Ernest Shackleton's expedition finds the magnetic South Pole**

First setting sail from Torquay, UK, in 1907, Sir Ernest Henry Shackleton organized and led the "British Antarctic Expedition" to Antarctica, also called the Nimrod expedition, named after the boat they sailed in.

After an initial exploratory voyage from New Zealand to Ross Island, the team set out on the 2,600-kilometer round trip to get to the South Pole. They set up camps and used ponies and sledges to aid them in their journey.

On January 16, 1909, three of Shackleton's team reached 88°23'S, only 180 kilometers from the South Pole. At the time, the South Magnetic Pole was located on land, but there was some doubt as to whether their location was correct. They only just made it back with dangerously low rations. Still, this was the furthest south anyone had been and would remain so until 1911 when Roald Amundsen reached the South Pole itself. The team also became the first humans to cross the Trans-Antarctic mountain range and set foot on the South Polar Plateau. Shackleton returned a hero and received a knighthood following the success of his expedition.

The magnetic South Pole is constantly shifting due to changes in the Earth's magnetic field, and is currently just off the coast of Wilkes Land, Antarctica. This is outside the Antarctic Circle and moving north-west by about 10 to 15 kilometers per year.

David Hall

Date 1909

Country Antarctica

Why It's Key This was the first trip to the magnetic South Pole. It laid the foundations for subsequent trips and for the first successful South Pole expedition by Amundsen.

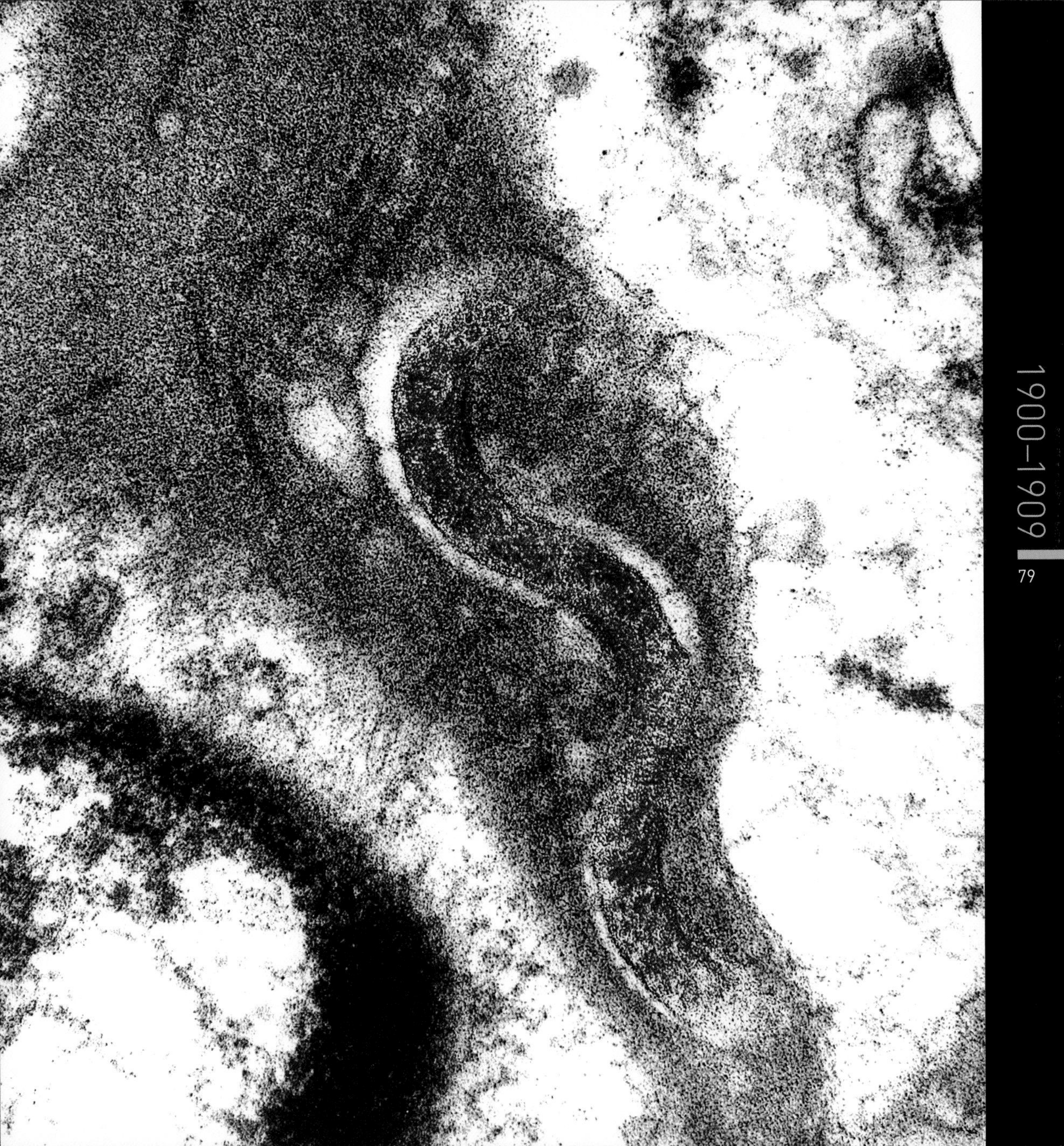

Key Publication **Dutch botanist coins the terms "gene," "genotype," and "phenotype"**

Now that words like "gene" and "genome" have become part of every day vocabulary, it is hard to imagine a time when the concept of genetics was abstract and unfamiliar. Amazingly, forty-four years before the structure of DNA was even identified, Dutch botanist Wilhelm Johannsen coined the terms "gene," "genotype," and "phenotype," and added important pieces to the genetic puzzle.

Gregor Mendel was the first person to introduce the concept of "factors" that caused inherited variations between different offspring, following his work with pea plants in the 1860s. This concept was refined slightly by Dutch botanist, Hugo de Vries, to "pangens," before Johannsen coined the term "gene" from the Greek "*genos*," meaning birth. Johannsen understood a gene to be a discrete factor that determines the inherited characteristics of an individual. With the visualization of the molecular structure of DNA by Watson and Crick in 1953, and subsequent advances, the term gene has been refined to mean a small piece of DNA that codes for a particular protein. Each human cell contains around 30,000 genes, arranged into double-stranded bundles of DNA called chromosomes.

Johannsen is also credited with coining the terms "genotype," meaning the complete genetic makeup of an organism; and "phenotype," meaning the outward characteristics of an organism, such as size, shape and color.

Hannah Isom

Date 1909

Scientist Wilhelm Johannsen

Nationality Dutch

Why It's Key Brought us one step closer to defining the principles of genetics.

Key Publication ***Inborn Errors of Metabolism*** Why first cousins don't marry

Archibald Garrod was already a prominent physician at St. Bartholomew's Hospital in London when he published *Inborn Errors of Metabolism*. Although his father, the distinguished physician Alfred Baring Garrod, encouraged him to study business, Archibald instead followed in his father's footsteps.

Following medical school, he was appointed assistant physician at the Hospital for Sick Children on Great Ormond Street in London. There, he noticed that a three-month-old boy, Thomas, had deep reddish brown urine and diagnosed alkaptonuria – an inherited condition that causes urine to turn black when exposed to air. Conventional wisdom at the time was that Thomas, like other children with alkaptonuria, had a bacterial infection. Garrod noted that the dark urine appeared in children of healthy parents and thus did not appear to be contagious. When two of Thomas' siblings were also noted to have dark urine, Garrod had an epiphany: it must be genetic. Confirmation came when he learned that the children's grandmothers were sisters.Using the ideas of heredity as developed by Gregor Mendel, which were then only just becoming widely known in England, Garrod realized that first cousins reproducing created conditions under which rare, recessive genes appeared in the offspring.

Garrod studied other recessive disorders such as albinism, and believed that such disorders arose from the genetic lack of an enzyme that led to a premature block in the chemistry of normal metabolism. He published his landmark paper in 1909 and described an important class of diseases that were genetic in nature.

Stuart M. Smith

Publication *Inborn Errors of Metabolism*

Date 1909

Author Archibald Garrod

Nationality British

Why It's Key Highlighted the importance of genetics in disease, and showed why inbreeding isn't such a great idea.

Key Person **Guglielmo Marconi** The first person to go wireless

The son of an Italian father and Irish mother, Guglielmo Marconi was born in Bologna, Italy in 1874. Growing up, he was privately educated, and developed a special interest in the science of electromagnetism, particularly radio waves. The telegraph equipment of the day needed to be tethered by cable to another station in order to be able to transmit a message, but Marconi was convinced that he could find a way to transmit signals over long distances using the waves. Working largely by himself on the family estate, Marconi was soon able to transmit a signal, wirelessly, over a few kilometers.

Immigrating to England in 1896, he began working with William Preece, engineer-in-chief of the British post office, and soon thereafter patented the first wireless telegraph system. It wasn't long before Marconi had established several permanent wireless stations around the British Isles and even placed his apparatus on some of the ships of the British Navy. His ultimate goal, however, was to transmit a radio signal wirelessly across the Atlantic, a feat which he accomplished late in 1901.

Marconi continued to refine his process, and eventually established the first commercial enterprise transmitting wireless signals from Europe to North America. He was subsequently awarded the Nobel Prize for Physics in 1909, but continued to improve his inventions until his death in 1937.

B. James McCallum

Date 1909

Nationality Italian

Why He's Key Marconi set the stage for modern radio wave transmission.

Key Person **Andrija Mohorovičić**

Andrija Mohorovičić was born in 1857 in Croatia. A meteorologist and seismologist, his work contributed to significant developments in the understanding of earthquakes and weather phenomena.

Mohorovičić's interest in meteorology started while teaching in the Croatian town of Bakar, where he founded a meteorological station, designed weather-measuring instruments, and undertook a measuring program. Throughout the 1890s, he produced a variety of meteorological gadgets, including a network of monitoring stations to trace the paths of thunderstorms. His studies also included the possibility of utilizing the strength of the strong bura winds of the Adriatic to produce energy. In the early 1900s, his main focus became seismology. Mohorovičić discovered that there are two different types of seismic waves that occur during an earthquake, longitudinal and transversal, which travel at different velocities. He also identified that the direction and velocity of seismic waves can be altered as they reach different materials and from this deduced that the Earth is made up of layers of different materials, surrounding a core. The boundary between the crust and the mantle has become known as the Mohorovičić discontinuity.

He made estimations of the depth of the Earth's crust to be 54 kilometers. Although we now know the depth of the crust to be much less under the ocean – as little as 5 kilometers – under the continents it is deeper and varies, an averaging 30 kilometers deep, making Mohorovičić's an impressive initial estimation.

Emma Norman

Date 1909

Nationality Croatian

Why He's Key His studies provided the foundations for many meteorological and seismological developments. This includes the causes and epicenters of earthquakes, and harnessing wind power.

Key Event **Channel crossing** The Blériot XI

Between 1907 and 1919, the *Daily Mail* newspaper offered various prizes for achievements in aviation, but it was the £1,000 prize for crossing the English Channel that convinced Louis Blériot to attempt his historic flight.

Blériot was an inventor and pilot who built a number of aircraft of his own design, the most famous of which was the Blériot XI. It was this 25-horsepower monoplane he took down to the shores of Les Barraques, near Calais, in July 1909, to take a shot at the £1,000 prize. But he was not alone. The daring Hubert Lathan, an Englishman living in France, had also been tempted to take the challenge; and joining them was Charles de Lambert, a Russian aristocrat with French roots who was one of Wilbur Wright's best students. The race was on.

Lathan tried first, but ran into engine trouble and landed in the sea. During testing, Lambert had a major crash and was forced to withdraw, while Blériot broke a petrol line and burned his foot.

Lathan and Blériot regrouped and repaired, waiting for favorable weather. At dawn on July 25, despite blustery winds and a badly burnt foot, Blériot set off. By the time Lathan's team had realized they had been caught napping, it was too late. Flying for thirty-seven minutes with no compass, Blériot made it to Dover to take the prize for the first Channel crossing in a heavier-than-air craft. Gaining worldwide fame, he went on to run a successful aircraft manufacturing company, and made planes for the Allies in World War I.

Douglas Kitson

Date 1909

Country France

Why It's Key The first flight across the channel in a heavier-than-air craft, delighting the French and worrying the British by making them vulnerable to attack by air.

opposite Louis Blériot in his aircraft during the first powered flight across the English Channel.

Key Discovery **Chlorine cleans water**

Medical research in the U.S. Army at the beginning of the twentieth century was beginning to focus on preventative medicine. In 1910, Major Carl Darnall, professor of chemistry at the Army Medical School, Washington DC, was concerned with finding a way of ensuring troops in the field had safe water to drink. He was experimenting with different methods for treating water supplies when he made the discovery that has, arguably, prevented more disease than any other medical breakthrough in history.

Darnall had found that using liquefied chlorine gas made water from potentially dangerous sources safe to drink. This was because the introduction of chlorine performed two important tasks. First it kills many kinds of bacteria and other organisms which cause diseases such as cholera, typhoid, and dysentery. The second important function of chlorine is that it leaves a residual level in the water, preventing contamination from harmful organisms in the transport, storage, and distribution of the water. Another army scientist, Major William Lyster, further improved on Darnall's technique by inventing a convenient system by which the chlorine-containing solid, sodium hypochlorite, could be added to water in a cloth bag. The use of "Lyster bags" became standard practice for troops in the field. The idea of adding such a toxic gas to the drinking water supply was understandably initially met with resistance, but eventually it became the standard method of treatment the world over.

Simon Davies

Date 1910

Scientist Carl Darnall

Nationality American

Why It's Key Throughout the world, communities have clean, safe drinking water as a direct result of Darnall's work.

Key Publication ***Principia Mathematica***
The mathematical principles of natural philosophy

Unlike many leading figures in the field of mathematics, Bertrand Russell knew very little of his own mathematical abilities before studying at Trinity College, Cambridge. He was orphaned at a young age and had been educated by a governess in virtual isolation. Despite his grandmother attempting to train Russell to become Prime Minister, his talents lay in mathematics and philosophy.

Generally recognized as one of the founders of analytic philosophy, it was Russell's desire to reduce mathematics to logic.

At times, he abandoned philosophy for politics, and become one of the world's most influential critics of nuclear weapons and of the American war in Vietnam. However, his political campaigns sometimes resulted in fines and even imprisonment.

It was while serving a six-month sentence for a pacifistic article he had written that Russell wrote the book *Introduction to Mathematical Philosophy*, later published in 1919. This was his last significant work in mathematics and logic, and largely formed an explication of his previous work, in particular *Principia Mathematica*, which was first published nine years previously. Russell had written the *Principia* with his former lecturer, Alfred North Whitehead, focusing on the view that mathematics is in some significant sense reducible to logic. Their work played a major part in developing and popularizing modern mathematical logic, and today remains one of the most influential books on logic ever written.

Sarah Watson

Publication *Principia Mathematica*

Date 1910

Author Bertrand Russell

Nationality British

Why It's Key Russell's work played a major part in developing and popularizing modern mathematical logic.

Key Person **Joseph John Thomson**
A founder of particle physics

Joseph John Thomson (1856–1940) spent the majority of his life investigating the atom, fascinated by what was inside. He was a key player in identifying the fourth state of matter – plasma – consisting of atoms stripped bare of electrons. This makes him partly responsible for the plasma television screens we use today.

One of Thomson's key achievements was the discovery of the electron. Whilst experimenting with electric charge inside an empty glass tube, known as a cathode ray tube, he considered how the unidentified rays generated inside could be the result of particles – electrons – flowing from one end to the other. This led Thomson to speculate about the structure of the atom and the conception of his "plum pudding model"; a positively charged nucleus, surrounded, and balanced by, a negatively charged "cloud" of electrons. The model was later disproved but he received a Nobel Prize in 1906.

In 1910, Thomson became the first to prove the existence of isotopes – different forms of various elements. He found that when he sent a stream of ionized neon into a magnetic and an electric field, the rays produced patches of light at different wavelengths. This showed him that two forms of neon must be present with different molecular weights.

Thomson's discovery set the stage for the use of isotopes in many different areas. Now known to contain different numbers of neutrons, isotopes are used in smoke detectors, to date fossils, and even to identify counterfeit money.

Nathan Dennison

Date 1910

Nationality British

Why He's Key Thomson made great strides in understanding the structure of the atom, both its positively charged core and negatively charged outer electron cloud.

Key Event
Henri Coanda makes first flight in jet aircraft

In 1910, Romanian inventor Henri Coanda became the first person to build and fly a plane that used a hybrid jet engine rather than a propeller. He was way before his time. While nowadays most modern commercial planes use similar, although usually much larger, forms of propulsion, it took more than three decades for jet aircraft to realize their potential: "A jet engine? It'll never catch on!".

One of the reasons why it took so long for jets to become popular might have been that, during its first and only flight, Coanda's prototype plane – the snappily named Coanda-1910 – destroyed itself. The inventor had placed his revolutionary engines, each containing a four-cylinder engine powering a compressor, too close to the fuselage. To his horror, the flames from the engines didn't shoot directly backwards, but licked down the plane's body, hugging the contours of the machine, torching it irreparably. The plane was a write-off and Coanda gave up his experiments soon after due to a lack of interest and support from scientists and the public in general.

Coanda's efforts weren't entirely in vain, however. His ill-fated flight illustrated, in dramatic style, the peculiar phenomenon where gases moving at high speed will wrap themselves around objects instead of carrying on in a straight line. Produced by the differences in air pressure on either side of the stream of gas, it is now known as the Coanda effect.

Logan Wright

Date 1910

Country Romania

Why It's Key Jet engines are the preferred form of air propulsion now. Almost all air vehicles, and even some radical land vehicles, use some type of jet engine.

Key Event **Through the keyhole**
A bright future for surgery

Although the technique of keyhole surgery has only been widely used since the 1980s, it was developed almost seventy years before, by Swedish surgeon Hans Christian Jacobaeus.

Born in Skarhult, Sweden in 1879, Hans Jacobaeus later became professor at the Karolinska Institutet in Stockholm. His main area of work was laparoscopy, also known as minimally invasive surgery, and he was a great believer in training his staff to use endoscopic techniques when performing internal thoracic and abdominal examinations.

Laparoscopic, or keyhole, surgery is also characterized by the use of very small incisions (usually 0.5–1.5 centimeters), and is usually carried out in the abdominal or pelvic cavities. Thoracic keyhole surgery is usually called thoracoscopic surgery, although now almost any kind of surgery may be carried out in this way. Both laparoscopic and thoracoscopic surgery belong to the broader field of endoscopy. Here, a fiber optic cable, a lens attached to a camera, and a cold light source, are used to view the internal space. This type of surgery can be carried out under either general or local anesthetic; leaves far less scarring; and requires much shorter recovery times.

Jacobaeus performed the first thoracoscopic diagnosis on a patient using a special endoscope in 1910. However, it was not until 1985 that computerized camera technology and optic fibers allowed the techniques to become widely used and expanded to other forms of surgery, such as gall bladder and gynecological operations.

David Hall

Date 1910

Country Sweden

Why It's Key Revolutionized surgery. Operations which no longer require large incisions mean less scarring and shorter stays in hospital.

Key Person **Howard Taylor Ricketts**
One dies so that that millions may live

One of Howard Ricketts' (1871–1910) lasting legacies is our understanding of the disease-causing bacteria bearing his name – *Rickettsia*. This skilled and imaginative pathologist discovered the unlikely causes of some very important diseases, his main interest being in disease vectors – the plants and animals that pass germs on to people. He often flouted popular research methods, and even injected himself with germs to see what would happen.

Ricketts' controversial methods worked. He proved that a fungus living in rotting vegetation by riversides caused blastomycosis – an inhaled fungal infection – and, after four painstaking years, he proved that bacteria from what is now known as the *Rickettsia* genus, carried by a biting wood-tick, causes Rocky Mountain spotted fever.

In 1909, Ricketts went to Mexico City, suspecting that a deadly outbreak of typhus ravaging the crowded, rodent-infested parts of the city might involve an insect carrier. Having infected monkeys in his makeshift laboratory with the disease, he was able to identify the insect culprit as the human body louse. He couldn't show which infective bacterium it was carrying, but he had enabled the city to deal with its sanitation problems.

In May 1910, just days before being reunited with his wife – his childhood sweetheart – and young family after their long separation, he was struck down by the very disease that he had returned from studying. Ironically, this remarkable man, who devoted his life to researching disease, fell victim to one of his own study subjects.

S. Maria Hampshire

Date 1910

Nationality American

Why He's Key Howard T. Ricketts was one of the first medical scientists to determine the ecological and environmental conditions behind transmission of vector-borne diseases.

Key Person **Paul Ehrlich**
The king of chemotherapy

By the time the nineteenth century turned into the twentieth, Paul Ehrlich (1854–1915) had forged himself a reputation as one of Europe's leading medical scientists. Born in 1854 to Jewish parents, he had developed an interest early on in his career in how dyes could selectively stain different biological samples. In particular, he developed methods to stain tuberculosis-causing bacteria, which, with a couple of improvements, are still used today. He was also the first man to show the existence of the blood-brain barrier – the membrane which protects the brain from harmful substances the blood.

It wasn't long, however, before Ehrlich realized that some of his dyes were killing bacteria, and started to wonder if there was a way that these colored substances could be used to treat disease. It was a question that set him on the way to the discovery of a cure for syphilis, and fame as the father of chemotherapy – the chemical fight against disease.

Apart from an early bout of tuberculosis, Ehrlich's lifestyle of working tirelessly, eating little, and smoking twenty-five strong cigars a day never impacted on his health. The outbreak of World War I, however, distressed him so much that he suffered a slight stroke in 1914. Having never fully recovered, he died the next year. His career was later dramatized in the 1940 film, *The Magic Bullet*, which took its name from Ehrlich's conception of chemical cures being targeted directly at disease-causing bacteria. He also has the dubious honor of having had an entire family of disease-causing bacteria – the *Ehrlichiaceae* – named after him.

Stuart M. Smith

Date 1910

Nationality German

Why He's Key Developed a treatment for the sexually transmitted disease syphilis, and coined the phrase "magic bullet."

opposite Paul Erlich in his laboratory

Key Event
Halley's Comet photographed for the first time

Halley's Comet is possibly the most observed 15-kilometer-wide lump of ice, dust, and rock in the history of human existence. Comets themselves are like massive dirty snowballs, which can pass very close to the Sun, although they spend most of their orbits in the far reaches of the Solar System. When comets do get close to the Sun, they heat up, leaving a characteristic trailing tail of gas and dust.

Halley's Comet is particularly special because it has a short orbital period of around seventy-six years and is visible even with the naked eye. Other larger comets are more visible but can take thousands or even millions of years to orbit the Sun. Halley's Comet was even visited by the European Space Agency's Giotto satellite during its recent 1986 appearance.

Halley's Comet was named after Edmond Halley, who correctly predicted the orbital period of the comet, and its return in 1757. Throughout history it has been regarded as an omen, including in 1066 before the battle of Hastings.

Despite scientists' best efforts, the 1910 appearance spread panic thanks to news that the comet contained cyanide gas. There was of course no danger to the public, but many salesmen made a quick buck out of "comet pills" designed to protect the nervous onlooker from speeding ice balls. Despite causing no fatalities, Halley's Comet did coincide with the death of American writer Mark Twain, author of *The Adventures of Huckleberry Finn* and *The Adventures of Tom Sawyer*, and born when the comet was last visible in 1835.

Jim Bell

Date 1910

Country International

Why It's Key One of the Solar System's best known snowballs.

opposite Optical photograph of Halley's comet taken on May 25, 1910 at Helwan, Egypt

Key Person **Srinivasa Aiyangar Ramanujan**
Mathematical genius

Srinivasa Ramanujan (1887–1920) was one of India's greatest mathematical geniuses. He showed an interest in mathematics at an early age and, in 1900, aged just thirteen years, he began to work on his own mathematics, summing arithmetic and geometric series. By the time he started high school Ramanujan was teaching himself mathematics from the book *Synopsis of Elementary Results in Pure Mathematics* by G.S. Carr. Unfortunately, a scholarship awarded to Ramanujan was not renewed beyond his first year of college, since his devotion to mathematics meant other subjects were neglected. Without a formal education he struggled to gain recognition for his work, until 1911, when he had a paper published in the *Journal of the Indian Mathematical Society*; this was the first step toward Ramanujan being recognized as a mathematical genius. Ramanujan sought to find work in order to "earn a pittance on which to live" so that he could continue his research. In 1913, he wrote to G.H. Hardy, an English mathematician, enclosing a list of about a hundred unproved theorems. In his list, Ramanujan had worked out the Riemann series, the elliptic integrals, and the hypergeometric series. Hardy's response was encouraging and, in 1914, he invited Ramanujan to Trinity College, Cambridge, to begin an extraordinary collaboration which was to lead to a number of important results.

Following Ramanujan's death, one mathematician alone published fourteen papers under the title "Theorems stated by Ramanujan," and a further thirty papers inspired by his work.

Sarah Watson

Date 1911

Nationality Indian

Why He's Key Ramanujan made substantial contributions to the analytical theory of numbers and left behind 4,000 original theorems.

Key Invention **The first chromosome maps**

Humble fruit fly provides a key moment in genetics

Chromosome mapping is a way of assigning genes to specific positions on a chromosome. The American geneticist Alfred Sturtevant, working with Thomas Morgan, thought of the idea of mapping around 1911 and produced the first genetic map of a chromosome in 1913.

Working with one of the most famous experimental animals, the humble fruit fly (*Drosophila*), Sturtevant and Morgan spent their days in the "fly room" at Columbia University. It was here that Morgan had first shown that genes are carried on chromosomes and are involved in the inheritance of certain physical features – an idea suggested by the founder of genetics, Gregor Mendel, nearly fifty years earlier. Chromosome maps are based on the idea that genes are arranged in order on a chromosome, like beads on a necklace, and that the gene for any particular feature is in a fixed position on the chromosome. By studying the way genes "cross over" when cells divide, Sturtevant and Morgan were able to work out which genes were closer to each other on the chromosome; the closer they were the more likely they were to cross over together.

Having maps that show where the gene for a particular physical trait is positioned has led to the development of techniques such as screening for genetic diseases. Scientists still use Sturtevant's method for chromosome mapping today, and school students still study genetics using fruit flies, more than a century after the first, and most famous, "fly room" was created.

Richard Bond

Date 1911

Scientists Alfred Sturtevant, Thomas Morgan

Nationality American

Why It's Key Producing chromosome maps for more and more complex organisms, including human beings, has led to significant advances in identifying and treating genetic disorders.

Key Person

Marie Curie does the double

Maria Sklodowska-Curie is undoubtedly one of the most awe-inspiring figures in the history of science. A Nobel laureate twice; and still, almost a century since she claimed her second, the only person to have won Nobel Prizes in separate science disciplines.

Curie was born Maria Sklodowska in Warsaw, Poland, in November 1867. The youngest of five children born to two teachers, the importance of learning was instilled in Marie at a young age. She quickly developed a strong work ethic, which helped to see her graduate top of her class at the age of fifteen. Despite her obvious ability she was denied a place at university in Poland, moving instead to the University of Paris, where she became the first woman ever to be awarded a French doctorate.

Curie's first Nobel Prize, in physics, was shared with her husband Pierre Curie and Antoine-Henri Becquerel, for their ground-breaking research into radioactivity. She was the first woman to be awarded a Nobel Prize. When, in 1906, Pierre was tragically run down and killed by a horse-drawn cart, Marie was devastated. But her work ethic and mental toughness saw her through. She was offered Pierre's teaching job, becoming, in yet another "first" for feminism, the first woman to teach at the Sorbonne.

Her second Nobel Prize came a few years later in 1911, this one for chemistry, following her discovery of two new elements, polonium and radium. Even in death Marie Curie remains unique – the only woman to be interred at the Pantheon in France.

Barney Grenfell

Date 1911

Nationality Polish

Why She's Key Marie Curie stands out as a singular figure in science, not only for the ground breaking research that saw her claim Nobel Prizes in both physics and chemistry, but also as a champion for women in science.

opposite Marie Curie

Key Person **Thomas Hunt Morgan**
Fruit flies form basis of genetics

From Watson and Crick's discovery of DNA to, more recently, the identification of the human genome, the ever expanding science of genetics has almost unparalleled importance both for biology and, through practical application in helping cure disease, humanity itself. But where did it all begin?

Thomas Hunt Morgan (1866–1945) was born into a wealthy American family – "Southern aristocracy" – with an impressive lineage. The young Morgan expressed a keen interest in nature and the world around him, an interest that eventually developed into a Bachelor of Science degree from the State College of Kentucky, and postgraduate work in zoology at Johns Hopkins, where he gained his PhD in 1891. Around the end of the nineteenth and beginning of the twentieth century Morgan was occupied with lecturing and laboratory research, as well as publishing his first book (on frog's egg development), in 1897. It was not until the end of the first decade of the 1900s that he began the work that he would be remembered for, and which would, arguably, change the face of biological science.

A resurgence of interest in Mendelian theory prompted Morgan to begin experimenting with fruit flies, mutating them through various means and then attempting to breed the mutated versions. Success was slow in coming, but his hard work did eventually pay off. Morgan's team provided key insights into the arrangement of genes on the chromosome, and into a phenomenon called "linkage" – when traits are inherited jointly due to their physical positions within the genome.

Barney Grenfell

Date 1911

Nationality American

Why He's Key Thomas Hunt Morgan's work laid the foundations for a modern unified theory of genetics, which incorporates evolutionary theory and heredity.

Key Discovery
Rutherford describes the shape of the atom

In 1906, Ernest Rutherford had noticed that when he passed a beam of alpha particles through a sheet of the mineral mica, the particles scattered and the beam went fuzzy. Fast forward three years to 1909, and Ernest Marsden and Hans Geiger, under the direction of Ernest Rutherford, performed an experiment which fired alpha particles through an extremely thin sheet of gold foil. Their job was to measure where the particles were going when they were scattered. What they found was one of the most significant results in physics.

Most of the particles whizzed through happily, as everyone had expected them to, but a very small number pinged off to the side, and in some cases actually rebounded. This was an astonishing result. Rutherford famously described it: "As if you fired a fifteen-inch shell at a piece of tissue paper and it came back and hit you."

In 1911, Rutherford published a paper that explained why this happens. Instead of being spread out evenly as they'd thought, most of the mass of the atom actually sits in a positive nucleus in the middle with the electrons orbiting around it. The repelled particles in Marsden and Geiger's experiment had collided with the nuclei of the gold atoms.

This was the birth of the nuclear model of the atom, a discovery that showed atoms not as tiny singular particles as everyone had thought, but as structures made up of even smaller sub-atomic particles.

Douglas Kitson

Date 1911

Scientist Ernest Rutherford

Nationality New Zealander

Why It's Key The birth of nuclear physics, giving us atomic energy, nuclear bombs, and an understanding of the way the Universe works at a fundamental level.

Key Experiment **Flying high for genetics**
Genes are located on chromosomes

For many people, flies are little more than a nuisance, but for geneticists they are a valuable pieces of laboratory kit. Fruit flies share nearly 60 per cent of human genes and show similarities in their physiology, for example getting "drunk" or addicted to cocaine. They are also cheap to breed, have a short life cycle, and can be kept in limited space.

In 1904, Thomas Hunt Morgan became a professor at the University of Columbia. When a colleague introduced Morgan to fruit flies in 1907, the famous "fly lab" was born. Morgan was dissatisfied that much research at the time referred to Mendel's "hereditary factors" without anyone having determined what they were. So, the fly lab aimed to find evidence for one of the contemporary theories at the time: that changing the flies' environment would cause mutations.

Although unsuccessful in this sense, the experiments confirmed the already known finding that inheritance was something to do with chromosomes. Reported in 1910, the results focused Morgan's attention on the chromosomes as the source of heredity.

That same year, a white-eyed mutant male appeared among the normal red-eyed flies. Experiments on this mutant eventually led the researchers to conclude that eye color was a sex-linked trait and that genes – the Mendelian "hereditary characters" that Morgan had found so unsatisfying – were located on chromosomes. It was this physical link to Mendelian theory that finally made Morgan a believer, and founded the science of genetics.

Christina Giles

Date 1911

Scientist Thomas Hunt Morgan

Nationality American

Why It's Key Finally pinned down Mendel's hereditary factors to material objects

Key Discovery **Really cool science**
The first superconductor

The brilliant Dutch theoretical physicist Heike Kamerlingh Onnes liked the cold. He originally worked on trying to cool gases to temperatures as low as possible to see how they behaved. His rival, James Dewar, had been the first to liquefy hydrogen and Onnes followed suit in 1908 by successfully liquefying helium, for which he received a Nobel Prize in 1913.

In 1911, at Leiden University, Netherlands, the "gentleman of absolute zero" began studying what happened to metals at these frigid extremes. This really was frontier science, illustrated by the fact that some other physicists believed that, as the temperature dropped, electrons passing through metals would simply come to a halt – in other words electricity simply wouldn't conduct through the extra-cold metals.

Onnes decided to concentrate on liquid mercury, because he could make extremely pure samples of it, and feared any impurities in other metals would spoil his results. He placed the liquid mercury in a U-shaped tube with a wire at both ends. Passing an electrical current through it allowed him to watch what happened to the electrical resistance of the mercury as he cooled it using liquid helium.

As he predicted, its resistance dropped with temperature, but he was shocked when, at 4.19 degrees above absolute zero, mercury's electrical resistance completely vanished. By accident, he had discovered superconductivity, and later went on to demonstrate the same phenomenon with other metals such as tin and lead.

David Hawksett

Date 1911

Scientist Heike Kamerlingh Onnes

Nationality Dutch

Why It's Key Among other things, the use of Magnetic Resonance Imaging (MRI) devices in modern medicine is a direct result of the discovery of superconductivity.

Key Event
Eugene Ely lands plane on the USS *Pennsylvania*

You don't need anything more than a passing knowledge of what an airplane is, and what an airplane does, to appreciate that landing one on the deck of a ship is quite a tricky task; landing safely on a runway is complicated enough. But this is exactly what Eugene B. Ely, a self-taught pilot from Iowa who flew exhibition flights, did on the instruction of Captain Washington Chambers.

Captain Chambers was in charge of the U.S. Navy's fledgling aviation programme, and the pair met at an international air convention in New York. Impressed by Ely's abilities, and looking for a pilot able and willing to carry out some important experiments, Chambers began to work with Ely.

On January 18, 1911, in front of thousands of shore-bound spectators, Ely took off from Tanforan racetrack in San Francisco, California and flew out toward the USS *Pennsylvania*. The plan was for a hook attached to the plane's landing gear to catch itself on a series of ropes attached to sandbags (known as a tail hook system), the weight of which would bring the craft to a rapid enough halt to land along the 36.6 meter length of the deck. It worked an absolute treat, and the provisional safety measures set up onboard the ship went completely unchallenged.

Sadly, later that year, Ely crashed his craft flying at an exhibition in Georgia and was killed. He was posthumously awarded the Distinguished Flying Cross for his contribution to naval aviation.

Chris Lechery

Date 1911

Country USA

Why It's Key As a result of these successes, naval forces then began to use similar tail hook systems in their warships, and integrate aviation machines into the force.

opposite Eugene Ely takes off from the USS *Pennsylvania*

Key Publication *Heredity in Relation to Eugenics*
Good intentions with ugly consequences

The term eugenics was coined in 1883 by English scientist Francis Galton, from the Greek, *eugenes* ("well-born" or "of good stock"). It was presented as a "science" that could be used to predict the traits and behaviors of humans, possibly even control human breeding so that only people with the "best" genes would reproduce, thereby improving the species.

It was a popular idea at the time; widespread concern existed that America was being swamped by immigrants who might displace the essentially European populace and erode the racial quality of the American people. Based in part on Mendel's laws of heredity and focusing on the idea that biological characteristics were determined by single elements (later associated with genes), eugenics research claimed that not only could physical characteristics such as eye color and disease be explained in a simple hereditary manner, but also mental and behavioral traits, mechanical skills, even a talent for music. The theory was promoted heavily by Charles Benedict Davenport, a prominent American biologist and head of a biological research laboratory at Cold Spring Harbor on Long Island, New York. In 1911, he published his famous work, *Heredity in Relation to Eugenics*, in which his related research, views, and inferences on how eugenics might be applied, were discussed.

Whilst many simplistic claims about human heredity were made under such titles, some eugenic investigation proved to be worthy, revealing, for example, that Huntington's disease results from a dominant gene and albinism from a recessive one.

Mike Davis

Publication *Heredity in Relation to Genetics*

Date 1911

Author Charles Benedict Davenport

Why It's Key By linking various human traits with specific ethnic and racial groups, the book impacted greatly on social attitudes and government policy.

Key Invention **Charles Wilson's cloud chamber**
A most original and wonderful instrument

The view from the top of Ben Nevis is spectacular; clouds that roll past Britain's loftiest peak combine with the Sun's rays to produce heavenly glowing edges to shadows. While working at the summit in 1894 the Scottish physicist Charles Wilson was struck by the beauty of these rainbow colored lining, known as glories. Early in 1895, he left his mountaintop observatory to recreate these conditions in the laboratory.

It was not until 1911 that Wilson completed his cloud chamber, using it to track the paths of alpha- and beta-radiation particles and electrons which he would photograph and describe as "little wisps and threads of clouds." He continued to improve on his chamber and, in 1923, published two papers on the tracks of electrons, prompting others to replicate his design.

When charged particles pass through air saturated in water, they ionize adjacent particles, forming condensation in their wake, and therefore revealing their trajectories. Wilson's invention would generate supersaturated air by expanding and therefore cooling it at a steady rate. An electric charge would remove old ions and therefore only the most recent ones would remain, allowing for clear photographs to be taken.

Arthur Compton's research into the particle-like behavior of electromagnetic radiation was one of the main beneficiaries of Wilson's work and this was recognized in 1927 when Compton and Wilson shared the Nobel Prize in Physics. Ernest Rutherford fittingly described the chamber as "the most original and wonderful instrument in scientific history".

Christopher Booroff

Date 1911

Scientist Charles Wilson

Nationality British

Why It's Key The Compton Effect, the positron, and electron-positron annihilation were some of the many discoveries made or confirmed using Wilson's cloud chamber.

opposite Charles Wilson's cloud chamber

Key Event
South Pole is finally conquered

Roald Engelbregt Gravning Amundsen was a Norwegian explorer who, in 1911, became the first man to reach the South Pole at a time when polar exploration was all the rage.

The expedition started from Framheim across the Ross Ice Shelf on the October 19, 1911 with a team of four others, four sledges, and fifty-two dogs. At the same time, Robert Scott was leading a similar team in a race for to reach the goal first. After eight weeks, and hundreds of kilometers, on December 14, 1911, the team arrived at the South Pole, 90°00'S. More importantly, they arrived thiry-five days before Scott's team. They only had sixteen dogs left, but all of the team were alive.

Amundsen was admired for his meticulous preparation and experience, which was a major factor in why the trip was largely uneventful in terms of drama, unlike Scott's. In total, the trip took 99 days and covered about 3,000 kilometers.

Unlike most polar expeditions of the time, Amundsen's was almost entirely focused on simply reaching the pole and very little to do with conducting scientific research. His trip, however, along with the failed effort of Scott, left a lasting scientific legacy. The lid had been lifted on questions of how to get to the continent, how to survive there, and even how to think in such extreme conditions.

Amundsen, who was also the first person to reach both the North and South Poles, disappeared during a rescue expedition in June 1928.

David Hall

Date 1911

Country Antartica

Why It's Key This is amongst the great feats of exploration, conquering one of the harshest environments on the planet for the first time.

Key Person **Carl Jung**
The other grandfather of psychology

In early twentieth century Europe, the science of psychoanalysis was emerging. Sigmund Freud helped develop the theory of the subconscious and the repression of thoughts by stifled sexual instincts, and gained a band of devoted followers and collaborators. The work of one acolyte of Freud's, a Swiss doctor by the name of Carl Jung (1875–1961), would come to follow quite a different course from that of his one time friend and mentor, resulting in a public split between the two in 1912.

Jung's believed that the unconscious not only held repressed sexual desire, but also hidden hopes and fears. Jung was also more influenced by society and spirituality than Freud, highlighting the need to treat people in a historical and cultural context. He believed Freud's mechanistic approach was too rigid, and that Freud himself was dogmatic and unreceptive to new ideas. Some of Jung's most influential and famous ideas include the concepts of the collective unconscious and of personality types. This latter idea was later developed by Katharine Cook Briggs and her daughter, Isabel Briggs Myers, and is used widely in psychometric testing. Many people will have undergone the Myers Briggs test during job interviews.

Although not one to shun controversy – allegations of being a Nazi sympathizer and his dalliances with spirituality and the occult have been well-documented – Jung has remained an influential figure in psychiatry and psychology. To this day he is considered, alongside Freud, as one of the greatest figures in the field.

Neal Anthwal

Date 1912

Nationality Swiss

Why He's Key Founded analytical psychology and developed many influential theories of the psyche.

opposite Carl Jung

Key Discovery **Variable stars**
How to identify galaxies

By the late nineteenth century, computers had become available to astronomers. At that time, however, these "computers" were not machines, but women – employed to carry out the drudge tasks that the main researchers didn't want to do themselves. Henrietta Leavitt worked in the Harvard astronomy department from 1895 until her death in 1921, during which time she discovered a vast array of variable stars – stars which changed their brightness at a fixed rate.

Through years of close study, Leavitt found that the brighter stars pulsed more slowly and, what's more, a star's pulse rate could be used to calculate how bright it actually was. For the first time astronomers could work out whether a star was dim because it was a long way away or because it was... well, just not very bright. It was the yardstick that allowed other astronomers, notably Edwin Hubble, to make their important contributions to our knowledge of the Universe.

Leavitt was discouraged from researching her findings further. But the relationship later proved so important that she was considered for a Nobel Prize. Unfortunately, being dead by that time, she couldn't be awarded one.

Measurements derived from variable stars have since provided evidence for the changing shape and size of the Universe. Pretty good going for a deaf woman being paid half wages to work on the boring jobs.

Kate Oliver

Date 1912

Scientist Henrietta Leavitt

Nationality American

Why It's Key Finally, a way of deciding whether stars are dim or just far away. And one up for female scientists.

Key Discovery
A head for heights reveals cosmic rays

Science in the early twentieth century was not for sufferers of vertigo. In 1910, German physicist Theodore Wulf climbed the Eiffel Tower with an electroscope, hoping to show that radiation decreased with height. He was on a mission to prove that radioactive materials in the Earth caused gases to ionize – that is, lose electrons. Unfortunately his results showed the opposite; radiation increased with height. Wulf doubted his results over a height of 300 metres.

To get more conclusive results, Victor Hess took to the skies two years later in a hot air balloon, planning to ascend to a height of more than 5 kilometers. Measuring radiation levels as he ascended, Hess found that although they initially decreased with height, at about 600 meters the levels started to increase again. The higher he went, the more radiation there was.

Since the radioactive elements in the Earth were getting further away, they couldn't explain the effect. Hess' conclusion was that "radiation of exceptionally high penetrative capacity" was entering Earth's atmosphere from outer space. This radiation was later to be called "cosmic rays".

Cosmic rays are actually high-energy particles – protons, electrons and atomic nuclei – which arrive at nearly the speed of light from all over the galaxy. They most probably originate in the remnants of supernovae. Some, however, have too much energy to have been produced this way. We still don't know the origins of these ultra-energized particles; they could even be caused by defects in the very structure of the Universe itself.

Kate Oliver

Date 1912

Scientist Theodore Wulf, Victor Hess

Nationality German, Austrian/American

Why It's Key Provided proof that conditions in Space affect conditions on Earth, and gave us a fantastic sci-fi phrase.

opposite Theodore Wulf

Key Publication **Accessory factors**
The many men of vitamins

In 1912 the English biochemist Frederick Hopkins published a paper with the catchy title "Feeding Experiments Illustrating the Importance of Accessory Food Factors in Normal Dietaries." In it, he postulated that the human diet required "accessory factors," in addition to proteins, carbohydrates, and fats. Hopkins would later share a Nobel Prize with Dutch physician, Christiaan Eijkman, for the discovery of what would become known as "vitamins."

Much of Hopkins' work was based on that previously carried out by the Scottish surgeon James Lind, who had noted that citrus foods helped prevent scurvy, a particularly foul disease which causes poor wound healing, bleeding of the gums, severe pain, and eventually death. It was the use of limes to prevent scurvy that led to British sailors being called "limeys."

Working separately from Hopkins, Eijkman discovered that eating brown rice helped to prevent the disease beriberi, whereas white rice did not. "Beri" is Sinhalese for "I cannot", as the disease can lead to profound weakness, pain and death.

But it was down to yet another scientist to christen these new chemicals. Around 1912 the wonderfully named Polish biochemist Casimir Funk proposed that such nutritional compounds be named "vitamines" for "vital amines". Hopkins' "accessory factors" now had a slightly snappier name, and by the time it was shown that not all vitamins contained amine groups, the word was already ubiquitous. The final "e" was later dropped to lessen the "amine" reference.

Stuart M. Smith

Publication *"Feeding Experiments Illustrating the Importance of Accessory Food Factors in Normal Dietaries."*

Date 1912

Author Frederick Hopkins

Nationality British

Why It's Key Improved the healthcare and treatment of rickets, beriberi, pellagra and a host of other conditions.

Key Experiment **MDMA**
Teutonic love drug created in the lab

Methylenedioxymethamphetamine is a nondescript, colorless oil that boils at around 155 degrees Celsius. Yet this substance has become one of the most controversial chemicals of our time, for it forms the base for the illegal drug ecstasy.

Now a "Class A" drug, ecstasy has found extensive usage in the latter quarter of the twentieth century. Its effects include feelings of euphoria and empathy, which have led to it being dubbed the "love drug."

The history of MDMA is shrouded in mystery. Until recently, very few authenticated facts were known about its development, and myths abounded about how the drug had been developed as an appetite suppressant. One thing is known for certain, the original patent was filed by the German pharmaceutical company Merck, in 1912. Anxious about their role in its development, Merck recently commissioned extensive research to unveil the history of MDMA. Their version of events is this: MDMA was indeed synthesized at Merck, by a little known chemist called Anton Köllisch, who was trying to develop an alternative to the blood-clotting medicine hydrastinine.

Köllisch died in World War I, unaware of the impact his discovery was to have in the century to come. MDMA was used, for a while, as a chemical treatment to help some patients in areas of psychotherapy, but is best known for having a substantial impact on the illegal drug scene – the global trade in ecstasy is estimated to run into billions of dollars every year. Despite the real, and well publicized, health risks, there is little sign that its popularity is waning.

Barney Grenfell

Date 1912

Scientist Anton Köllisch

Nationality Germany

Why It's Key MDMA was, for a while, used as a treatment for patients in some areas of psychotherapy, but is best known for its impact on the illegal drugs scene.

Key Discovery
X-ray crystallography

Since its discovery in 1912, X-ray crystallography has become an experimental technique that is used to analyze samples from all science disciplines. It is the essential tool that allows us to gain information about the structure of everything, from biological samples to minerals.

However it was only discovered when Max von Laue attempted to prove that X-rays were part of the electromagnetic spectrum. To show that X-rays were effectively a kind of light he predicted that when these mystery waves passed through the crystal they would show a diffraction pattern.

The X-ray diffraction pattern illuminates the properties of the crystal, since crystals are made up of repeated units of atoms in layers which have the same dimensions as X-rays. This property allows them to interact: as a wave enters a crystal, some of it is reflected from the first layer of units, and some from the second. Where these reflections have overlapping peaks, they produce a dot on the photographic plate, thereby building up a "diffraction pattern" from the dots. It was a phenomenon that allowed scientists to measure the wavelength of X-rays.

Having figured out the size of the X-rays, it wasn't long before scientists were using them to investigate a large variety of crystals; effectively von Laue's experiment in reverse. The technique is now a prerequisite in drug design and is used extensively throughout modern science.

Leila Sattary

Date 1912

Scientist Max von Laue

Nationality German

Why It's Key X-ray crystallography is now an essential laboratory tool and has been involved across the scientific disciplines.

opposite X-ray crystallography model of a crystal of human serum albumin (each side of a square represents one molecule)

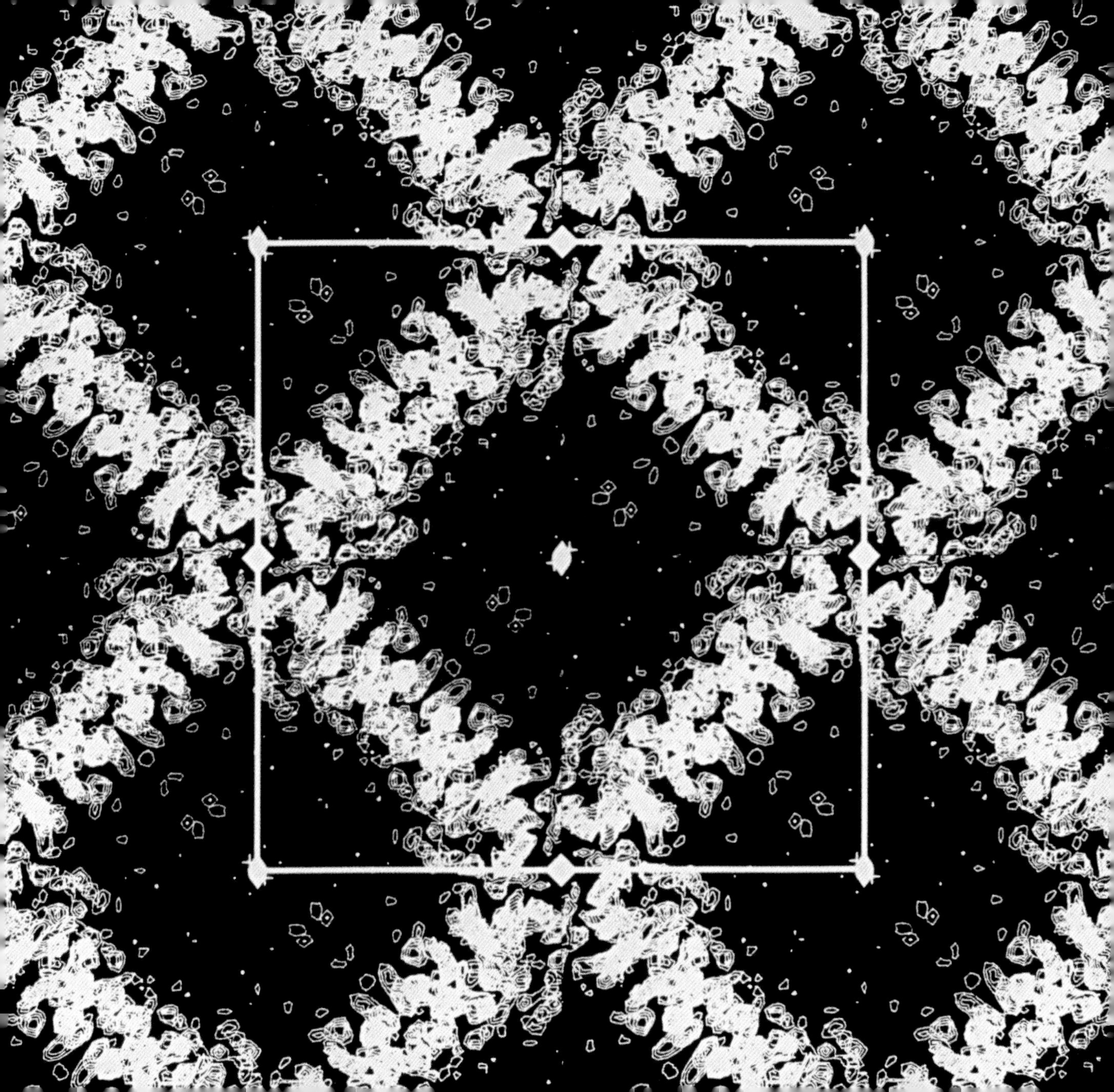

Key Discovery
Diffraction unlocks the secrets of the atom

Henry Moseley was born in 1887 and studied in Manchester under the father of nuclear physics, Ernest Rutherford. After only a few years' work on atomic structure, he joined the army and was killed in 1915. His groundbreaking research may well have won him a Nobel Prize, but unfortunately they are not awarded posthumously.

Moseley studied various elements using X-ray diffraction. He fired a beam of electrons at a pure sample of an element, which excited electrons in the sample to a higher energy state. They then decayed back to the original energy level, emitting the excess energy as X-rays, which Moseley passed through a crystal before recording the lines produced on photographic film. The position of the lines was unique for each element, and related to the element's atomic number. Prior to this, atomic number referred only to an element's position on the Periodic Table. The table was mostly sorted by atomic weight, but with variation to match observed chemical properties. Moseley's research showed the atomic number is the charge on the nucleus – the number of protons it contains.

Electrically speaking, an atom is made of positively-charged protons surrounded by negatively-charged electrons. As the nuclear charge changes, the energy needed to excite an electron – and hence the wavelength of the X-ray released – changes too. It was later shown that isotopes – elements with identical nuclear charge but different mass – were responsible for the variations in atomic weight.

Bill Addison

Date 1913

Scientist Henry Moseley

Nationality British

Why It's Key Provided undeniable physical evidence for Rutherford's model of atomic structure, and showed exactly why the Periodic Table worked so well at predicting behavior.

Key Publication **Bohr's atomic theory**
Orbital quantification

At the start of 1913, the general consensus was that atoms were put together very similarly to our Solar System. You have a massive nucleus in the middle and electrons orbiting it like tiny planets, trapped by their electric charge rather than by gravity.

But there's a problem with this theory. Classical mechanics tell us that a moving charge will release electromagnetic radiation. If an electron is giving all its energy out as radiation it won't have any left to keep orbiting, meaning it'll crash into the nucleus. All of this would happen in milliseconds, which means the fact that there's an age-old Universe full of non-collapsed atoms here is a fairly compelling counter argument to this view. But experiments have proven the shape of the atom, which means the problem lies somewhere else.

In 1913, Niels Bohr published the answer. By using the ideas of quantum theory, Bohr was able to find a way for the shape of atoms to match up with their existence. He quantified their orbits, meaning that rather than being like planets, free to fall into the Sun and through all the space in between, electrons can only follow certain paths. The only times electrons release their energy is when they jump between these paths, so their orbits are safe and atoms can exist.

For his services in the investigation of the structure of atoms and of the radiation emanating from them, Niels Bohr was awarded the 1922 Nobel Prize in Physics.

Douglas Kitson

Publication "On the Constitution of Atoms and Molecules"

Date 1913

Author Niels Bohr

Nationality Danish

Why It's Key It put the finishing touches to the model of the atom that has survived to this day.

Key Publication **Robert Millikan measures the fundamental unit of electric charge**

American physicist Robert Millikan was the son of a preacher, and taught himself physics as there was no one at his college qualified to do it for him. His physics career was varied and significant, but it was his paper "On the Elementary Electric Charge and the Avagadro Constant", published in 1913, which eventually earned him a Nobel Prize. In 1909, along with his student Harvey Fletcher, he began work on the famous oil drop experiment. It consisted of an atomizer that turned oil into a very fine spray of droplets, some of which were electrically charged. A small hole would allow a charged droplet to occasionally enter a chamber with metal plates on the top and bottom. A voltage placed across the plates was able to halt the charged droplet, and it was watched with a microscope. By illuminating the droplet with X-rays, the charge on the droplet could be altered and Millikan and Fletcher found that the measured charge was always a multiple of 1.602×10^{-19} Coulombs.

Millikan has been accused of cheating his student out of the credit for the historic 1913 paper, as well as "cooking" the data it contained – an approach frowned upon by many scientists. The momentous significance of the discovery stands though, and the experiment and its results are still considered among the most important in scientific history.

David Hawksett

Publication *"On the Elementary Electric Charge and the Avagadro Constant"*

Date 1913

Author Robert Millikan

Nationality American

Why It's Key The charge on the electron of 1.602×10^{-19} is now regarded as one of the fundamental constants of the Universe.

Key Event **Ford installs first conveyor assembly line**

Henry Ford is the man that put wheels under the world, beginning with his Model T, which was made between 1908 and 1927. Sturdily constructed and cheaper than other automobiles, it ran well on the many dirt roads of the time. The Model T looked expensive but was actually very simply made. It sold in quantities previously unimaginable. The secret of its success? Mass production.

The first car to be mass produced in the USA was a 1901 model, the Curved Dash Oldsmobile, designed by American car manufacturer Ransome Eli Olds. He was responsible for inventing the basic concept of the assembly line – his cars were placed on small carts for ease of movement during assembly. He is generally credited with starting the Detroit area automobile industry. He produced 425 Curved Dash Olds in 1901, and was America's leading auto manufacturer from 1901 to 1904.

American car manufacturer Henry Ford perfected the assembly line technique and, in 1913, installed the first conveyor belt-based assembly line in his car factory in Ford's Highland Park, Michigan plant.

The assembly line drastically cut production costs for cars by decreasing assembly time. Ford's famous Model T could now be put together in ninety-six minutes. After installing the moving assembly lines, Ford became the world's biggest car manufacturer. By 1927, 15 million Model Ts had been manufactured.

Mike Davis

Date 1913

Country USA

Why It's Key Perfected and confirmed the assembly line technique as the standard for mass-produced vehicle manufacture, thereby reducing costs and making cars readily available to the average person.

Key Invention
The development of sonar

It is often the way that some of the greatest achievements of man are born out of disaster and tragedy. When the RMS *Titanic* sank in 1912, killing around 1,500 people, it spurred a number of scientists into action to prevent similar sorts of atrocity from occurring again. One field that saw some great developments in the following weeks and months was "Sound Navigation and Ranging" – more commonly known as sonar.

The German physicist Alexander Behm decided that he would work on some sort of device that could detect icebergs in the sea. The device he eventually produced turned out to be almost entirely useless for this proposed purpose, as the icebergs didn't reflect the sound waves he was firing at them at all effectively, but he had stumbled across something very important nonetheless. Inadvertently, Behm had created echo sounding. By pointing these waves downwards toward the ocean floor – a surface that reflected them much better – ships and submarines could use it to measure the depth of the water beneath them.

The icebergs didn't go undetected for long however. Reginald Fessenden (famed for being the first to transmit voice by radio wave) created an electromagnetic moving-coil oscillator that was able to detect an iceberg under water from a distance of two miles. Because it made use of low frequencies – the wavelengths it emitted measured three meters – it was not a particularly precise contraption, but it was instrumental in the development and design of submarine detection units.

Chris Lochery

Date 1913

Scientists Alexander Behm, Reginald Fessenden

Nationality German, Canadian

Why It's Key Took theoretical ideas and concepts regarding the potential of sound waves, and created useful, adaptable inventions with them.

opposite Reginald Fessenden pictured on a boat, using his invention, the electric oscillator

Key Discovery
Vitamins A and B

In 1907, the American biochemist Elmer McCollum was studying the effects of different single-grain diets on dairy cattle. It was assumed that similar grains (corn, oats, and wheat) would have similar dietary effects. However, only the corn-fed cows remained healthy; animals fed wheat or oats did not thrive. McCollum reasoned that there must be some undiscovered difference in the structures of the grains.

He decided that nutritional studies would benefit from using small animals like mice or rats rather than cows for experiments – they ate less, took up less space, and reproduced rapidly. He persuaded his university's Dean to allow him to purchase twelve albino rats. This colony of rats was the first established in the United States for nutritional studies

By 1913, McCollum reported that when laboratory rats were put on diets in which lard or olive oil was the only source of fat, they eventually stopped growing. But these same rats quickly resumed normal growth when fed butter or extracts of egg yolk. He had discovered a fat-soluble factor in certain foods that was essential for growth and survival. His findings were backed up when a similar study by Thomas Osborne and Lafayette Mendel using cod-liver oil was published just a few months later. By 1917, McCollum had discovered a water-soluble factor in milk whey. These became "fat-soluble factor A" and "water-soluble B."

The water-soluble factor was later determined to be identical to the "anti-beriberi factor," previously discovered by Christiaan Eijkman and Gerrit Grijns. They had found vitamins.

Stuart M. Smith

Date 1913

Scientist Elmer McCollum

Nationality American

Why It's Key Since the first vitamins were found, they have revolutionized our understanding of nutrition.

Key Invention
Stainless steel

In Sheffield, the home of steel, Harry Brearley, the son of a steel melter, set up Brown Firth Laboratories to analyze and experiment with steel. In 1912, he was asked to help solve the problem of erosion of the internal barrels of rifles, and began to experiment with adding chromium to steel mixtures.

Over the previous century, several scientists had studied alloys of iron with chromium, and had shown them to have high melting points and to be resistant to acids. Brearley was able to use new melting techniques and purer metals to produce, on August 13, 1913, an alloy with a low carbon content containing 12.8 per cent chromium. The next stage in his investigation was to etch the surface of the alloy to examine the grain structure. Etching was carried out using highly corrosive nitric acid, so when the new alloy showed itself to be extremely resistant to this etching reagent, Brearley realized he had found a very special product.

He called the alloy "rustless steel" and commissioned a local cutler's to make knives from it. This was a great step forward for knife manufacture because previously they had either been plated with silver or nickel, or made from carbon steel, which was prone to rust. The new steel, renamed stainless steel, would go on to revolutionize the entire industry.

Simon Davies

Date 1913

Scientist Harry Brearly

Nationality British

Why It's Key A new iron-based alloy which was cheap but extremely resistant to corrosion.

Key Person **Henry Ford**
The original mass producer

A prolific inventor and the holder of 161 patents, Henry Ford (1863–1947) is one of those rare characters who seems to have an inhuman capacity for achievement.

Most famously, he is the Ford behind the Ford Motor Company. As founder and sole owner of the company, he gave his name to the Ford Model T, the revolutionary vehicle that is generally considered to be the world's first affordable automobile. This was due, in part, to his methods of production. Ford is also credited with fathering the modern mass-production assembly line; the T Model line could have a car completed in as little as ninety-six minutes.

As if this were not enough, he also has a selection of social theories and economic philosophies named after him. "Fordism" was a phrase coined to describe his idea of paying his workers a wage relative to the cost of the product they were putting on the market. He reduced their working week and doubled their daily wage – to US$5 – meaning the workers could afford to buy his autos for themselves, creating both a satisfied workforce and a whole new customer base in one swift move.

His Midas touch was used in the war effort, when Ford turned his expertise to aircraft production to boost the Allied forces. Ford managed to work a system whereby B-24 bomber planes could be produced at twenty times their original rates, assembling six hundred of them in just one month.

Chris Lochery

Date 1913

Nationality American

Why He's Key Ford didn't just revolutionize the automobile industry, he revolutionized industry itself.

Key Event
The start of the Behaviorist School of psychology

Behavioral psychology was founded by John Watson and was outlined in his work *Psychology as the Behaviorist Views It*, published in 1913.

Watson's work began with home-based rat studies. He investigated their behavioral responses to various stimuli, building on Pavlov's classical conditioning model. Pavlov showed that dogs could be conditioned to expect food by ringing a bell whenever they were fed. Eventually they salivated whenever they heard a bell, regardless of whether food was presented or not.

Watson believed that any animal's response to a specific situation was dependent on their previous experience. He stated that our reactions as humans were, basically, no different to a rat's. His work brought clarity and objectivity to the previously fuzzy world of psychology, and provided a solid basis for understanding facets of human behavior.

He began to test his theories in humans in 1919. One infamous series of experiments involved a baby known as Little Albert, whom Watson conditioned to have a fear of white rats. Albert was not originally upset by rats, but gradually learned to fear them with repetition of an associated loud noise made whenever rats appeared. After repetition, no noise was needed; Albert became distressed at the sight of not only white rats, but also any furry white object.

Katie Giles

Date 1913

Country USA

Why It's Key Marked the beginning of a more scientifically sound psychological school.

Key Discovery **Nervous beginnings**
The first neurotransmitter

Early in the twentieth century, British pharmacologist Sir Henry Hallett Dale was intrigued by the fungus that causes ergot – an infection of grasses. For centuries, people who had eaten ergot-contaminated wheat had suffered a range of puzzling symptoms such as hallucinations, burning sensations in their limbs, and sometimes even gangrene – caused by the tightening of blood vessels, restricting the flow of oxygen to the tissues.

In 1914, Dale isolated from extracts of ergot one of the substances involved – acetylcholine. His work offered the first clues about how nerves send signals around the body, triggering muscles to contract and organs to function. Seven years later, German-born Otto Loewi published the first of a series of papers showing that nerve signals depend upon the release of chemicals, for example, to stimulate the heart to beat. Loewi's key experiment was to show that a chemical substance is released from the heart upon stimulation via the vagus nerve. He did this by placing two frog hearts in separate chambers, connected in such a way that chemicals could move between the chambers. Loewi called his chemical "Vagusstoff." Dale promptly proved Vagusstoff and acetylcholine to be one and the same.

We now know that there are tens of different neurotransmitters, allowing different types of nerve signaling to take place in different tissues and organs of the body. Dale and Loewi received joint accolades when they shared the Nobel Prize in Physiology or Medicine in 1936.

Julie Clayton

Date 1914

Scientist Otto Loewi

Nationality British, German

Why It's Key The discovery of the first neurotransmitter opened the door to a whole field of research into nerves and how they trigger action throughout the body in a fraction of a second.

Key Invention
Fluoride first added to toothpaste

Toothpaste hasn't always been available in a range of minty fresh flavours. Throughout history its ingredients are thought to have included ox hooves, eggshells, crushed bones, and charcoal. Fluoride as an ingredient is a relatively modern invention. In 1914, Cecil Rudolph Lidgey of Middlesex, England filed a patent for a toothpaste – then called "dentifrice" – that contained "one or more fluorides amongst its components." Fluoride was added to try and prevent decay on the surface of teeth. Its inclusion, however, proved to be a contentious issue.

Some fifteen years later, a representative from the Silica Products Company registered a patent in the United States for a "dental preparation" containing sodium fluoride. However, this was not well received by the American Dental Association (ADA). In 1937 the ADA's Council of Dental Therapeutics noted several objections, stating: "The use of dentifrices is unscientific and irrational, and therefore should not be permitted."

The ADA changed its position some years later when it approved Procter & Gamble's Crest toothpaste, saying it was "an effective anti-caries" agent. This process took ten years from the first time Procter & Gamble approached the ADA, who insisted that the company conduct detailed research into the effectiveness of the toothpaste in preventing tooth decay.

Today, the vast majority of toothpastes sold contain fluoride. As for the hooves, bones, and charcoal, that's a different story...

Christina Giles

Date 1914

Scientist Cecil Rudolph Lidgey

Nationality British

Why It's Key Produced toothpastes that, for the first time, combined therapeutic benefits with cosmetic ones.

opposite Fluoride ions combine with enamel to produce protective fluorapatite crystals

Key Person **Joseph Goldberger**
Fighting a cereal killer

Joseph Goldberger (1874–1929) was an epidemiologist who worked tirelessly for fifteen years investigating pellagra, a disease which caused diarrhea, dementia, dermatitis, and in some cases death, throughout much of the Southern United Sates.

Goldberger grew up in New York, after emigrating from Hungary at the age of six with his parents and five siblings when the family's sheep flock was ravaged by disease. Intent on becoming an engineer, his ambitions changed after sitting in on a medical lecture. He became a doctor in 1885 and, after working in private practice for a few years, joined the US Marine hospital service. Goldberger investigated many infectious diseases, contracting yellow fever himself along the way. His work on pellagra started in 1914, at the personal invitation of the surgeon general, and was considered controversial in that it challenged the common belief that pellagra was an infectious disease. Despite injecting himself and a colleague with infected blood and eating the scabs of a sufferer, Goldberger could not convince other colleagues that pellagra was non-infectious. He ran trials in prisons, showing that a balanced diet stopped pellagra developing, but the implication that poor diet and poverty were responsible offended some Southerners, who perceived his criticism as a threat to their incomes.

Goldberger's search for the elusive "pellagra preventative factor" was never completed. Sadly, he died of cancer in 1929, eight years before Conrad Elevjhem confirmed niacin (vitamin B3) deficiency as the cause of pellagra.

Katie Giles

Date 1914

Nationality Hungarian-American

Why He's Key An uwavering belief in his own knowledge led to a discovery that saved many lives.

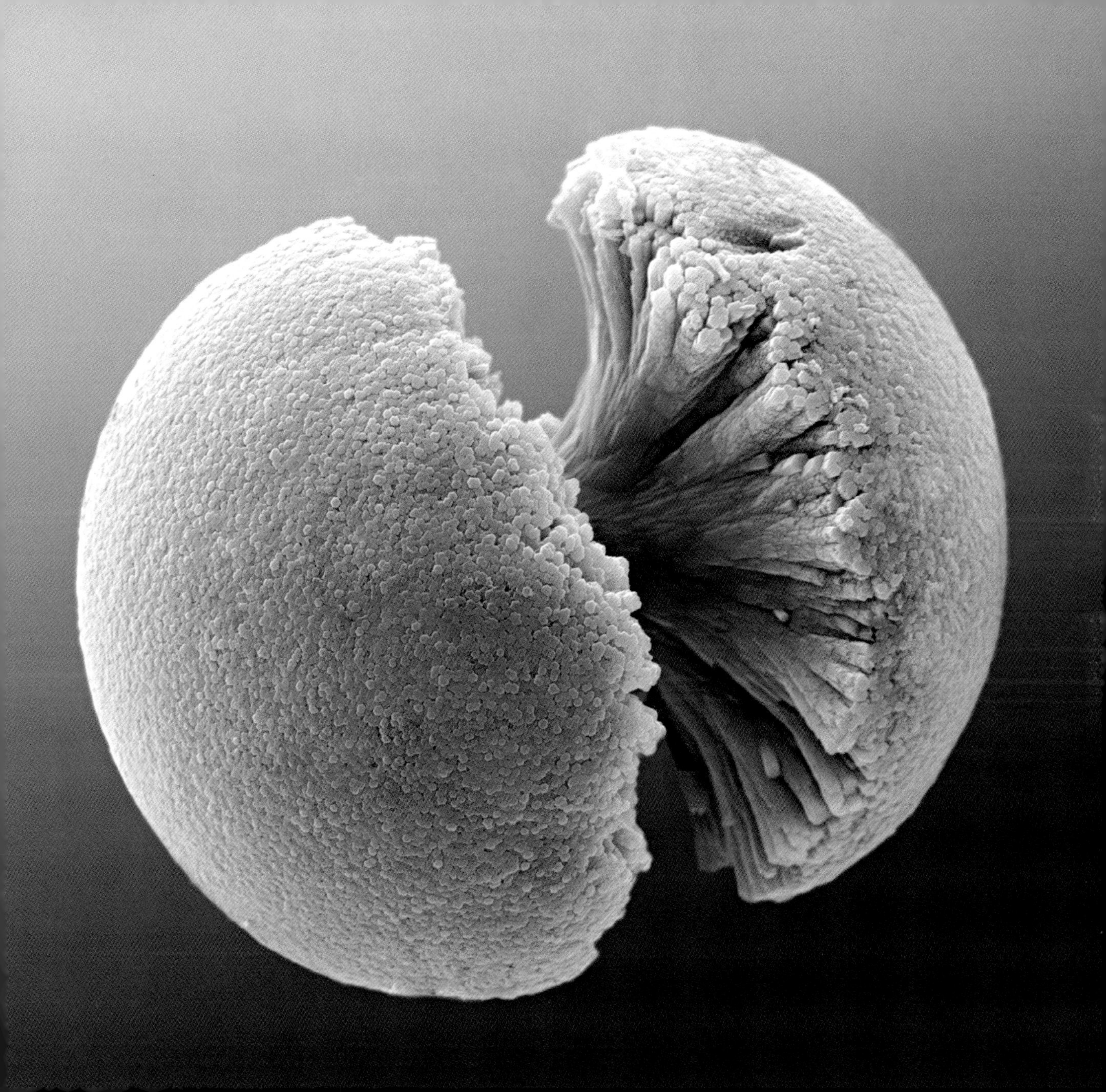

Key Event
Passenger pigeon goes extinct

Once the most abundant bird in North America, it is thought that there were around five billion passenger pigeons alive just five hundred years ago. Flocks up to a mile wide and 300 miles long were recorded during their annual migration from Canada and Northern USA to Mexico; each flock would take several days to pass.

And then, over the course of just a hundred years, the passenger pigeon went from being one of the most abundant birds in the world to being extinct. Following the European colonization of America, large areas of land were deforested, destroying much of the pigeons' habitat. Pigeon meat was then introduced as a hugely abundant source of high-protein food. These two factors led to a catastrophic decline in their population.

In 1857, a bill was brought to the Ohio State Legislature seeking protection for the passenger pigeon. A select committee responded: "The passenger pigeon needs no protection. Wonderfully prolific and no ordinary destruction can lessen them, or be missed from the myriads that are yearly produced."

Despite this optimism, by 1890, numbers were beyond rescue. A few remaining birds were kept in captivity with a view to captive breeding. However, the passenger pigeon was known to be a highly sociable species that would only breed when in large colonies.

The last remaining passenger pigeon, Martha, died in Cincinnati Zoo, Ohio on September 1, 1914. Her remains are in the archived collection at the Smithsonian Institution.

Vicky West

Date 1914

Country USA

Why It's Key Humans drive one of the most abundant species in the world to extinction.

Key Discovery
How to make a hormone

The condition we know as hypothyroidism was first described in 1874 by Dr William Gull, physician to Queen Victoria (and unlikely Jack the Ripper suspect). The afflicted suffer from fatigue, muscle cramps, and other unpleasant symptoms.

It was soon discovered that the condition could also be caused by removing the thyroid gland from someone's neck. In an early, if unrefined, effort to correct the problem, Victorian doctors tried implanting sheep thyroids, with some success. Thyroid "extract" from animals was shown to be effective, and the search was on for the active ingredient. The first clue came in 1895, when large protein-bound concentrations of iodine were found in the gland. It took another nineteen years before Edwin Kendall, working at the Mayo Clinic in Minnesota, hit upon the source. He isolated an iodine-containing compound from thyroid extract. The molecule, which he dubbed "thyroxine," acts as a hormone, circulating round the body, controlling metabolism. Kendall was never able to pin down the chemical structure of his compound, however, which took others another twelve years to find.

Thyroxine soon entered the clinician's armory and is used to treat hypothyroidism. Knowledge of the hormone also led to treatments for hyperthyroidism, a disease in which the thyroid gland is overactive.

Matt Brown

Date 1914

Scientist Edwin Kendall

Nationality American

Why It's Key Led to effective treatment for several diseases and greatly increased our knowledge of hormone action.

opposite TEM of a follicle from the thyroid gland. A layer of epithelial cells (pink) producing the hormones tri-iodothyronine and thyroxine, surround a central chamber where they are stored as the glycoprotein thyroid colloid (brown).

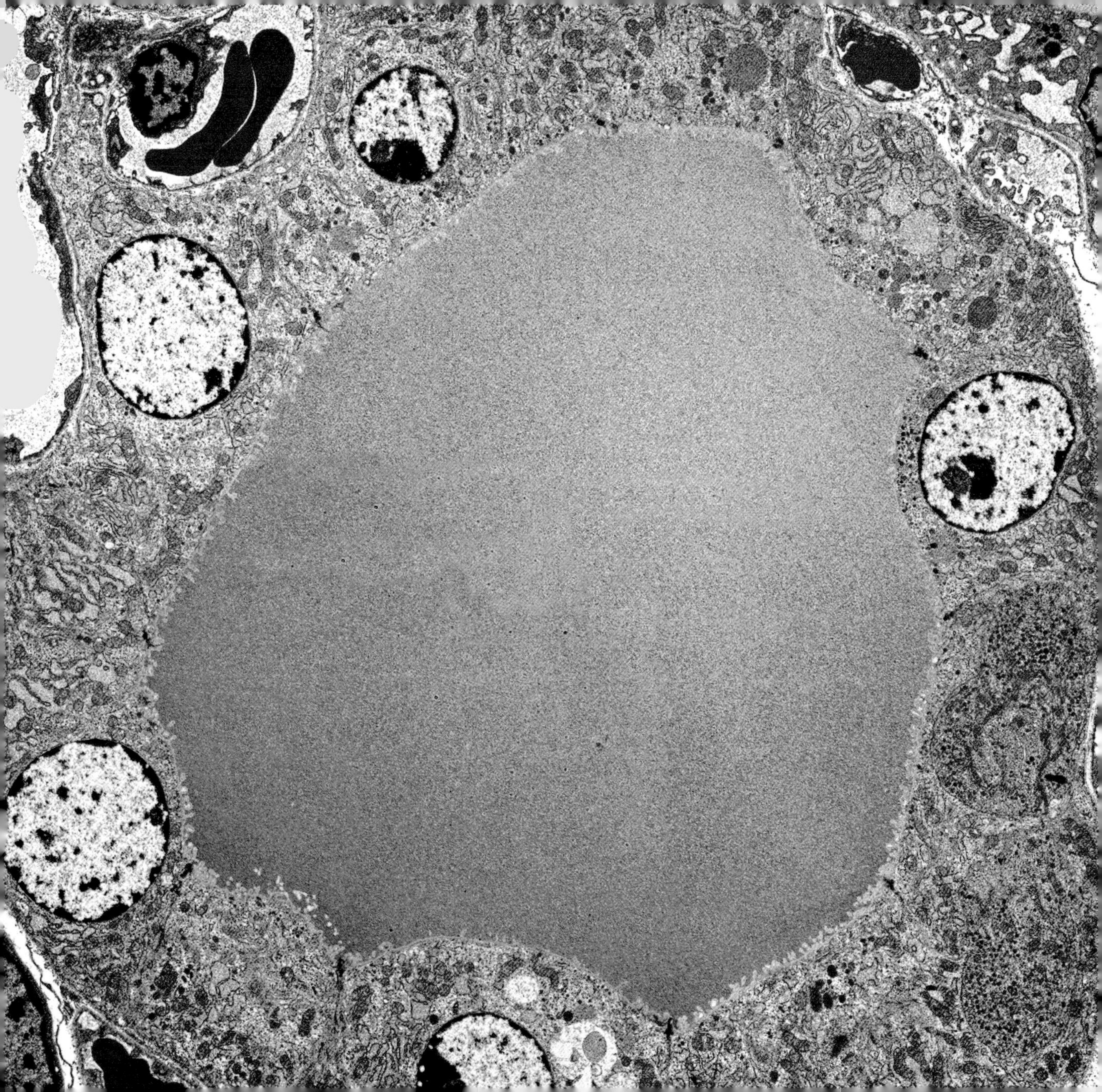

Key Discovery
Bohr's quantum theory of the electron confirmed

In 1913, the Danish physicist Niels Bohr introduced his radical new model of how an atom works. It had been known previously that the positively charged nuclei of protons and neutrons were surrounded by a cloud of negatively charged electrons, but what Bohr proposed was that the electrons could only exist in certain "allowed orbits," each with a distinct associated energy level.

One year later, two scientists were ready to perform an experiment that would test Bohr's discrete, rather than continuous, "quantized" energy states for electrons – quantum theory. German physicists James Franck and Gustav Hertz built a tube containing mercury vapor. An electrode at one end fired electrons into the vapor. These electrons were accelerated through a wire mesh in the middle of the tube, using a voltage. At the far end of the tube was a collector that measured electrical current.

As Franck and Hertz increased the voltage across the mesh, the current measured at the far end started to increase as anticipated, since more and more electrons were reaching the collector. But when the voltage reached 4.9 volts, the current dropped suddenly. What had happened was that the mercury atoms in the tube had been fed enough energy by the free electrons to allow the electrons in the mercury atoms to move into a higher orbit, and the free electrons were not left enough energy to reach the far end of the tube. It was enough to prove that electrons orbit an atom in discrete orbits, rather than a continuous cloud.

David Hawksett

Date 1914

Scientists James Franck, Gustav Hertz

Nationality German

Why It's Key Proof that Bohr's model of an atom was correct.

Key Person **Richard Martin Willstätter**
From cocaine to chlorophyll

"For his researches on plant pigments, especially chlorophyll," reads the citation for Richard Martin Willstätter's 1915 Nobel Prize for Chemistry. By the date of this award, the German chemist had also invented paper chromatography and revealed the structure of cocaine.

Born in Karlsruhe in Baden on August 13, 1872, Willstätter attended Technical School in Nuremberg. At eighteen, he went to the University of Munich, where he studied science, receiving his doctorate in 1894. He then continued his research into the structure and synthesis of related plant alkaloids as an assistant to Adolf von Baeyer.

In 1905, he accepted a professorship at the University of Zurich and commenced his research on chlorophyll. He determined its structure and showed that the porphyrin compound found in it bears a structural resemblance to that of the blood pigment, heme.

During his time as honorary professor of chemistry at the University of Berlin and director of the Kaiser Wilhelm Institute, his investigations disclosed the structure of many of the pigments of various fruits and flowers. When, in 1918, this work was interrupted by the war, he turned his attentions, at the request of Fritz Haber, to developing a gas mask. His glittering career came to an end when, as a gesture against increasing anti-Semitism, he announced his retirement in 1924.

Mike Davis

Date 1915

Nationality German

Why He's Key Willstätter's discoveries in the area of photosynthesis and into the nature and activity of enzymes were the precursors of modern biochemistry.

Key Event **Einstein announces the General Theory of Relativity** Everything is warped

By 1915 Einstein had already stated that no matter how fast you and everything around you is moving, the laws of physics should be the same. He had also equated the action of gravity with plain, straightforward acceleration. Finally, Einstein knew that the speed of light was a constant, regardless of the motion of the person measuring it. With these three pieces of information in hand, Einstein put together a description of the Universe that was like nothing seen before.

Since everything in the Universe is moving relative to everything else, there can be no absolutes. And as simply as that, Einstein trashed the idea of one absolute time and absolute space. Everyone measures their own time and their own distance, and these can only be true for them. As if this wasn't bizarre enough, one of Einstein's earlier ideas now comes into play; anything that can be described as due to acceleration can also be described as due to gravity, leading to the disturbing realization that a gravitational field can distort space and time.

Einstein's theory was arguably the single most incredible leap of understanding one man has ever made. If Einstein hadn't produced the theory of special relativity, someone else would have; if he hadn't produced the general theory, wrote C.P. Snow in 1979, we would still be waiting for it.

Kate Oliver

Date 1915

Country German-Swiss

Why It's Key The Universe became bendier, and a lot weirder, than we previously thought. Space, time, and measurements were no longer sacred but at the whim of acceleration and gravity.

Key Publication **General relativity stands up to scrutiny** Schwarzchild solves the unsolvable

The sixteen equations accompanying Einstein's mathematical description of general relativity looked so complicated that he doubted a solution would be found. But just a month later, German physicist Karl Schwarzchild submitted two exact solutions to the equations, which Einstein then verified. Still more surprisingly, Schwarzchild had submitted them from the front line, where he was fighting the Russians in World War I.

The Schwarzchild solutions for the distortion of space-time were for two special cases, one describing the distortion in an object of uniform density that looks the same from all directions, and another describing the distortion in the empty space around a large body.

The latter description was accurate for describing the gravitational field of the Solar System, caused by the Sun, and the predictions it made about the orbit of the planet Mercury fitted the observed data perfectly – something no previous theory had been able to do. This provided more support for the principle of general relativity.

But Schwarzchild's description also held true for another object – one we now know as a kind of black hole – that drew more and more matter inside it, gathering it all at one point. At the time, Schwarzchild regarded his solution as "physically meaningless", having no representation in the real world, but it was later extended to describe the radius of an event horizon – the nearest distance to which something can approach a black hole before being sucked in.

Kate Oliver

Publication "On the Gravitational Field of a Mass Point According to Einstein's Theory"

Date 1915

Scientist Karl Schwarzchild

Nationality German

Why It's Key A physicist solves an unsolvable equation, describing the shape of space-time, while at war with Russia. Twice. What more do you want?

Key Invention **Pyrex** The breakthrough glassware

Eugene Sullivan and William Taylor of the Corning Glass Works, New York, probably never expected their names to become part of history. They established a research lab at the glass works in 1908, with the aim of making a heat-resistant glass for the lenses of railroad lanterns. Normal glass, which is not only a poor heat conductor, but also expands greatly when it gets hot, tended to break under the fierce glare of the lanterns. In their attempt to improve on these qualities, they came up with borosilicate glass, which they called Nonex. This material found many uses, including as battery jars for railroad telegraph systems.

When Corning colleague Jesse Littleton took home a cut-off battery jar to his wife and asked her to bake a cake in it, the rudimentary culinary device showed the first signs of how massive a development it would become. It was Littleton who worked to refine Nonex's properties, and eventually Pyrex was trademarked and patented in 1915.

The timing could not have been better. The outbreak of World War I meant that supplies of high-quality glassware from Germany were cut off and Pyrex began to take over the new gaps in the market. Borosilicate glass takes its name from the element boron, which likes to form a variety of strong chemical bonds with other elements. It was Sullivan and Taylor's work in exploring the new properties of their tough glass that made Pyrex the most important breakthrough in glassmaking for more than three thousand years.

David Hawksett

Date 1915

Scientists Eugene Sullivan, William Taylor, Jesse Littleton

Nationality American

Why It's Key Just look in your kitchen cupboards.

Key Invention **The neon lamp**

Neon gas, along with krypton and xenon, were discovered by William Ramsey and Morris Travers in London in 1898, but it took another seventeen years for one of its most useful properties to come to light.

French engineer and chemist Georges Claude was the first person to apply an electrical charge to a glass tube of neon, creating a lamp which he patented in 1915. Claude's company subsequently sold neon signs to a Packard car dealership in Los Angeles for US$24,000, and the popularity of neon, or "liquid fire" signs spread throughout the world. Not only visible during the day, neon provides a spectacular advertising display at night.

Neon signs can be made in almost any shape and color although most are continuous, hollow glass tubes, which are heated and shaped. Most of the air is removed from the tube and an electrical charge is applied, then the remaining air is removed to create a vacuum, which allows the tube to be back-filled with argon or neon at high pressure.

Different types of glass are used depending on the size and voltage of the sign. Neon glows even at room temperature. All true neon signs are red in color although over 150 different colors are available by adding other gases – blue (mercury), white (carbon dioxide), and gold (helium), can be mixed in to create further colors, and the glass tubes can be coated with phosphor to produce the ultraviolet spectrum.

Vicky West

Date 1915

Scientist Georges Claude

Nationality French

Why It's Key Neon lamps revolutionized the advertising industry.

opposite Four of the earliest neon lights including a neon lamp of about 1922 (far right); an Osram "Glimmelamp" (top); and a beehive-type neon lamp

neon

Key Publication ***The Origin of Continents and Oceans*** Wegener's Pangaea

When Alfred Wegener proposed that there had once been a giant super-continent that later split into the land masses we recognize today, he was met with a fair dose of scepticism. The problem he faced was finding a convincing mechanism that would explain the movement of the continents, to back up the extensive evidence he had collected.

It was in 1911, whilst teaching at the University of Marburg, that Wegener first became struck by the idea that the continents could once have been joined. Whilst browsing the University library, he came upon a list noting identical fossil types and animal species, which were separated by vast oceans. From there he began collecting data to support his hypothesis of continental drift; he noticed how the coastline of Africa fitted the coastline of South America; found evidence of matching distinct rock strata in South Africa and Brazil; and came across arctic fossils of plants usually found in tropical regions.

In 1915, Wegener published his findings in *The Origin of Continents and Oceans*, and named the super-continent Pangaea, from the Greek for "all the earth." Updated editions were published – in 1920, 1922 and 1929 – to include the mounting evidence.

Unfortunately, it wasn't until after his death, in 1930, that scientists found the mechanism Wegener was searching for. We now know that the continents and the ocean floor form plates that "float" on the asthenosphere – the upper part of the liquid-like mantle – and remain in motion to this day.

Faith Smith

Publication *The Origin of Continents and Oceans*

Date 1915

Author Alfred Wegener

Nationality German

Why It's Key Provided the first evidence that the continents were once all part of the same super-continent.

Key Event **First full-scale deployment of chemical weapons** The second Battle of Ypres

During World War I, the small Belgian town of Ypres was the scene of three vicious battles. During the first, the German forces lost 135,000 soldiers against the Allied army, represented by French, Algerian, and Canadian troops. So in the second battle, the Germans switched tactics. After an initial bombardment, they released large quantities of chlorine gas; it was the first ever large-scale deployment of chemical weapons. The gas formed a yellow-green cloud which drifted slowly toward Allied trenches. The soldiers thought it was simply a smoke-screen and the sweet smell of pineapple and pepper gave little warning of its toxicity. Chlorine is denser than air; it worked its way down into the trenches, causing the hideous devastation later immortalized in Wilfred Owen's poem "Dulce et Decorum est."

Soldiers initially complained of pain in their chests and a burning sensation in their throats. Those who had inhaled a lot of gas had their respiratory systems destroyed and suffered choking attacks; a slow and painful death ensued by asphyxiation. The soldiers soon realized that they had been gassed and retreated as fast as they could.

After these chemical attacks, Allied troops were supplied with cotton pad masks that had been soaked in urine; the ammonia in the urine neutralized the chlorine. The pads were held over the face until the soldiers could escape from the poisonous fumes.

Riaz Bhunnoo

Date 1915

Country Belgium

Why It's Key The release of chlorine gas represented the first ever large-scale deployment of chemical weapons.

Key Publication
The Respiratory Exchange of Animals and Man

August Krogh's expertise covered a wide number of areas of animal biology, and throughout his prolific career he contributed a great deal of important work to the field, including previously under-studied areas of physiology. His greatest and most renowned work was that for which he picked up a Nobel Prize in 1920, a study he entitled *The Respiratory Exchange of Animals and Man*.

Krogh and his colleagues looked at respiratory action in muscles during "work," and discovered that blood flow increased greatly during activity. They began to realize that there must be some other action, besides blood pressure, working to increase the passage of oxygen into the muscle. Krogh concentrated on the actions of capillaries – tiny, one cell-thick blood vessels – and in particular, those in frogs' tongues. Under the microscope, he observed the appearance and disappearance of the hard-to-spot vessels as they seemed to contract and relax. He suggested that they opened and closed via a contraction mechanism, increasing and decreasing their ability to diffuse oxygen.

Not one to concentrate on just one group of animals, Krogh later went on to show that similar exchange mechanisms occured in the tracheal systems of insects. Thanks mainly to this work, he is now considered by many to be one of the founders of exercise science.

Katherine Ball

Publication *The Respiratory Exchange of Animals and Man*

Date 1916

Author Schack August Krogh

Nationality Danish

Why It's Key Uncovered the method which increases the passage of oxygen to the muscles, and laid the groundwork for showing that respiratory action occurs in a wide range of species.

Key Experiment **Crystallizing ideas**
The Czochralski method

One evening in 1916, a young Polish chemical engineer called Jan Czochralski was at work in his laboratory in Berlin. He was studying the crystallization of metals, and had just melted some tin in a crucible. He placed it to one side while he wrote up some notes, before absent-mindedly dipping his pen in the tin instead of his ink pot.

Immediately, his mind was ignited by what he saw. A very thin thread of solid metal led from his pen to the crucible as he pulled it away. He set to work to repeat the experiment using narrow capillary tubes and seed crystals, and his analysis showed that what he was producing were single crystals of metals. They were just one millimeter thick and up to 150 millimeters long.

Czochralski published his work, but his method of producing large single crystals was forgotten until after World War II. It was only then that the need for crystals of silicon and other semiconductors led scientists to this method again. Now named the "Czochralski method," this way of making single crystals is used throughout the electronics industry and is especially important in solar panels.

Czochralski didn't fare quite as well as his technique following World War II. Although later found to have committed no wrongdoing, he was stripped of his professorship at the Warsaw University of Technology for having worked with the Nazis. He saw out his days running a small cosmetics shop in his hometown of Kcynia.

Simon Davies

Date 1916

Scientist Jan Czochralski

Nationality Polish

Why It's Key A foundational technique for producing semiconducting crystals for the electronics industry, and important components necessary for harnessing clean solar energy.

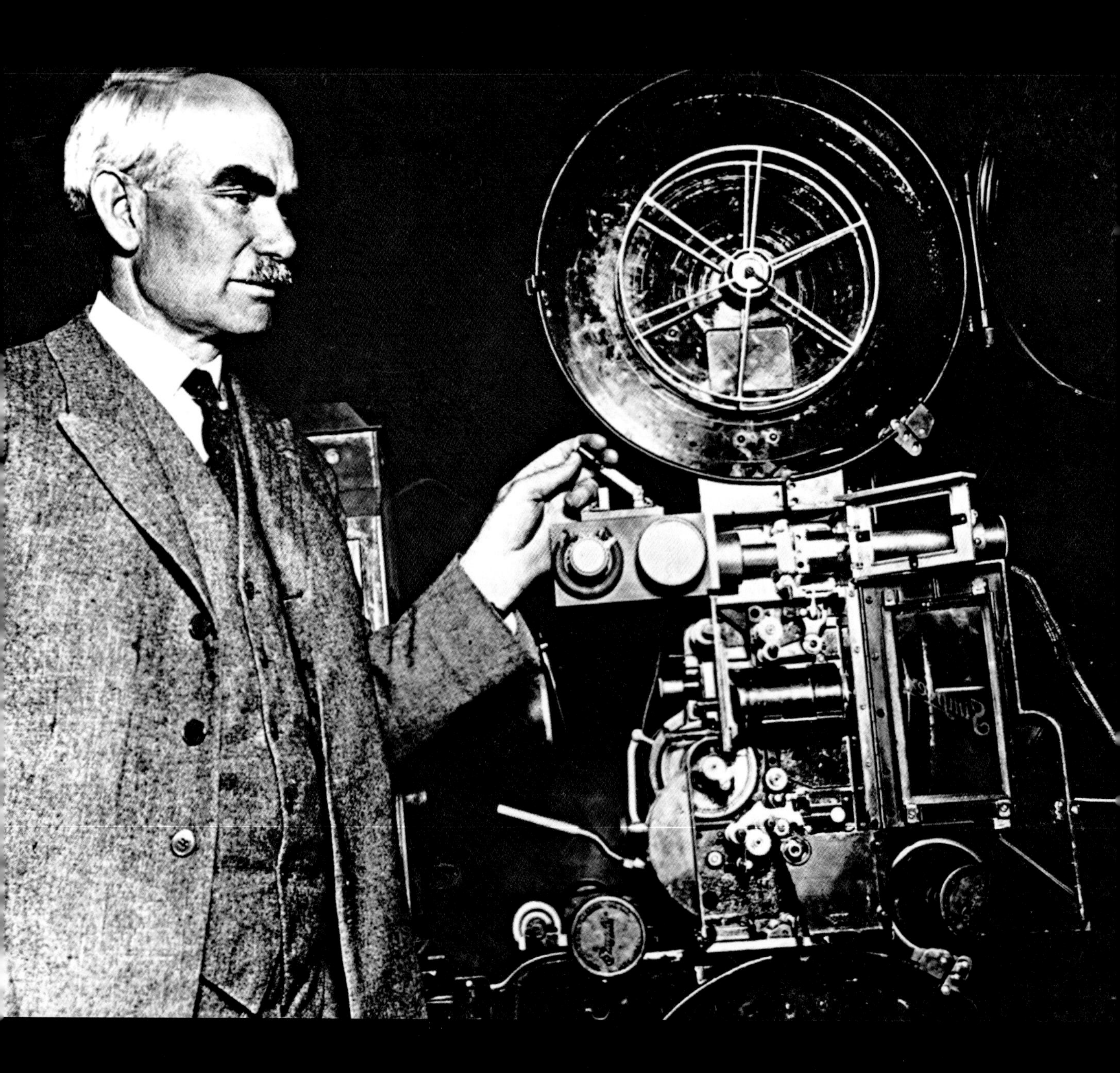

Key Person **Lee De Forest**
Inventor extraordinaire

Lee De Forest (1873–1961) was born in Iowa, USA and pioneered wireless telephone transmission as an alternative to telephones using cables. This was intended for the U.S. Navy but later evolved into transmitting radio programmes for the general public. The basis for these and similar developments followed from his invention of the triode valve, which converts radio frequencies to audible frequencies and vice versa. Voice and music signals could be sent over long distances without the use of wires, clearly of great value to ships at sea.

The radio telephones for the Navy were tested by playing gramophone records into a microphone and checking how well they could be heard at the remote receiver. But owners of radio-telephones could also hear the transmissions and listened to them simply for enjoyment. This led De Forest to the idea of developing music and voice for the general public and he started to transmit general-interest programmes. A unique "first" was his coverage of the Hughes-Wilson presidential election of November, 1916. He also invented ways of adding sound signals to motion pictures, heralding the age of the "talkie." But even as his inventions shaped the future, Forest wasn't known for his powers of foresight. "While theoretically and technically television may be feasible, commercially and financially it is an impossibility," he said in 1926.

Lee De Forest was a controversial figure with many lawsuits to his name, but also many patents – about three hundred; his contribution to radio technology is universally recognized.

Fabian Acker

Date 1916

Nationality American

Why He's Key Paved the way for modern radio with his technical inventions and innovative programmes.

opposite Lee De Forest and his thermionic amplifier

Key Discovery **The secret of succession**
How grassland becomes meadow becomes forest

We might think of meadows and forests as still, peaceful places, but ecologists see them as constantly changing worlds where entire alliances of species eventually replace each other. This concept, known as "plant succession," is the brainchild of two ecologists, Henry Cowles and Frederic Clements.

In Cowles' time, ecology was in its infancy, but he was so taken by it that he learned Danish just to be able to read an important study. He became fascinated by plants on the sand dunes bordering Lake Michigan and realized that walking inland from the lake was like walking through time. He saw grasses give way to flowers, and eventually to trees and argued that these physical changes mirrored changes over time. Each plant community creates conditions that allow it to be replaced by the next. He published his theory in 1899 and, even though it was only his PhD thesis, it remains one of ecology's most influential works.

A few years later, Clements took up Cowles' ideas and ran with them. He imagined vegetation to be a sort of super-organism, where different individuals and species are akin to cells and tissues. The species that make up this super-organism change over time in predictable ways, eventually settling in a "climax community," that persists until an external influence like fire or climate change pushes the reset button. He published his theories in 1916 and they dominated plant ecology for decades, although his climax community idea later fell out of favor. Theories too, it seems, must yield to succession.

Ed Yong

Date 1916

Scientists Henry Cowles, Frederic Clements

Nationality American

Why It's Key Revolutionized ecology by advancing the idea of ecosystems as dynamic entities that change over time in predictable ways.

Key Discovery **Heparin**
Putting a stop to blood clots

The identity of the first person to discover heparin is richly disputed. Since the 1940s, heparin has been used during various surgical and medical procedures – such as in open-heart surgery and in treating patients following heart attacks – to prevent blood from clumping and blocking small veins. Medically speaking, it is an "anti-coagulant," and distinct from clot busters which break down existing blood clots.

The controversy centers on Jay McLean, who in 1916 was a medical student working in the laboratory of William Henry Howell. McLean discovered an anti-coagulant in a fatty extract of dog liver – which Howell acknowledged in 1917. They called the substance "heparphosphatide" from the Greek *hepar*, for liver. But in the early 1920s, Howell isolated an anticoagulant from dog liver that was apparently chemically different from McLean's isolate – it contained carbohydrate – and named it "heparin." McLean devoted much of his life to arguing that heparin was present in the 1916 extracts. Either way, the extracts were crude and poisonous to humans, so not of much use, until Connaught Laboratories purified heparin ready for clinical trials in 1935.

We now know heparin to be found throughout the animal kingdom, including in creatures with no blood clotting system, such as lobsters and clams. Some scientists suggest therefore that its true purpose is as yet unknown, and that the prevention of blood clotting is a side effect.

Julie Clayton

Date 1916

Scientists Jay McLean, William Henry Howell

Nationality American

Why It's Key The discovery of heparin has made much of modern medicine possible, through its use in preventing blood clotting during surgery and other medical procedures.

Key Person **Gilbert Lewis**
The man behind the chemical bond

At the beginning of the twentieth century, physicists related the electronic structure of atom to two phenomena: the chemical bond and the valence, or number of bonds that an atom will form. Noble gases such as helium were found to be stable and it was believed that this was because they have eight electrons in their outer shell. Gilbert Lewis (1875–1946) termed this the "group of eight" and theorized in 1902 that an atom was cubic with electrons at its eight corners. This explained the eight groups of the periodic table and backed up the idea that chemical bonds formed by transfer of electrons.

Fourteen years later, Lewis published a theory that a chemical bond is formed when a pair of atoms share two electrons so that each can have eight electrons in their outer shell and remain stable. This became known as the octet rule, and enabled Lewis to propose structures for the halogens, such as chlorine. His rule explained both covalent bonds, where the pair of electrons sits in between two atoms, and ionic bonds, where the pair of electrons resides much closer to the more positive atom.

Lewis went on to describe the behavior of acids and bases using this theory. A Lewis acid has only six electrons so will look for a Lewis base, a chemical with a spare pair of electrons, to form a covalent bond. Each species will have eight electrons in its outer shell and will therefore be stable.

Riaz Bhunnoo

Date 1916

Nationality American

Why He's Key Lewis revolutionized the theory of chemical bonding. His "group of eight" work was fundamental, and today the octet rule underpins the whole of modern chemistry.

Key Publication **Birth Control**
Tackling a taboo subject

Birth control nowadays comes in many shapes and forms, but imagine the hostility to the idea back in the prudish 1900s. It took the pioneering work of two women from very different backgrounds to put it firmly on the agenda. Margaret Sanger was an American public health nurse; Scottish-born Marie Stopes was a highly successful botanist and geologist. Their paths crossed due to a mutual ambition: to develop safe birth control methods for women.

Sanger's then radical views on female sexuality resulted in her deportation from the United States, while Stopes discovered that her first husband was impotent and had their marriage annulled on the basis it had never been consummated. With the backing of her second husband, Stopes published *Married Love*, the UK's first sex manual. She was not, however, medically trained, and consulted with Sanger before publishing a guide to contraception, *Wise Parenthood*. Sanger's own book, *What Every Mother Should Know*, and *Wise Parenthood* were well-received by the female public, but birth control in any form other than abstinence was forbidden by the Church of England and the Catholic Church. Both women were tarnished by supposed links to eugenics movements.

Nevertheless, Sanger and Stopes opened birth control clinics across the United States and UK, sowing the seed in women's minds that they had a right to sexual fulfillment and safe birth control. Sanger was an elderly woman by the time the contraceptive pill, which she helped develop, was being marketed. Stopes didn't live to see it.

Ceri Harrop

Publication *Married Love, Wise Parenthood, What Every Mother Should Know*

Date 1917

Authors Margaret Sanger, Marie Stopes

Nationality American, British

Why It's Key It finally became acceptable for women to embrace their sexuality and control the size of their families, eventually leading to nationalized family planning in the late 1970s.

Key Event
Guided missiles tested for first time

Solid fuel rockets date back to Chinese firecrackers in 300 BCE, but by the early 1900s, scientists were starting to get a little more adventurous. One outlandish concept involved launching rockets into clouds and exploding them to prevent hailstorms, but unfortunately the rockets weren't the only things to explode; so did the hailstorms.

Starting in 1914, the first studies into guided missiles – rockets whose trajectories could be pre-programmed or altered in-flight – were started. Under the direction of Prof Archibald Low, the British army began a secret missile project, code named "Aerial Target" (AT), with the remit to find a way of remotely controlling an unmanned aircraft – thus effectively producing a guided missile. The project name was meant to confuse the enemy into thinking that they were merely developing targets for anti-aircraft guns.

On March 21, 1917, in front of an audience of generals, the AT was launched from the back of a lorry, using compressed air. The team managed to demonstrate controlled flight before the aircraft crashed due to engine failure. Despite receiving a "thumbs up" for feasibility, it was considered that the project still had limited value in its current state, and was later scrapped; but this was just the start.

The Nazis were first to use guided missiles operationally during World War II, including the famous V2s, whose flight path was programmed prior to launch. These days, advanced computer technology allows missiles to be fired across thousands of miles and landed to within a few feet of their intended target.

Andrew Impey

Date 1917

Country UK

Why It's Key Changed the face of modern warfare.

Key Discovery
Tesla's primitive radar units

Nikola Tesla was brilliant. He invented the twentieth century. From robots to radios, and not forgetting alternating current electricity, the great Serb either invented it himself or gave the foundations for others to build on.

Among his lesser known ideas was something that sounds remarkably like radar. In the early 1900s Tesla was working at his Wardenclyffe Tower research station on Long Island, USA, where he had moved in order to undertake experiments with high voltage and high frequency electromagnetic waves. During this remarkably productive period, when he also started to work on a directed energy weapon – or "death ray" – Tesla made a conceptual breakthrough by conceiving of a system for using transmitted high frequency signals to detect metal objects at a distance. Following the outbreak of World War I, Tesla described one such system for the detection of submarines in *The Electrical Experimenter* magazine in 1917.

These ideas, though flawed by Tesla's failure to account for the attenuation of radio waves traveling though water, demonstrated the great man's foresight and inventiveness; it wasn't until 1935 that the first true RADAR systems began to be developed by Sir Robert Alexander Watson-Watt of the UK Air Ministry. Just like the use of radio waves for signal transmission, however, these systems depended upon Tesla's theories and experiments relating to the properties of high frequency electromagnetic waves.

Neal Anthwal

Date 1917

Scientist Nikola Tesla

Nationality Serbian

Why It's Key Tesla was among the first to conceive of the use of radio waves to detect distant objects, leading to the later development of radar.

opposite Nikola Tesla in his laboratory

Key Event **Spanish flu**
The deadliest disease you've barely heard of

What killed more people than World War I, in the space of less than two years, but is only very casually mentioned in the history books? Why the 1918 flu pandemic, of course. What started out as relatively benign flu season in 1918 blossomed into one of the most virulent and deadly outbreaks ever seen. Though often referred to as the Spanish flu, it is thought to have actually originated in China and infected some 40 million people worldwide. As a direct result of this virus, the average life span in the United States dropped by ten years.

The 1918 flu was unique because it targeted the section of society diseases tend to leave alone. Influenza typically poses the greatest risk for individuals at the extreme ends of life: the elderly and the very young. The 1918 flu was different; it typically killed young, healthy, strong adults, and it did so with speed. In many cases, it was spread by soldiers, mobilized in great numbers for the Great War, often bringing the virus with them as they traveled home.

What made this flu so virulent was the fact that it had recently mutated in form and, as a result, humans had very little resistance to it. Many scientists see a strong parallel between the 1918 flu epidemic and the current strain of avian flu. Both influenzas most likely originated in animals other than humans and subsequently infected humans – which means that another global epidemic of 1918 proportions is not out of the question today.

B. James McCallum

Date 1918

Country Global

Why It's Key The 1918 flu pandemic killed more people than World War I, and there is no reason why we couldn't have a similar pandemic today. Though seldom spoken of, it was the twentieth-century equivalent of the bubonic plague.

Key Discovery **Our galactic center**
It's not Earth

In the history of astronomy, there have been many chances to redefine our Universe and to change our place in it. But whereas in the past, changes have met with dangerous opposition from religion, the modern world has been much more open to these new ideas.

In 1918, an astronomer named Harlow Shapley made a discovery that followed in the footsteps of Galileo and Copernicus, moving humanity's perceived place in the cosmos far from where it had been before. Where Copernicus found the Earth to be far from the center of our solar system, Shapley discovered the Sun to be far from the center of our galaxy.

Shapley came to this grand conclusion by looking at bright yellow pulsating stars called Cepheids, inside globular star clusters. What makes Cepheids so useful to astronomers is that they flash regularly, and, by using equations, it's possible to tell exactly how bright they are.

Once Shapley had measurements for the brightness of his Cepheids, he was faced with the relatively simple task of comparing these to how bright they look from Earth. Using the differences, he could tell where they were. Instead of surrounding us, they were gathered around a place in space that's actually thousands of light years away, in the direction of the constellation Sagittarius. It was this spot that he correctly labeled as the center of the galaxy, and changed the shape if the Universe as we know it.

Douglas Kitson

Date 1918

Scientist Harlow Shapley

Nationality American

Why It's Key Changed our perception of our place in the Universe, from the center of the galaxy to a truer position nearer the edge.

Key Person **Max Planck**
Quantum pioneer

In the final year of World War I, an elderly physicist from the German Empire stepped up to receive the Nobel Prize in Physics "in recognition of the services he rendered to the advancement of physics by his discovery of energy quanta."

Max Karl Ernst Ludwig Planck was born in Kiel, Germany, on April 23, 1858. An extremely gifted musician, he took singing lessons and spread his talents across a number of instruments. But he opted against a career in music, deciding instead to study physics, matriculating at the University of Munich.

Planck's first lecturing position was in Munich, but he went on to lecture at Kiel, New York City (briefly), and Berlin. It was while at Berlin that he began to study black body radiation and, in looking for a theoretical proof that supported experimental evidence, came to express energy as packets called "quanta." This revolutionary idea wasn't recognized as incredibly significant at the time. But it soon helped to pave the way for a paradigm shift – a significant change in fundamental ideas in science – that would see other scientific luminaries such as Einstein, Bohr, and Heisenberg developing the new quantum physics, which forms the basis for most of our contemporary physical theories.

Barney Grenfell

Date 1918

Nationality German

Why He's Key Max Planck's discoveries formed the basis of quantum theory – the most important scientific theory of the twentieth century.

opposite Max Planck

Key Publication **How evolution works** The birth of Neo-Darwinism

Although Darwin introduced the idea of evolution by natural selection in the mid nineteenth century, it took another seventy years before the underlying mechanisms began to be understood. Gregor Mendel laid the foundations of what we now call genetics in the 1890s, but it was a group of population biologists in the 1930s and 1940s who combined Darwin's ideas with those of Mendel and other scientists to produce the modern theory of evolution.

A key moment came in 1918 when Ronald Fisher published "The Correlation Between Relatives on the Supposition of Mendelian Inheritance." It had already been shown that genetic mutations caused variation in a species but this didn't seem to be enough to explain how one species evolved into another. Fisher took a mathematical approach, working out the statistical effects of variable biological features in whole populations. Fisher proposed that this variation was the result of the interaction of lots of different genes, not just one or a few as Mendel had described. Using clever statistics, he showed that all these small variations across a whole population could bring about new species by natural selection. Along with J.B.S. Haldane and Sewall Wright, he was able to combine Mendel's idea of inheritance with Darwinism to help explain how evolution worked. This became known as Neo-Darwinism.

Fisher's contribution to genetics was profound. Being a keen smoker, however, he later argued against the idea that smoking caused lung cancer, proving that you can't always be right.

Richard Bond

Publication *"The Correlation Between Relatives on the Supposition of Mendelian Inheritance"*

Date 1918

Scientist Ronald Fisher

Nationality British

Why It's Key Our modern understanding of evolution owes much to Fisher's mathematical explanation of how natural selection works.

Key Invention **Mass spectrometer**

One of the pieces of equipment that has contributed most to the advancement in scientific knowledge in the twentieth century is undoubtedly the mass spectrometer. It is used to identify molecules, essentially by smashing them into pieces and measuring the mass of each fragment.

The first machines that can truly be called mass spectrometers were built in 1918 and 1919 by the American Arthur Jeffrey Dempster and British-born Francis Aston respectively. These were, however, essentially modifications of similar machines that Aston had been working on with the great Joseph (J.J.) Thompson in the Cavendish Laboratories at Cambridge University.

Thompson had been working on "positive rays," which were produced when gases were bombarded with X-rays, removing electrons. He found that these "positive rays" were actually positively charged particles which were much heavier than electrons. The atoms and molecules in gases, by losing a negatively charged electron, had become positively charged ions which could be deflected by magnetic fields.

Aston used this phenomenon to show that the particles were deflected by different amounts, depending on their mass and charge. He made his mass spectrometer based on this idea, using photographic plates to detect the particles. His first big discovery using the machine was to show that there were not one, but two types of neon atom – each one a different "isotope" with a different mass.

Simon Davies

Date 1918

Scientists Arthur Jeffrey Dempster, Francis Aston

Nationality American, British

Why It's Key A new addition to the toolkits of chemists, who were now able to analyse in finer detail the substances they were working with.

opposite A modern mass spectrometer

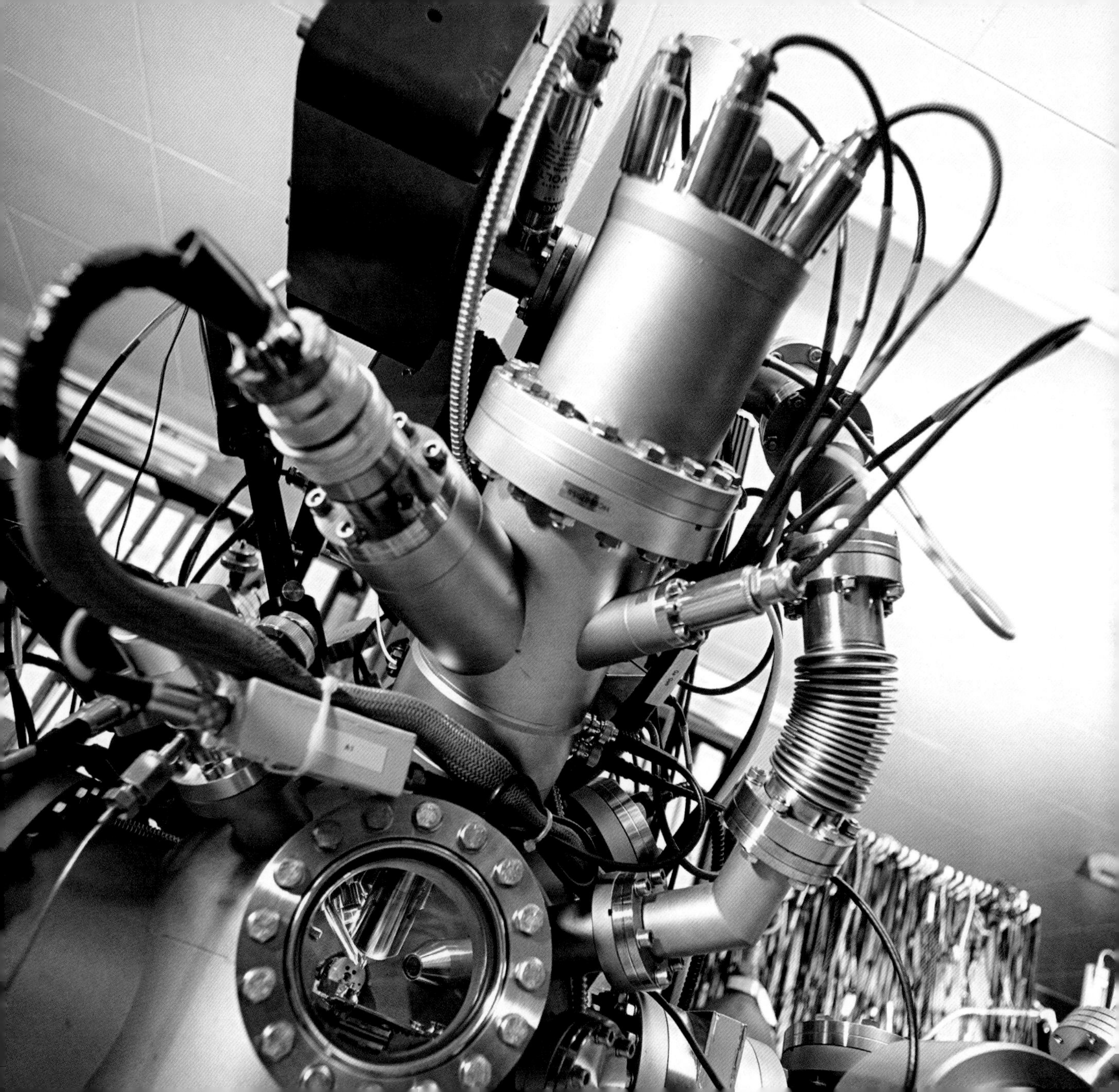

Key Invention **The rotor machine**
How a horse-stealing lumberjack began the computer age

How a middle-aged sawmill worker-come-carpenter made the biggest breakthrough in encryption for around four hundred years, whilst doing time for stealing a horse, nobody really knows. Why, at thirty-eight years of age, a man with no previous convictions should choose to steal a horse is quite another matter, but there is much about Edward Hugh Hebern that remains something of an enigma.

Before the rotor machine, codes were usually created by substituting letters of the alphabet for different letters; A might become W, B could become T, and so on. But monoalphabetic ciphers like this are easy to break. Polyalphabetic encryption, by comparison, is the substitution of a letter depending on the encryption of all previous letters and was invented back in 1466 by Leon Battista Alberti. It was a complex and very time consuming process, so was not widely used. The process meant that the first A in a message might be encrypted as a W, but the next A could be a completely different letter.

Whilst languishing in jail, Hebern devised a machine that mechanized this process, using typewriters connected to rotating wheels. Each wheel had twenty-six positions – one for each letter of the alphabet. Added complexity was achieved by having many cogs set up together, which made the code extremely hard to break.

Although Hebern was the first to apply for a patent for the machine, it was rapidly superseded by other more complex versions, including one that later became the famous "Enigma" machine of World War II.

Jim Bell

Date 1918

Scientist Edward Hugh Hebern

Nationality American

Why It's Key The earliest of a number of increasingly complex coding machines used throughout World War II and beyond, and a first tentative step toward the creation of computers.

Key Person **Ernest Rutherford**
Not a fan of stamp collecting

Ernest Rutherford was born in New Zealand; a tall man with a booming voice and bright blue eyes, he had ambitions to become a teacher. After receiving double first-class honors in mathematics and mathematical physics, as well as in physical science, Rutherford studied geology and chemistry, before trying his luck in the classroom. His teaching career never quite got off the ground however, and in 1895, Rutherford left New Zealand at twenty-three, with three degrees and a formidable research reputation. He went to Cambridge University's Cavendish laboratory, where he worked with J.J. Thomson, the man who discovered electrons.

In 1898, Rutherford discovered two different emissions coming from radioactive atoms, which he named alpha and beta particles. Having famously said, "All of science is either physics or stamp collecting", it was ironic then, that he received the Nobel Prize in Chemistry, for his work on radioactivity.

He went on to roughly determine the age of the Earth from the decay rate of uranium, before discovering, in 1911, that atoms weren't solid; but mostly nothing, with a nucleus in the middle. Three years later, Rutherford said that this nucleus must contain what we now call protons and, in 1920, figured out that neutrons are what keep it from flying apart.

In 1919, by a process he described as "playing marbles," he broke a nitrogen nucleus apart, becoming the first person to split the atom, after which he said that, "anyone who expects a source of power from the transformation of the atom is talking moonshine."

Douglas Kitson

Date 1919

Nationality New Zealander

Why He's Key As well as describing the pieces that make up an atom's nucleus, he discovered two out of the three types of radioactive emission, and was the first person to split an atom.

opposite Ernest Rutherford (right) in the Cavendish Laboratory at Cambridge University

Key Invention
Superheterodyne receiver invented

Before Edwin Armstrong invented the superheterodyne receiver, radio listeners had to constantly adjust settings on their wireless just to stay tuned in to the programs. Previous inventions tackling radio amplification, also by Armstrong, worked but were prone to drifting and loss of signal. His new idea, formed while he was a major in the army, involved mixing an incoming radio signal with a locally produced one of lower frequency. This also made it much easier to lock on to the frequency you wished to listen to by filtering out unwanted frequencies. Making use of these resonant "beats" produced by mixing two frequencies is essentially the same technique used when tuning musical instruments.

The superheterodyne system made picking up broadcast signals much more reliable and easy to handle – the age of radio and tv communications could begin in earnest.

Complex legal issues surrounding the highly technological nature of his inventions meant Armstrong spent more time in court than a judge, defending his patents against corporate giants. Eventually, on January 31, 1954, he wrote a two-page letter to his estranged wife, dressed smartly in an overcoat, hat, scarf, and gloves, and jumped from a thirteenth floor window – an end to a remarkable life and career that can only be described as "death by lawyers."

Today virtually all radios and televisions which receive broadcast signals with an aerial still use superheterodyne receiver technology.

David Hawksett

Date 1918

Scientist Edwin Armstrong

Nationality USA

Why It's Key Picking up radio signals became reliable for the first time.

Key Invention
Radio crystal oscillator

The piezoelectric effect was first demonstrated by the brothers Pierre and Jacques Curie in 1880. They had discovered that certain crystals generate an electrical charge when put under mechanical stress. Critically, the process is reversible and the crystals can be made to change shape by applying a voltage to them. The best crystals for this effect were found to be quartz and Rochelle salt, which is easy to grow into large crystals in the lab.

It took nearly forty years for this scientific curiosity to be turned into something useful by Paul Langevin in France who, in 1917, applied it to a technique for detecting sound waves in water – the forerunner of sonar.

This work inspired engineers Walter Cady at Wesleyan University and Alexander Nicholson at Bell Telephone Laboratories. Nicholson got there first to patent the idea in 1918. He was able to attach electrodes to a Rochelle salt crystal and make it oscillate at a very precise frequency, using the crystal's natural tendency to change shape with a voltage. By reading the voltage across the crystal, amplifying it, and feeding it back into the crystal, he was able to turn this vibration into an electrical signal with the same frequency precision as the vibration. This allowed radio transmitters to sustain a precise frequency of transmission for the very first time.

Today more than two billion quartz crystals are made each year globally for electronic circuits that need very precise frequency control, including mobile phones and computers.

David Hawksett

Date 1918

Scientist Alexander M. Nicholson

Nationality American

Why It's Key Revolutionized the way radio signals are transmitted

Key Event **First non-stop flight across the Atlantic** John Alcock and Arthur Whitten Brown

In June 1919, John Alcock and Arthur Whitten Brown became the first people to fly non-stop over the Atlantic Ocean, having taken up the challenge set as a *Daily Mail* newspaper competition. The prize for completing the flight was £10,000.

Their plane was a Vickers Vimy, originally designed as a bomber during World War I. The pair achieved this momentous flight only fifteen and a half years after the first ever powered flight by the Wright brothers. Alcock and Whitten Brown took 16 hours 27 minutes and flew across 1,890 nautical miles of open sea. A month earlier a U.S. Navy Flying Boat had completed the first flight across the Atlantic; however, it made multiple stops throughout the journey, taking nineteen days. Alcock and Whitten Brown took off from St John's in Newfoundland and landed in Clifden, County Galway in Ireland, in a swamp. They brought with them a bag of letters from Newfoundland, making them the fastest Atlantic postal service the world had ever seen.

The pair encountered many problems caused by fog, sleet, and snow. At times the fog was so dense that they had to fly only 300 feet above the sea. The fuel inlet into the engines kept freezing, threatening to freeze up the fuel and the plane's progress. A number of times during the flight, Brown had to climb out onto both wings to clear them of snow. It was a brave and daring feat which rightly placed the two in the history books.

Emma Norman

Date 1919

Country Canada-Ireland

Why It's Key A record-breaking flight that proved that long distance air travel was possible, enabling people to travel much quicker between continents.

Key Publication **Electron arrangements** The theory of covalent bonding

One of the fundamental concepts of chemistry is the way atoms join together to form molecules. A significant step forward was made in this area by chemists, Gilbert Newton Lewis and Irving Langmuir.

Lewis used the results of studies aimed at understanding the Periodic Table, to propose that atoms were made up of a nucleus surrounded by concentric shells of atoms. The first of these holds two electrons, and subsequent shells hold eight. Elements with the same number of electrons in their outer shells exhibit similar reaction behaviors, which explains the existence of eight distinct groups in the Periodic Table. The stable character of the noble gases is attributed to the fact that they have full outer shells (he called them octets). It also explains why atoms with few electrons in their outer shell (sodium or potassium) and atoms with almost full shells (chlorine) are so reactive – they willingly give up or accept electrons to form "octets". Langmuir went a step further than Lewis in his 1919 article on electrons. He suggested some atoms could share electrons in their outer shells to complete their octets forming "non-polar" bonds, which he called covalent bonds. So a carbon atom, with four electrons in its outer shell, could share electrons with four hydrogen atoms, forming four covalent bonds: a methane molecule (CH_4).

The most common type of covalent bond is the single bond, where two atoms share one pair of electrons. Double bonds exist where atoms share two pairs. The largest covalent bonds, sextuple bonds, can form between two tungsten atoms.

Simon Davies

Date 1919

Publication "The Arrangement of Electrons in Atoms and Molecules"

Author Irving Langmuir

Nationality American

Why It's Key A key concept learned by all chemistry students to elegantly explain chemical bonding.

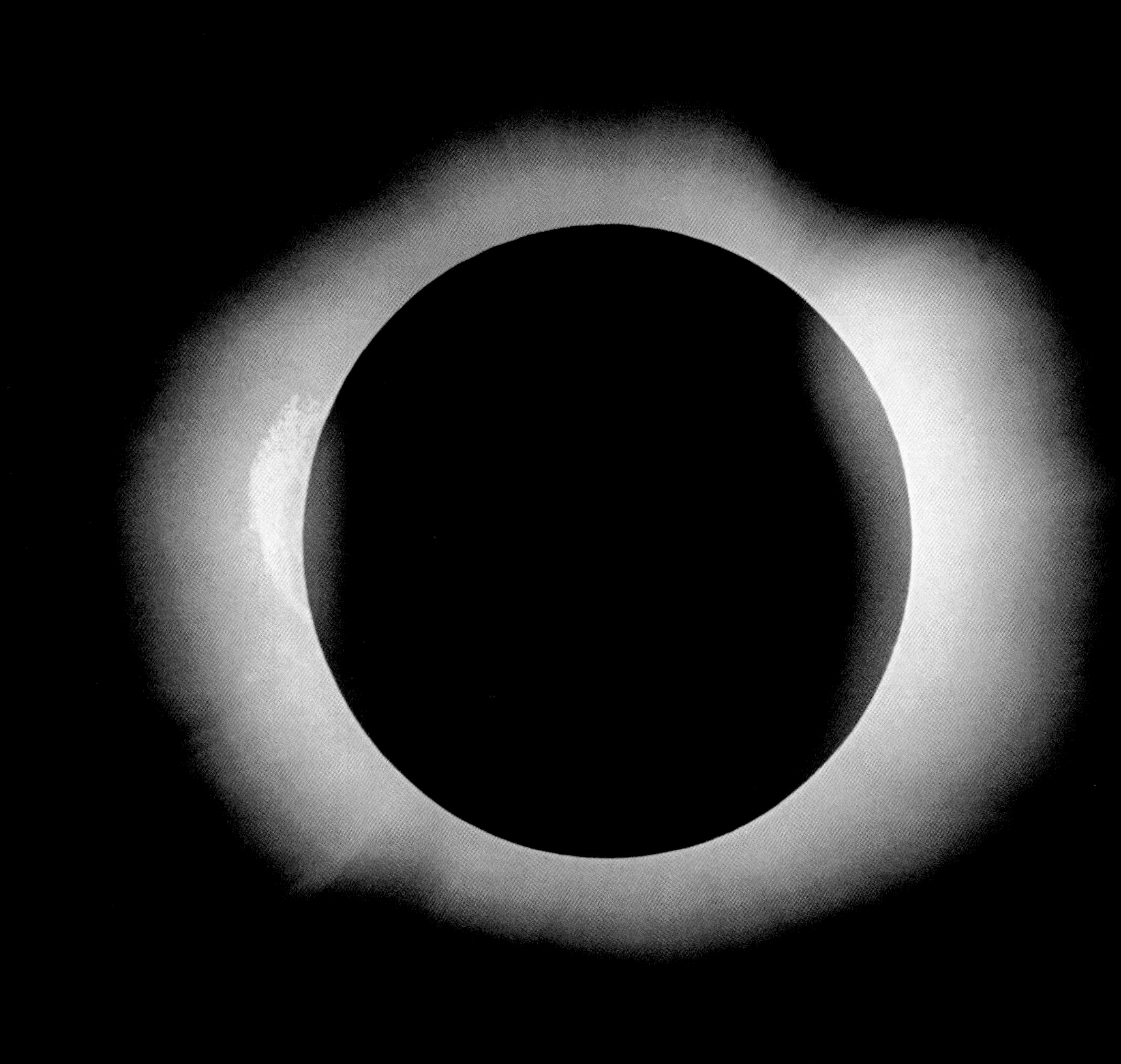

Key Experiment **Arthur Eddington proves relativity works!** The sun bends starlight

Einstein's theory of general relativity was published in 1915 in the middle of World War I. It made some revolutionary claims, including the fact that light could be affected by gravity.

One of the proposed ways to test this theory was to see whether light from stars was bent by the Sun's gravitational field. Practically, this was difficult to prove – the sun is so bright it is usually impossible to see stars behind it, so the only time this is possible is during a solar eclipse.

In 1919 the formidable English astrophysicist Arthur Eddington was dispatched to the island of Príncipe off the coast of West Africa to observe the eclipse and, hopefully, verify Einstein's theory. Eddington, having refused to fight in the war, had been saved from potato-peeling duties in an army camp by a letter from the Astronomer Royal. In return, he agreed to make the trip.

As is often the case, the only phenomenon to be relied on was bad weather; it rained non-stop for the nineteen days leading up to the eclipse and for the first 400 seconds of the eclipse, cloud obscured the sun. During this time Eddington did as any sensible scientist would and prayed. Fortunately the cloud lifted and he was able to take several photographs which appeared to show the effect that he had been hoping for. The first experimental evidence for Einstein's grand theory had come to light.

Anne-Claire Pawsey

Date 1919

Scientist Arthur Eddington

Nationality British

Why It's Key First experimental validation of the Theory of General Relativity

opposite Image taken by Arthur Eddington, confirming Einstein's Theory of General Relativity

Key Experiment **Starry eyed** Michelson and Pease measure a star

The diameter of a star is one of the most difficult things to measure. Although stars are very easy to observe, measuring their exact size can be a complicated affair, not least because the images we see through our telescopes are obscured by the fuzziness caused by light traveling across millions of light years and through our own atmosphere.

The first thing you would think you might need when trying to measure a star is a really high resolution telescope. But building a telescope powerful enough to make fine-scale measurements is expensive; astronomers use a different method. Their solution is to use a device called an interferometer, which creates that higher resolution. In its simplest form, an interferometer consists of two telescopes that combine their readings into one.

The first scientists to use an interferometer to "accurately" measure the diameter of a star were Albert Michelson and Francis Pease in 1920. At the Mount Wilson Observatory in California, they measured the diameter of Betelgeuse, a red giant in the Orion constellation. They estimated the star's diameter was 240 million miles – some 275 times greater than the Sun. The Universe had suddenly got a whole lot bigger. Even these readings, however, were an underestimate. Current measurements hold that Betelgeuse is bigger still, maybe 670 million miles in diameter.

The technique of interferometry is still in use today, with very large arrays of radio telescopes being used to observe some of the most distant objects in the galaxy.

Anne-Claire Pawsey

Date 1920

Scientists Albert Michelson, Francis Pease

Nationality American

Why It's Key The truly awesome size of some of the Universe's astronomical bodies is revealed.

Key Person **Albert Einstein**
Man of the century

He is quite possibly the most brilliant man ever to have lived, but if you had met Albert Einstein (1879–1955) as a boy, it's unlikely you would ever have predicted his future greatness.

Einstein showed little promise as a youngster; he didn't learn to talk until he was three and failed his college entrance exams. Bright, but not outstanding, he eventually graduated from the Zurich Polytechnic Institute and got a job as a patent examiner in 1902.

Then something amazing happened. During the "Annus Mirabilis" of 1905, Einstein published no fewer than five groundbreaking papers, aged just twenty-six. In these papers, he proved atoms existed, explained what light was, and presented the simple, now iconic, formula $E=mc^2$. Not bad for a man with no formal links to a university or access to a laboratory. It wasn't until after World War I, however, that his brilliance was generally recognized. His fame blossomed, helped in no small measure by that even greater feat of intelligence, his General Theory of Relativity.

By the early 1920s, Einstein was fast becoming an international superstar. Fame didn't stem the flow of revolutionary research, but it did highlight Einstein's qualities as a humanitarian. A passionate and vocal pacifist, he was described by Mahatma Gandhi as "a role model for the generations to come," and just days before his death in 1955 Einstein put his name to Bertrand Russell's manifesto calling for all nations to ban nuclear weapons. The Russell-Einstein Manifesto formed the cornerstone of all major proposals to abolish nuclear weapons.

Mark Steer

Date 1920

Nationality German-Swiss-American

Why He's Key "Most people say that is it is the intellect which makes a great scientist. They are wrong: it is character." Einstein

Key Discovery **Dendrochronology**
Wood you care to guess my age?

It's an age old story: astronomer finds key botanical discovery that revolutionizes archeology. Sounds bizarre, but it's actually the way that the science of dendrochronology – tree ring dating – came in to being.

In the early 1900s, an Arizona astronomer named Andrew Douglass theorized that sunspots had an impact on the earth's weather. He was also familiar with a theory that proposed that weather patterns were reflected in the growth of the rings on trees.

He began collecting samples of trees rings hoping to show a relationship between weather and sunspot cycles. He soon realized that differing periods of drought and abundant rainfall were reflected in the amount trees had grown. Trees develop a characteristic pattern of narrow and wide rings over a period of years – effectively a fingerprint for the time period – with a wide ring denoting a year of abundant rainfall, and a narrow ring, a year of drought. Douglass soon realized that by matching the pattern of rings in a piece of wood to the rings in a piece of wood of known age from the same area, he could determine the age of the unknown piece.

While this discovery was not particularly useful for astronomical research, it did allow for the precise dating of wooden structures, initially in the southwestern United States, and then worldwide. All that an archeologist now had to do was take a sample from a wooden beam in a structure and match the pattern to a ring pattern of known age.

B. James McCallum

Date 1920

Scientist Andrew Douglass

Nationality American

Why It's Key It allows the relatively precise dating of archeologically significant wooden structures and trees.

Key Publication
A new spin on ice ages

In the mid-nineteenth century, scientists looking at geological and fossil records concluded that the Earth had undergone prolonged periods of glaciation. But they couldn't work why these ice ages occurred.

In 1920, Serbian astrophysicist Milutin Milanković provided a solution to the problem. He proceeded to elaborate his theory and in 1941 published a more concrete version of his 1920 paper with his *Canon of Insolation of the Earth and Its Application to the Problem of the Ice Ages*. Building on previous astronomical theories of climate variation proposed by Joseph Adhémar and James Croll in the nineteenth century, Milanković spent many painstaking years measuring the slow changes in Earth's orbit around the Sun, by calculating gravitational effects and the positions of stars and planets in relation to the Earth. His results showed that our planet travels across space according to three cyclical variations.

First, the shape (eccentricity) of Earth's orbit varies from being almost circular to being elliptical, taking around 100,000 years to complete one cycle. Second, the tilt (obliquity) of Earth's axis, which dictates the seasons in the different hemispheres, fluctuates between 22.1° and 24.5° in 41,000 year cycles. And third, Earth's axis of rotation acts like a spinning top, causing Earth to "wobble" (precession) in cycles of 24,000 years. All this explains glaciation in terms of seasonal and latitudinal variations of solar radiation reaching Earth, and it is a significant theory for understanding climate change. It is thought, however, that additional factors are required to trigger ice-ages.

James Urquhart

Publication "Mathematical Theory of Thermic Phenomena Caused by Solar Radiations"

Date 1920

Author Milutin Milanković

Nationality Serbian

Why It's Key First theory that could adequately explain why Earth periodically experiences dramatic long-term changes in climate.

Key Publication **The ionization equation**
Seeing inside stars

Stars are balls of gas, superheated until the atoms are ripped apart. This gives a sea of charged particles – electrons and the positively charged remainder of the atoms that are left when an electron is knocked or ripped out – all moving very fast because they are so hot. This state of matter is called plasma, and the charged atoms are known as ions.

Megh Nad Saha, a Bengali Indian astrophysicist, formulated an equation that, taking into account how hot the star is, how dense it is, and how much energy it takes to rip each electron out, is able to tell you the percentage of atoms in the gas that are ionized compared to those left intact. This was important.

When ions and electrons recombine to make a neutrally-charged atom, the energy the electrons lose is emitted in the form of a photon of light. Quantum mechanics says that this photon can only have certain amounts of energy, depending on the type of atom. The light from the star, therefore, should be made up of only these energies. The brightness of the light representing each energy level depends on the number of ions and electrons recombining, so the Saha equation can be used to learn about a star's internal composition from its recorded light emissions.

Harvard astronomers had previously observed that different types of stars emitted characteristic light spectra, but there was no theory to explain why until Saha published his equation.

Kate Oliver

Publication "The Ionization Equation"

Date 1920

Author Megh Nad Saha

Nationality Indian

Why It's Key An explanation of why stars have characteristic light emissions, and what they mean.

Key Experiment **Little Albert**
A study in conditioning, a lesson in ethics

One of the most notorious and misrepresented psychological experiments of the last century was a study involving the child known as "Albert B." Its importance to the development of psychology has arguably been over-shadowed by the experiment's lack of good ethical practice.

John Watson was a behaviorist who believed that our emotional responses to things were conditioned – that is, learned through exposure to stimulation – rather than innate. This was a controversial stance at the time.

Watson, along with his assistant Rosalie Rayner, set about trying to prove that fear responses in particular were a result, at least partially, of learned behaviors. Their methodology was fairly simple. When a variety of stimuli were shown to poor little Albert for the first time, he reacted without fear. Watson then presented the same things accompanied by a loud noise, made by banging a steel bar with a hammer. Over time Albert developed a fear response to the animal/object in question without the presence of the loud noise, proving Watson's theory correct – emotional behavior could be learned.

But it was the ethical aspects that hit the headlines. Besides the age of the child (just eleven months old) there was the fact that the mother did not know of the experiments and that Albert received no rehabilitation or counseling because he had to leave one month into the experiment – a fact allegedly known by Watson.

Barney Grenfell

Date 1920

Scientist John Watson

Nationality American

Why It's Key Watson's experiment arguably proved that emotional behaviors could be learned, but today, his unethical experiment would be illegal.

Key Event **Advent of commercial radio broadcasting**
Through the ether to our ears

For many years, receiving speech, let alone music, by radio was the exclusive domain of amateur radio hobbyists ("hams") and engineers. Radio itself was essentially a point-to-point communication system requiring bulky, complicated equipment. "Broadcast" transmissions intended for general reception were deemed a novelty, and most radio frequencies were thought to have no commercial or practical application. Nevertheless, wartime advances and postwar experimentation meant that, by 1919, exploitation of the previously-ignored "shortwave" radio frequencies (2-30 Mega hertz) had begun.

Attitudes toward the use of frequencies were also changing. By August 1920, the first radio news program had been broadcast by Michigan-based station, 8MK. On October 17, a license to broadcast was given to radio station KDKA of Pittsburgh and on November 2, they began what was arguably the world's first regular, commercial broadcast service. The original test broadcasts from the garage of Westinghouse engineer Frank Conrad were mentioned in a store's newspaper advert, triggering a sell-out of their receivers and prompting Westinghouse boss Harry P. Davis to take notice. Funded by advertising and the sale of the US$10 "Music Box" receivers, the programming included historically important news items and innovative roof top concerts.

In the UK, in November 1922, the "BBC" was formed and commenced broadcasting with low power transmitters in major British cities, principally "2LO," a London station and "5IT" in Birmingham.

Mike Davis

Date 1920

Country USA

Why It's Key These events were the foundations of the modern broadcast media services we all take for granted today.

opposite An early edition of *Popular Wireless Weekly*

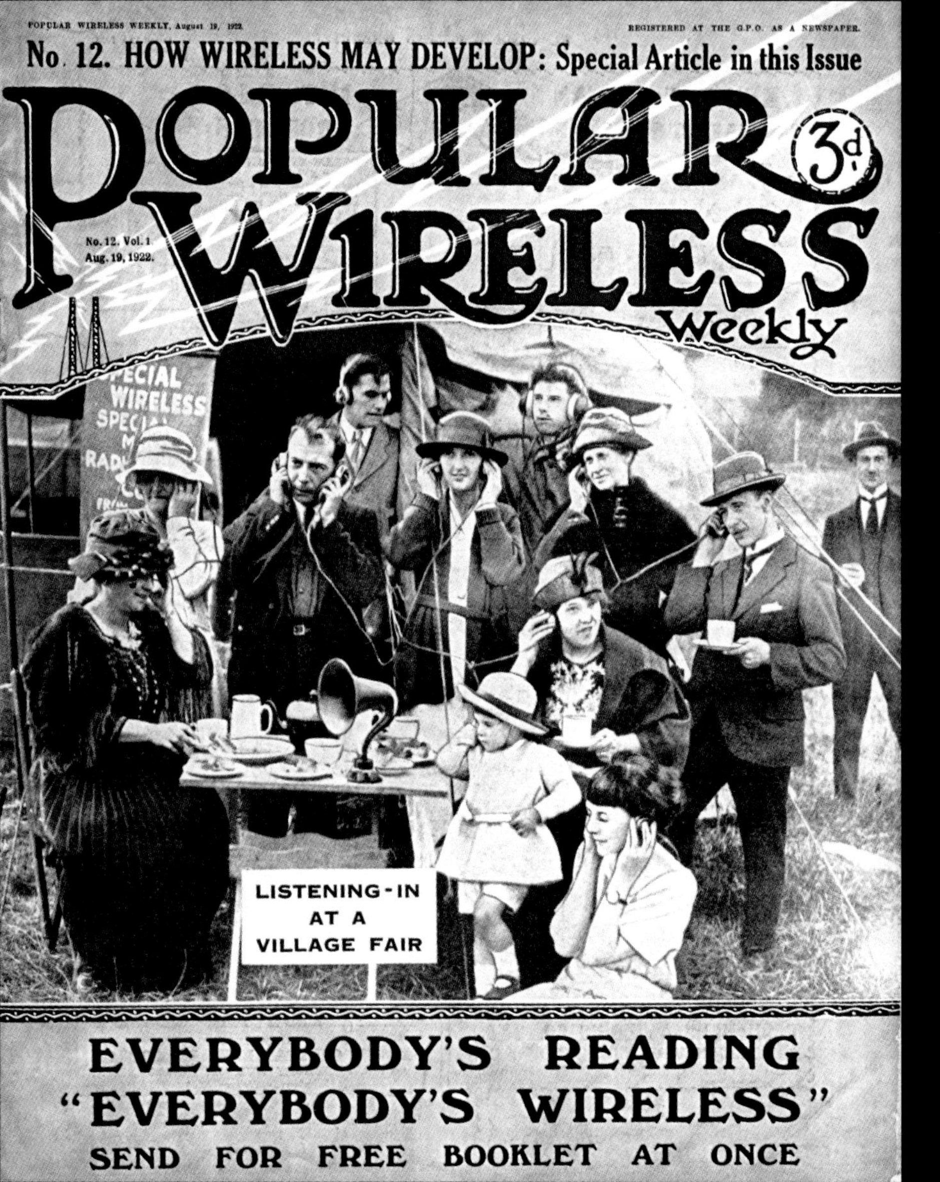
POPULAR WIRELESS WEEKLY, August 19, 1922.
REGISTERED AT THE G.P.O. AS A NEWSPAPER.
No. 12. HOW WIRELESS MAY DEVELOP: Special Article in this Issue
POPULAR WIRELESS Weekly
3d.
No. 12. Vol. 1
Aug. 19, 1922.
LISTENING-IN AT A VILLAGE FAIR
EVERYBODY'S READING
"EVERYBODY'S WIRELESS"
SEND FOR FREE BOOKLET AT ONCE

Key Event
The dawn of the electric age

In 1912, less than 15 per cent of American homes had electricity; within twenty years, that number stood at 85 per cent. The deaths attributed to World War I and the Great Flu Pandemic had sharply decreased the labor pool for domestic servants, but many other reasons could be made to account for the sudden sky-rocketing demand for electric appliances: the introduction of assembly line manufacturing, the development of improved trucking, rampant consumerism, or increased radio advertising and magazine sales.

Similar developments were taking place in Europe, and the Electrical Association of Women in Great Britain soon began to promote the benefits of electricity to overworked housewives. Englishman Hubert Booth had introduced his horse-drawn, gas-powered suction cleaner in 1901. James Spangler, an asthmatic custodian, improved upon this cumbersome design with a vacuum cleaner made out of a soap box, a fan, and a pillow case. With refinements, this became the first Hoover vacuum cleaner in 1908. Also introduced that year was "Thor," the first electric-powered clothes washing machine. Commercial refrigeration had been around since the 1850s, but by 1910 the units, although often prohibitively expensive, were small enough to fit in the average home.

Although all of these machines had been knocking around for a while, it took the particular combination of social factors of the 1920s to kick start their popularity among those that could afford them. Electrification had begun in earnest.

Stuart M. Smith

Date 1920

Country USA, Europe

Why It's Key Quite simply, electricity transformed modern life.

Key Discovery **Neutrons from nothing**
The solution to everyone's problems

The discovery of subatomic particles, such as electrons and protons, led to the hypothesis that there was plenty going on inside atoms. One particular theory was based around the plum pudding model, which envisaged electrons as negatively charged plums mixed into a positive pudding. But this couldn't be right – Ernest Rutherford showed there must be a nucleus full of positive charge and electrons flying round the outside, attracted to the positive innards. However, if all the positive charge is crammed in the middle, what is there to stop it flying apart like when the north poles of magnets are pushed together? There needed to be something else there.

Furthermore, there was an issue with atomic weights. Atoms are most stable when the numbers of protons and electrons are equal, so the charge is balanced, but the further up the Periodic Table you go the more the weights seem to point to there being more protons than electrons.

Two men were working separately on these problems and, in 1920, reached the same conclusion – there was another particle in the mix.

Ernest Rutherford and American physicist William Draper Harkins both predicted the existence of a neutral particle inside the nucleus. It was later named the neutron and helped explain many of the problems with the new atomic model. The protons are kept together by the attractive nuclear force of neutrons, which were of course causing the mysterious weights in the atoms. Twelve years later, in 1932, James Chadwick confirmed the existence of neutrons.

Douglas Kitson

Date 1920

Scientists Ernest Rutherford, William Draper Harkins

Nationality New Zealander, American

Why It's Key The discovery of neutrons explained how the atomic nucleus stays together and why atomic numbers and weights are different.

Key Publication
Polymers and their constituent parts

When Hermann Staudinger first began work on polymers, it was widely believed that these materials were made up of small molecules held together by weak forces. He had other ideas.

Staudinger first studied botany but, after taking advice from his father, later took up chemistry. In 1910, he was working for the chemical firm BASF on the synthesis of isoprene. Staudinger had a hunch that isoprene was the basic unit that made up the polymer rubber. He published a theory in 1920, with little proof at the time, that polymers were large molecules (macromolecules), made up of smaller sub-units called monomers, linked by covalent bonds.

Covalent bonds are strong chemical bonds between atoms of molecules, for example hydrogen and oxygen in water. Other chemists of the time ridiculed Staudinger, not believing that covalent bonding could be involved. They still believed polymers were small molecules being held together by other weaker forces. Some believed that there was a limit to the number of atoms that could be joined together in a single molecule, and rejected his macromolecule idea. He even suffered verbal attacks in lectures.

Staudinger stuck to his guns, however, and carried out a series of experiments to prove his theory. These included making polymers from formaldehyde and styrene. It wasn't until 1928 that chemists were finally convinced, when the X-ray crystal structure of a polymer revealed that it was made up covalently-bonded monomers.

Riaz Bhunnoo

Publication "Concerning Polymerization"

Date 1920

Author Hermann Staudinger

Nationality German

Why It's Key Staudinger's pioneering work provided the theoretical basis for polymer chemistry. His work underpins the multibillion-pound nylon, plastic, and rubber industries.

Key Person **Sir Arthur Stanley Eddington**
Relativity's biggest ally

Someone once remarked to Sir Arthur Stanley Eddington that, "It is common knowledge that you are one of the three persons in the world who understand Einstein's theory [of relativity]." Eddington (1882–1944) appeared perplexed and annoyed by this statement, but when questioned, replied, "Don't worry. It is merely that I could not imagine who the third person might be."

Eddington was a highly respected astrophysicist whose greatest accomplishment was, almost certainly, bringing the work of Albert Einstein to the English-speaking world. As there was an embargo on German work after World War I, the work of Einstein was almost unknown outside of Germany and Austria. Eddington helped both to popularize the theory by lecturing about it, as well as providing observations of a solar eclipse that helped support it. Eddington also produced several books aimed at a non-scientific audience. Possibly the best known of these was *The Nature of the Physical World* in which he used "infinite monkey theorem" to explain statistical concepts of infinity and probabilities. Essentially it implies that an infinite number of monkeys, sat at an infinite number of typewriters, given an infinite amount of time, would be able to produce any great literary work. "All the books in the British Museum" as Eddington stated. The book itself explained many new scientific concepts of the time, from relativity, to life on other planets, to evolution.

Jim Bell

Date 1920

Nationality British

Why He's Key The theory of relativity has been the basis of many of the major breakthroughs in science over the last 100 years, and without Eddington, it would not have influenced the work of so many scientists.

Key Invention **The toaster**
The best thing since sliced bread?

What's so brilliant about sliced bread anyway? Aside from the fact that fashioning your own rustic-chic style sandwiches with that simple implement we call "a knife" can actually be rather rewarding, the bread slicer can hardly be regarded as an invention that merits its iconic status. It should by now have been superseded by some far more exciting piece of technological wizardry.

Exhibit number one: the toaster. Perhaps not the twentieth century's finest invention, but a first class example of one that outstrips the bread slicing machine by miles. The advantages of a toaster over antiquated toasting methods – which revolved principally around fireplaces – are obvious: vastly reduced toasting and cleaning time as well as lower risk of charring and fire hazard. How useful, by contrast, is saving yourself three seconds not having to cut your own slice of bread?

The first electric toaster that enjoyed any real commercial success was reputedly the "D-12," patented by one Frank Shailer in 1909. But it wasn't until a few years later that a pop-up version appeared on the breakfast scene. The brainchild of Charles Strite, it didn't make it onto the shelves until 1926, in the form of the "Toastmaster," so called for its remarkable ability to "Make perfect toast every time! Without turning! Without burning!".

Not only was Strite's ingenious device capable of spitting out toast when a timer cut off the electricity supply, it was the first to use heating elements that toasted both sides of a slice simultaneously.

Hayley Birch

Date 1921

Scientist Charles Strite

Nationality American

Why It's Key Revolutionized the morning routine and made more time for lie-ins.

Key Event
BCG jab first given

The BCG vaccine for *Mycobacterium tuberculosis*, the bacterium that causes tuberculosis (TB), was first administered in 1921. Earlier attempts at vaccines had been fraught with disaster. After Jenner successfully vaccinated against smallpox using the cowpox virus, it seemed reasonable to assume that a TB vaccination with the form of the disease which infects cows (*Mycobacterium bovis*) should protect against the human strain. It rapidly became apparent, however, that bovine tuberculosis was every bit as infectious and deadly as human TB.

French bacteriologist Albert Calmette and his associate Camille Guerin grew *Mycobacterium bovis* in a series of over 200 cultures until it was weakened enough to summon an immune response without causing tuberculosis. They had produced a vaccine that worked. It now has a good safety profile, but an early incident almost put an end to the BCG jab. From December 10, 1929, until April 30, 1930, 251 of 412 infants born in Lubeck, Germany were given the vaccine. Of these 412, 207 contracted TB and 72 died of the disease; the BCG had become contaminated with full strength TB. Some countries – notably the USA and the Netherlands – still do not advocate the vaccine as it can complicate screening tests for the disease.

The success of the BGC vaccine varies greatly. It is more effective at preventing severe forms of TB, such as meningitis, than the more common pulmonary TB. Its efficacy also wanes with age. Most puzzlingly, its effectiveness varies with latitude: the further from the equator, the better the protection.

Stuart M. Smith

Date 1921

Country France

Why It's Key Decreased the incidence of tuberculosis.

opposite TEM of phagocytosed (engulfed) *Mycobacterium bovis* bacteria (yellow) in a macrophage (blue) white blood cell

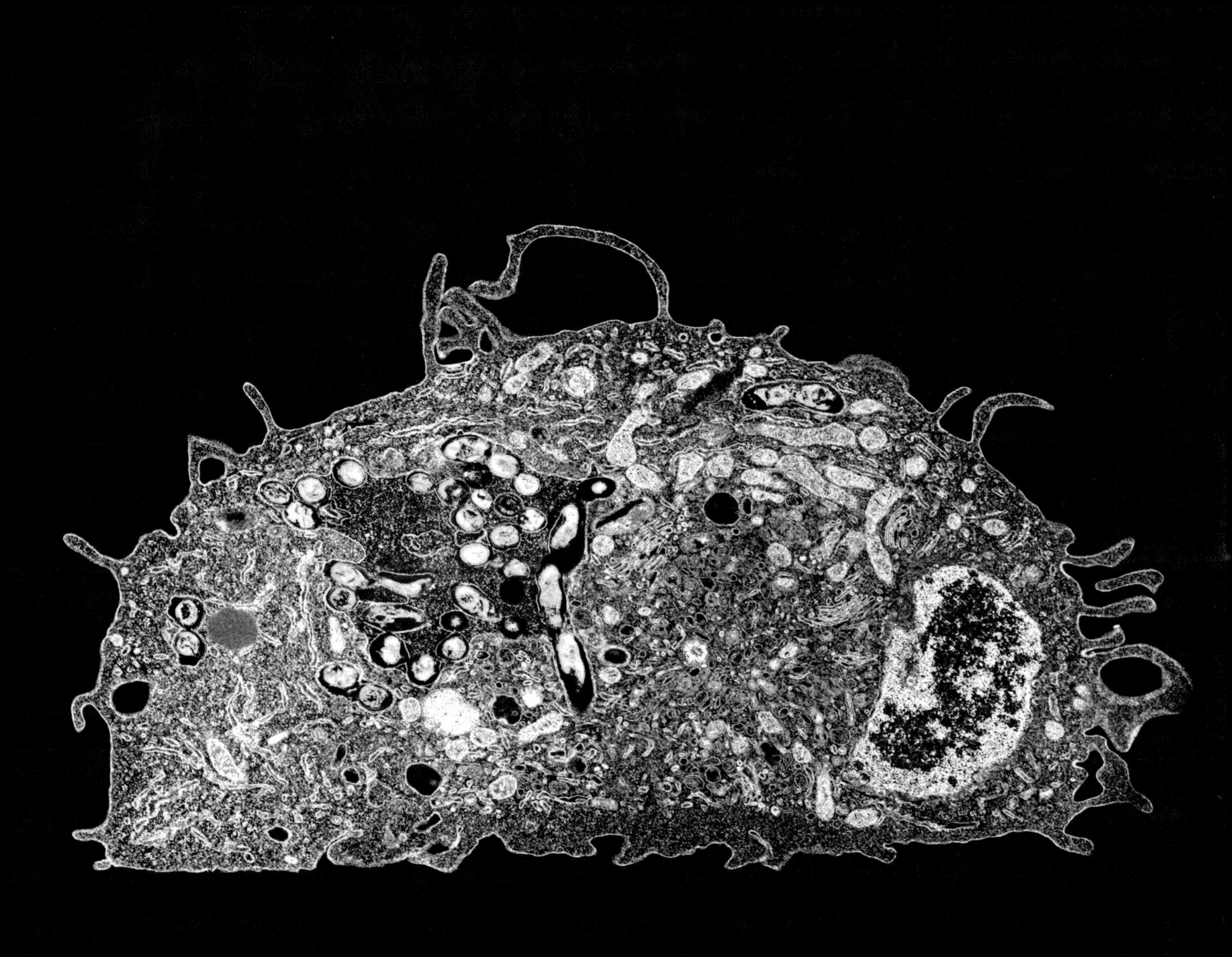

Key Event
Caltech formed

The California Institute of Technology, known universally as "Caltech," is one of the world's most famous and influential scientific research and educational facilities.Its origins, however, are modest, grown as it has from a small, nineteenth-century vocational college in Pasadena, California.

It was established in 1891 by Chicago philanthropist, universalist, and politician, Amos Throop "to furnish students of both sexes and all religious opinions a liberal and practical education." Originally called Throop University, it was altered to Throop Polytechnic Institute in 1893 when Charles H. Keys – a man who shared Throop's belief in the school motto, "Learn By Doing" – was hired as its president. It would not become "Caltech" until 1921, and much would happen in the interim.

In 1907, a new era dawned for Caltech when astronomer George Ellery Hale, the first director of the nearby Mount Wilson Observatory, joined the board of trustees. With scientific research in the United States still in its formative stages, Hale perceived an opportunity to turn the Institute into a major center for both education and research into the natural sciences. His enthusiasm and status attracted gifts of land and funds, enabling the construction of modern laboratory facilities. Noted chemist Arthur Amos Noyes, and physicist Robert Andrews Millikan joined the faculty at this time and began to steer the project toward its position of scientific excellence. Caltech has managed the famous Jet Propulsion Laboratory for NASA since the late 1950s.

Mike Davis

Date 1921

Country USA

Why It's Key Since its inception, a veritable "who's-who" of the world's greatest scientists have worked at Caltech; their research and discoveries affecting nearly every aspect of our lives.

Key Invention **Lie detectors**
The truth can set you free

There are some people who can spot a liar a mile off; they notice the telltale signs that you give under anxiety: increased heart rate, blood pressure, breathing rate, and perspiration. William Marson invented the first machine to measure the blood pressure of an individual under interrogation. Incidentally it was Marson who also created the action hero "Wonder Woman" arming her with the "Lasso of Truth."

It was, however, the medic, and later forensic psychologist, John Larson who came up with the first true polygraph which continuously measured breathing rate, heart rate, and blood pressure, on a revolving drum of smoked paper. Larson began a close association with California's Berkeley Police Department, using the test in hundreds of cases.

Many U.S. states still use more recent versions of the polygraph test in police investigations. In the UK and Europe, results from polygraphs cannot be used as evidence but can be used to monitor the supervision of sex-offenders. In 2003, the National Research Council, part of the U.S. National Academy of Sciences, set up an expert panel that concluded that "national security is too important to be left to such a blunt instrument," adding, "no spy has ever been caught by using the polygraph."

Results depend on the polygraph operator, and rely on an emotive response to a question, not necessarily a lie. Additionally, if the subject doesn't feel guilty, or just doesn't care, it won't be able to catch them out.

Umia Kukathasan

Date 1921

Scientist John Larson

Nationality American

Why It's Key Larson's lie detector became the first to be used by the police. It is an indicator of anxiety, not necessarily guilt, but it is still used in the USA.

opposite A lie detector in use in the USA, 1920s

Key Person **Thomas Midgley**
One man, two environmental catastrophes

Until the day that someone cracks the problem of climate change, Thomas Midgley (1889–1944) will be remembered as the man who "had more impact on the atmosphere than any other single organism in earth history".

The engineer-turned-scientist first made his name in 1921, when he discovered that adding lead tetraethyl (known as ethyl) to gasoline could make car engines more reliable and more powerful. Lead's toxicity was well known but this didn't stop Midgley happily demonstrating ethyl's "safety" in public. In private, however, he avoided it at all costs – he was already suffering from the poisonous effects of lead that were killing off workers at his company's chemical plant.

Having played his part in one great environmental calamity, Midgley went on to create another. Whilst working for General Motors in 1928 he produced Freon, the first of the CFCs – chemicals that would later leave a gaping hole in the ozone layer. The discovery was lauded at the time and earned our antihero a plethora of medals and honorary degrees.

Thomas Midgley wouldn't live to see the effects his inventions had on the Earth. At the age of 51 he contracted polio which left him seriously crippled. An innovator to the end, he designed a complex array of ropes and pulleys with which he could haul himself out of bed. It was the last of his disastrous inventions. Four years later he slipped, accidentally becoming entangled in the ropes; in 1944 Midgley's own invention strangled him to death.

Mark Steer

Date 1921

Nationality American

Why He's Key From chemical king, to ozone destroyer, Midgley's presence on the planet was felt long after his demise.

Key Discovery **Knock knock – who's there?**
Leaded gasoline prevents "knock" in car engines

Engine knocking occurs in an engine when a pocket of fuel mixes with air and ignites away from the normal combustion front. It causes a pressure spike that wears away at the engine walls and, back in 1921, the problem of knocking was holding up the development of higher efficiency engines.

Thomas Midgely and his co-workers at General Motors discovered that the highly toxic compound tetraethyl lead could be added to fuel to prevent knocking; leaded gasoline was born. It was added to gasoline as a mixture known simply as "ethyl" that also contained lead scavengers such as dibromoethane and dichloroethane to mop up any traces of the metal in the engine, and a red dye. General Motors and Standard Oil set up the Ethyl Corporation to market the additive.

The use of tetraethyl lead was controversial from the beginning, particularly in light of its toxicity. During the 1920s, workers at the Ethyl Corporation died from exposure to tetraethyl lead while others went insane, even Midgely himself fell ill with lead poisoning.

Leaded gasoline was eventually phased out in the United States from the 1970s, and was finally withdrawn from European sale in 2000 due to concerns over air pollution. Ironically, leaded gasoline is still used for aviation fuel as no suitable unleaded alternative has been found.

Helen Potter

Date 1921

Scientist Thomas Midgely

Nationality American

Why It's Key Enabled development of better engines, but with toxic side effects.

Key Person **George Washington Carver**
On top of the crops

George Washington Carver was a renowned agronomist and botanist who dedicated his work to educating struggling farmers during a decline in the Southern United States' agricultural industry.

Born in 1864, he was an African-American who lived with slavery and then racial segregation. Once slavery ended, Carver grew up on a farm in Missouri, where his love and knowledge of nature and plants developed. From a young age he determinedly set out to gain an education, attaining both bachelor's and master's degrees at Iowa State Agricultural and Mechanical College, before joining the faculty in charge of systematic botany. Carver moved, in 1896, to the Tuskegee Institute in Alabama to educate Southern farmers struggling with poor soil fertility resulting from monoculture production of cotton. He taught farmers about crop diversification and rotation, which would conserve their farmland's soil fertility, and reduce their vulnerability to pests and market fluctuations by reducing dependence on any one crop type. He also promoted the planting of legumes, such as peanuts, that would replenish soil nitrogen levels, and provided widespread education on sustainability and productivity.

With the public support of President Theodore Roosevelt, in 1921, Carver developed many uses for the new crops that farmers were producing, in order to stimulate these markets to grow. Most famous were his range of peanut products but he also developed new foods, beverages, cosmetics, dyes, paints and stains, animal feeds, and medicines.

Emma Norman

Date 1921

Nationality American

Why He's Key Overcame the restrictions in place during racial segregation, and the benefits of his work to agriculture were recognized throughout the United States and abroad.

Key Person **J.B.S. Haldane**
"Queerer than we can suppose"

The world of science has thrown up its fair share of eccentrics, but few have been as barmy – or as brilliant – as John Burdon Sanderson Haldane (1892–1964), known as Jack to his friends and J.B.S. to posterity. Haldane embarked early on his science career, assisting his equally idiosyncratic scientist father, John from the age of three. By his teenage years, father and son were experimenting with gases and gas masks, taking it in turns to see how long it took each other to pass out.

Remaining a keen scientist and experimenter throughout his life, Haldane was instrumental in marrying Darwin's theory of evolution to Mendel's theory of genetic inheritance – founding the study of population genetics. He also made further advances in comparative anatomy, the study of enzymes and was the first person to suggest using hydrogen-generating windmills as an alternative to mining fossil fuels.

Haldane's major interest, however, remained in helping divers avoid the bends. Having acquired a decompression chamber, he managed to persuade many people, including members of his own family and even an ex-Prime Minister of Spain, to take part in a variety of dangerous, but insightful, experiments. Haldane exploded all the fillings in his teeth during one episode and in another induced a fit so severe that he crushed some his vertebrae.

Not just a researcher, Haldane was a talented writer and keen popularizer of science. He inspired his friend, Aldous Huxley, to pen the massively influential *Brave New World*.

Mark Steer

Date 1922

Nationality British-Indian

Why He's Key A remarkable intellect, an ardent Marxist, an eccentric visionary; Haldane's lasting influence is felt across many scientific disciplines.

Key Person **Frederick G. Banting**
Sweet success for diabetics

Frederick G. Banting (1891–1941) was a Canadian scientist who spearheaded the discovery of insulin, bringing hope to millions of people with diabetes. Born in 1891, he originally studied medicine at the University of Toronto. After graduating, he served in World War I, and was awarded the Military Cross for bravery.

After the war, Banting returned to medicine and became fascinated by diabetes after reading about it in a journal article. At that time, diabetes was a devastating disease. Sufferers faced blindness, loss of limbs, and premature death, and the only way to treat the disease was with a strict starvation diet. It was known that the cause was a lack of the hormone insulin, which converts sugar into energy.

Over the summer of 1921, Banting and his colleagues, Charles Best and John J.R. Macleod, began experimenting on dogs to find a way of isolating insulin. In 1922, with the help of biochemist James B. Collip, they managed to purify the hormone and inject it into their first human subject, fourteen-year-old Leonard Thomson. Thomson, who had been within weeks of death, made a miraculous recovery, and the modern treatment for diabetes was born.

The following year, Banting and Macleod were awarded Canada's first Nobel Prize in Physiology or Medicine. Banting also refused to patent the drug, to ensure that it could be produced cheaply and be made available to all. He was killed in service during World War II, but his discovery is used in medicine to this day.

Hannah Isom

Date 1922

Nationality Canadian

Why He's Key Banting gave diabetics across the world a way to control their symptoms and prolong their lives.

opposite Frederick Banting with Charles Best

Key Publication **Proposing an expanding Universe**
Friedmann creates a Big Bang

Imagine the objects of the Universe – the planets, the stars, etc. – stuck to a piece of cling film. Now imagine that the film is being contorted and stretched and that the objects themselves don't move, but the space between them changes as the piece of film is stretched. This is a crude and simple explanation of what Russian scientist Alexander Friedmann's equations describe.

His equations were the solution to one of Einstein's biggest problems. Einstein's original Theory of General Relativity had predicted a Universe that was either expanding or contracting, but this was incompatible with the idea, prevalent at the time, that the Universe was static.

Friedmann's work, in which he described how an expanding Universe was mathematically possible, was published in 1924 and earned a prompt reply from Einstein himself. The Russian's work, Einstein thought, seemed "suspicious." Friedmann responded by sending his calculations to Einstein who, after a careful look, agreed that he had solved the problem – the Universe could indeed expand. The solution itself is named after Friedmann and three other scientists who all worked independently on the problem.

The expanding Universe theory predicted a red-shift in the light coming from distant galaxies, a prediction confirmed by the Hubble telescope in 1929. Friedmann's work paved the way for the Big Bang theory of how our Universe began – a theory that is widely accepted and agreed upon by astronomers and physicists around the globe.

Helen MacBain

Publication "On the Curvature of Space"

Date 1922

Author Alexander Friedmann

Nationality Russian

Why It's Key Showed that the Universe could be expanding after all.

Key Invention **Predicting the future**
The birth of weather forecasting

Vilhelm Bjerknes began his successful career in science at an early age, assisting his father's study of hydrodynamics – the movement of water. After graduating from university, Bjerknes worked with Heinrich Hertz, the famous German physicist. Together, they carried out research that would later be invaluable in the development of the radio and radar.

Bjerknes then began to apply hydrodynamics and thermodynamics on a global scale. He developed theories about how air masses move to create weather, and crafted formulas that, given sufficient information about the atmosphere's current state, could predict its future state. The technology of the day, however, nullified his work; the calculations could not be carried out quickly enough for the results to still be predictions.

Undaunted, Bjerknes was joined by his son, Jacob, and several colleagues in an ambitious project that eventually yielded the polar front theory. The theory explained how warm and cold air fronts interact like World War I battlefronts. The interaction produces cyclones – sizable areas of low atmospheric pressure usually involved in any noticeable weather condition. The polar front theory, on which current weather forecasting is based, has not been changed in the slightest since its publication in 1922.

Seeing radar's potential for weather observation, scientist David Atlas harnessed its power to further cement our meteorological might. Weather radars are now extensively used to collect the data necessary to predict future weather patterns.

Logan Wright

Date 1922

Scientist Vilhelm Bjerknes, Jacob Bjerknes

Nationality Norwegian

Why It's Key Before you complain about the times when weather forecasters get it wrong, consider how many times they're right and how engrained weather forecasts are in our society.

opposite Weather balloon is launched with a meteorograph at the end of a string

Key Person **Niels Bohr**
Unlocking the structure of the atom

Niels Bohr (1885–1962) is one of the giants of modern physics. After receiving his doctorate in 1911, he then went on to study with scientific heavyweights J.J. Thomson and "father of nuclear physics" Ernest Rutherford.

It was during his time with Rutherford that Bohr carried out his most famous work. Rutherford had already discovered that atoms were made of a positively charged nucleus surrounded by negatively charged electrons. Along with later refinements, Bohr's breakthrough resulted in the model of the atom we have today. In 1913 he published his atomic model, in which the electrons orbit the nucleus of the atom in rigid orbits or shells. Critically, his model included the ability of electrons to drop to lower energy orbits. Bohr proposed that the change in energy results in a photon of light being emitted as the electron makes an instantaneous "quantum leap" from one discrete orbit to another, contradicting the theories at the time which expected energy levels around the nucleus to be continuous. This work formed the foundation of quantum theory.

Three years later, he was named director of the new Institute for Theoretical Physics in Copenhagen – now called the Niels Bohr Institute. In 1922, the Nobel Prize Committee awarded him their physics prize for his work on atomic structure.

Receiving his Nobel Prize was far from being the end of his career, however. During World War II, he fled from Nazi-occupied Denmark and worked for the Allies on the Manhattan Project to develop a nuclear bomb.

David Hawksett

Date 1922

Nationality Danish

Why He's Key Bohr's work on the atom was one of the key steps in the foundation of modern physics.

Key Experiment **The Stern-Gerlach experiment**
Electrons do the twist

The Pauli Exclusion Principle tells us that all particles of matter in a system must be in different energy states, and so have different quantum numbers. An electron orbiting a nucleus, for instance, has a variety of quantum numbers denoting how far it is from the nucleus, how fast it's orbiting it, and at what angle. It also has another quantum number called spin, which can be thought of as describing how the electron is spinning on its own axis while orbiting the nucleus, just as the Earth spins on its own axis while orbiting the Sun.

Our electron can be represented as a little bar magnet. In a regular magnetic field, both ends of the "magnet" experience the same force and it travels in a straight line. However, if the magnetic field isn't constant, our electron gets deflected off course.

In 1922, Otto Stern and Walter Gerlach observed that, after passing through the wonky magnetic field, electrons landed in just two distinct groups. If the electrons had been spinning in any old direction, they'd simply have seen a big smear of electrons – two distinct groups showed there are just two spin values for electrons, which we now call "spin up" and "spin down."

The experiment had a huge impact on the scientific world, and paved the way for a multitude of advances as diverse as MRI scanners and atomic clocks.

Kate Oliver

Date 1922

Scientists Otto Stern, Walter Gerlach

Nationality German

Why It's Key A rare physical demonstration of quantum weirdness, confirming the idea of electron spin.

Key Discovery **Banishing bendy bones**
Vitamin D cures rickets

Rickets – the softening of bones that leads to bowed legs and an increased risk of fractures – was first described by British doctors in the late nineteenth century. As more families moved from the countryside to smoggy industrial cities during the Industrial Revolution, cases of rickets increased dramatically. German physicians later found that one to three teaspoons of cod liver oil daily could reverse the effects of rickets, and by the early twentieth century, experiments showed that exposure to sunlight could also act as a cure.

In 1921, Edward Mellanby adopted a different, if rather cruel approach, to studying the disease. By feeding dogs exclusively on porridge he found he was able to induce rickets. He then cured the mutts by dosing them with cod liver oil, indicating that there was a nutritional cause for the disease. Mellanby attributed the missing nutrient in the porridge to his recently discovered vitamin A.

It wasn't until American scientist Elmer McCollum followed up Mellanby's experiments, by testing for other nutrients in cod liver oil, that a different version of events emerged. On heating the oil, he found that it no longer cured night blindness, indicating that its vitamin A had been destroyed. But it still cured rickets. He dubbed the new substance vitamin D.

Further work over the next forty years established that vitamin D was vital for the incorporation of calcium to form strong, healthy bones, and that the "vitamin" was in fact synthesized in the skin on exposure to ultra-violet light.

Helen Potter

Date 1922

Scientist Elmer McCollum

Nationality American

Why It's Key A cure for one of the most prevalent nutritional deficiencies in the western world.

opposite Colored X-ray of the legs of a child with rickets

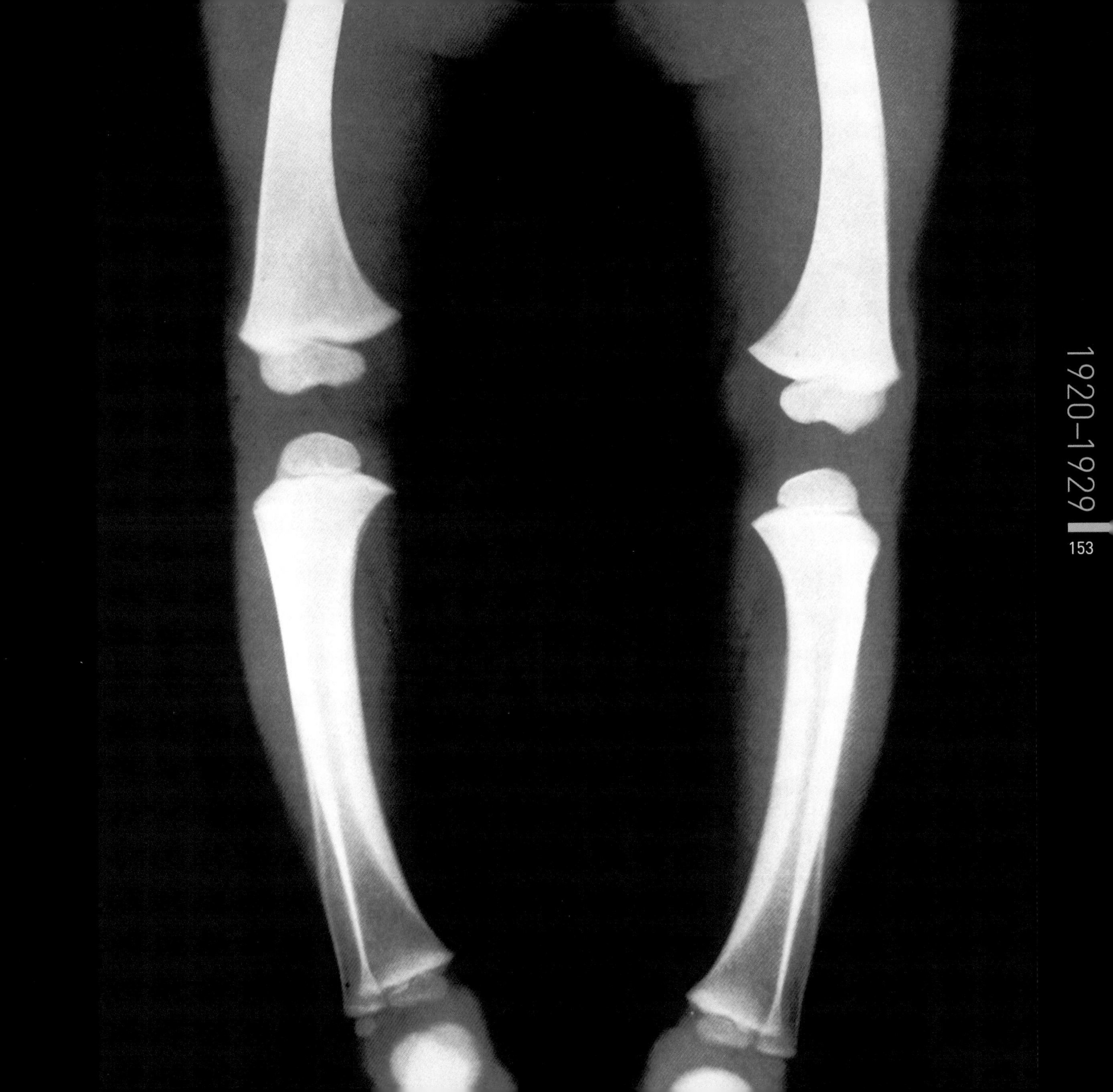

Key Publication
Tractatus Logico-Philosophicus

Ludwig Wittgenstein's major work, *Tractatus Logico-Philosophicus*, nearly didn't see the light of day. Wittgenstein wrote it while he was on the frontline in World War I and when he was taken hostage in 1918 the original copy of the book went with him. Luckily for modern philosophy both he and the book survived, only to be rejected by virtually every publisher Wittgenstein approached. However, when it was finally put in print in 1922 it received high acclaim from his contemporaries.

In the *Tractatus*, Wittgenstein investigated the relationship between language and reality. The book deals with how propositions create pictures of reality. Written musical notes do not seem at first sight to be an accurate description of a piece of music and yet the language is pictures, even in the ordinary sense, of what they represent. Something that cannot be pictured cannot be described by language and he alludes that such things are mystical and if we try and describe them then we will end up talking nonsense. Wittgenstein famously summarized the book in the following words: "what can be said at all can be said clearly; and what we cannot talk about we must pass over in silence." If his logic is right then much of philosophy, which deals with the mystic and cannot be pictorially described, is twaddle.

Tractatus captured the imagination of a generation of philosophers and clearly shows that philosophy and logic have many limitations on how they can describe the world around us.

Leila Sattary

Publication *Tractatus Logico-Philosophicus*

Date 1922

Author Ludwig Wittgenstein

Nationality Austrian-British

Why It's Key Key investigation into the relationship between language and reality.

Key Event **Thar she blows, blows, blows**
The first 3D movie

Although demonstrations of various stereoscopic (3D) film techniques and at least one public 3D presentation had been made by this date, the first 3D feature film release proper is generally acknowledged to have been Nat Deverich's "Power of Love," which premiered at the Ambassador Hotel Theatre, Los Angeles, in September 1922. The film revolved around the adventures of a young sea captain in California in the 1840s.

A Perfect Pictures production, the film used an "anaglyphic" 3D process developed by Harry K. Fairhall and Robert F. Elder. It was shot with a custom camera, designed by Fairhall, which combined two films in one camera body. The camera recorded two slightly different viewpoints of each scene which were then printed in two different colors – red and green. Two projectors are used to show the film. In the words of a review from the time: "...This stereoscopic method is obtained by projecting two positive prints on the screen simultaneously and super-imposing them.

The 3D effect created is not however, resolvable by the naked eye. Instead, the film is viewed through a pair of special glasses which have one red lens, which filters out the red image, and one green lens, which filters out the green image. So each eye sees one of the two slightly different images; these are interpreted by the brain as a single, three dimensional image.

Mike Davis

Date 1922

Country USA

Why It's Key Renewed interest amongst film-makers and movie-goers in the 3D phenomenon.

Key Event **Tutankhamun's tomb unearthed**

The real tomb-raiders

When British archaeologist Howard Carter discovered Tutankhamun's tomb and the forgotten treasures therein, he would set in motion a sequence of events that continue to captivate the world to this day. The grave would tease researchers with discoveries that provoked as many questions as they answered.

Tutankhamun's chamber was like no other tomb discovered before. Buried deep under ground in the Valley of the Kings, on the western bank of the Nile, it was overlooked by robbers and archaeologists alike until Carter, and Lord Carnarvon, his patron, entered it in late November, 1922. Amazingly, Tut's remains and the treasures surrounding them were intact and would thus provide a fascinating insight into ancient Egyptian burial practices and customs.

It was soon discovered, however, that as such sites go, this one was unusual. Modest in size compared to the typical royal chambers and, apparently, decorated at short notice; paint-splashed walls and second-hand treasures. Several actually bore the hastily erased names of other deceased people. Even the embalming had been spoiled by buckets of unguents dumped unceremoniously over the mummy. It all seemed very strange; had this been a rite or an attempt to conceal something?

It wasn't until 2005 that anyone had a definitive answer. Reanalyzing X-rays taken in 1968, a team of researchers concluded that the Pharoah, who died in 1352 BCE, had been killed by an infected broken leg, which was most likely sustained in an accident.

Mike Davis

Date 1922

Country Egypt

Why It's Key Revealed much about the ancient Egyptian civilization, their culture, their beliefs, and burial techniques.

Key Discovery **Childbirth drug, scopolamine, declared to be "truth serum"**

In the early 1900s forensic science was gaining importance in the criminal justice system, and doctors were becoming more involved with the police. One American physician, Robert House, was concerned about hostile police interrogations. He proposed the drug scopolamine as a non-violent way to reveal someone's innocence or guilt.

He had given it to women to relieve the pain of childbirth, by inducing a special "twilight sleep." In this artificial unconsciousness, people could still hear and speak, and responded to questioning with child-like simplicity and honesty. House believed scopolamine rendered people incapable of lying. His first results with convicted criminals were published in 1922. They were received with overwhelming enthusiasm by legal professionals, police departments, and the general public. Scopolamine was hailed as a true "truth serum." Hundreds of scopolamine interviews followed. Some suspects confessed to vile murders, some to impossible scenarios, and some were set free.

The drug's popularity could not last. It was a well-known poison (the henbane that killed Hamlet's father) with side effects including violent sickness. It was increasingly viewed as a form of torture. Also, any "truths" were likely to be distorted by the vivid hallucinations it induced. It came to be disregarded in the courts as a truth serum by the 1950s. It still has dark associations, being popular in the grim interrogations of fiction and used in "date rape". However it still has value as an anticholinergic, primarily in treatments that act on the parasympathetic nervous system .

S. Maria Hampshire

Date 1922

Scientist Robert House

Nationality American

Why It's Key The ability to conceal truth was considered the germ of social disorder, and truth serum was seen as the vaccine. For a while, scopolamine promised this.

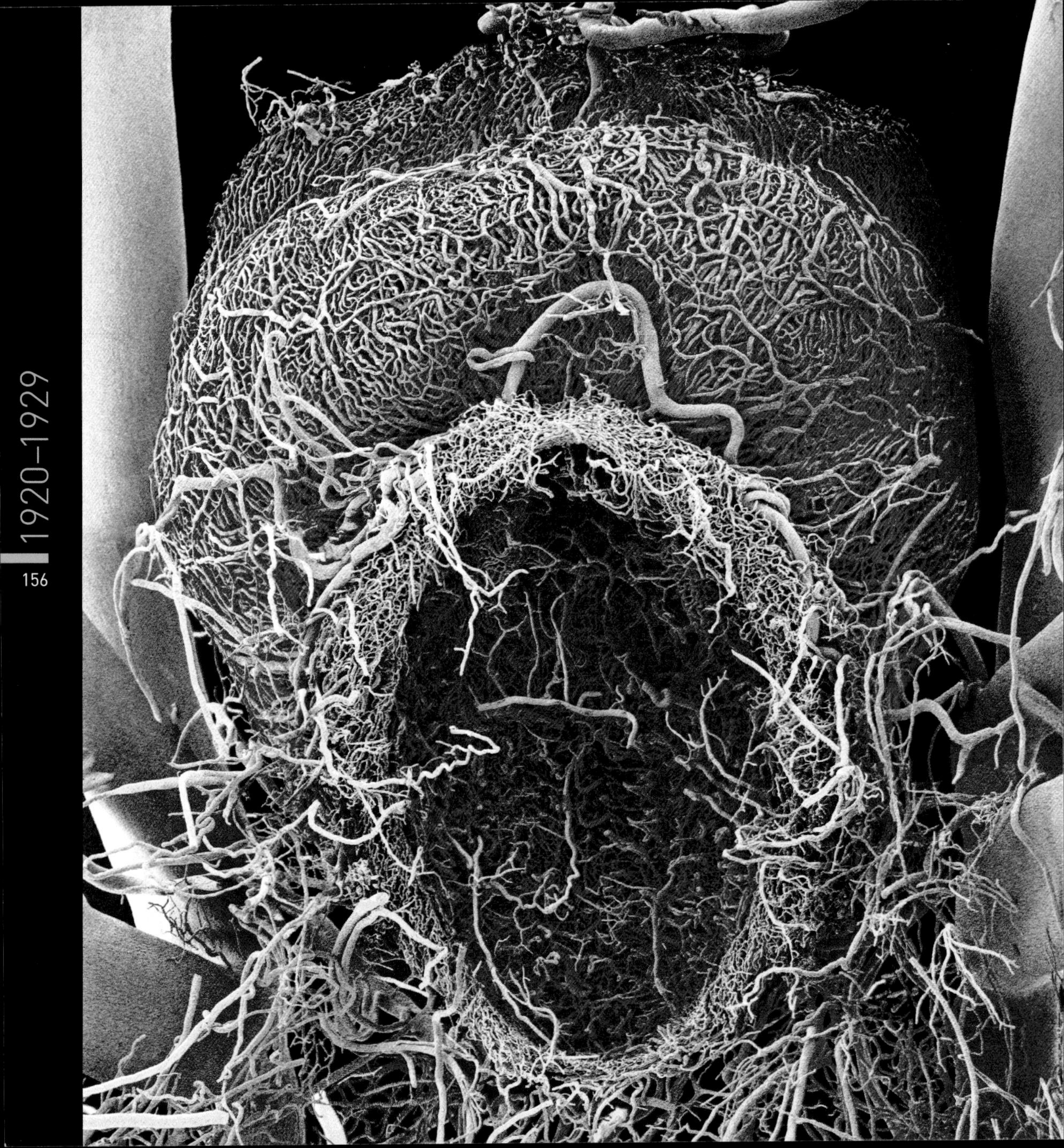

Key Experiment
A cure for stunted growth

Growth failure is a problem that affects thousands of children worldwide. It was a major problem at the beginning of the twentieth century, when Herbert Evans was making his way through medical school. His father had hoped he would take over his surgical practice but he chose anatomical research instead.

Evans was looking into how rats reproduce when he chanced upon the growth hormone. In an extraordinary experiment, he submerged the back part of a rat's pituitary gland – located in the brain – in various liquids, and then spun it in a centrifuge. This released a fluid that he injected into other rats' abdomens. He monitored the rats over the following days and was surprised to find the rats that had been injected with the fluid had grown a lot more than the rats that hadn't. It later became clear that this fluid contained a growth hormone. Human Growth Hormone (HGH) was first isolated in 1956, and its chemical structure was identified in 1972.

Since 1959, it has been helping stunted children by stimulating the growth of tissues. Morbidly, back in the 1960s and 1970s, the only source of this hormone was from dead bodies. The pituitary glands were removed from corpses and, after processing, the fluid was injected into patients with growth hormone deficiency. Its production represented one of the first applications of genetic engineering in the1980s and it is now synthesized chemically in the laboratory.

Riaz Bhunnoo

Date 1922

Scientist Herbert Evans

Nationality American

Why It's Key Without treatment, growth hormone deficient boys would only grow to a height of 130 centimeters by age 18. With the HGH medical breakthrough, they can now expect to reach a normal height of around 180 centimeters.

opposite Colored resin cast of blood vessels in the pituitary glands

Key Invention
Portable electric hearing aids

In the UK alone, there are nearly nine million people who are deaf or hard of hearing. Hardly surprising then, that so many devices and inventions throughout the ages have been designed to improve the ability of deaf and hard of hearing people to communicate.

Hearing aids have ranged from a basic ear trumpet – possibly used since the time of the ancient Greek and Roman empires – to direct stimulation of the auditory nerve by cochlear implant. Surgical implants are by no means a modern idea; as long ago as 1640, a German physician called Marcus Banzer created an artificial ear drum out of pig bladder and elk hoof.

A hearing aid, as it would be understood today involves converting sound into an electrical signal, which is then amplified to allow the sound to be heard more easily. The first electrical hearing aids began to become available in the 1920s, but as these devices, such as the 1923 Otophone, were based on vacuum tube technology they tended to be rather bulky – about 7k kilograms in the case of the Otophone.

Smaller transistor-based hearing aids were the first "wearable" aids, and became available in the 1950s. The Medical Research Council estimates that today there are two million hearing aid users in the UK. There are, however, thought to be four million more people who could benefit from using the devices, but are put off either by the expense or the stigma that is still attached to wearing an aid.

Jim Bell

Date 1923

Scientists Marconi Company (the Otophone)

Country UK

Why It's Key Hearing aids help millions of people across the world lead normal, independent lives.

Key Event
"Oil reserves are running out"

The early twentieth century was seeing a shift in fuel use. The rapid increase in motorized transport was pushing oil towards the forefront of energy production; it would go on to overtake coal – a mainstay of human life since the 1700s – as the largest source of primary energy in the 1960s. But there was a problem – where was all this oil going to come from?

In the 1920s, geologists were beginning to worry about whether they would be able to find enough oil to power the booming Industrial Revolution. A 1922 survey of known oil reserves in the USA estimated that just nine billion barrels remained – enough to last just twenty more years. The pressure was suddenly on to find more oil.

A worldwide search was launched to find new oil reserves, and by the mid 1930s, the oil industry – and everyone who relied on it – was able to breathe a sigh of relief. Large reserves had been uncovered, firstly in Texas (1930) and in 1935, following a frustrating five-year search, prospectors struck black gold in Saudi Arabia, now the single largest oil producer in the world.

The search for oil has since intensified along with mankind's dependence. A vast array of techniques is used to scour the subterranean world for the black lakes, including "thumpers", which bounce sound waves off underground deposits. Current estimates for the world's reserves differ depending on who you ask, but it's likely that there's at least another 400 years worth of oil left underground.

Mark Steer

Date 1922

Country USA

Why It's Key Stimulated the burgeoning oil industry to intensify the search for black gold. Oil has shaped the face of the modern world.

Key Invention
The autogyro

The autogyro is best described as a cross between a helicopter and an airplane, having a propeller at the front, but also rotor blades on the top. Its inventor was a Spanish aeronautical engineer named Juan de la Cierva and on January 9, 1923, his "windmill plane" made its maiden flight.

Cierva was obsessed with flight from an early age and as a teenager he was always busy building planes and messing around with gliders. He started experimenting with rotor blades in 1919 but his initial attempts were unsuccessful until he struck upon the idea of mounting the blades on hinges so that they would flap. This equalized the lift in the rotor while the craft was in forward motion and this discovery later proved to be a crucial aspect in the development of the helicopter.

His early models were known as compound aircraft – possessing wings and a forward mounted engine, in addition to the rotors mounted above the pilot's cockpit. Having proved that his autogyro was capable of flight, Cierva moved to England in 1925, forming his own autogyro company where he continued to manufacture and develop his aircraft. By the 1930s they were being used for both civil and military use.

In a cruel twist of fate, Cievra tragically died in a plane crash in 1936 but his autogyro lives on and with futuristic two-seater models already in pre-production, it looks set to reach new heights.

Andrew Impey

Date 1923

Scientist Juan de la Cierva

Nationality Spanish

Why It's Key The technical breakthrough in hinged rotors was crucial to the future development and success of the helicopter.

opposite The autogyro

G-EBY

Key Discovery
First vaccine for diphtheria

In the early 1900s, diphtheria was a major killer; as many as one in ten people in the UK – mostly children – caught it, and many died from suffocation, paralysis, and heart failure. By 1913, it had become apparent that the disease was caused by toxins produced by a bacterium called *Corynebacterium diphtheriae*. Early work by Emil von Behring had produced long-lasting immunity in guinea pigs and monkeys, using a mix of the bacterial toxin and antitoxin from these animals.

When, in 1923, A.T. Glenny and Barbara Hopkins discovered a way to detoxify the poison using formaldehyde, it opened up the possibility of a human vaccine against the disease. Their discovery allowed the production of an effective vaccine against diphtheria, without relying on difficult-to-produce antitoxins.Following this leap forward, a widespread immunization program was started in the 1940s which has led to the virtual elimination of the disease in countries like the UK and the United States. Success hasn't been seen worldwide, however; following the breakdown of the USSR, vaccination rates plummeted, leading to an explosion of cases in 1998. As a direct consequence, diphtheria was named the most resurgent disease in the Guinness Book of Records that year.

This discovery was so important that, to this day, the inactive diphtheria toxoid used in vaccines is still made via formaldehyde deactivation, using essentially the same technique as discovered in 1923.

Catherine Charter

Date 1923

Scientists Alexander Thomas Glenny, Barbara Hopkins

Nationality British

Why It's Key A way of producing inactivated diphtheria toxin, which saved thousands of lives in the form of a vaccine.

opposite Inoculated horse has blood taken from its exterior jugular vein, which is then used to treat diphtheria

Key Discovery **Acids and bases**
It's a matter of give and take

While the idea of acids and bases has intrigued scientists since the Middle Ages, it wasn't until the early twentieth century that an explanation at the atomic level came about.

Dutch scientist Johannes Bronsted, working as a professor at the Polytechnic Institute and University in Copenhagen, began developing his theory of acids and bases. His explanation of the properties of acids and bases dramatically changed the way scientists perceived acid-base chemistry.

Until this point, most chemists were using Swedish chemist Svante August Arrhenius' theory, which described acids and bases as chemicals that produce certain types of ions when they dissociate in water. Bronsted took the concept one step further. An acid, according to Bronsted's theory, is any substance that is able to donate a hydrogen ion, also known as a proton. A base, on the other hand, is a hydrogen ion acceptor. The explanation was as simple as that. And Bronsted's theory wasn't limited only to water as the solvent of the acid or base; it worked for all types of solvents.

Simultaneously to Bronsted's publication, British scientist Thomas Lowry had independently come up with essentially the same theory about acids and bases. As a result, this simple model, still an essential concept in basic chemistry classes, is now known as the Bronsted-Lowry theory of acids and bases.

Rebecca Hernandez

Date 1923

Scientist Johannes Nicolaus Bronsted

Nationality Dutch

Why It's Key Described one of chemistry's basic ideas at its most stripped-down level.

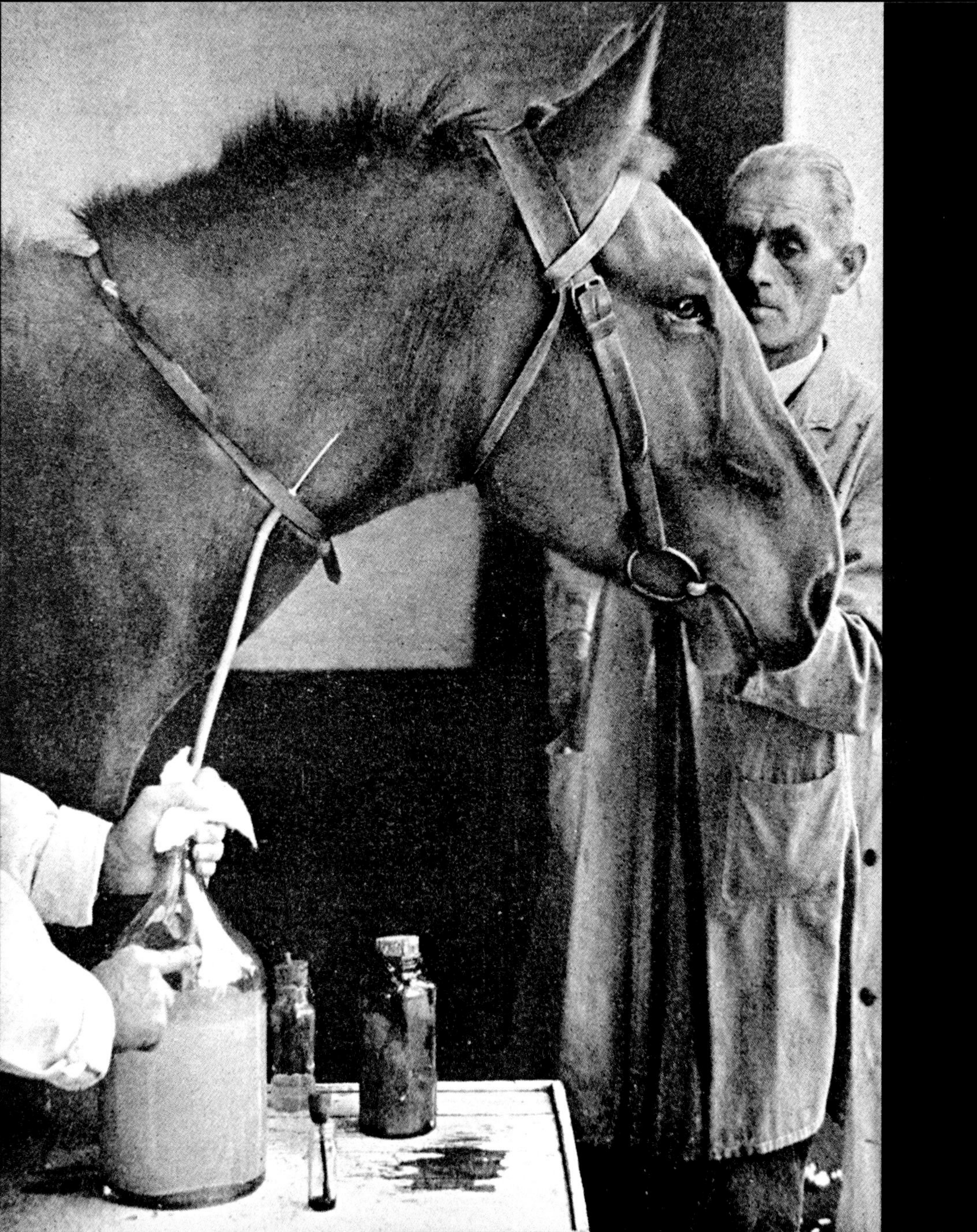

Key Person **Jean Piaget**
Development in stages

Jean Piaget (1896–1980) is probably best remembered as "that stage guy," but his true legacy is in the field of "Genetic Epistemology" – the study of the development of knowledge. His earliest forays into science were in biology. He published an account of his sighting of an albino sparrow at age ten, and later published papers on mollusks. He said his mother was neurotic, accrediting to this his interest in psychology. Following his graduation from the University of Neuchâtel, Switzerland, Piaget met Théodore Simon, co-author of the Simon-Binet intelligence scale. Simon secured work for him in Binet's laboratory, standardizing tests on children. Here Piaget began to notice that children of similar ages made similar types of mistakes; he thought the key to understanding human intellectual development may lie not in what children get wrong, but how they get it wrong. He postulated children reason differently to adults.

While observing his three children, Piaget constructed his four stages: the sensorimotor stage begins at birth, and lasts until age two. Intelligence develops through motor interactions with the environment. The preoperational stage lasts until the child is six or seven, when children are able to make mental representations of unseen objects, but cannot use deductive reasoning. The concrete operations stage follows, lasting until the child is eleven or twelve: they are able to use deductive reasoning, demonstrate conservation, and differentiate their perspective from that of other people. Formal operations is the fourth stage: at this stage children now think abstractly.

Stuart M. Smith

Date 1923

Nationality Swiss

Why He's Key Piaget's work was revolutionary; it transformed our thinking about cognitive development.

Key Person **Roy Chapman Andrews**
The original Indiana Jones

Roy Chapman Andrews was born in 1884 in the United States. An avid explorer, he was renowned as much for his brave adventures into hazardous territories and near death experiences, as for the discoveries his expeditions led to.

Andrews was a keen explorer as a child, and even taught himself taxidermy. After graduating, he pursued his interest in the natural world and worked his way up the staff ladder at the American Museum of Natural History in New York, starting off by sweeping their floors. He had an exciting and varied career at the museum, pursuing his passion for exploring under the guise of numerous scientific quests. Initially he started researching marine mammals off the east and west Pacific coast. His trips focused first on zoology, then anthropology, leading him eventually to paleontology. Regarded more as an expedition leader than as a scientist, Andrews' explorations significantly advanced the museum's research.

His most famous expeditions took him to the Gobi desert in Mongolia and China between 1922 and 1930, where he and a team of scientists ventured into previously uncharted territory. Having gone in search of evidence of early humans and their evolution, the team instead discovered the first known dinosaur eggs, and fossils of mammals that lived amongst them.

Fieldwork in the Gobi desert had to stop in 1930 due to political instability in the region. When he returned to America, Andrews became the museum's director before retiring in 1941 to write numerous books about his explorations and discoveries.

Emma Norman

Date 1923

Nationality American

Why He's Key Andrews explored new areas of the world and unearthed many exciting discoveries. Rumor has it that the character of Indiana Jones was based on him.

opposite Roy Chapman Andrews

Key Discovery
Particle nature of photons confirmed

The discovery that the X-rays scattering off a material weren't the same as those being directed at it was an interesting phenomenon, but seemed no cause for concern when it was first discovered.

But, investigating further, physicist Arthur Holly Compton measured the rays given off after scattering. He discovered that while a beam of the original wavelength was present, another beam with longer wavelength (and so a lower frequency) appeared as well. The difference was present regardless of what material the X-rays were scattered off, but the magnitude of the difference varied depending on the angle that the X-rays were scattered by. This was neatly explained if the X-rays were considered not as waves, but as particles. Compton's theory was that, when traveling through the material, X-ray particles (photons) are knocked off course in collisions with electrons. During the collisions, they transfer some momentum – and therefore some of their kinetic energy – to the electrons they collide with. This loss of energy manifests as a drop in frequency of the ray, because lower frequencies carry less energy. When electrons recoiling from the collisions were detected, as predicted, the particle nature of X-rays was confirmed.

This was a shock. Theories about how everything could be described as a wave had explained perfectly all physical phenomena so far observed. But here was the distinct possibility that X-rays, and maybe even light itself, were particles: how was the whole science of optics to redesign itself based on particles rather than waves?

Kate Oliver

Date 1923

Scientist Arthur Holly Compton

Nationality American

Why It's Key Compton succeeded in explaining how light and matter interact, paving the way for Einstein's theory of wave-particle duality.

Key Discovery
Dinosaurs laid eggs

In 1859, geologist Jean Jacques Pouech uncovered what may possibly have been the first recorded discovery of dinosaur eggshell, in Southern France. However, at the time it was believed giant birds probably laid the eggs. Subsequent fossilized eggs unearthed by Philippe Matheron in 1869 were also of uncertain origin – reported as either a giant crocodile or tortoise. It wasn't until 1923 that the first confirmed discovery of fossilized dinosaur eggs occurred.

The American explorer Roy Chapman Andrews was leading an expedition to the Gobi desert in Mongolia and, while exploring the Flaming Cliffs, his team from the American Museum of Natural History made their amazing discovery. Of a similar structure to those of a modern-day bird, the eggs had presumably been buried to protect them from predation. They had not hatched, possibly due to a change in weather conditions, but only the shell itself had been fossilized with no contents preserved.

At the time, the eggs were identified as belonging to a plant-eating species called *Protoceratops*. Scientists have now questioned this, and new evidence suggests that they in fact belonged to a dinosaur from the *Oviraptor* family. These meat eaters, which resembled small ostriches, have very distinctive skulls, and fossilized embryos found close to the original nest site showed these characteristic features.

Subsequent studies of the site also found the fossilized remains of an adult *Oviraptor* lying on top of the nest, suggesting that the parent may have either been incubating or guarding the nest before its death.

Andrew Impey

Date 1923

Scientist Roy Chapman Andrews

Country Mongolia

Why It's Key The first confirmed evidence that dinosaurs laid eggs, found at a site which has since taught us much about dinosaur behavior.

Key Event **Heroin**
From wonder-drug to illegal substance

In 1898, German drug manufacturer Bayer started producing heroin from morphine. It was aggressively marketed as a non-addictive cure for a variety of physical and mental ailments, and as a treatment for morphine addiction. Within four years, heroin comprised 5 per cent of the global pharmaceutical industry's entire turnover. Concern grew about its use, especially in relation to morphine addiction. Heroin is twice as potent as morphine and is transformed in the liver to morphine. Heroin addiction subsequently became a major problem in the USA; by the 1920s, 98 per cent of New York's addicts were hooked on it.

In 1909 and 1911, two international conferences were held, prompted by disputes over opium exportation worldwide. Recommendations from these conferences resulted in the 1914 Harrison Act being passed in the USA. This was a tax law requiring registration of drug importers, manufacturers, and prescribers. Although medical prescription of heroin remained legal if it was "in the course of professional practice," addiction was not regarded as an illness, so doctors faced arrest for supplying addicts with drugs. As a result, addicts – who were from all social classes – had to find drugs on the black market.

The deteriorating behavior of addicts, and the replacement of morphine by heroin, led U.S. authorities to believe heroin was more dangerous than morphine. In 1924, the sale, importation, and manufacture of heroin was banned. There are still some people who argue that legalizing and regulating drug use is more appropriate than criminalizing addicts.

Katie Giles

Date 1924

Country USA

Why It's Key An embarrassing u-turn regarding heroin use. Addicts were treated as criminals and problems with drug abuse continued.

Key Event **IBM formed**
From scales to supercomputers

In 1911, the Computer Tabulating Record Company (CTR) was formed. Incorporated from the merger of the Tabulating Machine Company (which made its money by producing tabulating machines for the 1890 U.S. Census), the Computing Scale Company of America, and the International Time Recording Company, their combined might spawned an enterprise that sold everything from commercial scales to cheese slicers. Never heard of them? In 1924, after expanding their product range to include the first complete school time control system and the electronic accounting machine, the company underwent an identity change, and emerged as the International Business Machines Corporation – IBM.

Over the twentieth century, IBM's operations have shifted from tabulating machines to bombsights, rifles, and engine parts during World War II; and computers, servers, and business solutions since then. IBM took its first steps toward computers in 1944 with the automatic sequence controlled calculator. At 19.7 meters long and 3.2 meters high, it took six seconds to complete a division problem. Currently IBM has a yearly revenue of over US$90 billion and employs over 350,000 people in 170 countries. In January 2006, the US.. Patent Office reported that IBM had earned more patents than any other company for the thirteenth year running, resulting in a total of over 31,000 patents. The company is the world leader in supercomputers, with 237 of the top 500 systems, including the world's most powerful, BlueGene/L, used to study everything from climate modeling to protein folding.

Helen Potter

Date 1924

Country USA

Why It's Key A world-leading IT company gets off the ground.

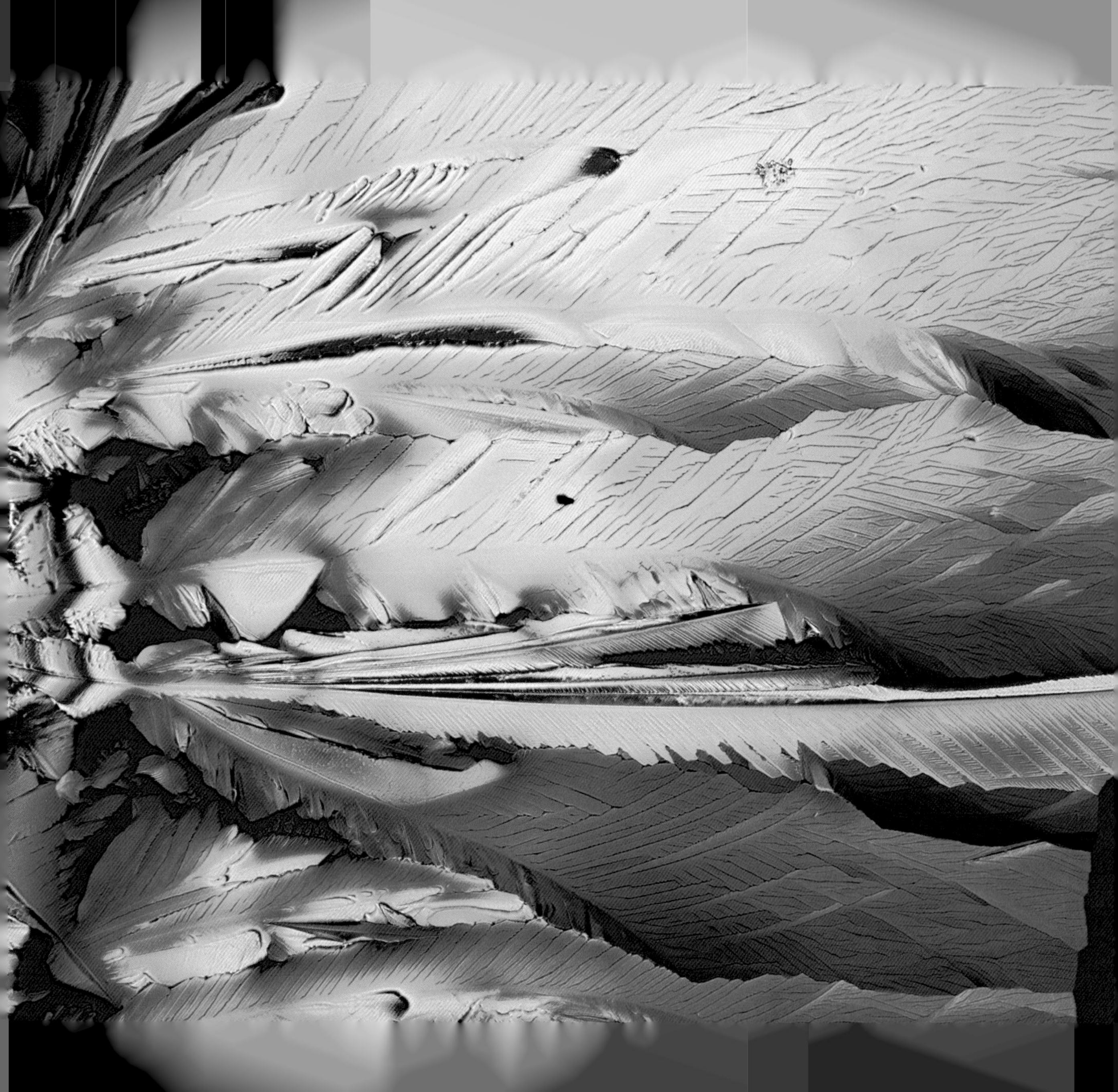

Key Experiment **Shedding light on rickets** UV radiation increases vitamin D in food

In 1924, Harry Steenbock received a telegram from his colleague urging him to publish his work without delay; rumors were rife that someone might underhandedly take his glory. He did as he was advised, and his work went on to have significant consequences for the treatment of rickets, a disease that leads to unnaturally soft, twisted bones.

Steenbock had read about an experiment where rats that had been kept in glass jars and irradiated with UV light grew more than those that hadn't. He couldn't think of a rational explanation for this, so he conducted a series of experiments to establish the cause.

In one experiment, he fed rats food that had been irradiated with UV light, and was surprised to find that they grew much bigger than those that had not been fed irradiated food. He also found that their bones were stronger and richer in calcium. He attributed this to the conversion of an inactive type of cholesterol in the food to vitamin D, using energy from the UV light. This reaction also occurs in the skin, which is why the rats that were irradiated directly had grown so much more than their counterparts. In fact, most people get the majority of their vitamin D through UV rays in sunlight.

The application of this irradiation process to milk led to the virtual elimination of rickets in the United States by 1945.

Riaz Bhunnoo

Date 1924

Scientist Harry Steenbock

Nationality American

Why It's Key Steenbock's discovery provided a simple method to increase the vitamin D content of food, eventually leading to the virtual elimination of rickets.

opposite Polarized light micrograph of crystals of vitamin D

Key Event **Pip pip: hourly time signals first broadcast from Greenwich**

Hourly time signals broadcast from the Royal Greenwich Observatory have been crossing the airwaves since February 5, 1924. They are more commonly referred to as "the pips," a term coined by the BBC.

In 1924, the Astronomer Royal, Sir Frank Dyson, and head of the BBC, John Reith, came up with the idea of an hourly signal that would be broadcast by the BBC from the observatory. Before this, the BBC's disk jockeys had indicated the hour by playing a few notes on tubular bells or a piano.

Until 1971, the pips were calibrated to coincide exactly with Greenwich Mean Time (GMT), but, after an international conference, it was decided that all time keeping should be in line with international atomic time. These two time frames were slightly different; GMT being determined by the spin of the Earth, which fluctuates over time, and atomic time being kept by more stable atomic clocks. In reality, atomic time was only about one second faster than the original GMT method over the course of a year, but this still had to be accounted for.

In the late 1980s, the Royal Observatory underwent major changes, and there was no longer the capacity to broadcast the pips. The decision was made to move the equipment to the BBC's Broadcasting House, where it resides to this day.

Josh Davies

Date 1924

Country UK

Why It's Key Nowadays, the pips are seen as a British institution, and they have preceded many of the late twentieth century's most important news announcements.

Key Publication **The de Broglie hypothesis**
Light and matter become the same

When French physicist Louis de Broglie earned his Nobel Prize in 1929, the achievement was pretty special. It was the first Nobel Prize awarded for a PhD thesis. De Broglie, the son of a French aristocrat, built upon the work of Niels Bohr in explaining the behavior of electrons in the atom. While revolutionary, Bohr's model did not explain why electrons emit or absorb energy when they jump between orbits around a nucleus. In de Broglie's thesis, he suggested that matter actually consists of waves, like light or sound. He assumed that any particle of matter had a wavelength which could be found by dividing Planck's constant by the particle's momentum. It had already been shown that light waves had this property, and that photons of light sometimes seemed to behave like waves and sometimes like particles. This neatly explained why an electron gives off a photon as it drops to lower atomic orbits; because its frequency and therefore energy has changed, and so the excess energy must be lost through photon emission.

In 1927, Clinton Davisson and Lester Germer at Bell Labs proved de Broglie's ideas fully. In an accidental discovery they found that a beam of slow-moving electrons, fired at a crystalline nickel surface, scattered back with the same pattern as X-rays did – showing for the first time the wave-like behavior of matter.

David Hawksett

Publication The de Broglie Hypothesis

Date 1924

Author Louis de Broglie

Nationality French

Why It's Key It married together the studies of radiation and matter.

opposite Electrons from an electron gun pass through graphite and hit a luminescent screen, producing patterns of rings associated with diffraction as de Broglie hypothesized

Key Invention **The ultracentrifuge**
Chemicals in a spin

Anybody who has ever swung a ball on the end of a string has seen centrifugal forces causing the ball to move outwards. Theodor Svedberg (who was also known as "The Svedberg" which sounds very self important in English, but is just a nickname in his native Swedish) turned that principle in to the 1924 Nobel Prize-winning invention.

Surprisingly, Svedberg's first love wasn't spinning stuff around really fast. His early work involved proving that Einstein's theory about the origins of Brownian Motion were correct – by slightly smaller pieces of matter. While this work helped to provide the first concrete proof for the existence of molecules, it was his later work that landed him the big prize.

Svedberg realized that if you crank up the revs of a centrifuge to a very rapid rate, substances of different masses and shapes will separate out at different rates. He was therefore able to prove that complex molecules of certain substances, such as hemoglobin, were of a distinct size and shape, making them easy to identify, and he was subsequently able to determine just how heavy these molecules were. He continued to refine the role of the ultracentrifuge, building machines that were capable of generating forces up to one hundred thousand times the force of gravity. The same basic techniques are still used today to determine molecular weights.

B. James McCallum

Date 1924

Scientist Theodor Svedberg

Nationality Swedish

Why It's Key The Svedberg's ultracentrifuge established a way to accurately measure the weight of complex molecules.

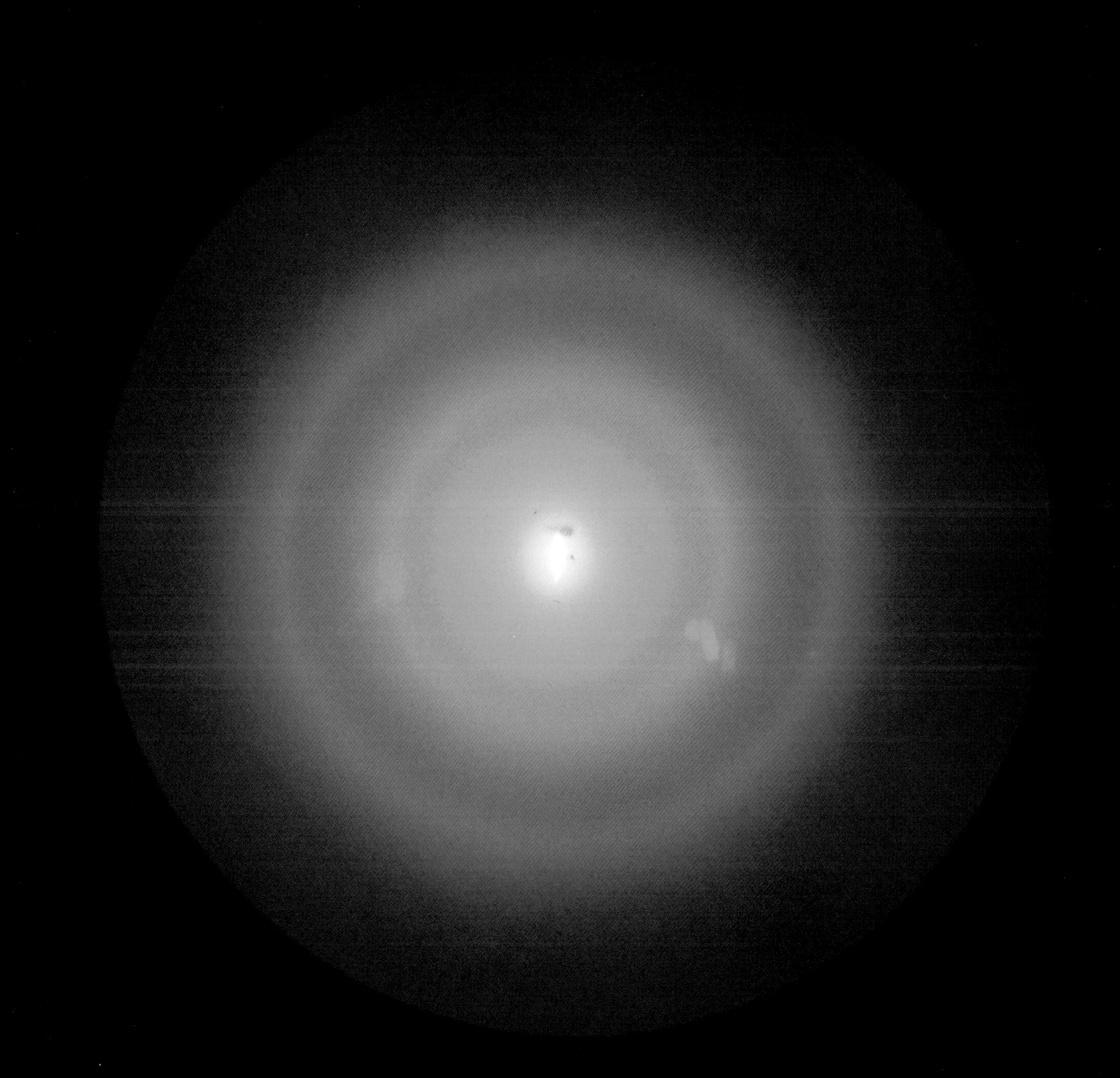

Key Invention
Fast freeze keeps food fresh

Freezing food to keep it fresh for later seems obvious now, but it's a trickier feat than you might imagine. The acknowledged "father" of the frozen food industry is American, Clarence Birdseye who gave his name to Birds Eye, Inc., a company which is still going strong today.

Birdseye was working as a field naturalist for the United States government in Labrador, Canada, when he noticed that fish frozen by icy winds still taste fresh when thawed out afterwards. The secret, Birdseye realized, lay in the speed of the freezing: a rapid freeze means that only small ice crystals have time to form, causing less damage to the food's tissue cells than large ones.

It took all of Birdseye's skills as a businessman and experimenter to perfect rapid freezing back in New York City – and to extract the full value of what he knew could be a lucrative discovery. In 1924, after a series of experiments, Birdseye patented a system whereby fresh food was packed in cartons between two refrigerated surfaces, and frozen under pressure.

In 1930, the first public sales of quick-frozen fish, meat, fruit, and vegetables took place in Springfield, Massachusetts, under the trade name Birds Eye Frosted Foods. By the 1950s, frozen food sales exceeded US$1 billion a year. Thanks to the astute Mr Birdseye, we can now enjoy fresh peas all year round.

Richard Van Noorden

Date 1924

Scientist Clarence Birdseye

Nationality American

Why It's Key Birdseye's quick-freezing technique underpins the modern frozen food industry.

opposite Frozen food comes out of the packaging process

Key Event
Around the world in 175 days, by biplane

On April 4, 1924, four Douglas World Cruiser biplanes left Seattle, on America's west coast, and headed to Alaska. The four planes were named Seattle, Chicago, Boston, and New Orleans, and this was the start of a 175-day trip that would make history. The biplanes were variations on DT-2 torpedo bombers, but were fitted with huge fuel tanks for the long journey ahead. Fifteen spare engines, fourteen pairs of pontoons, and enough spare parts to build two new planes were sent along the route the planes were to follow.

The Seattle had to stop for repairs and, while trying to catch up, crashed into a mountainside on the Alaskan peninsula in dense fog. The crew hiked out of the wilderness to safety, but the plane was destroyed.

Avoiding the Soviet Union, which they didn't have permission to cross, the remaining planes flew to Japan, the coast of China, Hong Kong, Thailand, and Burma, before flying up through the Middle East and into Europe. They arrived in Paris for Bastille Day, flew on to London, and then north before crossing the Atlantic.

While making the crossing, the Boston was forced down and then capsized while it was being towed. The two remaining planes crossed Iceland and Greenland into Canada, where they were met by the Boston's crew, flying a test aircraft named Boston II. The three crews flew to Washington, receiving a hero's welcome before flying west to complete the first circumnavigation of the world by air.

Douglas Kitson

Date 1924

Country International

Why It's Key Although it wasn't non-stop, it was a major feat of aviation, succeeding where others had failed by packing huge amounts of spare parts.

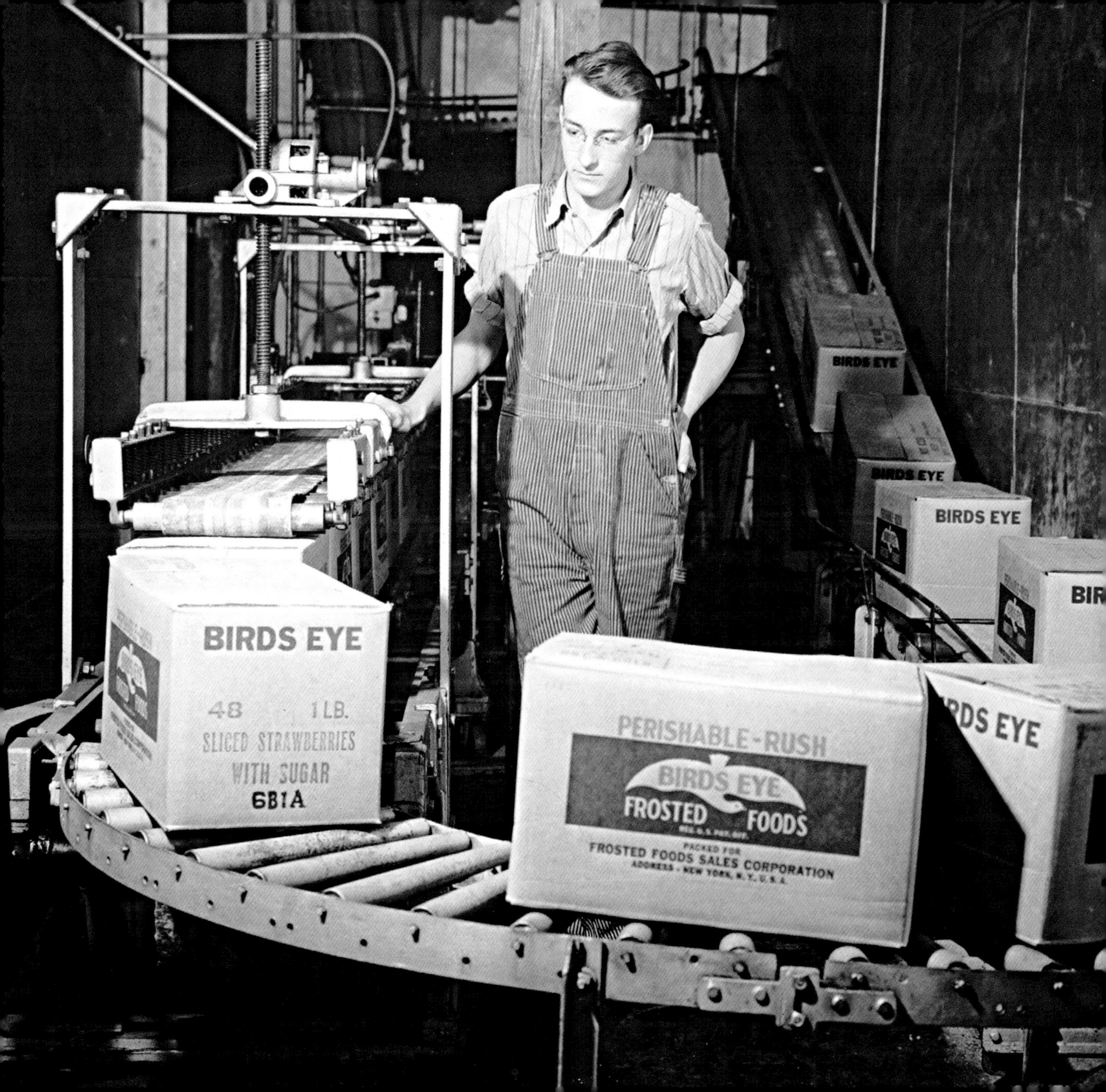
BIRDS EYE
BIRDS EYE
BIRDS EYE
BIRDS EYE
48 1 LB.
SLICED STRAWBERRIES
WITH SUGAR
6B1A
PERISHABLE-RUSH
BIRDS EYE
FROSTED FOODS
PACKED FOR
FROSTED FOODS SALES CORPORATION
ADDRESS - NEW YORK, N.Y. U.S.A.

Key Person **Arthur Holmes**
Geo-genius

In 1911, aged twenty-one, Arthur Holmes (1890–1965) developed a reliable and accurate method of calculating the age of rocks – the process of uranium-lead radiometric dating, which uses known facts about the half life of the radioactive element uranium, and the amount of uranium in the rock. Even at this young age, he had established his life's research objectives, which were to formulate of a geological timeline and to find similarities between Precambrian rocks (the oldest of all) taken from different parts of the world.

Using the methods he and Bertram Boltwood had devised, Holmes dated the Earth to at least 1.6 billion years old. This age was initially highly controversial within the geology community – the Earth simply couldn't be that old. Holmes couldn't afford to remain as a university researcher, and took a job as chief geologist for a company in Burma. It was a disaster; the company folded and he returned to the UK penniless. Fortunately he was able to return to research in 1924, when he set up the geology department at Durham University. By this time, his Age of the Earth theory had been more widely accepted.

Another hot topic at the time was Wegener's theory of continental drift. Holmes was the first to propose that convection currents under the Earth's crust were causing the continents to break up and move. Once again, his ideas faced much opposition. Holmes died in 1965, just as theories of plate tectonics were being seriously developed; at last he finally began to receive recognition for his pioneering theory of how continents move.

Emma Norman

Date 1924

Nationality British

Why He's Key Holmes pioneered a number of major geological developments. His ideas often came under criticism from his peers, yet time and again his work proved to be ahead of its time.

Key Discovery **Existence of other galaxies**
Hubble proves there is a whole lot more out there

In the 1920s Einstein's Theory of General Relativity was getting everyone excited, partly because it suggested the universe may be expanding. The so called "Big Bang" theory of the creation of the Universe was popular within physics circles, but badly needed some observations that supported it. Proof of other galaxies outside of the Milky Way would be a good start, especially if they were moving away from us.

Enter Edwin Hubble, an ex-barrister who was bored of law and had decided to study the stars. Plucked straight from his doctorate at the University of Chicago (after a quick stint in the armed forces during World War I) by none less than George Ellery Hale, director of the Mount Wilson Observatory, Hubble went on to prove the existence of other galaxies.

Using the largest telescope in the world at the time, the Hooker telescope, Hubble developed his techniques for measuring the distances to what were called "nebulae." Back then a "nebula" was essentially anything that was seen in space that wasn't a star. By analyzing the light emitted by objects such as Andromeda, which we now know as our closest galaxy, he proved that they were in fact galaxies in their own rights.

Although the Big Bang theory was derived from Einstein's General Relativity theory, Einstein himself thought the Big Bang rather far-fetched. Hubble's findings soon changed his mind and Einstein referred to his rejection of the Big Bang as "the biggest blunder of my life."

Jim Bell

Date 1924

Scientist Edwin Hubble

Nationality American

Why It's Key Proved there were galaxies outside the Milky Way and made sure astronomy was recognized as a "proper" area of physics.

opposite Edwin Hubble

Key Invention
Whooping cough vaccine

Though it has one of the more amusing names in the world of infectious disease, whooping cough, or pertussis, is nothing with which to be trifled. The characteristic cough is the most common symptom associated with the disease, which is caused by the bacterium *Bordetella pertussis*, but it can also lead to pneumonia and even death – especially in very young children. As a result, when the first vaccine for whooping cough was developed in 1925 by Danish physician Thorvald Madsen, it was quickly introduced across the globe and the incidence of whooping cough plummeted. But then reports of rare side effects began to circulate – some claimed the vaccine caused seizures and other neurological problems. It was even blamed for two deaths in Japan. In truth, the original vaccine was probably fairly safe and only perceived as dangerous in the public eye – at any rate there was a very real risk of dying from whooping cough if one was not vaccinated. Due to the well publicized controversy, however, parents soon began to refuse the vaccine. As a result, the number of immunizations dropped steadily and whooping cough staged a comeback.

Scientists realized that the problem with the vaccine stemmed from the fact that the whole *Bordetella pertussis* cell was used in the vaccine to generate the immune response. Soon efforts were being directed at finding a suitable replacement. The resulting acellular pertussis vaccines produce comparable results and have fewer side effects.

James McCallum

Date 1925

Scientist Thorvald Madsen

Nationality Danish

Why It's Key The case of the whole-cell pertussis vaccine showed that medical intervention not only needs to be effective and safe, but also recognized by the public as safe. The best vaccine in the world will not save lives if parents are afraid to give it to their children.

Key Discovery ***Australopethicus africanus***
Digging up human history

Part man, part ape – this was the original observation of Raymond Dart when he discovered what he would later refer to as *Australopithecus africanus* in a 1925 *Nature* paper. Dart, then a professor at the University of Witwatersrand in South Africa, was working at the Taung site in South Africa with his students, when a skull was discovered, originally presumed to be that of a primate.

However, upon closer inspection the skull appeared more human-like. Over seventy painstaking days later, the skull was reported by Dart to be that of a child with an ape-sized brain, but human dental and postural qualities such as bipedalism. Many powerful scientists in the field disagreed with Dart's assumption that this was a new hominid species; they believed the skull instead to be that of a young chimpanzee or gorilla. Dart also faced dissent from scientists who believed that humans originated not in Africa, but rather Asia or Europe. All of this resistance proved too much for Dart, who did not continue further excavations after his discovery.

Further discoveries of the same type emerged years later, and it was agreed upon by many paleontologists that this type of Australopithecine, which dated to 2.5 to 3 million years ago was indeed bipedal and worthy of its own species, although the position of *Australopithecus africanus* on the hominid family tree is still debated today.

Rebecca Hernandez

Date 1925

Scientist Raymond Dart

Nationality South African

Why It's Key The first clues that the earliest human history lay in Africa.

opposite Fossilized skull of the primate, *Australopithecus africanus*

Key Publication **Heisenberg vs Schrödinger**
When particle physicists collide

In the early part of the twentieth century, the physics world was in turmoil. It had lurched from a previously homogenous view of Newtonian physics into the world of quantum physics, peopled by a multitude of rival theories. In 1925 to 1926 two complete – and completely different – versions of quantum theory were developed independently.

Werner Heisenberg's revelation was called matrix mechanics, which reinterpreted Bohr's model of the atom. In his 1925 paper "Quantum-Theoretical Re-Interpretation of Kinematic and Mechanical Relations" he argued that matrix mechanics was built on a mathematical interpretation of observable physical results, with no conceptual image of what an atom "looked" like. Many of Heisenberg's contemporaries were uncomfortable with this idea, including Erwin Schrödinger. He had developed an alternative, more classical, theory of quantum physics based on Louis de Broglie's idea that matter could be described as a wave – a departure from Bohr's orbiting electron model of the atom. Going head to head with Heisenberg, he published the paper "Quantization as a Problem of Proper Values" in 1926.

While the two men debated each other's work ("it's crap" Heisenberg wrote of wave mechanics), it fell to two other physicists, Max Born in Germany and Paul Dirac in England, to prove the two theories were equivalent. Via wildly different routes, Schrödinger and Heisenberg had arrived at the general principles of quantum mechanics – the subject that was set to dominate physics for the rest of the century.

Barney Grenfell

Date "Quantum Theoretical Reinterpretation of Kinematic and Mechanical Relations"

Date 1925–6

Authors Werner Heisenberg, Erwin Schrödinger

Nationality German, Austrian

Why It's Key Quantum mechanics has completely changed our understanding of physics and chemistry.

Key Discovery **Raw liver and red blood cells**
Iron is essential for health

Anemia was a huge problem throughout the 1800s and many people died of the disease during this time. The illness results from a shortage of hemoglobin in the blood cells, usually brought about by a diet lacking in iron. Hemoglobin is the part of a red blood cell that carries oxygen as it travels around the circulatory system, but a deficiency in iron can reduce the body's ability to synthesize this essential substance, leading to lethargy, dizziness, and fainting.

George Whipple's discovery in 1925 – iron's hugely important role in the production of red blood cells – changed the way anemia is treated and has prevented many people from falling ill in the first place. Whipple's work was concerned largely with the liver and his early studies focused on anemia caused by parasites attacking the organ.

It was while Whipple was studying anemia in dogs that he noticed that liver cells had almost limitless powers of regeneration. Diet seemed to have a profound affect on the anemic condition of the liver, and he discovered that feeding raw liver to dogs cured or eased their symptoms. He soon realized that it was the iron in the liver which was alleviating their symptoms and went on to propose the importance of the mineral as a component of red blood cells.

Katherine Ball

Date 1925

Scientist George Whipple

Nationality American

Why It's Key It provided a successful treatment for anemia and other blood diseases.

opposite Colored SEM of human red blood cells (erythrocytes)

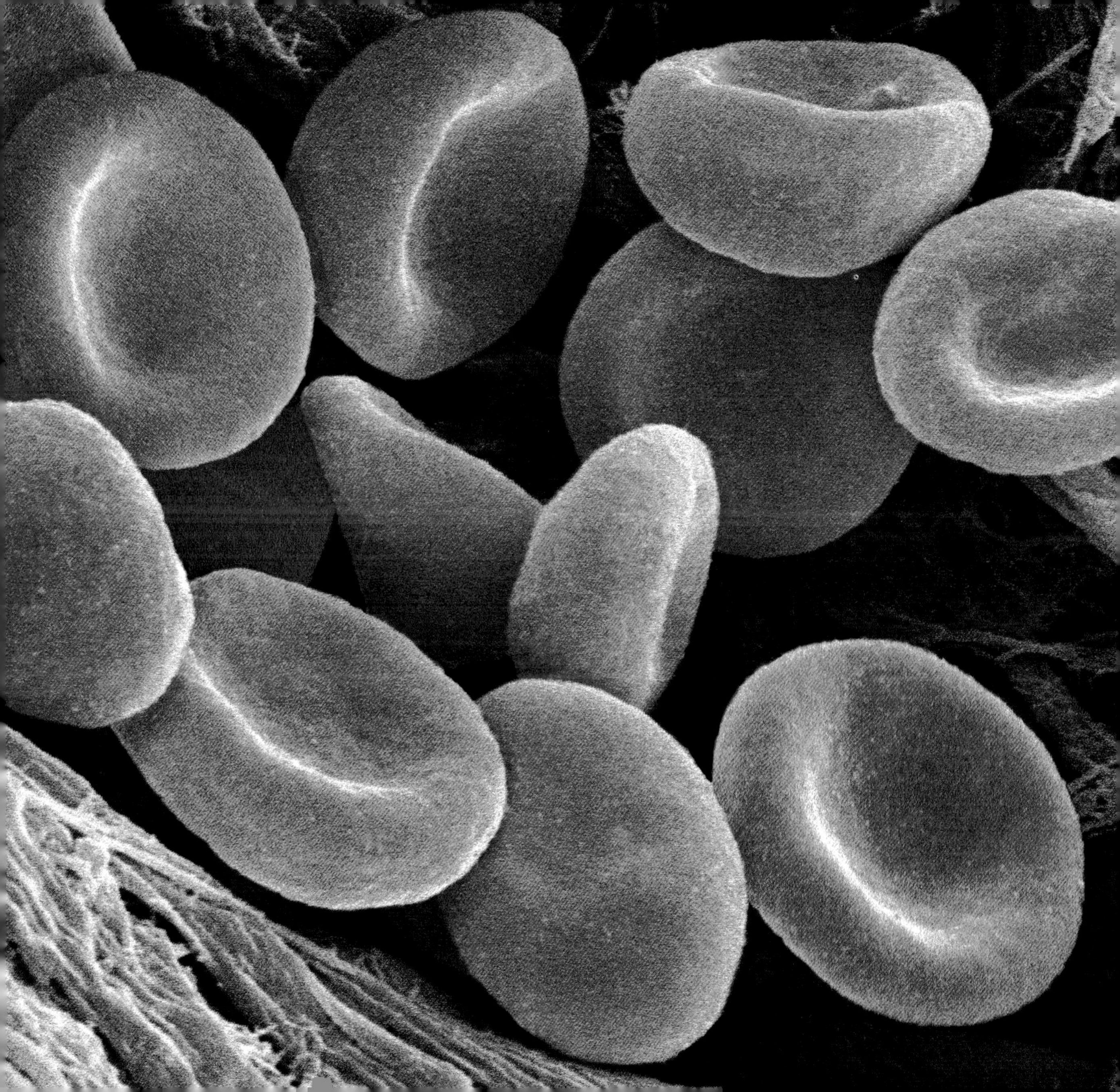

Key Event **Evolution in education**
The trial of Thomas Scopes

On May 25, 1925, in a courtroom in Dayton, Tennessee, high school teacher John Thomas Scopes was found guilty of teaching evolution.

Eager to challenge the constitutionality of the Butler Act – a law prohibiting the teaching of evolution in Tennessee schools – the American Civil Liberties Union (ACLU) had decided to fund a case which would dominate the national press for weeks. Scopes, a football coach and substitute teacher at Rhea County High School, was chosen as the face of the defense.

Led by a group of businessmen set on getting publicity for their town, the defense team argued that, while prohibited to teach evolution, teachers were still required by the state to use the set textbook *Civic Biology*, which featured a chapter on the subject. In other words, the state required them to break the law.

After a week of heated debate, the jury found Scopes guilty and he was fined US$100. The ACLU offered to pay his fine.

It wasn't until after this so-called Monkey Trial that Scopes revealed he might actually have been innocent of the crime he admitted to committing. He told a reporter that he skipped the lesson on evolution and that his lawyers had coached his students to lie on the witness stand. It later emerged that he was offered US$50,000 to lecture on evolution, but had to turn down the sum as he felt he didn't know enough about the subject.

Faith Smith

Date 1925

Country USA

Why It's Key This trial led to a string of cases arguing for the equal footing of evolution and creationism to be taught in American schools.

Key Publication ***Elements of Physical Biology***
The rise and fall of predators and prey

In his *Elements of Physical Biology* (1925), American biophysicist Alfred Lotka set out a pair of differential equations to describe how predators and prey species interact. The Italian mathematician Vito Volterra was working on the same problem and wanted to model the rise in predatory fish in the Adriatic, following a decline in fishing during World War I. He independently derived the same equations as Lotka and published them a year later.

When solved, the "Lotka-Volterra" equations paint a picture of predator and prey populations rising and falling in periodic cycles, with the predators lagging behind the prey. Thriving prey feed more predators, but as the predators get more numerous, they outstrip their own food supply and prey numbers fall. With fewer prey, the predator population also crashes and this declining threat gives the prey the opportunity to thrive again.

The model is a simplistic one. It assumes, for example, that the prey have an unlimited food supply and are the predators' only food source. Nonetheless, several real world examples have matched its predictions. Populations of lynx and snowshoe hare in Canada follow the boom and bust cycles predicted by the equations, and their relationship is now one of the most famous in the biological world.

Lotka and Volterra's work was critical for the development of more sophisticated models of population dynamics. With so many of the world's species in jeopardy, these have never been more important.

Ed Yong

Publication *Elements of Physical Biology*

Date 1925

Author Alfred Lotka

Nationality Italian

Why It's Key Showed us how populations of predators and prey interact in a natural setting.

Key Event **Geneva Protocol**
Humanity gets the upper hand

The Geneva Convention consists of four treaties signed in Geneva, Switzerland, that set international laws for the treatment of prisoners of war. Other conventions included laws on the state of refugees, on war crimes, and on the use of gas and biological weapons.

The latter was called the Geneva Protocol; it was signed on June 17, 1925, and was entered into force on February 8, 1928. This prohibited the use of chemical and biological weapons, and the use of "asphyxiating gas, or any other kind of gas, liquids, substances or similar materials." The Geneva Protocol, however, says nothing about the production, storage, or transfer of biological or chemical weapons.

At the time, chlorine gas had been used in trench warfare, and was being replaced by phosgene and mustard gas. However, growing realization that the use of these agents was inhumane, even in wartime, led to the treaty's genesis.

It failed at first due to individual countries not wanting to put their names to total cessation of the chemical's use, with many saying that they would still use chemical agents if they were used against them first. After more deliberation, the French suggested a protocol for non-use of poisonous gases, to which Poland suggested adding bacteriological weapons. This was agreed, and eventually signed, by the international committee.

David Hall

Date 1925

Country Switzerland

Why It's Key The use of chemical and biological weapons was officially prohibited by international agreement.

Key Event **Picture this**
"Radiovision"

There are several great inventors credited with developing and promoting the transmission of synchronized pictures and sound that we now call television. Of these, Charles F. Jenkins, born August 22, 1867, in Dayton, Ohio, is best known as the main North American contender.

A keen, although initially amateur, inventor, by 1892 he had developed a machine that could project small moving pictures onto a wall or screen. But the images were too small to be viewed by more than a few people at once. In 1895, Jenkins, with a partner, Thomas Armat, demonstrated an improved version to a paying audience at the Cotton States Exposition in Atlanta, Georgia, and, by default, invented possibly the very first movie theater. In 1925, Jenkins unveiled "Radiovision." Using a form of mechanical television based on a modified film projector and specially adapted radio receivers ("Radiovisors"), he was able to transmit cloudy, 40- to 48-line images, viewable on a 15-centimeters square mirror. This was publicly demonstrated on June 13, when Secretary Wilbur and other members of the U.S. Navy saw in his laboratory, in Washington, what was taking place several miles away at the Anacostia Naval Air Station.

On July 2, 1928, Jenkins' television station, the first in the United States, commenced broadcasts. Its first program consisted of a ten-minute segment of a revolving windmill; in all likelihood, a satellite channel is repeating that show even now.

Mike Davis

Date 1925

Country USA

Why It's Key Spurred interest in, and contributed to, the development of television as we know it today.

Key Publication **Pauli's Exclusion Principle** You just have to be different, don't you?

Physicist Wolfgang Pauli was attempting to explain variations in energetic stability between elements. Having observed the lifetime of different elements, he wondered if the existing description of how electrons move was incomplete.

It was thought that three quantum numbers were enough to describe an electron orbiting a nucleus – one describing the distance from the nucleus, one describing the rate it was spinning at, and a third describing the angle of orientation of its orbit. In his 1925 paper "On the Connexion between the Completion of Electron Groups in an Atom with the Complex Structure of Spectra", Pauli put forward a new number, the spin quantum number, which described how the electron spun on its own axis, like a planet around the Sun. With four numbers, the ordering of the elements could be explained; as long as each electron had a different set of numbers to all the others.

With some mathematical development, this idea became the statement we now know as the Pauli Exclusion Principle. It excludes two identical particles from occupying identical states, and defines each state in terms of its quantum numbers. The resistance of nature to having two electrons in the same energy state is what keeps neutron stars from collapsing in on themselves, and gives objects their rigidity. This force is totally unique to quantum mechanics and nothing like it can be found in the ordinary world. But it has been found experimentally that all particles of matter, not just electrons, obey the exclusion principle, refusing to all act the same.

Kate Oliver

Date 1925

Author Wolfgang Pauli

Nationality Austrian

Why It's Key Another example of quantum weirdness permeating up to the macroscopic scale, the Pauli Exclusion Principle is a fundamental part of our knowledge about the sub-atomic world.

opposite Wolfgang Pauli (left) and Niels Bohr

Key Discovery **Stellar structure revealed**

The Sun is the source of all life on Earth, yet for centuries we remained ignorant as to how it shone. Finally, in 1926, Arthur Stanley Eddington enlightened us, when he developed the first true understanding of the structure of stars.

Astrophysicists began modeling the Sun in the early twentieth century. Just after World War I, Eddington – an early advocate of Einstein's Theory of Relativity – photographed a solar eclipse. He used the images to show that the light of the stars around the eclipse had been shifted by the Sun's gravity, hence providing the first experimental proof of Einstein's claim that light could be bent by the presence of a gravitational field.

Eddington went on to develop a model for "stellar equilibrium." He showed that the pressure within a star, coupled with the radiation of light and energy from it, is perfectly balanced by that of its inward gravitational pull, creating a stable structure.

We now have a good understanding of our own Sun's structure and evolution. A star like ours is composed of shells, with the core containing the reactions that power it. As the star ages, the hydrogen-based nuclear reactions form helium debris at the center, and the shell of reactivity moves out to enclose this core. Eventually, as the star burns up all the hydrogen, it swells to form a red giant, before the outer layers detonate. Its remnants eventually shrink to form a white dwarf; the same fate that will befall our sun in 5 billion years.

Steve Robinson

Date 1926

Scientist Arthur Eddington

Nationality British

Why It's Key An understanding of how the Sun will change over time is critical to foreseeing the evolution of our Solar System.

Key Invention **Vinyl revolution** PVC into Plastic

The music industry has seen a succession of new formats come and go over the last century, but for those who frequent old-fashioned record stores, there's no greater pleasure than taking home a 12" LP. It's almost inconceivable, then, that this well-loved format only exists due to chance.

PVC (polyvinyl chloride), or vinyl, was discovered several times – at least twice before the turn of the twentieth century – purely by accident. In its naturally rigid state, however, it is difficult to put it to any commercial use. It wasn't until Waldo Semon stumbled across a way to "plasticize" it, that the real potential of PVC became apparent, not least because of the low cost of its production.

PVC combines two ingredients we're all fairly familiar with – chlorine, from salt water, and ethylene, from natural gas or petroleum. A two-step process turns these into vinyl polymer, which is then modified to create everything from water pipes to Wellington boots. Plasticizers make the material more flexible, meaning it can be used in cables and packaging.

Despite being one of the world's most widely used and versatile plastics, PVC's reputation is by no means untarnished, and not just because it forms the basis of some appalling fashion trends and oddball fetishes. PVC has long been at the center of an environmental row over waste disposal. The economic weight of this multi-billion dollar material, however, means that it's unlikely to be replaced in a hurry.

Hayley Birch

Date 1926

Scientist Waldo Semon

Nationality American

Why It's Key Turned a previously unusable chemical curiosity into one of the most useful materials of the modern world.

Key Event **TV history** Logie Baird shows off his television

In the 1920s, British people were introduced to one of the most influential and ground-breaking inventions of the technological age – fully functional television sets. The first set was actually demonstrated by Scottish inventor John Logie Baird in 1923, but this initial model was somewhat primitive. Large rotating discs acted as lenses to scan across images, sending out a meager thirty lines of picture to viewers.

Having improved upon this design, Baird began giving demonstrations of his televisions in the windows of London's Selfridge's department store in 1925. Shoppers flocked to see the new machine, which was now capable of transmitting images of letters and silhouettes.

This success, however, was incomparable to Baird's real breakthrough, which came in January 1926. Here, he showcased a new and even more improved television set to a small audience of Royal Institution members at his laboratory on Frith Street, in London. The machine was now able to transmit pictures of living human faces, complete with shading and light gradations; silhouettes were a thing of the past.

To a generation in which photography was a relatively new and exciting phenomenon, the idea of moving images being broadcast through a machine was completely revolutionary. Baird and his television set went on to drive the twentieth century into its technological age of maturity.

Hannah Welham

Date 1926

Country UK

Why It's Key Where would Friday nights be without a television to curl up in front of?

opposite John Logie Baird using one of his transmitting inventions

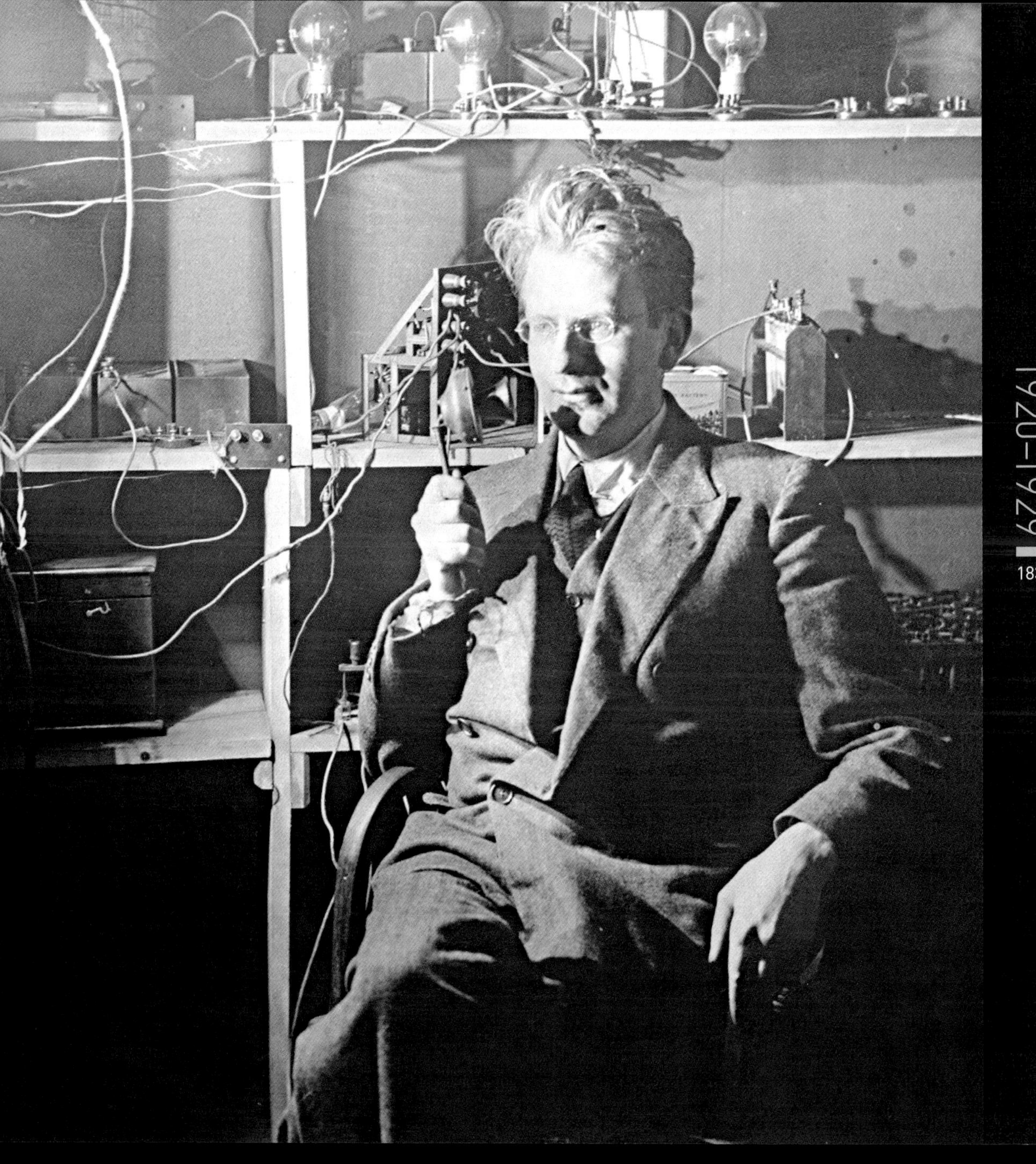

Key Discovery **Mutant X**
X-rays cause genetic disruption

Working at the University of Texas in Austin in 1926, American geneticist Professor Hermann Muller became fascinated by the simple fruit fly of the genus *Drosophila*. He noticed that when his fruit flies were exposed to X-ray radiation, they were almost certain to develop mutations. Following a series of ingenious experiments, Muller was able to show that the X-rays were literally breaking apart the *Drosophila* genes and, in some cases, completely knocking them out.

Muller's work was the starting point for a wealth of further research into genetic mutations, much of which now underlies our understanding of genetic disease. His theories about how genes cross over between chromosomes and how they make copies of themselves are now taken as fact. In addition, he suggested that a slow accumulation of mutations could not be avoided by asexual animals – mutation avoidance, in Muller's eyes, was the whole point of inter-gender sex. On the back of Muller's work, *Drosophila* shot to fame as one of the most useful test subjects of modern biology.

His X-ray discoveries earned Muller the Nobel Prize for Physiology or Medicine in 1946. His work also raised concerns about the dangers that supposedly helpful radiation, such as X-rays, could pose to humans and animals alike. Thanks to his research, the modern world can now take simple, life-saving precautions when it comes to using X-rays.

Hannah Welham

Date 1926

Scientist Hermann Muller

Nationality American

Why It's Key Muller exposed the dangers of X-rays and the world of genetic mutations.

Key Invention
First liquid-fuelled rocket launched

On March 16, 1926, on his Aunt Effie's farm in Auburn, Massachusetts, American physicist Robert Goddard launched the first ever liquid-fuelled rocket. The flight lasted just 2.5 seconds, with the 3-meter-long rocket reaching a height of 12.5 meters before crash-landing 56 meters away in the snow. But Goddard had just demonstrated the basic technology that would later see space exploration take off.

Beforehand, rockets had only used solid, gunpowder-like propellants. Inspired by a vision of sending humans to Mars, that he'd had as a teenager, Goddard spent years experimenting with solid fuels until realizing only liquid fuels could achieve the thrust required to escape the Earth's gravity. Following the successful launch, he continued to master the technical issues posed by liquid-fuelled rockets.

Goddard's achievements went largely unnoticed at the time due to his secrecy, but by 1927, physicist Hermann Oberth had independently come to similar conclusions and was developing liquid-fuelled rockets with Verein für Raumschiffahrt (Germany's Society for Space Exploration). In the 1930s, the German Army recruited several members of the society, including Wernher von Braun, who created the V-2 ballistic missile. They reached similar engineering solutions as Goddard, such as gyroscopic control and power-driven fuel pumps. After World War II, the U.S., Russian, and European governments sought to develop the technology, ultimately sending satellites and humans to space. With a manned mission to Mars on the horizon perhaps Goddard's dream will soon be realized.

James Urquhart

Date 1926

Scientist Robert Goddard

Nationality American

Why It's Key An invention which has since enabled scientists to explore space and further understand Earth, and to put satellites into orbit which have profoundly changed the way we live, work, and communicate.

opposite Dr. Robert H. Goddard with the world's first fuel rocket

Key Publication *Animal Ecology*
Energy in the food chain

It is startling to think that virtually all of the world's energy comes from the Sun. Energy from sunlight is converted into glucose through photosynthesis in plants. The plant stores glucose by converting it to starch, thereby storing the Sun's energy. If a rabbit eats the plant, then the energy stored in the plant is transferred to the rabbit. A fox may eat the rabbit, therefore transferring the energy from the rabbit to the fox. The energy therefore passes up the chain. Food is the common factor and it was Charles Elton that first described this as a food chain.

Elton published a book on animal ecology in 1927 in which he described the plant as the provider and those above it in the chain as consumers. The grass, rabbit, and fox are linked in a chain and if something affects one link, the rest will be affected too. For example, if the grass in a particular area was dug up and the rabbits starved because of a lack of food, this would have a knock-on effect for the foxes even though they don't eat grass directly.

Elton also noticed that one species of animal will only eat another within a certain size range. In addition, he found that the species at the bottom of a food chain is always greater in number than the species at the top. He called this the pyramid of numbers.

Riaz Bhunnoo

Publication *Animal Ecology*

Date 1927

Author Charles Elton

Why It's Key Elton identified the food chain principle which has important implications for the way we manage the Earth to sustain life on our planet.

Key Event
First vaccine for tetanus

Tetanus is a disease of the nervous system caused by soil-living bacteria, called *Clostridium tetani*. The bacterium enters the body through a wound in the skin (usually a cut or animal bite), and secretes a poisonous substance or "toxin" that causes muscle spasms. The first sign of tetanus is a characteristic spasm of the face and neck muscles called "lockjaw." The spasms then spread to other limbs and can eventually cause heart and lung problems, which may be fatal.

C. tetani was first isolated from a human in 1890 by Japanese biologist Kitasato Shibasaburo. He also found that if he injected this organism into rabbits they produced antibodies, which neutralized the effects of the harmful toxin. Then, in 1927, Gaston Louis Ramon – a French vet at the Pasteur Institute – discovered a way to create a vaccine consisting of the tetanus toxin, deactivated by formaldehyde and heat, which he named "anatoxin" (later called toxoid).

The vaccine was first widely used during World War II, when U.S. soldiers were routinely vaccinated against tetanus. It was so successful that only twelve cases of tetanus were reported among soldiers during the war, in stark contrast to the high number of cases among enemy troops. The tetanus vaccine was eventually combined with the vaccines for diphtheria and pertussis (whooping cough), and systematic vaccination of children began in the 1950s. In developing countries, where immunization programs are not in place, however, the disease continues to be a real threat, particularly to newborn babies.

Hannah Isom

Date 1927

Country French

Why It's Key Thanks to successful vaccination programs, tetanus in developed countries is now virtually non-existent.

opposite Colored SEM of a cluster of the bacteria *Clostridium tetani*, cause of tetanus

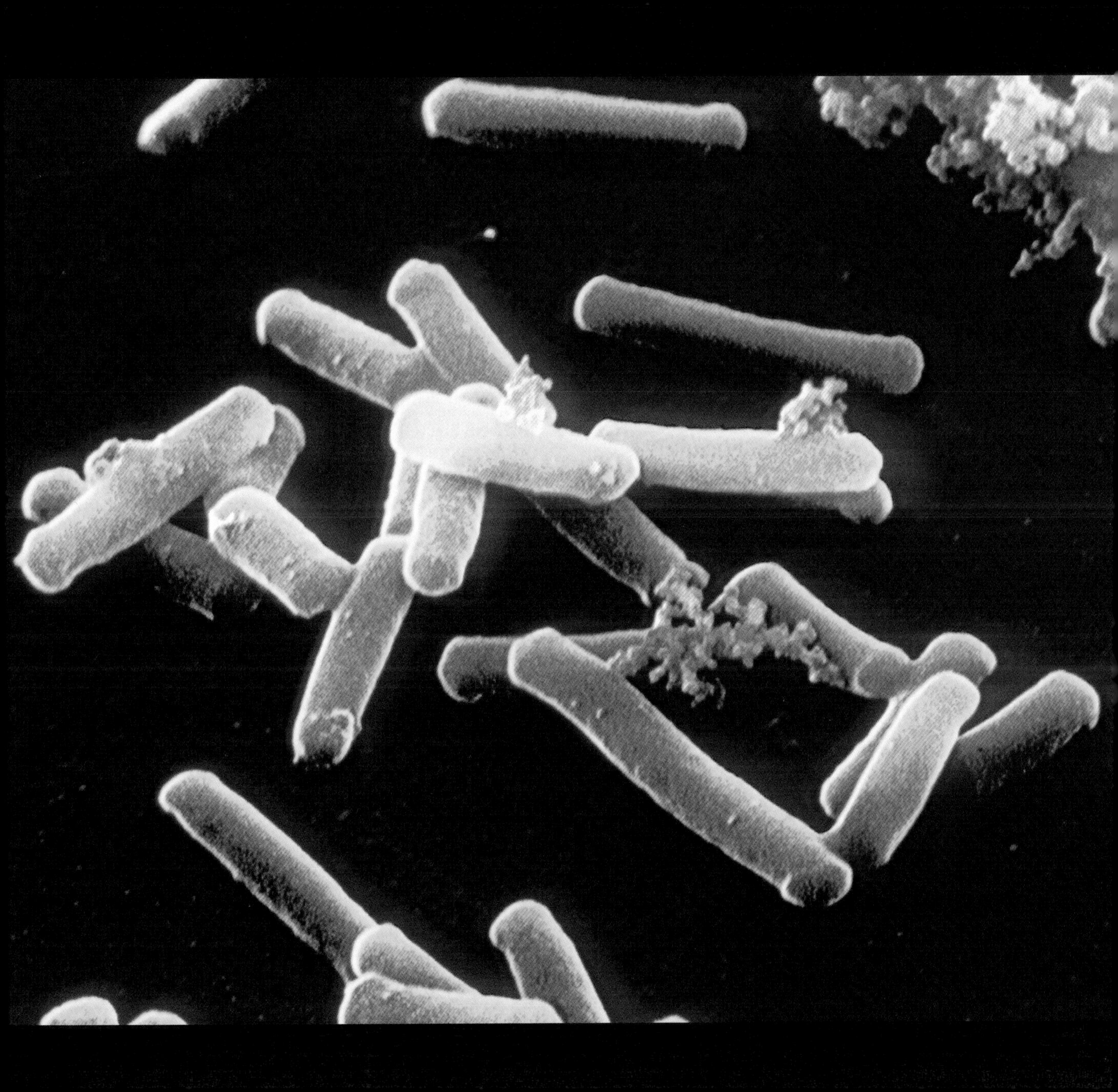

Key Publication **Heisenberg's Uncertainty Principle**
Do not attempt to determine your momentum

Say you have a particle and you want to know about it. If it's a very simple particle, there are only two things you can ask: where is it and which way is it going? Let's find out where it is. We shine light on it to get a good look; unfortunately the photons give the particle a knock, so now we don't know how fast it's going or which way. Or we could try and not disturb our particle by examining it very gently with dim light, but now we can barely see where it is. We seem to be unable to get both bits of information at once.

From ideas like this and a number of mathematical models of the quantum world, the German physicist Werner Heisenberg derived a simple relation called the uncertainty principle. This states that if you are absolutely certain about the location of the particle then you cannot have a clue about its momentum and therefore where it's likely to be in the future. Conversely, if you know exactly how fast your particle is moving and what its mass is, it is impossible to know exactly where it is.

The uncertainty principle was a crucial step in the development of quantum mechanics which showed that measurements had to be made as probabilities instead of absolute values.

Kate Oliver

Publication "The Actual Content of Quantum Theoretical Kinematics and Mechanics"

Date 1927

Author Werner Heisenberg

Nationality German

Why It's Key Heisenberg's uncertainty principle-laced new limits on what science could hope to discover, and totally buried the idea that you could, even hypothetically, predict the future if you had enough knowledge.

Key Event **Compulsory sterilization of mental patients in the USA**

By the late 1920s, support for the theory and practice of eugenics had become widespread. In much of Europe and the United States its application, it was believed, would prevent social degradation and stem the spread of various human pathologies because, as the American Eugenics Society declared, "unfit human traits such as feeblemindedness, epilepsy, criminality, insanity, alcoholism, pauperism, and many others run in families and are inherited."

Around twenty-five American states had by now passed eugenics sterilization laws; laws intended to prevent "defective" individuals from reproducing among themselves, thereby reducing the burden on the state. Many such laws had been passed in haste and not effectively tested. This would change on May 2, 1927 when representatives of one Carrie Buck sought to appeal against an order for sterilization delivered by an Amhurst County Circuit Court on the grounds that it was unconstitutional. Like her mother, "feebleminded" Carrie Buck had been committed to the Virginia Colony for Epileptics and Feeble Minded in Lynchburg, Virginia. Carrie's child, Vivian, was similarly labeled at seven months old. Hence, three generations of "imbeciles" became the ideal family for Virginia officials to use as a test case in support of the eugenic sterilization law enacted in 1924.

Tragically, the Circuit Court's decision was upheld. "Carrie Buck...potential parent of socially inadequate offspring, likewise afflicted...may be sexually sterilized without detriment to her general health, and her welfare and that of society will be promoted."

Mike Davis

Date 1927

Country USA

Why It's Key It gave the go-ahead to similar eugenics laws in other states. Ultimately, some 650,000 Americans would be sterilized against their will, without their consent or that of a family member.

Key Invention

The first spray can

While technically an aerosol is a suspension of liquid or solid particles in air, the more popular use of the word refers to the device that produces it: the aerosol can.

The concept of a liquid spray was not new by 1927, but it was in that year that Norwegian inventor Erik Rotheim made a real breakthrough. He was trying to find the best way of applying wax to his skis and discovered that a pressurized aerosol was perfect. He had tinkered in his own lab and eventually came up with a device that consisted of a metal can with a valve, containing pressurized gas that could propel liquids. This was the forerunner to the modern aerosol can.

At first, Rotheim's invention failed to impact the world significantly. It was not until World War II that it first became widely used. The U.S. military, serving in the Pacific, used aerosol cans to spray insecticide preventing the spread of deadly malaria amongst the troops.

Urban graffiti soon became the vandal's dream mode of expression after canned spray paint was invented in 1949. But by the mid 1970s aerosol cans had gained a more sinister reputation; they were partly responsible for the destruction of ozone in the upper atmosphere. The CFCs responsible for this were finally banned in aerosols in Europe in 1989.

David Hawksett

Date 1927

Scientist Erik Rotheim

Nationality Norwegian

Why It's Key The most efficient and hygienic method of dispensing liquids.

Key Event

Transatlantic phone calls go public

Thanks to a ollaborative effort between the British Post Office and Bell laboratories, the first commercial telephony service between New York and London was established on January 7, 1927.

Some fifty years after the telephone's initial invention, radio technology had made it possible for two people on opposite sides of the Atlantic Ocean – which if you need a slightly more tangible figure to work with spans a distance of 3,500 kilometers from London to New York City – to have a conversation.

Initially, it was only possible for a single call to be made at any one time and only between those two specific locations. Later that year, on October 3, a transatlantic service was set up between the UK and Canada – the first call being made between the two countries' Prime Ministers, Stanley Baldwin and William Lyon Mackenzie King. As the years went on, the facility spread across North America and Europe.

Of course, the use of such cutting edge and exclusive technology always comes at a price and, in 1927, that price was an eye-watering US$75 for the first three minutes of talktime – which puts even the most profitable of price plans for modern telecommunications to shame.

Chris Lochery

Date 1927

Country USA, UK

Why It's Key A huge milestone in international communications; a monumental step toward our current globally connected culture.

Key Publication **Covalent bonds' quantum corollary**
Because atoms just love to share

Walter Heitler's and Fritz London's research paper, published in 1927, is mentioned in every modern textbook on physical chemistry.

A covalent bond is something a chemistry student becomes acquainted with early on. It occurs between two non-metal atoms and involves the sharing of pairs of electrons in the atoms' electron shells. In non-metallic atoms, the outermost shell of electrons "wants" to contain eight electrons, as this makes it more stable. In the case of hydrogen, the smallest element, it "wants" to have two electrons in its outermost shell, although it naturally has one. Chlorine on the other hand, naturally has seven in its outermost shell. Neither the chlorine nor the hydrogen has strong enough powers of electromagnetic attraction to steal electrons from the other. The solution is that they share an electron – the chlorine is tricked into "thinking" it has eight and the hydrogen is tricked into "thinking" it has two. The two elements have formed a covalent bond and made a molecule of HCl – hydrochloric acid.

Max Planck proposed his quantum hypothesis in 1900, that energy is radiated and absorbed by matter in discrete packets or "quanta," and the early twentieth century saw many scientists attempt to apply this to science at large. What Heitler and London did was to examine the covalent bonds in molecular hydrogen, H_2. They applied the quantum equations of Schrodinger to covalent bonds and their results successfully made quantum mechanics compatible with chemistry for the very first time.

David Hawksett

Publication *Zeitschrift für Physik* (Journal of Physics)

Date 1927

Authors Walter Heitler, Fritz London

Nationality German

Why It's Key Heitler and London's work demonstrated that the new field of quantum mechanics did not violate known classical chemistry reactions.

Key Discovery **Evolutionary journey from Africa to Asia**
Peking Man

When ancient hominids first left Africa just under two million years ago, they headed to Asia. We know this because *Homo erectus* fossils dating back to this era were found in China.

Prior to the advent of scientific excavations, people had dug up fossilized bones from ancient caves. Locals were doing just this at a cave in the village of Zhoukoudian, China, when they stumbled across what they thought were dragon bones. What they had actually found was the remains of an ancient human, now called Peking Man. Researchers began a lengthy sweep of the area and found a rich collection of bones.

The initial discovery had been of two human-like teeth. This astonished scientists because *Homo erectus* bones had never been found in Asia. Two lower jaws were then found followed by the most significant find – an almost complete skull cap featuring a long, sloping forehead and thick brow ridge. In all, fourteen partial craniums, eleven lower jaws, numerous teeth, and some skeletal bones were found at the site. These findings proved that *Homo erectus* had lived in Asia and had evolved from the African species *Homo ergaster*.

In 1941, social conditions in this part of China were deteriorating due to World War II. To preserve the bones, researchers packed them for shipment to the United States. The shipment never arrived. Fortunately, many casts and descriptions remain of the original fossils, and more bones from *Homo erectus* have since been discovered across China.

Riaz Bhunnoo

Date 1927

Scientists Otto Zdansky, Johan Gunnar Andersson, Walter Granger

Country China

Why It's Key These fossils prove *Homo erectus* lived in Asia under two million years ago and evolved from the African species *Homo ergaster*. They also established the upright man in human evolution.

opposite Fossilized skull of Peking man (*Homo erectus pekinensis*)

Key Invention **Antifreeze**

Car engines get hot; no surprise there. Engine coolant is used to prevent the engine overheating and doing something inconvenient, like catching fire. Water is the natural choice for a coolant – it being cheap, plentiful, and safe – but is inconvenient for locations where the temperature drops below zero.

Antifreeze works by simply lowering the freezing point of water to below the minimum predicted outside temperature. Handily, it also raises the boiling point, making it useful in hot climates for preventing coolants from evaporating. The first antifreeze to be used was methanol. Whilst it did prevent the water from freezing, methanol had the unfortunate side-effect of decreasing the mixture's ability to actually cool things down. Despite first being prepared back in 1859 by the French chemist Charles Adolphe Wurtz, it wasn't until 1927 that ethylene glycol went into commercial production and became the antifreeze of choice. It was marketed as "permanent" antifreeze as its high boiling point meant that it did not evaporate during the summer months. Unfortunately ethylene glycol is also highly toxic, causing renal failure and death. Children and pets are particularly vulnerable as the chemical is dyed a bright color and has a sweet taste. Legislation is currently underway in the United States to add bittering agents to antifreeze in order to combat this problem.

Recently, however, propylene glycol has found favor as a safer form of antifreeze – it has similar properties to ethylene glycol but without the toxic side effects.

Helen Potter

Date 1927

Scientist Charles-Adolphe Wurtz

Nationality French

Why It's Key Enabled people to use their cars year round.

Key Person **Arthur Holly Compton** The X-ray examiner

Arthur Holly Compton (1892–1962) is best known for his work on the behavior of X-rays. His discovery that X-rays, like light, were capable of diffraction and refraction added weight to the argument that they were also electromagnetic waves. Compton's work on diffracting X-rays through gratings (which split electromagnetic beams according to wavelength) showed that they would fan out just as any other wave would. X-ray diffraction is now a common technique used to determine the internal structure of materials, and in fact, it was this method that elucidated the double-helix structure of DNA.

When not proving that X-rays were waves, Compton was proving that they were particles. His observations of the Compton Effect that bears his name showed that X-rays lose momentum in discrete amounts when hitting an electron or nucleus. This can only be explained if X-rays are made of distinct little packets of energy. Understanding the wave/particle confusion of photons, such as X-rays, is an integral part of modern quantum physics. For his contribution to this understanding, he received the 1927 Nobel Prize.

During World War II, Compton was a senior member of the committee responsible for atomic bomb production. He appointed Oppenheimer to run the new Manhattan Project, and his investigations led to the establishment of plutonium and uranium fission piles – the basis of nuclear weaponry and nuclear power. Compton has a NASA gamma ray laboratory named after him, and a crater on the moon named jointly for himself and his brother Karl.

Kate Oliver

Date 1927

Nationality American

Why He's Key The Compton Effect and its photon implications are crucial to modern physics, perhaps even more so than sustainable nuclear fission – the culmination of Compton's war work.

Key Experiment **Griffith's experiment**
Pneumonia, DNA, and the transforming principle

British Army surgeon, Frederick Griffith, was more interested in how to treat his pneumonia patients than in DNA. Nonetheless, he discovered that living cells could be "transformed," and some time later, DNA was revealed as the culprit.

Griffith noticed that two different strains of bacteria, which he called "rough" and "smooth" due to their respective appearances, were causing pneumonia in his patients, and he used the two to infect mice. Only the smooth bacteria turned out to be lethal. Intrigued, he tried heating the smooth bacteria before infecting the mice; this time they carried on running around their cages unaffected. But when he infected the mice with these heated smooth bacteria mixed with the rough, the little rodents were cut down in their prime. Something from the dead smooth bacteria had turned the rough bacteria into a lethal strain. Griffith had identified the transforming principle. He knew that something had been exchanged between the bacteria, although the nature of this mystery material eluded him. Later work by Oswald Avery, Colin MacLeod, and Maclyn McCarty, in 1944, identified DNA as the substance that could transform cells.

The results of Griffith's experiment can now be fully explained. The smooth bacteria had been killed by heat, but their DNA survived to be taken up by the rough bacteria. The DNA encoded an enzyme which enabled the smooth bacteria to create the capsules that allowed them to invade the mice, triggering fatal pneumonia.

Nathan Dennison

Date 1928

Scientist Frederick Griffith

Nationality British

Why It's Key The transforming principle revealed that DNA holds hereditary information. Transformation later played a part in the dawn of the genetic engineering era.

Key Publication
Coming of Age in Samoa

Cultural anthropology might not sound like the sort of area of study where there's much scope for being controversial, but you'd be very wrong to think that. In 1925, a twenty-three-year-old masters graduate named Margaret Mead set off for American Samoa (a small island in the South Pacific – just to the east of the Independent State of Samoa) to perform her first field work.

Her findings were written up as the classic text which presented the idea to the public for the first time that adolescence and sexual development could be shaped by various different cultural experiences and, as a result, could be more or less problematic for the individual. Mead suggested that the so-called "civilized" world could learn a great deal from the cultures that it considered to be "primitive."

She claimed that, when compared to the more permissive Samoan society, the traditional nuclear family set-up and pervasive Christian values regarding monogamy and exclusivity in "civilized" romance induced a great deal of stress and neurosis in the growing American adolescent – an opinion that disgusted and shocked readers in great measure.

Mead died in 1978 and, in the years following, her work underwent critical academic discussion, with fellow anthropologist Derek Freeman making the rather bold suggestion that the Samoan subjects had been winding Mead up. Opinion on the validity of his claims, however, is somewhat divided.

Chris Lochery

Publication *Coming of Age in Samoa*

Date 1928

Author Margaret Mead

Nationality American

Why It's Key A highly influential work that continues to generate discussion and debate within the psychological profession, eighty years after its initial publication.

Key Discovery
Alexander Fleming stumbles upon penicillin

In 1896, Ernest Duchesne, a final-year student at the Pasteur Institute, discovered penicillin. Unfortunately for him, this groundbreaking step forward seemed to go completely unnoticed in the world of science, until Alexander Fleming was credited with the discovery thirty-two years later.

Fleming was the first to admit that his discovery of penicillin was quite an accident; he is widely quoted as commenting, "one sometimes finds what one is not looking for." Known for his carelessness, he went on holiday for two weeks and left his lab in its usual state of chaos. On his return, he noticed that many of the culture plates that he had left to grow had developed a clear halo surrounded by a lawn of mould that had contaminated the plate. It had happened, quite by chance, that a spore of *Penicillium notatum* had contaminated his culture plates. He found the penicillin to have an antibacterial effect on a number of pathogens, including scarlet fever, meningitis, pneumonia, and diphtheria.

Unfortunately, Fleming had difficulty in finding a way to produce penicillin in any quantity, and became further convinced that its slow-working nature would prevent it from lasting long enough in the human body to kill bacteria.

It was here that he left his work, only for it to be picked up by Howard Florey and Ernst Chain nine years later. The duo succeeded in manufacturing large quantities of penicillin and, along with Fleming, they were awarded the Nobel Prize in their field.

Faith Smith

Date 1928

Scientist Alexander Fleming

Nationality British

Why It's Key Fleming discovered the first known antibiotic, indirectly saving millions of lives.

opposite Fleming's petri dish culture photographed 25 years after his amazing discovery

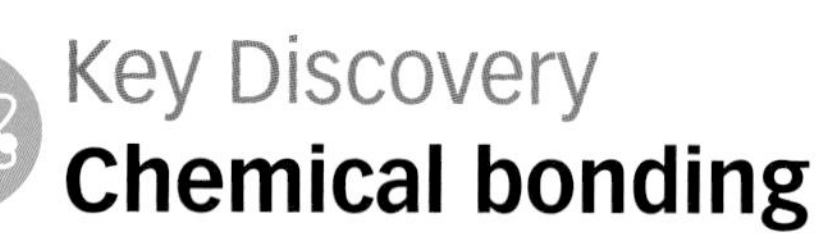

Key Discovery
Chemical bonding

By the late 1920s, physics and chemistry were rapidly advancing our understanding of the nature of the atom. The problem, however, was that the two different fields of research gave completely different results. This was especially true of the new area of quantum mechanics.

Carbon was one point of contention between the two disciplines. Both physicists and chemists knew that it had six electrons, two in an inner "shell" and four in an outer shell. Chemists had learned by experimenting that carbon formed four bonds with other atoms in a tetrahedral, triangular-based pyramid shape. These bonds must have been formed by each outer shell electron being shared with an electron from another atom. Physicists, however, had recently shown that two of the outer four electrons hung about at a slightly lower energy level, leaving only two electrons available for bonding. Current bonding theory didn't allow for this. Two chemists were capable of working through the complex mathematics needed to reconcile the two observations – Linus Pauling and Robert Mulliken. Working independently, using very different methods, they showed how electron orbitals were scrambled together, or "hybridized," to form molecular bonds.

Pauling, who first suggested the answer to this problem in a paper in 1928, later published an important text book on the subject, entitled *The Nature of the Chemical Bond and the Structure of Molecules and Crystals*. Both men were awarded Nobel Prizes for Chemistry for their contributions to the understanding of bonding.

Simon Davies

Date 1928

Scientists Linus Pauling, Robert Mulliken

Nationality American

Why It's Key This discovery demonstrated how the new physics of quantum mechanics could explain the fundamentals of chemical interactions.

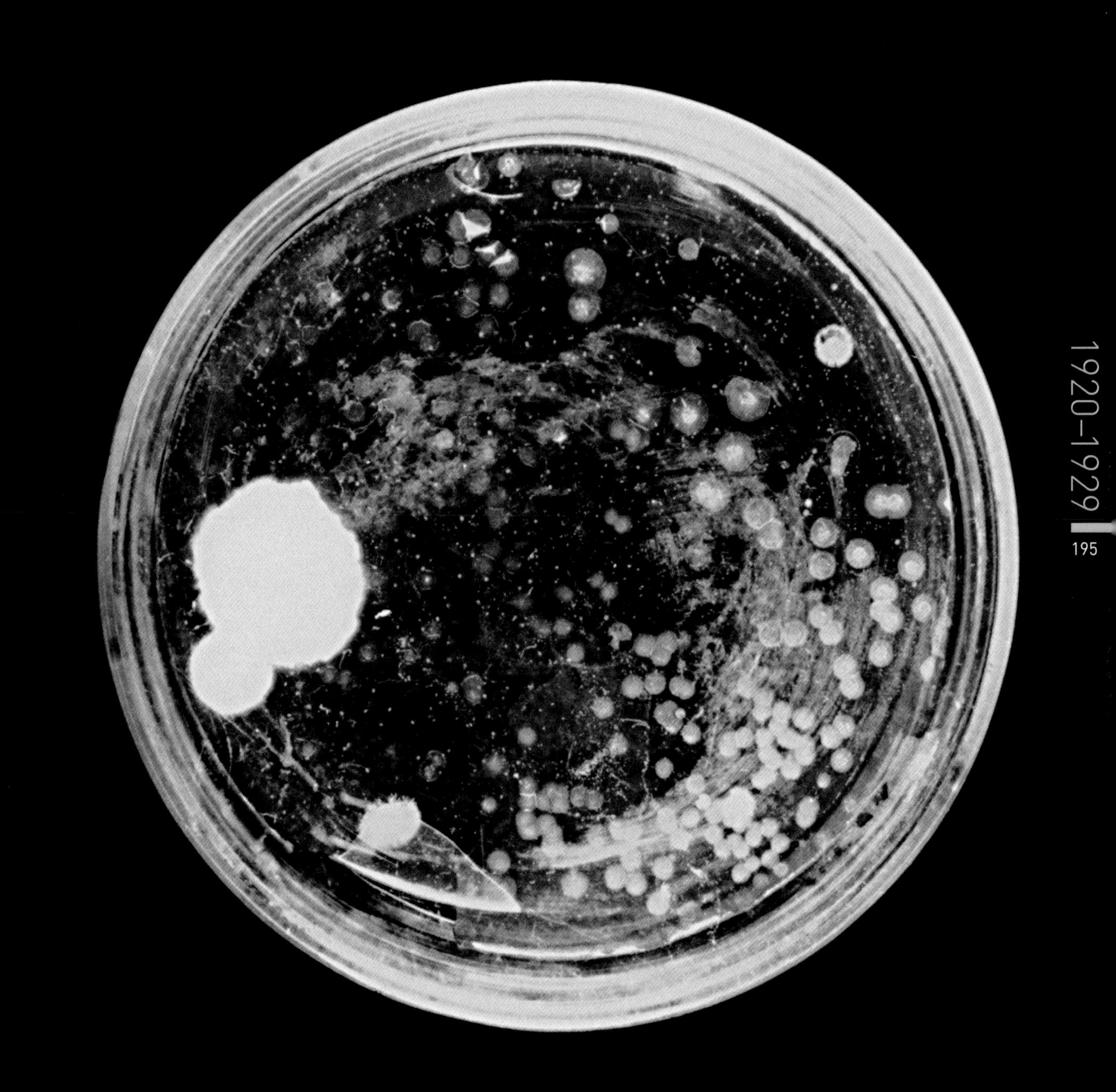

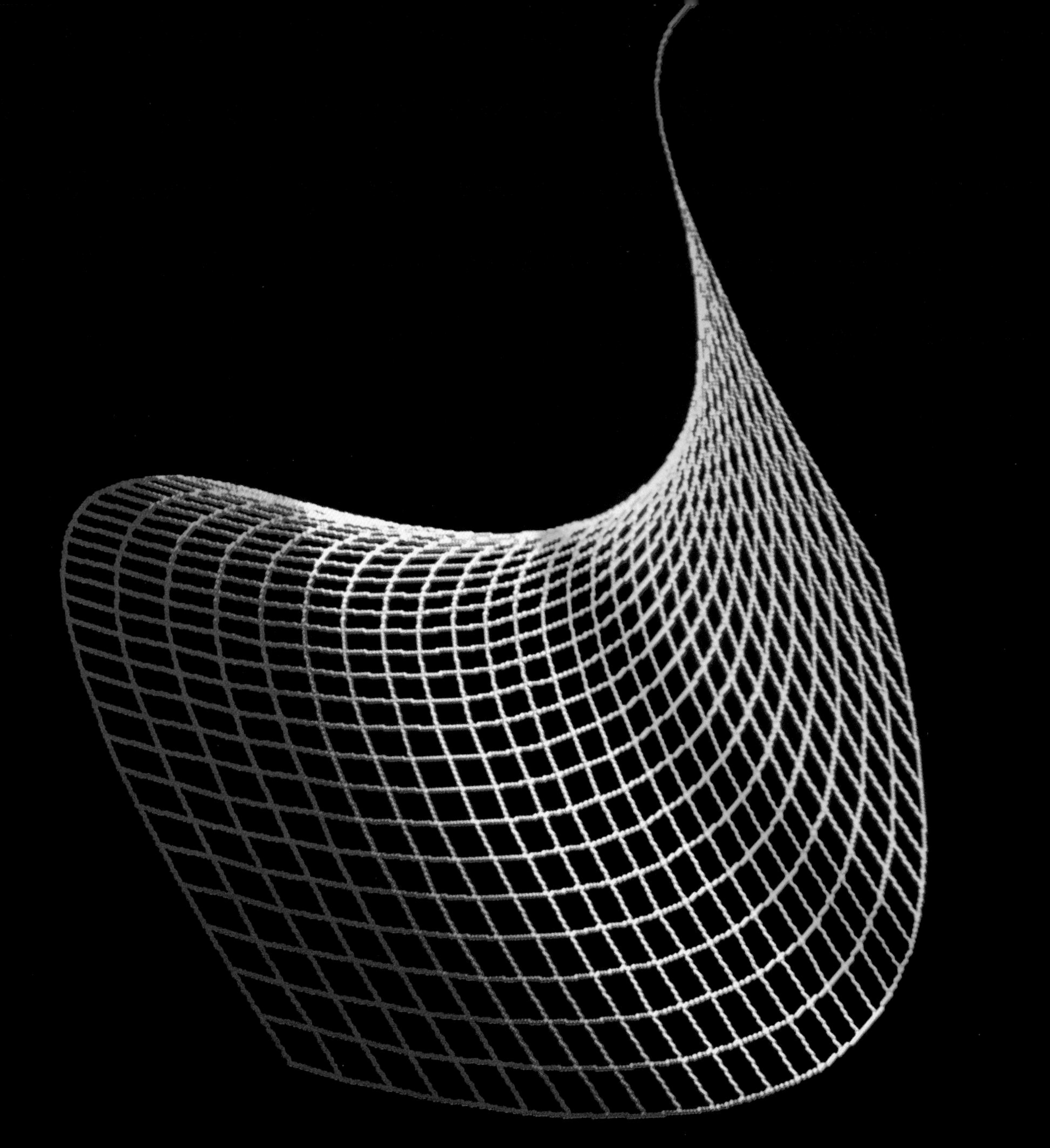

Key Discovery **On the scattering of light** The Raman Effect

Inspired by visual effects, such as the beautiful blue coloring of the seas, scientists at the University of Calcutta began investigating how molecules can scatter light. They quickly discovered, in 1928, that molecular light scattering was important in many materials, from crystals to gases.

Researchers found that light didn't just change direction after hitting a molecule, but it also changed color; indicating that the light waves had changed frequency. The research leader, Chandrasekhara Venkata Raman, and his colleagues already knew that a change in the frequency of the light indicated a change in its energy. This could only mean that the light being scattered off these molecules was transferring energy – but where to? There was only one place it could be going: the molecule itself. Molecules can vibrate, rotate, and shuffle their electrons around; all these processes require energy in well-defined, discrete parcels that depend on the internal structure of the molecules, so the exact amount of energy absorbed is a detailed indicator of what's going on inside. Since the changes in energy levels can be measured by looking at the changes in the frequency of scattered light, then measuring whether its scattered light has a higher or lower frequency gives invaluable information about a molecule's structure.

The phenomenon of the changing frequencies of light, named the Raman Effect after its discoverer, was designated a Historical Chemical Landmark in 1998, thanks to its significance as a tool for analyzing the internal composition of molecules.

Kate Oliver

Date 1928

Scientist Chandrasekhara Venkata Raman

Nationality Indian

Why It's Key The Raman Effect provides chemists with a vital method for analyzing the composition of solids, liquids, and gases.

Key Person **John von Neumann** The theory of games

As a child, von Neumann (1903–1957) entertained friends with his ability to recite pages picked from a phone book, using his photographic memory. The Hungarian continued to use this gift as an adult, memorizing an immense library of jokes, and using them to entertain at the many house parties he held.

By his mid-twenties, von Neumann was recognized worldwide within the academic community as a young mathematical genius; today, credit is given to von Neumann for developing and popularizing game theory, an organized system for playing games, first envisioned by the French mathematician, Emile Borel. Von Neumann's inspiration for his work on game theory was poker; as an occasional player who relied solely on probability theory in his game, he wanted to formalize the idea of "bluffing." This led to the publication of "Theory of Parlor Games" in 1928, discussing game theory and proving the famous Minimax theorem. Von Neumann was aware that his work would be invaluable to economists; however he was more interested in applying his methods to politics and warfare. At the start of World War II, he sketched a mathematical model of the conflict. Using game theory in his predictions, he deduced the Allies would win.

Von Neumann's postwar work included the development of the digital computer. Using his mathematical abilities, he improved the computer's logic design; he had the idea to base computer calculations on binary numbers and to store coded programs in the computer's memory instead of punchcards.

Sarah Watson

Date 1928

Nationality Hungarian

Why He's Key He created a theory now used widely in economics, politics, warfare, and recreational games.

opposite A mathematical "saddle" used to plot the outcome of a game between players whose interests are diametrically opposed. Should both players choose perfect strategies the game is said to have a "saddle point"

Key Invention **Freon**
A "safer" chemical coolant

Home refrigeration only became commercially available at the turn of the twentieth century. The first fridges were extremely large and consumed a huge amount of power. They primarily used ammonia, methyl chloride, or sulphur dioxide as coolants, and cost about twice as much as a car at the time. The use of such toxic chemicals as coolants, and a number of fatal accidents involving fridges leaking, spurred Charles Kettering and Thomas Midgley to invent a non-toxic alternative. Their answer to the problem was the supposed miracle chemical Freon (a chlorofluorocarbon, or CFC).

Freon is odorless, colourless, non-flammable, and non-corrosive; the perfect alternative to the gases used before. The ease of Freon production also helped to bring the price of fridges down, meaning that more people could afford them, though they didn't become a household essential until the end of World War II.

But in the late 1970s, the dangers posed by this "safe" chemical became apparent. If CFC gases escape into the atmosphere, they reach high elevations, where they are broken down by ultraviolet light and eat away at the protective ozone layer surrounding our planet.

Since its invention in 1928, the use of Freon had grown exponentially, expanding far beyond refrigeration, to propellants in aerosol cans and fire extinguishers. But, in 1994 the production of new Freon was stopped worldwide, potentially saving humans from generations of skin cancer. During a UN summit in 2007, it was agreed to completely rid the world of the chemical by 2030.

Josh Davies

Date 1928

Scientists Charles Kettering, Thomas Midgley

Nationality American

Why It's Key Toxic refrigerants were removed from fridges and replaced by the wonder chemical Freon (a type of CFC). At the time, scientists were oblivious to the devastating impact of CFCs on the atmosphere.

Key Invention **Pacemaker**
Get with the beat

Today, three million human hearts in the world rely on artificial pacemakers to keep a steady rhythm. But these people wouldn't be alive if it wasn't for the remarkable work carried out by a doctor from Sydney, who discovered a way of kick-starting a human heart without relying on artificial respiration or adrenalin injections.

With the support of physicist Edgar Booth, Dr Mark Lidwell from the Royal Prince Alfred Hospital in Sydney devised a portable device with which he could artificially pace a human heart. The device, which plugged into a light point, consisted of two poles. One connected to a pad soaked in salt solution and was placed on the patient's skin; the other consisted of an insulated needle with an exposed point inserted into the appropriate heart chamber.

In 1928, Lidwell was given the opportunity to test his apparatus when a stillborn baby was delivered at Crown Street Woman's Hospital. Using his equipment, he administered 16-volt impulses to the infant's heart for ten minutes. When the current stopped, the heart was able to beat on its own accord, and the child went on to lead a normal healthy life.

Lidwell, however, wished to remain anonymous at the time due to the controversy surrounding the issue of artificially extending human life. Nowadays, the extension of human life is very desirable, and scientists have developed tiny permanent pacemakers just 2.5 centimeters long, which can be placed within the fat of the chest wall to keep the heart pumping indefinitely.

Faith Smith

Date 1928

Scientists Mark Lidwell, Edgar Booth

Nationality Australian

Why It's Key Enabled the extension of human life without having to rely on artificial respiration and adrenalin injections.

opposite Colored chest X-ray showing a surgically-implanted heart pacemaker

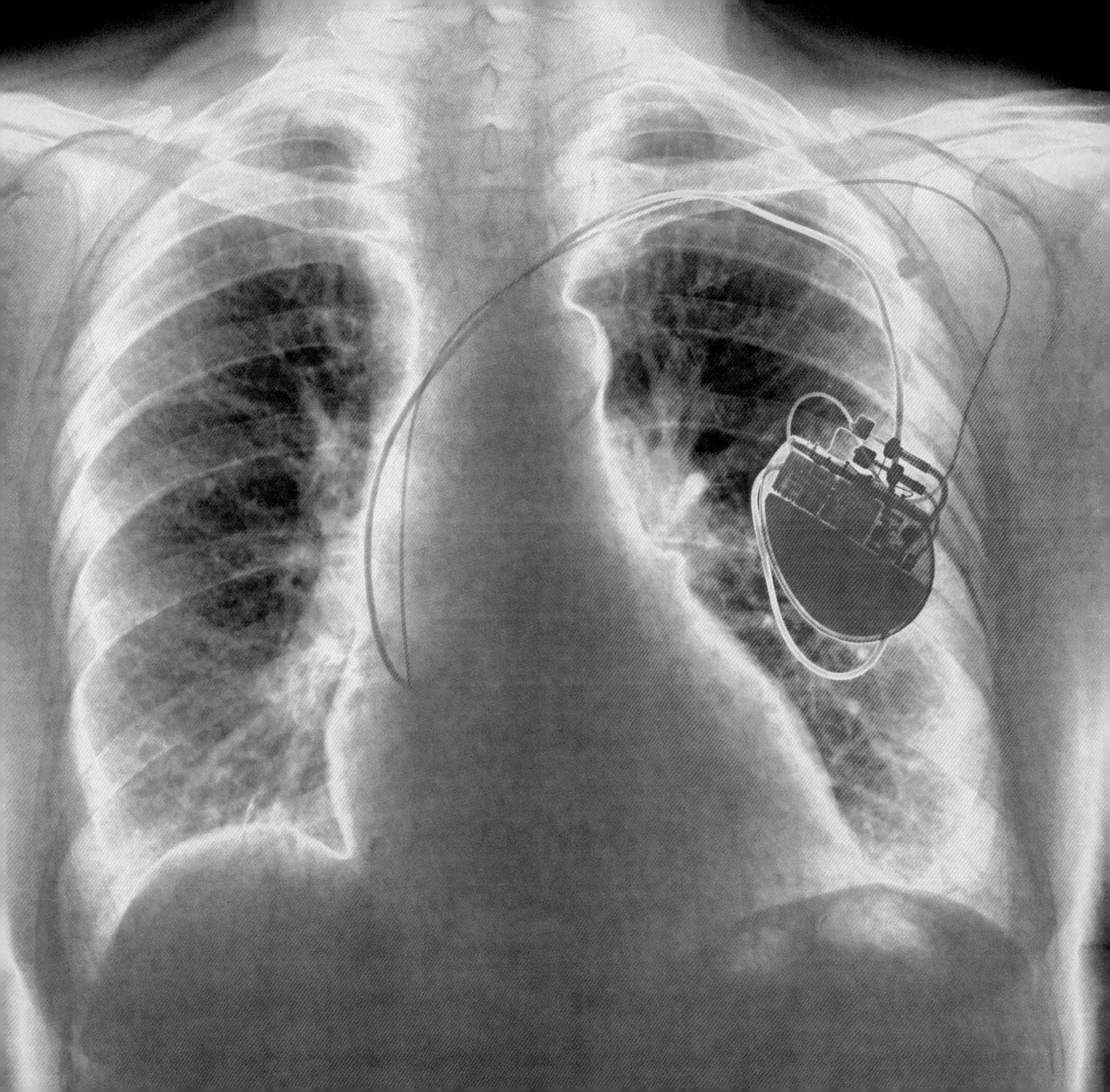

Key Invention
Urine pregnancy test

"The rabbit died" was a common saying in 1940s Britain, used by women who'd just found out they were pregnant. But how did they know and what exactly did it have to do with a rabbit?

In the early 1900s, there was a desire to improve the health of mothers and infants. The earlier a pregnant woman visited the doctor, the better the outcome. But most women had to wait for some obvious sign of pregnancy, which never appeared early.

Scientists across Europe began looking for a reliable way to detect pregnancy early, using microscopes to gather evidence. Two German scientists, Selmar Aschheim and Bernhard Zondek, identified a substance, found only in pregnant women, which affected a tiny structure in the ovary called the corpus luteum. They devised a test in which they injected young female mice with small amounts of the woman's urine. After a hundred hours the mice were killed and their ovaries carefully examined. If the woman was pregnant, then the substance was present and the corpus luteum of the mouse showed changes. Their highly accurate A-Z test was introduced to the world in 1928, and was followed by other, easier and quicker tests that used frogs and the infamous rabbit.

We now know this substance is the hormone human chorionic gonadotropin. Modern pregnancy tests can detect it in the first week or two of pregnancy, giving almost instantaneous results. Of course no rabbits need die these days. That phrase was misleading in any case – the rabbit always died.

S. Maria Hampshire

Date 1928

Scientists Selmar Aschheim, Bernhard Zondek

Nationality German

Why It's Key Early detection of pregnancy was not possible before this test. It meant that a woman and her unborn child could be given quality care much sooner.

Key Person **Frank Whittle**
Original jet setter

Frank Whittle has a medal named after him; he was knighted; and anyone who has ever flown to a far-flung place on holiday has benefited from his invention. In fact, he may just be the most famous person you've never heard of. Though he wasn't quite a child prodigy, Whittle was pretty close. Imagine patenting the idea behind what would become the jet engine in your early twenties. That's exactly what he did.

Born in Coventry in 1907, Whittle joined the Royal Air Force in 1923 as a cadet and, by 1928, he had attained pilot status. Soon after, he started developing an idea about how to propel a plane through the air at tremendous speed. At high altitudes, the air is thinner so planes should be able to move faster. The problem was, as aircraft flew higher and higher, propeller and piston-driven planes performed worse and worse. The standard mechanism of propulsion simply wasn't able to provide the thrust necessary to take advantage of high altitude flight.

Whittle postulated that placing a gas turbine inside a fuselage might make for a superior engine. He initially couldn't get any governmental backing for his project, so turned to private sources for financial support. He created a prototype of the progenitor of the modern turbo jet in the mid 1930s, and soon thereafter secured the financial backing of the British government. In 1941, his first jet powered plane, the Pioneer, flew successfully.

B. James McCallum

Date 1928

Nationality British

Why He's Key Frank Whittle's work led to the development of the jet engine.

Key Discovery
Oxygen isotopes

An important concept in chemistry is that of the atomic mass – how much an atom weighs. This is impossible to measure directly, so the scientific community have devised a method of defining the mass of atoms. If a standard element is used which has an experimentally determined mass, then the masses of other elements can be determined relative to the standard. At the beginning of the twentieth century this standard was oxygen. Oxygen, it was declared, had a mass of 16 units and there were no discrepancies.

In 1929, however, a group of scientists published some experimental results which threw doubt on this assertion. Led by Canadian-born chemist William Giauque, they were investigating oxygen and similar gases by spectroscopy – a technique that can be used to observe how the wavelength of light changes as it bounces off different molecules. Each chemical's pattern is different and depends on its mass. The team's results showed extra lines in oxygen's pattern – lines that shouldn't have been there if all oxygen atoms had a mass of 16.

This meant that a small proportion of oxygen atoms had a different mass. These "isotopes" – oxygen atoms which contained extra neutrons – had masses of 17 and 18. There were not many of these isotopes, but oxygen was no longer as reliable as a standard for atomic mass determination. This discovery eventually led the scientific community to adopt the mass of one isotope of carbon, carbon-12, as the standard for all other atomic masses.

Simon Davies

Date 1929

Scientists William Giauque, Herrick Johnston

Nationality American

Why It's Key The need to define a more accurate standard for the determination of atomic masses gave a more solid foundation for chemical calculations.

Key Discovery
Pole reversal

Rocks, it is sad to say, are generally considered by most to be nothing more than boring lumps of matter that clutter up an otherwise plush planet. Much like their candy counterparts, however, there is a message written right the way through each and every piece of rock and, if you know what you are looking for, they can make for fascinating historical documents.

The geologist Bernard Brunhes knew this only too well. Studying rocks from an ancient lava flow, he found that they were magnetized in a direction almost exactly opposite to the way the current magnetic field of the Earth would dictate. From this he reasoned that the polarity of the Earth's magnetic field must have been working in reverse when the rock was formed. In other words, at some point in history, the magnetic North Pole was situated somewhere in Antarctica.

Taking this further, Motonori Matuyama found that, far from being a freak occurrence, this chopping and changing of poles happens on a regular basis. It's nothing to set your watches by, of course, but the last reversal, now known as the Brunhes-Matuyama Reversal, is thought to have happened as recently as 780,000 years ago.

If this makes you think that we're long overdue another flip in polarity, you can rest easy. The Brunhes-Matuyama reversal took between 1,200 and 10,000 years to complete, so if you are the proud owner of a compass, you won't need to chuck it away quite yet.

Chris Lochery

Date 1929

Scientists Bernard Brunhes, Motonori Matuyama

Nationality French, Japanese

Why It's Key Changed our understanding of the Earth's geomagnetic properties, and paved the way for further discoveries and advancements such as sea floor spreading and Dynamo Theory.

Key Discovery **Hubble's law confirms the expansion of the Universe**

When astronomer Edwin Hubble began his work at the Mount Wilson Observatory, USA, in 1919, the Universe was considered to be much, much smaller than we think it is today. What we now know to be other galaxies, such as the Andromeda Spiral, were thought to be nebulous clouds within our own galaxy, rather than separate islands of stars much further away than anything in our own Milky Way. Hubble changed all that. Using the 100-inch (254-centimeter) telescope at Mount Wilson, he observed, within these "spiral nebulae," a certain type of star with known, consistent luminosity. The stars were so dim, however, that Hubble realized that they must be much further away than anything in our own galaxy. He had demonstrated, for the first time, that other galaxies exist outside of our own.

While classifying thousands of these galaxies into different types, Hubble paid particular attention to the colors of these objects. In the same way that the pitch of the engine of a passing car decreases once it is moving away from you, the wavelength of light from a distant galaxy also changes with speed. Hubble noticed that all the galaxies appeared slightly redder than they should – they were all moving away from our own galaxy. Critically, he realized that the speed of the galaxies was directly proportional to their distances from us. His groundbreaking paper, published in 1929, introduced Hubble's Law, allowing astronomers to accurately measure the distances across the Universe by examining the light from other galaxies.

David Hawksett

Date 1929

Scientist Edwin Hubble

Nationality American

Why It's Key Provided the first real evidence that the Universe was expanding and probably began in a Big Bang.

opposite Bubble-shaped shroud of gas and dust 14 light-years wide expanding at a rate of 2,000 kilometers per second

Key Invention **A stiff drinker** The iron lung

Polio epidemics began in Europe in the early nineteenth century and spread rapidly. The disease returned almost every summer and struck down children and young adults regardless of race, class, or location. In 1952, the worst epidemic hit the United States, with nearly 58,000 cases of polio reported; over 20,000 patients were left paralyzed. Death was often caused by paralysis of the respiratory muscles, although if a means of ventilation were employed, the patient could survive.

Many attempts at artificial respiration had been attempted over the years. Most of these used pressurized air ("positive pressure") to force air into the lungs, but these methods suffered from a lack of control. Back in 1670, however, British physicist John Mayow had demonstrated that air was normally drawn into the lungs by enlarging the thoracic cavity, creating a negative pressure system. It was this thinking that spawned the development of the first ventilators.

In 1929, pediatric clinician Philip Drinker and pediatrician Charles F. McKhann published an article on the successful clinical testing of the Drinker respirator, which became better known as the "iron lung." The device, developed by Drinker and Louis Shaw at Harvard, consisted of a tank in which the patient would lie with their head extending out through a rubber collar. The collar provided a seal so that when a partial vacuum – a negative pressure – was created in the tank, the chest and lungs expanded. For the first time, there was hope for parents whose children complained of the classic stiff neck and back of polio.

Stuart M. Smith

Date 1929

Scientists Philip Drinker, Charles F. McKhann

Nationality American

Why It's Key It gave hope to parents and saved thousands of lives.

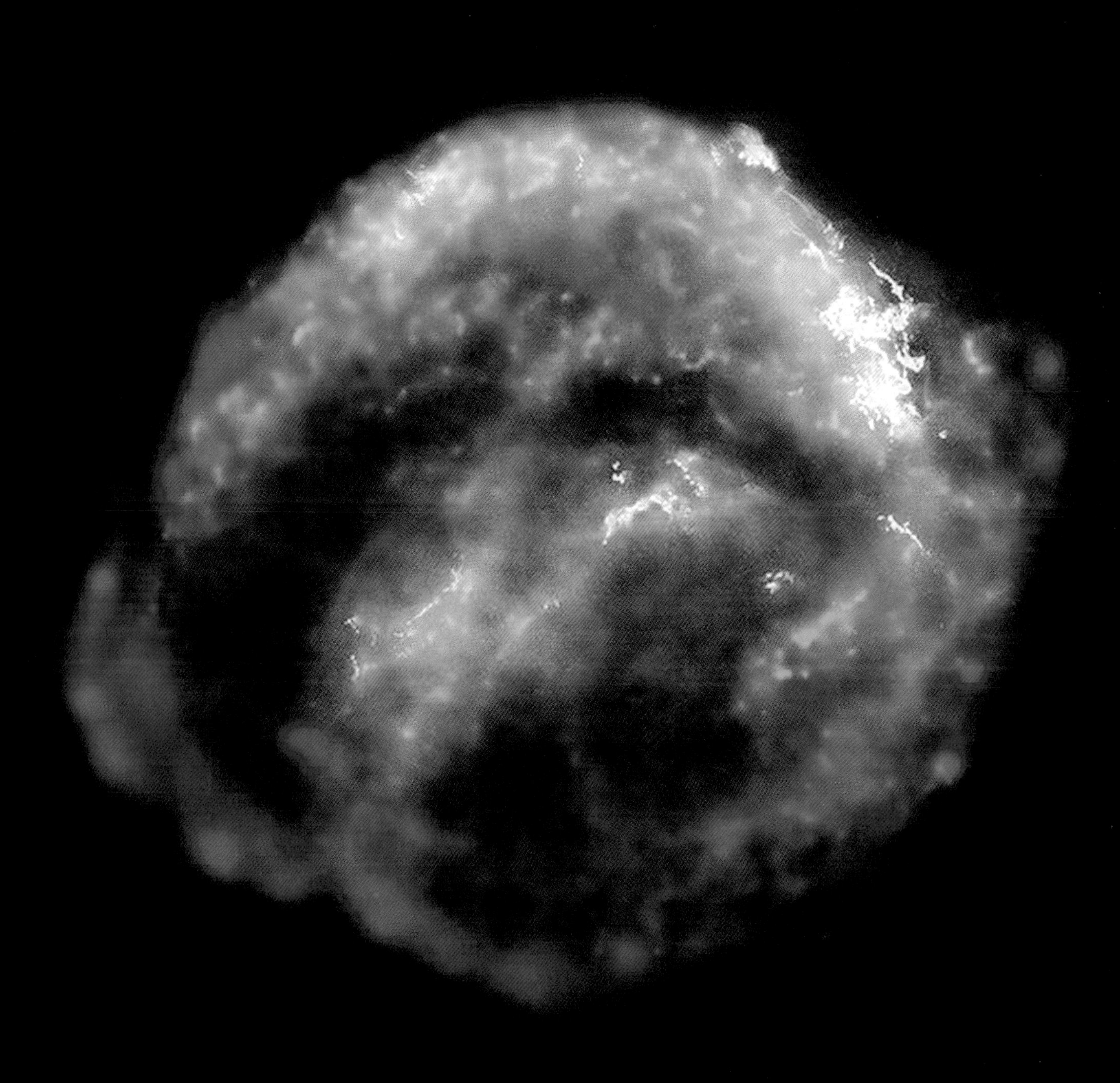

Key Discovery **Hydroponics**
Soil-free plants

Right now, there are crops growing near the South Pole, in extremely arid countries, and even in space. How? Through hydroponics, the growing of plants in soil-free environments, made possible by the ingenious work of one Professor William Gericke.

The principle of growing plants without soil was first demonstrated in 1859. The idea that the method could be used on a commercial basis, however, didn't emerge until 1929, during Gericke's stint at the University of California. Gericke announced that he believed water-grown crops could out-compete those from soil, and in 1937 dubbed the system "hydroponics," taken from the Greek words for water (hydro) and labor (ponos).

Gericke's work attracted huge interest, but initially he refused to reveal his secrets. This led, eventually, to his resignation from the post at the University of California and, in 1940, the publication of his book, *The Complete Guide to Soilless Gardening*.

Over the next seventy years, hydroponics made a huge mark on agriculture in some of the most unlikely places. Crops grown by this method do not suffer from weeds, over-watering, or soil borne diseases. They also require less pesticides, although the process can be expensive. The main benefits however, lie in hydroponics' potential to allow crops to flourish where no soil exists. Some agriculturalists in California have now switched to cultivating crops by hydroponics alone. Soil-free plants have also invaded the classroom, in the form of watercress grown by schoolchildren, and will be essential for future use in space.

Steve Robinson

Date 1929

Scientist William Gericke

Nationality American

Why It's Key Hydroponics allows agriculture to exist in the most unlikely of places, and may hold the key to feeding the world's growing population, both on Earth and in space.

Key Publication **Cosmic space trains**
Ticket to the stars

The Space Shuttle launch is perhaps the best known image of modern space travel, but the leap from fireworks to space ships was a big one, and would have been impossible without the ideas of Konstantin Tsiolkovsky.

There's a vicious cycle in rocketry. To launch your firework into space you need a lot of fuel, but the more fuel you add, the heavier you make your rocket. That extra weight needs more fuel, which is more weight, and this just keeps going round and round and round. Not only did Konstantin Tsiolkovsky realize this, he came up with a solution. In his book, published in 1929, he proposed the idea of "Cosmic Rocket Trains," which are known today as multi-staged rockets.

A cosmic rocket train is made up of separate rockets joined together. The first one gets the rocket off the ground and traveling pretty fast, but when the fuel tank has been emptied it drops off, and a second stage takes over.

The second stage isn't needed to get the rocket started; it only has to get it to go faster. With the dead weight of the launch rocket dumped, the remaining rocket moving, and the whole thing being rather high up already, the second stage has a much easier job. Finally, when it's high enough, the second stage comes off too, and the last stage boosts the leftovers into orbit.

Douglas Kitson

Publication *The Space Rocket Trains*

Date 1929

Author Konstantin Tsiolkovsky

Nationality Russian

Why It's Key Multi-staged rockets are the difference between fireworks that go quite high, and real space travel, people in orbit, and the Moon landings.

opposite Konstantin Tsiolkovsky Russian rocket pioneer, surrounded by models of his rocket designs

Key Invention **The Van de Graaff generator** More than just a physics experiment

Although the Van de Graaff generator has found fame by being the most memorable experiment during physics lessons at school, it has much more scientific significance than just making kids' hair stand on end.

American physicist Robert J. Van de Graaff produced the first generator that now bears his name in 1929. The machine used a silk ribbon driven by a small motor to create static electricity. The charge was collected on a terminal made from a tin can. From this simple construction, Van de Graaff was able to produce a massive 80,000 volts. By 1933, he had improved his design to include spherical aluminum terminals, which maximized efficiency. In order to increase the output of his generators, he began large scale constructions with columns reaching 13 meters tall, producing voltages 100 times larger than his original device. Scientists working with particle accelerators quickly took advantage of this new-found power supply which allowed them to crash particles into one another at higher speeds, opening up new possibilities for investigating the nuclear structure of atoms. In research today, Van de Graaff generators have largely been overtaken by solid state power supplies which have no moving parts. Instead, they have now taken on a different role in schools; as a very visual demonstration of the power of electricity. The largest air insulated Van de Graaff generator – built by Van de Graaff himself – is still in operation, and on display at the Boston Museum of Science.

Leila Sattary

Date 1929

Scientist Robert Van de Graaff

Nationality American

Why It's Key It was advanced nuclear and particle physics in its time, and is still a very useful and memorable class-room experiment for demonstrating physics today.

opposite One of the first Van de Graaff generators, built by Robert Van de Graaff at the Massachusetts Institute of Technology

Key Invention **Atom smashing with the cyclotron**

In the early part of the twentieth century, interest in the structure of the atom began to mount. Scientists started to literally smash them open to s ee what was inside. As Robert Van der Graaff first demonstrated, particles could be accelerated to near the speed of light and allowed to collide in order to break them open. Smashing open the nucleus of an atom releases a mess of radiation and subatomic particles and can reveal what the atom is made up of.

The first particle accelerators were linear, but Ernest Lawrence bucked this trend and used many small electrical pushes to accelerate particles in circles; starting off as a sketch on a scrap of paper, his first design cost just US$25 to make. Lawrence continued to develop his "proton merry-go-round," using parts including a kitchen chair and a clothes horse, until finally he was awarded the Nobel Prize in 1939.

Lawrence's idea used electricity to accelerate particles. In the "cyclotron," charged particles were accelerated inside two semi-circular "dees," whilst a magnetic force bent them into a circular path. As the particles gained speed, they spiraled outwards, the magnetic field no longer having such a controlling effect, and collided with a target particle.

Particle accelerators based on the first cyclotron are still used today for scientific research and commercial uses. In Lawrence's day, it meant that, for the first time in history, scientists were able to peer inside the atom.

Nathan Dennison

Date 1929

Scientist Ernest O. Lawrence

Nationality American

Why It's Key The invention of the cyclotron allowed scientists to discover how the atom is constructed.

Key Discovery
What heats the stars? A fusion of ideas

Two important areas of endeavor in physics – one concerning the very small, and the other the very large – converged at the end of the 1920s, thanks to the insights of the Russian-American physicist George Gamow.

Gamow's first brainwave related to the nuclei of atoms, which he suggested could approach one another closely enough to fuse together. He then used his famous imagination to envisage an atom's nucleus as a liquid droplet composed of protons and neutrons. This meant that it was possible to think of the nucleus changing shape, splitting, and joining with others. Using complex quantum-mechanical formulae he was able to prove that it was possible for two positively charged nuclei to come close enough together to fuse. This was named the "Gamow factor."

At the other end of the scale, Gamow turned his thoughts to the stars – and their source of energy. Building on speculations by Arthur Eddington in 1920, he suggested that the vast quantities of energy given out by stars could be explained by the fusion of hydrogen atoms, the commonest atoms in the Universe. These fusion reactions would produce lots of energy and result in helium atoms being formed.

Gamow's insights inspired later generations of physicists to develop more sophisticated theories of fusion in stars, as well as to produce weapons like the hydrogen bomb. Nuclear fusion remains one of man's best hopes for producing large quantities of clean energy.

Simon Davies

Date 1929

Scientist George Gamow

Nationality Russian-American

Why It's Key An inspired idea that allowed huge strides forward in our understanding of nuclear physics.

opposite Formations of molecular hydrogen and interstellar dust in the shape of pillars – gaseous star nurseries

Key Invention First practical coaxial cable for telephone network

As the United States began to get seriously wired in the early twentieth century, telephone wires began to crisscross the countryside. Telecommunication company AT&T's first network, built in the late 1800s, used simple copper wires on telephone poles to carry the signals. The system was cumbersome – the wires were thick and there had to be two of them, one for each direction a telephone signal traveled in. By 1918 engineers had worked out how to carry more than one call on the same wire at the same time by superimposing signals that had different, non-overlapping frequencies, but the system was still pretty inefficient.

By the 1920s one in three American homes had a phone and the increasing demand for connections led Lloyd Espenschied and Herman Affel to search for a much more efficient method. The result was their broadband coaxial cable for which they filed for a patent in 1929. Their revolutionary cable, which remains in use to this day, consists of internal layers. A conducting copper core is surrounded by an insulating layer which, in turn, is surrounded by another conducting layer of braided copper. An outer jacket then encases the whole assembly.

This new method meant that the electromagnetic signals carrying phone calls were confined within the cable itself and were much more resistant to signal degradation and interference. Their patent was granted in 1931 and, five years later, the first voice transmission was made using coaxial cable between Philadelphia and New York.

David Hawksett

Date 1929

Scientists Lloyd Espenschied, Herman Affel

Nationality American

Why It's Key Allowed thousands of calls to be made over the same cable simultaneously, paving the way for broadband global communications.

Key Event
Graf Zeppelin circumnavigates the globe

The Graf Zeppelin was a spectacular craft both in size and technology; it was 30 meters in diameter, and was only limited to this size by the dimensions of the hangar in which it was built.

Although later Zeppelins were produced, none was ever met with public adoration in the same way as the Graf. Named after airship pioneer Graf (German for Count) Ferdinand Zeppelin, the Graf flew for the first time in September 1928, and less than a year later it completed its first trip around the world.

The Graf was not the first aircraft to circumnavigate the globe, but it was the fastest. Previous airships had been powered by diesel or gasoline, but as the liquid fuels burn, the craft loses weight and hydrogen needs to be vented to stop the ship gaining height. This had limited the potential of airships to fly long distances. The Graf overcame the issue using specially developed gas cells to contain the hydrogen and stabilize the airship at altitude.

On August 8, 1929, the Graf departed New Jersey and set across the Atlantic to its home in Friedrichshafen, Germany, where it refueled before crossing Siberia to land in Tokyo. Many of the countries it crossed had not been seen from the air before. It completed the first non-stop flight of the Pacific Ocean, passing San Fransisco and Chicago. Eventually it returned to Lakehurst Naval Air Station in New Jersey with a total journey time of twenty-one days, five hours and thirty-one minutes. This marked the zenith for zeppelins; the Hindenberg disaster eight years later would spell the beginning of the end for airship travel.
Vicky West

Date 1929

Country USA

Why It's Key The fastest global circumnavigation by an airship.

opposite The LZ 127 Graf Zeppelin

Key Publication **"On the Electroencephalogram of Man"** Hans Berger discovers EEGs

When Hans Berger was two years old, Richard Caton, lawyer, surgeon, scientist, and one-time Lord Mayor of Liverpool, discovered electrical signals in the brains of animals. Almost fifty years later, Berger proved their existence in humans and opened up a new field of medicine. Berger initially studied astronomy at the University of Jena, Germany, but after one semester, transferred to medicine. In 1897, he received his doctorate, became Otto Binswanger's assistant at the university's psychiatric clinic, and began to work on cerebral localization. His early efforts focused on blood pressure in the brain, which was directly measured after trephination – where part of the skull is removed.

Berger then began to measure the brain's electrical activity. In his publication "On the Electroencephalogram of Man" (1929), he named July 6, 1924, the date of discovery of the human "Elektroenkephalogramm." He had made visual pictures of brain wave rhythms with electrodes attached to an instrument called an oscillograph. He reported the brain generates electrical impulses or "brain waves" which change dramatically when the brain is stimulated. His studies, repeated with brain-injured subjects, and techniques, would soon prove to be clinically useful. Berger became a victim of the Nazis, who forced him to retire and dismantle his lab. Having fallen into despair, in 1941 Berger entered the hospital in Jena and hanged himself. The EEG has revolutionized neurological and psychiatric diagnosis, as well as research in neurological sciences.
Stuart M. Smith

Publication "On the Electroencephalogram of Man"

Date 1929

Author Hans Berger

Nationality German

Why It's Key The EEG threw open a window onto the brain and revolutionized neurological and psychiatric diagnosis.

Key Invention **Polarizing filter**
Putting light to work

Like that other famous Harvard drop-out, Bill Gates, Edwin Land founded a hugely successful company that brought a new product into homes and businesses worldwide. And all before Gates was even born. A brilliant scientist, Land's singular focus on the subject of light polarization left little time for his college studies, but led to the polarizing filter and the creation of the Polaroid camera. His filters also gave birth to modern sunglasses, anti-glare safety glass and plastic, instant photography, and 3D movies.

Using synthetic compounds and his physics know-how, Land made the first polarizing sheet filter in 1929. Within three years, along with Harvard physics professor George Wheelwright III, Land had founded Land-Wheelwright Laboratories (LWL) focusing on finding uses for it. Cheap, flexible, and durable, his invention filtered out light traveling in certain planes, making it useful for cutting out glare and reflections through glass and lenses. Applications abounded, and soon polarizing filters were to be found in microscopes, automobiles, and airplanes, and were poised to revolutionize photography.

Photo giant Kodak began doing business with Land-Wheelwright Laboratories in 1932. By 1937, the lab had enough experience in the photo business to form the Polaroid Corporation. At first Polaroid was known mainly for safety glass, but by 1950, it would be synonymous with instant cameras. Now recognized as an essential tool in photography, the use of the Polaroid filter in safety glass and plastics means it helps us save lives and capture our fondest memories.

Raychelle Burks

Date 1929

Scientist Edwin Land

Nationality American

Why It's Key Polaroid filters gave birth to safety glass and revolutionized photography.

Key Person **Frederick Hopkins**
Discoverer of vitamins

Frederick Hopkins (1861–1947) wasn't really interested in science as a child, and was much happier reading books. However, all this changed when his mother gave him a microscope. By the age of seventeen, he had a degree in chemistry and had published his first scientific paper. After finishing medical school, he became the first Professor of Biochemistry at Cambridge, and worked on the living cell, thinking of it as a chemical machine.

Hopkins was the first person to isolate tryptophan, an important amino acid, and he went on to prove that this and other essential amino acids could not be manufactured by the body. Intrigued by work being done by Christiaan Eijkman on the importance of a proper diet, Hopkins took this further and suggested diet this was the only way that some essential nutrients could be supplied. He conducted an experiment to prove his theory, in which rats were fed only the known components of milk: fats, proteins, and carbohydrates. He found that these rats grew much less than rats that had been fed milk itself. His work proved therefore that there were unknown substances in the milk that were essential for normal development. These "accessory food factors" as he called them turned out to be vitamins.

Prior to his work, most researchers believed that diet-linked illnesses, such as the scurvy that sailors suffered from during long trips, were caused by a toxic substance in certain foods. Hopkins proved that such diseases were actually due to vitamins missing in the diet.

Riaz Bhunnoo

Date 1929

Nationality British

Why He's Key Hopkins first discovered vitamins, and the need to supply the body, through diet, with essential nutrients that it can't manufacture itself.

Key Event
The term “quantum leap” is coined

The common interpretation of the phrase “quantum leap,” coined in 1929, is far removed from its scientific definition. Although the term is generally taken to mean making a great stride, in reality, it has more to do with something on a much smaller scale – minute changes within atoms.

In an atom, the central nucleus, consisting of protons and neutrons, is surrounded by electrons. These electrons are able to change their position, or energy state, within the atom – changes that physicists call quantum leaps. The electrons are moving distances on a scale of less than a billionth of a meter, hardly a massive movement by any stretch of the imagination.

Electrons change their states discontinuously, meaning that their paths from one point to another are not seen. They are said to exist only in “quantum states” – they do not move in between. It could be compared to time travel; starting at one point, disappearing, and returning elsewhere, seemingly without movement.

The notion is hard to believe. Movement without moving; it doesn’t seem like it should happen. But it does. And in fact, the term “quantum leap,” in the sense of this particular unbelievable phenomenon, lends itself quite readily to the description of another: In the TV series of the same name, scientist Sam Beckett passes through time via a quantum leap accelerator, without experiencing the years in between.

Nathan Dennison

Date 1929

Nationality Argentinean

Why It’s Key The term “quantum leaps” is commonly used to describe a huge change when in fact it refers to the movement of electrons between energy states, one of the smallest movements imaginable.”

Key Discovery **The Chandrasekhar Limit**
A key to Life, the Universe and Everything?

Subrahmanyan Chandrasekhar was a man ahead of his time. So much so that he was awarded the Nobel Prize in Physics in 1983 and yet much of his work dated back to the 1930s. In particular, his application of special relativity to stellar bodies revealed much about the nature of dwarf stars, supernovae, and black holes.

Chandrasekhar applied special relativity, which suggests that the speed of light is the same to all observers regardless of their speed relative to each other, to prove that there is a point at which a dwarf star will inevitably collapse in on itself causing a supernova. The critical point comes at 1.44 solar masses – 1.44 times the amount of stuff in the Sun. Beyond this Chandrasekhar Limit any white dwarf star will be unable to sustain itself and so collapse to become a neutron star or black hole at the end of its life.

White dwarfs are the final state of the majority of all stars, and although they can contain the mass of up to 1.44 Suns, they are comparable in size to Earth. The star can no longer produce energy by nuclear fusion; it has essentially run out of fuel. White dwarfs slowly cool and become dimmer over billions and billions of years.

At the time this was viewed as little more than an interesting aside. Today however, it underpins much of what we believe about the workings of the Universe.

Jim Bell

Date 1930

Scientist Subrahmanyan Chandrasekhar

Nationality Indian-American

Why It’s Key Paved the way for the modern-day scientists such as Stephen Hawking to ask increasingly complicated questions about the nature of the Universe.

Key Discovery
Pluto and the search for "Planet X"

The search for a ninth planet in the Solar System, dubbed "Planet X" by American astronomer Percival Lowell, commenced as soon as the existence of the eighth, Neptune, was established in 1846. Curious anomalies in the orbit of the planet Uranus had led astronomers to pinpoint Neptune's precise position. But once it had been found, and its mass and orbit had been estimated, deviations in the predicted orbits of both Neptune and Uranus remained.

Astronomers around the world joined the hunt for "Planet X," but for nearly a century after Neptune's discovery it remained elusive. Lowell died in 1916, still hoping his observatory would find the missing planet first. It was not until a young American, Clyde Tombaugh, was invited to join the Lowell Observatory in 1929 that the search there resumed in earnest. The twenty-two-year-old overhauled the old-fashioned planet-hunting techniques that had previously resulted in false positives – wandering asteroids and comets – being taken for possible planets.

Using a device called a blink comparator to compare photographic plates taken of the same area of the sky at different times, Tombaugh's ingenuity and meticulous attention to detail were rewarded on the night of February 18, 1930. Another astronomer, W. H. Pickering, tried to lay claim to Tombaugh's discovery; as it turned out, Pickering had photographed the correct portion of sky in 1919, but Pluto had fallen just outside the area shown on his plates. Tombaugh's ashes are currently in space, aboard NASA's New Horizons mission, headed for Pluto and beyond.

Arran Frood

Date 1930

Scientist Clyde Tombaugh

Nationality American

Why It's Key A long sought-after discovery that improved on techniques for planet, comet, and asteroid hunting. Some of these methods are still used today.

opposite Clyde Tombaugh using his blink comparator

Key Person **Ronald A. Fisher**
"A great mind and a great friend"

As a child, Ronald Aylmer Fisher displayed great ability in mathematics. Due to severe short-sightedness and having to work without electric light, he learned to perform difficult calculations in his head.

Something of a mathematical genius, Fisher excelled at Harrow and achieved a First from Cambridge University, but he had also developed a passion for biology. The key to his success was in combining these two disciplines, by using complex mathematical formulae to explain fundamentals in evolutionary biology. He developed new statistical methods for analyzing biological data, and published fundamental texts in these areas. His 1930 book *The Genetical Theory of Evolution* is still considered a work of great influence and importance, and is described as the basis of modern population genetics.

In 1952, Fisher became one of the first men to be knighted by Queen Elizabeth, following her accession to the throne, but he wasn't without his idiosyncrasies. His deep distrust of puritanical tendencies, for example, led him to vehemently oppose Richard Doll's assertions that smoking caused cancer.

A close friend, E.B. Ford, described Fisher as a man whose appearance resembled King George V, with a fierce, pointed red beard, pale face, and often "challenging" expression. Ford recalls an afternoon of shooting where Fisher, unperturbed by his poor eyesight, took part: "Fisher – his finger on the trigger, waved the gun uncertainly while his friends dived for cover." A man who would try anything, he was regarded both as "a great mind and a great friend."

Katie Giles

Date 1930

Nationality British

Why He's Key A remarkable man who made major contributions to the worlds of mathematics and biology in the twentieth century.

Key Discovery **The neutrino**
Saving nuclear physics

A few years after Henri Becquerel discovered radioactivity in 1896, it was discovered that the tiny electron was one of the culprits. It was responsible for what is known as beta decay – electrons being fired off from an atom during radioactive decay. By 1914 observations of beta decay by different scientists had revealed a problem. It seemed that, when an electron was emitted as radioactivity, there was some energy that was completely disappearing. This flew in the face of all physics; energy could never just disappear, it could only be converted from one form to another.

The problem haunted science until 1930 when Austrian physicist Wolfgang Pauli thought of a solution. To explain the disappearance of energy he proposed that a new particle must exist. Unlike the proton and electron, the new particle must carry no charge; it must be electrically neutral. This particle would also be tiny and difficult to detect. His idea was that his new particle – a neutrino – was emitted along with an electron during beta decay, carrying the missing energy with it.

On December 4, 1930, he sent a bizarre letter to a group of fellow nuclear physicists. "Dear radioactive ladies and gentlemen... I have hit upon a desperate remedy to save... the law of the conservation of energy..."

His bold idea was not proven until 1956, when the neutrino was detected for the first time in an experiment monitoring the radiation from a nuclear reactor. Nuclear physics was saved.

David Hawksett

Date 1930

Scientist Wolfgang Pauli

Nationality Austrian-Swiss-American

Why It's Key Filled a critical missing link in our knowledge of radioactivity.

Key Person **Dirac**
Mastering the quantum world

The son of a hugely authoritarian school teacher, Paul Dirac (1902–1984) had an unhappy childhood. Growing up an introvert, he remained incapable of small talk and uncomfortable in the company of others throughout his life. It was during his long periods of solitary thought that he made many of the astonishing breakthroughs for which he is remembered today.

Having become fascinated with Einstein's theories of relativity while studying engineering at university, Dirac devoted the rest of his life to theoretical physics. By 1926 he had established himself as one of the world's leading scientists by combining the seemingly opposing theories of wave mechanics and matrix mechanics to produce the first complete mathematical formalism of quantum mechanics. His 1930 book *Principles of Quantum Mechanics* is universally regarded as a landmark publication; and that was just for starters.

Satisfied that he pretty much had quantum mechanics licked, Dirac set about trying to merge the, seemingly contradictory, special theory of relativity with quantum mechanics. In doing so he fathered a whole new field of enquiry, quantum electrodynamics (QED), which describes some aspects of how electrons, positrons and photons interact. QED is now often referred to as the "jewel of physics", because of the extreme accuracy of many of the predictions it has since produced. One of the first of these predictions was the existence of antimatter, the first particles of which would be discovered, much to the amazement of many physicists, within a few short months.

William Scribe

Date 1930

Nationality British

Why He's Key "Solved" non-relativistic quantum mechanics and fathered the field of quantum electrodynamics. The precision and clarity of Dirac's work has been said to verge on the poetic

Key Experiment
Unpicking pepsin

Enzymes are the substances that make things happen in life; they are biological catalysts, essential for virtually all biological reactions. But they also have important applications in industry, in everything from washing powder to cheese making.

The structure of enzymes was poorly understood in the early twentieth century. Scientists knew that enzymes existed, but were uncertain of their make-up. It was understood, for example, that saliva contained enzymes involved in breaking down dietary proteins, but their structure was not known.

John Howard Northrop was the American chemist who discovered that enzymes were in fact proteins. His work involved pepsin, a digestive enzyme that the stomach secretes to break up dietary proteins. By producing pure pepsin crystals, Northrop was able to analyze its structure. He used a series of experiments to break pepsin into its constituent parts, then sequentially analyzed each fragment's activity and compared it to the activity of whole pepsin, finding no major difference in the activity of the various fragments. This supported his hypothesis that pepsin was a protein.

Northrop's findings were published in 1930, and his techniques were then used to study other enzymes resulting in his fundamental text *Crystalline Enzymes*, in 1939. Northrop's work was recognized in 1946 when he was awarded the Nobel Prize in Chemistry, along with two of his colleagues.

Katie Giles

Date 1930

Scientist John Howard Northrop

Nationality American

Why It's Key Unlocked the secrets of nature's catalysts.

Key Publication **Evolution evolves**
Genetical Theory of Natural Selection

R.A. Fisher's *Genetical Theory of Natural Selection* married two fundamental theories of evolution: Darwin's theory of natural selection and Mendel's patterns of genetic inheritance. It is quite possibly the most important book in evolutionary biology, after Darwin's own *Origin of Species*.

Fisher's publication demonstrated his brilliance in two different areas: mathematics and biology. Using his expertise, Fisher devised complex formulas to explain previously poorly understood mechanisms of inheritance and provided evidence for processes his predecessors had described, but could not prove. His work was seen as the basis of population genetics, a key area in evolutionary biology. His eye-twitchingly mathematical tome also described other fundamental elements of modern biology, such as mimicry, where individuals of one species copy the appearance of another. Some species of butterfly, for instance, mimic others to feign toxicity and avoid predation.

Fisher's beliefs about eugenics were also a large feature of his text, and he applied his views on genetic selection to humans. This area of his work, however, is less celebrated; eugenics became much more controversial following Hitler's Nazi regime.

The publication of Fisher's book influenced biology for years to come, but it was by no means an easy read. Fisher himself alluded to this when the first edition was printed in a relatively large font; he wrote to his publisher: "Fairly large print is a real antidote to stiff reading."

Katie Giles

Publication *Genetical Theory of Natural Selection*

Date 1930

Author Ronald Aylmer Fisher

Nationality British

Why It's Key A key work which married Darwin's theory of evolution to Mendel's concepts of genetic inheritance.

Key Invention
The meteoric rise of the radiosonde

Measuring weather prediction data in the different levels of the atmosphere was a tricky business in the early twentieth century. Meteorologists initially had to make do with kites or airplanes. That was until Pavel Molchanov came up with the radiosonde – a small instrument box with radio transmitting capabilities that could be carried into the atmosphere by a helium or hydrogen balloon. He first launched the radiosonde in 1930 in Pavlovsk, Russia, and successfully recorded a radio signal from the stratosphere.

The instrument box can take meteorological measurements at different atmospheric levels as it ascends into the air. These measurements are then transmitted back to the ground via radio signals. The fully inflated balloon can measure up to 3 meters in height, and when released can remain in the atmosphere for around ninety minutes. The balloons are designed to burst when they reach a designated altitude, usually around 30,500 meters, and fall back to Earth via a parachute.

While in the air, radiosondes use a thermometer to measure temperatures, a hygrometer to take humidity measurements, and a barometer to measure air pressure. The flight path taken by the radiosonde can also be used to indirectly measure wind speed and direction. The biggest advantage is that they don't need to come back to Earth to have their data retrieved.

Meteorologists around the world release radiosondes twice a day, at midnight and midday. The data collected from these is correlated in adjacent areas to predict weather patterns over large regions.

Riaz Bhunnoo

Date 1931

Scientist Pavel Molchanov

Nationality Soviet

Why It's Key Radiosondes provide a cheap and effective way to take measurements in the Earth's upper atmosphere, transmitting the data via radio waves. These measurements underpin weather prediction.

opposite A weather balloon carries a radiosonde up into the atmosphere

Key Publication
Big Bang theory explodes into science

The Big Bang theory revolutionized our view of the creation of the Universe, and lies at the heart of modern astrophysics. Who takes credit for this theory? A Roman Catholic priest by the name of Father Georges Lemaître.

After serving in the Belgian Army during World War I, Lemaître enrolled as a graduate student at Cambridge University, where he began studies in cosmology. Edwin Hubble's observations in 1924 inspired Lemaître to turn his gaze to a particular galaxy that was moving rapidly away from Earth. He realized that if it was receding, then it must mean that the Universe was expanding. In 1927, he published these beliefs, which were contrary to those of his peer, Albert Einstein, who had claimed the Universe was static. Lemaître's work was, however, verified by Hubble.

Lemaître realized that if the Universe was expanding, there must have been a time when it had occupied a single point in space-time. Lemaître called this the "primeval atom," now known as a singularity. He deduced that the Universe must have spawned from this initial creation event, and published his theory in the scientific journal *Nature*, four years after his expanding Universe paper.

Though there was much skepticism initially – it was actually skeptic Fred Hoyle who inadvertently coined the phrase "Big Bang" – observational evidence has bolstered the theory. Lemaître's early ideas have evolved into the best explanation we currently have for the origin of our Universe.

Steve Robinson

Publication "The Beginning of the World from the Point of View of Quantum Theory"

Date 1931

Author Georges Lemaître

Nationality Belgian

Why It's Key Our best shot yet at describing the Universe's explosive beginnings.

Key Invention
Polystyrene first produced commercially

In the early part of the last century, polymer science was gaining momentum. Long-chain molecules, made up of many thousands of repeating units, were proving useful as a new type of material, in all sorts of contexts. There was one polymer in particular that had been synthesized, that was showing great potential as a material, but it was difficult to produce commercially: the polymer was called polystyrene.

This polymer had been accidentally discovered by the German chemist Eduard Simon, back in 1839. He had assumed that when his newly discovered chemical "Styrol" hardened it had oxidized. Other chemists had later proved that this wasn't the case and went on to show that the hard material was actually a polymer. The problem for those trying to produce polystyrene was that styrene (the single-unit chemical which forms the building block of the polymer) tended to react and turn into the polymer before it was supposed to.

Eventually, work by chemists Herman Staudinger and Carl Wolff for the German company BASF allowed the commercial manufacture of polystyrene to begin in 1931. The polymer is hard and can be cast into molds with very fine detail. In this form it is used for economical, rigid plastic items.

Its main use now, of course, is in its expanded form, also known as Styrofoam. First manufactured in 1954 by combining styrene with isobutylene, a volatile liquid, under pressure, this is used for many types of packaging where fragile material needs protection from impact damage.

Simon Davies

Date 1931

Scientists Herman Staudinger, Carl Wolff

Nationality German

Why It's Key Overcoming premature polymerization resulted in a cheap, hard plastic which is also used as a versatile packing material.

Key Person **The Leakeys**
The origins of man is a family business

Louis Leakey (1903–1972) was born to British missionaries working in Kenya. While bird-watching, aged twelve, he discovered some stone tools, sparking a passion for archaeology that never left him. After graduating with a degree in archaeology and anthropology from Cambridge, he began leading expeditions with the awesome task of proving Darwin's theory of an African origin for mankind.

In 1936, Louis married Mary Douglas Nicol (1913–1996), the illustrator of his book *Adam's Ancestors*. Louis and Mary continued their fossil hunting in Olduvai Gorge, Tanzania, and were rewarded in 1948 with their first major discovery – the jaw of pre-human creature called *Proconsul*. The excavations continued while Louis served as a spy and translator for the British government in Kenya during the late 1940s and early 1950s. Their son Jonathan continued the family tradition and, in 1959, the three Leakeys worked together to uncover *Homo habilis*, the "handyman" hominid who made the stone tools found at Olduvai.

From the mid-1960s, Louis and Mary led separate lives. Mary continued to excavate at Olduvai, while Louis funded studies in primatology. He died in London and Mary in Nairobi. Their son Richard and his wife Meave continued excavations at Koobi Fora on Lake Turkana, discovering some of the earliest known hominids. A third generation of the Leakeys is still actively seeking man's origins in the same area; Richard and Meave's daughter Louise now runs the Koobi Fora research project.

Helen Potter

Date 1931

Nationality Kenyan, British

Why They're Key A scientific dynasty that traced man's origins to Africa.

Key Invention **Cephalometrics**
The new face of orthodontics

In 1931, American orthodontist B. Holly Broadbent published an article in *Angle Orthodontist* entitled "A New X-Ray Technique and Its Application to Orthodontia." The method described a novel technique which would significantly contribute not only to orthodontic research, but to general office orthodontic practice as well.

Throughout the 1920s, Broadbent and his colleagues at Western Reserve University developed an instrument called a roentgengraphic cephalometer. This machine was devised from a craniometer – a device previously used for holding only dry skulls – and an X-ray machine. The craniometer was adapted for use with living patients, and the completed instrument was able to properly position the head relative to the X-ray source and the film.

Broadbent's cephalometric radiographer had a huge impact on the developing field of orthodontics. It allowed practitioners to measure changes in tooth and jaw positions due to growth and orthodontic treatment. The invention was used in a thirty-six year study called the Bolton-Brush Growth Studies in which dental and facial X-rays of 5,400 children were studied from birth to adulthood. Some of the early results were displayed at the Chicago World's Fair in 1933.

Cephalometrics is considered one of the greatest achievements in the history of orthodontics. It is a major part of training programs and is standard procedure in orthodontic practices.

Rebecca Hernandez

Date 1931

Scientist B. Holly Broadbent

Nationality American

Why It's Key Provided orthodontists with a way to study and measure changes in the jaw and teeth.

Key Discovery **Radio astronomy**
A window on the Universe

The science of radio astronomy has brought us closer to an understanding of the structure of the cosmos than ever before, thanks largely to the work of a young scientist named Karl Guthe Jansky.

By 1931, it was well established that electromagnetic radiation – such as radio waves and microwaves – could in theory be emitted by objects in space outside of the visible spectrum. But it was only when Jansky detected evidence of cosmic radio waves, purely by accident, that we had definitive proof.

At the time, Jansky was working for Bell Telephone Laboratories investigating interference on long-distance voice transmissions. He noticed a distinct hissing noise on the line, which peaked every day at roughly the same time. He concluded, after a few months of tracking the signal, that the hiss was caused by electromagnetic radiation emanating from a fixed point in space – the center of the Milky Way.

In 1933, he announced his discovery, and astronomy changed forever. Four years later his work was pursued by Grote Reber, who built the first radio "dish," and proceeded to scan the sky for radio waves, leading the way for larger, more detailed studies of the sky. As telescopes grew larger and more complicated – culminating in the Very Large Array in New Mexico, USA – so our understanding of astronomical features such as quasars, black holes, galaxies, and supernovae developed at an astonishing rate.

Steve Robinson

Date 1931

Scientist Karl Guthe Jansky

Nationality American

Why It's Key Opened a window on formerly unseen corners of the Universe, allowing us to see the things our eyes could not.

Key Invention
The electron microscope

Since their invention in the 1600s, microscopes have been crucial to our understanding of the world, but by the beginning of the twentieth century they had reached their limit. The resolving power of a light microscope – the smallest detail it can bring into focus – is roughly the same size as the wavelength of light itself. Visible light has a wavelength of around 500 nanometers (0.0000005 meters, or five millionths of a meter) – small enough to observe things like bacteria, but not viruses or minute cell structures. Scientists wanting to peer at the tiniest parts of life had a problem.

The solution was electrons. Due to a quirk of quantum physics, it is possible to get electrons to behave as waves, and therefore like light. They can be accelerated, given a precisely defined wavelength, and even focused in a beam. It was the discovery that a magnetic field could be used to focus electrons that led to the construction, in 1931, of the very first electron microscope by Ernst Ruska. While working on a project to create an improved oscilloscope, Ruska discovered that electrons could be focused by an electromagnet – by using two of these together, he was able to create a basic microscope. Its resolving power wasn't very high and it had a tendency to set fire to the samples being observed, but it was a useful proof of principle.

The first commercial microscopes, designed with resolutions of around 10 nanometers, were in production by the end of the 1930s. Finally, in 1986, Ruska won a Nobel Prize for his work.

Anne-Claire Pawsey

Date 1931

Scientist Ernst Ruska

Nationality German

Why It's Key Brought to light a new level of detail, allowing scientists to work on an astonishingly small scale.

opposite Ernst Ruska with his electron microscope

Key Publication **The Incompleteness Theorem**
Dashing hopes of universal proof

Legendary German mathematician, David Hilbert, dreamed of reformulating mathematics using a set of absolute proofs. He never found the proofs, and following his death, his dreams were dealt a hard blow by Kurt Gödel's findings in 1931. Using modern logic, Gödel showed that there will always be some mathematical statements that cannot be proven either true or false. You will never be able to reduce mathematics to the application of fixed rules; there will always be truths that will elude proof.

Gödel published his work in an ominous sounding paper, "On Formally Undecidable Propositions," in which he showed that it is possible to construct a sentence so that neither it nor its contrary is provable. Even if you were to decide that the sentence was true anyway, you would be adding to the system's axioms (statements that are so blindingly obvious they don't need a proof) and just create more equally undecidable sentences – basically you can never find enough axioms.

Gödel's theorem has been the catalyst for many philosophical controversies. It was used by Roger Penrose to argue that a computer can never be as clever as a human being since the extent of its knowledge is constrained by a fixed set of axioms, whereas humans can discover unexpected truths. The theorem is also arguably used to show that a Theory of Everything in physics could never be reached due to the inherent nature of mathematics. Gödel truly ended an age of innocence in mathematics by showing that everything is destined to remain forever incomplete.

Leila Sattary

Publication "On Formally Undecidable Propositions"

Date 1931

Author Kurt Gödel

Nationality Czech

Why It's Key A key concept in modern logic, the Incompleteness Theorem changed our viewpoint of the nature of mathematics.

Key Person
Nikola Tesla honored at seventy-five

Scientists rarely make the cover story of the American *Time* magazine. However, in 1931, when Serbian scientist Nikola Tesla (1856–1943) reached seventy-five years of age, the publication honored him in just this way. Readers were left in no doubt as to the significance of this strange pioneer: "... people who have forgotten there were an Alessandro Volta, an Andre Marie Ampere, a Georg Simon Ohm... or a James Watt, are reminded that there is still a Nikola Tesla."

The young inventor-scientist came to the USA in 1884 to work for Thomas Edison and quickly found fault with Edison's preoccupation with direct current (DC) electrical power. Lighting powered by Edison's power stations was weak and inefficient and needed stations every couple of miles, as DC could not be transported over long distances.

So began the "war of the currents," as Tesla developed alternating currents (AC) and Edison fought to save his DC empire. Edison, of course, lost. Tesla's method of transmitting electricity over long distances enabled the Technological Revolution and is still powering the globe today. Tesla's achievements are countless. He is credited as the father of radio and conceptualized methods for wireless power transmission. His discovery that the Earth could be used as a conductor was demonstrated spectacularly when he lit 200 lamps from 40 kilometers away without wires using his Magnifying Transmitter.

In 1997, *Life* magazine declared Tesla one of the hundred most famous people of the last thousand years.

David Hawksett

Date 1931

Nationality Serbian

Why He's Key Without Tesla's many innovations, our world would be very different indeed.

opposite The July 1931 edition of *Time* magazine that celebrated Nikola Tesla

Key Discovery Harold Urey makes light work of heavy hydrogen

Deuterium is known as "heavy hydrogen," and deuteriated water (D_2O) is called "heavy water." This is because deuterium (D or ^2H) is an isotope of hydrogen; a standard hydrogen (or "protium") nucleus has just one proton, whereas a deuterium nucleus has one proton and one neutron. This makes no difference to the electrical charge, but there is a difference in mass.

Frederick Soddy, an English radiochemist, came up with the idea of isotopes in the early twentieth century while studying radioactivity in Canada with Ernest Rutherford. He suggested that versions of the same element with different radioactive properties and different atomic weights were chemically the same. "Identical outsides but different insides." It wasn't until the 1930s, however, that anyone was able to prove it.

Hydrogen, being the simplest atom, seemed the obvious place to start. Harold Urey theorized that any heavier isotopes of hydrogen existed only in very small quantities, as the atomic weight of hydrogen is only just over 1. Also, due to its increased atomic weight, it would boil at a lower temperature. He collected four liters of liquid hydrogen and let it evaporate slowly down to one milliliter. Spectroscopic analysis of this sample detected deuterium for the first time.

Today, heavy water is routinely extracted from normal water by electrolysis, and is commonly used as a non-radioactive tracer molecule. Individual molecules can be followed through complex chemical and biological reactions, giving a detailed understanding of the process.

Bill Addison

Date 1932

Scientist Harold Urey

Nationality American

Why It's Key Urey's discovery gave the first experimental proof of one of the key chemical theories of the twentieth century.

TIME
The Weekly Newsmagazine
Keystone
NIKOLA TESLA*
All the world's his power house.
(See SCIENCE)
*From a portrait by Princess Lwoff-Parlaghy.
Volume XVIII
Number 3

Key Person **Werner Heisenberg**
Subatomic theorist

In June 1922, the famous physicist Niels Bohr was finishing up another lecture when someone stood up and questioned the great man. To everyone's surprise Bohr hesitated in his response. A precocious PhD student named Werner Heisenberg made the objection; a few years later he went on to produce the first formulation of quantum mechanics at only twenty-three years of age.

Born Werner Karl Heisenberg in 1901, little Werner had a seemingly unremarkable childhood, attending primary school in Würzburg before his family relocated to Munich in 1910. Following World War I, Heisenberg attended the University of Munich, completing his PhD just a few years later. Shortly after attaining his doctorate, he finalized the first formulation of matrix mechanics, an essential contribution to the burgeoning field of quantum physics – and one for which he received the Nobel Prize in 1932 – and described "the uncertainty principle," now a central tenet of quantum physics.

Controversially, Heisenberg had a lot of help in the formulation of matrix mechanics. Max Born and his student Pascal Jordan reformulated Heisenberg's mathematics; the end product of which was the first formulation of quantum mechanics. Yet neither Born nor Jordan was recognized by the Nobel administration.

And this was not the only controversy of Heisenberg's career; his role in the development of the Nazi's nuclear programme has been debated at length.
Barney Grenfell

Date 1932

Nationality German

Why He's Key Heisenberg's contribution to quantum physics, through his invention of matrix mechanics and description of the uncertainty principle, opened the door for new theoretical developments.

Key Experiment **The Kennedy-Thorndike experiment**
The speed of light is constant

Roy Kennedy and Edward Thorndike's experiment in 1932 was one of the key tests of Einstein's Theory of Special Relativity. Einstein claimed that motion depends on your frame of reference, for example you can regard yourself as stationary and observe the relative motions of everything around you when, actually, everything in the Universe is moving relative to everything else. The driver of a passing car has just as much right to think of themselves as stationary as you do standing on the pavement. Einstein stated that the laws of physics hold true for all frames of reference, and that the speed of light will be seen as constant from any reference point, however fast it is moving as seen by someone else.

In 1887, Albert Michelson and Edward Morley conducted an experiment which split and recombined light. As the peaks and troughs of the light waves combined, they produced a pattern which could be seen to change with any slight variation in the speed of light.

Kennedy and Thorndike's experiment was a modification of Michelson and Morley's but it went one step further to test Einstein's prediction that time slows down as speed increases. They watched the patterns in the recombined light over many months. If Einstein was wrong then the changing speed of Earth as it orbited the Sun would produce changes in the light pattern as the speed of light minutely changed. They found no such differences; their results favored Einstein. Even today, more accurate versions of the experiment continue to support Einstein's theory.
David Hawksett

Date 1932

Scientists Roy Kennedy, Edward Thorndike

Nationality American

Why It's Key One of the most important tests for Einstein's famous work.

Key Publication *The Causes of Evolution*
Haldane shows how evolution works in practice

J.B.S. Haldane (1892–1962), known to his friends as "JBS," was one of a group of mathematical biologists, including Ronald Fisher and Sewall Wright, who established the science of population genetics, laying the foundations for a modern theory of evolution.

In the 1920s, Haldane wrote a series of papers on the mathematical theory of natural selection, including the first studies of the speed and direction in which genes move through real populations. By far his most famous example – and one that has had to withstand substantial criticism in recent years – was the peppered moth. Haldane claimed in some English cities it had evolved, over a few generations, from pale to black, where the black form benefited by being camouflaged against soot-covered trees. Haldane summarized his work in *The Causes of Evolution* (1932), a landmark book in the synthesis of modern biology. It explained how the mathematical effects of Mendel's genetics on whole populations could be used to describe the way Darwin's idea of natural selection worked in practice.

Haldane was a colorful character; Scottish aristocrat, World War I veteran, popularizer of science, Communist, and environmentalist. In 1923, he suggested that hydrogen-generating windmills could replace coal as a source of power and, in 1963, he coined the word "clone" to describe genetically identical cells or organisms derived from a single cell or individual. When he died he left his body to medical science, noting that, "its refrigeration should be a first charge on my estate."

Richard Bond

Publication *The Causes of Evolution*

Date 1932

Author J.B.S. Haldane

Nationality British

Why It's Key By applying mathematics to the study of real populations, Haldane showed how evolution worked in practice.

Key Discovery **Is it a bird? Is it a plane? No, it's a neutron**

Irene and Frederic Joliot-Curie, daughter and son-in-law of Marie and Pierre, observed a curious effect when experimenting with the element beryllium: when the radiation it emitted hit paraffin, protons were ejected at high speeds. Their findings inspired English physicist James Chadwick to conduct further research, discovering that the beryllium radiation also liberated protons from other substances; though these moved more slowly.

He calculated that, in order to smash protons out of a substance, the mass of the particles in the beam of beryllium radiation had to be similar to the mass of a proton itself. But there was a major difference between protons and the particles from the beryllium – when aimed at a thick slab of lead, these passed straight through, while protons would be stopped after just a millimeter. Since the only thing that could be causing this was a difference in charge, Chadwick concluded that the particles making up the beryllium radiation must be chargeless. He named them neutrons.

Chadwick's discovery helped a lot of other questions fall into place. The process of beta radiation – where electrons are emitted – could now be explained while conserving charge: A neutron decays to a positively-charged proton and a negatively-charged electron. Neutrons also revolutionized the model of the atom, explaining how atoms of the same element could have different masses, but the same chemical properties. Finally, they became important experimental devices in their own right; neutrons can be used to bombard atoms and probe their structure.

Kate Oliver

Date 1932

Scientist James Chadwick

Nationality British

Why It's Key The discovery of a neutral particle simplifies the model of atomic structure and explains a number of pesky things about atoms.

Key Discovery **Positrons** The first antimatter particles

When British physicist Paul Dirac's latest equation gave him solutions with negative energies, he didn't let it bother him. He had successfully combined Schrodinger's idea of describing particles as probability waves with Heisenberg's matrix mechanics; it was too good to give up. But he still had to explain his puzzling results – negative energy just didn't make sense.

His solution was to declare that previously unknown particles must exist; particles that act just like electrons but have an opposite charge; in other words, antiparticles. Dirac, to the disbelief of some of his colleagues, reckoned that the occurrence of positrons – positively charged electrons – explained his negative energy conundrum.

Fast forward four years from Dirac's original work to 1932; physicist Carl Anderson is making observations in a cloud chamber, where lines of condensation mark the trajectories of particles passing through it. A curly line typical of a charged particle in a magnetic field appears. It looks like the path of an electron, but it's spiraling the opposite way to a negatively charged particle. More appear. To determine their direction of travel precisely, Anderson inserts a plate of lead into the chamber. After passing through the lead, the particles lose energy, travel more slowly, and make tighter spirals. The direction of their paths is now obvious.

Anderson's results were conclusive. They could only be explained in terms of a particle identical to the electron, but with a positive charge. Anderson had found a positron; Dirac's antimatter was real.

Kate Oliver

Date 1932

Scientist Carl Anderson

Nationality American

Why It's Key A new type of particle, the antiparticle, shows up and becomes available for investigation.

opposite Anderson's cloud chamber photograph shows the positively-charged particle's track. It loses energy and curves more in the magnetic field after traversing the lead plate in the middle.

Key Person **Wernher von Braun** The rocket man

The history books would tell a very different tale if it wasn't for Wernher von Braun (1912–1977), who nurtured a life-long desire to propel humans beyond the reach of Earth's gravitational grasp. His contributions to rocket science encompassed both the honorable and the more sinister applications of the discipline; as well as working on liquid-fuelled rockets required for space exploration, he helped to develop the Nazi's V-2 ballistic missiles.

Inspired by the science fiction of Jules Verne and H.G. Wells, von Braun's interest in space flight was sparked at an early age. By the time he was twenty, he had an aeronautical engineering degree and several years experience helping the German Society for Space Travel conduct tests with liquid-fuelled rockets. The German army became intrigued by the society's activities, and offered von Braun a job in 1932. With government funding to fuel the construction of bigger and better rockets – and pay for a PhD in physics – he could hardly say no. Despite von Braun's cosmic aspirations, he soon found himself developing rockets for the purposes of warfare, and, by 1937, was leading the team that created the V2 missile, which the Nazis used toward the end of World War II.

After the Nazi defeat, von Braun's team was employed by the American army to share the V2's secrets. Eventually, in 1960, the team relocated to work for the newly established NASA at the Marshall Space Flight Center. Von Braun became its director and the chief designer of the Saturn V rocket that propelled the first men to the Moon in 1969.

James Urquhart

Date 1932

Nationality German

Why He's Key Pioneered the development of liquid-fuelled rockets, leading to the creation of the V-2 rocket, the forerunner of those used for space exploration and modern guided missiles.

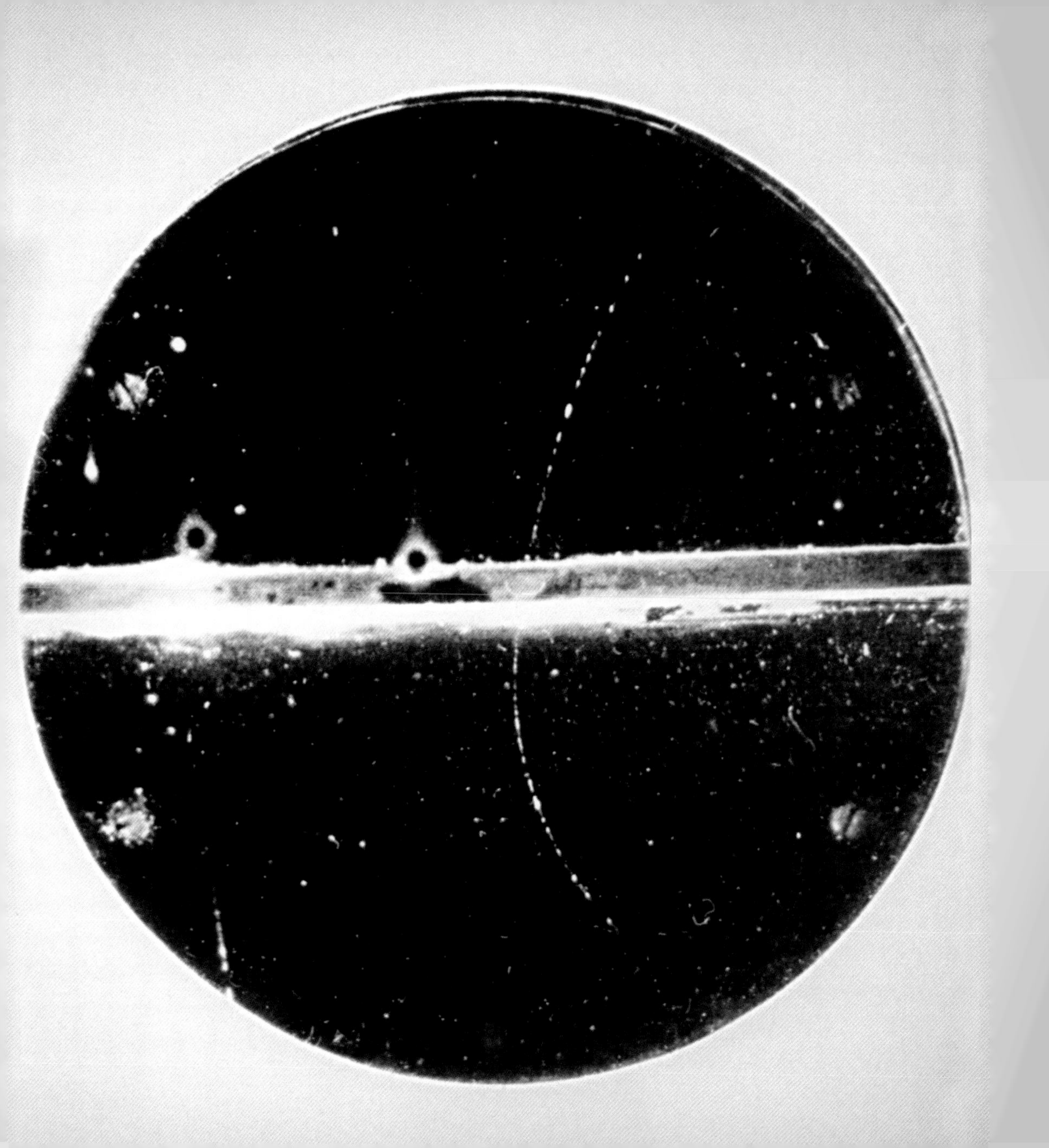

Key Invention
Jet engines developed – twice

Hans von Ohain and Sir Frank Whittle are jointly credited with inventing the jet engine, despite neither being aware of the other's work. Whittle, an English aviation engineer and pilot, received very little official support for his early work into jet propulsion. It was mainly through his own diligence and determination that, in 1932, he became the first person to be granted a patent for a jet engine. Having finally secured some much needed private funding, Whittle started work on his first jet engine in 1935, which he went on to successfully bench test in the lab two years later.

Meanwhile, unbeknown to Whittle, German scientists were hot on his heels with development of their own jet engine design. Hans von Ohain was just twenty-two years old when he developed his initial plans for a continuous cycle combustion engine – similar in design to Whittle's, but with a different internal arrangement. By 1939, Ohain's engine was incorporated in the first jet aircraft to see flight, some two years before Frank Whittle's.

During World War II, German scientists worked tirelessly in an effort to develop a jet aircraft that would give them an edge in the sky, however, due to heavy Allied bombings, it was not until mid-1944 that any jet fighters emerged.

After the war, jets progressed at blistering speeds – quite literally. The first commercial jet airliner, Britain's De Havilland Comet, was soon succeeded by faster models such as the Russian's Tu-104 and the American Boeing 707.

Logan Wright

Date 1932

Scientists Hans von Ohain, Sir Frank Whittle

Nationality German, British

Why It's Key Air travel has never looked back since jet engines were developed; propeller-driven planes are now largely a thing of the past.

opposite The first jet-plane

Key Discovery
Vitamin C gets the lime light

For centuries, sailors who went on long voyages – and had a limited diet – suffered scurvy: bleeding gums, bruising, and wounds that were slow to heal. And although by the fifteenth century Danish sailors knew that they could prevent scurvy if they consumed lemons and oranges, it was not until 1753 that the British medical establishment described the condition as being linked to a diet deficient in citrus fruit. By 1795, limes were standard rations for the British Navy.

The race was on to identify the curative substance, and it attracted the curiosity of Albert Szent-Györgyi, a man so determined to do science rather than fight in World War II that he shot himself in the arm to make himself unfit for military service. In 1932, Szent-Györgyi reported his success in showing that a substance isolated from citrus juice could prevent scurvy-like disease in guinea pigs. Some fifteen days earlier, the American Charles Glen King had also published his own discovery of the same substance – later called ascorbic acid, or vitamin C. But only Szent-Györgyi was awarded the Nobel Prize in 1937.

We now know that vitamin C is critical to health, particularly for maintaining the structure of collagen, and hence the integrity of the body's mucous membranes. It can be found in many fruits and vegetables, including broccoli and potatoes. Its popularity as a cure and treatment for a range of maladies, however, goes beyond the scientific evidence. Its usefulness against the common cold and cancer – advocated by Linus Pauling – remains controversial.

Julie Clayton

Date 1933

Scientists Albert Szent-Györgyi, Charles Glen King

Nationality Hungarian, American

Why It's Key Vitamin C is essential to good health in humans who, unlike primates and most other animals, lack a key enzyme needed to produce it in the body.

Key Discovery **Link between Vitamin A and night-blindness**

As urban legend has it, carrots help us to see in the dark. The claim has its roots in World War II propaganda. A heavy diet of carrots was claimed to aid British gunners' ability to shoot down German planes at night; the story was used as a cover for the newly-introduced radar technologies used in the Battle of Britain. But what is it about carrots that made this plausible? A link between eye diseases and poor diet has been noted throughout history. Around 400 BCE, raw beef liver was used as a cure for night-blindness – the inability to see in dim light. It wasn't until the early twentieth century, however, that we understood why this cure worked.

After the discovery of vitamins A and B in 1913, experiments on rats showed that the addition of butter fat, liver extract, and even alfalfa leaves to their diets could improve growth and alleviate blindness. The explanation, however, remained unclear.

In 1933, the work of George Wald and colleagues proved the link between vitamin A deficiency and night-blindness by isolating vitamin A from the retina of the eye. The human body needs vitamin A to make the chemical retinal, without which our eyes cannot absorb light. As light enters the eye, a series of chemical reactions occur which lead to our perception of color. Retinal is destroyed during this process, and in order to replenish it, we need vitamin A, meaning that it is an essential part of our diet.

Ceri Harrop

Date 1933

Scientist George Wald

Nationality American

Why It's Key The importance of vitamins was poorly understood, and their role in physiological processes unknown. Vitamins can now be chemically synthesized and are taken as dietary supplements to improve health.

Key Person **August Krogh**
A pioneer in physiology

August Steenberg Krogh (1874–1949) was a Danish professor at the University of Copenhagen where he specialized in zoophysiology from 1916 to 1945. He was something of a pioneer in his field and was awarded the Nobel Prize in 1920 for his research into capillary regulation in muscles. His main area of interest was in studying the mechanisms of physiological processes. He famously described the "Krogh Principle," which states that it is more useful to study a physiological mechanism in a large group of animals rather than a small, more limiting group, and that some organisms are more useful to study than others. His renowned work, *Osmotic Regulation*, was published in 1939 and made huge waves in the world of biology, as well as science as a whole. "Osmoregulation" – as it's more commonly known – is a method for regulating osmotic pressure by moving water from one solution to another through the process of osmosis. It allows animals to maintain correct levels of water, waste products, and solutes in their systems. Krogh's work described the mechanism in marine animals which allowed them to keep their internal water solutions at a constant state through changing pressures and salinities.

In addition to being a major breakthrough in marine physiology, Krogh's work made a significant contribution to understanding the methods of water regulation in all animals, including humans.

Katherine Ball

Date 1933

Nationality Danish

Why He's Key He helped us form a better understanding of regulatory mechanisms in animal physiology.

Key Discovery **A star is born** Congratulations, it's a neutron!

Only a year after the discovery of neutrons – uncharged particles in atoms – Swiss-born Walter Baade and Geman Fritz Zwicky, two eminent astrophysicists of the time, postulated the existence of neutron stars. They came up with the idea whilst investigating the existence of supernovae.

They suggested that when stars got old they would start to collapse from the inside out. As atoms in the core of the star become unstable, their outermost particles – the negatively charged electrons – collapse into the atoms' nuclei, causing positively charged protons to turn into neutrons. The removal of charge allows the atoms to pack together very tightly to form an unimaginably dense ball known as a neutron star, which spins rapidly as it comes into existence, and has an extremely strong gravitational pull. The outer layers of the star that escape this collapse explode in a spectacular display called a supernova.

Baade and Zwicky discovered this phenomenon while investigating the phenomenon of cosmic rays proposed by Victor Hess in 1912. They realized that many of the cosmic rays reaching the Earth's atmosphere could not be produced in our galaxy, and invoked the ideas of their production in supernovae, and the creation of neutron stars, to explain them.

In 1933, neutron stars and supernovae were considered pretty wacky and far-fetched. Today, however, Baade and Zwicky's discovery is considered one of the most influential in the field of astrophysics.

Josh Davies

Date 1933

Scientists Walter Baade, Fritz Zwicky

Nationality Swiss, German

Why It's Key The discovery of neutron stars was instrumental in the advancement of astrophysics in the twentieth century, and gave Hess' cosmic rays a source.

Key Discovery **Mysterious dark matter found in galaxies**

Swiss astrophysicist Fritz Zwicky was studying galaxies at the California Institute of Technology in 1933, when he noticed something strange about the Coma Galaxy Cluster, 320 million light years away. Galaxies at the perimeter of the cluster were moving much faster than they ought to be.

In 1937, Zwicky stumbled upon another problem. It emerged whilst he was trying to "weigh" the galaxies using the Virial Theorem, which allows scientists to gauge an object's mass from its speed. He found that the observed mass of the Coma Cluster was 500 times too small; the visible mass alone could not account for the huge gravitational pull of the cluster. This became known as the "missing mass problem." Zwicky then attempted to explain the paradox by an invisible presence he dubbed "dark matter." This mysterious substance, he said, did not emit or reflect light, and so was concealed from direct detection.

Zwicky's work was largely ignored, however, and he spent the next two decades trying to convince the skeptical scientific community that his "dark matter" existed. Only in 1974, the year of his death, was his work rediscovered and taken seriously.

Further evidence for dark matter has come from studies of the amount by which a galaxy bends light, which are useful for calculating their mass. Understanding dark matter is the aim of scientists at the forefront of current scientific research, as they strive to shed light on the origins of galaxies, and of the Universe itself.

Steve Robinson

Date 1933

Scientist Fritz Zwicky

Nationality Swiss

Why It's Key The first knowledge of a strange substance in space that would baffle scientists for decades to come and change our understanding of how the Universe is built.

Key Discovery **Shock horror**
Insulin therapy

In the early 1900s, mental illness was widely regarded as incurable. Patients with conditions such as psychosis and schizophrenia were routinely locked up in asylums, where they suffered horrendous conditions. The introduction of psychoanalysis by psychiatrists such as Sigmund Freud was the start of a changing era in the treatment of mental illness, and the analytical approach was eventually supplemented by physical treatments such as electroconvulsive therapy and surgery.

Insulin Shock Therapy was pioneered by Polish scientist Manfred Sakel, who studied neurology and neuropsychiatry at the University of Vienna from 1919 to 1925, becoming a researcher at the neuro-psychiatric clinic thereafter. Forced to emigrate to the USA in 1936, when the National Socialist Party came to power in Austria, he became an attending physiologist at the Harlem Valley State Hospital. It was here that he realized insulin-induced comas calmed schizophrenic and psychotic patients due to the lower level of glucose in the blood system. This "hypo-glycemic crisis" in patients causes reduced brain function due to a deficit of glucose in the brain cells, effectively calming the patient by reducing nerve cell activity.

Sakel's discovery, now known as the Sakel Technique, was publicly communicated in 1933, although he had already put the technique to use in 1927 in Berlin. In his tests, up to 70 per cent of patients responded positively to the treatment and the therapy was heralded as a breakthrough in psychiatric circles, changing the way mental illness is treated worldwide.

Katherine Ball

Date 1933

Scientist Manfred Sakel

Nationality Polish

Why It's Key Transformed the way mental illness was treated and changed the approach of psychiatry for ever.

Key Invention **A big wave hello**
The arrival of FM radio

The mass media is an almost inescapable part of our modern lives, and the technology used to transmit media messages is advancing at an ever increasing rate; almost as soon as you stop to recharge your MP3 player you will have fallen behind.

So how does the unassuming FM radio fit into all this? Frequency modulation (FM) radio was patented in 1933 by Edwin Armstrong as an improvement upon the existing amplitude modulation (AM or MW) radio. FM radio varies the frequency of the transmission waves rather than the amplitude, or height, of the wave. It is less likely to be affected by static and allows for many more radio stations than AM, but does require "line of sight" transmission, with a maximum range of around 80 kilometers. Although it has had its critics, FM is generally accepted to produce a much improved, high fidelity (hi-fi) sound – as long as the signal is strong.

FM became the preferred radio medium across the world, and is comparable to the "digital revolution" of today. It is notable for being one of the first developments of this kind; a technology aimed at consumer satisfaction within a global market. It is also remarkable that, in our digital age, analog radio is still a massively popular media source. Even in the UK, which has seen rapid take up of digital radio, the vast majority of radio receivers are still analogue, and, unlike the analog TV signal, there is as yet no date for a switch off of FM radio.

Jim Bell

Date 1933

Scientist Edwin Armstrong

Nationality American

Why It's Key Better radio broadcasts mark an early step toward the media hungry information age we live in today.

opposite One of the first FM radios – media to the masses

Key Person **Erwin Schrödinger**
The diverse discoverer

The name Schrödinger is most often associated with the invention of wave mechanics, as well as the quirky exploration of a problem for quantum physicists, using the poisoning of a cat as an analogy. On closer inspection, however, the life of the Nobel Prize-winning scientist reveals that his work and interests were, in fact, much wider than this.

Born in 1887 in Vienna, science was already inherent in Erwin Schrödinger's family. His grandfather was a renowned chemistry professor and his father was a keen botanist. Indeed botany was an area of science that Schrödinger himself also studied, publishing several papers on plant phylogeny. Before exploring plants, Schrödinger had studied Italian painting. His physics career began in his home city, but world conflicts played a part in the geographical path he took. He first read and was influenced by Einstein's new theories in 1915, while serving as a soldier, and later decided to move to Zurich, as Switzerland remained untouched by the war. It was here that he published his most famous work, his theory of quantum mechanics: "Quantization as an Eigenvalue-Problem." It is for this work that, in 1933, after another move across Europe to Oxford University – again due to world conflicts – Schrödinger received the Nobel Prize for Physics. His theories have helped solve many problems in atomic physics.

Schrödinger retired in 1955, but not before branching into yet another area of science. Some genetic biologists cite his book, *What is Life?*, published in 1944, as the start of molecular genetics.

Helen MacBain

Date 1933

Nationality Austrian-Irish

Why He's Key Pioneered quantum mechanics, describing how atomic particles behave.

Key Experiment **Magnetic refrigeration cooling**
Is it cold in here, or is it just me?

In one of those strange twists of fate, it turns out that one of the best ways to cool something down is to heat something else up. In 1933, chemists William Giauque and Duncan MacDougall were able to demonstrate that they could cool atoms down to previously unimaginable temperatures by using magnets to heat them up.

Known (rather unromantically) as adiabatic demagnetization, their technique uses strong magnetic fields to control the behavior of atoms which themselves resemble tiny magnets. These atoms spin about an axis which normally points in a random direction, but when a strong magnetic field is turned on, it forces the atomic spins to line up all in the same direction, producing magnetization. The energy contained within the atoms that made them spin along random directions now emerges from the atoms as heat, warming the sample up. This extra heat energy can be removed from the system.

When the magnetic field is later turned off, the atomic spins remain lined up, which corresponds to a much lower temperature. To return to the initial state, with random spins, the atoms would need to absorb energy from the sample; the grand result is that everything cools down.

The general technique Giauque and MacDougall pioneered has been used ever since to produce the lowest temperatures man has ever attained. In 2003 a team of physicists cooled atoms to just half a billionth of a degree above absolute zero (approximately -273 degrees Celsius).

B. James McCallum

Date 1933

Scientists William Giauque, Duncan MacDougall

Nationality American

Why It's Key Allows physicists to study how matter behaves at near absolute zero. It's also used to cool highly sensitive X-ray telescopes and may yet turn out to be an environmentally friendly way to keep food cold.

Key Publication ***Logik der Forschung***
Popper bursts empiricists' bubble

Karl Popper is arguably the most influential philosopher of science of the twentieth century; so famous, even scientists know who he is. Born in Austria in 1902 and educated at the University of Vienna, Popper attained a PhD in philosophy in 1928. The rise of Nazism caused him to leave Austria and emigrate to New Zealand and then on to England in 1946. There, he became reader in logic and scientific method at the London School of Economics, gathering more awards and honors than you could shake a sizeable stick at.

Popper's philosophical work covers a broad spectrum, including social and political philosophy, history, and metaphysics. But the work for which he is arguably best known is his first publication *Logik der Forschung* (*The Logic of Scientific Discovery*). In *Logik der Forschung* Popper reasons that the main problem in philosophy of science is that of demarcation – distinguishing between science, non- science, and pseudo-science. According to Popper, this latter category would include logic, metaphysics, and psychoanalysis. In order for something to be considered science it must formulate theories that are falsifiable, meaning that they can be disproved by the discovery of a counter instance. For Popper science starts with problems rather than observations. *Logik der Forschung* has had a massive impact on the way that people view the process of science, and as a bonus, has stuck one in the eye for the "science" of psychoanalysis – take that, headshrinkers!

Barney Grenfell

Publication *The Logic of Scientific Discovery*

Date 1934

Author Karl Popper

Nationality Austrian

Why It's Key Fundamentally changed people's perception of the way scientific theories work.

Key Discovery
Seizures for schizophrenia

By the early 1930s, Ladislas Meduna, a Hungarian neuropathologist, had noticed that the brains of patients with schizophrenia had a lower concentration of glia – the cells that support neurons – than the brains of epileptics. He had also heard reports of psychotic symptoms decreasing when patients developed seizure disorders and noticed that the prevalence of epilepsy in patients with schizophrenia was rare. He postulated that these two diseases must be opposing conditions, and sought to cure schizophrenia by causing epilepsy.

In 1934, Meduna induced seizures, using camphor dissolved in oil, in twenty-six patients who had been diagnosed with schizophrenia; he noted improvement in over half of his treatment group. Buoyed by this early success, he undertook further tests, this time using the cardiac stimulant Metrazol in 110 patients, and reported improvement in over 90 per cent of cases. He published his findings and, by 1936, Metrazol-induced seizures were being used worldwide.

It wasn't long, however, before Meduna's techniques were further improved upon. Drug-induced seizures were soon replaced by electroconvulsive therapy (ECT), which was considered safer. In 1952, the anti-nausea medicine chlorpromazine was noted to have anti-psychotic effects, and by 1954 the drug was approved for use as a treatment for schizophrenia.

Although the supposed antagonistic relationship between epilepsy and schizophrenia has long since been disproved, ECT remains a therapeutic option for some schizophrenic patients.

Stuart M. Smith

Date 1934

Scientist Ladislas Meduna

Nationality Hungarian

Why It's Key Ushered in new treatments for mental health ailments.

Key Invention **The electric guitar**
Born to rock

The twentieth century has undoubtedly been a century of innovation, within which a few inventions stand out above the rest: Random Access Memory, nuclear power, and most importantly of all, the electric guitar.

There is some controversy over who actually invented the electric guitar, but George Beauchamp, co-founder of Rickenbacker guitar makers, first filed the patent for a solid body guitar with electric pick-ups in 1934. Beauchamp's key innovation was the invention of those electric pick-ups.

What makes the electric guitar distinct from other stringed instruments is the manner in which the music is amplified. In a conventional stringed instrument, the vibration of the strings is amplified through the "body" of the instrument; a large hollow chamber. In an electric guitar the vibration of the steel or iron strings causes a small magnet in the guitar (the "pick-up") to vibrate, which creates an electric current. This is then amplified through a speaker system, creating the sound.

Beauchamp's design was technically and commercially successful; the electric guitar sold like hot-rock cakes. In the process, a whole new generation of music was born, a musical form which has gone on to dominate the popular music scene throughout the whole of the twentieth and early twenty-first century, elevating George Beauchamp's invention to the iconic status it now enjoys.

Barney Grenfell

Date 1934

Scientist George Beauchamp

Nationality American

Why It's Key Jimi Hendrix, Eric Clapton, Jimmy Page – how many more reasons do you want?

opposite An early 1930s Rickenbacker guitar

Key Person
Fritz Zwicky and his astronomical ideas

Fritz Zwicky was born in Bulgaria in 1898, and is considered by many as one of the most influential and unusual people ever to grace the field of astrophysics. His ideas varied from inspired to crackpot, and his people skills left a lot to be desired. This is probably the reason he isn't the household name today that his genius deserves.

In 1931, Zwicky was the first to explain cosmic rays as originating from supernovae, the catastrophic explosions of dying stars. After teaming up with Walter Baade in 1933, he took this theory further and postulated the existence of neutron stars – a highly controversial idea, especially since the neutron itself had only been discovered the previous year.

Also in 1933, while investigating the speeds of galaxies in clusters, he noticed that the masses of clustering galaxies were about ten times less than they should be. He explained this phenomenon by proposing the existence of particles within the galaxies that didn't emit any light – dark matter.

Alongside his brilliant ideas, Zwicky entertained the ridiculous. He believed that people could influence fission in the Sun by bombarding it with objects from Earth, and change the travel path of the galaxy in space. Zwicky considered that this would be a viable method of propelling ourselves through the Universe and postulated that we could visit Alpha Centuri, our closest star system, in 2,500 years.

Josh Davies

Date 1934

Nationality Swiss

Why He's Key Zwicky's research was instrumental in sculpting the science of astronomy in the twentieth century.

Key Discovery **See that noise?** The secrets of sonoluminescence

Meaning "sound light," the strange phenomenon of sonoluminescence was hypothesized upon in 1933 by Reinhardt Mecke, of the University of Heidelberg, based on observations that very high power sound from military sonar arrays could accelerate chemical reactions in water. It was first observed by Mecke's fellow Germans, H. Frenzel and H. Schultes, in 1934.

While attempting to speed up the process of photographic developing, Frenzel and Schultes placed an ultrasonic speaker into a drum full of developing liquid and noticed that, when the pictures were developed, small dots were visible. They concluded that the bubbles in the liquid were somehow producing light. It would be the late 1980s before technological developments made it possible to properly repeat and study the phenomenon. Recent experiments indicate sonoluminescence appears when a microscopic bubble suddenly collapses and its wall implodes at several times the speed of sound. This is believed to provoke the shock compression of any gas within the bubble, briefly heating it to a temperature higher than that of the Sun's surface. The compressed gas radiates both visible light and high-energy ultraviolet radiation. Temperatures generated by sonoluminescence are thought to reach 1 million degrees Celsius. If so, this is close to the temperatures needed for thermonuclear fusion. Real-life experiments in the area of sonoluminescent fusion continue in the hopes of finding a way of triggering reactions that would supply the world with endless clean energy.

Mike Davis

Date 1934

Scientists H. Frenzel, H. Schultes

Nationality German

Why It's Key Aside from the theoretical potential for fusion, sonoluminescence may be used in place of some expensive lasers to direct therapeutic drugs to specific sites in the body.

Key Person **Enrico Fermi** The original atom smasher

It is hard for us in this age of nuclear power to imagine a time when the atomic Pandora's box was still closed. But it was in fact only seventy or so years ago that the atom was split for the first time, by an Italian physicist called Enrico Fermi. Ironically, he didn't even realize what he had done.

Fermi was born in Rome in 1901. He became interested in physics as a teenager, and showed an immediate aptitude for the subject. The essay he wrote as part of his entrance exam for the Scuola Normale Superiore – where he was to get his undergraduate and doctoral qualifications – was judged worthy of a PhD. Fermi was just seventeen.

His first professorship was in Rome, a post that was created for him, in atomic physics. In this post Fermi also experimented with bombarding different elements, including uranium, with neutrons. The results he observed caused him to suggest that new elements were being created. For this work he was awarded a Nobel Prize in Physics, though it was not until 1938 that physicists Lise Meitner and Otto Frisch actually confirmed that the nucleus had in fact split.

Fermi's contribution to twentieth-century physics is astounding, not just for his work in particle physics, but also as co-author of Fermi-Dirac statistics, important in quantum theory. He even has a group of particles named after him – Fermions.

Barney Grenfell

Date 1934

Nationality Italian

Why He's Key Fermi was one of the last to be both a theoretical and an experimental physicist. For his role in the exploration of the structure of the atom (including splitting it), he should be recognized as one of the century's greats.

opposite Enrico Fermi

$-\frac{\hbar}{i}\frac{\partial}{\partial t} = \frac{p^2}{2m} - \frac{Ze^2}{r}$
$\alpha = \frac{\hbar^2}{ec}$
ϑ

Key Invention
Cats' eyes

Even in total darkness, gleaming roadstuds – better known as cats' eyes – guide drivers down the lanes of motorways across the world. Their glass spheres are backed with mirrors that reflect light shining on them from car headlamps. The same process occurs in a living cat's eye, where a reflective layer at the back of the retina works as a mirror – pushing light back into the eye and helping the cat see in low light.

British-born Percy Shaw patented the original self-cleaning cats' eyes in 1934. At the time, buses were making trams obsolete and British motorists were losing the reflecting tramlines they'd relied on to light the road ahead.

According to one romantic tale, Shaw's moment of inspiration came when he was saved from driving off a foggy road by the reflection of his car's headlights in the eyes of a cat preening itself on a boundary fence. Rather disappointingly, Shaw later told British journalist Alan Whicker that he simply saw a reflective sign one foggy night, and thought "we want those things down on the road, not up there."

Whatever the truth, Shaw's invention – after some tinkering with materials – quickly caught on. Production at his company, Reflecting Roadstuds Ltd, was boosted by enforced blackouts during World War II. Nowadays cats' eyes are ubiquitous on every road; like the very best inventions, they do their job so simply and effectively that we take them for granted.

Richard Van Noorden

Date 1934

Scientist Percy Shaw

Nationality British

Why It's Key A simple reflecting roadstud that saves lives across the world.

Key Event **The Great Plains turn to dust**
Dust, drought, and depression

When European settlers reached North America's Great Plains in the sixteenth century, they found vast tracts of fertile prairie land, perfect for ranching and subsequently growing wheat. It was North America's breadbasket.

During the early part of the twentieth century there was a dramatic intensification of agriculture, due in part to the demands of World War I. Enormous wheat fields stretched as far as the eye could see on the eastern plains, while in the west, livestock were packed onto the land in ever higher densities.

For a while everything worked, but it couldn't last. The topsoil was being exhausted of all of its nutrients and the organic matter that kept it bound together. And then disaster struck. In 1930, the year after the Wall Street crash, the rains failed. It was the start of a decade of droughts. Topsoil dried out and simply blew away in huge dust storms – black blizzards that swamped large areas of Kansas, Colorado, Oklahoma, Texas, New Mexico, and beyond. In 1935, a journalist described the whole area as a dust bowl and the name stuck.

That same year, the U.S. government passed the Soil Conservation Act, aimed at promoting farming methods that would combat erosion, including crop rotation and the planting of shelterbelts. While the methods the government espoused were eventually successful, they didn't stop the dust bowl exacerbating the Great Depression. The cost of this environmental disaster ran into millions of dollars and cost countless lives.

Mark Steer

Date 1935

Country USA

Why It's Key A stark warning about the perils of overusing natural resources. Many scientists now worry that global warming will dry the Great Plains out once more.

Key Invention **Vannevar Bush creates first useable calculating engine**

Devices to aid mathematical calculation have existed for millennia; the simple abacus has helped solve complex problems in arithmetic for so long we don't know how old it is. Toward the beginning of the twentieth century, another machine was created, to assist in the solution of tricky differential equations: the "differential analyzer."

Differential equations are extremely important equations in mathematics, economics, science, mechanics, and engineering. Newton's Laws of Motion, for example, allow the position and movement of a body in space to be calculated using differential equations. Machines that would aid in the solving of these complex equations had been created before; Charles Babbage's unsuccessful "difference engine" had been one attempt to remove the drudgery of calculating mathematical tables. Vannevar Bush set about creating a functional mechanical device to solve differential equations, at a time when others were beginning work on the first electronic computing devices. He produced the "differential analyzer," an analogue computer that, although powered by electricity, was a mechanical device which used a complex system of wheels and discs to perform the actual calculations. This monstrous calculator, the first practically usable analogue computer of its kind, filled a whole room, but it worked.

Unfortunately for Bush, another innovation, the digital computer, superseded the analogue computer fairly soon afterwards, rendering the mechanical calculating device obsolete.

Barney Grenfell

Date 1935

Scientist Vannevar Bush

Nationality American

Why It's Key The creation of a practically useful device that would solve differential equations with a high degree of accuracy.

Key Invention **Wired for sound** Tape takes over

In the 1930s, magnetic recording machines were predominantly used by broadcasters. Largely impractical and potentially lethal, machines such as the "Blattnerphone," used by the BBC in the UK and the "Steel Sound Machine" designed by Semi J. Begun for the C. Lorenz company in Germany, utilized rapidly spinning spools of steel which could be "edited" only with wire cutters and soldering equipment. If a spool snapped during its high-speed rotation, it could break apart like an exploding grenade. Despite this, such machines remained in service with various commercial broadcasters until the 1950s.

Begun, a pioneer of magnetic recording, also developed the first (wire-based) consumer tape recorder and the "Mail-A-Voice," which magnetically recorded audio onto one side of a paper disk. Practical magnetic tape recording originated with Austrian inventor, Fritz Pfleumer, who had created a process for striping gold-colored bronze bands on cigarette paper. He felt certain the technique could be used to make magnetizable recording paper to replace expensive and impractical wire systems. In 1932, German electrical company AEG hired Pfleumer to work with Theo Volk and chemist Friedrich Matthias to develop a magnetic tape recording system. By 1933, magnetic heads that created a magnetic field without touching the tape, and tough, readily-editable magnetic tape proper had been developed. Progress was then fast.

In August 1935, the "Magnetophon K1" recorder was unveiled at the Berlin Radio Fair. This was, effectively, the first s magnetic tape machine.

Mike Davis

Date 1935

Scientist Fritz Pfleumer

Nationality Austrian

Why It's Key Opened the door to smaller, less expensive tape recording systems that would form the basis of audio and video recording for many years to come.

Key Invention **Nylon**
The era of pantyhose begins

When the DuPont Company made the decision to focus on pure research, rather than the usual aim of companies – money-making – little did they realize that the decision would prove to be highly lucrative.

In the late 1920s, Wallace Carothers, an accountant-turned-chemist, was employed by the DuPont Company to run an organic chemistry laboratory. In particular, Carothers' work focused on polymers, large molecules made up of repeating chemical units – think beads on a string.

After some rather inspired chemical "cooking," Carothers' efforts led first to the invention and manufacture of synthetic rubber, or neoprene, in 1931. But his skills were soon to be needed on a matter of some urgency. As trade relations between Japan and the United States turned sour in the mid-1930s, the United States' supply of silk became limited. While today we may see silk as a luxury item, at the time it was essential for parachutes, tents, ropes, and other military supplies. During the first few months of World War II, finding a replacement for silk was a top priority for many research institutes.

It was Carothers' ingenious research that eventually led to the invention of nylon, a polymer that DuPont claimed was "as strong as steel, as fine as a spider's web." Commercial production of nylon began in 1939, and nylon stockings went on sale to the American public in 1940, selling 64 million pairs in the first twelve months. The rest, as they say, is history.

Ceri Harrop

Date 1935

Scientist Wallace Carothers

Nationality American

Why It's Key Previously non-existent, nylon rocked the organic chemistry world in the 1930s, and continues to feature heavily in the otherwise fickle world of fashion.

opposite Light micrograph of a piece of nylon stocking

Key Experiment
Structure of a virus revealed

In the early part of the twentieth century, viruses were only understood by scientists as infectious agents that cause disease. That view was to change forever in 1935 when American biochemist Wendell Meredith Stanley peered down the eyepiece of a new invention called the electron microscope, and saw the molecular structure of a virus for the very first time.

When Stanley began his research at the Rockefeller Institute in 1931, it was still not clear whether viruses were living organisms or some kind of chemical molecule. Although light microscopes had been available since the early seventeenth century, they were just not powerful enough to view objects as small as viruses; most of which range in size from about 20 to 400 nanometers (1 nanometer = 1 billionth of a meter).

The advent of the electron microscope, which uses beams of high-speed electrons to illuminate the specimen, proved key in finally deciphering the structure of a virus. In 1935, Stanley became a scientific celebrity when he succeeded in crystallizing the tobacco mosaic virus and studying it under an electron microscope. From what he saw, Stanley deduced that a virus was actually made from protein. These findings were later refined, concluding that viruses are made of genetic material enclosed in a protein shell called a capsid. Although Stanley's conclusions were somewhat inaccurate, his work on the tobacco mosaic virus is widely heralded to be the birth of molecular biology, and won win him the Nobel Prize for Chemistry in 1946.

Hannah Isom

Date 1935

Scientist Wendell Stanley

Nationality American

Why It's Key Understanding the structure of viruses was a first step for molecular biology, and electron microscopes gave scientists the power to view specimens never before visible.

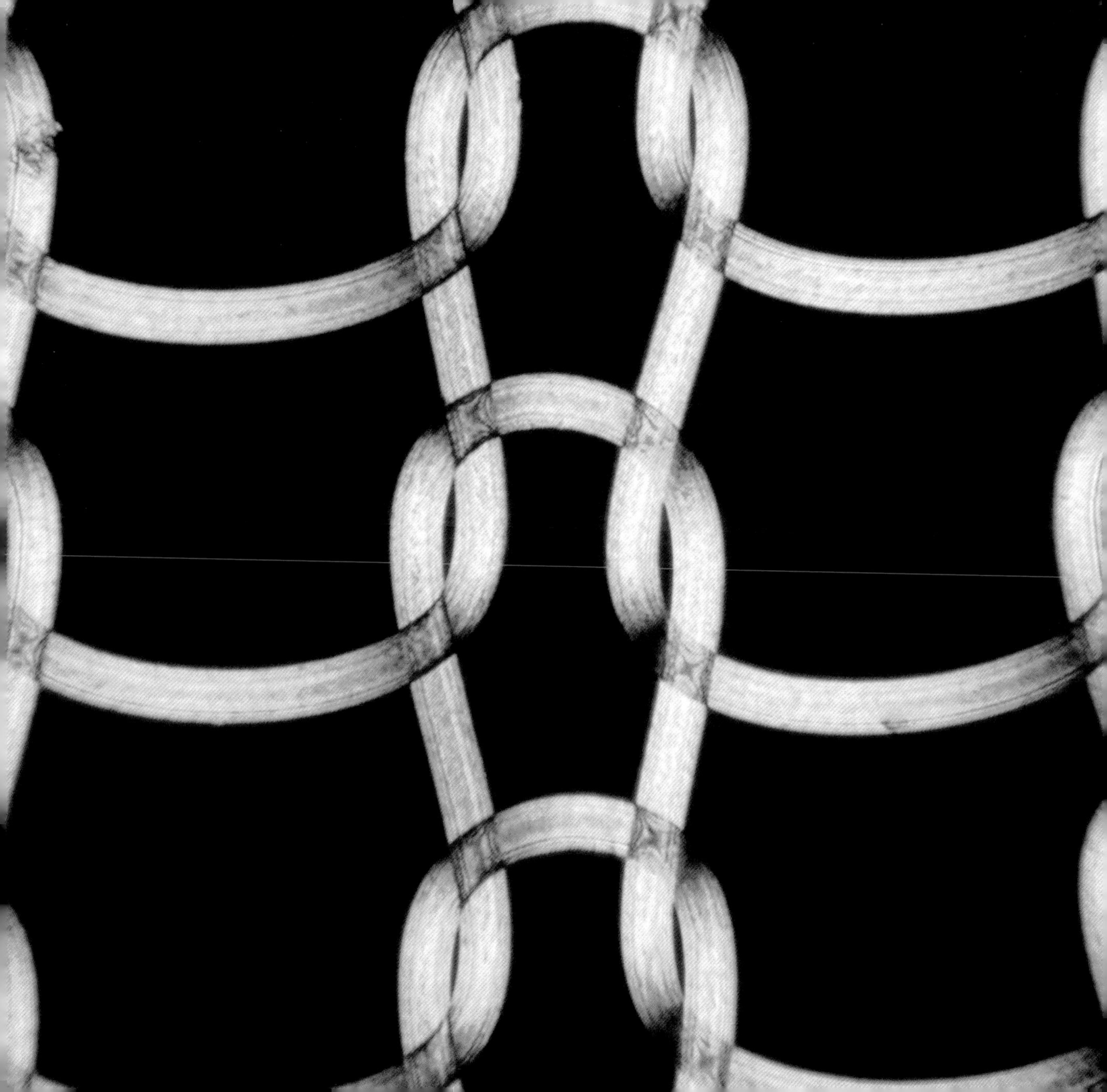

Key Invention
Heart-lung machines

At 8.00 am on October 4, 1930, in Cambridge, Massachusetts, an unconscious woman was rushed into emergency surgery to have a blood clot removed from her lung. The patient had been diagnosed as suffering from a pulmonary embolism and, although the clot was successfully removed, she never regained consciousness. A young surgeon training at Harvard had been assigned to monitor the patient's vital signs over the course of the previous night. To him, the tragedy was to prove inspirational. That surgeon was John Heysham Gibbon, inventor of the heart-lung machine.

John Gibbon reasoned that the woman's life could have been saved if only it had been possible to remove deoxygenated blood from her body, oxygenate it, and return it to circulation. Gibbon resolved to construct a machine for exactly this purpose. By 1933, following a series of experiments on small animals, Gibbon had built a functioning prototype. Assisted by his wife Mary, he undertook further experiments to perfect techniques for the extracorporal circulation of blood. Two years later, he was ready for a full trial.

On May 10, 1935 a cat became the first patient to have all of the functions of its heart and lungs carried out by a machine. The cat's heart was successfully restarted. In 1954, Cecilia Bavolek became the first human patient to successfully undergo the procedure, and thus heart-lung bypass surgery was born.

Jamie F. Lawson

Date 1935:

Scientist John Gibbon

Nationality American

Why It's Key Handing over the functioning of the lungs and heart to a machine has been vital to the development of modern surgical techniques.

Key Discovery **The birth of antibacterial drugs**
For the love of a daughter

In 1935, six-year-old Hildegard Domagk was gravely ill. She had pricked her finger on a knitting needle and developed a serious infection; the only known treatment was to amputate her arm. The little girl, however, was lucky. Her father Gerhard Domagk had been experimenting with a red dye known as "prontosil rubrum." Although the dye exhibited no antimicrobial activity on cultures, he elected to test it in mice anyway and found that the drug cured many bacterial infections. If it worked on mice, maybe it could work on his sick daughter too. Domagk had to give it a go. To his relief, Hildegard made a complete recovery ; the first antibacterial agent to be effective in humans had been discovered.

Prontosil was subsequently used to treat another famous offspring: Franklin Roosevelt, Jr. In late 1936, he developed a sore throat and sinus infection which was successfully treated with Prontosil and a related medication, Protolyn.

Following their early success, antibacterials continued to develop apace. A sulfa drug made by the British drug company, May and Baker, has even been credited with saving Winston Churchill's life. In December 1943, Churchill contracted pneumonia while traveling to Tunis to plan the D-Day landings. The infection was successfully treated with the sulfonamide drug, M&B 693, and some brandy.

Gerhard Domagk was awarded the 1939 Nobel Prize in Medicine. However, Hitler had forbidden German citizens from accepting the prize and Domagk was not able to claim his award until 1947.

Stuart M. Smith

Date 1935

Scientist Gerhard Domagk

Nationality German

Why It's Key Prontosil was the first effective antibacterial; it ushered in modern methods of treating many diseases.

Key Event **Watson-Watt's ministry memo outlines the radar system**

Although Sir Robert Alexander Watson-Watt can not be held entirely accountable for inventing the idea of radar – various prominent scientists had been experimenting with radio waves before him, most notably Heinrich Hertz and Nikola Tesla – in terms of creating the first workable radar system, he is definitely the man to thank.

Watson-Watt started his research working at the UK Air Ministry Meteorological Office, where he developed an idea for a short-wave radio device that could be used to detect thunderstorms. It worked by receiving the radio signals given out by bolts of lighting as they formed and ionized in the sky; this information was then used to warn pilots of oncoming storms.

Then, on February 12, 1935, Watson-Watt sent a memo entitled "Detection and Location of Aircraft by Radio Methods" to the Air Ministry. A 2,000-word document, it detailed how he envisioned reflected radio energy could be used to determine the distance, range, and direction of attacking aircraft from afar, and how urgent it was to start work on basic systems immediately. Watson-Watt's radar system later became key to the outcome of the Battle of Britain in World War II, and his memo has since been hailed as one of the most prophetic scientific documents ever produced.

Chris Lochery

Date 1935

Country UK

Why It's Key Led to the development, creation, and installation of the first working radar units. Not only were they a valuable, war-winning commodity during World War II, but they have since become an integral part of aviation technology.

Key Person **Hideki Yukawa** First Japanese scientist to win the Nobel Prize for Physics

In nature there are really only four fundamental forces: gravity, electromagnetic, and the strong and weak nuclear forces. The strong force is the strongest of all but only acts within the nucleus of an atom, binding together the protons and neutrons that form the nuclei of all matter.

It was not understood how the strong force worked until Japanese physicist Hideki Yukawa (1907–1981) proposed a method in 1935 which required the existence of brand new particles called mesons. These new particles would interact strongly, and provide the mechanism that overcomes the very strong natural repulsion between positively-charged protons, allowing them to bind together with neutrons and form atomic nuclei. With a predicted mass equivalent to around 200 electrons, it was hoped that a meson would be detectable in experiments, proving Yukawa's theory. A false alarm raised hopes in 1935 when a different new particle was discovered in cosmic rays, but this new particle, the "muon," did not show the strong interactions with nuclei.

It wasn't until 1946 that Yukawa's particle was finally seen for the first time by British scientist Cecil Powell, and named the pi meson, or "pion." It was the first major discovery in physics of the postwar era.

In 1949, the Nobel Committee awarded Yukawa their physics prize for his prediction of the pion. He was the first Japanese scientist ever to receive one.

David Hawksett

Date 1935

Nationality Japanese

Why He's Key He predicted the existence of mesons, which are now known to be key subatomic particles.

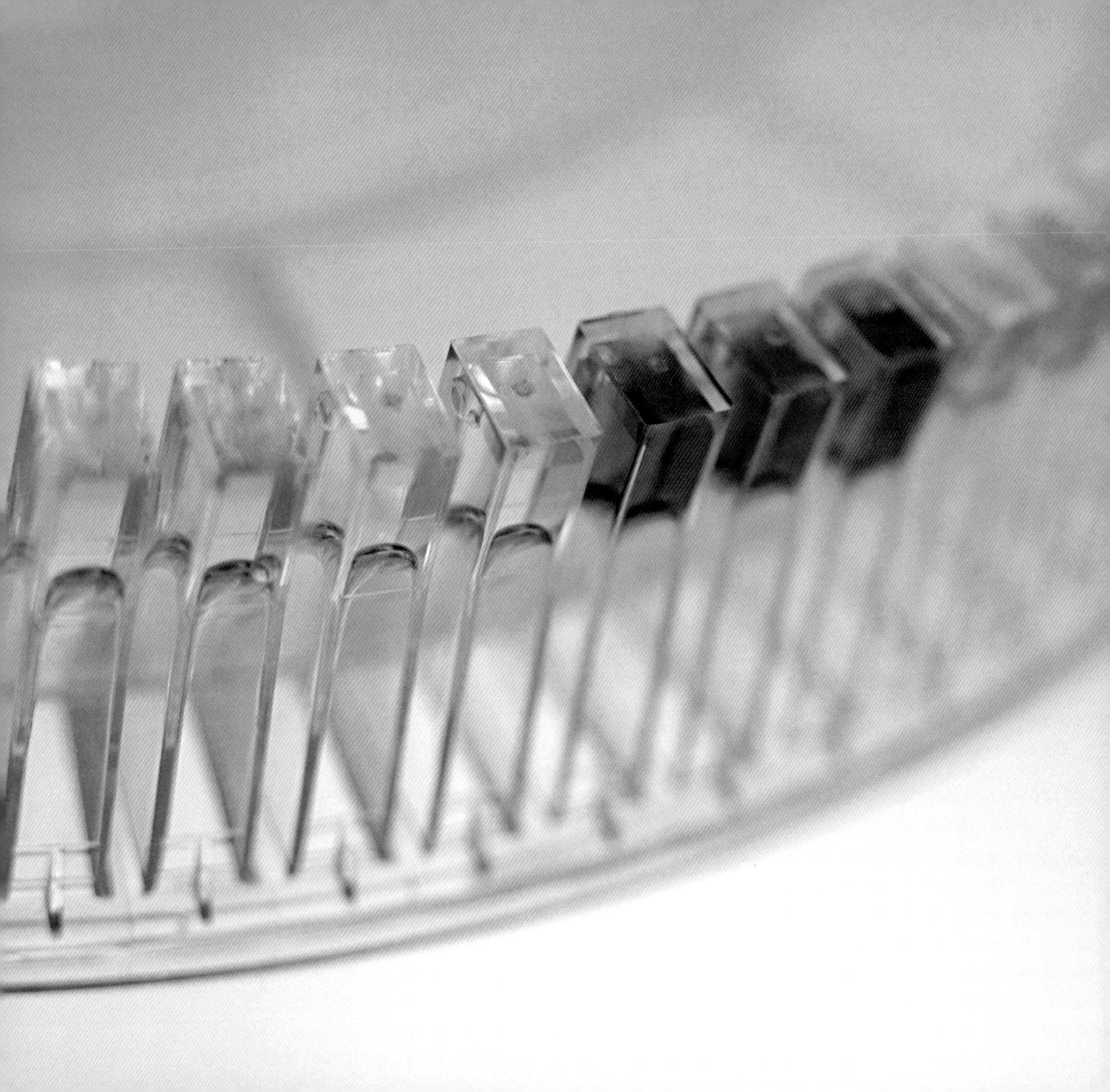

Key Invention **Spectrophotometers**
Seeing the light

Chemical elements absorb and emit light at different wavelengths. The human eye is sensitive to these wavelengths, and variations in them, resulting in the perception of color; we see light produced by sodium, for example, as the yellow-orange glow of streetlights. These different radiation and absorption characteristics form the basis of a method used to identify elements and molecules: spectrophotometry.

Spectrophotometry initially concentrated on the wavelengths of light absorbed by substances in the visible spectrum. A spectrophotometer could therefore be used to identify and compare colors by analyzing the wavelengths of light absorbed by a sample. Early devices required a human observer to decide when a reference color matched that of the sample, which meant that results were often very subjective.

In 1935, professor of physics Arthur C. Hardy patented a spectrophotometer that could distinguish many different colors and also record the results, removing the need for a human observer. This invention was first used in industries such as printing and dying to ensure that colors reproduced consistently, but became a useful tool for identifying biological molecules.

Improvements mean there are now spectrophotometers which can "see" outside of the visible spectrum, including infrared and ultraviolet light. The technology has been invaluable in many different fields of science – for instance, in proving that some birds have plumage that reflects UV light.

Anne-Claire Pawsey

Date 1935

Scientist Arthur C. Hardy

Nationality American

Why It's Key The first automatic and objective method for determining color.

opposite Spectrophotometer cells holding coloured solutions. The cells are placed in a spectrophotometer which measures the optical densities and hence the concentrations of the solutions they contain.

Key Publication
Ecosystems: a brand new concept

Founder of the British Ecological Society and two leading plant journals, Arthur George Tansley ranks among one of the twentieth century's most important ecologists. A lifelong writer and conservation enthusiast, he is probably best remembered for introducing the term "ecosystem" in his landmark scientific paper "The Use and Abuse of Vegetational Concepts and Terms."

The concept became popularized after World War II, and Tansley expanded the term to encompass the interactive system established between groups of living creatures and their environment. This interaction between the living biotic community and their non-living abiotic surroundings is central to the concept of an ecosystem. At a time when ecological studies were specific to individual species, the concept of viewing habitats as a whole was progressive, but the study of the ecosystem as a whole has since grown in importance.

The Convention on Biological Diversity (CBD), which is now supported by the governments of over 180 countries, defines "the protection of ecosystems, natural habitats, and the maintenance of viable populations in natural surroundings" as the basis for environmental protection strategies and conservation projects throughout the world. Looking at the bigger picture is now globally accepted as the best conservation strategy, so Tansley was way ahead of his time in predicting the shift to thinking in terms of ecosystems.

Vicky West

Publication "The Use and Abuse of Vegetational Concepts and Terms"

Date 1935

Author Arthur Tansley

Nationality British

Why It's Key The introduction of the term ecosystem as a concept shifted biologists' focus from single species to communities.

Key Publication **Schrödinger's cat**
Dead and alive at the same time

Two years after becoming a Nobel Prize-winning physicist, Erwin Schrödinger grabbed the attention of the physics world once again, with the publication of a paper now so famous it is merely known as "Schrödinger's cat."

Far from being, as the title might suggest, a piece on the scientist's feline friend, his 1935 essay (actual title "The Present Situation in Quantum Mechanics") is an analogy of a quantum mechanics problem. Schrödinger himself described the problem as a "ridiculous example." A cat is shut in a box with a bottle of poison and a small amount of radioactive material. If one particle of this material decays it will trigger a hammer that will smash the poison and kill the cat.

The question is: how do we know whether the cat is alive or dead without opening the box? The chances of the material decaying are 50:50, so the chances of the cat being alive are 50:50 – it is alive and dead in the box at the same time. As time passes, the probability of decay increases, hence the probability that the cat is dead increases, but we can never know for sure until the box is opened. It is the act of observation which collapses all possible realities into the one we see.

Even though this thought problem was published over seventy years ago, it is still discussed by students and professors alike and referenced in television programs and popular fiction. By linking a complex problem to an everyday animal, Schrödinger made thinking about quantum mechanics amusing and accessible.

Helen MacBain

Publication *The Present Situation in Quantum Mechanics*

Date 1935

Author Erwin Schrödinger

Nationality Austrian

Why It's Key Was and still is a much talked about thought problem concerning the nature of reality.

Key Invention
The Richter scale

The Richter scale is now a familiar term used all over the world. Whenever an earthquake is reported, it is measured in terms of Richter factors, with a step up on the scale being proportional to an increase in seismic wave size. The larger the seismic wave, the more disastrous the quake.

Despite not being published until 1935, the Richter scale was actually invented by Charles Richter in 1933. Working at the California Institute of Technology during that year, he was able to experience first-hand the infamous Long Beach earthquake, which hit along the Newport-Inglewood fault line. The quake, which was also witnessed by Einstein, was recorded by an accelerograph installed earlier that year by the U.S. Coast and Geodetic Survey. Having observed this astonishing event, Richter set about constructing a scale to record future earthquakes. He started by taking the logarithm of the seismic wave height (or amplitude), with an increase of one scale point representing a thirty-fold jump in amplitude. By 1935, his scale was ready for publication, and he soon became a household name.

Thousands of instruments around the world are now calibrated to the Richter scale and used to monitor seismic activity, helping to predict major earthquakes before they happen.

Hannah Welham

Date 1935

Scientist Charles Richter

Nationality American

Why It's Key Provided a means of measuring and monitoring earthquakes, one of the world's most unpredictable natural disasters.

Key Invention
The first canned beer

On January 24, 1935, the first canned beer (Kruger Cream Ale) went on sale in the United States, in a rather flashy looking cone-topped can. It is unclear who purchased the first canned beer, but one would imagine it was very satisfying, as a first beer usually is.

Beer is thought to have been the first ever alcoholic drink, yet there have been relatively few innovations in the design of beer containers over its thousands of years of history. Simple tarred pots were initially replaced with the sealable wooden beer barrel, a mainstay of the beer industry for more than a thousand years. Barrels, however, are not perfect for storage; they need to be kept cold as they are prone to spoiling. Beer bottles have been used for over three hundred years, but bottling is an expensive process.

Nowadays, the beer can, along with its big brother, the beer keg, are now the most common forms of beer container. Canned beer is cheap and easy to mass produce, and has a very long shelf life.

The beer can has also so far resisted any attempts at modernization in the form of plastic containers or cartons. Most contemporary methods of packaging liquid are, in fact, unsuitable for containing beer, as modern beers are generally carbonated and produce too much pressure. Plastics allow too much light and oxygen in, so, for now, the can remains king.

Jim Bell

Date 1935

Scientists Kruger Brewing Co.

Nationality American

Why It's Key It might not have quite the same taste as from a keg, but a canned beer is the cheapest, most portable, and most reliable way to enjoy beery goodness. Cheers!

Key Event **Cane toads introduced into Australia**
From predator to pest

Cane toads were introduced into Australia to control beetles that were destroying large tracts of sugar cane. In hindsight, it was one of the most disastrous attempts at biological control ever conceived.

A hundred cane toads were imported to Queensland from Hawaii in 1935. They were bred in captivity and released into the sugar cane plantations in the north. It wasn't long, however, before farmers noticed that the toads couldn't jump very high and were unable to reach the higher parts of the canes, which were inhabited by the beetles they were supposed to be catching.

Instead, the toads proved to be remarkable pests themselves, for a number of reasons. As well as happily munching their way through many native species, they were also virtually immune to predation. Cane toads have glands on their backs which produce a deadly toxin capable of killing any predator that tries to eat them. Released from the pressures of being eaten while surrounded by a bounty of food for themselves, the Australian cane toad population rocketed. They have now spread across Queensland, most of the Northern Territory, and south along the New South Wales coast. And they show little sign of stopping.

There is a small glimmer of hope, however – some of the local species appear to be learning to deal with the toads. Black kites, for instance, have learned to attack them belly-first, avoiding the poison glands.

Emma Norman

Date 1935

Country Australia

Why It's Key An important lesson about the damage that can be caused by introducing non-native species.

Key Person **The Curies continue** Nobel prizes four and five

Pierre Curie and his wife, Marie, were both awarded the Nobel Prize for Physics in 1903. After her husband's death, Marie was also awarded the Nobel Prize for Chemistry in 1911, becoming the first person twice honored and the only person ever to win prizes in two different sciences. While most families would be sufficiently chuffed with three Nobels between them, there was more to come for the Curies.

Their daughter, Irene, was born in Paris in 1897. She was educated in Paris, and in 1918 joined her mother working at the University of Paris' Institute of Radium, where she met Frédéric Joliot. Joliot, born in 1900, also in Paris, was educated at the School of Industrial Physics and Chemistry. In 1925, he became an assistant to Marie Curie. Irene was asked to teach him the precise laboratory techniques required for radiochemical research, and during this time they fell in love. They married in 1926, formed a scientific super team, and hyphenated their names to Joliot-Curie.

The Joliot-Curies worked on natural and artificial radioactivity – bombarding boron with alpha particles to produce radioactive nitrogen. For their synthesis of new radioactive elements, they received the 1935 Nobel Prize in Chemistry. Irene was appointed Undersecretary of State for Scientific Research, director of the Institute of Radium, and was recipient of many awards. Frédéric was the president of a resistance movement during World War II, and became the director of the Institute of Radium after Irene's death.

Douglas Kitson

Date 1935

Nationality French

Why They're Key Pierre and Marie are the recipients of the fourth and fifth Nobel Prizes in the Curie family. Irene is the only daughter of a Nobel Prize winner to win one herself.

opposite Frédéric and Irene Joliot-Curie at work in a laboratory

Key Person **Edwin "The Ego" Hubble**

Edwin Powell Hubble (1889–1953) was born in Missouri in 1889. At school he displayed no special talent for science; his early academic career was marked instead by sporting prowess. In a single event in 1906, he won seven first places and a third. Hubble studied at the University of Chicago before traveling to Oxford as one of the first Rhodes scholars. He left with a slightly peculiar English accent and mannerisms that would later combine with his infuriating good looks and over inflated ego to exasperate his work colleagues.

After returning to America, Hubble worked as a high school teacher, served in World War I and eventually gained a PhD. Soon after he was offered a position at the Mount Wilson Observatory, where in 1925 he showed that the Andromeda nebula was nearly a million light years away, making it a separate galaxy rather than part of our own. Not content with fundamentally changing our view of the universe just once, Hubble then moved on to the redshift of light observed from distant galaxies.

This suggested that the galaxies were moving away but Edwin noted that the further away a galaxy was, the greater the red shift. The Universe, he concluded, was expanding. This had a profound effect on Einstein, who was able to discard the "corrections" he'd made to his General Theory of Relativity. In 1931 Einstein paid Hubble a visit at the observatory to thank him in person. Hubble spent the last years of his life campaigning to have astronomy considered eligible for the physics Nobel prize. He died in 1953; in accordance with his wishes, no funeral was held.

Helen Potter

Date 1935

Nationality USA

Why He's Key Gave us Hubble's Law, Hubble's constant, the proof of separate galaxies and the concept of the expanding universe. He had good reason for being a bighead.

Key Person **Wallace Carothers**
The father of man-made polymers

Wallace Carothers (1896–1937) was a brilliant chemist. As an undergraduate he was made head of the chemistry department at Tarkio College, Missouri; although, admittedly, this was partly because of a lack of personnel during the war. After becoming a professor at Harvard, he was lured into industry by chemical company DuPont, who had opened a science laboratory to find new products.

In 1928, he headed up a team looking into artificial materials. Carothers was interested in the hydrocarbon acetylene and its family of chemicals. Together with his boss Elmer Bolton, he developed the first synthetic rubber polymer, which DuPont mass-produced as neoprene in 1931.

Shortly afterwards, he turned his attention to developing a man-made synthetic fiber that could replace silk. Japan was the United States' main source of silk, and trade relations between the two countries were breaking down. In 1935, he made significant steps toward a silk-like fiber by reacting the monomers hexamethylene diamine and adipic acid to form a polymer. The reaction formed strong bonds between the individual chemical units, producing water as a by-product. Removing water from the reaction as it formed led to an even stronger fiber. This fiber is known as nylon, introduced to the world in 1938.

Carothers suffered from severe depression. A co-worker once remarked that he could name all of the chemists that had committed suicide. In April 1937, Carothers added his own name to the list, when he took a deadly dose of cyanide.

Riaz Bhunnoo

Date 1935

Nationality American

Why He's Key Carothers developed the synthetic fiber nylon, which underpins the multi-billion pound textile industry. During the war, it replaced silk in parachutes but afterwards found a lucrative market in nylon stockings.

Key Discovery
The structure of vitamin B

The story of vitamin B, also known as thiamine, began with investigations into the cause of the disease, called beriberi, which causes inflammation of the nerves. Scientists originally thought beriberi was caused by toxic bacteria. Following the work of Dutch scientists Christian Eijkman and Garrit Grijns in the late nineteenth century, however, beriberi was found to be cured by a natural factor found in certain foods.

In 1911, Casimir Funk at the Lister Institute in London tried to crystallize the compound, naming it "vitamin," short for "vital amine." The chemical that he isolated, however, turned out not to be the anti-beriberi factor. The substance was finally crystallized from rice bran by Dutch chemists B.C.P. Jansen and W. Donath in 1926, who dubbed it "aneurin." However, Jansen and Donath missed one important element of the substance, a sulfur atom, and their published formula caused confusion for many years.

Around the same time as Jansen and Donath's studies were being performed, American chemist Robert Williams was working toward the same goal of isolating the anti-beriberi factor. In 1934, he successfully isolated one-third of an ounce of the same substance from one ton of rice hulls, in a very expensive and labor-intensive technique. Williams also discovered the sulfur atom in the structure, and the correct chemical structure was finally published in 1936, the same year his laboratory was able to chemically synthesize the material. It was subsequently named "thiamine" in order to reflect the sulfur (thiol) present in the compound.

Rebecca Hernandez

Date 1936

Scientist Robert Williams

Nationality American

Why It's Key Finally solved the mystery of one of man's most important dietary supplements.

opposite Polarized light micrograph of crystals of thiamine, also known as vitamin B1

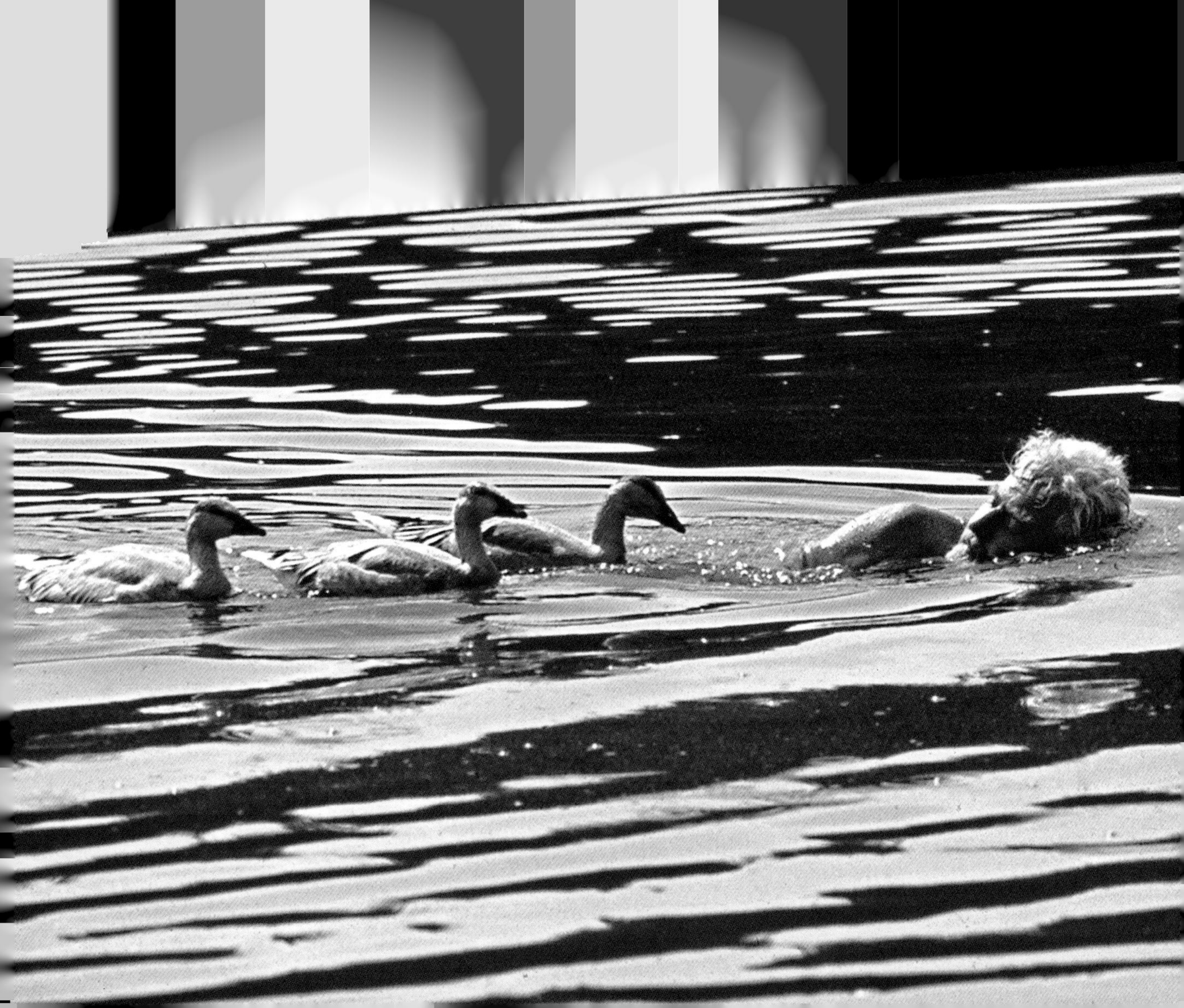

Key Invention
Beckman's pH meter

The pH scale is the universal measure used to quantify the relative acidity and alkalinity of different solutions. From stomach acids to industrial alkalis, it is a hugely important gauge for many everyday processes and reactions.

The pH meter was first invented in 1936 by American Arnold Beckman, and was originally known as "Beckman's acidimeter." Ironically, its development was something of an offshoot from his main project at the time which was to aid the California Fruit Growers Exchange to measure the acidity of their lemon juice.

Prior to 1936, the chemical industry relied on inaccurate color-changing litmus papers (like those you used in school) and complicated measurements of hydrogen ions to get an indication of acidity and alkalinity. The advent of the pH meter provided a much easier method for measuring pH, based on some simple electrochemistry. To use Beckman's original pH meter, a charged probe was simply placed into the solution being tested, where it attracted positive hydrogen ions. The pH meter then measured the amount of these ions. The more hydrogen ions there are in a solution, the more acidic that solution is.

In 2004, the pH meter was named as a National Historic Chemical Landmark by the American Chemical Society. In addition to his hydrogen probe, Beckman is also famous for giving rise to Silicon Valley, having funded the first silicon transistor company.

Hannah Welham

Date 1936

Scientist Arnold Beckman

Nationality American

Why It's Key Revolutionized the accurate and reliable measurement of acidity and alkalinity.

Key Event **Animal instincts**
Tinbergen and Lorenz's new science

Ethology, the study of animal behavior, is scarcely over seventy years old. Although its roots date back centuries, the seminal event in the founding of ethology (from the Greek "ethos," meaning "character") as a discrete field occurred in 1936. It was at the Instinct Symposium in Leiden, Netherlands, that two ornithologists, Konrad Lorenz and Nikolaas "Niko" Tinbergen, met and agreed to collaborate.

Lorenz, an Austrian, had noticed the role instinct had in the development of young animals. Newly hatched goslings, for example, would "imprint" on the first moving object they saw. In some instances, this was Lorenz himself, whom the goslings then treated as though he were their mother.

Tinbergen, a Dutchman, focused on methodology; his "Four Questions" are used by ethologists today as the underpinning of all of their studies. These are:- Which stimuli cause a behavior to occur? How does behavior change with age? How has the behavior evolved? And how does the behavior help an animal survive or reproduce?

Lorenz and Tinbergen's work was interrupted by World War II. Tinbergen spent much of the early 1940s in a German prisoner of war camp, and Lorenz was drafted into the Wehrmacht to serve as a medic, until he was captured and held as a prisoner of war in the Soviet Union. The two reconciled in 1949, after Lorenz's release, and later shared the 1973 Nobel Prize in Physiology or Medicine with Karl von Frisch.

Stuart M. Smith

Date 1936

Country Netherlands

Why It's Key Lorenz and Tinbergen's first meeting led to a collaboration that would revolutionize the study of behavior and pose questions that plague the minds of ethologists to this day.

opposite Zoologist Dr Konrad Lorenz acting as mother to three goslings as they follow him

Key Discovery **Molten Earth**
A hardcore planet

Traveling to the Earth's core is somewhat problematic. The distance and amount of tunneling involved is one thing, but it's the crushing pressures and sweltering temperatures that really put it out of the question. So how on Earth did Inge Lehmann, a Danish scientist, discover in 1936 that the planet's center comprises a solid inner core surrounded by a molten liquid outer core? The answer is: she studied waves.

Seismic waves, generated by earthquakes and explosions, ripple through Earth, and can be measured and recorded using the aptly named seismograph, invented in 1880.

There are two particular kinds of wave that seismologists depend on in order to understand the internal structure of Earth: P (primary) waves, which can penetrate fluids and solids; and S (secondary) waves, which can only pass through solids. Both types bend or reflect, and change speed as they travel, indicating that Earth's interior consists of layers with different densities, both liquid and solid.

Before Lehmann's discovery, seismologists had proposed a molten liquid core, encompassed by a solid mantle and an outer thin crust, all separated by density changes called discontinuities. But when Lehmann studied the shock waves of an earthquake in New Zealand in 1929, she noticed weak P-waves being detected in areas of Europe, which the liquid core theory couldn't explain. She hypothesized that Earth must have a solid inner core with an outer liquid core, separated by what is now called the Lehmann discontinuity.

James Urquhart

Date 1936

Scientist Inge Lehmann

Nationality Danish

Why It's Key Bettered our understanding of Earth's interior, which was necessary for learning more about its formation and the internal processes that govern its evolution.

Key Discovery **"Insecticides kill"**
Shock report

Insecticides are a vital part of today's industrialized world but the synthetic organic compounds used today were only developed in the 1930s, and they have been put to much darker uses than eliminating leaf mites. The most well-known insecticides today are organophosphates. These are derived from phosphoric acid and disrupt the nervous system, interrupting useful functions such as breathing.

Organophosphates are different from other organic insecticides such as DDT (Dichloro-Diphenyl-Trichloroethane) in that they are only weakly soluble in water and don't get stored in body fat. They also don't cause problems with bioaccumulation – the increasing concentration of toxins moving up the food chain – and so are safe to use on crops. In 1936, Gerhard Schrader was looking for new insecticides based on fluorine compounds. He discovered Tabun, which is so poisonous it killed all the insects in his studies. Schrader and his assistant also observed its effects after exposure; these included sweating and tremors .

The discovery was demonstrated to the Nazi government, who saw the military potential and 12,000 tons were produced before the end of World War II. At least ten workers died, including a pipe-fitter who had two liters of Tabun poured on him and died in minutes, despite his protective clothing. Schrader and colleagues went on to produce around 2,000 other organophosphate nerve gases, including Sarin and Soban. Along with Tabun, these are known as G-agents. Organophosphate insecticides are still routinely used across the world as insecticides.

Bill Addison

Date 1936

Scientist Gerhard Schrader

Nationality German

Why It's Key An example of a scientific discovery that was meant to benefit mankind being appropriated by the military to cause harm.

Key Publication **Turing's "Universal Machine"** The precursor to the computer

In his 1936 paper "On Computable Numbers," Alan Turing set out to address the Entscheidungsproblem – the "decision-problem" in mathematics, which asks if there's always a way of telling whether a statement is true or false.

Although the Entscheidungsproblem was a key question, and Turing's paper provided an answer (basically, no, you can't always figure it out), it was the way Turing got to his answer that had the most impact on the real world. Instead of using calculus to reach his solution, Turing described machines which could calculate mathematical answers. "Turing machines" were theoretical devices that would read symbols from a paper tape, and follow their instructions, moving the tape forwards or backwards, erasing symbols, and writing new symbols on the tape.

Turing showed that for a given mathematical problem, a Turing machine could be built to work out the answer. Any question could be asked, even one – such as "what's the value of π?" – that would take for ever to answer, as the hypothetical paper tape was infinitely long.

Turing proved that you could build a "universal machine" capable of doing the job of any other Turing machine, effectively inventing the modern computer. Although Turing machines were theoretical – it's difficult to get hold of an infinitely-long piece of tape – his work laid the foundations for programmable computers. Within ten years of the paper, the first truly programmable computers were built, and Turing's ideas remain the cornerstone of computer science.

Matt Gibson

Publication "On Computable Numbers"

Date 1936

Author Alan Turing

Nationality British

Why It's Key The most important step forward in computer science since Charles Babbage planned his Analytical Engine.

Key Event **First commercial catalytic cracking plant built**

"Cracking" is the name given to the process of breaking up complex strings of molecules, such as the hydrocarbons found in crude oil. In its crude form, oil extracted from the Earth is largely useless to us, until it is refined. Cracking is used to break a chemical chain of, say, twenty carbon atoms into a series of smaller and more reactive chains, altering both its physical and chemical properties.

Catalytic cracking is a variation on this process whereby zeolites – complex, negatively-charged molecules made up of aluminum, silicon, and oxygen atoms – are introduced to the hydrocarbons at a temperature of about 500 degrees Celsius. Using a zeolite catalyst in the cracking process gives a much higher yield of hydrocarbons with chains of between five to ten carbon atoms – perfect for use in gasoline.

The process was invented by a French mechanical engineer named Eugene Houdry, working in collaboration with the American oil companies SoconyVacuum and Sun Oil. A year later, in 1937, the first full-scale Houdry unit was opened at a refinery in Marcus Hook, Pennsylvania, and the timing could scarcely have been better. When war broke out two years later, the demand for high-octane aviation gasoline soared. Compared to other cracking methods, the catalytic cracking process meant that, compared to other cracking methods, twice the amount of useful gasoline could be extracted from each barrel of crude oil.

Chris Lochery

Date 1936

Country USA

Why It's Key Effectively doubled the entire supply of usable gasoline on the planet.

Key Event **Last thylacine dies** Tasmanian Tiger lost forever

The thylacine, also called the Tasmanian tiger or Tasmanian wolf, was a large carnivorous marsupial that went extinct in the twentieth century. They bore a striking resemblance to dogs, with sharp teeth and powerful jaws, despite having totally separate evolutionary paths. They had characteristic dark stripes along the lower half of their back, rump, and at the base of the tail. Being marsupials, females had pouches in which to rear their young.

They could open their mouths to an incredible extent, up to 120 degrees, and their prey included kangaroos, wallabies, wombats, birds, and small animals. They also ate the sheep which had arrived with Western farmers in the woodland and coastal areas of Tasmania they inhabited. This trait did little to endear thylacines to their new neighbors.

Although they went extinct on the Australian mainland thousands of years before European settlers arrived, they survived on Tasmania until the 1930s. Hunting by humans and the introduction of dogs, however, led to a rapid decline in numbers. The last specimen was captured in 1933 and sent to Hobart Zoo, where it lived for three years. It died on September 7, 1936, and was thrown out with the rubbish – even when the last one had died, thylacines still had a rotten, but largely undeserved, reputation in Tasmania.

Many sightings have been made since the 1930s, but despite these, no conclusive evidence has been found to change the "extinct" status. However, many people believe the animal still exists.

David Hall

Date 1936

Country Australia

Why It's Key The largest marsupial carnivore in modern times, and a symbol of Tasmania, was forced into extinction by humans.

opposite A female Tasmanian tiger – the last one to be captured – that died in the old Hobart Zoo, now closed

Key Event **Blood, sweat and tears** Fantus opens the first "blood bank"

The twentieth century was a big century for blood. One of the key discoveries made at the start of the century was that blood could be stored by adding a chemical called sodium citrate – which blocks the blood's natural clotting mechanisms – and putting it in the fridge.

Soon after, the first depots for stored blood were established by the British during World War I. In the early 1930s, several "blood centers" opened in Russian cities, and doctors there published their experience of giving transfusions of stored blood taken from cadavers. Progress in blood transfusion was also m ade in the United States. Among the pioneers was Hungarian-born Dr Bernard Fantus, who established the Cook County Hospital Blood Preservation Laboratory in Chicago in 1935. He spent two years investigating the best way to preserve blood taken from living donors. In 1937, he managed to preserve blood for up to ten days, by using a 2.5 per cent solution of sodium citrate and keeping the blood at 4 degrees Celsius.

On March 15, 1937, Fantus opened his laboratory to the public. He also coined the phrase "blood bank," emphasizing the importance of people not only taking blood from the laboratory, but donating some too if possible. He said: "Just as one cannot draw money from a bank unless one has deposited some, so the blood preservation department cannot supply blood unless as much comes in as goes out."

Christina Giles

Date 1937

Country USA

Why It's Key By coining the term "blood bank," Dr Fantus emphasized the importance of giving and taking in blood transfusion.

Key Discovery **The citric acid cycle**
Powering life

In 1937, Hans Krebs was a well established scientist working in Sheffield, having being kicked out of Germany by the Nazi purge of Jewish scholars. He had already elucidated the urea cycle, but was just about to make the discovery that won him the Nobel Prize for Physiology and Medicine in 1953.

Krebs was measuring the amounts of organic acids generated in pigeon liver and breast muscle, when sugars are burned to yield carbon dioxide, water, and energy. From his observations, he was able to show that there is a cycle of chemical reactions which underpin the production of ATP – the chemical which powers life at the cellular level.

During the citric acid cycle, carbon compounds, derived from the food we eat, interact with an enzyme catalyst called CoA to form acetyl-CoA, an important molecule in metabolism. This is transported in the blood to all the body's cells, where it is combined with an acid, producing citric acid and kicking off a cyclical chain of reactions. One molecule of citric acid ultimately generates twelve molecules of ATP and two molecules of carbon dioxide, which exits the body in the air we breathe out.

Also produced, however, are molecules of the exact same acid that acetyl-CoA reacted with in the first place to form the citric acid. As long as enough carbon is brought in to keep the cycle stocked, the chain of reactions is sustained, and ATP is produced indefinitely.

Catherine Charter

Date 1937

Scientist Hans Krebs

Nationality UK

Why It's Key Finding out how energy was made from food in higher organisms paved the way for many new discoveries in cell metabolism, biochemistry, and molecular biology.

Key Discovery **Cooler than cool**
Superfluidity is discovered

Russian scientist Pyotr Kapitsa, and Canadians John Allen and Donald Misener discovered the strange properties of liquid helium at around the same time in 1937. The prestigious journal *Nature* published their articles back to back a year later.

It had only been thirty years since helium had first been liquefied by Heike Kamerlingh Onnes, and he had already found hints that this strange liquid had some unusual properties. Kapitsa, who had been working on magnetic fields at Cambridge with Ernest Rutherford, became interested in low temperature physics and developed a new method for making large quantities of liquid helium, allowing him to perform a variety of experiments with it.

He found something completely unexpected when he cooled liquid helium down to a chilly 2.17 K (equal to -270.83 degrees Celsius, just 2.17 degrees above absolute zero). At this temperature, atoms have no thermal energy and helium undergoes a phase transition into a liquid with bizarre properties. Viscosity is a liquid's natural resistance to flow – very low for water; much higher for syrup – and ultra cold helium has zero viscosity and infinite thermal conductivity. Kapitsa and his stateside rivals had discovered a "superfluid." If placed in any sealed container, superfluid helium can flow against gravity and coat the container walls in a layer of helium just one atom thick.

Superfluids are now used as quantum solvents and in precision instruments such as gyroscopes. They have even been used to slow the speed of light down to just 17 meters per second.

David Hawksett

Date 1937

Scientists Pyotr Kapitsa, John Allen, Donald Misener

Nationality British-Canadian

Why It's Key Liquids that flow against gravity – magic.

Key Invention **Metal detector**
Beaches no longer just for bathing

A lonely middle-aged man walks a solitary path up and down a wind-swept beach, peering at the ground beneath his feet. But he hasn't gone completely mad, at least not yet; he is merely using his metal detector. This humble piece of technology has become ubiquitous in our modern lives, and can be seen everywhere from airport security checks to schools in downtown Los Angeles.

The inventor of the modern metal detector, Gerhard Fischer, already held the patent for radio direction finders in aircraft. It was while investigating some odd anomalies reported by pilots, in the results of these radio direction finders, that Fischer began to realize the possibility for creating a device that would detect metal. Previous attempts at creating such devices had been made, most notably by Alexander Graham Bell at the end of the nineteenth century, but had met with limited success.

Fischer's was the first metal detector to make use of a radio frequency to find its target. The "metallascope" was, by all accounts, an unsightly looking assemblage of coils and vacuum tubes, but it worked, and found extensive usage among hobbyists and professional organizations.

The use of the metal detector in World War II helped to seal the deal and set "Fisher Lab" up as one of the main manufacturers of metal detecting devices.

Barney Grenfell

Date 1937

Scientist Gerhard Fischer

Nationality German

Why It's Key The metal detector is used in a wide variety of ways; for security, mine-sweeping, and giving geeky dads something to do on a Sunday afternoon.

Key Invention
Xerox machine

In 1937, Chester Carlton was working for a patenting company. He was often troubled by the difficulty of replicating documents, the only two options available to him at the time being expensive and time consuming photography, or re-drawing and re-typing entire documents. Spurred on by a childhood interest in inventing, Carlton tried to find a more efficient way.

The method he developed involved applying an electrostatic charge to a bed of sulfur, which was placed below a document. A bright incandescent light was shone through the document, causing the charged sulfur to be attracted to the ink. When the document was removed an exact replica of the manuscript remained in the sulfur.

As soon as Carlton discovered the process he patented it, worrying that others would try to steal such a useful and easy-to-build machine. His fear was, however, unfounded; Carlton spent the next ten years of his life refining his invention and trying to sell it to companies. Eventually, in 1944, Battelle Memorial Institute became interested in the idea and agreed to sign a contract with Carlton. Three years later they came to an agreement with a small photo-paper company called Haloid – now known as Xerox – who further developed the machine.

The photocopier they created was so successful that in the first six months more units had been sold than they had expected it to sell in its entire lifetime.

Josh Davies

Date 1937

Scientist Chester Carlton

Nationality American

Why It's Key The invention of the photocopier revolutionized administration practices and provided a new source of entertainment at office parties worldwide – photocopying body parts.

Key Event **The Hindenburg tragedy** Death of a behemoth, birth of a mystery

The German Zeppelin airship LZ 129 Hindenburg, three times longer than a Boeing 747 and with a diameter of 41.15 meters, remains the largest aircraft ever to have flown; and the controversy over the cause of its loss is one of the longest running.

On May 6, 1937, at 7.17 pm, as radio reporter Herbert Morrison narrated Hindenburg's arrival through storm-filled skies at Lakehurst, New Jersey, its supply of hydrogen lifting gas ignited. Flames engulfed the ship as passengers and crew leapt for their lives. Thirty-six people died. Morrison's eye witness description of the disaster, which included his despairing (and often parodied), "Oh, the humanity!" has become an iconic piece of audio history.

The cause of the disaster has been repeatedly examined by theorists: Sabotage? Lightning? Were supplies of the safer helium withheld by the U.S. government for political reasons? Did a static spark ignite the unusual (and, it is claimed, violently flammable) painted surface of the airship's skin?

Recent evidence and testimony suggest a leak. The airship became heavier at the back and required aggressive maneuverings to keep steady. This broke an internal bracing wire, which in turn ruptured a hydrogen bag, releasing gas and creating a "fluttering" sound. Minutes later, landing ropes became rain-soaked and conductive. The static electricity caused by the storm now had a route to earth, and the resulting spark was all that was needed to destroy the mighty craft.

Mike Davis

Date 1937

Country USA

Why It's Key The disaster marked the end of public confidence in rigid airships.

opposite One of the victims of the Hindenburg tragedy is carried away as the wreckage burns

Key Person **Hans Krebs** The man who greased the wheels of biochemistry

Hans Krebs was one of the true giants of biochemistry, teasing out details of the most important chemical pathways in living organisms. The son of a surgeon, he was born in Hildesheim, Germany, on August 25, 1900. Krebs followed in his father's footsteps, studying surgery at the universities of Gottingen and Freiburg. After working his way around various German institutions, he settled back in Freiburg where he was to make his first significant discovery.

The urea cycle is a series of biochemical reactions that convert toxic ammonia molecules into the relatively harmless urea, a process that occurs in the livers of all mammals. Krebs worked out the major steps with Kurt Henseleit in 1932.

With the rise of the Nazi Party, Krebs, who was Jewish, decided to leave Germany for England. He worked in Cambridge with Sir Frederick Hopkins who had won the Nobel Prize for Physiology or Medicine in 1929 for his discovery of vitamins. Krebs went on to gain the same distinction in 1953, for discovering many of the reactions and components of the citric acid cycle, commonly called the Krebs cycle.

The Krebs cycle describes how cells generate energy by converting derivatives of sugars, fats and proteins into carbon dioxide and water. Krebs went on to work at Sheffield University, and was knighted in 1958.

Matt Brown

Date 1937

Nationality German

Why He's Key Krebs' insights into cellular pathways are central to our understanding of biochemistry.

Key Person **Franz Boas**
A real "people person"

Franz Boas (1858–1942) was born in Germany to liberal parents who encouraged free-thinking, and he is now generally considered the founder of modern anthropology. He enjoyed the natural sciences, studying mathematics, physics, and geography, and in 1881, completed his doctorate before taking his first expedition to the Arctic. His initial interest in geography, however, lessened after spending time with the indigenous people; from here on in, his attentions turned to anthropology.

Boas emigrated to the United States in 1887 so he could study more freely. His work was fundamentally different to other anthropologists of his day, in that it used a scientific approach to people and their societies. In contrast to the often unproven and racist ideas of his time, he believed understanding of cultures required in-depth study of all aspects of "society."

He did not believe that Western cultures were superior, instead observing that other cultures had become "civilized" in their own way. Boas' ethnocentric analysis involved examining cultures in their own terms, a fundamental change in anthropology. He worked in Columbia University's anthropology department for forty-one years, becoming Professor Emeritus in 1937.

In the United States, he was an ardent campaigner against racism. He celebrated the diversity within different races, and defended fellow scientists who left Nazi Germany. After his death, Boas' work in anthropology lived on, and his scientific principles and methods guide students of the discipline to this day.

Katie Giles

Date 1937

Nationality German

Why He's Key Founder of modern anthropology, and changed the way the West viewed the world.

Key Experiment **The accidental hallucinogen**
Hofman discovers LSD

The process of pharmaceutical research involves taking a chemical which is known to be medically active in some way and making other chemicals of similar structures. A few of these structures produce effects similar to that of the original one. Others produce no effects at all. Some produce totally unrelated effects and research spins off at a tangent.

This was the case for Dr Albert Hofmann, a Swiss chemist working for Sandoz Chemical Company in Basel. He was working on derivatives of a substance called lysergic acid and had produced compounds which were used to induce labor in pregnant women and others which helped with certain geriatric conditions.

In 1938, he produced a compound known as LSD-25 (from the German for lysergic acid diethylamide). This compound was soon discarded because it showed low pharmaceutical activity. Five years later, however, Hofmann decided to make this compound again, and somehow absorbed some through his fingertips. He soon found he had to go home from the laboratory suffering from strange hallucinogenic experiences including fantastic, colourful images.

He decided to undertake a self-experiment to see if the experience was repeated. He swallowed a very small quantity dissolved in water and experienced a similar but more intense episode. Hofmann had discovered a powerful hallucinogen which would go on to be used in psychotherapeutic treatments, for artistic inspiration, and for recreational use, especially during the hippy movement of the 1960s.

Simon Davies

Date 1938

Scientist Albert Hofman

Nationality Swiss

Why It's Key The production of a drug which kick-started the explosion in neuroscience that continues to this day.

Key Discovery
Electroconvulsive therapy

The treatment of severe mental health problems has been transformed over the past hundred years. One crucial, but highly controversial, discovery was electroconvulsive therapy (ECT); the application of an electric current across the scalp to induce a seizure or convulsion. The origins of this treatment lie with a Hungarian doctor. In 1933 Ladislas von Meduna believed that schizophrenia and epilepsy were antagonistic or opposite conditions, and reasoned that inducing convulsions – as seen in epilepsy – could provide a cure for schizophrenia. He induced seizures with chemicals, but the results were unpredictable and the procedure very unpopular with patients.

In 1938, two Italian doctors made a shocking discovery that would transform the treatment of mental health problems. Ugo Cerletti and Lucio Bini realized electricity could be used to induce a convulsion. After extensive animal tests they tried out ECT on newly diagnosed schizophrenics. Not only was ECT much safer and easier to control than chemical convulsants, it caused retrograde amnesia in patients, meaning they didn't remember the events prior to or during the procedure, alleviating fear and anxiety.

ECT is highly controversial and its use has varied throughout the years. A huge backlash against ECT followed its depiction as a form of cruel punishment in the film *One Flew Over the Cuckoo's Nest*, forcing the medical fraternity to tighten restrictions for its use but it is still performed under strict supervision with the use of anesthetics, and only as a last resort.

Katie Giles

Date 1938

Scientists Ugo Cerletti, Lucio Bini

Nationality Italian

Why It's Key An alternative but controversial treatment for individuals with severe mental health problems.

Key Discovery **Echolocation**
Listening in the dark

In 1793, the Italian scientist Lazzaro Spallanzani discovered that blinded bats could avoid obstacles. How they managed to do so remained a mystery until 1938, when Harvard student Donald Griffin was able to listen to the ultrasonic sounds produced by the winged mammals as they flew around. Working with his colleague Robert Galambos, Griffin determined that the bats were echolocating – "seeing" with sound. The discovery was initially treated with profound skepticism.

We now know that dolphins, as well as bats, use echolocation to sense their environment, producing sounds and listening for echoes that bounce off nearby objects. By comparing the outgoing signal with the returning echo, images of the surroundings are produced in the animal's brain. Echolocation allows these animals to move freely in situations where vision is ineffective. Bats use echolocation for orientation and also to detect, localize and identify prey at night: dolphins echolocate in murky conditions underwater.

Echolocation is precise; bats can perceive the position and even texture of objects in three dimensions. Distance is determined from the time delay between sound production and echo reception, while projections of tissue in the ear create interference patterns that allow the bats to calculate elevation. Horizontal angle (azimuth) is computed from differences in the loudness of sounds received by the two ears. Echo strength gives a measure of target size, and peaks and troughs in the echoes' sound waves reveal surface texture.

Gareth Jones

Date 1938

Scientists Donald Griffin, Robert Galambos

Nationality American

Why It's Key Echolocation's signals are exquisitely shaped for sensing the environment – an amazing example of adaptive design by natural selection.

Key Discovery **Scientists hit the evolutionary jackpot** The coelacanth resurfaces

A few days before Christmas 1938, trawler boat captain Hendrik Goosen returned from South Africa. He invited his friend Marjorie Courtenay Latimer, a local museum official, to have a quick look through his catch for anything that might interest her. Lo and behold, she spotted a strange blue fin poking out of the top of the net and quickly realized that she was looking at a fish that ought to have been extinct for 60 million years.

The coelacanth is a remarkable fish, with strange lobed fins that resemble terrestrial animals' limbs. It can grow up to two meters in length, live for a hundred years and, most importantly, has been on the planet for over 400 million years. The fish is often referred to as a "living fossil," because it has remained remarkably unchanged throughout this time. On the whole, species tend to last about two million years, barring any major catastrophic interventions, so to have lasted for over 400 million is quite simply amazing.

But there is much more than longevity to this story. The lobe-finned ancestors of our modern day coelacanth, or "old four legs," may well have been the ancestors of all limbed vertebrates, so the arrival of living examples of these fish gives us an unprecedented insight into evolutionary history.

It took until 1952 for another coelacanth specimen to be caught, but since then many have been found, including a population off the coast of Indonesia, nearly 9,650 kilometers away from the first catch.

Jim Bell

Date 1938

Scientist Hendrik Goosen, Marjorie Courtnay Latimer

Nationality South African

Why It's Key Provided an astounding window onto evolutionary history.

opposite Researchers Looking over a coelacanth

Key Person **Katherine Blodgett** Pioneering invisible glass

In 1926, Katherine Blodgett (1898–1979) became the first woman to be awarded a PhD in Physics from Cambridge University. She was also the first woman employed in General Electric's Research Laboratory in Schenectady, New York, where she worked with Nobel laureate Irving Langmuir on applying single molecule layers to a range of surfaces, including metal and glass.

Despite having up to 35,000 layers one on top of each other, these incredibly thin oily coatings were only as thick as a piece of paper. By developing techniques to lay down thin films on glass, Blodgett was able to radically reduce glare from the reflective surface. In 1938, this led her to develop and patent the process for producing the first totally non-reflective or "invisible" glass. The process became crucial in the development of spectacles, microscopes, telescopes, and camera lenses, and the coating itself became known as a Langmuir-Blodgett film. In World War II, her glass was used in periscopes, range finders, and aerial cameras, and the coatings were used to make airplane wings resistant to ice.

Blodgett also invented the "color gauge" to measure the thickness of these coatings on glass to the nearest one millionth of an inch. The Langmuir-Blodgett technique continues to be used widely, in applications ranging from solar energy conversion to integrated circuit manufacturing. Among her many awards was the designation of June 13, 1951 as "Katherine Blodgett Day," in the city of Schenectady, USA, in commemoration of her scientific and civic contributions.

Richard Bond

Date 1938

Nationality American

Why She's Key A true pioneer who developed techniques that have become standard tools for applying thin films in a wide range of important applications.

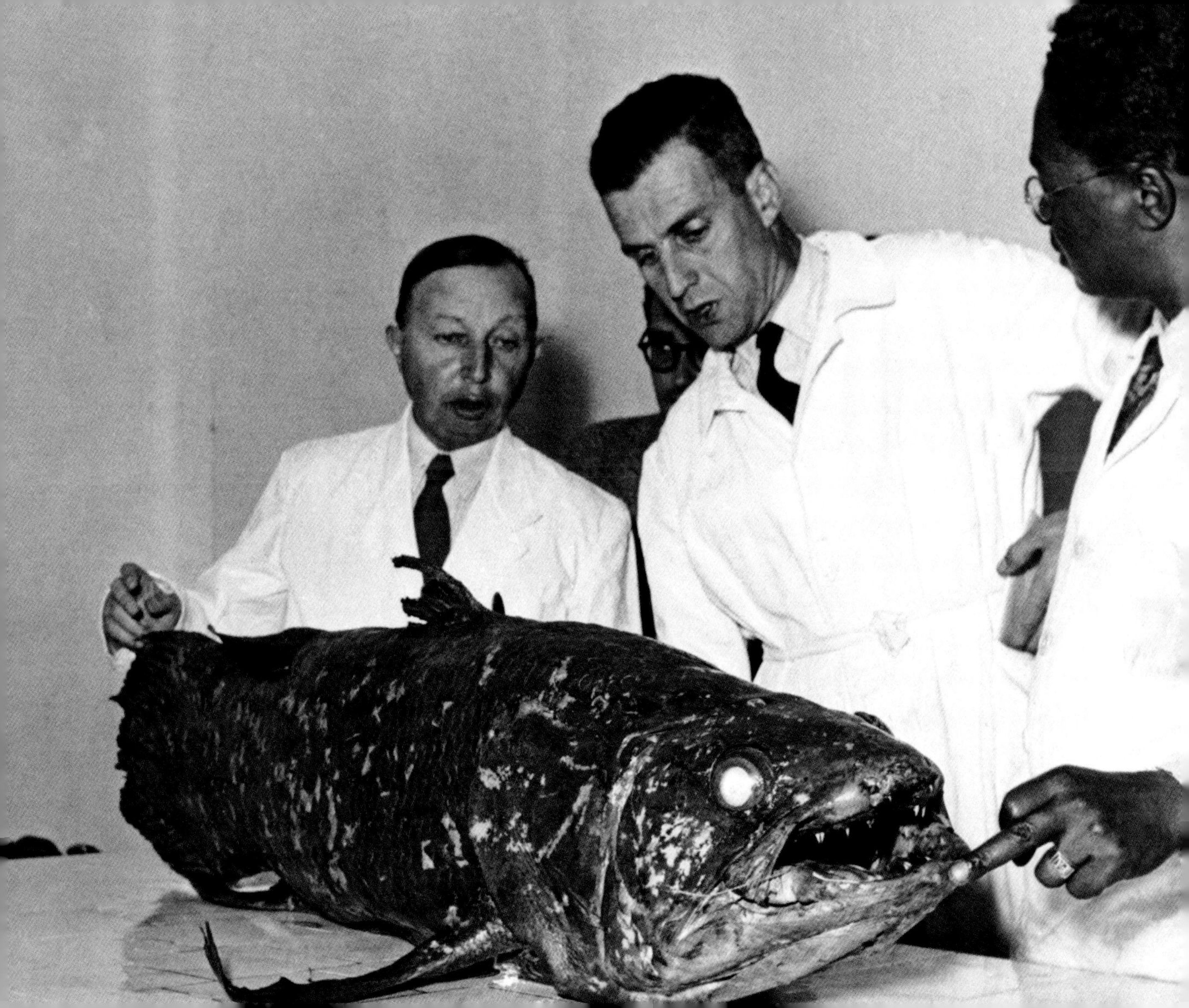

Key Discovery
Teflon discovered by chance

Cooks across the world love it; the space shuttle couldn't fly without it; and even bicycles work better because of it. But Teflon, the non-stick, non-melting, non-reactive lubricant was discovered by one of the great staples of scientific discovery: blind luck.

In 1938, Roy Plunkett was working methodically, synthesizing new forms of the DuPont refrigerant, Freon. During a routine experiment, one cylinder of the gas TFE seemed to have nothing in it. They put it aside, but Plunkett's assistant, Jack Rebok, noticed it was too heavy to be empty. Despite the risk of explosion, they cut open the pressurized cylinder and found an unexpected waxy white powder. Investigation showed this new substance was much more slippery than anything else ever seen, as well as being immune to virtually any other chemical reaction, and having a very high melting point. High pressure had polymerized the gas. In other words, many gas molecules had joined together to form giant molecules; Teflon, in fact, has a molecular weight of around 30,000,000 – one of the heaviest molecules known. Teflon's unique properties were a result of shields of fluorine atoms, which arranged themselves around core strings of carbon atoms, making the material repellent.

Putting aside his other work, Plunkett worked on reproducing this happy accident and by 1941, DuPont had patented the name and the manufacturing process. Today Teflon coats millions of frying pans across the world, and has led to a range of polymers used in everything from airplanes to drug manufacturing.

Bill Addison

Date 1938

Scientist Roy Plunkett, Jack Rebok

Nationality American

Why It's Key Because fried eggs just wouldn't be as (over) easy without it.

opposite Roy Plunkett (right) reenacts the 1938 discovery of the fluorocarbon polymers that led to Teflon.

Key Invention
Sunburn's savior

Whilst climbing Piz Buin – a mountain on the Swiss-Austrian border – in the 1930s, young Austrian chemist Franz Greiter got himself a hefty dose of sunburn.

At the time, little was known about the cancerous dangers of the ultraviolet (UV) radiation absorbed by the skin during exposure to the sun. In fact, pain was the main worry associated with the unfortunate burns. Returning home in an irritated state, Grieter decided to tackle the problem of sunburn head-on, by creating a protective barrier between the skin and the sun.

Whilst developing his protective cream, named Gletscher Crème, Greiter also came up with the idea of the sun protection factor. This scale is now used as the benchmark for different strengths of sun cream. Greiter's original product only managed to reach a factor of 2, while most modern products range from factors 5 to 50, but its popularity grew by the year. By 1946, the cream had formed the basis of the company Piz Buin, named after Greiter's site of inspiration. Since then, his product has gone on to inspire the many different brands of cream we know and use today.

Greiter's invention effectively brought about an optional end to the pain that a day in the sun can cause, and with the dangers of sunburn now fully understood, his work is all the more noteworthy.

Hannah Welham

Date 1938

Scientist Franz Greiter

Nationality Austrian

Why It's Key Protects us against skin cancer and looking like a lobster.

Key Event **The Mallard sets speed record**
The peak of steam

A century after the first section of Isambard Kingdom Brunel's Great Western Railway opened, forever revolutionizing long distance travel, steam trains reached the pinnacle of their powers. On July 3, 1938, the "Mallard" reached 126 miles per hour (200 kilometers per hour) – the fastest speed ever attained by a steam train.

Designed by engineer Nigel Gresley (a keen ornithologist) the Mallard was one of only thirty five Class A4 passenger locomotives ever to be built. Class A4 machines differed from their predecessors with a much more efficient combustion chamber and aerodynamic design. During its record-breaking run, the train, driven by Joseph Duddington, picked up speed down a shallow incline at Stoke Summit in Lincolnshire, hitting 126 miles per hour for a just a few seconds before the speeds caused the engine to start overheating. The attempt was symbolic as, in a time of heightening tensions in Europe, it eclipsed the previous record of 124.5 miles per hour – set by a German train.

The advent of steam locomotion had changed the face of the world, allowing the industrial revolution to power into action as well as making long-distance travel much quicker and much more accessible to a wider range of society – all of a sudden holidays at the coast were no longer the preserve of the wealthy. By the time of the Mallard's historic run, however, the reign of steam was beginning to draw to a close. Diesel engines started to appear on mainline services in the 1920s and gradually pushed steam into the sidings. By the 1970s steam trains were obsolete.

William Scribe

Date 1938

Country UK

Why It's Key This was the pinnacle for steam locomotion, the driving force behind the industrial revolution.

Key Discovery **Giant nuclear reactors**
The fire at the heart of the Sun

Stars like the Sun burn at a temperature of 16 million degrees Celsius for about nine billion years before exploding. This is both hot and long, so they need a pretty efficient source of fuel. In 1938, the physicists Carl von Weizsäcker and Hans Bethe discovered how it is our Sun keeps burning.

In small stars, such as the Sun, the most important process is known as the p-p chain reaction. It's a form of nuclear fusion, where atomic nuclei are fused together. By fusing and fusing and fusing again, nuclei change from hydrogen, to heavy hydrogen, to light helium and finally to full blown helium, kicking out huge amounts of energy along the way. This is the kind of reaction fusion power stations might use in the future.

The other big burning process in the Sun is the CNO (carbon, nitrogen, oxygen) cycle; a process that becomes more common the bigger a star is. As before, hydrogen nuclei are burnt, but this time carbon is used as a catalyst, to help it along. Starting with carbon, adding four hydrogen nuclei, and releasing all sorts of subatomic bits and bobs en route, stars burn through nitrogen, to oxygen and back to carbon, producing a helium nucleus and massive amounts of energy as a result.

Between these two processes our Sun and every other star in the sky are kept burning. It's an understanding of this that has led to many other discoveries about the stars and our Universe.

Douglas Kitson

Date 1938

Scientists Carl Friedrich von Weizsäcker, Hans Bethe

Nationality German, German-American

Why It's Key These are the processes that fuel the star, which produces light, which makes plants grow, which keep everyone on Earth alive. Also, we get suntans.

opposite A solar prominence – cooler dense solar plasma suspended in the hotter thin corona – is seen at the Sun's right

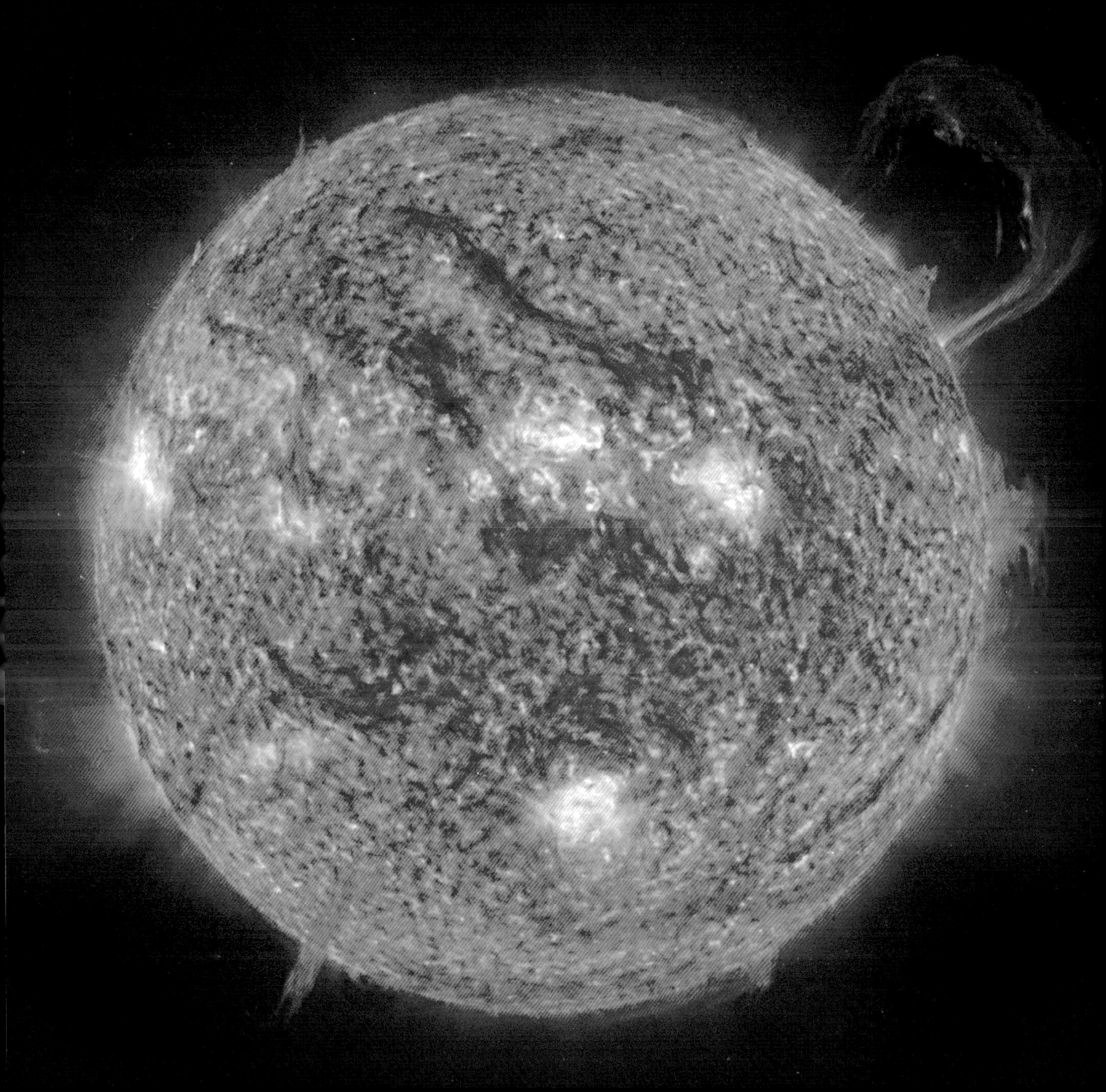

Key Discovery
Meitner, Frisch, and nuclear fission

Just a year before World War II erupted in Europe, Otto Hahn was working on an experiment with Lise Meitner in Germany. They were trying to create elements with heavier atoms than could be found in nature, by firing neutrons at uranium in the hope that they would stick.

In the middle of the experiment, Hitler's purges forced Meitner, who was Jewish, to flee to Sweden. When Hahn came to analyze the results on his own, he found that, far from producing heavier elements they had produced lighter ones – radium and barium. Hahn was puzzled; radium is only a little lighter, but barium has exactly half the atomic mass of uranium. Creating these much lighter atoms would require the neutron to knock around a hundred particles out of the nucleus of uranium, which seemed impossible. Baffled, Hahn sent the results to Meitner. Meitner discussed the puzzling results with her nephew, physicist Otto Frisch. It didn't take them long to work out what was happening. When a neutron was fired at a uranium atom, it was absorbed by the nucleus, making the atom unstable. The easiest way for the atom to become stable again was to split in half, creating two barium atoms. These two atoms had less mass in total than a single uranium atom – the lost mass was converted into energy as described by Einstein's famous equation, $E=mc^2$.

Inadvertently, their experiment had set the wheels in motion for the production of the atomic bomb, and changed the face of war for ever.

Anne-Claire Pawsey

Date 1938

Scientists Otto Hahn, Lise Meitner

Nationality German, Austrian-Swedish

Why It's Key The discovery of nuclear fission led to the development of the atomic bomb and nuclear energy.

opposite Neutrons in a cloud chamber knock on protons and other nuclei, creating short tracks. One neutron has induced fission in a nucleus of uranium coating a thin gold foil. Two fission fragments shoot off sideways, creating long tracks.

Key Discovery **Chain reactions**
DNA to RNA to protein

In the early twentieth century, biologists were baffled by the transfer of genetic material. DNA had been discovered, but was thought to be in short strands only a few base pairs long and with a regular pattern. Its function as the genetic blueprint we know it to be today was unrecognized; scientists had instead cast proteins in its role.

In the 1930s, the Swedish scientist Torbjörn Caspersson was using an ultraviolet (UV) microscope to study the distribution of RNA and DNA in cell structures. The two molecules absorbed UV light of specific frequencies, enabling him to calculate the amount of each molecule in different cells. Collaborating with Jack Schultz, he found that cells making proteins were rich in RNA, implying that RNA must be necessary for protein synthesis.

Together with the independent work of Jean Brachet, his discovery established the movement of RNA from the nucleus to the outer reaches of the cell, more than fifteen years before it was confirmed that DNA in the nucleus carried the genetic information and that RNA was responsible for transferring this information into functional proteins. We now know RNA to be a mixture of "messenger" RNA, which codes for the correct amino acid sequences and "transfer" RNA, responsible for attaching amino acids to the growing protein chain. The trio's results not only successfully applied physical measurements to biological processes, but set molecular biology on the way to its fundamental dogma – "DNA makes RNA makes protein."

Helen Potter

Date 1939

Scientists Torbjörn Caspersson, Jean Brachet, Jack Schultz

Nationality Swedish, Belgian, American

Why It's Key Established the link between RNA and protein in cells, and laid the groundwork for modern molecular biology.

Key Publication
The beginnings of black hole science

Black holes have dominated science fiction more than any other scientific phenomenon. They have provided us with terrifying deaths, opportunities for time travel, and portals to alien worlds. They have also provided scientists with plenty of quirks to busy them.

The concept of a star that trapped light had been discussed since the early nineteenth century. Much like a demanding science fiction fan, however, the scientific community had always remained skeptical. But on September 1, 1939, with the publication of their paper entitled "On Continued Gravitational Contraction," Robert Oppenheimer and Hartland Snyder provided a major breakthrough in understanding the phenomenon that would later be described as a "black hole."

Although the shape and size of a star is determined by the balance between the gravitational force that pulls matter inwards and the thermonuclear energy pushing outwards, Oppenheimer and Snyder explained that "when all thermonuclear sources of energy are exhausted a sufficiently heavy star will collapse." In other words when fuel levels are sufficiently low and the star is sufficiently heavy, gravity will win the battle, collapsing the star into a body so dense not even light can escape its gravitational field.

In 1916, Karl Schwarzschild had already revealed a solution to Einstein's Theory of Relativity which indicated that black holes were theoretically possible. What Oppenheimer and Snyder offered the world was an explanation of how this state could arise. The notion of gravitational collapse of massive stars fuelled a new era in both science and science fiction.

Christopher Booroff

Publication "On Continued Gravitational Contraction"

Date 1939, *Physical Review*

Authors Hartland Snyder, Robert Oppenheimer

Nationality American

Why It's Key Understanding how black holes could be formed provided astronomers with an excellent hint as to where they should be looking for them.

Key Discovery **Plastic fantastic**
The birth of polythene

The chemical giant ICI coined the word plastic in 1927, but it was actually by mistake that polythene was discovered. It followed a laboratory accident in 1933, during an experiment to see how reactions took place under extreme pressure. The new polymer was created when ethylene, a very light gas prepared from petroleum, was placed under high pressure in the presence of oxygen, leaving a waxy solid behind. The event was actually serendipitous for ICI as, following the Global Financial Crash of 1929, they had been forced to let 10 per cent of their staff go. Profits had subsequently fallen dramatically.

Initially, polythene was used to insulate radar cables in submarines. As demand soon began to exceed supply, a production plant was opened in 1939, coincidentally on the same day the Germans invaded Poland at the start of World War II. The discovery of polythene aided the Allied Forces as it was used in improved airborne radars, which helped defeat the U-boats that threatened Britain's food supply in the Battle of the Atlantic.

While polythene was kept top secret during the war, soon afterwards it went into commercial production and today polythene has thousands of commercial and domestic uses; from saran wrap to electrical cables. Polythene was later modified to withstand higher temperatures using the Ziegler catalyst, developed in 1953. This increased the density of the final product; modern-day polythene is now referred to as High Density Polythene.

Katherine Ball

Date 1939

Scientist ICI

Nationality British

Why It's Key Polythene has made a huge difference to the world; aiding the war effort and creating an ultra-usable commercial polymer.

Key Discovery **Francium**
The last of the natural elements

Francium isn't the most famous of elements. Indeed, after astatine it's the second scarcest in nature. Little wonder then, that its discovery took until the eve of World War II, making it the last of the natural elements to be found.

The radioactive species was finally isolated in France (hence the name), in 1939, by Marguerite Perey. She'd taken a technician's position as Marie Curie's personal assistant in 1929, and soon became an expert on the little understood element actinium.

It was while attempting to find the half-life of this element that she made her discovery. A highly pure sample of actinium-227 showed an unexpectedly rapid decay to a particle that behaved chemically like an alkali metal. Perey had spotted the long sought-after element, which had been missing from the foot of Group I on the Periodic Table. After some internal squabbling among her superiors, she was given full credit for the discovery. The element – the 87th in the table – was given its official name in 1949.

Because it is rare and unstable, francium is a useless element as far as practical applications go. Instead, it pulls its weight in the laboratory. Francium has a relatively simple atomic structure, making it a useful testbed for probing electron energy levels. Our understanding of atomic structure and interactions has been much improved thanks to this Gallic latecomer.

Matt Brown

Date 1939

Scientist Marqeurite Perey

Nationality French

Why It's Key It filled the last gap in the Periodic Table, and refined our knowledge of the atom.

Key Discovery **DDT**
A killer and a cure

Ten million civilians died in World War I, most from disease. In 1943, civilians in World War II were not faring much better. Louse-borne typhus had broken out in Naples and the customary measures were proving inadequate. But when the new synthetic insecticide DDT was introduced to take care of the lice, the epidemic was quelled within three weeks.

Initially produced in 1873, it was not until 1939 that Swiss chemist Paul Müller, working on a contact insecticide to protect textiles from moths, rediscovered DDT. Müller found that a wide variety of pests were killed by trace amounts, and the insecticide lasted for weeks. DDT also saw off other pests during the war years. In Asia, mosquitoes were causing malaria; General MacArthur reported in the early stages of the war that two thirds of his troops in the South Pacific were sick with the disease. Unless something was done, MacArthur complained, it would be "a long war." DDT was that "something."

In 1948, Müller won the Nobel Prize for his work with DDT and his discovery became the centerpiece of one of the most ambitious public-health campaigns in history: the Global Malaria Eradication Program. Between 1945 and 1965 the World Health Organization effort led by Fred Soper helped eliminate malaria from the developed world and saved tens of millions of lives.

Due to decreased funding, however, development of resistance, and concerns of the environmental impact of DDT, the World Health Organization formally abandoned global malaria eradication in 1969 and by the mid-1970s, most of the world had rejected DDT.

Stuart M. Smith

Date 1939

Scientist Paul Muller

Nationality Swiss

Why It's Key Saved millions of lives; may have inadvertently started a modern environmental movement.

Key Invention **Helicopter**
Sikorsky achieves da Vinci's great vision

The Sikorsky VS-300 was the first machine that would be recognizable to us today as a helicopter. By the late 1930s, vertical-flight machines had been at the cutting edge of aeronautical research for three decades or so, but the production of a practically useful helicopter was proving difficult.

The idea for a vertical-flight machine is thought to date back to an ancient Chinese child's toy, which inspired Leonardo da Vinci to create what he called an "aerial screw." Sadly the design was not practical for powered human flight, and it was not until the advent of the internal combustion engine that a suitable power source became available. The first autogyro, similar to a helicopter but with a single un-powered rotor, was built by Juan de la Cierva in 1923. It was, however, unable to take off vertically.

To prevent early helicopters from flipping over and crashing, the designs typically incorporated two rotors to balance the machine. A functional helicopter with two sets of rotors was developed by the Germans during World War II, but it was Igor Sikorsky who developed the first functional single rotor helicopter. The VS-300 flew tethered in 1939, and untethered in 1940, and was developed into the first ever mass producible helicopter, the R-4.

As well as being a pioneering designer, Sikorsky was a pilot, and carried out test flights of many of his inventions personally. Despite wearing only a protective bowler hat and overcoat during these flights, he lived to the ripe old age of eighty-three.
Jim Bell

Date 1939

Scientist Igor Sikorsky

Nationality Soviet-American

Why It's Key Helicopters are incredibly useful machines for reaching remote areas that aircraft cannot access. Vertical flight was a defining moment in aeronautical engineering history.

opposite Igor Sikorsky making a helicopter flight

Key Event
Einstein writes to President Roosevelt

On August 2, 1939, Albert Einstein made what he later described as the "greatest mistake" of his life, for it was on this day that he signed and sent a letter to United States President Franklin Roosevelt.

The letter, composed mostly by Einstein's Hungarian colleague Leo Szilard, informed Roosevelt of recent research which had revealed the possibility of creating a nuclear chain reaction with a large amount of uranium. The letter went on to explain that the chain reaction would lead to the production of new elements and also generate a vast quantity of power which, if incorporated into a bomb, would cause widespread devastation. Furthermore, the letter described how Germany, who had discontinued exporting uranium mined from occupied Czechoslovakia, might also be developing atomic weapons.

The content of the letter prompted Roosevelt to form a committee dedicated to nuclear chain reaction research. Out of this committee, the Manhattan Project was born, which focused on weapon development. Ultimately, the weapons led to the deaths of 220,000 people when the "Little Boy" and "Fat Man" bombs, containing the technology discussed in Einstein's letter, were dropped on the Japanese cities of Hiroshima and Nagasaki. Einstein later regretted sending the letter to the President, forever feeling responsible for causing such a catastrophe six years after his communication.
Gavin Hammond

Date 1939

Country USA

Why It's Key Einstein's letter prompted the development of atomic weapons by the U.S. government, leading to massive destruction in Japan and the eventual end of World War II.

Key Discovery **Neptunium** A new addition to the Periodic Table

One institute stands out in the creation of new elements. This is the Berkeley Radiation Laboratory at the University of California. The Laboratory has discovered more than a dozen transuranic elements and, in the process, created a staggering eleven Nobel Prize winners. No great surprise then that it was here that the first transuranic element was discovered.

Elements 1 to 92 on the Periodic Table are all, with a few exceptions, easily found in reasonable quantities on Earth, but the Periodic Table is awash with other elements – those with atomic numbers higher than that of uranium (atomic number 92). These are called the transuranic elements. Of these elements only two occur naturally on Earth, and these in very small amounts. Any significant quantities of these elements have to be synthesized. In 1940, two young men from the recently relocated Berkeley Radiation Laboratory, Edwin McMillan and Philip Abelson, were experimenting with bombarding uranium with slow-moving neutrons. The product of this process was a new element with atomic number 93. It was named neptunium, Neptune being the next planet in line after Uranus.

A radioactive, silvery, metallic element, neptunium's most significant practical application is as a precursor in the production of plutonium, the next transuranic element in line. plutonium was first synthesized in the same year as neptunium, just a few short months later. It was for this that McMillan – with Glenn Seaborg – won the Nobel Prize.

Barney Grenfell

Date 1940

Scientists Edwin McMillan, Philip Abelson

Nationality American

Why It's Key McMillan and Albertson's discovery, and the process by which they made it, paved the way for the production of many more transuranic elements such as plutonium.

Key Event **Geneticist Nikolai Vavilov is sentenced to death**

In August 1940, the eminent botanist and geneticist Nikolai Vavilov was sentenced to death for his belief in a scientific theory. Vavilov's research career was dedicated to the improvement of crops. He aspired to eliminate the famine that often affected his country – the Soviet Union. His research brought him acclaim all over the world, including the Soviet Union where, in 1930, he became director of the Institute of Genetics.

During the 1930s, governmental support for his research was withdrawn due to the Communist state's backing of the biologist Trofim Lysenko. Lysenko rejected the theories of modern genetics that formed the firm foundations of Vavilov's work, and turned his own ideas into a powerful scientific movement – Lysenkoism. Lysenko was a master of spin, offering fast practical solutions to agricultural problems which he claimed academics would procrastinate over. He failed to deliver on his promises, largely because his assertions were not based on sound scientific investigations. His success continued, because, unlike most scientists at the time Lysenko was from a peasant family. Embodying the Soviet ideal of the "peasant genius," he was the propaganda machine's dream. His popularity afforded him a platform from where he denounced genetic theory and its proponents, labeling them "wreckers of agriculture" and orchestrating their imprisonment.

Vavilov remained an opponent of Lysenko's pseudo-science, a decision that resulted in his death sentence. Although commuted to twenty years behind bars Lysenko starved to death in prison 1943 aged 55.

Becky Poole

Date 1940

Country USSR

Why It's Key The ultimate illustration of how politics can affect the scientific process.

Key Invention
Automatic transmission

Though the purist will claim that it gives you less feel for the road; the frugal will claim that it is an unnecessary luxury; and everyone's father will claim that you have to know how to drive a "stick shift", for a lot of people, the invention of the automatic transmission was a fantastic thing.

In 1940, General Motor's Oldsmobile Division introduced the first mass produced automatic transmission with great fanfare, making it easier for drivers everywhere to drive and drink coffee at the same time. The device, called the Hydra-matic, gave the driver a choice of neutral, reverse, drive, and low, and meant that they no longer had to worry about using their left foot to work the clutch pedal.

Rather than having to shift gears manually, drivers need only had depress the accelerator. As the speed of the car increases, the car automatically shifts gears upwards. To help this along, the engine shaft and drive shaft are mechanically linked through a fluid-filled coupling. The transmission automatically attains the appropriate gear ratio and establishes a smoother and more efficient ride. As the car slows down, the Hydra-matic performs just the opposite way, shifting from high gear back down to low.

And how much did this "automatic, systematic, Hydra-matic" wonder cost when it was first introduced? Just US$57.

B. James McCallum

Date 1940

Scientists General Motors

Nationality American

Why It's Key The automatic transmission made driving much more convenient, and arguably meant that you only need one hand to do it.

Key Invention The cavity magnetron
Radars and ready meals

The cavity magnetron might sound like something straight out of a science fiction film, but it is an invention that changed the course of World War II and is now used in almost every home, to heat food.

When war broke out in 1939, the British were desperate to develop their radar technology. Radar at this time was based on long radio waves and although it worked, it used very large antenna and so was not portable. Scientists needed to find a way to produce waves of a much shorter wavelength but with just as much power.

John Randall and Harry Boot were scientists at the University of Birmingham, UK. They were working with a piece of apparatus called a magnetron, based on the idea that electrons emitted by a hot filament wire could be accelerated by a magnetic field. They built a new type of magnetron with sets of holes called cavities. When electrons flowed over the cavities, short wavelength radio waves mere produced. These waves were called microwaves.

Randall and Boot's new fangled magnetron was soon being used by the Allied forces to detect the approach of airplanes, ships, and even submarine periscopes at a range of almost ten kilometers. Since then, other uses of microwaves have been developed, including the ubiquitous microwave oven.

Simon Davies

Date 1940

Scientists John Randall, Harry Boot

Nationality British

Why It's Key An invention, now used in the war on fresh food, which gave the Allies the upper hand in the Battle of Britain and other turning points of World War II.

Key Person **Kurt Gödel**
Living logic to the last

Born in 1906 in what is now the Czech Republic, Kurt Gödel (1906–1978) was one of the many scientists who left for America to avoid the Nazi regime. Gödel's work included his Incompleteness Theorem which showed that there will always be some mathematical statements that cannot be proven either true or false; taking away mathematical naivety in a way. This was only a theory and therefore, by his own rationale, he couldn't prove it. He also elaborated on documenting proof of the existence of God from original work by Leibniz.

A close friend of Einstein, Gödel even caused the great man to doubt his Theory of General Relativity by raising the issue of paradoxical solutions that could lead to time travel. Late in Einstein's life he said that he only went to the Institute of Advanced Studies to have the privilege of walking home with this mathematician.

His work on logic had a great impact on scientific and philosophical thinking in the twentieth century and in later life, Gödel became almost distrustful of common sense as a way to arrive at the truth. He was paranoid of being poisoned and would not eat a thing unless his wife had tasted it first. When she was taken ill and unable to do this he refused to eat until he eventually starved to death. For a man who had such a fundamental grasp of logic, it was a very sad way to go.

Leila Sattary

Date 1940

Nationality Czech

Why He's Key His work had large impact on scientific and philosophic thinking, particularly his theory that not all theories can be proved.

Key Discovery **Plasma transfusions**
"Blood for Britain"

Blood transfusions have been performed in some form or another for hundreds of years, but the practice has always been riddled with problems. Two major difficulties with transfusions were troubling Dr Charles Drew during the late 1930s. Firstly, the red cells perished quickly, so blood could only be stored for about a week. Secondly, the unique markers on the surface of the red cells in blood needed careful matching between donor and recipient – unless a match was found, the patient could die, but finding a match took up precious time.

Drew wanted to find a substitute for whole blood. In 1940, he devised a method for separating the red cells from the watery plasma, by spinning whole blood extremely quickly in a centrifuge. The red cells sank to the bottom, leaving straw-colored plasma above, which could be tapped off and stored. This solved problems in many of the cases where whole blood transfusions had previously been needed. Plasma lasted longer so could be transported over greater distances, and the lack of red cells meant there was no need to match blood types, saving time. Plasma could be transfused into anyone, therefore providing instant help for people suffering massive blood loss who might go into fatal shock.

During World War II, Drew spearheaded the "Blood for Britain" campaign, in which huge quantities of plasma were sent in response to the blood shortage. Plasma banks were even set up on battleships. Countless lives were saved as a result, proving that sometimes less is definitely more.

S. Maria Hampshire

Date 1940

Scientist Charles Drew

Nationality American

Why It's Key The production of plasma – blood without the red cells – revolutionized the collection, storage, and distribution of blood products, eventually leading to the elaborate system used worldwide today.

Key Publication
One gene, one enzyme hypothesis

In 1941, Edward Tatum and George Beadle answered the question that had occupied biologists since the turn of the twentieth century: what do genes do?

At first it might not seem an obvious choice, but to answer this question Tatum and Beadle turned to a bread mold. From their previous work on fruit flies they had an idea that genes were involved in metabolism and this mold was the perfect subject in which to investigate the link further. Bread mold could metabolize a very simple food source to produce all of the nutrients it required to grow and methods to look at its genetics were already established.

Assuming X-rays would cause mutations in genes, they bombarded the mold with X-rays and produced a mutant that could no longer grow on the simple food source. However, with the addition of just one nutrient, the mutant mold was happy again and grew normally. They demonstrated the mutation had occurred in one gene, which had caused a change in just one enzyme. From this they proposed the "one gene, one enzyme" hypothesis in their 1941 paper "Genetic Control of Biochemical Reactions in Neurospora." This hypothesis forms a tenet upon which molecular biology is founded, despite its having evolved in the intervening years.

The magnitude of this achievement cannot be overstated. They proved that genes are distinct units that carry the information to make enzymes at a time when there wasn't even a consensus about the physical nature of a gene. That they were able to do this is simply mind boggling.

Becky Poole

Publication "Genetic Control of Biochemical Reactions in Neurospora"

Date 1941

Authors Edward Tatum, George Beadle

Nationality American

Why It's Key This physical definition of a gene was a massive leap forward for science and opened the door for molecular biology, without which gene function would still be a mystery.

Key Discovery **Rhesus factors**
Are you positive you know your blood group?

Most people know their blood type, be it A, B, O, or AB. Most people remember that there is a plus sign or negative sign associated with that letter, but many do not realize that the positive or negative sign has nothing to do with the rest of the designation.

Karl Landsteiner revolutionized blood transfusion in the early 1900s when he figured out that only transfusions between people with like blood types would result in a good outcome for the patient. But that wasn't the end of the story and, in 1940, Landsteiner and Dr Alexander Weiner announced that they had discovered a second important factor in the blood.

They had found that antibodies produced by rabbits, in response to an injection of monkey blood, caused some people's blood to clot. The two scientists concluded that these people must have an antigen in their blood similar to one found in the monkeys, whilst other people lacked the chemical. Since they had taken blood from Rhesus macaques, they dubbed this new finding "Rh," or the "Rhesus factor." People whose blood clotted were Rh-positive whereas the others were Rh-negative.

The discovery was good news for anybody needing a transfusion, but it was especially great news for babies. As it turns out, the new blood antigen was responsible for a condition known as erythroblastosis fetalis, a disease in which an Rh-negative mother produces antibodies which attack her Rh-positive child. Recognition of this antibody as a cause has led to effective treatments for this disease.

B. James McCallum

Date 1940

Scientist Karl Landsteiner

Nationality American

Why It's Key The discovery of the Rhesus factor further refined the science of blood transfusion, and saved countless babies' lives.

Key Invention
The Atanasoff-Berry computer

It seems almost poetic that the two people who would come to invent the world's first electronic digital computer be named John Vincent Atanasoff and Clifford Berry, because it means that the digital revolution started, like so many in things in life, with the ABC.

Built at the Iowa State University, the Atanasoff-Berry Computer was a calculation device capable of solving systems of up to twenty-nine simultaneous linear equations, with twenty-nine variables and one constant. Though this may not seem like much of a computer to our advanced sensibilities, it introduced the ideas of binary arithmetic, regenerative memory, parallel processing, and separation of memory and computing functions – all of which place it firmly atop the family tree of all of today's personal computers.

Though the machine was prototyped in 1939, Atanasoff wasn't fully credited with the idea until much later, in 1973. Instead, two other scientists, J. Presper Eckert and John W. Mauchly, invented a device known as the ENIAC, which drew heavily on Atanasoff's research. They claimed to have come up with the idea over ice-cream and coffee in a restaurant in Philadelphia.

According to a decision made by a U.S. District Court, however, the truth was that Atanasoff had come up with the idea over bourbon and water in a tavern in Illinois, but was called away at the start of World War II to help the Naval Forces, before he had chance to file a patent for it.

Chris Lochery

Date 1941

Scientists John Vincent Atanasoff, Clifford Berry

Nationality American

Why It's Key Marked the birth of modern digital computing as we know it today.

opposite An early version of the Atanasoff-Berry computer

Key Discovery
Plutonium back on the scene

Plutonium hadn't been seen on Earth since our planet first formed over 4.5 billion years ago. Produced in stars and thrown out in supernovae, all plutonium had decayed into more stable elements long before humans came into existence. But chemists knew that, if element 94 could be produced again, it should be able to start a chain nuclear reaction for use in weapons and nuclear reactors. One kilogram of plutonium would be able to generate 22 million kilowatt hours of energy, or the equivalent of 20 million kilograms of chemical explosives.

Producing plutonium involves bombarding uranium-238 with deuterons, a combination of a proton and a neutron. This process was first successfully carried out in 1941, and the new radioactive element was given the designation "Pu" as a joke on the part of lead researcher, Glenn Seaborg, who suggested it as reminiscent of the "Pee – yew!" noise we sometimes make when smelling something bad. To his surprise, the code was accepted and no comment was made.

With World War II raging, the discovery was kept secret while chemists and physicists frantically studied its properties. Plutonium is chemically very complicated and highly radioactive; it's not as toxic, however, as it's sometimes reported to be – no more so than lead. When the element finally made its appearance on the world stage, it was a dramatic event to say the least: It was a plutonium bomb that detonated above Nagasaki in August 1945.

Kate Oliver

Date 1941

Scientist Glenn Seaborg

Nationality American

Why It's Key A newly recreated element which offered novel ways to blow people to bits, irradiate things, or power millions of houses; take your pick.

Key Event **The Manhattan Project**
Destroyer of worlds, bringer of light

Born in 1941, when President Franklin D. Roosevelt gave the U.S. Army Corps of Engineers' Manhattan Engineering District unit supervisory control over the construction of an atomic bomb, the "Manhattan Project" is an epic tale of human endeavor. It was developed at numerous locations, including a 428,000-acre complex in Los Alamos, New Mexico.

Following a British report on the feasibility of building an atomic bomb, and fearful that the Nazi regime in Germany had already begun such a project, the USA, UK, and Canada recruited scientists, technical experts, and skilled workers from around the world to secretly develop such a weapon. Ultimately there would be two: One uranium-based, and one plutonium-based. Under the direction of General Leslie R. Groves, Deputy Chief of Construction of the U.S. Army Corps of Engineers, and physicist Robert Oppenheimer, research and testing took place at breathtaking speed. To maintain this momentum, unprecedented levels of funding were made available – some US$2 billion in total.

But as the scientists' understanding of nuclear fission developed, so concerns grew. Would such a weapon actually work? Would it, as another project physicist, Enrico Fermi, suggested, set fire to the Earth's atmosphere and create huge conflagrations around the world? On July 16, 1945, at 5.29 am, the world's first atomic bomb test – code named "Trinity" – took place at Alamogordo in the New Mexico desert. A month later, atomic bombs would be dropped on Hiroshima and Nagasaki.

Mike Davis

Date 1941

Country USA

Why It's Key Accelerated developments in the use of nuclear energy for both military and peaceful purposes and, arguably, shortened World War II.

Key Experiment **ATP proposed as cellular currency**
An energy transferring molecule

ATP, or adenosine triphosphate, packs a powerful punch for a collection of just a few dozen atoms. This small molecule is the universal fuel that keeps all organisms ticking over at the molecular level.

It's formed from a couple of chemical rings known as adenine, a ribose sugar group, and a trio of phosphate groups. The molecule is relatively unstable and will readily lose one of its phosphate groups to form adenosine diphosphate (ADP). Now, breaking a bond releases energy. And the phosphate bond is easily broken. Thus ATP is an efficient source of energy that can be channeled into many cellular processes from muscle contraction to protein synthesis. The molecular battery is recharged with a new phosphorous group when energy is put back in through photosynthesis or glucose breakdown.

German-American biochemist Fritz-Albert Lipmann was the man who first proposed ATP as the main energy transferring molecule of the cell. In 1939, with the growing threat of Nazi fascism, Lipmann moved from Copenhagen to Cornell, USA, where he continued studies on the breakdown of glucose by bacteria. He noticed that certain oxidation reactions would not occur unless phosphate was present. He hit upon ATP, previously suspected as an energy source for muscle, as the active agent in cellular energy transfer. His suggestion was soon verified by biochemists.

Lipmann won the 1953 Nobel Prize in Physiology or Medicine for his later discovery of coenzyme A, another important constituent of the cellular economy.

Matt Brown

Date 1941

Scientist Fritz-Albert Lipmann

Nationality Dutch

Why It's Key Understanding how ATP powers the cell was an important step in untangling the complex web of biochemical processes.

Key Event **Yellow fever vaccine causes hepatitis in thousands**

Viral hepatitis causes an estimated 1.5 million deaths each year worldwide. Although its symptoms have been known for centuries, prior to the 1940s no one was certain how it was spread. When some 50,000 U.S. soldiers who had been vaccinated for yellow fever developed "icteric hepatitis," British physician F.O. MacCallum realized that the vaccine must somehow be to blame.

The jabs had been developed by Max Theiler in the 1930s at the Rockefeller Institute, and had been used with great success. Theiler had developed a weakened strain of yellow fever, which he grew in eggs. However, human serum had been added to the vaccine to stabilize it for transport. MacCallum realized that it was the human serum that must have contained another virus, which he later named hepatitis B.

In the 1960s, Baruch Blumberg – along with Harvey Alter, Alfred Prince, and others – isolated the virus and developed a way of testing for it.

The discovery that blood and blood products could spread hepatitis revolutionized transfusion medicine. In the early 1970s, the risk of contracting viral hepatitis from blood was as high as 8 per cent, per unit transfused. Today, with the advent of widespread testing, the risk of contracting hepatitis B is 1 in 250,000, and the risk of contracting hepatitis C is 1 in 2 million, per unit transfused.

Stuart M. Smith

Date 1942

Country UK

Why It's Key Identified the cause of a newly identified disease and led to safer blood transfusions.

Key Experiment **Chain reaction**
Enrico Fermi and the dawn of the nuclear age

When Enrico Fermi left Italy in 1938 to receive a Nobel Prize for his work on radioactivity, he never returned, moving instead to America, to escape Italy's rising fascism. Fermi was one of the scientists who realized that the splitting of the atom in nuclear fission would release colossal amounts of energy.

In 1939, US President Franklin D. Roosevelt received a letter from Albert Einstein and Leo Szilard warning that the Nazis were potentially working on a bomb using nuclear fission. Roosevelt was alarmed enough to create the Advisory Committee on Uranium. This committee not only led the Manhattan Project to develop a nuclear bomb, but also funded Fermi's work at the University of Chicago.

In a squash court under the university's football field, Fermi built the world's first nuclear reactor. Chicago Pile-1 was a massive "pile" of uranium fuel and graphite bricks. Inside, neutrons were fired at uranium atoms, which would split apart ejecting neutrons of their own, which could then go on to split other atoms. Left to its own devices, this process would cause a chain reaction and a nuclear explosion. But Fermi discovered that rods of cadmium could slow down the process by absorbing the neutrons and controlling the reaction.

On December 2, 1942, the experiment was ready. If it went wrong, they could destroy half of Chicago. Fermi's team began gradually removing the control rods and the reaction began. They found that not only could they control the rate of reaction, but that it was self-sustaining; the nuclear age was born.

David Hawksett

Date 1942

Scientist Enrico Fermi

Nationality Italian

Why It's Key Chicago Pile-1 provided the Earth with its first controlled flow of energy from a source other than the Sun.

Key Invention
Kodak click for color photography

In order to produce a photo from a camera film, negatives must first be processed. In these negatives, dark areas appear light whereas light areas appear dark – a so-called tonal inversion.

Nowadays, both black-and-white and color negatives are available, but up until 1942, color photography was one of those technologies that was only whispered about. At the time, photos were all produced in black-and-white or sepia monochromes, far from the digitally enhanced colors we're used to seeing today.Then Kodak, one of the most successful American film development businesses of the twentieth century, produced the world's first color photographs from a color negative film.

These color negatives are made of stacks of black-and-white negatives, one on top of the other. A chemical is applied to each layer, to make the negatives color sensitive. Like light in a black and white negative, color is also reversed in a color negative, so that blue might appear yellow, and green appear magenta. When it comes to processing a color negative, a chemical coupler present on the negative simply reacts with an applied development chemical to produce color photographs.

Kodak's color negative technology, "Kodacolor Film for Prints," was soon made available *en masse*, with photographers flocking to produce still images in color. The advent of digital cameras has seen the popularity of film photography wane in recent years, but it was Kodak who first brought color into our cameras.

Hannah Welham

Date 1942

Scientist Kodak

Nationality American

Why It's Key Color photography becomes a painless process.

Key Invention **Napalm**
The devastating sticky brown syrup

Louis Fieser was a highly successful and award-winning chemist, recognized for his research into the chemical causes of cancer; developing a way of making the hormone cortisone; and pioneering the production of vitamin K, the body's blood clotting agent. However, despite all this acclaim, he received some very angry letters in his lifetime. The reason for this is that he invented napalm.

Flamethrowers were being used in combat in World War I, however, they were not effective because the petrol in them burnt out easily and splashed everywhere, instead of sticking to the target. In World War II, latex from rubber trees was used to thicken petrol; but supplies of rubber were sparse. The government launched a competition to find a replacement thickener, and Harvard University, DuPont, and Standard Oil all took part. Fieser and a team of chemists from Harvard managed to get there first with the invention of napalm – a mixture of the powdered aluminum salts of naphthenic and palmitic acids, found in crude oil and coconut oil respectively. When mixed with petrol, it forms a sticky brown syrup that travels further when thrown, sticks better, and burns more slowly than petrol itself.

Napalm bombs resulted in immense fire devastation and temperatures of up to 1,200 degrees Celsius. Incomplete combustion of the napalm mixture resulted in the release of large quantities of carbon monoxide, and asphyxiation of those in the vicinity. Those unfortunate enough to be splashed with the tar-like substance suffered severe third-degree burns.

Riaz Bhunnoo

Date 1942

Scientist Louis Fieser

Nationality American

Why It's Key Napalm was cheap, easy to manufacture and significantly improved the impact of incendiary weapons. Its devastating effects however, helped turn the tide of public opinion against the Vietnam War.

opposite Napalm bombs burning fiercely through Brunei Town during American air raids on Papua New Guinea

Key Publication
The biological species concept

For centuries, scientists pondered how best to define a species. Many thought it essential to have a standard definition, thus allowing for accurate measures of biodiversity, and consistency in the biological world. Even so, some, including Darwin, thought the issue trivial: "I look at the term species as one arbitrarily given for the sake of convenience to a set of individuals closely resembling each other."

Ernst Mayr disagreed and, using the "biological species concept" outlined in his 1942 book *Systematics and the Origin of Species*, he defined a species as "a group of actually or potentially interbreeding natural populations which are reproductively isolated from other such groups."

Along with his definition, Mayr finally answered the question of how a species originates. He proposed that an isolated population could evolve new traits over a period of time, leaving them unable to interbreed with the main group.

Mayr's concept isn't the only one used to define a species. For one, the "phylogenetic species concept" eliminates the constraint of reproductive isolation, and, if widely used, would spectacularly increase the number of defined species. What's more, the "morphospecies concept" defines species on the basis of morphological (physical) similarities. Even so, it is Mayr's "biological species concept" that has become widely accepted, with environmental laws framed around this model.

Faith Smith

Publication *Systematics and the Origin of Species*

Date 1942

Author Ernst Mayr

Nationality German

Why It's Key This new definition of species allowed accurate measures of biodiversity.

Key Person **Edward Teller**
National defence scientist

Hungarian physicist Edward Teller (1908–2003) was born into a Jewish family. In 1935, he fled the anti-Semitism of Nazi Germany to become a key player in U.S. national security. As a physicist, world events had spurned Teller to focus on science as a form of defense. "I wanted to work on theoretical physics, not on weapons," Teller once said. His focus changed upon hearing President Roosevelt speak on the same day that Hitler invaded Belgium and Holland: "He said that if scientists in the free countries won't work on weapons, freedom will not survive." Moved by Roosevelt's speech, Teller dedicated the rest of his career to weapons research.

One of the architects of the atomic bomb, Teller went on to design the hydrogen bomb, having been invited to join the Manhattan Project in 1942. He also co-founded another premier national security lab, Lawrence Livermore, serving as its director until his retirement in 1975. Rather than taking things easy in his twilight years, Teller continued research and championing the use of science and technology for national defense. In the 1980s, Teller became involved in the Strategic Defense Initiative, better known as "Star Wars." His role in high-tech weapons brought him enormous attention; not all of it positive. Alternately hailed as one of the world's brightest scientists and burned in effigy as a warmonger, Teller's impact is undisputed. Like his contemporaries Albert Einstein, Enrico Fermi, Hans Bethe, and Robert Oppenheimer, Teller advanced the field of quantum mechanics, bringing theoretical physics to practical use.

Raychelle Burks

Date 1942

Nationality Hungarian-American

Why He's Key Dedicated to fusing science and defense, Teller helped bring about the most powerful weapons of our time.

Key Experiment **The Delbrück-Luria Experiment** Identifying mutations in bacteria

Nearly a hundred years after Darwin's theory of evolution was published, the question of how evolution worked on organisms had still not been fully answered. In the blue corner were the proponents of adaptation theory: organisms evolve due to direct interaction with selective pressures exerted upon them by the environment, such as predation or exposure to disease. In the red corner, champions of mutation theory: random mutations occur independently of selection pressures and are the driving force of evolution. It was like asking why rabbits can outrun foxes – because those rabbits that can outrun foxes pass on this ability to their offspring, or because some rabbits are just born fast runners and tend to survive?

German Max Delbrück and Italian Salvador Luria solved the problem with the famous Delbrück-Luria experiment, musing in the paper that followed, "It may seem peculiar that this simple and important question should not have been settled long ago."

Scientists had noted that bacteria sometimes became resistant to bacteria-killing phage viruses. But the only way to demonstrate the resistance was by exposing bacteria to the virus. Delbrück and Luria sidestepped the problem by allowing small bacterial cultures time to mutate. Delbrück then developed a mathematical model to prove that the variation in the number of resistant bacteria emerging was too great to be due to exposure to the virus. Mutations must occur spontaneously. We now know the selection of random, beneficial mutations to be the driving force behind evolutionary change.

Arran Frood

Date 1943

Scientists Max Delbrück, Salvador Luria

Nationality German, Italian

Why It's Key Answered a long-standing question in evolution and contributed to the understanding of the later problem of antibiotic resistance.

Key Discovery **How enzymes work**

Taking a fairly reductionist view of life, living organisms are just large bags of water containing an assortment of proteins, nucleic acids, carbohydrates, and fats. These chemicals are packaged into billions of cells, where they carry out vital processes, whether that's making cells divide – for example to create sperm and eggs or grow new bones – or triggering muscle contraction. But what controls these processes? What makes sure that they don't occur too fast or too slow? The answer is enzymes – specialized proteins that bind to substances called substrates, to control the pace of chemical reactions.

In 1930, American John Howard Northrop was the first to crystallize an enzyme – pepsin – and did the same for several other enzymes, providing biochemists with pure forms with which to investigate enzymes more precisely than was previously possible. But it was not until 1943 that biochemist Britton Chance revealed exactly how enzymes work. He designed a special apparatus with which he could mix together a peroxidase enzyme from the horseradish plant, the substrate hydrogen peroxide, and a dye that changes to the color green when oxygen is released. He revealed that enzymes work by sticking tightly to their substrate for a brief moment, and defined the dynamics of the process in unprecedented detail, providing the foundations for modern enzyme biochemistry.

As if scientific achievements were not enough, Chance had also developed a passion for sailing, and won an Olympic gold medal in 1952.

Julie Clayton

Date 1943

Scientist Britton Chance

Nationality American

Why It's Key Enzymes are the engines of life and understanding how they work has enabled scientists to develop all manner of biological and industrial processes, including the production of modern drugs.

Key Invention
The ball-point pen rolls into action

The staple of every desk-tidy, pencil case, and bureau drawer the world over – and the one item that any self-respecting scientist will always have sticking out of the breast pocket of their lab coat – the humble Bíró is actually the product of a great deal of research, effort and development.

As early as the 1880s, there were a number of inventors all working on a basic alternative to the somewhat messy and fiddly fountain pen. All this was still decades before László Bíró, the invention's eventual creator, had even been born.

After taking up a position at the Hungarian newspaper, *Elôtte*, Bíró found himself plagued by a number of pen-related problems, which clearly made his professional life so tricky he felt obliged to provide a solution. Having noticed that the ink of the printing presses dried much quicker than the ink in his fountain pen, he decided one day to fill his pen with printer ink. However, the printer ink was too thick to flow through the nib, so he enlisted the help of his brother Georg, a trained chemist.

In 1939, László and Georg fled Hungary, fearing for their lives, and emigrated to Argentina where they continued their work. They filed a patent for the device in 1943 and by 1945, the Eterpen, as it was then marketed, was already being sold in large numbers all over Argentina. Its popularity hasn't waned, and to day there are, on average, fifty-seven Biros sold worldwide every second.

Chris Lochery

Date 1943

Scientists László Bíró, Georg Bíró

Nationality Hungarian-Argentinian

Why It's Key Changed the nature of day-to-day writing and created a smudge-free society.

opposite László Bíró

Key Invention
The Aqualung

Since the dawn of civilization, humans have been fascinated with the mysteries of the sea. With two thirds of the planet made up of the blue stuff, it was only a matter of time before someone found a way to explore what lies beneath the surface. That someone was Jacques-Yves Cousteau, who, with the help of fellow Frenchman Émile Gagnan, invented the revolutionary Aqualung in 1943.

Fairly early on in the quest to breathe underwater, it was established that the two key elements of success were an oxygen supply, and a way to regulate the increased pressure exerted on the chest and lungs when diving in deep water. Before the Aqualung, the closest anyone had got to tackling this problem was Christian Lamertson, who invented a breathing system for the U.S. military's SCUBA program (self-contained underwater breathing apparatus). However, this design was far from safe, and divers were often killed or injured from oxygen toxicity.

The breakthrough came when Gagnan, an engineer, drew inspiration from a valve used in gas generator engines, to create a device called the demand regulator. By using a series of valves, the demand regulator releases oxygen slowly, at an equal pressure to the surrounding water, allowing the diver's lungs to inflate. Cousteau and Gagnan had succeeded in creating a device that was safe, reliable, and easy to use, opening up sub-aqua exploration to the mainstream. By 1952, the Aqualung was being sold across the world, and the international phenomenon of recreational diving had been born.

Hannah Isom

Date 1943

Scientists Jacques-Yves Cousteau, Émile Gagnan

Nationality French

Why It's Key The aqualung allowed humans to explore the un-charted depths of the deep blue sea for the first time.

DRUCCI

Key Discovery **Streptomycin**
TB's biggest enemy is unmasked

Unlike the discovery of penicillin by chance by Alexander Fleming, the discovery of streptomycin was the result of a much more systematic and deliberate process. Selman Waksman, a Russian-born scientist working at Rutgers University, began looking for antibiotics in soil microbes during the 1930s. His work, funded by the Merck pharmaceutical company, initially produced antibiotics that were effective but too toxic for human use.

Finally in 1943, Waksman's student Albert Schatz discovered an antibiotic that appeared to inhibit the growth of penicillin-resistant bacteria. Waksman's group soon realized that the antibiotic was active against several types of bacteria. One of these was *Mycobacterium tuberculosis*, the bacteria which causes tuberculosis (TB), which was still widespread and deadly throughout the United States in the 1940s. Streptomycin, as it was named, quickly moved out of the laboratory and into the clinic. After promising results in guinea pigs infected with TB, pathologist William Feldman and bacteriologist H. Corwin Hinshaw of the Mayo Clinic pushed the drug into human trials in 1945, with astounding results: streptomycin was safe and effective in the treatment of tuberculosis.

By the late 1940s, streptomycin was being produced by eight pharmaceutical companies, and the drug was later found to be effective against seventy penicillin-resistant bacteria. Waksman was awarded the Nobel Prize for his work in 1952, much to the chagrin of Schatz, who felt slighted that his work had not been acknowledged properly.

Rebecca Hernandez

Date 1943

Scientist Albert Schatz, Selman Waksman

Nationality American

Why It's Key A miracle drug in every sense of the word, streptomycin gave hope to patients with diseases that had proven to be incurable by penicillin.

opposite Polarized light micrograph of crystals of Streptomycin

Key Invention **Colossus**
Secretly changing the world

Even though the computer built by Thomas H. Flowers in 1943 epitomized British ingenuity, Winston Churchill demanded that it should be smashed to pieces once World War II was over; such was the fear that the world's first electronic digital computer, Colossus, might fall into enemy hands.

By the end of the war, Britain had ten of the machines in operation, each the size of a room, and each one eavesdropping on the highest level of enemy transmissions. During the war German communications were usually encrypted using Enigma machines. These were frequently decrypted, however, so the most important messages were sent using bulkier and heavier stream cipher teleprinters, which the Nazis believed to be totally secure. The Colossus computers used logic to decipher intercepted communications – which were then rewritten in order to disguise their origin and prevent Hitler from discovering that his messages were being understood. In some cases it was decided to ignore the intelligence on the basis that it would be too obvious where the inside knowledge had come from.

The secrecy surrounding this awesome military invention is evocative of a comic-book superhero, who protects their identity from the world in a bid to prolong the effectiveness of their powers. Thomas H. Flowers was an unlikely superhero of the 1940s, but with the computers he designed, he saved hundreds of thousands of lives and dealt one of the most telling blows to Nazi Germany.

Christopher Booroff

Date 1943

Scientist Thomas H. Flowers

Nationality British

Why It's Key Colossus was instrumental in defeating Germany in World War II, and instigated the computer age.

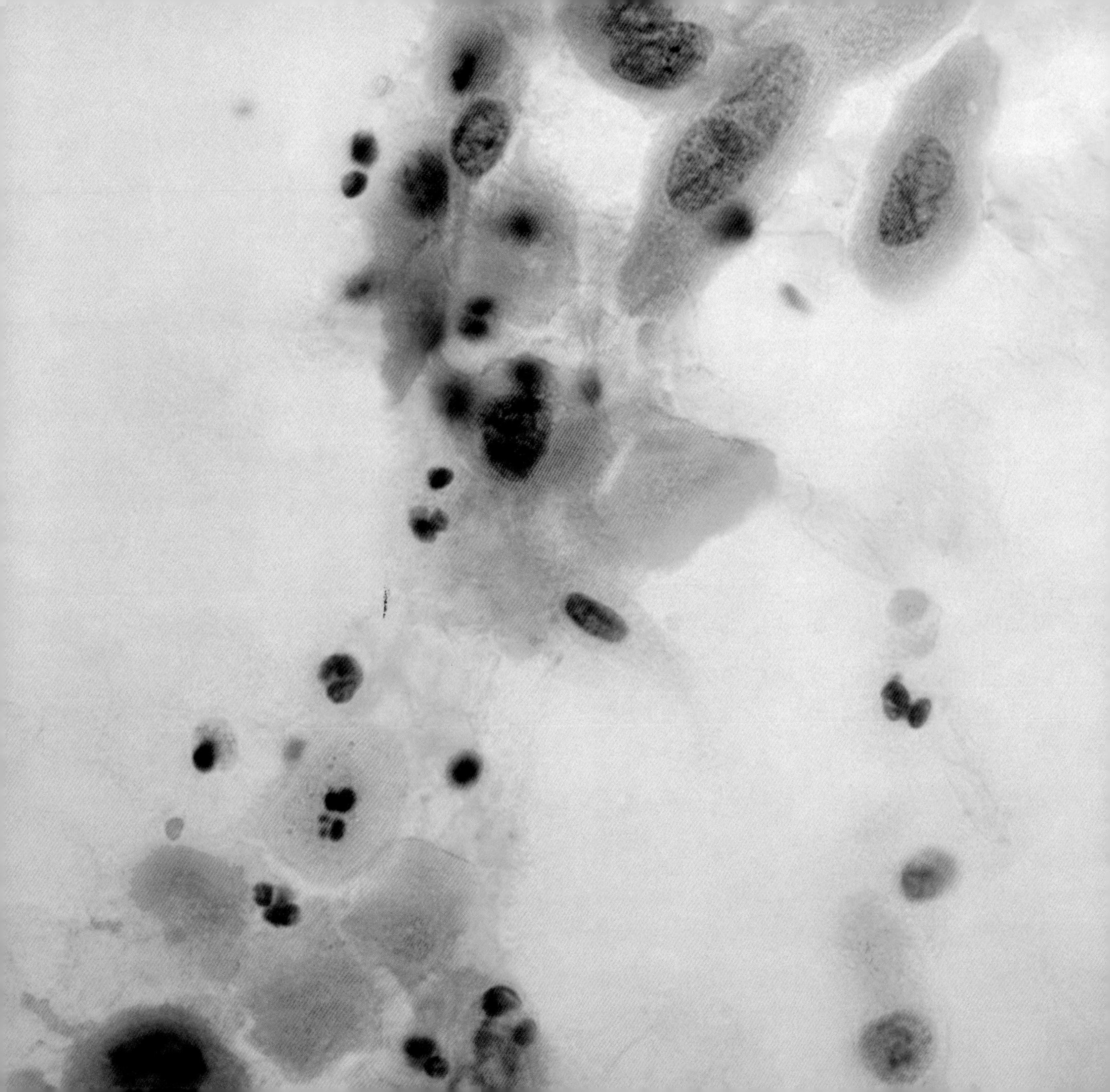

Key Publication **The hierarchy of need**
All you need is love?

Human beings have just a few basic needs – shelter, water, and food, right? Not according to Abraham Maslow. In 1943, he postulated that human behavior could only be explained by a complex series of needs – usually represented by a pyramid.

At the lowest level are the physiological needs – for example, food, sleep, and water. The next level up includes safety, securing a job, financial security, and having good health, among other things. The third level up consists of social needs. It is this level that incorporates love, friendship, and a sense of belonging. The fourth level, called esteem, involves self respect and achievement. The fifth level is self actualization, which consists of more esoteric qualities: things like truth, justice, and morality. Maslow called the first four levels deficiency needs – in that people could not be satisfied until these needs could be met. In other words, people were motivated to meet these needs. The highest level of the pyramid, that of self actualization, is different. It can only be attained after the lower levels have been fulfilled. In contrast with the first four levels, it is characterized by personal growth. Also unlike the lower levels, it can never be fully "conquered." Nevertheless, people who achieve this level of functioning, per Maslow, are quite happy.

Maslow's ideas won initial favor amongst many psychologists, but this has since waned. However, while some subsequent studies have failed to find evidence of people having a mental hierarchy, the theory remains influential.

B. James McCallum

Publication "A Theory of Human Motivation"

Date 1943

Author Abraham Maslow

Nationality American

Why It's Key Maslow's Hierarchy of Needs attempts to explain complex human behavior, and the motivation behind it.

Key Invention **The Pap smear**
Accidental eureka

The Pap smear test has become the most successful and widely used cancer screening method in history. Not many people realize, however, that the first ever smear test was carried out not on a willing human guinea pig, but on a real life fur ball of a guinea pig.

The smear test, which detects early warning signs of cervical and uterine cancers, was invented in an "accidental" eureka moment. Its creator was Greek scientist George Nicholas Papanicolaou, who in 1916 was studying the menstrual cycle of guinea pigs at his lab in New York. He took samples from the vaginal lining of the guinea pigs, using a small ear swab, and then "smeared" the sample onto a glass plate to observe the cells under a microscope.

Encouraged by his findings, Papanicolaou tried to replicate the results in the vaginal smears of women. Surprisingly, in one sample he noticed some abnormal cells that he identified as uterine cancer cells. He soon realized that if he could detect early signs of cancer before a woman even had symptoms, there could be a much greater chance of a curing her.

In 1928, Dr Pap presented his findings to his peers, who were skeptical about the test. In fact, it wasn't until 1943 that his research was taken seriously at all. The test was given the name "Pap smear" in 1953, when large-scale screening programs first started. Their widespread use has resulted in a drop in cervical cancer deaths of almost 80 per cent.

Hannah Isom

Date 1943

Scientist George Nicholas Papanicolaou

Nationality Greek

Why It's Key The most successful and widely used cancer screening method in history.

opposite Cervical Pap smear showing pre-cancerous cells

Key Event

The bouncing bomb raid

Barnes Wallis was a British inventor, scientist, and engineer. Shortly after the Germans invaded Poland in 1939, he recognized the potential value of strategic bombing of enemy infrastructure in slowing down or halting their hostile activities. By 1941, he was working on a bomb that bounced on water, based on observations by the Royal Navy in the nineteenth century that cannonballs occasionally bounced on the surface of the ocean, increasing their range.

The RAF had identified German hydroelectric dams as critical targets, but due to their size, they were immune to conventional bombing, and the Germans had already placed anti-torpedo nets upstream as protection. Wallis's solution was the bouncing bomb – a large cylindrical mine weighing over four tons, three tons of which was Torpex explosive. The principle was ingenious: a modified Lancaster bomber would carry the bomb, which would be spun backwards at 500 revolutions per minute, before being dropped. A precise bombing run was needed with a forward speed of 390 kilometers per hour, a height of 18.3 meters, and a range of 400–500 meters from the target. Striking the water, the bomb would skip like a stone and, upon encountering the dam, its backspin would make the bomb hug its face while also driving downwards to the dam's base. A hydrostatic fuse would detonate the explosive at a depth of 9 meters.

On the night of May 16, 1943, Operation Chastise took place, in which six bouncing bombs were deployed against German dams, destroying two and damaging four.

David Hawksett

Date 1943

Country UK

Why It's Key The Dambusters raid – along with Wallis' ingenious invention – helped bring the war one step closer to its end.

opposite Barnes Wallis, designer of the ten-ton bomb, and the bouncing bomb

Key Publication

Theory of Games and Economic Behavior

John von Neumann was one of the greatest mathematicians of the twentieth century; Oskar Morgenstern was an economist and professor at Princeton University. Together, in 1944, they published the book *Theory of Games and Economic Behavior*, a classic work on which modern day game theory is based.

The book introduces modern logic into economics, and focuses on a mathematical theory, based on the theory of games, which has been described as a "radical new approach to economic theory." The publication of *Theory of Games and Economic Behavior* revolutionized economics and yielded game theory as a new field of scientific enquiry. It has been said that if von Neumann and Morgenstern had never met, it is unlikely that game theory would ever have developed. Game theory is interested in how someone will behave when their choices are affected by other people, the most famous example being the Prisoner's Dilemma game devised in 1950.

The work of von Neumann and Morgenstern was intended solely for economists, but its applications in many other fields including psychology, sociology, politics, warfare, and recreational games soon became apparent. Since their book was published, game theory has been used to analyze real-world phenomena such as the arms race, optimal policies of presidential candidates, vaccination policies, and even major league baseball salary negotiations.

Sarah Watson

Publication *Theory of Games and Economic Behavior*

Date 1944

Authors John von Neumann, Oskar Morgenstern

Nationality Hungarian-American, Austrian

Why It's Key Provided a new approach to economic theory, and has since been used to analyze real world phenomena in various other fields.

MÖHNE DAM

Key Experiment
DNA is the hereditary material

In 1944, three biochemists were trying to solve an interesting problem. Sixteen years previously, Fred Griffith had found that extracts of a virulent strain of *Pneumococcus* bacteria could transform a harmless strain into a pathogenic one. At the time, accepted wisdom had it that protein was the genetic material, containing the information being transferred between bacteria. Griffith, however, claimed this was not the case; that it was in fact DNA which contained the genetic material. He had been ridiculed for his unconventional theories, and his results discounted.

Intrigued by his findings, Oswald Avery, Colin McLeod, and Maclyn McCarty decided to revisit Griffith's experiments. To avoid claims that their DNA was contaminated with protein – which everyone knew as the "real" information transfer molecule – they added in a few extra clever experimental techniques.

The trio treated their bacterial extracts with two enzymes: protease, which breaks proteins into its building blocks, amino acids; and DNAse, which degrades DNA into nucleotides. They found that while samples treated with protease still performed the transformation of the bacteria, pre-treatment with DNAse stopped any transformation from occurring. In other words, once DNA had been broken down, there was nothing to transport genetic information between bacteria. This was strong proof that the hereditary material in cells was DNA. Griffith's experiments were vindicated and over the next few years, it was proven beyond all doubt that DNA was the genetic material.
Catherine Charter

Date 1944

Scientists Oswald Avery, Colin Macleod, Maclyn McCarthy

Nationality Canadian-American, American

Why It's Key Before the experiments of these three shrewd scientists, no one would believe that DNA could provide the template for life.

Key Event
First operational use of the V-2 rocket

September 1944 saw the start of "Operation Penguin," a terror campaign that marked the first operational use of a technological triumph and deadly new weapon – the V2 ballistic missile. Lasting until March 1945, this was Hitler's final desperate attempt to avoid defeat and witnessed over 3,000 V-2 rockets with their 1,000 kilogram warheads launched against Allied targets across Europe.

With a range of approximately 320 kilometers and the ability to cut through the air at speeds exceeding 5,600 kilometers per hour – several times faster than the speed of sound – the V2 quickly and silently reached its target. No defense, no warning. It took five seconds from its launch in the Netherlands to reach the city of London. Only after it exploded could the ghostly whine of whistling air, followed by the thunderous roar of the rocket be heard. The weapon was developed for the Nazis by Wernher von Braun and his "rocket team" during the late 1930s and early 1940s. They pioneered four essential rocket technologies: the immense 25-ton thrust liquid-fuelled engine, the aerodynamic shape, the innovative guidance system, and the radio transmission system.

The V2 caused around 5,000 deaths, but at least 20,000 prisoners of war died as a result of the poor work conditions, starvation, and beatings whilst being forced to manufacture the weapon in underground facilities. Under pressure from the Allies, the Nazis eventually retreated beyond the rocket's range, and their surrender came soon after, in May 1945.
James Urquhart

Date 1944

Country Netherlands

Why It's Key The USA, the Soviets, and other nations acquired the V2 technology following World War II, creating a revolution in warfare and marking the beginnings of space exploration projects.

opposite The deadly V-2 rocket missile

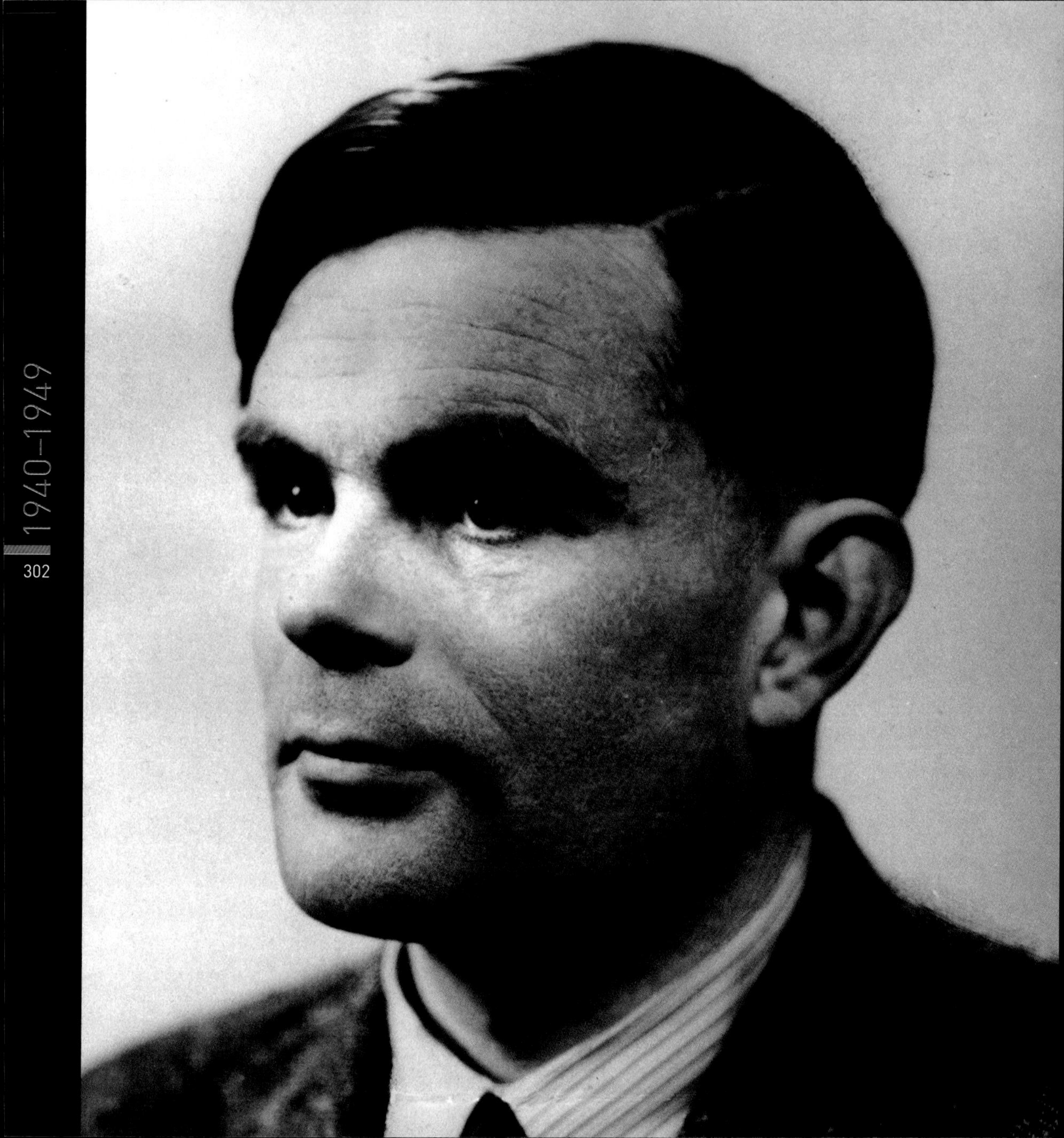

Key Event **The Green revolution**
Flour to the people

The green revolution – far from being the invasion of tree huggers – was a revolution with its origins in 1940s Mexico which led to massive increases in the global production of wheat and rice.

The Cooperative Wheat Research and Production Program, a joint venture between the Mexican government and the Rockefeller Foundation, was initiated in 1944. One of the key figures in this cooperative was Norman Borlaug, a microbiologist from the United States who was appointed to direct the operation. Borlaug remained with the project for sixteen years. During this time, he bred a series of remarkably successful high-yielding, disease-resistant, semi-dwarf wheat plants which doubled the amount of flour that could be produced in a year. In part thanks to Borlaug's hard work, Mexico had become self-sufficient in wheat production by 1951, and began to export wheat thereafter. In 1970, Borlaug received the Nobel Peace Prize for his work on wheat breeding.

This was only the beginning, however; in 1964 there was another major step forward. New strains of rice were developed by the International Rice Research Institute in the Philippines. These doubled the yield of earlier strains if enough fertilizer was used.

This was the peak of the green revolution. Increased food supplies across Asia meant that an ever increasing population could be fed, and to a better standard, underpinning the social development, population growth, and urbanization that has characterized Asia and the rest of the world over the last 30 years.

Catherine Charter

Date 1944

Country USA

Why It's Key Without the green revolution, the human population simply wouldn't have been able to expand the way it has.

Key Person **Alan Turing**
An apple for the teacher

Turing's contributions to the world as we know it are not incalculable, thanks in part to the role he played in the development of computers, software, and artificial intelligence. Inextricably linked with the British World War II "Enigma" decoding projects, Turing himself was a man with many complex and often hidden traits.

Born on June 23, 1912, in Paddington, London; Turing attended Hazelhurst Preparatory School; continuing his education at Sherborne, Dorset; and ultimately King's College, Cambridge, where he read mathematics. University for Turing may not have been as automatic as we might imagine. A Sherborne English master noted on his progress, "I can forgive his writing, though it is the worst I have ever seen, and I try to view tolerantly his unswerving inexactitude and slipshod, dirty, work." The headmaster of the school commented, "...bound to be a problem for any school or community." Turing was attracted to men.

With homosexuality then unlawful, this visionary, who would one day reveal the secrets of a nation, fought to keep his own secrets well hidden. In wartime England, employed by the government, and of great value, his sexuality could be overlooked, but by the 1950s his behavior had become more overt and, in 1952, he was convicted of "gross indecency." Turing's security clearance was revoked and he had to resign from his post at the Government Communications HQ.

On June 8, 1954, Turing's body was discovered by his cleaner; beside him, and later found to be laced with cyanide, lay a half-eaten apple. Following a long bout of depression, he had poisoned himself.

Mike Davis

Date 1944

Nationality British

Why He's Key Turing's work formed the basis of modern computer science.

opposite Alan Turing

Key Invention **Paper chromatography**
Divide and conquer

Paper chromatography is one of the first techniques that any budding scientist learns. It's an unsophisticated way of checking the purity or composition of a mixture.

The experiment takes place on a sheet of filter paper, which is called the "stationary phase." The sample under study is applied to the paper, in the form of a concentrated spot. One end of the paper is then immersed in a suitable solvent, which creeps up the paper and is known as the "mobile phase." The sample is carried along with the solvent. Its various constituents will travel different distances depending on how strongly they are absorbed in to the paper and dissolve in the solvent. They can then be isolated and studied. Amazingly, this simple analytical technique was unknown until the mid-1940s, when it was developed by Archer Martin and Richard Synge at the Wool Industries Research Association in Leeds. The pair needed a way to separate amino acids, the building blocks of proteins.

After experimenting with more complicated techniques, involving columns of ground up silica gel, they discovered that a simple paper-based method could be used to separate the complex mixtures. The pair picked up the 1952 Nobel Prize in Chemistry for their efforts. The technique has now largely been superseded by more sophisticated methods, but is still widely used in teaching laboratories.

Matt Brown

Date 1944

Scientists Archer Martin, Richard Synge

Nationality British

Why It's Key From separating simple mixtures, to finding the composition of whole proteins, this simple technology revolutionized analytical techniques.

opposite A paper chromatography strip showing the separating colored inks

Key Invention
The Tetra Pak

The Tetra Pak company was built on the sustainable concept that each package should save more than it costs. For its founder, Dr Ruben Rausing, it also turned out to be a highly lucrative business enterprise. When he died in 1983, he was Sweden's richest person.

Rausing started working on ways to revolutionize food packaging in 1943 and, just a year later, he released his first product – Tetra Classic. The initial designs were folded into tetrahedrons, four-sided cartons which maximized storage capacity and could be easily opened. In 1963, the design was revamped and the company released the Tetra Brik, a rectangular carton resembling the Tetra Pak we are familiar with today. Milk, juices, smoothies; they all come in Tetra Paks. Some Californian wine producers have even started selling their vino in the four-sided packets. Fully airtight materials (plastic coated paperboard), and a merciless bacteria-beating sterilization process make the Tetra Pak just about impenetrable to food bugs. Hydrogen peroxide is used to kill anything secreting itself in the creases before the package comes into contact with food.

Ruben Rausing's son Hans took the reins from 1954 until 1985, and managed to turn Tetra Pak into one of the most successful businesses in the world. Today the company is still working to become more environmentally responsible; to offset pollution, Tetra Pak has created the world's first packaging recycling plant in Brazil.

Vicky West

Date 1944

Scientist Ruben Rausing

Nationality Swedish

Why It's Key A safer way to store food, which improved shelf-life, condition, and distribution of food products. It also made its inventor a mint.

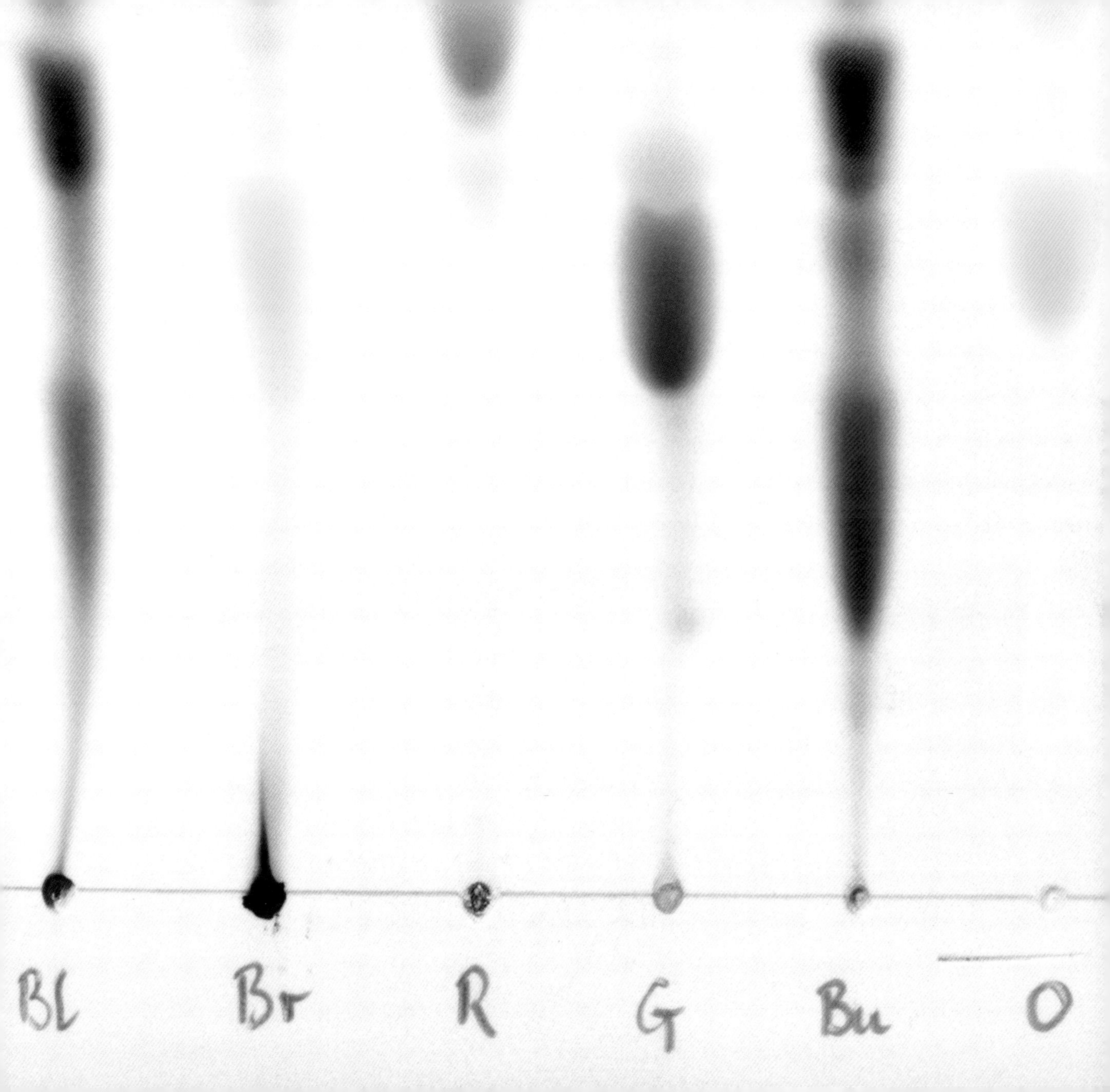

Bl
Br
R
G
Bu
O

Key Person **Lise Meitner**
Austria's Marie Curie

Austrian born Lise Meitner (1878–1968) was part of the team that discovered nuclear fission, a discovery for which her colleague Otto Hahn won the Nobel Prize in Chemistry in 1944. Referred to by Einstein as "our Marie Curie," she was the first person to give a theoretical explanation for the nuclear fission process.

Meitner worked on developing the basis of nuclear theory before its worrying applications were known. She realized that heavy nuclei could break apart and undergo a chain reaction to give out lots of energy, but she never expected her research to lead to the development of nuclear weapons.

When Adolf Hitler came into power in 1933, Meitner fled Germany because of her Jewish background. Hahn gave her a diamond ring in case she needed to bribe her way across borders, but she made it to Sweden without difficulty. Only a few years later, Hahn published their work but could not add her as an author due to the political situation. She continued her research in Sweden and corresponded with her colleagues in Germany, as well as establishing a working relationship with the great Niels Bohr. In his Nobel Prize speech in 1944, Hahn gave full credit to Meitner for creating the foundational theory of nuclear fission.

Arguably, this was the biggest scientific achievement the Nobel committee ever overlooked. Perhaps some small consolation for having been denied the Nobel Prize was being immortalized in the Periodic Table as the element "meitnerium."

Leila Sattary

Date 1944

Nationality Austrian

Why She's Key A great talent in science, and a woman whose work carried huge political implications.

Key Person **George Gaylord Simpson**
Marrying fossils and evolution

In the early twentieth century, the newly accepted Darwinian theory of evolution came up against a sticking point – genetics. It didn't seem as if natural populations of organisms had enough genetic variation to lead to the creation of new species. Instead alternative explanations were entertained, such as evolution toward a specific goal, or sudden large mutations that led to new species.

American paleontologist George Gaylord Simpson (1902–1984) was the man to reconcile Darwin and genetics. Through his work in the 1930s and '40s, he applied the concepts of genetics to evolution, claiming that the geneticists' microevolution could lead to the paleontologists' macroevolution. His 1944 book *Tempo and Mode in Evolution* revolutionized the field by marrying biology, paleontology, and many other fields into what became known as Modern Evolutionary Synthesis. His key example was that of the modern horse, which was seen as a directed evolution toward its current specialized form. Simpson instead saw it as an evolutionary tree with branches that had led to extinction for the other species created.

Simpson applied the new science of population genetics to show that evolution occurred within a gene pool rather than within individual organisms, and also showed that evolution occurred at different rates during the fossil record. He had enabled a thorough scientific approach to fossils and evolution, and dispelled the theories of directed evolution. As he put it, "Man is the result of a purposeless and natural process that did not have him in mind."

Helen Potter

Date 1944

Nationality American

Why He's Key Applied the science of genetics to the world of fossils, enabling a deeper understanding of evolutionary processes.

Key Discovery **Antihistamine**
Quick relief for allergies

During the "golden years of pharmacology," Daniel Bovet, a Swiss native, embarked upon a career in drugs. His first appointment was at the Pasteur Institute in Paris, with his wife Filomena working at his side. Here, among his many achievements, he discovered pyrilamine, the first antihistamine – a substance that counteracts the painful swelling, itching, and breathing problems that occur during an allergic reaction. Antihistamines fight against histamine, a chemical produced naturally by the body that causes these reactions.

Bovet went on to work out the chemical formulas for what became a whole range of drugs to control the symptoms of allergies such as hay fever, asthma, and hives. In 1947, he left Paris for the Istituto Superiore di Sanita in Rome, where he challenged traditional hierarchy with his informal and friendly manner toward junior colleagues, and eventually became an Italian citizen. When he was awarded the 1957 Nobel Prize in Physiology or Medicine, Bovet was lauded as the first Italian to do so since 1906. Remarkably, Bovet did not take out any patents on his discoveries, or profit personally from their commercial use, insisting his success had as much to do with chance observations as it did with ingenuity.

Today, antihistamines are still the some of the most widely used medicines for relieving symptoms of allergies such as hay fever.

Julie Clayton

Date 1944

Scientist Daniel Bovet

Nationality Swiss-Italian

Why It's Key Antihistamines are one of the most widely used classes of drugs. Allergic reactions need no longer be so debilitating.

Key Person **Wilhelm Heinrich Walter Baade**
Banished into the darkness

Walter Baade (1843–1960) was born in Germany, but moved to America in 1931 to take up a position at the Mount Wilson Observatory near Pasadena, California. Baade demonstrated his passion for his work by working past his formal retirement, up until his death, aged sixty-seven. As much as he loved astronomy, however, he detested bureaucracy; a trait that would prove hugely influential in his career.

Having lost papers regarding his application for citizenship during the move, Baade did not bother to repeat the process, and so remained a foreigner. A German foreigner. When war broke out, and Baade's colleagues were being utilized in the war effort, he was banished to the telescopes of Pasadena, where he was treated as an alien, albeit one with special privileges to use the observatories.

It was during this period that Baade took spectacular advantage of the black-out in Los Angeles, and the underused telescopes, to produce his most significant work. The dark sky improved the quality of his observations to such a degree that in 1944 he was able to divide stars into two discrete populations. His results led to a better understanding of astronomical distances and, in the blink of an astronomer's eye, he had doubled the estimated size and age of the Universe.

Noted for his kindness to others and enthusiasm for young astronomers, Baade may have been in the right place at the right time, but he was also the right man to take advantage of his fortune.

Christopher Booroff

Date 1944

Nationality German

Why He's Key Baade made huge contributions to astronomy for almost half a century, earning both the affection and respect of others in the process.

Key Event **My, what pretty teeth you have**
Grand Rapids, Michigan, gets fluoride in the water

The story behind one of the greatest examples of preventive medicine in world history started when a small-town Colorado dentist noted that his patients' teeth were covered with brown spots. That dentist, Fredrick McKay, realized that the spotting was occurring during the early development of a child's teeth; if someone made it to adulthood without the discoloration, then there was no further risk.

Through meticulous observation, he also discovered that the brown spotted teeth were curiously resistant to decay. McKay eventually determined that something in the water supply was turning the teeth brown. He joined forces with the chief chemist of a large aluminum producing company who was desperately trying to prove that aluminum was not causing the dental problem. Through the use of chemical analysis, they determined that the water contained large amounts of fluoride.

The next big step was when a certain Dr Trendley Dean was placed in charge of investigating the fluoride levels in the water of the United States. By the early 1940s, he was certain that, in small doses, fluoride could be used to prevent dental cavities. The city of Grand Rapids, Michigan, was convinced to fluoridate their water for fifteen years. The results were astonishing – the incidence of dental cavities dropped by over half. Fluoridation of the water supply was rapidly adopted across the world. Millions now have healthier teeth – a prime example of how preventive medicine can stop disease before it has even started.

B. James McCallum

Date 1945

Country USA

Why It's Key It's why we all smile with our mouths open these days.

Key Person **Archibald McIndoe**
The plastic man

While almost everybody knows of the heroic exploits of the RAF pilots during the Battle of Britain, a great deal fewer know of the exploits of the man who dedicated himself to saving their lives when things went wrong.

Archibald McIndoe was born and raised in New Zealand. He pursued his medical studies at Otago University, and then later in the United States. He arrived in London in 1930, and here he became a celebrated plastic surgeon. At the outbreak of World War II, he was appointed as the Royal Air Force's consultant in plastic surgery.

The RAF would soon be in need of McIndoe's talents. Fighter planes burned quickly when damaged, and even if a pilot survived a crash landing, he was often severely burned. McIndoe became a pioneer in the treatment of just such burns, and injured pilots were frequently sent to him at the Queen Victoria Hospital in West Sussex. "The boss" or "the maestro," as McIndoe became known, wasn't just performing difficult surgeries; he was literally learning from trial and error as his techniques evolved. To be operated on by McIndoe placed one in the "Guinea Pig Club."

McIndoe realized that he had to treat the burned pilots' medical difficulties, but he was also quite cognizant of the isolation these young men felt because of their now disfigured bodies. He worked tirelessly to help his patients rejoin society.

B. James McCallum

Date 1945

Nationality New Zealander

Why He's Key Prior to McIndoe's efforts, there was no treatment for severe burns, and most victims died.

opposite The Guinea Pig Club

Key Person **J. Robert Oppenheimer**
The man who thought he destroyed the world

Experiments, equations, and epiphanies tend to define scientific careers. At 5:30 am, July 16, 1945, 480 kilometers south of Los Alamos, (Julius) Robert Oppenheimer's (1904–1967) career was defined with the detonation of the world's first atomic bomb.

Oppenheimer was the civilian leader of the Manhattan Project at the time of the blast, and later compared the experiment to a line from the Bhagavad-Gita, an ancient Hindu text in which Krishna attempts to convince the Prince that he should do his duty: "I am become death, the destroyer of worlds."

Oppenheimer often pondered whether responsibility for the bomb belonged to the politicians who controlled it, or to the scientists who devised it. However, his most poignant quote on the matter came when he lamented that "physicists have known sin; and this is a knowledge which they cannot lose." One thing Oppenheimer did lose was his security clearance. The Atomic Energy Commission took away his privileges in 1954, on the basis that previous associations rendered him a threat. This was seen by many as the victimization of a brilliant, honest scientist.

Ironically, Oppenheimer also produced papers on the gravitational collapse of stars, an occurrence which could one day destroy our world; however he certainly did not cause global destruction and, in 1963, he was presented with an award for his "outstanding contributions to theoretical physics, and his scientific and administrative leadership." Like the mushroom cloud from an atomic blast, however, the bomb inevitably always overshadowed the rest of his work.

Christopher Booroff

Date 1945

Nationality American

Why He's Key
Oppenheimer's work impacted the outcome of World War II and sculpted the future of the human race through the development of the atom bomb.

Key Person
Alexander Fleming

Famous for discovering penicillin, Alexander Fleming was a Scottish biologist and pharmacologist – but he never actually intended to study bacteriology. In fact, if it wasn't for an unusual sequence of events, he may never have gone on to discover the antibiotic that was to revolutionize the way infected wounds were treated.

Fleming literally stumbled into the medical profession. After an unfulfilled career in the shipping industry, he used a well-timed inheritance gift of £250 to change direction, using the money to pay his way through medical school. After passing the entrance exam with distinction, he was offered a place at any school of his choosing. He chose St Mary's in London – on the basis that he'd once played water polo there – and specialized in surgery, but he soon switched to bacteriology when he learnt he would otherwise have to leave St Mary's to take up a position as a surgeon.

When World War I broke out, Fleming served as a captain in the Army Medical Corps. After witnessing the needless deaths of hundreds of patients from what he saw as minor infections, he began his search for an antiseptic. Back in the lab, he discovered a natural antiseptic found in many bodily fluids, which he named lysozyme. He continued his search for a more powerful substance that could fight stronger infections, and in 1928, returning from holiday to find his culture dishes infected with mold, discovered, quite by accident, what he was looking for: penicillin.

Faith Smith

Date 1945

Nationality British

Why He's Key He discovered penicillin and also the natural antiseptic, lysozyme.

opposite *Time* magazine cover, May 15, 1944 of Dr Alexander Fleming

TIME
THE WEEKLY NEWSMAGAZINE
DR. ALEXANDER FLEMING
His penicillin will save more lives than war can spend.
(Medicine)

Key People **Ho[illegible]rd [illegible]lorey and E[illegible]st Chain**

Penicillin pioneers

In 1929, Alexander Fleming ceased his investigations into the effects of penicillin when he began to doubt the feasibility of making sufficient quantities of the drug to test. Nine years later, Ernst Chain and Howard Florey formed a partnership to challenge this notion, and succeeded in isolating and concentrating penicillin.

After extensive testing on mice, they were finally given a chance to test the effect of penicillin on humans in 1940. They managed to prevent Albert Alexander from succumbing to a deadly septicemia infection for five days before their supply of the drug ran out.

At the time, all of the factories in Britain were focused on the war effort, but Florey made use of contacts in America to help produce penicillin *en masse*. They used corn steep liquor in huge fermentation tanks, but it was clear, even at this early stage, that *Penicillin notatum*, the mold first discovered by Fleming, would never yield enough of the drug to be commercially viable.

After a worldwide search for better penicillin molds, Florey produced a strain with a yield a thousand times that of the original, and by the end of the war, American factories were producing 650 billion units of the drug each month.

Florey, Chain, and Fleming shared the Nobel Prize for Physiology or Medicine for their research into penicillin. After being awarded his prize, Howard Florey noted: "I don't think it ever crossed our minds about suffering humanity. This was [just] an interesting scientific exercise."

Faith Smith

Date 19

Nationa
German

Why Th
Manufa
known
on to sa

Key Discovery

The Jet Stream

Jet streams were discovered after World War II pilots noticed it was sometimes quicker to fly from one country to another than it was to fly back again. It soon became apparent that this was thanks to ribbons of fast-flowing air which meandered their way, west to east, through the atmosphere, at some ten kilometers above the surface of the Earth.

The streams form when large bodies of warm and cold air – which differ in atmospheric pressure – meet. Pressure differences usually would cause air to move from the area of high pressure to the area of low pressure (wind). However, a phenomenon known as the Coriolis Effect, caused by the spinning motion of the Earth in space, causes the wind to travel along the edge of the two air masses. The discovery of jet

Meteorologists noticed that rain-bearing depressions often form around these currents, and that measured changes in wind speed and pressure within them are useful in predicting the development of storms. In fact, jet streams play an important role in shaping global weather patterns. Shifting north and south with the seasons, but not always predictably, these ribbons of wind transport weather systems great distances across the globe.

The travel industry has since made good use of jet streams in international flight. These breezy boosts give planes a friendly push in the right direction, not only allowing airlines to reduce flight times, but also saving them fuel and therefore oodles of cash.

Nicola Currie

Date 19

Scientis

Nationa
America

Why It's
an indic
storms,
meteoro
unstable
and haz

opposite
photo sh
over Egyp

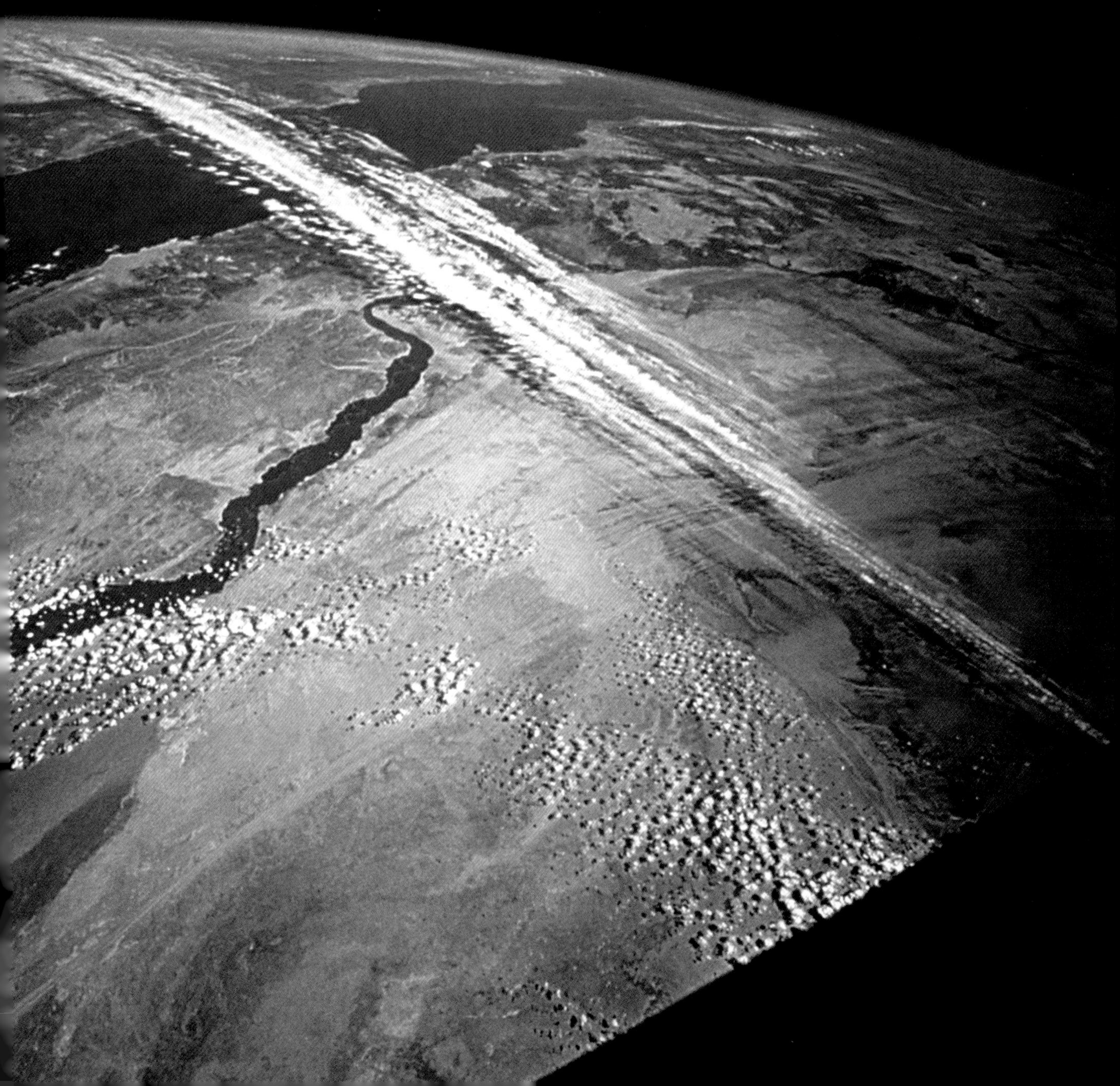

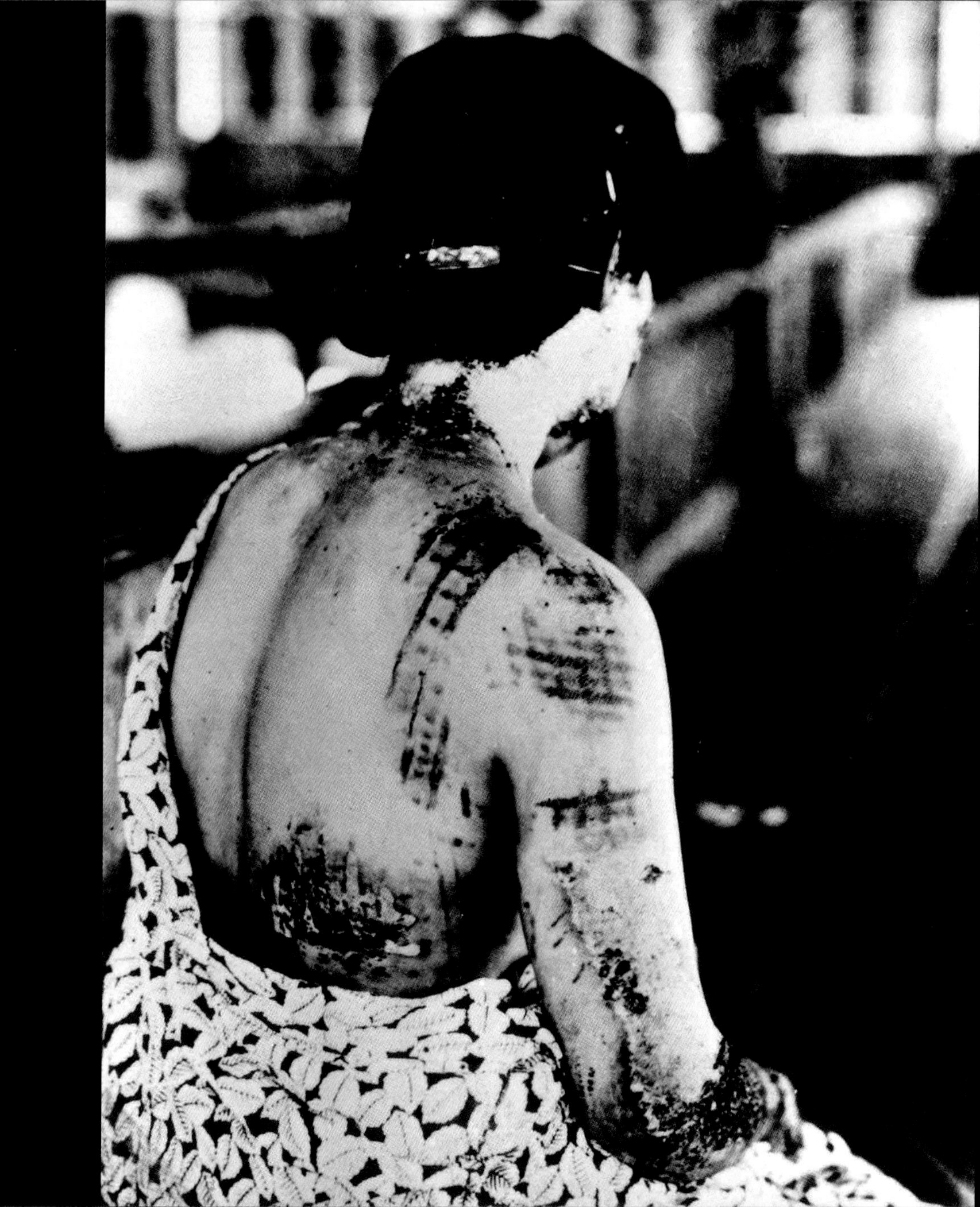

Key Discovery **Promethium**
Element 61 finally found

In 1915, during an escalating World War I, twenty-six year-old Henry Moseley, a British physicist whose research had already influenced the final order of elements in the Periodic Table, died in combat in Gallipoli, Turkey. A year earlier, he had confirmed John Casper Branner's 1902 prediction that another element – element 61 – should exist between the rare elements neodymium and samarium. Ironically, chemical proof would come some thirty years later, as a spin-off of an attempt to conclusively end yet another war.

In 1945, chemists Larry Glendenin and Jacob Marinsky – working at the Graphite Reactor, Oak Ridge National Laboratory (home of the Manhattan Project) under the leadership of Charles Coryell – finally produced element 61, both by uranium fission and by bombarding neodymium with neutrons. Working in the so-called "hot" laboratory and chemistry building, they made the first chemical identification of element 61, using a technique known as ion-exchange chromatography. Their breakthrough was announced at the 1947 American Chemical Society meeting. In 1948, they suggested "promethium" (an idea of Corvell's wife, Grace Mary) as the name for element 61; named for Prometheus, the Titan in Greek mythology who stole fire from Heaven for humans. The name was adopted by the International Union of Chemistry in 1949.

Promethium is a radioactive metal not naturally found in the Earth's crust. Among other possibilities, the element is used as a portable X-ray source, in nuclear-powered batteries, and can be absorbed by a phosphor to create light.

Mike Davis

Date 1945

Scientist Charles Coryell

Nationality American

Why It's Key A significant hole in the Periodic Table was finally plugged.

Key Event **Hiroshima**
Dawn of the nuclear age

At 08:15 local time on August 6, 1945, under United States President Truman's orders, an American B-29 bomber called Enola Gay released its payload above the unsuspecting Japanese city of Hiroshima, marking the dawn of the nuclear age as well as one of the gravest moments in scientific and in human history. After plummeting for forty-three seconds, the first operational atomic bomb, named "Little Boy," detonated, unleashing unimaginable horror.

Hiroshima was strategically important for the Japanese army, so it was a prime target for the Americans. It was also populated by over 300,000 civilians. The explosion caused by the 4,400 kilogram bomb was almost a thousand times more powerful than any other kind of weapon on Earth, vaporizing those in the immediate vicinity, and generating a massive mushroom cloud. It is estimated that around 70,000 people died as a result of the initial blast, but the final death toll exceeded 200,000, after the effects of radiation were taken into account.

America justified the A-bomb's employment by arguing that it would force Emperor Hirohito to surrender swiftly as opposed to a full-scale ground invasion that would have cost considerably more lives and, of course, money. Japan did in fact surrender eight days later bringing World War II to an end, but not before a second A-bomb was dropped on Nagasaki killing a further 70,000 people. The moral issue of the weapon's use remains a subject of debate; some have suggested America dropped the bomb to show the world how powerful it had become through science.

James Urquhart

Date 1945

Country Japan

Why It's Key Demonstrated, for the first time, the immense power and destructive potential of nuclear weapons developed as part of the Manhattan Project, sparking a nuclear arms race between America and the USSR.

opposite A Hiroshima survivor with the pattern from her clothing burnt onto her skin

Key Discovery **Cooking with Microwaves**
The radar range

The contents of Percy Spencer's pocket changed cooking as we know it. Spencer was a self-taught engineer who dropped out of school when he was twelve, before joining the navy to work on wireless telegraphy.

In 1945, he was working with magnetrons, using microwaves to build a better radar. Standing next to the machine, he noticed the chocolate bar in his pocket had melted. Other people had noticed a similar effect, but Spencer wanted to know more. He sent a boy out to purchase popping corn, which he promptly popped using the power of microwaves.

The next day he put the machine next to an egg. As its insides heated up, the egg started to shake and one of Spencer's colleges stuck his head in for a closer look. This was the second that the egg exploded, leaving the man covered in hot egg.

Realizing the potential, Spencer made a box to trap the microwaves in. He patented the "Radar Range," which at the time stood nearly 1.8 meters tall, weighed more than 340 kilograms, and cost around US$5,000. The microwave generating magnetron inside had to be water cooled, so the machine needed a plumber to install it. Since then, the magnetron has become air cooled and smaller, and the microwave oven has ingratiated itself as an essential part of the kitchen. It just goes to show what you can discover if you have pockets full of chocolate and no health and safety to worry about.

Douglas Kitson

Date 1945

Scientist Percy Spencer

Nationality American

Why It's Key Revolutionized cooking and helped hurry up dinner by speeding up the heating process.

Key Invention
The weather radar

Seen as a transitional period for all areas of scientific research, 1945 signaled the end of Word War II, and with it, the refocusing of technological investigation toward peacetime objectives. Significant among this would be the use of previously secret radar techniques. By 1943, radar operators had noticed ghost, or phantom echoes on their screens, but had not initially realized their significance. Once the link with weather patterns was deduced, they experimented with, and developed methods for weather detection, paving the way for today's meteorological techniques.

Originator, innovator, and driving force behind such efforts, David Atlas is generally acknowledged to be one of the founding fathers of radar meteorology. Not only did he help revolutionize our understanding of atmospheric processes, but he has also been central in developing methods to detect and predict them.

Among the first of the Army Air Corps' radar meteorologists, Atlas subsequently spent eighteen years as chief of the U.S. Air Force's Weather Radar Branch. There he foresaw the vital role of Doppler radar in weather prediction and tracking. At the NASA Goddard Space Flight Center, he steered development of space-based, atmospheric, cryospheric, and oceanic monitoring technologies. Atlas has always taken pains to explain how and why an existing technique might be improved whilst acknowledging the great value of others' research. He holds more than two hundred patents on related equipment and, although retired since 1994, he still continues his research at Goddard.

Mike Davis

Date 1945

Scientist David Atlas

Nationality American

Why It's Key They might not enable us to get it right all the time, but the use of radars revolutionized weather forecasting.

Key Invention **Tupperware**
From chicken frames to chicken soup

With a reputation for being smelly, greasy, and pretty unreliable, the fate of commercial plastic products in the mid-1940s was far from firmly sealed. New-Hampshire-born Earl Silas Tupper, however, was to change this. A farmer's son and hopeful inventor of numerous items, from chicken-dressing frames to creaseless trousers, Tupper, after various career moves, found himself in 1937 working at the DuPont factory in Leominster, Massachusetts. It was here that he developed the skills and techniques he would need to pursue his new-found love – plastic. Keen to experiment on his own with this new material, just a year later, in 1938, he founded his own company.

The defining moment of his life arrived in the shape of a lump of stiff, black polyethylene slag, a waste product from oil refineries. Tupper developed a method for transforming the slag into a plastic that was flexible, translucent, and odor-free. Realizing the potential for a new type of food container, he designed a snap-fitting lid and, in 1946, "Tupperware" was born.

Unfortunately the world was not particularly impressed. Tupper needed an innovative selling strategy. When, in 1948, he learned that salespersons – specifically one Ms Brownie Wise – from the Stanley Home Products company were selling lots of his products to potential buyers assembled at a "party," held at a "hostess'" home, he saw the way forward. Remembering his own boyhood success with doorstep sales, Tupper combined forces with Brownie Wise and other Stanley distributors, marking the arrival of the Tupperware Home Party.

Mike Davis

Date 1946

Scientists Earl Silas Tupper

Nationality American

Why It's Key Revolutionized food storage methods and took off thanks to Tupperware parties.

Key Discovery **Dynamo Theory**
Why is the Earth a magnet?

A problem that has puzzled some of science's greatest minds, including physics' poster boy, Albert Einstein, is how the Earth – that beautiful planet known to most of us as home – became, and continues to be, a massive dipole magnet.

It is a mystery that is still far from being fully understood and, at present, there is still no real answer. A number of theories, however, have been developed, the most prominent of which is Dynamo Theory. As the center of the Earth is much too hot to be a permanent ferromagnet (the type of magnet you would use to play with iron filings), our planet's magnetic field must come from another source.

In the mid 1940s, Walter Elsasser – one of the men credited with making instrumental developments in this particular train of thought – was interested in eddy currents that occurred in the liquid core of the planet. Eddy currents occur when a conductor is passed through an existing magnetic field. The movement naturally induces an electrical current, which has a magnetic field of its own; when this field reinforces the original field (by working in the same direction as it, and not against it), a dynamo is created.

Edward Bullard (not the eponymous eddy of eddy currents, in case you're wondering) later elucidated this theory, stating that the existing magnetic field could be a product of the movement of molten iron – a conducting fluid – in the hot outer core. Work is still being carried out on Dynamo Theory to try to solve the mysteries of the inner Earth.

Chris Lochery

Date 1946

Scientists Walter Elsasser, Edward Bullard

Nationality German, British

Why It's Key Provided a solid base on which to build our knowledge of geomagnetism.

Key Invention **ENIAC**
The not-so-personal computer

Arguably the most significant of all the important scientific breakthroughs spawned by World War II was the genesis of electronic computing. The U.S. Ballistic Research Laboratory was struggling with the many calculations necessary to work out how far shells would travel when fired from a gun. They employed the talents of Dr John W. Mauchly and Dr J. Presper Eckert, from the University of Pennsylvania, to develop a way of making the calculations faster.

Mauchly was a theorist who had recently seen a newly developed electronic machine working on mathematical applications; Eckert was the engineer who put Mauchly's ideas into practice. Together they designed and built the Electronic Numerical Integrator And Computer (ENIAC), which was officially unveiled on February 14, 1946.

The computer filled one of the few air-conditioned rooms at the university. It was made up of over 19,000 vacuum tubes, 1,500 relays, and hundreds of thousands of resistors, capacitors, and inductors. Programming the machine meant manually wiring different components of the different units together, a task performed by teams of women under the direction of engineers and mathematicians.

Running almost continually until it was decommissioned in 1955, ENIAC was used in the Manhattan Project development of the hydrogen Bomb. It was also used in 1949 to calculate the value of "Pi" to 2,037 decimal places. The next important development was the production of the EDSAC, the first stored-program computer, in 1949.

Simon Davies

Date 1946

Scientists John Mauchly, Presper Eckert

Nationality American

Why It's Key A huge step forward in the development of electronic computing, a technology which transformed the world in the late twentieth century.

opposite Technicians connecting the wiring of the ENIAC device

Key Discovery **From murder to medicine**
Chemical weapon becomes the first chemotherapy

When Louis Goodman started researching chemical weapons, little did he know that his work would save thousands of lives sixty years later. At the start of World War II, the U.S. Government asked Yale to study chemical weapons. Goodman began to work on nitrogen mustard, a derivative of the blistering mustard gas used in World War I.

Goodman knew from autopsies of exposed soldiers that nitrogen mustard could destroy lymphatic tissue. After successfully using it to treat lymphomas in mice, he decided to do a human trial and injected the chemical into a patient, JD, with terminal, radiation-resistant lymphosarcoma. It worked, albeit temporarily, and JD became the first person in history to be treated with chemotherapy. In further secret trials, Goodman successfully treated sixty-seven other leukemia and lymphoma patients and, in 1946, the government allowed him to publish his results. Nitrogen mustard is still used to treat cancers today and it kick-started the discovery of many other classes of cancer drugs.

Important as that was, Goodman insisted that his main contribution to medicine was a textbook he wrote with Alfred Gilman in 1942. The 1,200-page *Pharmacological Basis of Therapeutics* cost a small fortune, and the publishers promised the duo a case of scotch if their small print run had sold in four years. All were snapped up within six weeks. The highly readable book revitalized the dying science of pharmacology. It is still used by today's students, in its ninth edition, and is edited by Gilman's Nobel Prize-winning son.

Ed Yong

Date 1946

Scientist Louis Goodman

Nationality American

Why It's Key Created an entire line of life-saving cancer treatments and revitalized the field of pharmacology.

Key Experiment **Crystal therapy**
Dorothy Hodgkin maps penicillin

Sir Alexander Fleming's 1928 observation of the antibacterial properties of *Penicillium notatum*, and the subsequent isolation and purification of penicillin at the University of Oxford, by Drs Chain and Florey, ushered in one of the greatest advances in medicine. Bacteria, however, rapidly grew resistant to penicillin, and it was difficult to make any alterations to the penicillin molecule because its structure was unknown. If only there were some way to work out how penicillin was put together.

X-ray diffraction was a new technique which could determine the structure of simpler compounds. Scientists could break larger molecules into smaller parts, akin to a jigsaw puzzle, but the task of fitting the puzzle pieces back to "reassemble" the large compound was extremely laborious. It was going to need someone special to take on the task. Born in Cairo, Dorothy Crowfoot Hodgkin was initially interested in archaeology but soon turned her attention to chemistry and ultimately took a job at Oxford studying X-ray crystallography. It was she who was the first to decrypt penicillin's labyrinthine structure, for which she received the 1964 Nobel Prize in Chemistry. The knowledge of the structure of penicillin would help scientists develop semi-synthetic penicillins, and related antibiotics such as cephalosporins, which remain major treatments of bacterial infectious diseases.

In 1964, Professor Hodgkin announced the structure of vitamin B12 and, in 1969, she finally solved the puzzle of the structure of insulin.

Stuart M. Smith

Date 1946

Scientist Dorothy Hodgkin

Nationality British

Why It's Key Knowing the structure of penicillin allowed variations to be manufactured that could outflank resistant bacteria's defenses.

opposite Professor Dorothy Crowfoot Hodgkin at work in her laboratory

Key Event **SI units**
Europe gets the measure of... everything

France in the late 1700s: The French Revolution was in full swing and the country was in the throes of violent political turmoil. An unlikely time to begin establishing the system of measurement that would become *Le Système International d'Unités*, known elsewhere in the world as the International System of Units. But it was in eighteenth-century France that the standardized system that we now call the metric system, for measuring weight and length, was created.

Over the next 150 years, the metric system grew in stature as its utility became apparent. Carl Friedrich Gauss, a major proponent of the system, made the first proper measurements of Earth's magnetic force based around millimeter, gram, and second measures.

The International Committee for Weights and Measures officially recognized a system of four standardized measures, with amperes – the yardstick for electrical current – joining meters, kilograms, and seconds, in 1946. The system wasn't officially named the International System of Units until 1960.

Since the initial agreement, three more measures have been added to the international standards; Kelvin and candela were introduced as measures of temperature and luminosity respectively, and in 1971, with the addition of moles as a measure of the amount of substance, we finally arrived at the system that is still used by scientists across the world today. All of which shows that whatever else you want to say about the International Committee for Weights and Measures, they certainly don't believe in rushing anything.

Barney Grenfell

Date 1946

Country Europe

Why It's Key Introduction of a standardized system allowed greater accuracy in scientific measurement, and meaningful discussion of scientific results across the world.

Key Discovery **The age of the Earth**
Not millions, but billions

During the early part of the century, the American physicist Bertram Boltwood developed a technique of measuring the radioactive decay of uranium which could be used to calculate the age of rocks. Using these measurements, he tentatively suggested that the Earth might be staggeringly old, somewhere between 500 million and 2.2 billion years. Unfortunately his results were largely ignored by the scientific community and Boltwood himself didn't feel the need to follow up his research.

One young Brit, however, was fascinated. While still an undergraduate, Arthur Holmes set to work improving Boltwood's technique and, by 1913, had proposed the first geological time scale.

By the 1920s, Holmes was reaching the peak of his career just as scientific interest in geology was hitting an all time low. At Durham University he was, for many years, the entire geology department; funding was virtually non-existent and he had to attempt to make his extremely precise measurements with old, patched up equipment. It is a testament, then, to Holmes' will and persistence that in 1946, he was able to announce to the world that the Earth was at least three billion years old, and quite possibly a lot older.

Even these results, however, failed to find favor in many of the higher echelons of the science world. But Holmes was to have the final say. By the time he retired in 1957, his measurements had been confirmed by others' work – the Earth was around 4.5 billion years old, and no one could now deny it.

Emma Norman

Date 1946

Scientist Arthur Holmes

Nationality British

Why It's Key The ability to date rocks and the creation of an absolute geological timescale enabled fossils to be dated and led to an increased understanding of the Earth's history.

Key Event **Cybernetics**
A small word for a huge field of study

According to the American Cybernetics Society, "The history of cybernetics is a tangled story whose contents and significance are subject to multiple interpretations." The word cybernetics itself comes from the Greek *kybernetes*, which means steersman or pilot. Cybernetics as a concept, or as an underpinning philosophy in the development of certain systems of engineering, has existed for thousands of years. For these reasons, untangling the complex history of cybernetics is difficult.

The fact that, conceptually, cybernetics can be used in so many areas – the concept has been applied to the government of people, engineering systems, artificial intelligence, and evolution – makes it difficult to define, but it is also what makes it so useful as a term.

Norbert Weiner was a brilliant polymath, attaining a BA in mathematics at the age of fourteen, before going on to study zoology and philosophy. He received his PhD from Harvard at the age of eighteen, for a dissertation on mathematical logic.

Following work in World War II on the automation of anti-aircraft guns, Weiner worked at MIT, where he coined the term cybernetics in 1946. It is defined as "the science of control and communication in the animal and the machine." He went on to publish a book by the same name in 1948, which covered the themes of control and communication within systems. He popularized cybernetics as a concept and, through the cognitive science team established at MIT, performed groundbreaking work in contemporary cybernetics.

Barney Grenfell

Date 1946

Country USA

Why It's Key Cybernetics is a mathematical system or philosophy of thought which has relevance in a vast array of fields including biology, technology, computing, engineering, philosophy, psychology, learning theory, physics, and mathematics.

opposite Professor Norbert Weiner, American mathmematician who founded cybernetics, in a classroom at MIT

Key Event

Radio waves make a round trip to the Moon

The months following the Japanese surrender in World War II should have been quiet ones for the U.S. Army. It was during this time, however, that a small group of army radar scientists awaiting discharge achieved what many thought was impossible: to make radio contact with the Moon.

Project Diana, as the mission was officially called, was led by former amateur radio operator Lieutenant Colonel John DeWitt. The project was based at the Signal Corps Laboratories in Fort Monmouth, New Jersey. The radar equipment consisted of one large antenna made up of many smaller ones, pointing toward the Atlantic Ocean and the rising moon. The equipment also comprised an extremely sensitive receiver, which DeWitt called the "real trick" of the experiment.

On January 10, 1946, the scientists beamed radar signals from their transmitter toward the moon and were able to detect a faint signal echoing back. The round trip took about two and a half seconds. The signal had penetrated the Earth's ionosphere, hit the Moon, and made its way back to Earth.

This 2.5-second event was hailed by the media as a significant scientific achievement. Now that it was possible to penetrate the Earth's atmosphere with radio waves and reach celestial bodies, the dream of contact with outer space, which had only been imagined in terms of pure science fiction until this point, had become a reality.

Rebecca Hernandez

Date 1946

Country USA

Why It's Key Paved the way for future communication between Earth and outer space.

Key Event

First working transistors trump vacuum tubes

Unless you're seriously into guitars or stereo amplifiers, you've probably never seen a happily glowing vacuum tube. Large compared to modern electronic components, vacuum tubes started to come into use around 1915. They generated a lot of heat, were expensive to produce, and prone to failure. But the thing vacuum tubes were good for, was their ability to modify or amplify an electric signal. In this manner a small broadcast radio signal could be made powerful enough to drive a speaker system. Early radio apparatus relied heavily on this method.

After World War II, Bill Shockley, John Bardeen, and Walter Brattain were working at Bell Labs in the United States on a project to design a replacement for vacuum tubes. In 1947, they produced the first working transistor.

Still a fairly large device, it relied on two strips of thin gold foil which tapered down in a "v" shape, with insulating plastic filling the v. The gold foils did not meet at the tip of the v but were separated by a distance of fractions of a millimeter, where they were lodged on the surface of a germanium plate. Due to the electrical properties of the germanium, when no signal was present at one end of the apparatus, nothing would happen at the other. But if a tiny signal was present, then it would be amplified through the germanium and produce a larger effect on the other side of the circuit.

Andrey Kobilnyk

Date 1947

Country USA

Why It's Key Transistors were essential to provide reliability in electronic equipment. They also allowed for far greater miniaturization than possible with vacuum tubes.

opposite Bardeen, Shockley, and Brattain at work in the laboratory

Key Person **Jay Forrester**
Memories and modeling

In 1936, just three weeks before enrolling in agricultural college, Jay Forrester decided that cattle weren't going to be his destiny, and signed up for engineering instead. That path led him all the way to the Massachusetts Institute of Technology. There, throughout World War II, he worked on control systems for radar antennas and guns, making them better at tracking their targets.

Having survived an excursion to fix some systems on an aircraft carrier in Pearl Harbor, Forrester took charge of MIT's new Digital Computer Laboratory in 1947. It was when he began developing a digital computer called Whirlwind – later to become America's air defense system – that he realized he needed a fast computer memory, and made his greatest contribution to computer hardware. Building on work by Harvard's An Wang, his crucial improvements made "magnetic core" computer memory the industry standard for the next twenty years. Later, Forrester took his knowledge of complex electrical systems and computers and applied it to humans. Creating a new field of study, called System Dynamics, he showed that any complex system, a city for example, could only be predicted by looking at how the individual parts interact and affect the system as a whole. Forrester used computer models of cities to show how small changes in one system can ripple through and change everything else in the city in ways that are difficult to predict.

This might sound familiar to anyone who's played the computer game "SimCity." Will Wright, its creator, based his game on Forrester's theories.

Matt Gibson

Date 1947

Nationality American

Why He's Key Revolutionized computer memory, allowing bigger and more powerful computers than ever before and applied computer power to solving social problems.

Key Event ***Kon-Tiki***
The journey of a lifetime

The *Kon-Tiki* was a famous boat named after the Inca Sun God, Viracocha, who apparently went by the same name in times gone by. Despite the rather grandiose moniker, the boat was little more than a balsa wood raft designed to copy early South American boats. The Norwegian explorer, Thor Heyerdahl, believed that, despite their modest structure, these vessels could have been used by early settlers to sail to Polynesia in the Pacific. Heyerdahl set sail on the *Kon-Tiki* on April 28, 1947, eager to prove that with only the same materials and technology available to the early South Americans, he could get to Polynesia from Peru, thus proving that the islands were populated with settlers rather than indigenous people.

Accompanied by five equally daring crew members, Heyerdahl and the *Kon-Tiki* battled the ocean waves for 101 days and 6900 kilometers. After what must surely have been a rather rocky ride, the team smashed into a reef at Raroia in the Tuamotu Islands, proving that six men and some soft wood could indeed travel a vast distance across the open ocean. The voyage also suggested that the people of Polynesia probably descended from South Americans, and that long journeys of this type could have taken place, contrary to popular belief at the time.

Soon after, a film documentary celebrated the incredible journey made by the crew and promptly won an Oscar in 1951.

Katherine Ball

Date 1947

Location Pacific Ocean

Why It's Key It proved that pre-Columbian civilizations could have traveled from South America to settle in Polynesia.

opposite Thor Heyerdahl and his balsa raft *Kon-Tiki* crossing the Pacific Ocean

Key Invention **Holography**
Through the looking glass

Hungarian-born Dennis Gabor was trying to improve his electron microscope in the 1940s, when he came up with the concept of holography – a way of producing 3D images. Gabor coined the term hologram from the Greek words *holos*, meaning "whole," and *gramma*, meaning "message."

The most well known use of holography is to produce an image of an object in three dimensions by splitting a laser beam so that half falls on a photographic plate unaltered and half reflects off the object before causing an interference pattern with the first half. After developing the image, much in the same way as a conventional photograph, a 3D image can be seen when light is shone on the plate. Interestingly, if the plate is broken in half, then each half still shows the whole image.

Today, holograms are found on day-to-day objects from credit cards and official documents, to football shirts and toys from cereal boxes. Holography is still an active research area with a variety of applications, for example in medical imaging, microscopy, navigation, data storage, and even art. Despite all these however, a functioning holodeck – made famous in *Star Trek: The Next Generation* – is yet to be developed.

Gabor was awarded the Nobel Prize in 1971, following the advent of lasers. Holograms were very difficult to create up until the discovery of lasers, which were the first example of a coherent, single-frequency light.

Leila Sattary

Date 1947

Scientist Dennis Gabor

Nationality Hungarian

Why It's Key New theory of light which has played a part in a wide range of science applications.

opposite Holography – inference patterns formed by laser beams used to record the shape of an object

Key Event **The Shelter Island Conference**
Just what is quantum mechanics?

In 1947, theoretical physics was just beginning to recover from having been focused almost exclusively on the atomic bomb throughout World War II. In order to get things going again, a series of small conferences was organized. The first was held at the Ram's Head Inn, on Shelter Island, New York.

The conference, scheduled to begin on June 2, gathered together young and promising American scientists, along with more established physicists, to debate the key problems in quantum mechanics. Participants included Edward Teller, Hans Bethe, Richard P. Feynman, and J. Robert Oppenheimer – head of the Manhattan Project.

Among the key findings presented at the conference – of which there were many – were those presented by Willis Lamb. Dubbed the Lamb shift, his results showed that the energies of electrons in hydrogen atoms weren't exactly what they were predicted to be. A few days later, on the train ride home, Bethe succeeded in producing a theoretical explanation for this on the back of an envelope. Lamb went on to win the Nobel Prize in 1955 for his work.

The conference can be seen as laying the foundations for relativistic quantum electrodynamics (QED), a hugely complex theory that describes how individual charged particles like electrons interact with photons (light particles). This theory provides the basis for all subsequent theories of quantum physics.

Anne-Claire Pawsey

Date 1947

Country USA

Why It's Key Kick-started the post war era of physics research.

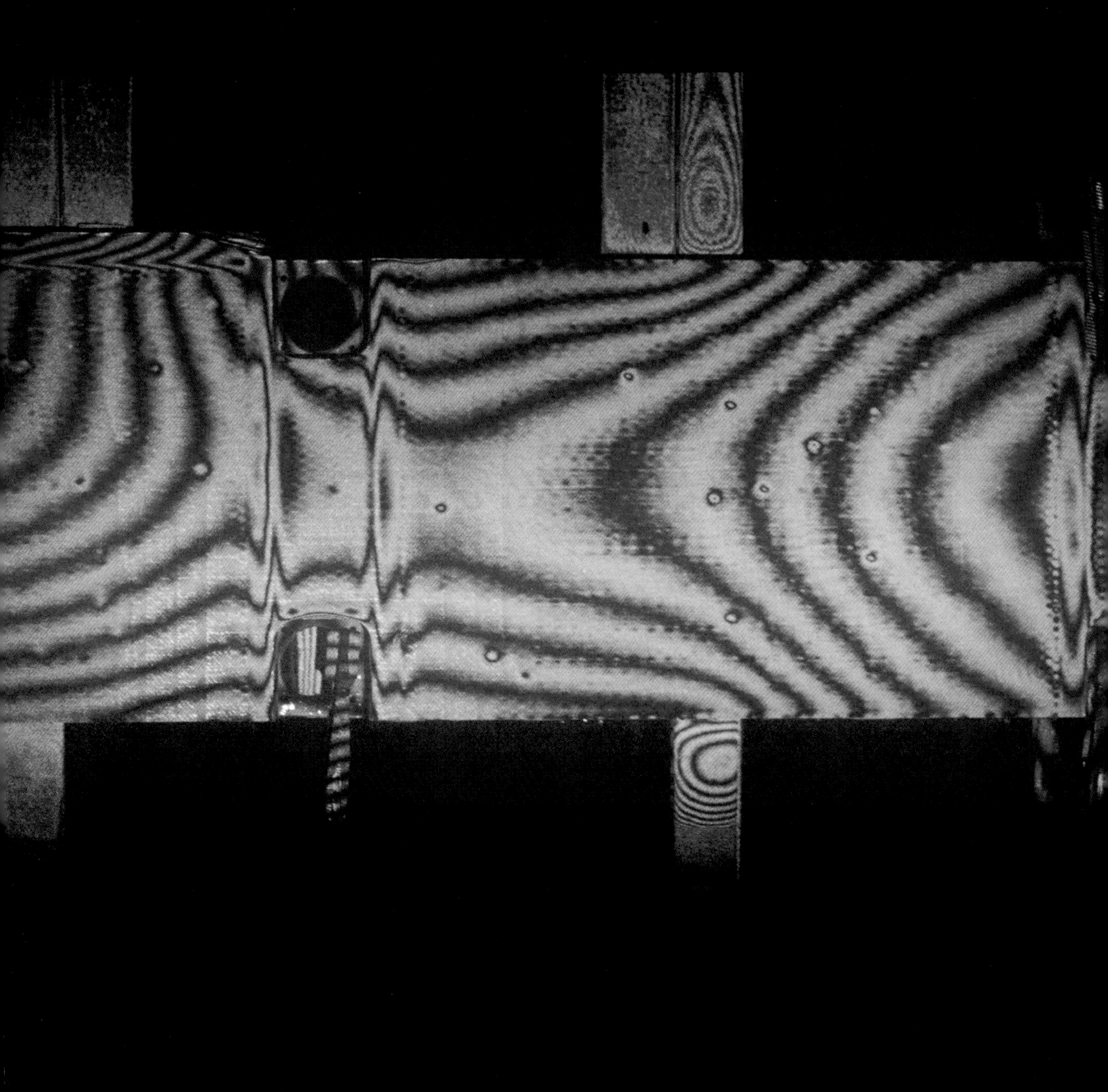

Key Invention **AK-47**
Symbol of war and icon of liberation

The AK-47, created by Mikhail Kalashnikov in 1947, was not the first assault rifle. These were developed in the late 1890s and first saw service in WWI, notably the German Sturmgewehr 44 – used with ruthless efficiency on poorly equipped Russian troops in World War II. Red Army tank commander Kalashnikov had witnessed first hand the devastation wrought on Russian troops in WWII by the Germans' newly invented automatic rifles and resolved to create a weapon to allow his people to fight to reclaim their country.

His assault rifle combined the rapid fire rate of a machine gun with the lightweight flexibility of a submachine gun to deadly effect. The AK-47 is simple but highly effective, cheap, and reliable. With both automatic and semi-automatic settings, it can fire up to 600 rounds per minute with reasonable accuracy. The gun has become one of the most common (an estimated 70 million exist worldwide) and deadly guns in history. Some analysts contend it has killed more people than any other weapon ever created. The characteristic weapon of rebels, guerrilla fighters and terrorists, the AK even features on the national flag of Mozambique.

The "success" of the AK was never more obvious than in the Vietnam War, when US troops regularly took AK-47's from their enemies to use instead of their technologically more advanced, but unreliable M-16 assault rifles. This most distinctive weapon help to create one a superpower, now torments another and shows no sign of being superseded.

Jim Bell

Date 1947

Scientist Mikhail Kalashnikov

Nationality Soviet

Why It's Key The gun itself was not the first assault rifle but it has become symbolic of many of the struggles in the latter half of the last century.

Key Event **Faster than the speed of sound**
Chuck Yeager crashes through the sound barrier

On October 14, 1947, Chuck Yeager became the fastest test pilot on Earth.

As a young man he had enlisted in the U.S. Army Air Corps, on graduation from high school, to serve in World War II. Shot down over enemy territory in 1943, just days after his first kill, Yeager managed to make his way to Spain, with the help of the French Resistance. He appealed to General Eisenhower and was allowed to fly combat missions again, completing a total of sixty-four missions and downing thirteen of the enemy's aircraft.

After the war, Captain Yeager continued to serve in the U.S. Air Force as an instructor and test pilot, until 1947, when he was chosen to test the rocket-powered X-1 plane in an attempt to break the sound barrier. At the time, experts considered the barrier impenetrable, and were not even sure if a pilot could survive a supersonic flight.

That October, after a summer of test flights, and just days after a horseback riding accident had left him with cracked ribs, the X-1 was launched from a B-29 bomber in flight at 23,000 feet. Yeager accelerated rapidly upwards, leveling out at 42,000 feet at a speed of Mach 0.92, before relighting the third chamber of his engine. At 43,000 feet the needle on his Machmeter jumped off the scale – the task was complete.

Helen Potter

Date 1947

Country USA

Why It's Key Tested the limits of the day's fastest aircraft, and of human endeavor.

opposite The world's first supersonic aircraft, the Bell X-1 in flight

6062

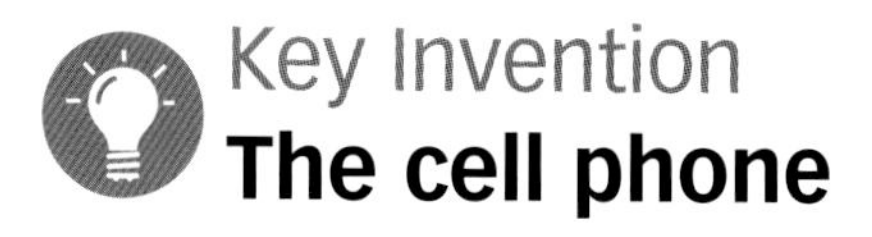

Key Invention
The cell phone

The concept of the cellular phone has been around since 1947, when Douglas H. Ring at Bell Labs proposed hexagonal "cells" for mobile phones in an internal memo. The cells referred to the hexagonal zones transmitted to by phone masts, arranged to ensure 100 per cent ground coverage. The idea was that people would be able to walk within the different cells while maintaining the same phone call. This idea was far beyond its time as the technology to create mobile phones wasn't invented until nine years later.

Land-based mobile phones had been in place since 1946 when the first commercial system was installed in St Louis, Missouri. Unfortunately it was limited to having manual connections via an operator. Despite this, twenty-five cities within the United States had installed the technology within a year.

The first fully automated mobile phone became available in 1956. The device used a single frequency (160 Hertz) and was subscribed to by 125 customers. At 40 kilograms, it wasn't exactly what you'd call portable, and its inability to sustain a signal across a cell boundary meant that users' conversations were often left hanging.

During the late 1950s and early 1960s Bell Labs continued to work on the cell phone idea. The problem of traveling between cells while still on the phone was not successfully solved until the 1970s, when economical computers and microprocessors became readily available. With these initial problems overcome, the mobile phone industry was primed to grow into the giant that it is today.

Josh Davies

Date 1947

Scientists Douglas Ring, Rae Young

Nationality American

Why It's Key The invention of the cellular network laid the groundwork for the multi-billion pound mobile phone industry that exists today. Douglas Ring and Ray Young were way ahead of their time with their ideas.

Key Event
Celsius becomes the temperature standard

Science's first breakthrough in reliably measuring temperature was by German physicist Daniel Gabriel Fahrenheit, who invented the mercury thermometer in 1714, and introduced his Fahrenheit temperature scale ten years later.

But it was Swedish astronomer Anders Celsius who proposed the most reliable and sensible method of temperature measurement. With a natural talent for mathematics, Celsius became a professor of astronomy in Uppsala in 1730, at the age of just twenty-nine. His job included making geographical and meteorological observations; it was for these weather records that he needed a precise modern method of measuring temperature. Any temperature scale must be based on two fixed, known points, with the freezing and boiling points of water being excellent and logical candidates. Celsius realized the complications and conducted careful research into how these values can vary with location, purity of water, and other factors. Using precise measurements, he discovered that while someone's position on the globe had no effect on boiling point, local atmospheric pressure has effects.

Celsius' scale originally had zero degrees as the boiling point of water and 100 degrees for the freezing point. While this may at first sound illogical, it served its purpose just fine, and avoided the use of negative numbers. The scale was reversed by Swedish scientist Carolus Linnaeus in 1745, one year after Celsius' death.

In 1948 Celsius' scale, originally called Centigrade, was formally adopted as the world standard by an international conference on weights and measures.

David Hawksett

Date 1948

Country International

Why It's Key Confirmed as a standard the Celsius scale as a reliable, and global, method of temperature measurement.

Key Publication **Going digital** *A Mathematical Theory of Communication*

Claude Shannon founded the subject of information theory when he published *A Mathematical Theory of Communication* in 1948. In information theory, "information" refers to a degree of order or non-randomness that can be measured and treated mathematically. Shannon focused his work on developing a method of expressing such information in a quantitative form – a measurable physical quantity – in an attempt to develop the most efficient way to transmit information.

In 1948, communication was thought of as requiring electromagnetic waves to be sent down a wire, but Shannon noticed a similarity between Boolean algebra – a system for logically manipulating 0s and 1s developed by George Boole – and telephone switching circuits. Shannon determined the fundamental unit of information as a yes-no situation and considered this expressed in Boolean two-value binary algebra so that 1 means "on" (the switch is closed and the power is on) and 0 means "off" (the switch is open and the power is off). A unit of information could therefore be expressed as a "binary digit" or "bit" (i.e. 1 or 0), with a combination of "bits" being used to express more complicated information.

The idea that information could be transmitted by sending a stream of 1s and 0s down a wire, was at the time fundamentally new; today it is taken for granted that pictures, words, and sounds can be transmitted using this method – it's called digital.

Sarah Watson

Publication *A Mathematical Theory of Communication*

Date 1948

Author Claude Shannon

Nationality American

Why It's Key Spelt out the concepts of the digital communications which have revolutionized the way we watch TV, use the phone and... well, communicate.

Key Experiment **Virus grown in culture for the first time** A breakthrough for vaccine production

In 1948, three scientists in Boston, USA, became the first to grow viruses in cell culture, leading to a breakthrough in the race to find effective and safe vaccines against infectious diseases.

John Franklin Enders, Thomas Huckle Weller, and Frederick Chapman Robbins had been researching how the immune system fights disease, but had become frustrated with inadequate microscopes and the inability to grow viruses outside of a living organism. This frustration led them to pioneer a technique called cell culture, where living cells are grown in a laboratory under conditions that mimic the environment of the organism they come from.

As they refined their technique, using antibiotics to stop bacteria contaminating the cultures, they were able to grow many generations of polio virus in cells from monkey kidneys. They could then select mutant strains of the virus that had become well-adapted to living under culture conditions, but were no longer able to thrive inside a human host. Viruses like this are called "attenuated" and form the basis of so-called "live vaccines," due to their ability to induce immunity to a disease, without causing any of the associated symptoms.

The work by Enders and his colleagues eventually led to the development of the modern-day polio vaccine, and the three received the Nobel Prize in 1954 in recognition of their achievement. Enders later went on to develop the first vaccine for measles in 1962.

Hannah Isom

Date 1948

Scientists John Enders, Thomas Weller, Frederick Robbins

Nationality American

Why It's Key Vaccination has become an extremely important tool for controlling – and even eradicating – disease, and the work of Enders has formed the basis of vaccines for diseases like polio, measles, and diphtheria.

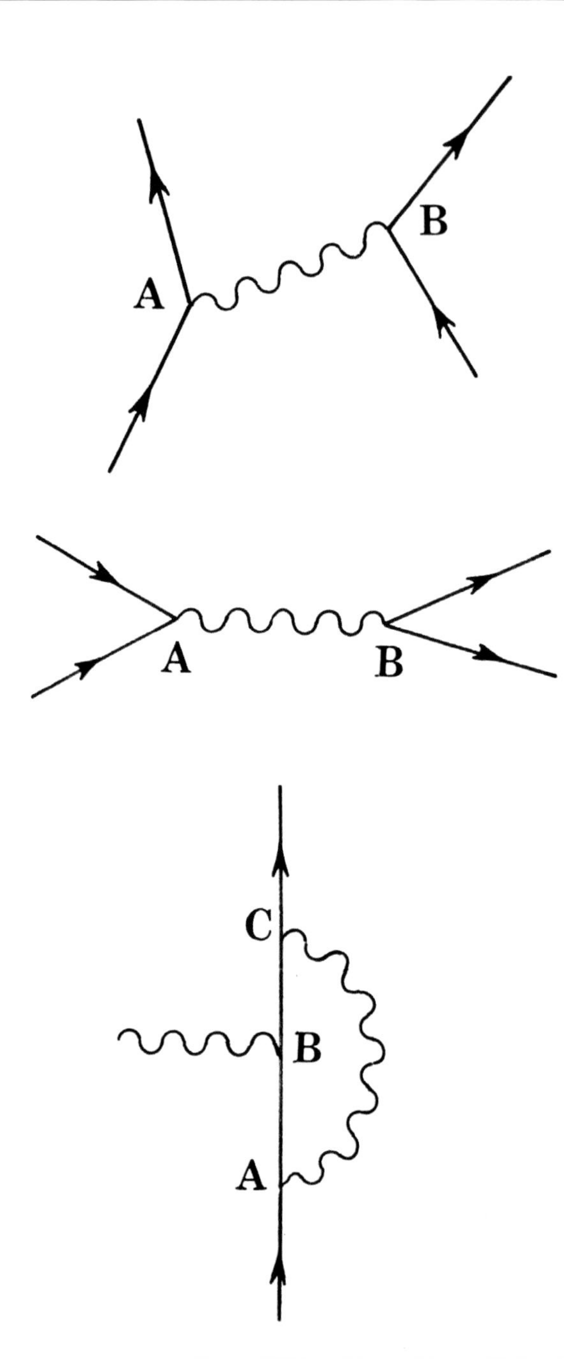
A
B
A
B
C
B
A

Key Discovery **Quantum electrodynamics**
The jewel of physics

In 1965, Sin-Itiro Tomonaga, Julian Schwinger, and Richard Feynman shared the Nobel Prize in Physics for their work on quantum electrodynamics. Feynman was a charismatic character and enjoyed interacting with his students and the public. There is a story though that asserts that even Feynman, when approached by a reporter and asked to explain the subject so that anyone could understand it, retorted, "Hell, if I could explain it to the average person, it wouldn't be worth a Nobel Prize."

Quantum electrodynamics is mathematically complicated, dealing with events at the subatomic level which are so counter-intuitive that they seem completely alien to the way in which most of us view the world. Much of this is due to the fact that the theory merges quantum mechanics with electromagnetism, field theory, and Einstein's relativity. It takes up the problem of how particles, particularly electrons, produce electromagnetic fields, and how they are influenced by these fields.

While the groundwork for quantum electrodynamics was set in the 1920s by Heisenberg, Dirac, and Pauli, some of the results produced by the early mathematical model resulted in meaningless numbers – infinities. The trio who won the 1965 Nobel Physics Prize expanded on these earlier efforts, and by 1948, had produced refinements to quantum electrodynamics which, when applied to results from experiments, were accurate to many decimal places. "Q. E." is still widely viewed as the most successful and experimentally validated theory in physics.

Andrey Kobilnyk

Date 1948

Scientists Sin-Itiro Tomonaga, Julian Schwinger, Richard Feynman

Nationality Japanese, American

Why It's Key Q.E. has been validated as an accurate model for the interaction of particles and fields, and was a stepping stone for the physics which followed

opposite Three Feynmann diagrams to help calculate electromagnetic interactions of charged subatomic particles

Key Event **WHO founded**
Healing the world

The World Health Organization (WHO) was officially established on April 7, 1948 – now World Health Day – as one of the original agencies of the fledgling United Nations. It inherited the tasks of epidemic control, drug standardization, and quarantine measures, from its predecessor, the Health Organization of the League of Nations, as well as from the International Office of Public Health in Paris. But the WHO also concerned itself with a much larger mandate.

The WHO defines health as "a state of complete physical, mental, and social well-being, and not merely the absence of disease or infirmity," and is committed under its constitution to promoting "the attainment of the highest possible level of health" to all people. This means that, as well as monitoring outbreaks of infectious diseases such as malaria, SARS, and AIDS, the World Health Organization also initiates treatment and prevention programs for many diseases, distributes safe and effective vaccines and drugs, and educates member states in their use.

One of the WHO's greatest achievements came in 1979, when it announced that smallpox had been eradicated worldwide, making it the first disease to be conquered purely by human endeavor.

In addition to education and disease prevention, the organization also carries out its own research into health risks such as smoking, ionizing radiation, and healthy diets. Although these results can sometimes be controversial, most of the health guidelines by which we live our lives today are the result of work by WHO.

Helen Potter

Date 1948

Country Global, based in Switzerland

Why It's Key Founding of the global organization responsible for world health.

Key Publication **"The Origin of Chemical Elements"**
As easy as Alpher, Bethe, Gamow

Why do only two elements make up 99.9 per cent of the atoms in the Universe? Ralph Alpher and George Gamow answered this question in a landmark scientific paper, to which they added another author, Hans Bethe, as an April Fools' joke.

A total of 92 per cent of the atoms on the surface of the Sun are hydrogen, almost eight per cent are helium, and only one in a thousand is anything else (oxygen, carbon, iron, and so on). In 1946, Gamow proposed that it might be possible to explain these abundances if the elements had been formed in rapid nuclear reactions as the Universe cooled off, shortly after the Big Bang. Alpher – Gamow's doctoral student – performed the calculations and discovered that there should be ten atoms of hydrogen for every atom of helium; almost exactly the ratio astronomers were observing in stars. Alpher and Gamow wrote up the results as a letter to the *Physical Review*. When Gamow discovered it would be published on April 1, he added prominent physicist Hans Bethe to the author list, to make it the Alpher-Bethe-Gamow (or α-β-γ) paper.

While being vital, the research was not able to explain the abundances of elements beyond helium, which were later found to arise through nuclear reactions in stars and supernovae. Fittingly, Hans Bethe played an important role in these discoveries.

Eric Schulman

Publication "The Origin of Chemical Elements," *Physical Review*

Date 1948

Authors Ralph Alpher, George Gamow

Nationality American, Soviet

Why It's Key Explained why there is so much hydrogen and helium in the Universe, and provided important support for the Big Bang theory.

Key Invention **Velcro**
A burr in our side

One could say Velcro was invented by nature; that it is a product of evolution. If it were not for a Swiss electrical engineer, however, Velcro would still be a natural nuisance, and not the wonder fastener it is today. In 1941, George de Mestral examined some burrs that had become stuck to his dog's fur. Inspired by their simple mechanism, he spent the next eight years developing Velcro, named from the French word *velours*, meaning "velvet," and *crochet*, or "hook."

The concept of Velcro is relatively straightforward. One side contains many miniscule, loose loops – the fuzzy side – and the other, many small hooks – the rough side. When the two sides meet, the hooks catch on to the loops and the hundreds of hook bonds hold the sides together firmly. Thanks to de Mestral's observance and dedication, we are now able to enjoy Velcro in all avenues of closure and fastening, not to mention the musical sound of hooks ripping from the fuzz. Stainless steel Velcro is even used in the car manufacturing industry to attach parts such as bumpers. As opposed to bolts, the vibration of the car actually reinforces the strength of the bond, since any hooks that come loose are more likely to re-attach again due to the motion.

Most children, struggling to tie their shoes, have probably wondered aloud as to why we use other fasteners at all.

Logan Wright

Date 1948

Scientist George de Mestral

Nationality Swiss

Why It's Key An effective product useful in a wide variety of applications, Velcro acts as a testament to the value of an inquiring, scientific mind.

opposite A false color scanning electron microscope close-up of a Velcro hook

Key Discovery **Paracetamol**
The forgotten painkiller returns

For a person with mild pain, like a headache or sore throat, modern chemists stock entire shelves of products that provide relief. People in the late nineteenth century weren't so spoiled for choice.

In 1886, chemists hunting for drugs to reduce fever hit upon acetanilide, which turned out to be a potent painkiller. However, it also caused a blood disorder called methemoglobinemia, associated with blue lips and darkened skin. In the hit-and-miss hunt for alternatives, two related chemicals were soon created. The first, phenacetin, became widely used but the second, paracetamol, was not, largely because German scientist Joseph von Mering wrongly thought that it too caused methemoglobinemia. Ignored, paracetamol languished in obscurity for decades. It was effectively rediscovered in 1948 by Julius Axelrod and Bernard Brodie, who had seen a report that found paracetamol in the urine of people taking acetanilide. Together they showed that that our bodies convert acetanilide and phenacetin into paracetamol, and that it was this third drug that accounted for the pain-killing and fever-reducing qualities of the other two.

Axelrod, Brodie, and others soon recommended paracetamol over other contemporary painkillers. It had all of its parent drugs' benefits but none of their weaknesses, and, unlike aspirin, long-term use didn't cause ulcers. Paracetamol was first marketed in the UK as a children's medicine in 1955, and as over-the-counter tablets called Panadol in 1956. For unearthing this safe pain remedy from the vaults of history, our heads owe great thanks to Axelrod and Brodie.

Ed Yong

Date 1948

Scientists Julius Axelrod, Bernard Brodie

Nationality American

Why It's Key Led to the widespread production of a safe and effective painkiller.

Key Person **Alfred Charles Kinsey**
"Dr Sex" gives it to us straight

Alfred Kinsey (1894–1956) wasn't always a sex fanatic. He started his career as an assistant professor of zoology at Indiana University, and it wasn't until the students called for a marriage course to be established that Kinsey discovered his fascination with sexual behavior. He took it upon himself to organize the course, and, with little knowledge of the subject, began to gather case histories.

Soon after the establishment of the course, the university's president gave Kinsey an ultimatum: to continue with his research, or the course. He chose the research. However, he was soon dismissed as a sex education lecturer due to his open support for contraception. This setback didn't stop Kinsey. In 1942, he set up the Institute for Sex Research and published many controversial books, including *Sexual Behavior in the Human Male* in 1948, which concluded that homosexual acts were much more common than anyone at that time imagined.

Although his book was a best seller, the subject of sex was still taboo in 1948, and religious critics believed his work to be immoral. When his sequel, *Sexual Behavior in the Human Female*, was released in 1953, America was afraid the book was a Communist plot to corrupt the moral norms and Kinsey was wrongly accused of being a supporter of the Reds. Consequently, his funding was withdrawn. He died of a heart attack a few short years later. Despite this, Kinsey remains a popular figure in America, and a musical based on his life, entitled *Dr Sex*, premiered in Chicago in 2003, and has since won many awards.

Faith Smith

Date 1948

Nationality American

Why He's Key Revolutionized attitudes toward sex and sexual behavior.

Key Discovery
Lobotomy developed for mental illness

In 1949, the Nobel Prize for Medicine was awarded not for an antibiotic, transplant, or X-ray technique, but for the frontal lobotomy. In the 1930s, a Portuguese neurologist by the name of António Moniz found out that, by severing the connections between the frontal lobes and the rest of the brain, it is possible to improve certain mental illnesses. The patients who underwent the operation were largely schizophrenic, and some dramatic improvements were noted – delusions and hallucinations all but disappeared in some cases. Hailed as a medical miracle, the lobotomy found favor worldwide, but it wasn't long before side-effects were being noted as well.

Those who had undergone the procedure appeared to lack emotion, drive, and even personality. Coupled with the fact that that new drug thorazine seemed to better treat schizophrenia, the lobotomy soon lost its popularity.

So how did such a seemingly barbaric procedure – early procedures involved drilling holes in the skull a nd inserting an ice pick-like device into the cerebral hemispheres – gain medicine's most prestigious prize? Well, one has to consider the alternative. Prior to the invention of the lobotomy, there were almost no treatment options available to physicians trying to treat their most desperately mentally ill patients. As a result, any procedure that would improve severe psychosis, no matter how outlandish it might seem, was Nobel Prize-worthy.

B. James McCallum

Date 1949

Scientist António Moniz

Nationality Portuguese

Why It's Key The lobotomy represented the first therapy for mental illness, even if it does seem cruel by today's standards.

Key Experiment
Discovering the cause of sickle cell anaemia

In 1930, Linus Pauling – famous for his work on chemical bonding – turned his attention to organic chemistry and in particular the protein hemoglobin. Haemoglobin molecules contain iron atoms and are employed by the body to transport oxygen around in the bloodstream.

A switch in the direction of Pauling's research led him to begin thinking about hemoglobin in terms of immunology – the study of the immune system, which plays a vital role in our bodies' defense against disease. This paid off when in 1945 he first heard of sickle cell anemia, a disease of the blood in which red blood cells bend into a crescent shape when devoid of oxygen. Pauling said that when he heard of the disease it took him no more than two seconds to understand what was happening.

He already knew so much about the structure and function of hemoglobin, and about chemical bonding, that he realized immediately the sickle shape must be due to the haemoglobin molecules. He knew that when hemoglobin is carrying oxygen, iron in the molecule forms a certain type of bond – a covalent bond – with the rest of the molecule. When it is deoxygenated, the iron forms a different type of bond – an ionic bond, which is due to the attraction of opposite charges.

In 1949, Pauling's research team performed experiments showing that chemical bonding is markedly different in the hemoglobin of sickle cell anaemia sufferers. This led to the conclusion that the disease was caused by molecular differences which must have been inherited.

Simon Davies

Date 1949

Scientist Linus Pauling

Nationality American

Why It's Key A major breakthrough in the understanding of genetically inherited diseases.

Key Publication ***The Production of Antibodies***
What's virus and what's me?

Sir Frank MacFarlane Burnet known as "Mac" to his friends – was a shy but opinionated Australian who modestly called himself "the last of the great amateurs." Yet he made a prediction in 1949 that was to inspire a new avenue of research into the body's immune system, and win him worldwide recognition.

An avid reader, Burnet had found reports describing how viruses could infect fetuses in the womb (in mice), and remain long after birth without triggering the production of antibodies or being destroyed. Burnet realized that fetal life is a special time when the body's defenses accept whatever is present as "self." In *The Production of Antibodies* he predicted that it should be possible to trick the immune system artificially into accepting foreign matter by placing it in the body during fetal life.

Burnet was too busy with other projects to follow up his idea. Nonetheless, its originality earned him the Nobel Prize for Physiology or Medicine in 1960, shared with Sir Peter Medawar, who carried out animal experiments showing the prediction to be true. Their work laid the foundation for understanding how the immune system tells the difference between the body's own tissues and foreign material – including organs transplanted from another person – and has inspired research into new treatments for improving the survival of transplant patients.

Julie Clayton

Publication *The Production of Antibodies*

Date 1949

Author Frank Macfarlane Burnet

Nationality Australian

Why It's Key Discovering how the immune system tells the difference between foreign material and the body's own "self" revolutionized transplant medicine and our understanding of autoimmune diseases, like rheumatoid arthritis.

Key Discovery **Radiocarbon dating**
Measuring history

You may not realize it, but in every fiber of your being there is a radioactive stopwatch just waiting to start a countdown. As a concept it sounds fairly terrifying, but there is a perfectly natural explanation.

Carbon appears in great abundance in our atmosphere and a trace amount of this carbon – one part per trillion – is an unstable, radioactive isotope known as carbon-14. Every living thing is made up of carbon, and their survival is dependent upon its constant intake; therefore organisms contain a concentration of the rare isotope that matches that of the surrounding atmosphere.

As soon as a plant ceases to photosynthesize, or an animal ceases to eat and breathe – widely considered to be foolproof symptoms of death – so too does their intake of carbon-14. The isotope is no longer being replenished from the surrounding atmosphere, and whatever carbon-14 is present in the organism's body at the time of death begins to decay.

Carbon-14 has a very slow decay rate. Its half-life (the time taken for the initial amount of a sample to reduce by half, due to radioactive decay) is 5,730 years, give or take the odd half century. So by measuring the amount of the isotope present in old organic samples, it is possible to trace them back to the time when their decay began. The process allows us to date organic remains from as long as about 60,000 years ago. The discovery, made by Willard Libby and his colleagues at the University of Chicago, was significant enough to win a Nobel Prize in 1960.

Chris Lochery

Date 1949

Scientist Willard Libby

Nationality American

Why It's Key Gave us a greater and more detailed insight into the organic remains recovered from archaeological sites, resulting in a more accurate understanding of history and chronology.

opposite Carbon dating has been used to calculate the date of the famous Turin Shroud

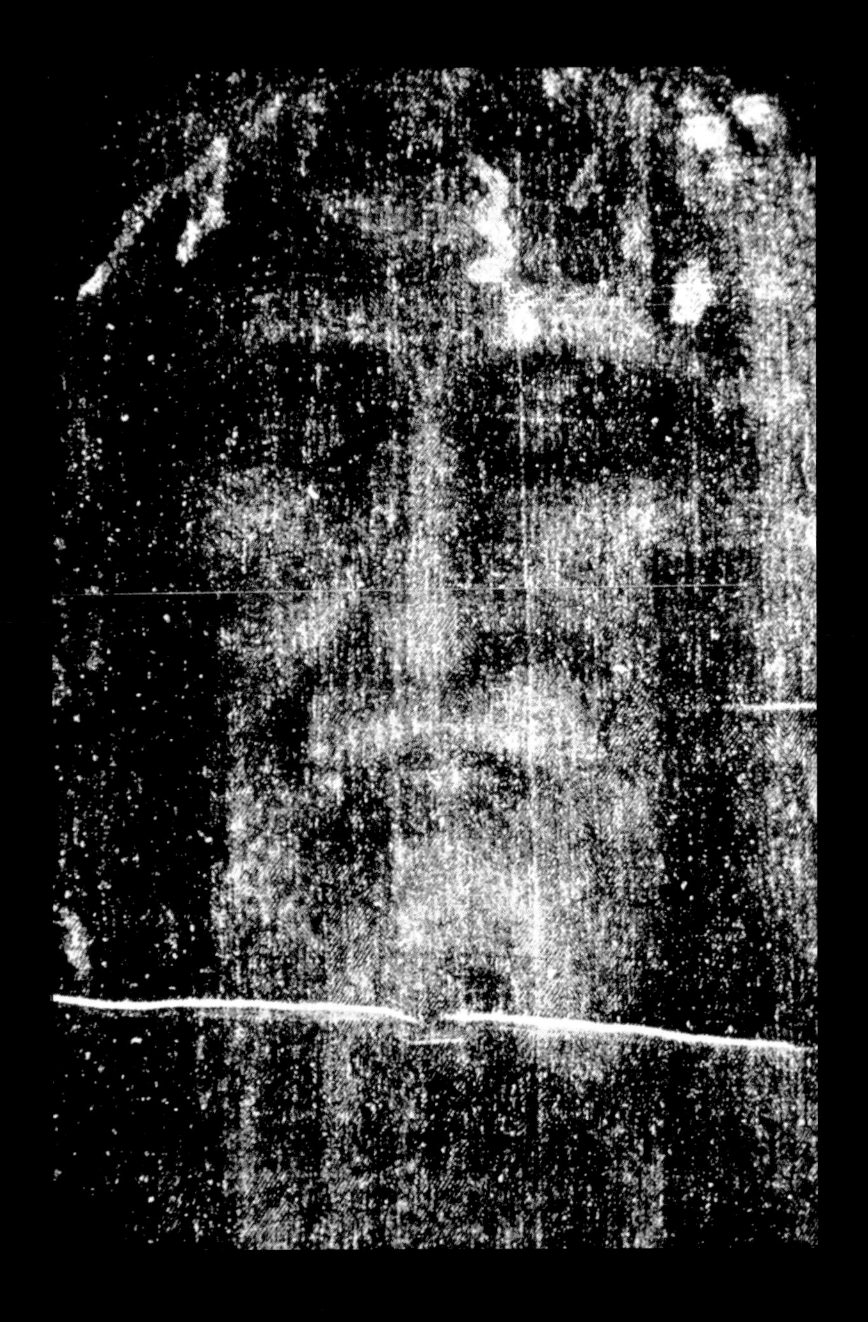

Key Event **Ahead of his time**
Sir Harold Ridley implants an intraocular lens

Following World War II, British ophthalmologist Sir Harold Ridley observed that British Royal Air Force pilots who had shards of plastic canopy in their eyes did not have any signs of a foreign body reaction. Years earlier, he had been struck by the comments of a medical student, who had remarked that it was a pity that they could not replace the opaque cornea they were removing with a clear one. At that time, the only treatment for cataracts – the clouded lens caused by accumulations of dead cells – was removal, which left the patient with clear, but unfocused vision. Thick glasses were required to see more clearly. Prior attempts at lens replacement were hampered by the body rejecting the implant.

Ridley's search for a material that would not be rejected led him to John Pike, an optical scientist at the lens manufacturers Messrs Rayners, who was able to obtain high-quality plastics to design and make the artificial lens.

On November 29, 1949, at St Thomas' Hospital, during a procedure shrouded in secrecy, Ridley implanted a lens made of Perspex, the same material as the aircraft canopies. His lens was bulky by today's standards and the surgical instruments were archaic, but the lens was tolerated, and vision was restored.

Today, over five million people receive intraocular lens transplants every year.

Stuart M. Smith

Date 1949

Country British

Why It's Key Ridley revolutionized the treatment of cataracts and heralded the birth of biomedical implants.

Key Event **EDSAC**
Stored-program computer put to work for first time

EDSAC – the Electronic Delay Storage Automatic Calculator – wasn't, in fact, the very first stored-program computer. That honor goes to the Manchester Small-Scale Experimental Machine, the "Baby" of Manchester University, which ran its first program in 1948. The EDSAC, though, was the first programmable computer that was put to work in the real world. It wasn't an experiment in pure computing; it was designed from the beginning to be a practical tool.

At a time when the transistor was so new that it had yet to be named, EDSAC was a valve-based computer, and took up a whole room at Cambridge University's Mathematical Laboratory. Programs were inputted on punched paper tape. The "electronic delay storage" of EDSAC's name comes from its memory system, which was based on mercury delay lines; the computer remembered its instructions as physical waves riding from one end of a tube of mercury to the other, over and over again.

EDSAC, built by Maurice Wilkes and his team, first ran a successful program on May 6, 1949, calculating a table of squares. Later, EDSAC was opened up for use by other departments. With EDSAC's help, geneticist Ronald Fisher was the first scientist in the world to publish a paper using computer calculations to support its conclusions. Another first may have come from graduate student Sandy Douglas, who made the EDSAC play noughts and crosses on a cathode ray tube screen, creating what is believed to be the world's first video game.

Matt Gibson

Date 1949

Country UK

Why It's Key It proved that computers could be truly useful in an academic environment. It also inspired business computing – the Lyons Electronic Office, the first real business computer, was based on EDSAC.

opposite The EDSAC computer

Key Person **Jacques Cousteau**
From Manfish to Captain Planet

In 1936, a serious car crash cut short Jacques-Yves Cousteau's (1910–1997) training as a pilot. No longer able to explore the skies, he turned instead to the sea; that same year he took a pair of goggles and went swimming in the Mediterranean. It was a revelation that changed his life. It has since been said that most of what the public knows of life under the ocean is thanks to the films and campaigns of Captain Cousteau.

Having discovered a fascination for the underwater life, Cousteau soon became frustrated by the diving equipment of the day, which didn't allow him freedom to move around. So, with engineer Émile Gagnan, he invented the first SCUBA equipment (Self-Contained Underwater Breathing Apparatus), earning himself the nickname Manfish.

In 1950, Cousteau purchased Calypso, an ex-minesweeping boat, fitted her with a slew of scientific equipment, and embarked upon a four-decade-long mission to study and film the underwater world. Through his 115 television films and fifty books, Cousteau brought the secrets and wonders of the ocean into millions of people's homes.

He was the first to understand and highlight the effects of marine pollution and, in 1974, set up the Cousteau Society. In 1990, he launched a successful petition to block mineral extraction in the Antarctic. A further petition calling for a protection of "Rights for Future Generations" – signed by over five million people – was presented to the UN in 1992.

Mark Steer

Date 1950

Nationality French

Why He's Key Pioneered the investigation of the seas, and became a leading environmental campaigner whose influence was felt across the globe.

Key Experiment
A Prisoner's Dilemma

The Prisoner's Dilemma describes the following hypothetical situation: Two criminals are captured by police. The police suspect they are responsible for a murder but they do not have enough evidence to prove it in court. However, they are able to convict them of a lesser charge, for example carrying a concealed weapon. The prisoners are put in separate cells with no way of communicating with each other. Each is offered the opportunity to confess.

If neither prisoner confesses, both will be convicted of the lesser offence and sentenced to one year in prison. If both confess to murder, both will be sentenced to five years. If, however, one prisoner confesses while the other does not, the prisoner who confessed will be granted immunity but the prisoner who did not confess will be sentenced to twenty years.

What should each prisoner do? In 1950, puzzles with the structure of the Prisoner's Dilemma were discussed by the mathematicians Merrill Flood and Melvin Dresher. The Flood-Dresher Experiment involves running the Prisoners Dilemma game one hundred times between two players. Playing the game multiple times, as in this experiment, results in it being beneficial for both players to stay silent; however on the hundredth move, both players would confess since there can be no fear of retaliation from their opponent after the final move. Work on the Prisoners Dilemma by Flood and Dresher formed part of the RAND Corporations investigation into game theory, pursued due to the possibility of its application to the global nuclear strategy.

Sarah Watson

Date 1950

Scientists Merrill Flood, Melvin Dresher

Nationality American

Why It's Key The Prisoner's Dilemma has applications in social sciences (economics, politics, and sociology) as well as the biological sciences (ethnology and evolutionary biology).

Key Person **John Nash**
A beautiful mind

John Nash (b. 1928) has not always been devoted to developing mathematics; he began by studying chemical engineering and then changed to chemistry, before finally settling on mathematics, having been assured that it was not impossible to make a good career as a mathematician in America. Nash's mathematical ability soon became apparent when his letter of recommendation to graduate school consisted solely of the line: "This man is a genius."

It was early on while studying at Princeton University that Nash decided against learning second hand, from lectures or from books, and instead began developing his own mathematical theories. He showed an interest in a range of pure mathematics topics including game theory, logic, algebraic geometry, and topology; and went on to publish a substantial number of papers from his work. In 1958, at the age of twenty-nine, Nash began to experience the symptoms of paranoid schizophrenia. His condition worsened over the following years, but despite the involuntary periods spent in hospital, he continued his mathematical work.

By the 1990's Nash had recovered from the schizophrenia that had tormented him. His ability to produce high-quality mathematics had not left him, and he returned to the Department of Mathematics at Princeton University, where he continues his research in logic, game theory, cosmology, and gravitation to this day. John Nash's story is widely known outside the mathematical community from the film *A Beautiful Mind*. The film is based on Nash's mathematical career and struggles against mental illness.

Sarah Watson

Date 1950

Nationality American

Why He's Key A key figure in the development of many mathematical theories.

Key Publication ***Foundations of the Theory of probability***

The mathematical field of probability theory had been constructed using a common point of view for some time prior to 1933, but what was lacking was a complete and concise presentation. This was resolved in 1933 by Andrei Nikolajevich Kolmogorov's publication of *Grundbegriffe der Wahrscheinlichkeitrechnung* which was later translated into Russian and then, in 1950, into English as the *Foundations of the Theory of Probability*.

Kolmogorov's earlier mathematical achievements gave him the authority to define the axioms, or rules, of probability which were to become the foundation of the modern theory. Although *Foundations of the Theory of Probability* is now out of print, its structure survives in many modern probability books. It brought together the work of many leading mathematicians in the field of probability and molded them into a cohesive unit.

A great deal of research on probability theory has been generated and influenced by Kolmogorov's approach to the subject, predicting everything from market forces and business performance, to weather patterns and the outcomes of horse races. Probability theory has even been used to debunk claims of extra-sensory perception. Such were the far-reaching natures of the theories and proofs supplied in his book, that Kolmogorov was able to establish his reputation as one of the world's leading experts in the field. He is regarded as possibly the most influential mathematician of the Soviet era.

Sarah Watson

Publication *Foundations of the Theory of Probability*

Date 1950

Author Andrei Kolmogorov

Nationality Soviet

Why It's Key Established the foundations of modern probability theory.

Key Publication **Smoking can seriously harm your health** Original evidence presented

The keen observations and dogged persistence of one man in the 1950s revolutionized medical understanding of today's most common type of cancer – lung cancer.

Sir Richard Doll joined the British Medical Research Council in 1948, as part of a project investigating an increase in mortality due to lung cancer. Eighty per cent of men were regular smokers at that time; smoking was considered a completely harmless habit. More cars on the roads and more roads being built meant that workers in particular were exposed to much higher levels of fumes and tar than we're used to today. Doll's research partner, Austin Bradford Hill, first attributed the increase in lung cancer to industrialization.

After interviewing hundreds of lung cancer patients, however, the real culprit was revealed; in 649 cases of lung cancer, only two were non-smokers. But despite overwhelming evidence collated by the pair, and a key paper outlining the evidence in the British Medical Journal, it took another four years for the UK government to accept their findings.

The effects of smoking were to become Doll's life's work. Just twelve months before his death in 2005, Hill published the results of a fifty-year study on mortality associated with smoking, concluding that even regular smokers, on quitting the habit, could prolong their lives. By the summer of 2007, the risk smoking posed to public health was so well-recognized that it was banned in all public places in many European countries, and in several states in America.

Ceri Harrop

Publication "Smoking and Carcinoma of the Lung"

Date 1950

Author Richard Doll

Nationality British

Why It's Key Made history, not only by proving smoking caused lung cancer, but also by putting epidemiology – the study of disease in populations – firmly on the medical map.

opposite Sir Richard Doll

Key Publication **Cosmic clouds** Oort's comet theory

Ernst Öpik, an Estonian astronomer, proposed in 1932 that all comets entering the near vicinity of our planet have a common origin. He believed that they originated from a dense field of comets orbiting our Sun at the edge of the Solar System.

The Dutch astronomer Jan Oort arrived at the same conclusion in 1950, when he calculated that there must be a source of comets – a "cloud" – just at the edge of the Sun's domain. He was working with the theory that comets have been around since the start of the Solar System, but are destroyed by multiple passes through it. He suggested that the cloud could contain trillions of comet nuclei which could be forced into the inner Solar System at any point by the gravitational forces of passing stars. The edge of the Oort Cloud is thought to be the extent of the Sun's gravitational influence. Despite the fact that both scientists came up with the same idea, and although Oort acknowledged Öpik's work in his paper, it was Oort who received all the credit, which is why we now refer to these comet fields as "Oort clouds."

There have still been no confirmed observations of the Oort cloud, but two comets have been found which have trajectories suggesting they originate from that area.

Josh Davies

Publication "The Structure of the Cloud of Comets Surrounding the Solar System and a Hypothesis Concerning its Origin"

Date 1950

Author Jan Oort

Nationality Dutch

Why It's Key Oort's hypothesis concerning the cloud at the edge of the Solar System explained the source of the anomalously large number of inner Solar System comets.

Key Invention **Electrophoresis**
Using gel and electricity to separate proteins

To study important biological molecules like DNA and proteins, scientists have to be able to separate them out. Gel electrophoresis is a really useful technique for doing exactly that.

A gel is a polymer substance which forms a solid but porous matrix. In gel electrophoresis, the protein is added to the gel, and an electric current applied. The current "pushes" or "pulls" different molecules through the gel at different speeds, and in different directions, depending on the size of the electric charge on the molecules, and whether they are positively or negatively charged. At the end of the process, the molecules can be stained using a dye so they can be seen, sometimes under ultra-violet light.

The result is a series of bands in different "lanes" on the gel – bands that end up at a given distance from one end contain molecules that passed through the gel at the same speed, which usually means they are about the same size. This technique allows scientists to analyze biological substances and to purify them to be used in other experiments.

The British-born American geneticist Oliver Smithies is credited with inventing the modern method of gel electrophoresis in 1950. He is probably better known for his later work in medical genetics, especially stem cell technology, which led to a Nobel Prize in 2007. Gel electrophoresis is still widely used in molecular biology, genetics, and biochemistry and has important applications in forensic science; in producing DNA "fingerprints" for example.

Richard Bond

Date 1950

Scientist Oliver Smithies

Nationality American

Why It's Key Separating complex biological substances like DNA and proteins is crucial to enable scientists to study and use them in a whole range of ways.

opposite Examining an electrophoresis gel under UV light

Key Event
Myxomatosis introduced to Australia

After living in Australia for twenty-eight years, Englishman Thomas Austin decided he wanted to shoot rabbits. So, in October 1859 he had twenty-four shipped over from the U.K. and released them onto his Victoria estate… a bad move. Rabbits quickly became Australia's biggest agricultural pest; the country's wool industry was pushed towards breaking point as up to a billion bunnies decimated grazing lands and wiped out native plants. Something had to be done.

Conventional methods of control – trapping, shooting, and poisoning – just weren't effective against the "grey blanket," so the Australian government looked to a different method: disease. Myxomatosis, first discovered in Uruguay in 1896, infects only rabbits. But, boy, does it infect them. It was first released into Australia in 1950, and two years later the rabbit population had plummeted from 600 million to 100 million. To this day it remains an important control measure in Australia, even though the rabbits there are becoming increasingly resistant to the virus.

While myxomatosis was ravaging rabbits Down Under, it was also gaining a generally less welcome foothold on the other side of the world. Unintentionally released into France in 1952, the disease spread rapidly throughout Europe, killing off 95 per cent of rabbits, with unfortunate consequences. For example, the Iberian lynx, a spectacular species of cat, and rabbit-hunting specialist, was left with nothing to eat. Its numbers also dropped precipitously, and there is still a very real danger that it will become the first wild cat species to go extinct for at least 2,000 years.

Mark Steer

Date 1950

Country Australia

Why It's Key Controlling pests using diseases is a tricky business. Myxomatosis gave Australia back much of its farmland, paving the way for its use to control rabbits in other countries. But its introduction had unforeseen negative impacts on many species in Europe.

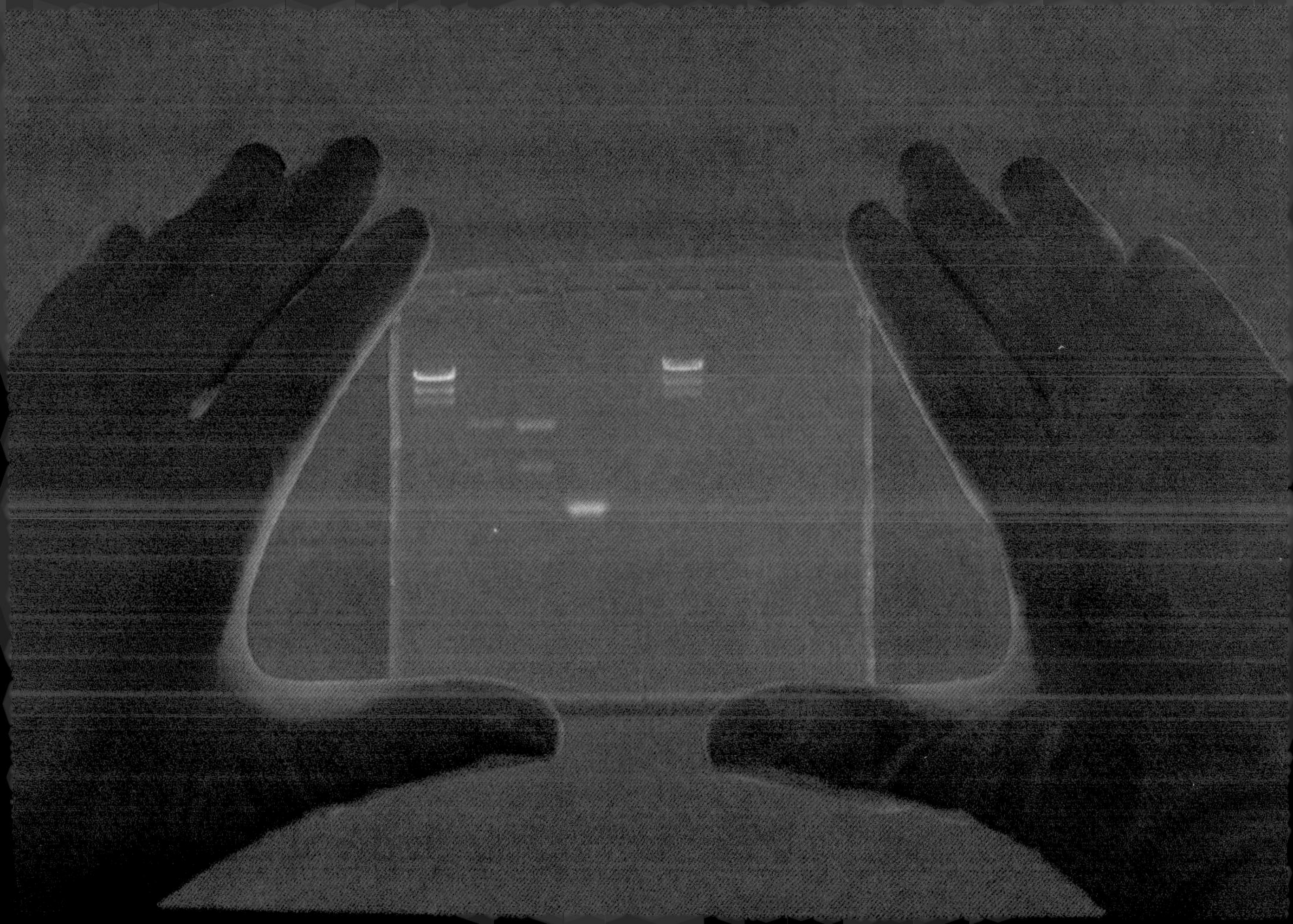

Key Discovery **First leukemia drugs**
Designed not stumbled upon

Modern drugs are designed carefully and methodically, but sixty years ago, they were more haphazardly stumbled upon by tinkering with the structures of natural chemicals. In 1948, all that changed when George Hitchings hired a bright, young chemist called Gertrude Elion, and started a decade-long partnership.

Hitchings knew that bacteria, parasites, and cancer cells depended heavily on nucleic acids, the building blocks of DNA, in order to grow. He tasked Elion with working out how cells produce these nucleic acids and how to stop it from happening. By 1948, she had discovered a chemical that stopped experimentally induced leukemias from taking hold, although it produced toxic side effects in human trials. Undeterred, Elion and Hitchings soon developed the first successful anti-leukemia drugs – thioguanine in 1950, and 6-mercaptopurine (6-MP) in 1951. Although both drugs sent leukemia patients into remission, they later relapsed and died. Elion continued her research and now 6-MP is used to treat leukemia, along with a combination of other drugs that cure most cases.

In 1988, the pair were awarded the Nobel Prize for their work. While the drugs they discovered saved many lives, the rational approach they pioneered was arguably more important. By founding their discoveries on basic biochemistry, they went on to develop drugs for gout, malaria, auto-immune disorders, and herpes, that all block nucleic acid synthesis without harming healthy cells. Their method also led to discovery of the best known AIDS drug, azidothymidine (AZT).

Ed Yong

Date 1950

Scientists George Hitchings, Gertrude Elion

Nationality American

Why It's Key Saved the lives of thousands of leukemia patients, and set a precedent for rationally designing drugs.

opposite Melvin Calvin with his illustration of the carbon dioxide assimilation cycle of plants

Key Discovery **The Calvin Cycle and photosynthesis**
Tracing chemistry

Melvin Calvin was the son of Russian and Lithuanian immigrants to America. Having settled in Detroit, Michigan, Calvin attended the Michigan College of Mining and Technology and became the school's first chemistry major. He also studied at the University of Manchester and the University of California, Berkeley. His work began to focus on photosynthesis in the early 1940s, and he used radioactive tracers to look at the chemical pathways in the reaction. It was during this research period that he went on to describe the Calvin Cycle of photosynthesis in plants.

The Calvin Cycle (or more expansively – and fairly – the Calvin-Benson-Bassham Cycle) is a system to describe the series of biochemical reactions that take place in the chloroplasts of photosynthetic organisms. The cycle describes a light-independent reaction (meaning it does not need visible or ultraviolet light) where stored energy is used to convert carbon dioxide and water into organic compounds. The enzyme RuBisCo is key to this process of "carbon fixation."

What Calvin – and his adversaries – did was to describe each stage of the light-independent process using a carbon-14 tracer to explain the chemical pathway. By adding carbon dioxide with trace amounts of radioactive carbon-14 to a single-cell green alga, he was able to identify the path of the reaction. This was the first time that this method had been used, winning Calvin the Nobel Prize in 1961.

Katherine Ball

Date 1950

Scientist Melvin Calvin

Nationality American

Why It's Key It was the first time a chemical pathway was described using a radioactive tracer, and it paved the way for this to become common practice.

CARBON
DIOXIDE
SUGAR
FOOD
WATER
ACID
HYDROGEN
SUGAR
OXYGEN
STORAGE

Key Publication
The Turing Test

Can computers think? In his 1950 paper, "Computing Machinery and Intelligence," Alan Turing proposed a test to find out.

Imagine you're in a room with a computer terminal. Hidden in two other rooms are a human, and a computer that's pretending to be human. Your job is to talk to them both through your terminal, and try to figure out which is which. You can ask them anything – set them a chess problem, ask them about women's fashion, even get them to write a poem. If you can't tell which one of the two is human, then the computer has passed the Turing test.

In 1950, computers were far too primitive to stand a chance of passing. Turing intended his test as a thought experiment, to get people talking about computers and the nature of intelligence. Scientists and philosophers have been debating it ever since. If a computer passes the test, does that mean it's really thinking, or just faking it? Is there actually a difference?

HAL 9000 in *2001: A Space Odyssey* is one of several fictional computers that could pass, but we haven't seen it happen in real life yet. The Loebner Prize, which offers US$100,000 for a computer that passes a simplified version of the test, remains uncollected. Futurist Ray Kurzweil predicts that computers with the processing power of a human brain will arrive around the year 2020, but even if they can pass the Turing test by then, humans will still be wondering what – if anything – is really on their minds.

Matt Gibson

Publication "Computing Machinery and Intelligence"

Date 1950

Author Alan Turing

Nationality British

Why It's Key Raised the question of what is intelligence and what happens when computers attain human levels of processing power.

Key Event **Embryo transplants**
Farmers can beef up their herd

Ever wondered if you could implant an embryo from one rabbit into the body of another? Well, apparently Walter Heape did. In the 1890s, using what would be viewed as very crude techniques by today's standards, he transplanted two Angora rabbit embryos into the womb of a pregnant Belgian Rabbit. The results were a brood of four Belgian rabbits and two live Angora rabbits; but more importantly, the first successful embryo transplant in a mammal. Despite this seemingly ground-breaking discovery, nothing practical came of this new technology for almost sixty years.

The first successful cattle embryo transfers were actually preformed by Elwyn Willett in 1950. Prior attempts had all ended with the spontaneous abortion of the cattle line. The technique was laborious and involved surgical implantation of the embryo into the cattle, not at all practical for commercial application, but it did prove that bovine embryo transfer was possible.

All this begs the question: Why would one want to transplant an embryo of one cow into the womb of another? Well, what this allows is the genetic enhancement of a herd or cattle. You can harvest embryos of cows with more desirable traits and transplant them into the womb of less desirable cows – or even take embryos from rare breeds of cattle and put them into any old common heifers – leaving the desirable cows free to become pregnant again, rather than be tied up with gestating. The result is a greater population of commercially viable cows.

B. James McCallum

Date 1950

Country USA

Why It's Key Bovine embryo transfer allows for one genetically desirable cow to produce many more offspring than she normally would.

Key Experiment **DNA**
As easy as A, C, G, T

The discovery of the double helical structure of DNA, credited to Watson and Crick, was one of the most fundamental scientific breakthroughs of the twentieth century. Without the work of other scientists, however, including Erwin Chargaff, they might never have made their discovery.

Chargaff became fascinated by the search for the secrets behind DNA in 1944, when it was found that the double helix itself carried genetic information within the cell. He could sense that this was the start of a major turning point for science, and set about his work with vigor.

He found success with new techniques for analyzing DNA samples. These allowed the quantification of the constituent parts of the DNA molecule, the four bases. His work (Chargaff's rules) revealed that DNA structure varied between different organisms and that, regardless of species, the amount of adenine (A) always equaled the amount of thymine (T); the same was true for cytosine (C) and guanine (G).

The significance of these findings was not clear initially. It wasn't until a discussion with Watson and Crick that their importance became apparent. This meeting of minds was not a great success on a personal level, but Chargaff's findings were of great interest. Watson and Crick used his rules – realizing that the pairing of A with T, and C with G, was responsible for Chargaff's findings – along with other fundamental work on the X-ray structure of DNA, to build the model of DNA structure so familiar to us today.

Katie Giles

Date 1950

Scientist Erwin Chargaff

Nationality Austrian-American

Why It's Key The unsung hero who helped fit the pieces together in one of nature's biggest puzzles.

Key Discovery
Corticosteroids

The discovery of corticosteroids and their biological functions during the 1940s was a major step in both understanding the significance of these chemicals and developing drug treatments for numerous medical conditions. The importance of the discovery led to the award of the Nobel Prize for Medicine in 1950 to Swiss scientist Tadeus Reichstein and two American colleagues, Edward Kendall and Philip Hench.

Corticosteroids are a type of steroid hormone involved in a range of important bodily functions. These include the way the human body responds to stressful situations, immune and inflammatory responses, and even behavior. Corticosteroids are made from cholesterol in the adrenal glands, on top of the kidneys. Once isolated, it became possible to develop corticosteroid-based drugs to be used in a wide range of medical treatments, for conditions including brain tumors, arthritis, asthma, allergic reactions, hepatitis, and various skin diseases. Early treatments promised a "cure" for diseases like arthritis until it was realized prolonged use of steroids had damaging side effects.

Given their huge significance, and potentially profitable use in medical treatments, the production of corticosteroids became a prime target of drug companies. Early expensive and complex production methods involved the use of ox bile, but more recently, Mexican yams have proven to be a cheaper and more convenient source. Despite the nasty side effects, steroid hormones like corticosteroids have also been misused, particularly in sport, by those illegally trying to boost their muscle strength and athletic performance.

Richard Bond

Date 1950

Scientists Tadeus Reichstein, Edward Kendall, Philip Hench

Nationality Swiss, American

Why It's Key Has led to the development of important and effective drugs for a range of medical conditions and diseases.

Key Experiment
Measuring the speed of light

The lower case letter "c" is not the most used letter in the English language or the highest scoring tile in a Scrabble set. However, for many, the letter "c" represents possibly one of the most important and indeed most used constants in science – the speed of light.

Scientists first began trying to measure the speed of light using the planets in the seventeenth century. But it was much later, several centuries later, in fact, that English physicist Louis Essen found a much more accurate value of "c": one that is still used today.

In 1946, Essen, together with Albert Gordon-Smith, set up an experiment using microwaves. Essen had been working at the National Physics Laboratory on radars when he realized he might be able to carry out the measurement.

Using a cavity with well defined dimensions, he measured the frequency of the normal patterns of motion of microwaves within the cavity. Because he knew the dimensions of the cavity, he was able to deduce the speed of the waves in the vacuum using electromagnetic theory, and thus had his value for the speed of light. It is almost as if he had used a simple microwave oven to find one of the key constants of science.

The value he came out with was much higher than previous predictions, so was met with skepticism. But after he conducted another experiment in 1950 and came out with almost exactly the same value, his figure of 229,792.5 kilometers per second was officially "adopted" as the correct one.

Helen MacBain

Date 1950

Scientist Louis Essen

Nationality British

Why It's Key First truly accurate measurement of the speed of light.

opposite Jack Parry and Louis Essen (right) with the Caesium resonator that counts oscillations of caesium atoms between energy levels

Key Publication **Jumping genes**
One woman's corny story

While working on maize at Cold Spring Harbor in 1944, Barbara McClintock noticed something strange about her kernels. Random patterns of coloration, which couldn't be explained, were appearing in her corn. She developed a theory around lengths of DNA she called "transposons," her idea being that the genes for coloration in the maize were being disrupted by these rogue pieces of DNA "jumping" around the genome. Corn kernels' characteristic yellow color, she realized, was produced by a transposon jumping into the middle of a gene involved in color; otherwise, they remained purple.

In 1950, McClintock published a paper which was the culmination of over ten years of groundbreaking work on transposons. Her complex theories were met with skepticism and hostility from her peers. But she persevered, continuing her work in this area and even going on tours of universities to promote her ideas. By 1953, however, there was such a body of resistance to her work that she felt she had to stop publishing to avoid alienating the scientific mainstream.

It was only following a series of experiments by other researchers that McClintock's theory of jumping genes was finally accepted. Since then transposons have become exceedingly important in genetic studies, giving scientists a way of altering the DNA of a living organism. Total vindication for McClintock came in 1983 when she was awarded a Nobel Prize.

Catherine Charter

Publication "The Origin and Behavior of Mutable loci in Maize".

Date 1950

Scientist Barbara McClintock

Nationality American

Why It's Key After a shaky start in research circles, transposons are now used in all manner of genetic studies, including in the search for new cancer treatments.

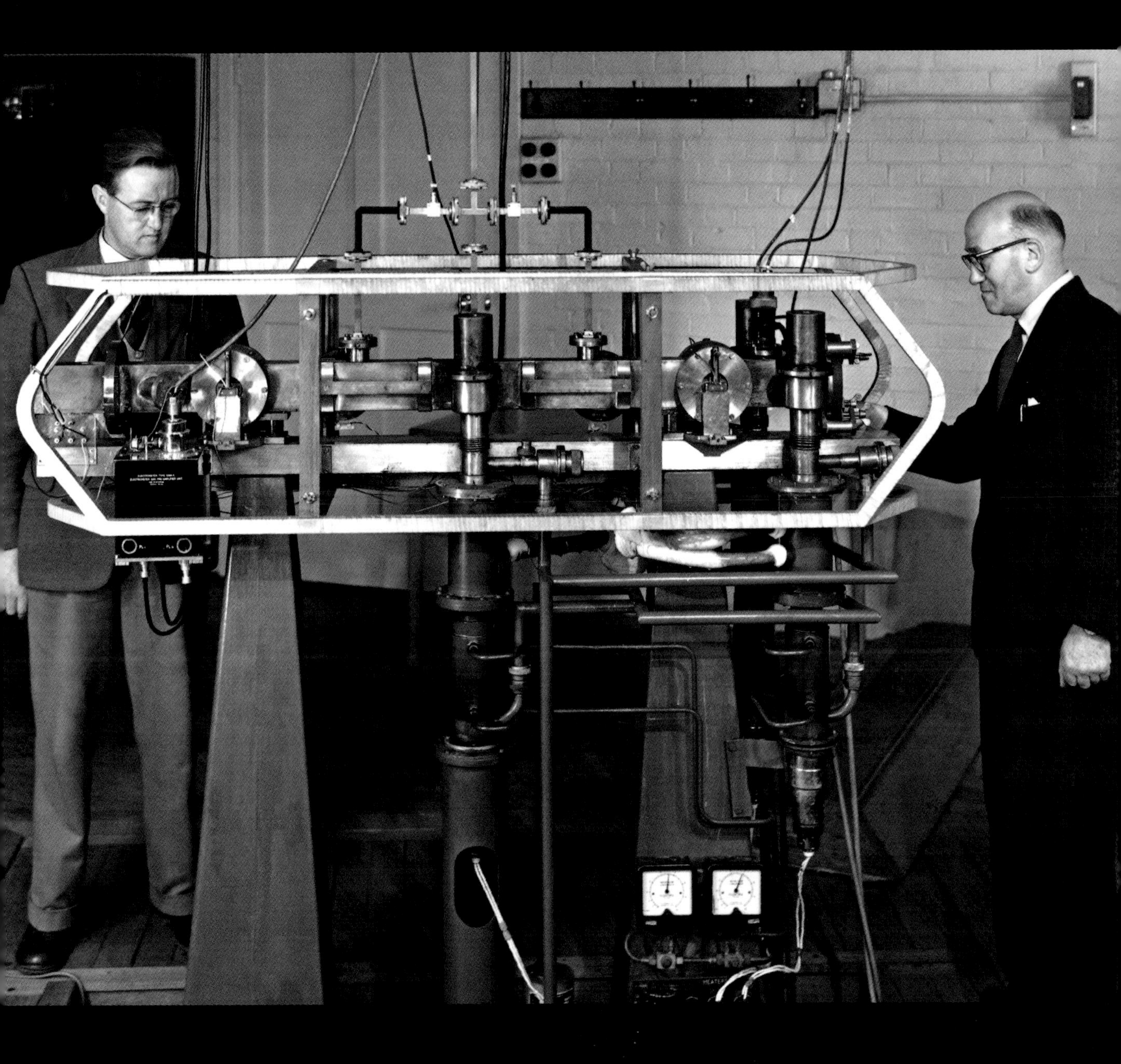

Key Person **Rachel Carson** Environmental pioneer

Born in 1907 on a farm in the rural Pennsylvanian town of Springdale, Rachel Carson fed her love of nature early in life by exploring the forests and streams of her hometown. After a brief stint as an English student at Pennsylvania College for Women, she switched her focus to biology, and eventually earned her master's degree in zoology at Johns Hopkins University.

After being hired by the United States Fish and Wildlife Service in 1935 to write seven-minute radio spots about marine life, entitled "Romance Under the Waters," Carson went on to publish numerous freelance articles on the wonder of nature and man's place among it. Her poetic, yet scientifically enlightening style, was revealed in her first book *Under the Sea Wind* (1941), and more famously in *The Sea Around Us* (1952). The latter placed her on the New York Times Bestseller List and won her the coveted National Book Award.

In 1962, she published the controversial book *Silent Spring*, which exposed the public to the dangers of pesticides used by the government and agricultural industries. The chemical industry and other opponents blasted Carson personally and professionally, however, the new concern over pesticides led to a Congressional hearing in 1963, and a subsequent ban on widespread use of a particularly dangerous chemical known as DDT. Carson died in 1964, after a battle with breast cancer, but left in her wake an enduring awareness of environmentalism, the likes of which had never before been observed in America.

Rebecca Hernandez

Date 1951

Nationality American

Why She's Key Carson is considered to be one of the pioneers of modern environmentalism.

Key Publication ***Epilepsy and the Functional Anatomy of the Human Brain***

When Wilder Penfield's sister died following an operation – which he himself had performed – to remove a brain tumor, he resolved to devote his life to studying the human brain. His brilliant collaboration with Herbert Jasper, and enduring commitment to discovering the mysteries of the mind had a lasting impact on neuroscience.

It all began with a chance encounter in 1937. Jasper, by then a respected neurophysiologist who had been pioneering the use of Hans Berger's brain-tracing electroencephalography (EEG) machine, bumped into Penfield at Brown University. Penfield was soon able to confirm, by operating on two of Jasper's patients, that he had identified the area of the brain responsible for epileptic seizures. Within two years of their meeting, Penfield had found the means to open a new EEG department at the Montreal Neurological Institute in Canada. There they enjoyed years of skiing sessions and the insight that thousands of patients' brains would bring. Penfield developed what became known as the "Montréal procedure" for treatment of epilepsy, in which he delved into the brains of conscious patients, probing gently whilst they described the effects. Through this method he was able to successfully cure epilepsy in many of his patients by removing the affected pieces of tissue.

By 1951, Penfield and Jasper had gained enough experience with epileptic patients to publish their landmark text *Epilepsy and the Functional Anatomy of the Human Brain*, which made major contributions to the understanding of the brain – not just in epilepsy.

Hayley Birch

Publication *Epilepsy and the Functional Anatomy of the Human Brain*

Date 1951

Authors Wilder Penfield, Herbert Jasper

Nationality Canadian

Why It's Key Penfield and Jasper's insight and pioneering techniques brought major steps forward in understanding the mysteries of the human brain.

Key Invention **Concept of microprogramming developed**

The 1940s and '50s were a key era in the development of modern computer technology. The Colossus Mark 1, delivered to Bletchley Park in England in 1943, was the first programmed electronic computer in the world. Five years later and Baby was born – the first operating stored program machine.

But early pioneers faced various limitations, including the size of available memory. One of the innovators of the age was Maurice Wilkes who, in the late 1940s began working on a computer called the EDSAC – Electronic Delay Storage Automatic Calculator. At this time, the simple computers available to the world had their programs hard-wired into their circuitry. This meant that computers were very different to one another depending on their function, and it could be difficult for an expert to switch between machines. During his work on the EDSAC, Wilkes realized that the sequencing control signals in his new computer were similar to the sequencing actions in a computer program. This was critical, as it meant he could use a rudimentary computer program instead of hard wiring everything directly into the machine.

He named the result microprogramming, where the microprogram controls the different parts of a computer's main central processing unit (CPU), which beforehand was hard-wired in. His new technique made it easier to find bugs, allowed computers to emulate other machines by changing their microprograms, and made it possible to add functions to a computer without having to rip out and replace its guts.

David Hawksett

Date 1951

Scientist Maurice Wilkes

Nationality British

Why It's Key Microprogramming revolutionized how computers were designed and operated.

Key Event **HeLa cancer cells** Henrietta Lacks leaves legacy for the world

Some people seek immortality through noble or nefarious deeds. Henrietta Lacks (1920-1951), a humble black woman from Virginia, achieved it by dying of a particularly aggressive form of cervical cancer.

Today, cells from her cancer are still grown in labs all over the world. They are used for studying a wide range of biological processes, and have contributed to more than 50,000 scientific publications. They were even used in the development of the polio vaccine. But how did this come about?

In the 1950s, medic George Otto Gey was working in the new Tissue Culture Laboratory at Johns Hopkins University, in the United States. He was given a sample from Henrietta's cancer, which was spreading rapidly through her body. The cancer cells grew merrily in the lab, and were dubbed "HeLa," to preserve Henrietta's anonymity. This led some people to think that they were from a woman called "Helen Lane," until their true identity was revealed in 1971.

Over the years, HeLa cells have been a source of controversy, both sociological and scientific. Some believe that Henrietta Lacks was exploited by the researchers, who eventually patented the cell line. There are also big questions about the purity of the cells; they are still mutating as they grow in the world's incubators. It's widely thought that HeLa cells are now wildly different from Henrietta's original cancer cells, and have contaminated many other cell lines. Some scientists have even proposed that the cells should be classified as an entirely new species.

Kat Arney

Date 1951

Country USA

Why It's Key Cells from one woman's cancer are now used by researchers all over the world.

Key Invention
The electric brain that predicted the president

We live in a world dominated by computers; they affect every aspect of our lives from the home to the workplace. But the first commercial computer was not available until as late as 1951, when the UNIVAC-1 (Universal Automatic Computer 1) was created by J. P. Eckert and John Mauchly. Previous computers, such as the U.S. military's ENIAC and those used in universities, were one-offs, whereas UNIVAC-1 was mass-produced.

A year later, with the whole world was hanging on the outcome of the U.S. presidential election, UNIVAC-1 hit the headlines. Broadcaster CBS used it to correctly predict the victory of Dwight D. Eisenhower, despite many claiming the result was "too close to call."

At a hulking 13 tons, taking up over 35.5 square meters and costing a million dollars, the UNIVAC-1 packed only a fraction of the computing power of the mobile phone in your pocket. Nonetheless, it was snapped up by U.S. governmental and military agencies such as the Census Bureau, and later by commercial outfits such as General Electric and the Prudential Insurance Company, with a total of forty-six machines being built.

Unlike many other early computers, UNIVAC-1 used magnetic tapes rather than punch cards, which had until then been used to store and transfer data. Processing of both numerical and alphabetic characters was carried out by the use of 5,200 vacuum tubes, piezo-electric crystals (which change shape slightly when an electric current is applied across them), and columns of mercury, allowing around 1,905 operations per second.

Neal Anthwal

Date 1951

Scientists John Presper Eckert, John Mauchly

Nationality American

Why It's Key UNIVAC-1 pioneered the era of commercial computing, and successfully predicted the future.

opposite Operators of the first UNIVAC computer installed at the US Censor Bureau

Key Experiment
Humans: 33 per cent sheep, half the time

Are you a sheep when it comes to lines? This is the question that the social psychologist Solomon Asch wanted to answer and, in 1951, he carried out a beautifully simplistic experiment. Asch had faith in people's ability for independent thought, believing they'd stand up against group pressure if they believed something was true.

Under the guise of testing visual perception, straight black lines drawn on card were shown to small groups of students around a table. Each was asked, in turn, to find the line of the same length as the reference line. Easy.

However, in reality, for each genuine volunteer, the rest of the students in the room were Asch's lackeys, (those sneaky social psychologists). The lackeys were instructed to initially give correct answers, and then later, to all give the same incorrect answer. Hardly peer pressure, nothing's at stake, you're just stating the length of a line. That's what Asch thought, so he was surprised that when it was their turn, only a quarter of the true volunteers consistently gave the correct response in defiance of the rest of the group. A third of volunteers agreed at least half of the time with the rest of the group's wrong answers.

Asch's experiment went on from hinting at human weakness to showing that if just one lackey gave the correct answer (against the majority), the volunteer was much happier to do so too. It seems it's easier to think for yourself if someone else thinks the same as you.

Umia Kukathasan

Date 1951

Scientist Solomon Asch

Nationality American

Why It's Key Asch showed how influential group pressure is, and how it is greatly reduced, with even one colluding "dissenter." This questions notions of individuality, perception, and democracy.

Key Discovery
Hydrogen line radiation

In the 1930s, astronomers noticed a "hiss" on radio frequencies emanating from the center of the galaxy. However, it wasn't until 1951 that Harold Ewen and Edward Purcell discovered the source of the hiss to be a constant frequency radiation emission – known as the hydrogen line. This discovery later allowed scientists to map our galaxy as never before.

A neutral hydrogen atom consists of one proton and one electron, both of which "spin" either clockwise or anti-clockwise. If the spin of one of these subatomic particles suddenly changes – extremely unlikely, but not impossible – then energy can be released by the atom in the form of radiation. This emission is the hydrogen line – known as "'21 centimeter radiation" after its wavelength. Dutch astronomer Jan Oort and his student van de Hulst discovered the properties of the hydrogen line in 1944. They realized that if this frequency line was emitted from points across the galaxy, then it could be detected by radio telescopes. This would allow astronomers to cut through the galaxy's dust – which had hampered efforts to map the galaxy with optical telescopes – as if it weren't there. In 1951, having built a setup designed to detect galactic radiation at very high sensitivity, Ewen and Purcell successfully detected the hydrogen line, a discovery confirmed by Oort.

The hydrogen line has since become an extremely important galactic constant, used to measure the galaxy's structure and rotation. It has provided an insight into the early stages of the Big Bang, and contributed to the search for life outside earth.

Steve Robinson

Date 1951

Scientists Harold Ewen, Edward Purcell

Nationality American

Why It's Key The hydrogen line allows astronomers to study the structure of the Milky Way without the interfering dust layer that sits in the galactic plane.

opposite The Milky Way and constellation Cygnus. The reddish area is North America nebula which glows read due to ionization of hydrogen gas

Key Event **Oil Rocks rig starts pumping**
A city on the sea

In order to access the huge oil deposits deep below the sea, large structures called platforms have to be built to house the workers and machinery needed to drill and then transfer oil to the shore.

The first oil platform in the world was the appropriately, and deliberately, named "Oil Rocks," built near Baku in the Soviet Union – now Azerbaijan – about 50 kilometers offshore in the Caspian Sea. Construction of the platform began in the late 1940s, and the first well was drilled in 1949. In 1951, the first tanker left Oil Rocks and headed for the shore, heralding a new era of massive oil production. At its peak in the mid-1960s, it was producing 21 million tons of crude oil a year.

Oil Rocks was a fully functioning city with 5,000 inhabitants and 200 kilometers of streets built in shallow seas on top of piles of rocks, sand, and landfill. It included nine-storey buildings, shops, a school, a hospital, a library, and even a park with trees.

Oil Rocks still produces oil, although it is now in a state of disrepair, with kilometers of roads, and many buildings, submerged. Only about two thirds of its 600 oil wells are still working, with a remaining population of some two thousand people, but it still produces about half the total crude oil of Azerbaijan, and remains the world's largest oil platform. In 1999, this strange "city on the sea" featured in the James Bond film *The World Is Not Enough*.

Richard Bond

Date 1951

Country USSR (now Azerbaijan)

Why It's Key Heralded an era of massive oil production and consumption, fuelling industrial and economic growth that is now proving unsustainable.

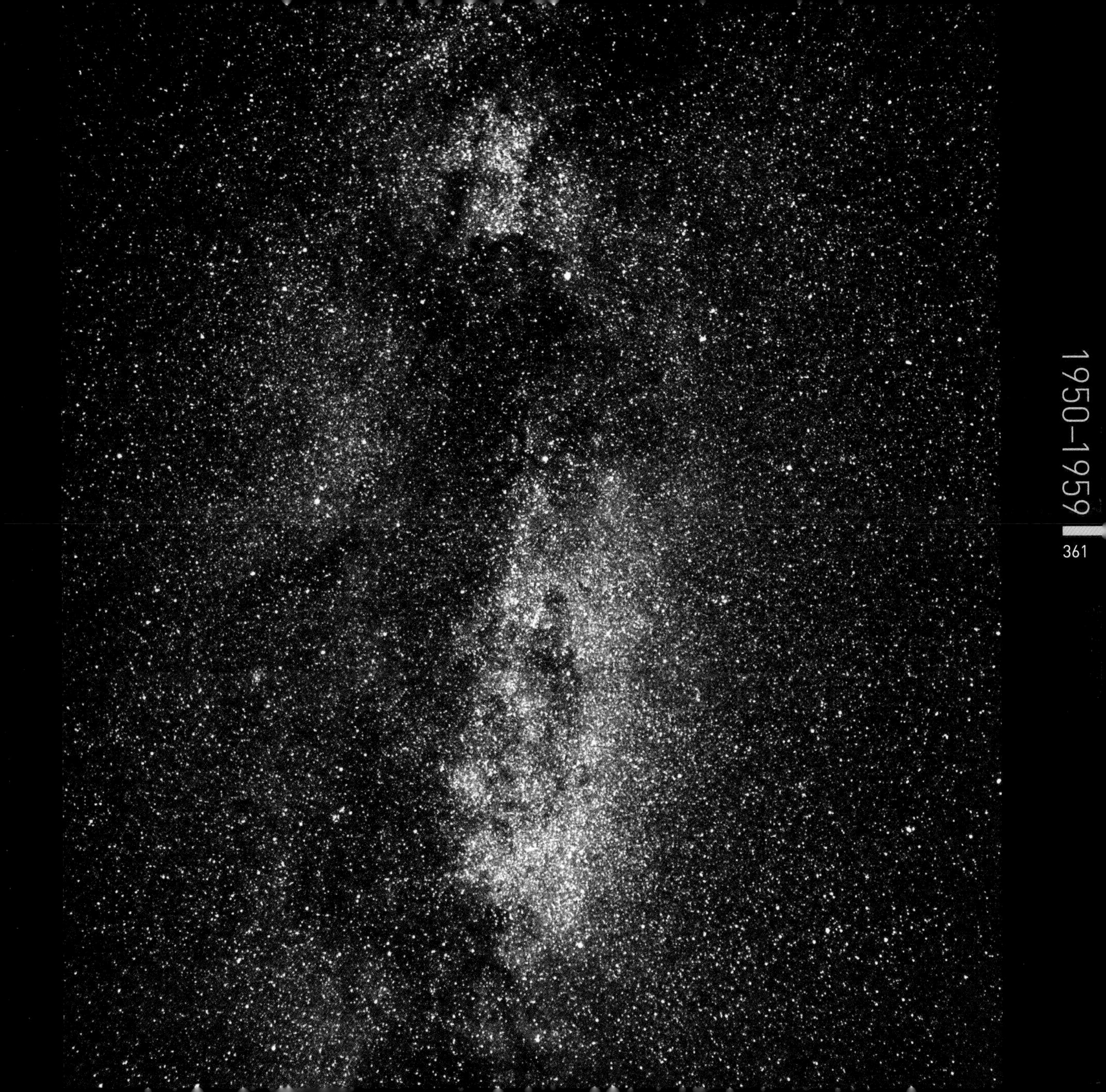

Key Invention
The airbag

John W. Hetrick was an engineer by trade, but it was a freak accident whilst out driving that proved the catalyst to him patenting a key invention in the automobile industry. Having swerved to avoid a rock in the middle of the road, Hetrick, together with his car and his family, ended up in the ditch.

This got him thinking about developing some form of safety cushion to protect passengers involved in an accident. Hetrick successfully patented his nascent idea in 1952 and then the race was on to develop his idea to its full potential. It was clear that any device would need to deploy very quickly; must accurately sense a collision, and must not cause any secondary injuries.

Following calls from U.S. President Johnson in the 1960s that "we can no longer tolerate unsafe automobiles," an American mechanical engineer named Allen Breed developed a crash sensor that would be the key to modern airbag development. The design we know today has three key elements. Firstly, the bag itself which is folded within the wheel and made of thin nylon, secondly, a sensor switch which is flicked to complete an electrical circuit thus inflating the bag, and thirdly, the inflation system which involves sodium azide reacting with potassium nitrate to produce nitrogen – hot blasts inflate the bag.

The trigger for all this to occur is a mass change in the car equivalent to running into a brick wall at 16 kilometers per hour with the inflation occurring in less than 40 milliseconds.

Andrew Impey

Date 1952

Scientist John W. Hetrick

Nationality American

Why It's Key Modern statistics show that airbags reduce the risk of dying in a head-on-crash by 30 per cent.

opposite The airbag is tested using a crash-test dummy

Key Invention
Superglue

Super glue was invented by accident during World War II, when Harry Coover at the Eastman Kodak Company was looking for a way to make gun sight lenses. He produced a new substance called cyanoacrylate which stuck fast everything it touched. By the early 1950s Coover and his colleague Fred Joyner had realized its potential, and it was marketed as a commercial product in 1958 in adverts with people being suspended at a height by a single drop of glue.

Cyanoacrylate is a generic name for the family of adhesives all sharing the same underlying chemical structure. They are very powerful, especially when used to stick materials that are non-porous or that contain tiny amounts of water. They are also very good at bonding skin and body tissue, hence their common use in surgery and medicine. These uses in medicine were first tested on a large scale during the Vietnam War in 1966, and the results were impressive. Cyanoacrylate could not only hold wounds closed but also stopped bleeding vessels. A slightly different formulation from normal super glue is used in the medical industry to prevent skin irritation, and this is now used in liquid plasters, which you can spray on to skin wounds.

Cyanoacrylate is highly effective because it produces long, strong bonds when it sets, and the setting time is very quick - usually within one minute. It reaches full strength in about two hours. Acetone, or nail varnish remover, can be used to soften hardened superglue, while freezing it makes it brittle enough to break apart.

David Hall

Date 1951

Scientist Harry Coover, Fred Joyner

Nationality American

Why It's Key Super glue was revolutionary, not only in industry but also in medicine, where it has been used extensively in surgery.

BREED
BREED

Key Invention
Birth of artificial intelligence

In 1951, a year after Alan Turing wrote "Computing Machinery and Intelligence," the first computer programs to play games like chess appeared. Turing himself had been interested enough in chess to write a computer program to play the game, but only on paper. American Claude Shannon's 1950 paper "Programming a Computer for Playing Chess" described the strategies a computer might use to win, but Shannon, too, was more interested in the theory than the practice.

In 1951, Dietrich Prinz, a colleague of Turing's, created the first practical computer chess program. Using Manchester University's Ferranti Mark I computer, his program couldn't play a full game of chess, but it could solve standard chess puzzles of the "mate in two moves" type. Prinz's program laboriously ran through every possible combination of moves until the right answer was found. While this worked for limited games, it would be impossible for a full game of chess – Shannon's paper estimated had that there are more potential games of chess than there are atoms in the Universe, and even today's computers have problems with numbers that big.

As computer chess developed, programs became more sophisticated, using smarter strategies, and databases of memorized games, to bring their play up to human standards. Although the term "artificial intelligence" wouldn't be coined until 1956, programs were already taking the first steps toward beating humans at their own games. In 1997, IBM's Deep Blue became the first computer to beat the reigning world chess champion, Gary Kasparov, in a six-game match.

Matt Gibson

Date 1951

Scientist Dietrich Prinz

Nationality German

Why It's Key The development of programs that could play chess made people wonder how proficient machines might become at this very human game, and more.

Key Person **James Bonner**
The plant polymath

James F. Bonner (1910–1996) could be regarded as something of a genius when it came to plants. His achievements in botany, and cellular and molecular biology were so far-reaching that one of his greatest discoveries – that of tiny powerhouse particles in plants called mitochondria – earns little more than a footnote in the biography of his life.

Bonner was highly skilled in separating out the cellular components of plants; he isolated mitochondria in mung beans in 1951. By this time, critical energy-producing metabolic pathways in animals had already been uncovered, but his discovery showed that plants were also employing the particles to create energy in their cells.

This was just one of his many successes. He had already elucidated key factors in the timing of flowering; pinpointed vitamins that act as root growth factors; and was later to make key discoveries about how genes regulate growth and development in plants. It was also Bonner who helped to show that plants create and utilize cellular energy in the form of ATP (adenosinetriphosphate) in the same way as animals.

So varied were his talents, that Bonner was even credited with doubling the production capacity of Malaysia's rubber trees, and inventing a method of harvesting oranges mechanically. He was, after all, the man who said "browsing in far-flung pastures is more fun."

Arran Frood

Date 1951

Nationality American

Why He's Key Bonner devoted his life's work to one of nature's often neglected kingdoms – plants.

Key Person **Amazing Grace** Murray Hopper creates first computer program

Grace Murray Hopper (1906–1992) was a New Yorker who wasn't going to let gender stereotyping get in the way of her career. After picking up a Master's degree and a PhD from Yale, she joined the Naval Reserve in 1943.

In 1944, assigned to duty at Harvard University, she saw her first computer. Working on the Mark I calculating machine – a huge mechanical computer with clacking relays and miles of wiring – she discovered a natural ability in this new field. It wasn't until she went to work for Remington-Rand in the 1950s, however, that she made her biggest contribution to computing: the compiler.

At its heart, a computer understands zeroes and ones. To tell a computer to count another bean, you have to say something like "01000010" to it – succinct, but hardly memorable. A compiler takes more understandable instructions, like "ADD 1 TO BEANS," and translates them into the machine language for you, saving hours of work looking up codes.

Grace developed compilers throughout the 1950s and was instrumental in creating COBOL, the COmmon Business-Oriented Language, which made computer programs easy for people in business to understand without years of training. Her work earned her the Data Processing Management Association's first "Man of the Year" award in 1969. On her retirement in 1986, at the age of eighty, Rear-Admiral "Amazing" Grace Hopper was the oldest active officer in the U.S. Navy, and she is probably the only mathematician to have had a warship named after her.

Matt Gibson

Date 1952

Nationality American

Why She's Key COBOL, with its origins in Grace Hopper's ideas, ran the computers of the business world for decades, and still runs many of them today.

Key Event **First mechanical heart implanted**

What do you get when you combine Wayne State University in Detroit, Michigan, the American Heart Association, and General Motors? In 1952, you got a mechanical heart.

The method of using a machine to pump blood around a patient's body while a surgeon operated was the brainchild of Dr Forest Dewey Dodrill, but he decided to use local builders instead of medical device manufacturers, asking General Motors' research division to help him build the machine. Surprisingly, they accepted. The American Heart Association played a vital role in funding the whole venture.

Soon they had fashioned a device that used a combination of air pressure and vacuum pumps to take the place of the left ventricle, the chamber of the heart that pumps blood throughout the body. The so-called Dodrill-GMR Heart Machine was used initially to replace a defective heart valve in a forty-one-year-old man. Considered a classic machine, it was only two years later, in 1954, that it was placed in the permanent collection of the Smithsonian Institute. In 1955, Dodrill's mechanical heart was combined with an oxygenator, and the heart-lung bypass machine was born.

The invention of this machine effectively ushered in the era of modern heart surgery. Heart valves are now replaced, blood vessels are bypassed, and hearts are even transplanted using heart-lung bypass machines that are descendants of the Dodrill-GMR.

B. James McCallum

Date 1952

Country USA

Why It's Key The Dodrill-GMR heart machine is the progenitor to the heart-lung transplant machine used in today's surgery. Countless lives have been saved by similar apparatus.

Key Discovery
The exciting world of nerve cells

One of the important models of neuroscience, now taught at high school level, is the basic theory of how impulses travel along nerve cells. Discoveries leading to this theory began prior to World War II and culminated in a Nobel Prize over twenty years later.

English physiologists Andrew Huxley and Alan Hodgkin began a collaboration to study nerve impulses in 1939. The two scientists first measured these impulses with electrodes inserted into the giant nerve fibres of squid. Huxley and Hodgkin later demonstrated that these transmissions were caused by the movement of potassium and sodium ions in and out of the nerve cells via proteins called ion channels. This discovery led to a fundamental understanding of how nerve cells work and transmit signals to each other.

Australian John Eccles' work, mostly performed in the late 1950s, focused on synapses, which are the points of contact between two nerve cells. Eccles, using tiny glass tubes inserted into nerve cells, was able to show that a change in a nerve cell's electrical charge will cause an inhibitory or excitatory effect through the synapse and into the surrounding nerve cells. His work also proved that nerve cells have an important impact on muscle cells, specifically the speed of muscle contraction.

The groundbreaking work of these three scientists now forms the basis of our understanding of the transmission of impulses between nerve cells, and their efforts were acknowledged in 1963 when they shared the Nobel Prize for medicine.

Rebecca Hernandez

Date 1952

Scientists Andrew Huxley, Alan Hodgkin, John Eccles

Nationality British

Why It's Key Advanced the understanding of how nerve cells communicate with one another.

opposite Colored transmission electron micrograph (TEM) of the DNA of a lambda bacteriophage, a virus that infects *Escherichia coli* bacteria.

Key Experiment **The Hershey-Chase Experiment**
DNA is hereditary material

DNA was first isolated in 1869 by Friedrich Miescher, but its exact function remained elusive for a number of years, until a series of experiments with viruses and bacteria showed that it was in fact the hereditary material of all life forms.

Bacteriophages are viruses that are able to invade and inject their DNA into bacteria, effectively taking them hostage. Originally, it was thought that the outer protein "coat" of a bacteriophage was the material that passed between the two, but the American biologists Alfred Hershey and Martha Chase proved this theory wrong.

Hershey and Chase exploited the infection cycle of a virus known as T2, to show how DNA is transferred. They labeled viral coats and DNA, by attaching radioactive molecules to them, and allowed them to infect *E. coli* bacteria. The viruses were then separated from the bacteria to look for the labels. They found that radiation marking showed up inside the bacteria when the DNA had been labeled, suggesting the viruses were causing infection by inserting their DNA. Offspring of the bacteria also contained radioactive DNA, showing that it could be passed on during replication. We now know these viruses attach to a bacteria cell surface and inject their DNA inside.

This initial 1969 Nobel Prize-winning work changed what we know about viral infections such as flu and the common cold, and became one of the inspirations for Watson and Crick in solving the mystery of DNA structure.

Nathan Dennison

Date 1952

Scientists Alfred Hershey, Martha Chase

Nationality American

Why It's Key Before this experiment it was believed that proteins were responsible for passing on all hereditary information; we now know that we have DNA to thank.

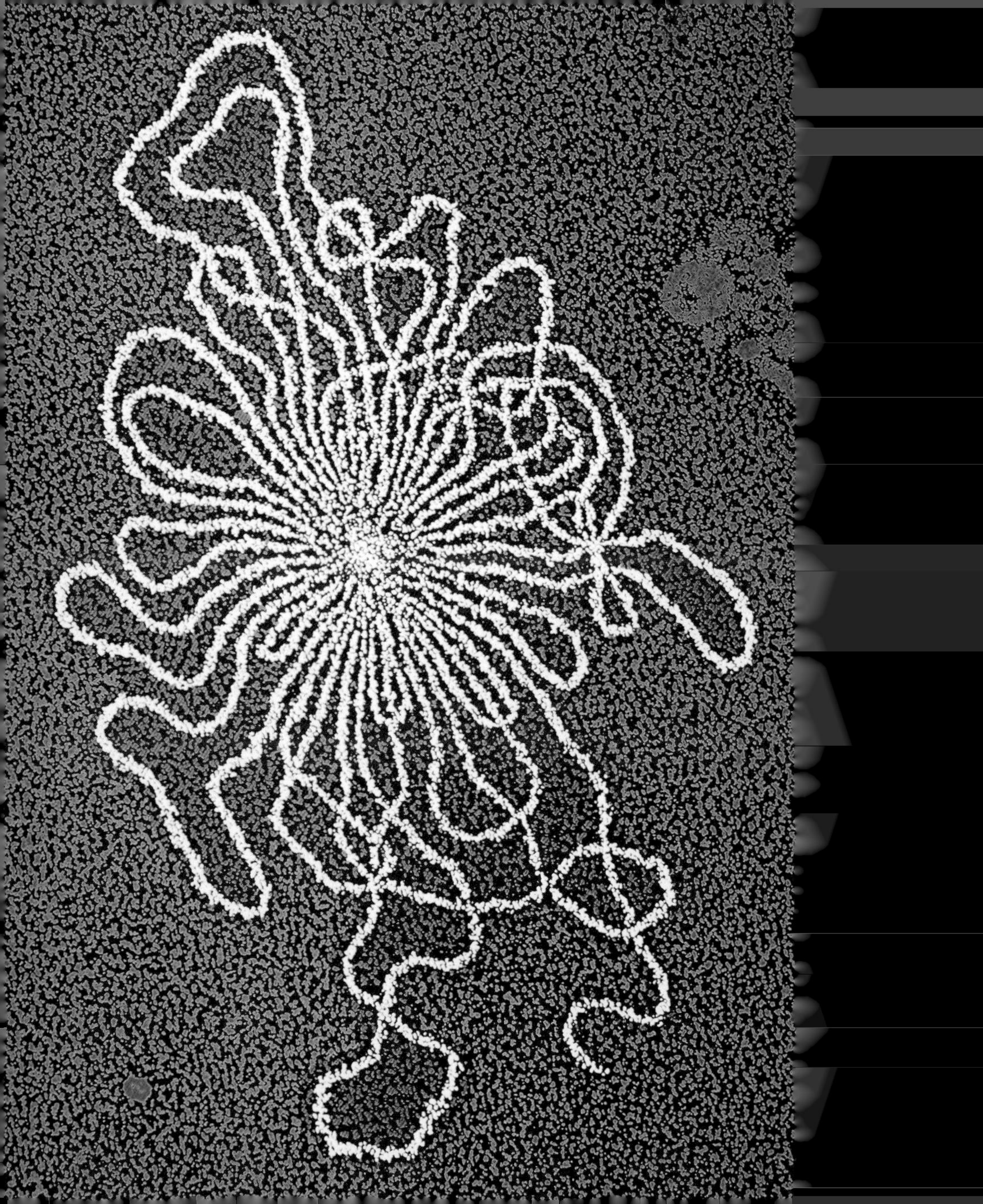

Key Person **Rosalind Franklin** Unsung heroine of the double helix

The scientists traditionally credited with uncovering the structure of DNA are James Watson, Francis Crick, and Maurice Wilkins. Their work in the field was recognized by the Nobel Committee in 1962 when they were jointly awarded the Nobel Prize for Physiology or Medicine. But traditional accounts of their discovery fail to acknowledge the contributions of one brilliant individual – Rosalind Franklin (1920–1958).

It is now generally accepted that the role played by Rosalind Franklin was substantial. Having spent several years in France using X-rays to make images of biological molecules, she had arrived in London in 1950 to begin work on DNA. But her skilful X-ray imagery, observations, and conclusions were trivialized – due to what might be termed a clash of personalities – by James Watson in his personal account of the discovery of DNA, *The Double Helix*. In his version, Franklin was presented as a subordinate of Wilkins at King's College. This myth persisted for some time.

In fact, it was the communication of Franklin's data and conclusions to Watson and Crick by Maurice Wilkins and Max Perutz that revealed missing pieces of the puzzle. These, along with data from other sources, allowed them to assemble their full description of DNA structure. Their model appeared in the journal *Nature* in April, 1953. Franklin was not informed about the publication, despite the fact that it was her photograph which had for the first time revealed the iconic double helix.

Mike Davis

Date 1952

Nationality British

Why She's Key Her detailed X-ray pictures and observational data facilitated the discovery of the molecular structure of DNA.

Key Event **Great Smog**

London and fog have never been strangers to one another – a glance at any Charles Dickens novel will tell you that. But the fogs of December 1952 were particularly serious, resulting in thousands of deaths and an act of Parliament.

On December 5, a fog settled over the city. Over the following days, it got thicker and thicker, trapping the pollution from burning coal and forcing around 2,000 tons of carbon dioxide, 1,000 tonnes of smoke particles and 140 tons of sulphur dioxide into the city air.

Visibility on the streets went down to almost nil, whilst the number of deaths rocketed, including all the ill-fated cattle at Smithfields market. Around 4,000 people died as a result of the fog and a further 8,000 in the weeks and months that followed. People couldn't drive, and even performances at Sadlers Wells had to be cancelled due to the fog permeating into the theatre.

The smog disaster caused a massive rethink in the way people used fuels. The 1956 Clean Air Act stipulated the use of cleaner coals, electricity and gas in order to reduce sulphur dioxide emissions. Power stations were moved to more rural locations and in urban areas, smokeless areas were set out, where only smoke-free fuel could be used. The environmentally-aware consciousness was beginning to emerge.

Fiona Kellagher

Date 1952

Country UK

Why It's Key Caused the premature deaths of thousands and sparked a suite of revolutionary measures to combat air pollution worldwide.

opposite A policeman on duty wears a smog mask issued to the force

Key Person **Konrad Lorenz**
The original Dr Dolittle

King Solomon's Ring, published in 1952, is one of the seminal texts in ethology, the study of animal behavior. Its title refers to the ring supposedly owned by King Solomon, a figure in Christian, Jewish, and Islamic scripts, which enabled him to talk to animals. Inspired as a child by the attentions of a day-old duckling and his nurse's tendency for nurturing animals, Konrad Lorenz (1903-1989) claimed it was possible to do this, albeit through an understanding of behavior, rather than a magic ring.

Elegantly written, but also practical and scientifically intriguing, Lorenz's book is popular science aimed at a lay audience. Accounts of his unusual and sometimes bizarre experiments make up a large portion of the text, but it also incorporates some cutting edge (for the time) scientific ideas, such as animal language, aggression, and imprinting.

Lorenz's sweeping generalizations and tendency to anthropomorphize have attracted much criticism. He often incorporates his own opinion and views as part of his writing, referring to house pets – including goldfish, tortoises, canaries, Angora cats, and lap dogs – as "dull animals" and all "true birds of prey" as "extremely stupid creatures." It's hard to find fault, however, with the man who was to become the joint receiver of the Nobel Prize in Physiology or Medicine in 1973, for his accomplishments in the area of animal behavior. In particular, his work on imprinting – when a young bird latches on to the first object it sees – was singled out for praise.

Jim Bell

Date 1952

Nationality Austrian

Why He's Key Lorenz was one of the fathers of the modern science of ethology – the study of animal behavior.

Key Event
The first accident at a nuclear reactor

Opened in 1944 as Canada's leading nuclear research and development laboratory, Chalk River became famous for its successful work throughout the 1940s. However, in 1952 disaster struck the plant, as a clumsy operational error combined with technical hitches to create the world's first nuclear accident.

Whilst working on a nuclear reactor at the laboratory, a technician mistakenly opened several valves leading to the equipment. This caused a sudden drop in air pressure at the top of the reactor, pulling its control rods straight out of their sockets. Luckily, alarms sounded and disaster was very nearly averted, as the valves were immediately closed.

However, the control rods did not fall straight back into their containers as they should have done. This caused a rapid increase in power output and heated the reactor's cooling water systems. Eventually, steam and helium began to leak from the reactor, causing its radioactive contents to ooze onto the floor.

Despite the potential severity of the accident, nobody died at Chalk River and there has been no direct evidence to suggest that any harm was caused by the exposure of workers to radiation. The event was followed by a similar accident at the UK Atomic Energy Authority's base of Windscale, in 1957, as a nuclear reactor caught fire, releasing radioactive material into the environment.

Hannah Welham

Date 1952

Country Canada

Why It's Key The first accident to occur at a nuclear power station.

Key Invention
Patent for barcodes granted

The barcode on the back of this book is there to help us and computers get along with each other. Computers love identifying things by numbers, but when cruising the supermarket aisles, we'd rather read "thick sliced bread" than "3488403275," and we especially don't like having to type numbers in on keypads all day long.

Barcodes, first patented in 1952, solve this problem by making it easy for computers to read the numbers for us. The numbers are encoded as a series of white and black stripes. These are a lot more readable for a computer than a string of normal numbers would be. They're read by shining a light – these days from a laser – across the stripes and analyzing the reflection with a light sensor or a camera.

Though they were invented for use in shops, barcodes didn't catch on until 1974, when the first retail barcode on a pack of Wrigley's Juicy Fruit gum, was scanned for real. Now, cheap and simple to produce, they're printed on anything that might need tracking or cataloguing, from cans in a supermarket to the trucks that carry them there.

Fifty years on from their invention, new uses are still being found for barcodes. If you order a concert ticket, it might not arrive by post; they're now being sent straight to our mobile phones as barcode picture messages, ready to be scanned at the door.

Matt Gibson

Date 1952

Scientists Joseph Woodland, Bernard Silver

Nationality American

Why It's Key For more than half a century, barcodes have streamlined every area of our lives, from global parcel deliveries to the queue at the supermarket.

Key Person **Selman Waksman**
The microbiology man

Selman Waksman was born in 1888, in the Ukraine but emigrated to the United States when he was twenty-two years old. He achieved a basic degree in agriculture and completed a PhD, becoming interested in the not-so-glamorous world of soil microbiology.

Waksman's expertise spanned many different aspects of microbiology. He made contributions to the war effort, investigating the fouling of the bottom of ships in marine microbiology. He also worked in industry, using his knowledge to advise in production of substances such as vitamins and enzymes.

Arguably, Waksman's greatest legacy was the discovery of certain antibiotics. The first antibacterial substance he isolated was from a bacterium found in the soil. This substance, actinomycin, was actually toxic when studied in animals, but this little setback did not discourage Waksman. He developed screening techniques and identified a number of antibiotic compounds produced by the bacterium, including streptomycin, the first substance active against TB.

As well as making major discoveries in microbiology, Waksman published twenty-eight books and four hundred papers. Despite his great ability, Waksman was a selfless man who donated 80 per cent of his earnings from the licensing and patenting of his antibiotics to establish the Institute of Microbiology, in association with Rutger's University. He also set up the "Foundation of Microbiology" to encourage publication and research in microbiology worldwide. In 1952, his efforts were recognized when he was awarded the Nobel Prize in Physiology or Medicine.

Katie Giles

Date 1952

Nationality Ukrainian-American

Why He's Key He made some major microbiological breakthroughs, finding valuable antibiotics from the soil.

Key Event **Operation Ivy**
The dawn of the thermonuclear age

It was the end of January and the beginning of the 1950s. A spy within the U.S. nuclear programs was revealed and the Soviet Union had tested a fission bomb, which put them equal with America, bomb-wise. That January, President Harry Truman announced the United State's plans to make a super bomb. Just a year later, Stanslaw Ulam and Edward Teller gave them the secrets of the hydrogen bomb.

The Teller-Ulam bomb has two stages. Like a traditional nuclear bomb, the trigger works by breaking up uranium and plutonium, in a process called fission, and releasing energy. But instead of just exploding, the energy from this trigger is used in the second stage.

The second stage contains deuterium, a type of hydrogen. When you heat deuterium and compress it – in the way that setting off a small nuclear bomb on top of it will tend to do – you get fusion. The nuclei of the atoms are stuck together and you're left with plenty of energy and some unpleasant nuclear fallout.

Another year on, and the Americans had a bomb that was ready to test, under the codename Operation Ivy. "Ivy Mike" was a 74-ton shed on Elugelab Island in the Pacific Ocean. Inside were cryogenic tanks and the 6-meter- tall bomb assembly nicknamed "Sausage."

The resulting explosion unleashed 10.4 megatons of power, more than 450 times the power of the bomb dropped on Nagasaki, and turned the island of Elugelab into an underwater crater, 1.9 kilometers wide and 50 meters deep.

Douglas Kitson

Date 1952

Country USA

Why It's Key Entering the thermonuclear age meant the possibility of being wiped out by nuclear bombs became much more likely, scaring the pants off pretty much everyone.

opposite Explosion of the first hydrogen bomb

Key Invention **Bubble chamber**
Watch the quark

Invented by American physicist Donald Glaser in 1952, the bubble chamber opened up a new world of possibilities in the field of nuclear and particle physics. This was recognized officially in 1960, when Glaser received the Nobel Prize in Physics for his work with elementary particles and bubble chambers.

When particles are fired into a bubble chamber – essentially a cylinder of liquid – it is possible to study them by observing the bubbles they create. The liquid is heated to just below its boiling point and, when a piston quickly decreases the pressure, high speed particles entering the liquid create bubbles. The density of the bubbles is used to determine various properties of the particle. As the bubbles expand over time, it is possible to capture the image of the tracks left behind. It is also possible to determine the charge of the particles by sticking a magnet on the side of the chamber and seeing if the particle responds.

In his original research, Glaser experimented with many liquids including water, soda, and even beer. He had his suspicions about these as useful media but thought it best to check before splashing out on proper chemicals. His experiments allowed him to calculate the lifetime and "spin" (a property that relates to particle structure) of particles made up of only two or three quarks (the smallest in existence).

Modern bubble chambers, which often use liquid hydrogen, are used to study these fundamental particles, and are an essential tool in the search for dark matter.

Leila Sattary

Date 1952

Scientist Donald Glaser

Nationality American

Why It's Key Was essential to the progress of nuclear and particle physics.

Key Experiment **Existence of the neutrino confirmed**
Detecting the undetectable

The existence of tiny particles called neutrinos was an ad hoc assumption, used to make the laws of particle physics work. In order for them to fudge the equations properly. For neutrinos to exist they would have to have an infinitesimal mass and interact very weakly with other particles. But these properties, which had enabled the neutrino to plug the gaps in the theory, meant it was virtually impossible to find them with experiments.

To some people, however, things that are virtually impossible cry out to be done and in 1953, Fred Reines and Clyde Cowan set out to track down this elusive little neutral thing. They had originally planned to detect the neutrino barrage from an atomic bomb blast, but in the end concluded a nuclear reactor, which emits a mere 10,000,000 million neutrinos through each cm^2 area per second, would suffice. To look for the particles, Reines and Cowan watched for a reaction that could only happen if the neutrino existed. They looked for a proton and a neutrino combining to give two different particles – a neutron and a positron. If this did happen, the positron would combine with an electron and annihilate, giving off a flash of light, which *could* be detected.

Even with the unimaginable number of neutrinos streaming through the $2m^3$ detector, only three flashes an hour were detected. But each of those three results was consistent with the result predicted for the theoretical neutrino. Finally it was concluded that if it walked like a neutrino and quacked like a neutrino, it was a neutrino.

Kate Oliver

Date 1953

Scientists Fred Reines, Clyde Cowan

Nationality American

Why It's Key An experiment which showed that neutrinos, previously thought to be either fictional or undetectable, really did exist. And that you're allowed the occasional ad-hoc assumption.

Key Event **Chilly start for a new science**
First calf bred with frozen semen

In 1953, a calf was born on a Wisconsin farm. This would have been a wholly unremarkable event were it not for the fact that it was the first calf to have been artificially fertilized by previously frozen semen.

For frozen cells to remain viable, you have to protect them during the freezing process. It was British biochemist Chris Polge who first worked out how to do this, with the help from a colleague, Audrey Smith, and a careless lab assistant. Polge had been trying to protect semen with sugars during freezing without much luck. One day, he found that the same bottle of fructose solution he had been using mysteriously started to work and he eventually realized that his lab assistant had mislabeled the bottles. He had actually been using glycerol and this made all the difference. Unfrozen semen starts to lose potency after just two days, but in glycerol it can be frozen at -78 degrees Celsius for over a year, and still be viable when thawed.

Polge's work meant massive improvements for the cattle industry. Armed with this new technology, breeders could use sperm from one high-quality bull to fertilize hundreds of cows throughout the year and anywhere in the world.

Even more importantly, he kick-started an entirely new field of science called cryobiology, which had immense implications for agriculture and medicine. The ability to freeze eggs, sperm, blood cells, and even entire embryos has allowed infertile human couples to have children and conservationists to breed endangered species.

Ed Yong

Date 1953

Country USA

Why It's Key The start of successful freezing of live tissues for later use, which revolutionized the cattle industry, and laid the groundwork for IVF (in-vitro fertilization) and other important medical applications.

Key Event **Piltdown Man is a fake**
Human evolution in turmoil

Far from being the discovery that would fill the missing evolutionary link between ape and man, the now infamous Piltdown Man is better known as one of the greatest scientific embarrassments to date.

The origin of mankind has forever fascinated us, and researchers have painstakingly tried to piece together the story of our evolution. Remains of early humans are, however, very rare. It's easy to see how, in 1912, when an amateur paleontologist discovered what appeared to be bones of an entirely new genus of hominid in Piltdown, southern England, the find was hailed as one of the most important scientific discoveries of the twentieth century. As time passed, other fossil remains were found and pieces of the human evolutionary jigsaw started to slot together. But one piece of the jigsaw was ill-fitting; that of the Piltdown Man. It was originally thought that the remains of the Piltdown Man were up to half a million years old. In the 1940s, however, this theory was thrown into doubt when the fluorine absorption test was developed at the British Museum. Results indicated that the skull and jaw bones were much younger than previously thought.

The forgery was officially revealed in 1953. The skull in question was no more than 50,000 years old, when modern *Homo sapiens* were already walking the planet. Furthermore the jaw bone wasn't even human; tampered with by filing and staining, it was in fact that of a modern ape.

Ceri Harrop

Date 1953

Country UK

Why It's Key One of the most famous scientific hoaxes in history was uncovered. Human evolution researchers could finally start piecing together the rest of the puzzle.

Key Publication ***Science and Human Behavior***
Science, Skinner style

Born in Pennsylvania in 1904, Burrhus Frederic Skinner became one of the most noted psychologists of the twentieth century. His work on the behavioral learning of various animals was considered revolutionary in that it established behavior as a valid scientific field. Until then, science had been about physics and little else.

Following on from these studies, Skinner went on to produce one of the most controversial publications of his age, in 1953. Called *Science and Human Behavior*, the book concentrated on the study of the behaviors of various animals. This may seem like little cause for talk, but in the 1950s, psychology was still a growing field. The "behaviors" and "concepts of self" dissected in *Science and Human Behavior* were thought of as vague and controversial topics.

To add insult to injury, Skinner attempted to draw comparisons between the behaviors of humans and other animals. He argued that, based on Darwinian evolution, we as *Homo sapiens* had little to separate us from other creatures. He wrote "I may say that the only differences I expect to see revealed between the behavior of the rat and man (aside from enormous differences of complexity) lie in the field of verbal behavior."

In a conservative and tight-lipped 1950s society, where man was thought of as highly superior to other animals, comments such as these were bound to cause a stir. However, Skinner's book is still in popular print today, with his theories becoming more and more widely accepted by the day.

Hannah Welham

Publication *Science and Human Behavior*

Date 1953

Author Burrhus Frederic Skinner

Nationality American

Why It's Key A key moment in the establishment of behavior as a valid scientific field.

Key Event **Sir Edmund Hillary and Tenzing Norgay reach the summit of Everest**

At 8,847 meters above sea level, Mount Everest is the highest of all the world's peaks, and for mountaineers it has always been the Holy Grail. In 1953, a British expedition led by Colonel John Hunt was planning an assault on the summit. The team comprised over 400 people carrying 4,500 kilograms of gear. By March they had set up base camp. The summit team comprised of two pairings: the British duo of Tom Bourdillion and Charles Evans, and the New Zealander Edmund Hillary together with the Nepalese Sherpa Tenzing Norgay. A year earlier, Tenzing had come within 243 meters of the summit but was forced to turn back because of bad weather.

On May 26, Bourdillion and Evans were forced to abandon their summit attempt due to oxygen equipment failure. Three days later, it was Hillary and Tenzing's chance. They were climbing in what is now commonly referred to as the "death zone" – above 8,000 meters the air is so thin that the body slowly dies from oxygen deprivation.

At 11.30 am, after battling the elements and risking their lives, the two men were literally on top of the world. Out of necessity, the pair only spent fifteen minutes on the summit – taking pictures and burying sweets in the snow as a Buddhist offering. In the intervening years, over 2,000 people have followed in the footsteps of Hillary and Tenzing; many, however, were not so fortunate, and their bodies still remain on the mountain.

Andrew Impey

Date 1953

Country Nepal

Why It's Key Far from being just an amazing feat of human endurance, this was the catalyst for medical research into the mysteries surrounding altitude sickness and oxygen debt in the human body.

Key Publication **"A structure for deoxyribose nucleic acids"** Unravelling DNA: the double helix is revealed

Arguably the most famous discovery of modern science began with the meeting of Francis Crick and James Watson in the Cavendish Laboratory at Cambridge, UK in 1951.

Interestingly, although the pair's best-known findings were related to DNA, they did not actually do any experimental work on this molecule. Instead, they combined their own expertise – for Watson, in viral and bacterial genetics, and for Crick, in X-ray diffraction – with the findings of other DNA researchers.

Two particular findings were critical for the pair's discovery. First was Austrian biochemist Edwin Chargaff's work in 1950, which showed that there were equal amounts of the chemicals adenine and thymine, and cytosine and guanine in DNA, suggesting they occur in pairs. Second was Rosalind Franklin's X-ray diffraction pattern of the "B" form of DNA, the most common form found in cells, which her colleague Maurice Wilkins showed to James Watson in January 1953. On April 23, 1953, Watson and Crick's famous paper detailing the characteristic "double helix" of DNA was published in *Nature*. In this paper, the authors recognized the potential importance of this helical structure in terms of inheritance. The paper concluded: "It has not escaped our attention that the specific pairing that we have postulated immediately suggests a possible copying mechanism for the genetic material."

In 1962, Crick, Watson, and Wilkins shared the Nobel Prize in Physiology or Medicine. Sadly, Rosalind Franklin died in 1958, before the prize was awarded.

Christina Giles

Date 1953 *Nature*

Authors James Watson, Francis Crick

Nationality British

Why It's Key Unlocked the secret of life and uncovered one of nature's most elegant and iconic structures.

opposite Crick and Watson with their DNA model, the double helix

Key Invention **Black box**
The truth of the matter

It was amid the aftermath of the mysterious crash of the world's first jet-powered plane that ideas for an early black box found form. Australian research scientist Dave Warren realized that it would have been most useful to know exactly what had happened before the airplane met its untimely end. He concluded that a miniature recorder placed in the cockpit, designed to continually record activity, was the answer.

Unfortunately, as with most technological and scientific advances at first, it failed to seize public interest and it was quite some time before the authorities realized its merits. One man with more foresight was former British Air Vice-Marshal, Sir Robert Hardingham, who immediately installed a black box on a flight back to England. It was a while before the rest of the world followed suit, but today black boxes are fitted as standard on aircraft worldwide.

Despite their name, black boxes are in fact orange. They are vital pieces of equipment in the event of an accident, and have aided crash reporters in their work since the early 1950s. The box is two components: the flight data recorder and the cockpit voice recorder. The former traces the forty-eight parameters of flight data, while the latter records audible sound in the cockpit. Amazingly, black boxes can withstand forces many times the magnitude of gravity; temperatures to rival any decent fiery inferno; and immense underwater pressure.

Katherine Ball

Date 1953

Scientist Dave Warren

Nationality Australian

Why It's Key Helped us understand the reasons for aircraft crashes and prevent future accidents.

Key Experiment **Amino acids**
The stuff of life made in a lab

Stanley Miller and Harold Urey, at the University of Chicago, weren't afraid of the big questions. They went in search of the origins of life, by trying to create its building blocks in the lab.

They demonstrated for the first time that organic molecules could be formed from inorganic precursors. Specifically, they were able to synthesize amino acids; small molecules that can aggregate to form proteins and therefore key building blocks for even the most rudimentary forms of life.

Miller and Urey decided to model the conditions thought to have existed on the early Earth. To simulate the atmosphere, they set up a sterile chamber containing water, methane, ammonia, hydrogen, and carbon monoxide. This was exposed to frequent electrical charges, to model lightning in the primordial skies. After a week of reflux, the resulting brown mess was found to contain a range of organic compounds. These included thirteen of the twenty-two amino acids present in proteins, as well as sugars, lipids, and nucleotides.

We don't know for sure whether Urey and Miller's experiment was a good model for the chemistry of early Earth. This and subsequent experiments, however, have shown that the stuff of life and the stuff of the universe are readily interchangeable – a message relevant beyond our planet as we seek out life elsewhere in the cosmos.

Matt Brown

Date 1953

Scientists Stanley Miller, Harold Urey

Nationality American

Why It's Key Miller and Urey's experiment touched on one of the fundamental questions of the universe: how did life arise?

Key Experiment
Making the love hormone oxytocin

Oxytocin is more commonly known as the hormone of love. It gets this reputation partly because it is released during orgasm, and partly because it regulates emotion and trust. However, its more significant roles include initiating muscle contraction during childbirth, stimulating milk release from the mammary glands of females, and encouraging maternal behavior. It is made in the hypothalamus – part of the brain – but is actually secreted from the front part of the pituitary gland.

In 1953, American chemist Vincent du Vigneaud decided to synthesize oxytocin in the lab. After painstakingly analyzing this protein, he found it was a made of nine amino acids. Synthesizing this would be difficult in itself, but the molecule presented further problems: the nine amino acids were linked together in a circle. Imagine making a daisy chain from nine daisies, it is possible to close the chain before using up all nine daisies - the same can happen in amino acid chemistry, and once the bonds have formed and the ring has closed, it is very difficult to undo it.

Du Vigneaud solved the problem by using a protecting group. This protecting group stopped the amino acid from reacting too early, so he was able, after joining the nine amino acids first, to close the ring by simply removing the protector using liquid ammonia. This protecting group method went on to be used in many hormone production methods. Today, oxytocin is used medically to induce labor in overdue pregnant women and to facilitate breast feeding.

Riaz Bhunnoo

Date 1953

Scientist Vincent du Vigneaud

Nationality American

Why It's Key Du Vigneaud's synthesis enabled the mass production of oxytocin for medical applications. The protecting group method also paved the way for the synthesis of many other hormones.

Key Experiment **X-ray vision**
Identifying the molecule of life

The discovery of the structure of DNA is mainly attributed to Watson and Crick, but many other scientists made breakthroughs that contributed greatly to that understanding. Rosalind Franklin and Maurice Wilkins were both scientists working at King's College London in the 1950s. Although they were not great friends, they worked in the same department developing X-ray chromatography to image DNA.

X-ray chromatography was used to gain information on crystalline structures too small to be seen with microscopes. With this technique, researchers aim X-rays at a substance, which "bounce off" the atoms they hit and produce a characteristic image on photographic film.

The X-ray images generated by the King's group were crucial in determining the double helical structure of DNA. One of Franklin's images, "photo 51," was shown to Watson and Crick without Franklin's knowledge. It showed an X-shaped diffraction pattern, suggesting that overlapping helical structures made up DNA. James Watson saw the image and felt a key question had been answered. He and Crick published their findings on the structure of DNA in 1953.

The work of Franklin was not formally acknowledged at this time. She died aged thirty-seven from cancer, and has since been verbally credited by Watson and Crick for her contribution. In 1962, Crick, Watson, and Wilkins were awarded the Nobel Prize in Physiology or Medicine – a prize that is not awarded posthumously.

Katie Giles

Date 1953

Scientists Rosalind Franklin, Maurice Wilkins

Nationality British

Why It's Key A major step in elucidating the structure of DNA.

Key Event **Space rock crashes into living room**

"I can't come into work; I've been hit by a meteorite."

Hollywood has made all manner of apocalyptic doomsday scenarios seem feasible, from lethal, zombie-making viruses to alien invasion, but no event could signal The End more urgently than a catastrophic meteorite impact.

Massive meteorite impacts are clearly something to worry about, but they are also very rare, occurring only every 50 to 100 million years, although the dinosaurs were wiped out by a large collision 64 million years ago, smaller impacts happen on a more regular basis. A 250-meter meteorite impact – packing the punch of fifty Hiroshima bombs – is estimated to occur every 10,000 years, whilst scientists have estimated that between 1,000 and 10,000 tons of meteorite material fall to Earth on a daily basis.

If this is indeed the case, then what is the chance of being hit by one? If the calculations are correct, on average a human should be hit by a meteorite weighing more than 100 grams once every fourteen years. Considering that the impact of a 500-gram meteorite is probably going to be fatal, a person is likely to be killed by a meteorite once every fifty-two years.

It's difficult to decide if Mrs Elizabeth Hodges, resident of Sylacauga, Alabama, was either very, very lucky or very, very unlucky. On November 30, 1954, a 3.9 kilogram meteorite crashed through the roof of her house and struck her left hip and arm. Despite being badly bruised, the impact was reduced by her roof and ceiling, and she survived to become the only human to be verifiably struck by a meteorite.

Jim Bell

Date 1954

Country USA

Why It's Key Meteorite impacts happen regularly and it is only a matter of time before a deadly impact comes our way.

opposite Mrs Hodges points out the hole in the ceiling where the meteorite came through

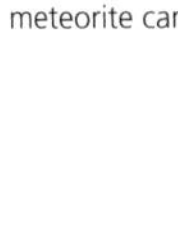

Key Experiment **Straight down the middle**

Roger Sperry and the split brain

The work of American neuropsychologist Roger Sperry gave a whole new meaning to the phrase "splitting headache." Starting life as a student of English, Sperry switched to zoology and gained a PhD from the University of Chicago. Here he and Ronald Myers began to unravel the mystery of the split brain.

The corpus callosum is a thick bundle of fibers connecting the two halves of the brain. Experiments in the 1930s and 1940s showed that severing this connection appeared to have no effect on laboratory animals. Sperry and Myers investigated these peculiar results. In an experiment with cats, they found that one hemisphere could "learn" simple tasks as fast as two. They concluded that a cat's brain had the ability to act as two separate brains, sharing information via the corpus callosum. This line of research then switched to humans. At the California Institute of Technology, Sperry and Michael Gazzaniga devised a series of ingenious experiments to be carried out on people who had undergone surgical cutting of the corpus callosum as a treatment for epilepsy. They showed that each hemisphere was "a conscious system in its own right, perceiving, thinking, remembering, reasoning, willing, and emoting, all at a characteristically human level," and that the two sides housed different areas of expertise. The ability to speak, for instance, lies almost entirely in the left hemisphere.

Sperry's work transformed neuroscience and philosophy. He is said to have remarked: "The great pleasure and feeling in my right brain is more than my left brain can find the words to tell you."

Christina Giles

Date 1954

Scientist Roger Sperry

Nationality American

Why It's Key Showed that the left and right hemispheres were conscious systems in their own right.

Key Event **The first kidney transplant**
A Christmas gift to remember

In December 1954, Ronald Herrick gave his twin brother Richard a unique and life-saving Christmas gift: one of his kidneys. In a five-and-a-half hour procedure, surgeon Joseph Murray and his team performed the world's first successful kidney transplant at the Peter Bent Brigham Hospital in Boston, USA.

Cornea and skin transplants had been performed before, but this particular transplant was important as it was the first successful solid organ transplant. Work earlier in the twentieth century had used part of a rabbit kidney to replace a diseased organ. This was successful only for a very short time, and ultimately proved to be fatal. In 1933, the first human kidney transplant resulted in failure when the recipient's immune system rejected the kidney, recognizing it as foreign. The secret to the twins' success was in their genes. Richard's immune system didn't reject Ronald's kidney because they shared the same genetic identity. This finding was key to the major advances made in transplantation medicine.

Thanks to his brother, Richard Herrick was able to live for a further eight years. Not only was this a medical first, but, more importantly, it offered a potential cure for patients dying of kidney failure. Joseph Murray continued his work and was rewarded with a Nobel Prize in Medicine or Physiology in 1990.

Katie Giles

Date 1954

Country USA

Why It's Key Unlocking the secret behind successful organ transplants.

Key Invention **Let there be light**
The first solar cells

The first primitive solar cell, although unrecognized as such at the time, was created in 1839. A nineteen-year-old French physicist, Antoine-Cesar Becquerel, noted that when certain combinations of metals and solutions were exposed to light, they could produce small amounts of electric current.

By 1877, Charles Fritts had developed the first "modern" solar cell, made of gold-coated selenium. It worked, but was plagued with inefficiencies and transformed less than 1 per cent of light into electricity.

The real breakthrough didn't come until 1954, when three Bell Laboratories scientists – Daryl Chapin, Gerald Pearson, and Calvin Fuller – demonstrated the first practical solar cell. They were seeking a replacement for the ubiquitous dry cell batteries, used in the Bell telephone system, which degraded too fast in tropical areas. Chapin considered wind power, thermoelectric devices, and steam power, along with photovoltaic cells. The small silicon solar cell they eventually designed returned an efficiency of 6 per cent in direct sunlight. Thirty years later, silicon cells with efficiencies of more than 20 per cent had been fabricated. In late 2006, Boeing-Spectrolab announced they had broken the 40 per cent efficiency barrier.

Another solution to the problem of dry cell battery decay involved radioactive isotopes. Atomic batteries functioned much like solar cells but emitted electricity when bombarded by highly radioactive gamma photons, rather than photons of visible light. Many environmentalists, understandably, consider atomic batteries to be a less than eco-friendly solution.

Stuart M. Smith

Date 1954

Scientists Daryl Chapin, Gerald Pearson, Calvin Fuller

Nationality American

Why It's Key The introduction of a viable "green" technology.

opposite A new solar cell is placed in a Van De Graaf accelerator to test its atomic endurance

Key Event **The Nile perch**
A big fish in a big pond

An alien species arrives in a sleepy backwater and starts consuming everything that stands in its way. Sounds like the plot to any number of science fiction movies, doesn't it? But this is not science fiction; this is the Nile perch, introduced deliberately into Lake Victoria, the largest freshwater lake in Africa.

The Nile perch is an impressive fish. A fully grown specimen can be up to two meters in length, and individuals have been reported as weighing in at over 200 kilograms. Being one of the world's largest species of freshwater fish, they are predictably voracious predators and will gobble up pretty much anything they come across.

Following its introduction to Lake Victoria, this massive predatory fish thrived in its new environment, eating hundreds of fish species, peculiar to the lake, into extinction. Though the fish has brought economic growth to the surrounding area – the Nile perch is the basis of a multimillion dollar fishing industry for both food and sport – it has come at a cost of more than just ecological diversity. Growth in the market for it has brought deforestation, pollution, and, with a dwindling average catch, the prospect of a collapse in the local fishing industries that many depend upon for their livelihood.

As one of the "100 Worst" species in a report published by the Global Invasive Species Database, the Nile perch ranks alongside the likes of the rabbit, the cane toad, and the crazy ant.

Jim Bell

Date 1954

Country Tanzania, Uganda, Kenya

Why It's Key The introduction of an alien invasive species is often ecologically damaging, but this fish now affects the lives of millions of people around the world.

Key Event
First atomic submarine launched

On January 21, 1954, First Lady Mamie Eisenhower christened the world's first nuclear powered submarine. Construction of the ship had been ordered by congress in 1951, and, a year later, the keel was laid by President Harry S. Truman. After a nineteen-month build, *Nautilus*, the sixth ship of the Navy fleet to bear the name, slid into the Thames River in Groton, Connecticut.

Nautilus was driven by a nuclear reactor developed by the Naval Reactors Branch of the Atomic Energy Commission. Nuclear power enabled the submarine to break a number of world records. It could stay submerged for weeks without having to return to the surface to recharge – unlike diesel-driven submarines – and therefore smashed many submerged speed and distance records. Due to its endurance abilities, *Nautilus* was chosen for "Operation Sunshine" – the first crossing under the North Pole. The operation succeeded on August 3, 1958 – a great boost to the United States after the Soviets' launch of Sputnik the year before.

Nautilus was able to outrun many of the old submarines and outmaneuver anti-submarine warfare weapons. This meant new tactics and better armaments needed to be developed to meet the threat posed by nuclear powered submarines.

After traveling half a million miles, *Nautilus* was finally decommissioned in March 1980. The famous submarine was later declared a National Historic Landmark and is now an exhibit at the Submarine Force Library and Museum in Connecticut.

Faith Smith

Date 1954

Country USA

Why It's Key Faster and more maneuverable than any other sub, the *Nautilus* led the way in nuclear submarine technology.

opposite The launch of Nautilus

571

Key Event **Georgetown**
The first language translation by machine

"Vladimir appears for work late in the morning." We don't know who Vladimir was, and it's not important that he was late. What is important is that this sentence is one of the very first to be translated by a computer. The original Russian words, *"Vladyimir yavlyayetsya na rabotu pozdno utrom"* were punched onto cards by an operator – who didn't speak a word of Russian herself – on January 7, 1954. They were then fed into an IBM Type 701 computer, and spat back out as their English equivalent a few seconds later.

The sentence was one of more than sixty Russian sentences successfully translated by the machine in a demonstration given at the New York headquarters of IBM. The experiment was a collaboration between IBM and the language department at Georgetown University, headed up by Professor Leon Dostert.

As one of the most advanced computers on the planet at the time, the IBM Type 701 took up an area around the size of a tennis court and cost US$15,000 per month to hire (more than US$100,000 in today's money).

Virtually everything about the project – the "Kitty Hawk of electronic translation," according to Professor Dostert – was a new challenge. In 1954, there wasn't even an established way of typing English into a computer, so to produce a machine capable of translating even a small selection of phrases from one language to another was remarkable.

Matt Gibson

Date 1954

Country USA

Why It's Key Up until the Georgetown experiment, computers and words really hadn't mixed. Georgetown demonstrated to the world that computers could do more than just math.

Key Discovery
First effective polio vaccine

In 1955, American scientist Jonas Salk made a breakthrough that would save thousands of lives: he developed the first safe and effective vaccine against polio.

Polio is a devastating disease with no cure, caused by a virus that spreads via the mouth and poor hygiene. It mostly affects young children, attacking the nerves and, in severe cases, causing paralysis. Public swimming pools were closed as summer epidemics swept through Europe and North America in the nineteenth and early twentieth centuries. Children and adults crippled by polio were a familiar sight, as were the "iron-lung" breathing machines used in hospitals to keep patients alive by helping them to breath. President Franklin D. Roosevelt, one of the disease's most famous victims, called for a massive scientific effort to create a vaccine to protect against infection.

Salk's achievement was to grow the virus in large quantities in the laboratory and then kill it using the chemical formaldehyde. This rendered it safe to inject into people without fear that it would cause the disease. Such was his confidence that Salk first tested his vaccine on his own family.

Luckily Salk's vaccine was a success, and was used widely around the world, slashing the number of cases. It is still used today, together with an oral polio vaccine licensed in 1962, created by Albert Sabin. Sabin's vaccine can be swallowed rather than injected, and so is the vaccine of choice for tackling outbreaks in countries where polio lingers, predominantly in South Asia and Africa.

Julie Clayton

Date 1955

Scientist Jonas Salk

Nationality American

Why It's Key There is no cure for polio. Salk's vaccine was the first safe and effective vaccine to protect against the disease and its crippling effects – slashing the number of cases worldwide.

opposite Jonas Salk in his office, inspecting polio vaccinations

Key Event **CERN, the world's largest science lab, is founded**

In 1949 Louis de Broglie proposed that a new European research center be founded to improve European science after World War II. Three years later, eleven European governments agreed to form the *Conseil Européen pour la Recherche Nucléaire* – CERN, and picked a site near Geneva for the laboratory.

After the rush to develop the atom bomb and nuclear power in the war, CERN adopted a different approach, bringing nations together after the conflict. No military research would be performed and any scientific results from the center would be publicly available. Once the founding governments agreed on the principle of CERN's operation, what was to become one of the world's largest and most respected science organizations was created on September 29, 1954. In 1957, CERN's first particle accelerator went online and it soon yielded results: the observation of a subatomic particle – a pion – decaying into an electron plus a neutrino. Bigger, better instruments were constructed and, in 1983, presidents of France and Switzerland were present at the ceremony to commemorate the construction project the laboratory is famous for: a vast circular tunnel with a circumference of 27 kilometers built underground to house the Large Electron-Positron Collider – the biggest scientific instrument ever built.

Today CERN is home to half the world's particle physicists, working toward the switch-on of the latest machine, the Large Hadron Collider, scheduled for 2008. Physicists hope that the LHC will reveal, for the first time, what happened fractions of a second after the Big Bang.

David Hawksett

Date 1954-5

Country Switzerland

Why It's Key CERN provided focus for collaborative European science after World War II.

Key Publication ***A Million Random Digits***

Random numbers are, as the name implies, random; they do not follow any regular or repetitive pattern so that any number in a random number series contains no information that can be used to determine the next, or the previous number. These numbers are important for solving problems in various kinds of experimental probability procedures; a process called the Monte Carlo method in which a random starting point is chosen for solving a problem.

Before the development of high-speed computers that allow the fast generation of pseudo-random numbers, random number tables were greatly relied upon for use in the statistics field.

A large supply of random digits required by, for example, statisticians and physicists; a need that was realized by RAND, a non-profit institution that conducts research on important and complicated problems. Consequently the book *A Million Random Digits and 100,000 Normal Deviates* was published by RAND in 1955, the first time such a large and carefully prepared table had ever been produced.

The book containing random number tables was produced by an electronic simulation of a roulette wheel attached to a computer, and still remains the largest published source of random digits. It is still routinely used in the work of statisticians, physicists, polltakers, market analysts, lottery administrators, and quality control engineers alike, and has become a standard reference in engineering and econometrics textbooks.

Sarah Watson

Publication *A Million Random Digits and 100,000 Normal Deviates*

Date 1955

Nationality American

Author RAND corporation

Why It's Key An important breakthrough in delivering random numbers. Such a large and carefully prepared table had never been available before the book's publication.

Key Invention **The hovercraft** Is it a bird? Is it a plane? Is it a boat?

Christopher Cockerell was a man with a fantastic imagination. When he fitted an empty cat food tin inside a coffee tin, connected them with an industrial air-blower and made them hover above a set of kitchen scales, he knew he had made an important breakthrough.

Cockerell's aim was to reduce the drag on sea vessels as they moved through the water by making them travel on a cushion of air. His new vehicle would be known as a "hovercraft." By trial and error in his small boatyard on the Norfolk coast, he finally managed to build a craft which floated above the water on a cushion of air from a special pump fitted on top. This was in 1955.

The hovercraft, however, had to wait a few more years to see the light of day. The delay was caused by the military who initially classified the idea as "secret," forbidding publicity so that potential enemies wouldn't develop the vehicle first. Finally in 1959, the first hovercraft was built and flown across the English Channel, from Dover to Calais, with Cockerell clinging to the deck as moveable ballast.

The hovercraft has since been transformed into the useful and versatile vehicle it is today. From huge passenger ferries and military vehicles to small craft able to access the upper reaches of rivers, the hovercraft can be seen zipping across land and sea the world over.

Simon Davies

Date 1955

Scientist Christopher Cockerell

Nationality British

Why It's Key A vehicle that hovers above water or solid ground, reducing drag and crossing the boundary between sea and land. It's the ultimate versatile carriage.

Key Invention **Atomic clocks** "The time is 11:42 and 3.1689785426 seconds"

Very accurate time measurement has applications in a whole variety of areas, both scientific and non-scientific. In science, many other measures fall back on the accurate measurement of units of time; these days, for example, distance is actually measured relative to how far light travels in a second. But previous definitions for the exact length of a second relied upon astronomical observations. This system lacked precision, as well as being pretty inconvenient.

Louis Essen had been working at the National Physical Laboratory in the UK, researching into ways of making more accurate time measurement. He had already developed the quartz ring clock – also called the Essen ring in his honor – which substantially improved the accuracy of conventional astronomical time measurements.

But Essen soon realized that advancements in this type of chronometer had reached their peak; the problem was that the standard they were using to determine the length of a second, at this point still astronomical, needed to be more accurate.

Spurred on by American research into atomic clocks, Essen set about developing an atomic clock, using the element cesium to generate a very accurate measure of an interval of time. Atoms of cesium leaving a microwave field are turned into charged particles, which are attracted to electrodes that register the charge. When amplified, these detected charges provide a timing mechanism accurate to one second in thousands of years.

Barney Grenfell

Date 1955

Scientist Louis Essen

Nationality British

Why It's Key Redefined the second, and now atomic clocks measure the passage of time 10,000 times more accurately than their astronomical counterparts.

Key Invention
Fibre optics offer a bright new future

It was in the 1840s when John Tyndall, an Irish physicist, demonstrated that light shone through a stream of water followed the path of the stream. Over a century later, physicist Narinder Singh Kapany invented the optical fiber, using Tyndall's work as a basis. On January 2, 1954, Kapany published an article in the journal *Nature* detailing his success at transmitting images through transparent rods made of glass and plastic.

The principle which both Tyndall and Kapany observed as they bent light is called *total internal reflection*. This means that if light hits a boundary of, say, a glass rod, at a particular angle, all of the incident light is reflected back inside the rod, with no light passing through the boundary. The light can propagate along the length of the glass rod with no transmission loss if the totally internally reflected light continues to strike boundaries at the correct angle and is totally internally reflected, time and time again.

Modern optical fibers can be as small as 9 microns (0.009 millimeters) in diameter, allowing them to be bundled together and placed in underground pipe networks. Despite their small size, optical fibers are extremely efficient at transferring information extremely quickly: globally transmitted television shows and telephone calls are possible, and electronic mail is received with almost no delay, thanks to a subterranean maze of optical fibers.

Gavin Hammond

Date 1955

Scientist Narinder Singh Kapany

Nationality Indian-American

Why It's Key Although the potential use of optical fibers was not truly recognized back in the 1950s, Kapany, the "father of fiber optics" is now considered to have revolutionized global telecommunications.

opposite The total internal reflecting beam of light is clearly visible in this length of fiber optic

Key Publication **Copy-bots**
Von Neumann's self-replicating machines

As a key player in quantum mechanics, a contributor to the Manhattan Project and co-creator of game theory, Hungarian John von Neumann had many areas of expertise. In the late 1940s, he turned his attention to robots.

Self-replicating robots are machines capable of producing copies of themselves from raw materials, without human intervention. They were proposed years before von Neumann formalized the idea in a scientific framework, but the idea wasn't popularized until 1955, when John Kemeny's article was published in *Scientific American* magazine.

Von Neumann devised a thought experiment in which he imagined a machine, surrounded by spare parts, that could produce copies of itself. This was feasible if the robot was able to manipulate its environment, and its memory – or program – could be transferred to this new copy, which could make another copy and so on. He called these robots Universal Constructors, although some still refer to them as "von Neumann machines."

The prospect of so-called "von Neumann probes" being employed in space exploration, is one that has attracted interest from scientists at NASA, who commissioned a detailed investigation in 1982. In theory, a self-replicating factory unit could be sent to mine an asteroid. Rather than one machine doing all the work, a myriad small drones could be created – some as miners, some as builders, some as repairers – and programmed to extract asteroid materials for delivery to Earth.

Steve Robinson

Publication *Scientific American*

Date 1955

Author John Kemeny

Nationality Hungarian-American

Why It's Key The future of space exploration could be based around robots able to set up base and reproduce themselves.

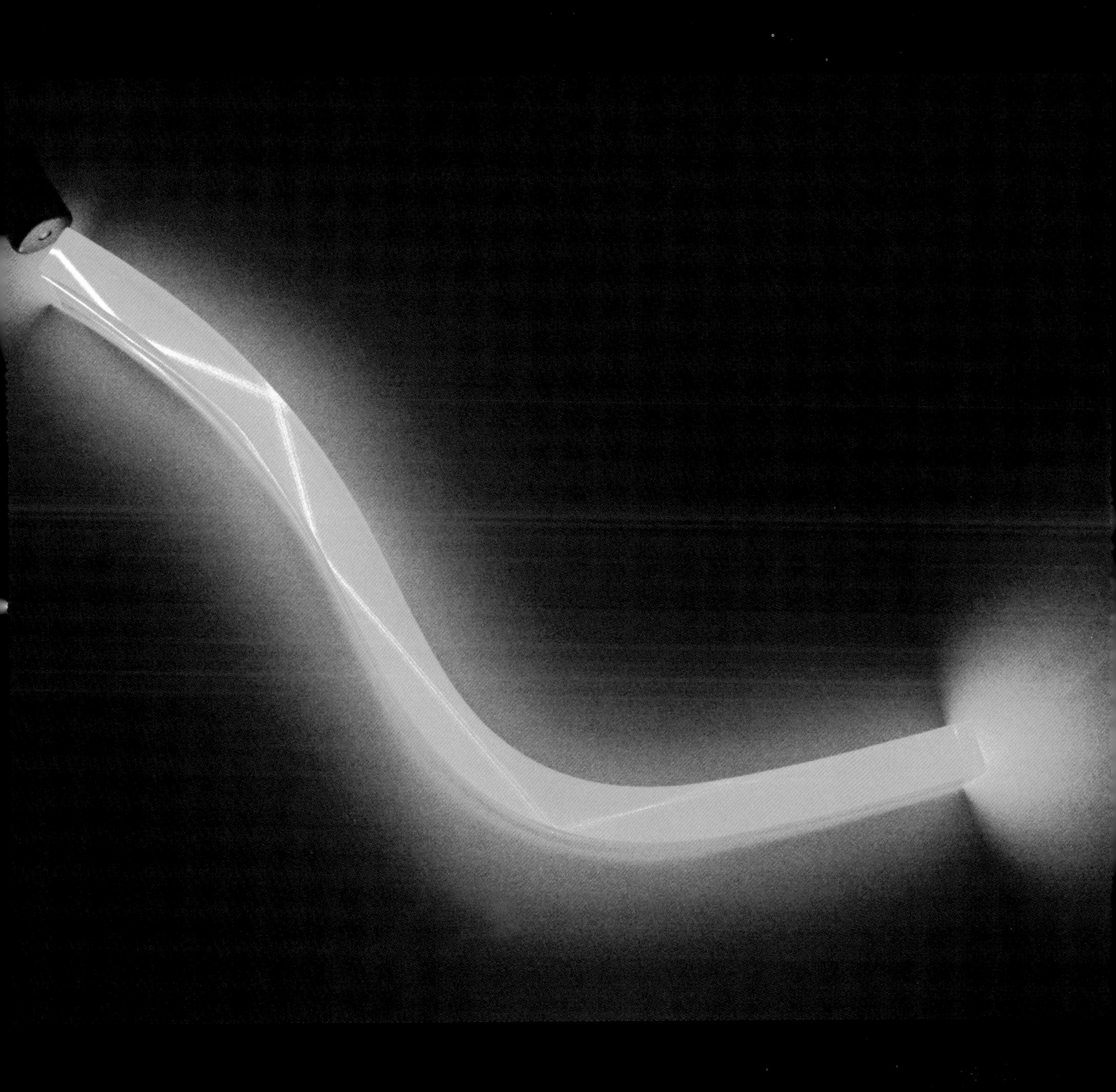

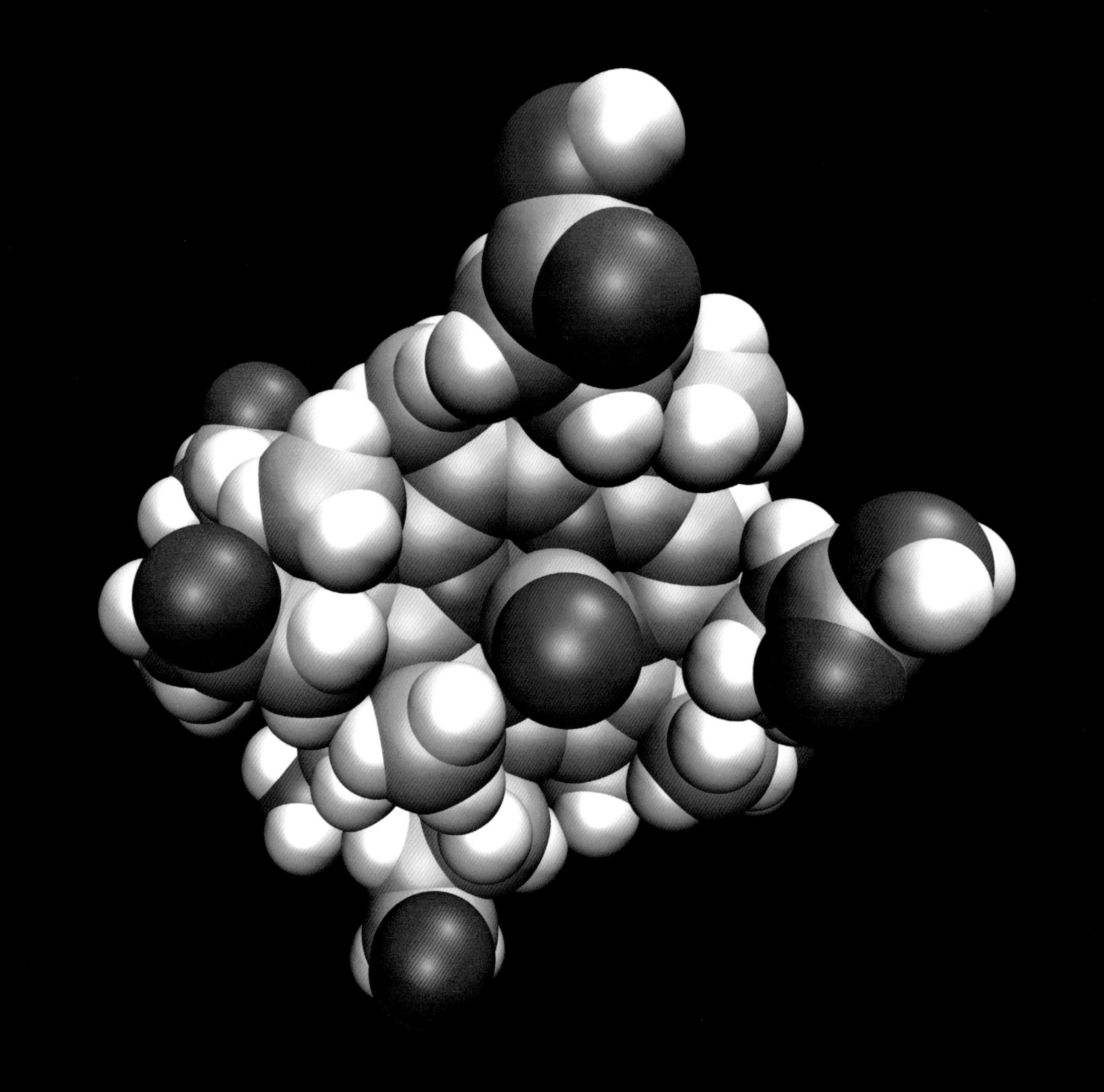

Key Person **Lawrence Whipple**
Dirty snowball theory

There isn't much that Fred Whipple (1906–2004) didn't know about comets and his work was been fundamental to the understanding of astrophysical research. Yet had it not been for a bout of polio at an early age, he might have been a professional tennis player.

It was a class in astronomy that ignited his passion for the mysteries of deep space and whilst still completing his doctorate he helped map the orbit of Pluto. Comet research was revolutionised in 1950 when Whipple published his theories on the physical nature of comets. In a stand against established theories, he argued that, far from being loose agglomerations, comets were in fact solid balls of ice containing ammonia, methane, and carbon dioxide mixed with dust. The media lapped up this revelation and proclaimed a new "dirty snowball theory." World War II interrupted his work but Whipple wasn't about to rest on his laurels. Researching radar countermeasures, he co-invented the chaff-cutter – a 3 oz piece of aluminum foil that acted as a miniature antenna. When dropped from bombers during Allied raids over Germany, they reflected radars, confusing the searchlights and anti-aircraft artillery. More advanced versions of chaff-cutter are still used today.

So many times throughout his career, Whipple was ahead of his time. He invented the meteor bumper to protect future spacecraft; he managed to view the Russian launch of Sputnik when the U.S. military failed and he was one of the first people to predict the potential benefits of satellites.

Andrew Impey

Date 1955

Nationality American

Why He's Key Whipple's dirty snowball theory is generally considered one of the most important contributions to Solar System studies in the twentieth century.

Key Discovery
The structure of Vitamin B12 revealed

In the 1950s, chemist Dorothy Hodgkin tried to work out the structure of vitamin B12 using the new computer technology and a technique called X-ray crystallography. This involved looking at how X-rays scatter in a crystal, as a way of determining the arrangement of atoms. She had already used the technique to study penicillin, in 1949, but vitamin B12 posed a whole new set of problems.

Vitamin B12 had been identified as an effective treatment for anemia – a condition where sufferers have a lack of red blood cells. Although well known as an often fatal disease, it wasn't until the 1920s that eating large amounts of liver was found to counteract the effects of anemia. And it wasn't until 1948 that Vitamin B12 was identified as the crucial ingredient in liver that gave it this useful property.

Working out the chemical structure of such an important and complex biological molecule like vitamin B12 had never been attempted before. At that time, both computers and X-ray equipment were laborious, difficult, and occasionally dangerous to use. So, when Hodgkin and her team at Oxford University successfully determined the structure of vitamin B12 in 1956, it was a major achievement.

Their success paved the way for further X-ray crystallography to work out the structures of other important molecules, and Hodgkin herself went on to determine the structure of insulin in 1969, another major milestone. Not surprisingly, Hodgkin's reward was a Nobel Prize, in 1964.

Richard Bond

Date 1956

Scientist Dorothy Hodgkin

Nationality British

Why It's Key Was important in understanding the role of vitamin B12 in the body, and in developing a new technique for working out the structure of complex molecules.

opposite Computer model of a Vitamin B12 molecule

Key Invention
The VCR

It is almost impossible to imagine that, when television broadcasting first began, there was no way to magnetically record television images for a further twenty years. Programs were broadcast live and had to be watched live; there was no such thing as video cassette recorders.

Sound recording on magnetic tape was fairly commonplace in the early fifties, but using the same technology to record video images required tape to go through the machines at fantastic speeds. Eventually, in 1956, a team from the Ampex Corporation headed by Charles Ginsburg developed a Video Tape Recorder (VTR) based on rotating head technology.

The groundbreaking technology was introduced in dramatic fashion. On April 14, William Lodge, engineering vice-president of the CBS network, gave a speech to the National Association of Radio and Television Broadcasters in Chicago. He briefly alluded to the appearance of exciting new technologies but made no further comment. Having concluded his speech, Lodge remained standing at the podium as the two hundred assembled delegates started to leave their chairs. Suddenly monitors around the room, on which images of the briefing had been relayed, began showing it repeated over and over again. Silence fell over the meeting for about fifteen seconds before they realized what was going on. Then a curtain was drawn back to show the Ampex VTR Mark IV prototype, and the crowd broke into sustained applause. The video recorder was born.

Simon Davies

Date 1956

Scientist Charles Ginsberg

Nationality American

Why It's Key Finally it was possible to record television pictures, and the video rental industry would follow close behind.

Key Invention **Tokamaks**
Fusion in a doughnut

For nuclear fusion to occur, hydrogen needs to be heated to millions of degrees – so hot that the atoms don't hold together anymore, and the gas becomes a plasma of free-floating protons and electrons. At this heat, protons travel fast enough to overcome the forces that make them repel each other; if they collide they can fuse together to create a heavier atom, giving off lots of energy in the process. But since protons are so small, to have a reasonable amount of collisions the gas needs to be very dense, which means it must be at high pressure.

So how do you compress plasma at a billion degrees? With magnetic fields. An electromagnet shaped like a doughnut creates a ring of magnetic field inside it which forces the charged particles inside the plasma to circle, following the direction of the field. This doughnut-shaped magnetic field generator is called a tokamak.

In 1968, Russian scientists at the Kurchatov Institute announced the best fusion results yet using a tokamak reactor. These were so impressive that, even at the height of the Cold War, British and American scientists immediately switched to using tokamaks in their own programs. Successful fusion has since been maintained for a period of several minutes, generating almost as much power as was needed to create the reaction. Current work aims to prove the viability of fusion as a long-term power source that can produce more energy than it uses.

Kate Oliver

Date 1956

Scientist Lev Artsimovich

Nationality Soviet

Why It's Key Tokamaks are our best hope yet for channeling the major power source of the Universe into a clean(er) nuclear power technology to fulfill Earth's energy demands.

opposite Interior of the JET (Joint European Torus) Tokamak device, Oxfordshire, England, the largest Tokamak in the world

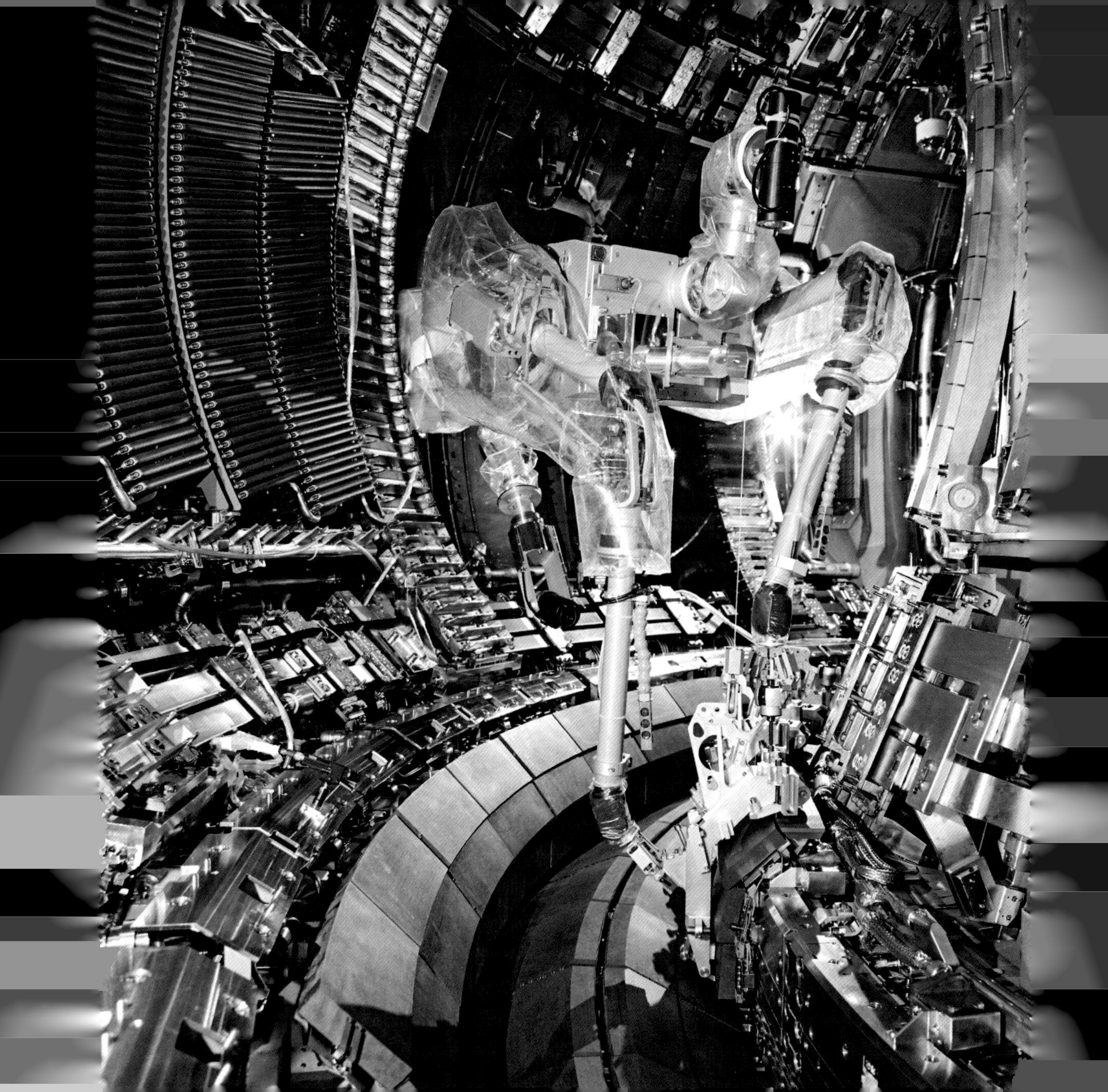

Key Event **Sellafield** World's first full-scale nuclear power station

On October 17, 1956, the Cumbrian town of Sellafield made the headlines as its nuclear power station was opened by Queen Elizabeth II. The full-sized station, named Calder Hall, was powered by uranium and was the first in the world to be opened. At 12.16 GMT, the Queen activated the reactor to supply electricity to the National Grid. By 16:00 that same afternoon, London had already started to receive power from the station.

It was only just over a decade since the bombings of Hiroshima and Nagasaki, and nuclear power was still a controversial topic, often associated with war and threat. The opening of the station at Calder Hall marked the advent of nuclear power as a useful alternative energy supply. The Queen said: "This new power, which has proved itself to be such a terrifying weapon of destruction, is harnessed for the first time for the common good of our community."

It wasn't long before the world caught on to the activities at Calder Hall, and nuclear power stations were built universally. At one time, a quarter of the UK's energy supply demands were being met by the technology. However, the phenomenon would soon be confronted by catastrophe as accidents and environmental concern began to obscure its advantages. Calder Hall was eventually closed in 2003 following the government's damning Energy White Paper, published in the same year.

Hannah Welham

Date 1956

Country UK

Why It's Key The 1956 opening of Calder Hall was a landmark in the use of nuclear power as an energy supply.

Key Person **Glenn Seaborg**

We all remember a good teacher. For Glenn Seaborg (1912–1999), it was his high school teacher who made science come alive and inspired him to achieve great things. So many great things, in fact, that Seaborg is in the *Guinness Book of Records* for having the longest entry in the "Who's Who in America" publication.

When Glenn Seaborg arrived at the University of California, Berkeley, the heaviest known chemical element was uranium with an atomic weight of 235. The search was on for heavier elements, and in 1941, Seaborg discovered plutonium-238. He went on to discover a number of other elements which found their place in the periodic table as the actinide series. Number 106, Seaborgium, was named after him as a tribute to his efforts. Seaborg's early work involved the discovery of isotopes – forms of the same element with different atomic masses. He was most famed for discovering that the isotope plutonium-239 was fissile – it could be split by neutrons to release large amounts of energy and could therefore be harnessed in a bomb.

After Japan attacked Pearl Harbour, Seaborg and his team joined the Manhattan Project. Their task was to devise a multi-stage chemical process for the isolation and concentration of plutonium-239 for the first atomic bombs. There was strong competition from the Germans, but after three years they produced enough plutonium to make a bomb. It was dropped on the city of Nagasaki. Seaborg spent the rest of his life campaigning for peaceful uses of fissile elements, such as nuclear power generation.

Riaz Bhunnoo

Date 1945

Nationality American

Why It's Key Seaborg's discovery of fission led to the atomic bomb, which marked the beginning of the end of World War II. It also enabled nuclear power generation, which provides a significant amount of our energy today.

Key Person **John Bardeen**
The brain of Bell

John Bardeen (1908–1991) was an American electrical engineer and physicist best known for his work on creating the first transistor. By the start of World War II, Bardeen was an assistant professor at the University of Minnesota. He became involved in military work in 1941 and, a couple of years later, was asked to participate in the Manhattan Project to build a nuclear bomb.

At the end of the war his old university did not appreciate the potential of his work in solid-state physics, so instead, Bardeen took up an offer from Bell Labs. It was here that he and his colleague Walter Shockley invented the transistor, which went on to replace vacuum tubes in radios and televisions.

On November 1, 1956, Bardeen was making breakfast when he heard on the radio that he had been awarded the Nobel Prize in Physics, along with Shockley and Walter Brattain at Bell Telephone Labs, for their work on transistors. During the Nobel ceremony in Sweden, King Gustav VI scolded Bardeen for not bringing his children to the event. Bardeen promised him that he would bring them all the next time he won a Nobel Prize.

Most accomplished scientists would have settled for that but by this time Bardeen was at the University of Illinois and working on his next major interest: superconductivity. The standard theory developed by Bardeen and colleagues earned him a second Nobel Prize, also for physics, in 1972. This time Bardeen did indeed bring his children.

David Hawksett

Date 1956

Nationality American

Why He's Key John Bardeen was the first person to receive two Nobel Prizes for Physics.

Key Discovery **Controlled cross-circulation in open heart surgery**

There was a day when being born with an abnormality of the heart meant almost certain death. Thankfully, due largely to the contributions of Dr C. Walton Lillehei, that's no longer the case.

The problem with performing open heart surgery is that the heart has to be beating in order to move blood throughout the body, but a moving organ is a very difficult thing on which to operate. Couple to that the fact that the inside of a heart is under high pressure and, as you can imagine, operating on it is not dissimilar to operating on a leaking garden hose.

Early open heart surgery was performed by inducing hypothermia in a patient and thus reducing blood flow. While this was a huge step forward, it proved less than ideal for the delicate operations needed to fix the tiniest of hearts.

Lillehei recognized that perhaps he could use another person to provide both the oxygenation and blood flow necessary to sustain a child's life during an open heart operation. This technique involved linking a child's arteries and veins to another person's with a compatible blood type – a process called controlled cross-circulation. The other person – usually a close relative – functioned as the heart and lungs of the infant, allowing any number of heart abnormalities to be fixed. Although it was eventually replaced with mechanical pumps and oxygenation devices, this technique saved the lives of many children in the 1950s.

B. James McCallum

Date 1956

Scientist Clarence Walton Lillehei

Nationality American

Why It's Key Cross-circulation saved the lives of many children who otherwise would have died from terrible birth defects of the heart. It also led to the creation of the heart-lung bypass machines utilized widely today.

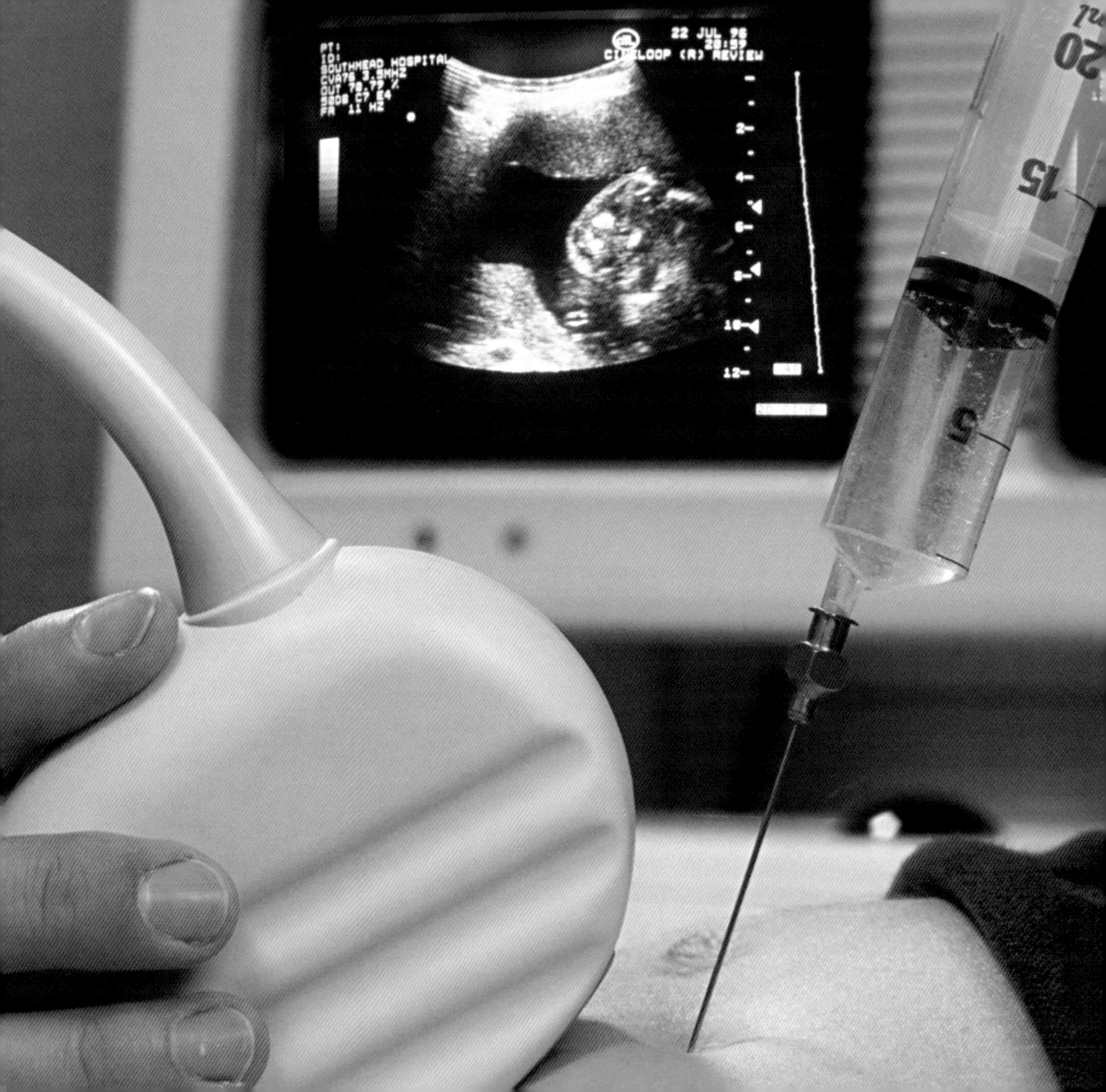
PT:
ID:
SOUTHMEAD HOSPITAL
22 JUL 96
REVIEW

Key Discovery **Amniocentesis** Monitoring the unborn bump

Amniocentesis aims to identify disorders in an unborn baby, by testing a small amount of the protective amniotic fluid from the womb. The practice of draining off this fluid through a long needle has been common since the first half of the twentieth century, in treating pregnant women who over-produce it, but the fluid was not used to diagnose the health of an unborn child until the 1950s.

Two Danish physicians, Fritz Fuchs and Povl Riis, wanted to determine sex before birth to help families affected by "sex-linked" genetic diseases. In hemophilia, for example, the defective gene is carried by girls and passed on to their children, but only boys suffer from the disease. The idea was to find out whether pregnant women known to have a defective gene were carrying a boy or girl.

It was already known that the developing fetus sheds cells into the amniotic fluid, so Fuchs and Riis examined samples under a microscope, looking for the tiny structures called Barr bodies, which only exist in female cells. This novel way to determine sex proved to be completely reliable, and in 1956 they published their findings. From this point on, sex-linked disorders could be predicted with certainty.

Amniocentesis is always risky, but is essential for revealing many serious problems before birth – from immature lungs and blood disorders, to neural tube defects. It is offered to women in whom these disorders are more likely, but many complex moral issues will always accompany the technology.

S. Maria Hampshire

Date 1956

Scientists Fritz Fuchs, Povl Riis

Nationality Danish

Why It's Key Amniocentesis is one of the most remarkable technological advances in monitoring pregnancy. Without it, our current understanding of clinical genetics would not have been possible.

opposite Taking a sample of amniotic fluid during an ultrasound

Key Discovery **Patterson dates the Earth**

In 1664, the creationist Archbishop Ussher famously calculated that the Earth was created in 4004 BCE. He believed this process had taken six days and that the Earth had remained relatively unchanged since. What Ussher didn't know, and certainly didn't want to know, was that the Earth was, and is, in a perpetual state of change; there was no way that it could have taken less than a week to form.

Over the next few centuries, scientists offered up estimates of the Earth's age ranging from seventy-five thousand, to a few billion years old. However, it wasn't until 1948, when Clair Patterson used a newly developed method in a dissertation project, that anyone got close to determining the figure scientists now generally accept as correct. Patterson spent years collecting specimens of igneous rock from the Canyon Diablo meteorite site, which he identified as having formed at the same time as the Earth. He then used a mass spectrometer to analyze lead isotope ratios, which allowed him to make a direct assessment of the age of the Earth. Armed with his results, Patterson rushed back home to his mother and asked her to check him into hospital, as he was sure he was about to have a heart attack. He had just aged the Earth at 4.55 billion years: a figure that has remained unchanged since 1956.

Fiona Smith

Date 1956

Scientist Clair Patterson

Nationality American

Why It's Key The first key use of mass spectrometry to age rocks.

Key Person **William Shockley**
The father of Silicon Valley

William Shockley (1910–1989) was born in London on February 13, 1910, to American parents. When he was three, his family moved to Palo Alto, California. After earning his BSc in physics from Caltech, and his PhD in physics from MIT, he worked at Bell Labs between 1936 and 1955.

John Bardeen and Walter Brattain were working for Shockley when they invented the transistor in 1947 and the following year, Shockley invented an improved version called the junction transistor. The next few years saw Shockley developing underlying transistor theory more fully and writing the fundamental book on the subject *Electrons and Holes in Semiconductors* (1950). Shockley, Bardeen, and Brattain were awarded the 1956 Nobel Prize in Physics "for their researches on semiconductors and their discovery of the transistor effect." In 1956, Shockley became Director of the Shockley Semiconductor Laboratory, which he chose to establish in Mountain View, close to where he had grown up. It was a choice that was to have a profound effect on the area – it would later become Silicon Valley. By 1957, eight of Shockley Semiconductor's researchers were increasingly unhappy with his management style. As Gordon Moore put it, "Working for Shockley proved to be a particular challenge." The eight left to found Fairchild Semiconductor Company, where they created the first silicon integrated circuit. Some of these would go on to found semiconductor company Intel, industrial conglomerate Teledyne, venture capital firm Kleiner Perkins, and many other Silicon Valley companies.

Eric Schulman

Date 1956

Nationality American

Why He's Key Instrumental in both the invention of the semiconductor and Silicon Valley's high-technology industries.

Key Event
Jodrell Bank goes online

The world famous Jodrell Bank radio observatory will always be linked with its founder, Sir Bernard Lovell. While working on radar development during the war, he thought that the sporadic blips in early warning systems might be due to high energy particles from the Universe. After the war ended, Lovell tried setting up some military equipment at the University of Manchester, but nearby trams caused too much interference. So he found a suitably quiet site on land owned by the Botany Department – Jodrell Bank – and set up his small receiver.

He quickly discovered that his radar "echoes" were actually due to the plasma trail left by meteors burning up in the atmosphere. After initial successes, Lovell began developing the site. He planned a huge receiving dish for the study of natural radio waves emitted by various objects in deep space. Initially, a wire mesh dish was planned, but the discovery of the natural radio emissions from neutral hydrogen in our galaxy made Lovell upgrade the design to a solid steel surface, which would enable it to study these emissions.

His new radio telescope was of colossal proportions. The solid dish measured 76.2 meters across, weighed 1,500 tons, and was fully-maneuverable. It was completed in 1957 and, within months, made the headlines as it tracked the rocket launch of Sputnik 1.

After thirty years of cutting edge research, the "Mark 1" was renamed the Lovell Telescope, in honor of its founder.

David Hawksett

Date 1957

Country UK

Why It's Key It is one of the greatest scientific instruments ever constructed in the UK.

opposite The construction of the huge satellite dish at Jodrell Bank

Key Event **Hewlett Packard**
From the shed to the shelves

William Hewlett and David Packard co-founded Hewlett Packard (HP) in 1939. Both from engineering backgrounds, they had become friends at Stanford University and had been encouraged to start a business by their professor and mentor Fred Terman. Twenty years later, they had a company employing over 2,000 staff and taking US$48 million annually in revenue.

HP's initial workshop was in Packard's garage, where the pair designed and built an audio oscillator, a machine used in the production of radios and stereos, which significantly undercut all competitors in terms of price, but was substantially more efficient. They soon branched out into other high-end electronics focusing on extreme accuracy and stability.

A year after founding their company, Hewlett and Packard moved from the garage to rented buildings and, in 1957, to a new site on an industrial park attached to Stanford University. During the 1960s, spurred on by their continuing success, HP started work on silicon semiconducting devices for use in personal computing equipment. Although not the first to discover silicon semiconductors, HP helped develop and promote them.

The Stanford Industrial Park is today considered to have played a key role in the founding of California's technology hub, Silicon Valley in San Francisco. HP's move was a step toward the company's future in high powered personal computing, a far cry from their humble beginnings in Packard's shed.

Josh Davies

Date 1957

Country USA

Why It's Key HP's move to what is now Silicon Valley turned a successful technology company into a multi-million-dollar empire operating in 170 countries.

Key Publication **Star dust**
The key to all matter

"We are all star dust" was Carl Sagan's summary of George and Margaret Burbidge, William Fowler, and Fred Hoyle's 1957 publication – the famous "B²FH" paper. This seminal paper, entitled "Synthesis of the Elements in Stars," explained how stars produce energy, and the processes by which they are responsible for creating (almost) all the elements in the Universe. The two subjects are closely linked; stars are powered by processes that manufacture heavy elements from lighter ones.

The process was first described in 1938 when Hans Bethe outlined the mechanisms by which stars produce energy – by fusing hydrogen atoms together to produce slightly larger helium atoms. Sticking hydrogen atoms together releases large amounts of energy, which is why stars give off lots of heat and light. But although Bethe's research explained how stars were powered, it didn't account for the formation of heavier elements.

The B²FH paper, however, described how elements such as iron are produced during the lifecycle of a star, starting with the fusion of hydrogen atoms. It also explained how heavier elements are formed by what is called neutron capture; uncharged particles called neutrons stick to the nuclei of atoms, increasing their atomic mass. This process takes place during the spectacular death of large stars. During its final phase of life, a star will explode, releasing enough energy to create new elements as atomically massive as uranium.

Anne-Claire Pawsey

Publication "Synthesis of the Elements in Stars"

Date 1957, *Reviews of Modern Physics*

Authors George Burbidge, Margaret Burbidge, William Fowler, Fred Hoyle

Nationality British, American

Why It's Key The so-called B²FH paper explains exactly how all the material that makes up the Earth was originally formed.

Key Person **George Evelyn Hutchinson**
In a niche of his own

At the tender age of fifteen, George Evelyn Hutchinson (1903–1991) was already starting to document field observations. How many children would notice a grasshopper swimming in a pond, let alone submit a detailed report to an academic journal?

British-born, Hutchinson joined Yale University in 1928 and for the next forty years he would prove to be a hugely influential figure in the world of ecology. Considered by many to be the father of modern limnology – the study of freshwater lakes – he exploited any excuse to don a pair of waders. Following his divorce from his first wife in 1933 Hutchinson was required to remain in the state of Nevada for six weeks. Some people would put their feet up, but not Hutchison. Instead he managed to produce important work on how lakes in arid areas remained as lakes.

He is probably best known for his theory of the ecological niche, detailing specific roles for certain organisms. Announced in 1957, this concept fundamentally changed the way ecologists study the environment. Ecology, he believed, should not be a purely descriptive science, but should analyze and predict. His work is now widely considered to be amongst the most important ever carried out in the field of ecology.

But there was more to Hutchinson's work than pure research. He was an extremely gifted and enthusiastic teacher. Many of his students, such as Robert MacArthur, would in their turn become the pre-eminent ecologists of the latter half of the twentieth century.

Andrew Impey

Date 1957

Nationality British

Why He's Key He developed the theory of niche that helped shape the way ecologists now view ecosystems.

Key Invention **FORTRAN**
Language of the ancients

In 1954, IBM announced plans for a computer language that would allow computers to handle mathematic formulas. It was named FORTRAN – the formula translator.

The system was kept as simple as possible; no long names for the variables, no long definitions for your functions, no complex numbers, and no apostrophes. Unlike other systems available at the time, FORTRAN used integers as well as floating points, meaning it could handle both whole numbers and those with decimals. It had an "if" statement in it, which could compare two numbers and give different outputs for less than, greater than, and equal to. It also had a "do" statement that could count from one number to another, following the commands attached to the numbers.

FORTRAN was released in 1957 for the IBM 704 and became so popular because it had no real competition. The only other languages around at the time were specialized and awkward, so while other languages have come and gone, FORTRAN stuck around through engineers, mathematicians, scientists, and students into the 1960s and 70s, becoming the first computer language good enough to be thought of as mainstream. It went on to be revised in 1977 and 1990 to add in modern features.

Although it may seem old-fashioned when you hold it up next to more modern languages, it does have vast libraries of usable code, which the younger languages haven't been around long enough to build up yet.

Douglas Kitson

Date 1957

Scientists IBM

Nationality American

Why It's Key The Daddy of programming languages. The first to be considered mainstream, and still taught in universities today.

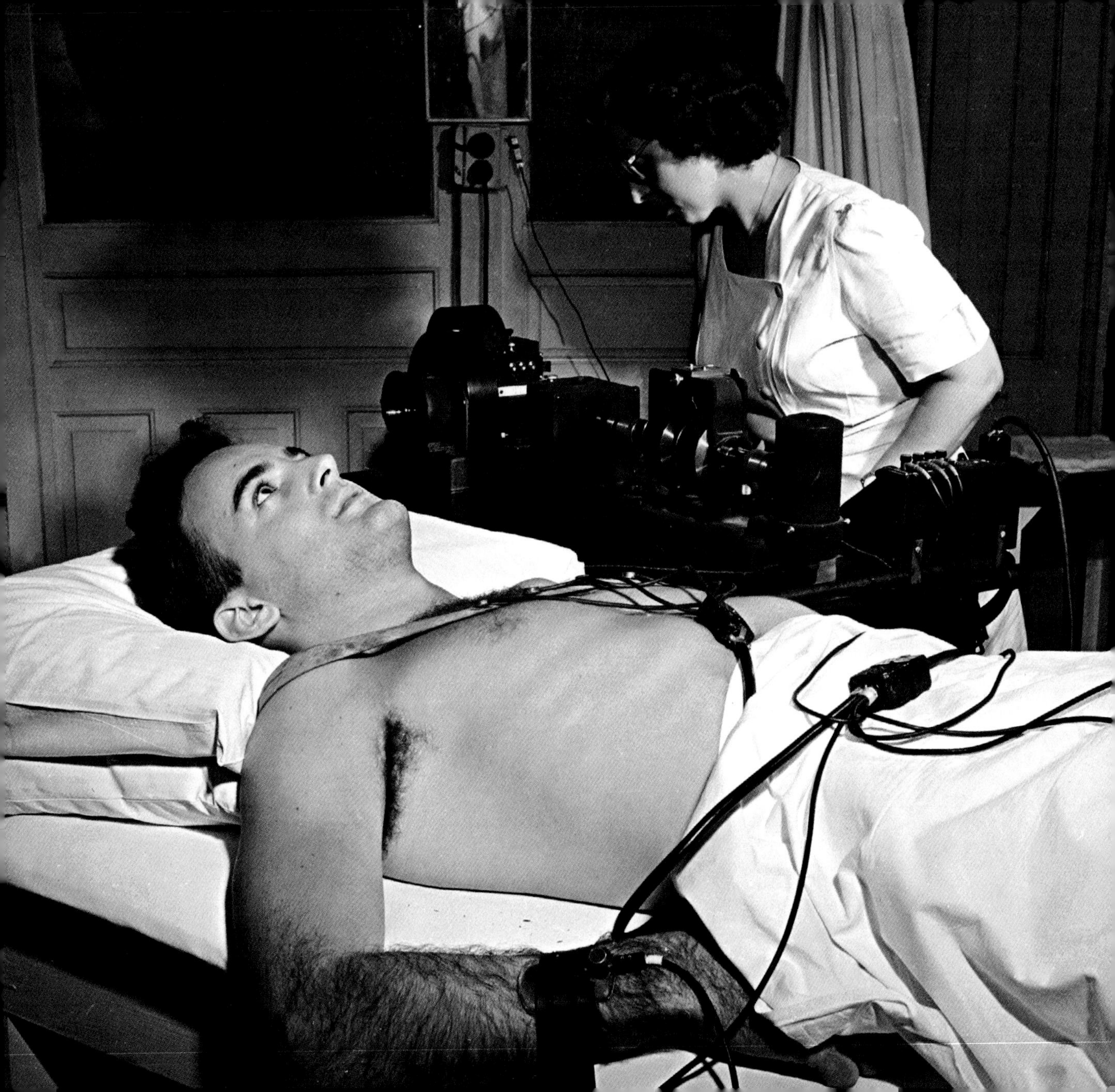

Key Invention **EEGs**
From rabbits to toposcopes

In 1875, lawyer Richard Caton, also known as the Lord Mayor of Liverpool, reported his observations of electrical impulses produced at the surface of the living brains of rabbits, dogs, and apes. These currents increased during sleep, responded to anesthesia and dissipated with the animal's death.

The German physician Hans Berger followed Caton's experiments and made the first recordings of human brain activity in 1924. He coined the term "electroencephalogram" (EEG) and described a number of normal and abnormal EEG phenomena. Many of the normal recordings were of his teenage son Klaus. He noted changes associated with attention and mental effort, and variations in the EEG read-outs of people with brain injuries. In the 1930s, William Grey Walter made further advancements. Using multiple electrodes and complex algorithms, he developed brain-wave topography. This allowed physicians to locate certain tumors and centers of epileptic behavior.

By 1957, he had added twenty-two cathode ray tubes and made the "toposcope," which could simultaneously display rhythms of activity present in different parts of the brain. Although advancements in CT and MRI scanning have since replaced EEG topography and toposcopy for localizing tumors, functional EEG is still used prior to surgery for epilepsy, and for research.

Stuart M. Smith

Date 1957

Scientists Hans Berger, WIlliam Grey Walter

Nationality German, American

Why It's Key It allowed the accurate location of tumors, before the advent of CT scanning, and is a vital part of surgery for epilepsy.

opposite An early cardiograph machine in use

Key Person
Tsung-Dao Lee and the particle puzzle

Aged just twenty-nine, Tsung-Dao Lee became the University of Columbia's youngest full professor. Born in 1926 in Shanghai, Lee had emigrated to the United States in 1946 and, despite having no undergraduate degree, managed to make quite a name for himself as a particle physicist.

By 1956, Lee's old friend and colleague Chen Ning Yang was at Princeton University, New Jersey and the two men continued their collaboration, working out a schedule that allowed them to meet once a week. Their focus was on the subatomic particle known as the K-meson, which had been discovered just a few years earlier. The K-meson was a puzzle; it seemed to be a single particle but it could decay in two different ways. Some physicists suggested that the K-meson was actually two different particles that had slightly different properties, but Lee and Yang were unconvinced.

A fundamental concept in particle physics was that of parity, which is said to be conserved if a particle and its mirror image have the same properties. There was so much evidence that the K-meson was a single particle, and not two, that Lee and Yang proposed that parity, in this case, was not conserved with one of the two types of decay. They announced their theory in 1956 and it took just six months for another scientist to experimentally prove them correct.

In 1957, Lee and Yang picked up a Nobel Prize for their efforts. Lee has since won countless awards and honors from universities worldwide.

David Hawksett

Date 1957

Nationality Chinese

Why He's Key Gave particle physicists pause for thought, and proved you don't have to be an ageing academic to win a Nobel Prize.

Key Discovery **Electrons being sneaky**
Escape artists

Erwin Schrödinger, using a straightforward equation, found a way of describing elementary particles, like electrons, as waves. The motion of an electron is nothing like the balls-in-space analogies we would usually think of. In fact, the little chaps travel around in a packet and behave, at the very smallest level, quite a lot like waves.

Wave-like behavior means the electron's position becomes a matter of probability. If you created a "potential well" – a place where electrons can be – surrounded by barriers that electrons are very much not allowed to cross, the probability exists that they will somehow tunnel through the barriers and emerge on the outside.

Scientific interest in this effect goes back to the late twenties, but conflicting results sent it out of fashion. While working on diodes and transistors at Sony in Japan, Leo Esaki made the discovery that brought it back. After finding unusual peaks in the graphs of currents for the diodes they were using, Esaki suggested the effect may be caused by tunneling. In the autumn of 1957, his team reported their discovery and in 1973, Esaki received the Nobel Prize in Physics for electron tunneling in solids.

As well as developing the Esaki diode, this discovery opened up research on tunneling in semiconductors. This became important because of the high sensitivity of tunneling and led to its use in scanning tunneling microscopes, showing objects too small to be seen with conventional microscopes.

Douglas Kitson

Date 1957

Scientist Leo Esaki

Nationality Japanese

Why It's Key Leo Esaki showed that electron tunneling in solids works. This jump-started tunneling research and gave us very sensitive microscopes.

Key Event **Sputnik 1**
First artificial satellite to orbit the Earth

"Listen now," said the presenter of American radio station NBC, "for the sound which forever more separates the old from the new." He was referring to the repetitive "beep beep beep" transmitted from Sputnik satellite launched on October 4, 1957, by the USSR; the sound heralded the space race.

In fact, the race had already begun. Since the mid-1950s, both the USA and the USSR had signaled intentions to launch artificial satellites as a part of International Geophysical Year (IGY), which spanned July 1957 to December 1958. The Soviet launch was timed for maximum political impact; Sputnik first orbited the Earth as scientists from the superpowers mingled at an IGY reception at the Soviet embassy in Washington. This led Dr Joseph Kaplan, Chairman of the American IGY committee, to congratulate his Russian rivals on their "remarkable achievement."

The launch was more than remarkable. At 83.5 kilograms, Sputnik dwarfed the Americans' satellite-in-waiting, Explorer I, which would not reach orbit until February of the next year. Worse still, the Soviets had the ability to launch Intercontinental Ballistic Missiles. What if the next launch contained an atomic bomb? To some influential American policy-makers, the launch of Sputnik was a second Pearl Harbor. The political fallout had far-reaching consequences for domestic science policy in the United States. Most significantly, this led to the birth of NASA, in July 1958, and funding flowed into scientific research and education – a reorganization that would eventually see the United States triumph in space.

Arran Frood

Date 1957

Country USSR (now Kazakhstan)

Why It's Key Not just the start of the space age, Sputnik caused a revolution in the way America invested in and organized the exploration of space.

opposite A postcard issued by the Russian government depicting Sputnik 1

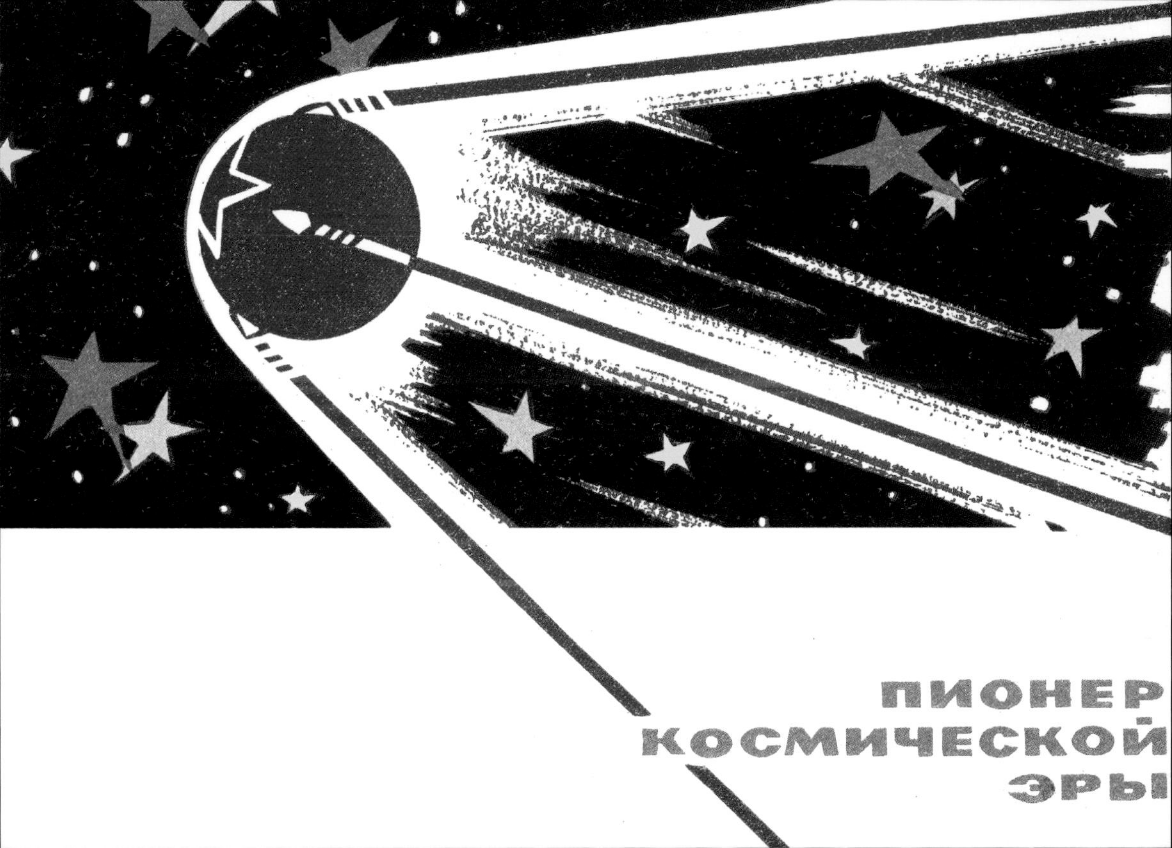

ПИОНЕР
КОСМИЧЕСКОЙ
ЭРЫ

Key Experiment **Project Orion**
A spacecraft powered by nuclear explosions

So here's your problem: You're living in the late 1950s, space flight's a big deal, but you can't afford a US$24 billion project like Apollo. If only a cheap alternative existed...

It's funny you should ask, because Project Orion could get you into space for much less. And though spacecraft powered by nuclear explosions may sound a little dangerous, there's no denying the savings.

Conceived by Ted Taylor and Freeman Dyson in 1957, Project Orion works on the "firecracker under a tin can" principle that many ten-year-old boys have proved to be sound. But instead of a tin can, imagine a 50-meter-tall bullet and replace the "firecracker" with "a large number of atomic bombs."

Dyson and Taylor had high hopes for Project Orion – its motto was "Mars by 1965, Saturn by 1970." They also had plans for something rather more spacious than a tin can. Instead of the cramped capsules endured by other space travelers, Orion would house 150 passengers, more like a battleship than the tiny weight-saving alternatives that other programs envisioned. The engines, known by the press-friendly name, "External Pulsed Plasma Propulsion," drop bombs out of the back and explode them 60 meters away. The rocket then works like the piston in a car engine, and is pushed away from the blast, with shock absorbers keeping the passengers safe from the heat, force, and radiation.

Project Orion could have blazed a trail to the outer reaches of the Solar System. If only it hadn't sounded so very dangerous.

Douglas Kitson

Date 1957

Scientists Ted Taylor, Freeman Dyson

Nationality American, British-American

Why It's Key Project Orion could have been a realistic alternative to the rockets we currently use, but for a few funding difficulties and nuclear testing bans.

Key Event
Laika, the first animal goes into orbit

On November 3, 1957, in a grand gesture to celebrate the fortieth anniversary of the Soviet revolution, a Russian dog called Laika became the first live animal to be launched into orbit around the Earth.

Sputnik 2, or "Muttnik" as it became known, was the second spacecraft to be launched into orbit. It was designed to transmit biological and engineering data, including measurements of solar radiation and cosmic rays. It was what would happen to a living organism in space, however, that was of most interest to a team that had ambitions to send a human into space.

The dog, a female terrier named Laika ("Barker"), was used to test the effects of weightlessness and high acceleration during launch, as well as systems for supplying a passenger with oxygen and food in space. Data transmitted live indicated that she showed signs of stress during the launch but calmed down once in orbit.

Sadly, it was never going to be possible to bring Sputnik back safely, so poor old Laika was doomed to posthumous fame. Although there was sufficient food and oxygen for a week, it is thought that she only survived for a few days, possibly only a few hours, because of overheating within the spacecraft caused by solar radiation. Although news of her fate gradually became known, the truth of her rather rapid demise was only confirmed by the Russians in the 1990s.

Richard Bond

Date 1957

Country USSR (now Kazakhstan)

Why It's Key It paved the way for the first manned spaceflight – just four years later – and to the first man on the Moon, in 1969.

opposite A model of Laika in Sputnik 2

Key Publication "Theory of Superconductivity"

The year 1957 was a good year for University of Illinois professor John Bardeen. Not only did he win the Nobel Prize for his work on semiconductors, he also published the 'Theory of Superconductivity' with his two students Leon Cooper and J. Robert Schrieffer. It was the first time anyone had produced a convincing theory for the strange behavior of some materials at low temperatures. And it would win Bardeen his second Nobel, in 1972.

Superconductivity was discovered in 1911, and although other theories had been proposed in the intervening forty-six years, BCS theory (named after the three authors of the paper) was the first complete explanation. It goes something like this: imagine an electron comes across a positive ion as it moves through the lattice of atoms and ions in a conducting material. The lattice distorts inwards, drawing another electron with opposite "spin" into the resulting positive trough behind. The two electrons become a "Cooper pair" held together by forces from the lattice. At low temperatures, this binding energy can overcome the influence of the thermal vibrations on the electrical conduction, and the Cooper pairs will not experience any resistance. While the discovery of high-temperature superconductivity in 1986 challenged this elegant explanation as BCS states superconductivity cannot occur above -243 degrees Celsius, the relative simplicity of the theory meant it could be applied to condensed matter physics, and beyond. It has been used to explain subjects as far-reaching as the core of neutron stars and the structure of atomic nuclei.

Kellye Curtis

Date 1957

Authors John Bardeen, Leon Cooper, J. Robert Schrieffer

Nationality American

Why It's Key BCS theory gave an explanation for superconductivity for the first time, and rekindled interest in the field. The theory later influenced all areas of physics.

opposite A demonstration of superconductivity

Key Event **Amundsen-Scott South Pole Station established**

Named for Roald Amundsen, who reached the South Pole in 1911, and Robert F. Scott, who made it there but not back again in 1912, the opening of the South Pole Station was a key moment in scientific history. It is the southernmost continually inhabited place on the planet and was built to celebrate International Geophysical Year (IGY) in 1957, an initiative to which sixty-six countries were signed up.

The South Station, as it is called, supports international scientific research into upper-atmosphere physics, meteorology, earth sciences, geophysics, glaciology, biomedicine, and even astrophysics. In 1975, a new station had to be built because the original was suffering damage from the severe weather conditions, as well as being buried under the constant heavy snowfall at a rate of four feet every year.

The international collaboration which established and promoted the IGY made an important political step forward in 1959. In December of that year, the twelve leading participating nations signed the Antarctic Treaty in Washington, DC. The treaty was framed as an agreement to use the continent solely for "peaceful purposes." It came into effect two years later and guarantees access and scientific research in all territory south of 60 degrees latitude, allowing continued research and collaboration between countries, and protection from development and other threats that would destroy the unique ecosystem and conditions of the Antarctic.

Catherine Charter

Date 1957

Country Antarctica

Why It's Key The South Pole station has led to many important scientific discoveries and was the catalyst for conservation of the Antarctic, via the Antarctic Treaty.

Key Invention **Doppler radar invented**

The Gatso is born

In 1842 Christian Doppler published his thoughts on an effect that was later to be named after him. Little did he know that this effect would one day provoke the use of more foul language among motorists than almost anything else – Doppler radar, used today in speed traps.

The Doppler effect occurs in waves of all kinds - sound, radio, light – but is most obvious to us with sound waves. Imagine the sound of a fast car racing by you, the pitch of the sound is high as it approaches and then abruptly gets lower as the car passes. This is because the sounds waves from the oncoming car are all bunched up – a high frequency – whereas they become stretched out as the car moves away – a lower frequency means a lower pitch.

Doppler radar utilizes radio rather than sound waves, but the principal is the same. Waves allow a speed trap to calculate the speed of a moving vehicle – bad news if you're exceeding the legal limit.

The first Doppler radar was made by a Dutch company founded in 1958 by racing driver Maurice Gatsonides, who wanted to know exactly how fast he was going on the racetrack. In some countries speed cameras are still referred to as "Gatsos," in honor of this.

Barney Grenfell

Date 1958

Scientist Maurice Gatsonides

Nationality Dutch

Why It's Key The Doppler radar in all its many forms is an important device that has contributed much to road safety all over the world.

Key Invention **Ultrasound scanners**

A safe way to peer inside

War has been the impetus behind many an innovation, but the field of gynecology is not one that we usually think of as benefiting from armed conflict. This was not the case, however, when it came to the development of the ultrasound scan.

Dr Ian Donald gained basic knowledge of both radar and sonar while serving in the British Royal Air Force during World War II. At the cessation of hostilities, he began to search for possible medical applications of some of the devices he had seen used during the war. Donald soon realized that sonar technology was almost perfectly adapted to the study of his patients.

Donald used a new device that capitalized on the fact that sound is reflected when it hits tissues of different densities. This new "ultrasound" machine generated a high frequency sound, and then "listened" for the echoes to be reflected back to the machine, which converted them into an image. He was soon able to safely and rapidly diagnose everything from uterine cysts to pregnancy.

Today, ultrasound scans are of high quality and provide two-dimensional images in real time. Using only sound waves, they are considered a safe alternative to radiation. They are also uniquely adaptable to showing the motion of a living thing – a heart beating for instance, or a baby moving around in the womb. As a result babies, as well as hearts, gall bladders, and a myriad other organs can now be viewed safely and effortlessly, leading to better medical care for all.

B. James McCallum

Date 1958

Scientist Ian Donald

Nationality British

Why It's Key Ultrasound allows a physician to see what is going on inside the body, without the risk of radiation exposure. It's a safe, cheap, and easy way to monitor health.

Key Invention **Mighty modems**
Connecting the world

Back in the 1960s, computers were just booting up. Almost all computers were time-shared; users would buy time and communicate with a (very) large computer located elsewhere by using a terminal. Terminals were made up of a keyboard, a screen, and a revolutionary new device called a modem, which allowed digital data to be transmitted and received over the analogue telephone system.

Modems not only modulate, they demodulate. Modulation is the process whereby electrical waves are modified so that they carry a message. Modems are essentially devices that modulate analogue signals, known as carrier signals, so that they can carry digital information. Receiving modems translate the signals back to the original transmitted data by decoding them – demodulation. The most common modems work by converting digital sequences of 1s and 0s into analogue sound waves, allowing digital data to be easily transmitted through analogue systems like telephone networks.

The first modems were used by the 1950s American SAGE air-defense system; they connected terminals at the system's many locations to the central director centers. Due to fears of communist spies, the SAGE modems used their own communication lines rather than public telephone lines. The technology was soon adopted for commercial use and, in 1958, AT&T launched the world's first commercial modem. Your granddad's snail-like 56k modem is a speed demon compared to that early modem, which transmitted data almost two hundred times slower.

Logan Wright

Date 1958

Scientists AT&T

Country USA

Why It's Key The various types of modems in existence are an integral part of the Internet and telecommunications, and have now become household items.

Key Discovery
Seeing proteins in three dimensions

Most of the complicated stuff that takes place in our bodies relies on proteins – long chains of amino acids that are able to bind small molecules and process them in some way. To understand how proteins work, you really have to get a feel for their shape and structure. And until X-ray crystallography came of age, there was no easy way of doing this.

X-ray crystallography is a technique that reveals the three-dimensional structure of molecules. Material must first be crystallized into a regular array of repeating units. When such a crystal is bombarded with X-rays, the regular spacing scatters the beam to form a diffraction pattern that is characteristic of the material's atomic structure. Scientists can study the pattern and, following the laws of geometry, infer the structure of the molecules. The technique was pioneered by Max von Laue on the eve of World War I, and was subsequently developed by the father and son team, William Henry and William Lawrence Bragg.

In the early years, X-ray crystallography was only useful for simple and highly regular inorganic structures, such as table salt or graphite. Small organic molecules followed in the 1930s. But the breakthrough into complex biological matter had to wait until 1958 when John Kendrew and Max Perutz first uncovered the atomic-level structure of a protein.

Perutz hit on the idea of using heavy metals to help interpret complex diffraction patterns. Armed with the new technique, the duo were able to work out the structure of myoglobin, a simple oxygen-carrying pigment found in muscle tissues.

Matt Brown

Date 1958

Scientists John Kendrew, Max Perutz

Nationality British, Austrian

Why It's Key Gave us a better understanding of how proteins work and why they are so important.

Key Event
NASA created

During the Cold War, US President Eisenhower commissioned the National Aeronautics and Space Administration (NASA) largely in reaction to the Soviet Space Program's successful launch of the first man-made satellite, Sputnik 1, the previous year. America put a satellite into orbit only a few months later and the space race began, capturing the imagination of the world.

When NASA formed in 1958, it employed eighty scientists and had four research laboratories. But it rapidly expanded as the U.S. Government put more emphasis on space exploration and particularly human spaceflight. Throughout the 1960s, they concentrated on developing rocket technology, experiments to find out if humans could survive a launch, and investigating what the effects of weightlessness on human health were likely to be. By the end of the decade, the Apollo program had made one of the most important steps in history: landing humans on the Moon. NASA had shown that celestial objects were within reach of the human race, and changed peoples' perceptions of space. Neil Armstrong and Buzz Aldrin returned to Earth as heroes, as well as with a wealth of new scientific data about our closest neighbor.

Since then, NASA has worked on a multitude of projects, including creating the first weather and communication satellites, establishing a permanent human base in space at the International Space Station, and many shuttle launches returning essential data about our Universe. In the future, NASA plans to land men on Mars and create the first Moon base.

Leila Sattary

Date 1958

Country USA

Why It's Key Historically, NASA has undertaken the large majority of space exploration, giving us essential insight into how our Universe works, as well as making human history by putting a man on the moon.

Key Event **Let's go exploring**
The first American satellite

Marking the start of a new era of scientific discovery and born out of numerous technological breakthroughs, the International Geophysical Year in 1958 was an effort to coordinate the global collection of geophysical information. This would be the year that saw the launch of the first successful U.S. orbital satellite: Explorer 1. Difficult as it is to believe now, prior to the late 1950s, space flight was the stuff of fictional books and movies.

Explorer 1 was launched from Kennedy Space Center, then known as Cape Canaveral, in Florida, USA at 10:48pm on January 31, 1958. It followed the deployment of the Soviet Sputnik 1 satellite by less than three months. The race for space was on; but so too was the Cold War. Sputnik was, to many in the West, a demonstration not only of technological advancement, but also the Soviets' ability to launch ballistic missiles at the United States. The Americans needed to respond quickly.

Through close cooperation between the Army Ballistic Missile Agency and the NASA/Caltech Jet Propulsion Laboratory, the existing Jupiter-C launch rocket was modified to carry the satellite; Explorer 1 itself was built in just eighty-four days.

Explorer 1 had onboard instruments, designed by James Van Allen, for the study of cosmic radiation. These instruments provided evidence that the Earth is surrounded by bands of intense radiation, now known as Van Allen belts. This was the first major scientific discovery of the space era.

Mike Davis

Date 1958

Country USA

Why It's Key Explorer 1 was the first successful US orbital satellite.

opposite The launch of satellite Explorer 1

Key Person **Burrhus Frederic Skinner**
The "rat man"

Burrhus Frederic Skinner (1904–1990) was one of the most influential psychologists of the twentieth century. he was a leader in a field known as "behaviorism"; the study of animal and human behavior and the way it is influenced by their environment.

Skinner was born in Susquhanna, a small town in America, and spent much of his childhood building and inventing things. After completing an English degree at Hamilton College, he tried his hand at becoming a writer, but soon decided it wasn't the career for him. He went back to University at Harvard in 1928 to study psychology, and it was here that he became fascinated with behavioral analysis.

Skinner is best known for coining the term "operant conditioning," a method of learning, where the animal in question learns by responding to – or "operating on" – its environment. He invented this theory after extensive research using his famous piece of apparatus called the "Skinner box," in which rats would learn to press a lever to retrieve food. He is also credited with developing the philosophy of "radical behaviorism" – the experimental analysis of behavior.

Skinner used the observations in his animal experiments in his philosophy of human behavior. He concluded that all behavior is a result of positive or negative reinforcement, and therefore that the idea of free will is merely an illusion. In 1948, he published his most famous novel, *Walden Two*, in which he describes a utopian community run on the principles of his operant conditioning.

Hannah Isom

Date 1958

Nationality American

Why He's Key Skinner was listed as the most influential psychologist of the twentieth century. His most famous contribution was coining the term "operant conditioning."

Key Event **Checkmate**
Human loses to computer

As electronic computers began to be used for applications outside the military, one direction they took was into the games market. The game of chess was used by researchers to determine what level of artificial intelligence could be attained by a computer.

At first, scientists could only teach the computer the moves. These early chess programs were easily beaten by novices. A great step forward, however, was made in 1958 by Allen Newell, Herbert Simon, and Cliff Shaw. They developed a program which used "algorithms" (step by step procedures) which analyzed all possible moves and outcomes and so decided the best move at any stage of a game.

This chess program was called NSS, and ran on a vacuum-tube computer. The programmers taught the basics of chess to one of their secretaries who had never played before. After an hour's tuition the secretary played the computer and was beaten. This clearly demonstrated that the computer was able to understand an hour's worth of human knowledge and put it into practice.

Although this seems to have no practical importance, this was the first time a computer had "outwitted" a human being – an important demonstration of the potential capabilities of electronic computers. Humankind was standing on the cusp of the computer revolution that would redefine almost every aspect of our society over the next few decades.

Simon Davies

Date 1958

Country USA

Why It's Key The initial steps of artificial intelligence demonstrated how computers could be taught to "think" and take decisions based on calculated outcomes.

opposite Early chess program in operation in the 1950s

DATA PROCESSING MACHINE TYPE 704
INDEX
STORAGE
ACCUMULATOR
MQ
INSTRUCTION

Key Invention **Microchips**
Good things come in small packages

The invention of the transistor in 1947 ultimately led to another electronic marvel: The microchip. The transistor itself certainly decreased the size of electronic equipment – no more clunky vacuum tubes were necessary – but this just wasn't enough miniaturization for some people.

The first steps on the road to the microchip revolution happened almost simultaneously in Texas and California. In late summer of 1958, a young researcher at Texas Instruments named Jack Kilby realized that he could use semiconductors to mimic all the common components of a circuit board – only at a fraction of the size. In short, he discovered that he could use silicon not only to make transistors, but also capacitors, resistors, and other components.

A few states away, Robert Noyce figured out the same thing at nearly the same time. Noyce's method of connecting components on his semiconductors base was distinctly different from Kilby's and, as a result, patents applied for by both men's companies were ultimately granted. Kilby went on to have a very successful career and Noyce went on to co-found international computing company Intel. Though the two are considered to be the co-inventors of the integrated circuit, or microchip, Noyce died in 1990, precluding him from sharing in the Nobel Prize granted to Kilby in 2000.

B. James McCallum

Date 1958

Scientists Robert Noyce, Jack Kilby

Nationality American

Why It's Key The microchip has now revolutionized the entire electronics industry. You probably have several of them in close proximity to you right now.

Key Experiment **Meselson-Stahl experiment proves DNA replication is semiconservative**

In 1958, a graduate student and a postdoctoral researcher proved that DNA underwent semiconservative replication and, in the process, proved that Watson and Crick's DNA structure was most likely correct. Matthew Meselson and Franklin Stahl had completed what has been described as the "most beautiful experiment in biology."

Meselson and Stahl very elegantly devised and carried out an experiment proving that during replication, the double helix of DNA unzips into two separate strands, and new strands are built along each of the former parent strands. The newly created daughter double helices therefore have one entire strand of DNA from the parent, and another strand entirely synthesized. Their ingenious experiment involved growing *E. coli* bacteria in a broth containing the heavy isotope of nitrogen. The first generation of *E. coli* DNA was then centrifuged and found to be heavier than that from *E. coli* not grown in the broth. This implied that the bacteria had taken up some of the heavy nitrogen. But had the parent DNA been dispersed equally amongst the two copies of the daughter DNA, or had one of the original helices stayed intact and an entirely new double strand been synthesized (semiconservative replication)?

The strands of first generation daughter DNA from the heavy nitrogen colonies were all the same weight, each double strand of DNA contained one old and one new strand – the mode of replication was clear.

B. James McCallum

Date 1958

Scientists Matthew Meselson, Franklin Stahl

Nationality American

Why It's Key In an elegantly simple experiment, Meselson and Stahl proved both the existence of the DNA double helix and its semiconservative replication.

Key Discovery
How bacteria change their genes

Although antibiotic resistance is often a bad thing, it has led to some breakthrough discoveries and advances in science. In war-torn Japan of the late 1940s and early 1950s, dysentery became rampant. Luckily a new champion emerged in the form of sulfonamide antibiotics. While these drugs initially enjoyed great success against the most prominent of the bacteria causing the disease, *Shigella dysenteriae*, it eventually became resistant.

What was even more disturbing was that the bacteria were becoming resistant to the other antibiotics of the day too – and doing so at an alarming rate. It seemed highly unlikely that the bacteria were spontaneously mutating and dividing fast enough to produce such rapid evolutionary changes. What was being witnessed was the transfer of genes between bacteria via DNA in the cytoplasm of the bacteria, rather than from the chromosome, where DNA is usually stored. In a series of elegant experiments, Japanese scientists showed that bacteria can transmit genetic information in the form of plasmids – circular strands of DNA separate from the main chromosome. These can subsequently become incorporated into their new bacteria's chromosome, where they become functional genes. Research showed certain viruses called bacteriophages could be used to insert genes into bacteria. Today, plasmids and bacteriophages are used to introduce genes into bacteria to produce substances they wouldn't normally manufacture – for example, bacteria are now used to make insulin for diabetics and human growth factor.

B. James McCallum

Date 1959

Scientist Susumu Mitsuhashi and colleagues

Nationality Japan

Why It's Key The discovery of extra-chromosomal genetic material that could be transmitted to other bacteria led directly to today's advances in genetic engineering.

Key Publication **"Searching for Interstellar Communications"** Is there anybody out there?

In 1959 Giuseppe Cocconi and Philip Morrison published a paper in the journal *Nature* explaining to the world why we should be searching for alien life. They believed that extraterrestrials might try to communicate with Earth using electromagnetic waves, which travel at the speed of light. They theorized that 1,420 megahertz would be the frequency used, as this is the frequency linked with the formation of hydrogen, the most abundant element in the Universe.

Their paper was met with skepticism by most governments, who considered it speculative fiction rather than "real" science. But it was regarded with some enthusiasm by academics and a handful of wealthy individuals willing to fund SETI (Search for Extra-Terrestrial Intelligence) projects. The first major breakthrough came in 1977 when a strong signal was recorded by two telescopes in North America. This signal has never been repeated. Some effort has been made to send a signal from Earth into space. The Arecibo message was broadcast in 1972 but was largely symbolic – it is expected to take 25,000 years to reach its destination. A conference in 2006 deemed it potentially dangerous to attempt further correspondence in case of a hostile reply.

Berkeley University now distributes a computer program called SETI@home which allows anyone's personal computer to search through electromagnetic data received from space for evidence of extraterrestrial communications. This project has created a network of computers which together are more powerful than the largest supercomputer.

Josh Davies

Publication "Searching for Interstellar Communications"

Date 1959, *Nature*

Authors Giuseppe Cocconi, Philip Morrison

Nationality Italian, American

Why It's Key Cocconi and Morrison established the scientific rationale for SETI, which has been in the public eye since the 1960s.

Key Experiment
Gravitational time dilation

Einstein's Theory of General Relativity predicted that the closer you are to a large mass, the slower time will move due to gravity. For a mass the size of the Earth, the difference between being near the surface and in empty space is tiny but measurable. Robert Pound and Glen Rebka succeeded in experimentally measuring this tiny difference in 1959.

To measure such a small shift they required a source with as well defined a wavelength as possible. They used a gamma emitter and put it on top of a tower in the Jefferson Physical Laboratory at Harvard and used the velocity difference between the emitter and detector at the bottom of the tower to obtain a wavelength shift.

If you imagine two atomic clocks, one at sea level and one on top of a mountain, eventually the sea level clock will drop behind the high altitude clock because the latter is affected by gravity less as it is further from the Earth's core. However, this only applies for an observer; in the clock's frame of reference time always passes at the same rate.

Many scientists have re-enforced time dilation by putting atomic clocks into planes and rockets and comparing their behavior at ground level and at various distances from the ground. Although gravitational time dilation effects are usually insignificant on Earth, they are essential for understanding black holes and the warping of space-time.

Leila Sattary

Date 1959

Scientists Robert Pound, Glen Rebka

Nationality American

Why It's Key Gravity slows time. Experimentally proving time dilation now allows a much greater understanding of some major theories of astronomy.

Key Event
Gerald Durrell founds Jersey Zoo

Conservationist and writer Gerald Durrell had always planned to open a zoo with a difference; one which didn't treat the animals as mere curiosities to be gawped at, but one that aimed to preserve vulnerable species from around the world. Unfortunately, however, Gerry had a reputation as something of a maverick and his early attempts encountered resistance from the zoological fraternity. Happily, a chance meeting with Major Hugh Fraser in the mid 1950s allowed him to explain his vision and resulted in the major agreeing to lease Les Augres Manor on the island of Jersey.

Opening officially in 1959, visitors flocked to see a variety of animals brought back from countries as far away as Africa and South America. Over the course of the coming years, Durrell pioneered the concept that the islands and highlands of the world needed to be protected as the "lighthouses" of the planet. His revolutionary menagerie, made more famous by his fantastically popular books and television series, helped alter the public perception of zoos. Shabby, cramped enclosures and inhumane conditions were no longer good enough – other zoos had to follow Jersey's lead or face ruin. His contribution to conservation was summed up by Sir David Attenborough, who commented: "Gerry was responsible for changing people's attitudes to zoology and changing their agenda. His work with endangered species was incredible in that he could persuade them to breed in captivity. He then returned them to the wild. He was a pioneer with a marvelous sense of humor."

Vicky West

Date 1959

Country UK

Why It's Key Now called the Durrell Wildlife Conservation Trust, it is still one of the world's foremost conservation organizations, and helped alter the role of zoos around the world.

opposite Gerald Durrell petting a South American Tapir in his zoo on the isle of Jersey

Key Event **Feynman dreams in nano**

Really big ideas about really small things

Once in a while, a scientist comes along who, through force of personality and originality of ideas, distinguishes themselves from all others, no matter how brilliant these others may be. Richard Feynman, in the second half of the twentieth century, became one such scientist.

Feynman was a brilliant theoretical physicist, and is well known for his work in particle physics and quantum electrodynamics, for which he was joint recipient of a Nobel Prize in 1965. Feynman was also the first person to introduce ideas that would form the basis of nanotechnology: technology on a really, really small scale.

Nanotechnology deals with materials on a molecular and atomic scale, and miniaturization, as it turns out, is big business. Materials at the nano-level behave very differently from materials on a larger scale, allowing for a much greater range of applications. These days nanotechnology is beginning to be applied virtually across the technological board, from computing to sunscreen to tennis rackets.

In his 1959 lecture, Feynman predicted many of the possibilities that nanotechnology might offer, including miniaturized computer components in "nano-computers." Remember, he was saying this at a time when high-power computers would fill a large room. Although he never used the word "nanotechnology," Feynman was almost clairvoyant in his prediction of the possibilities of this yet-to-exist field.

Barney Grenfell

Date 1959

Country USA

Why It's Key Richard Feynman's seminal lecture recognized much of the potential of nanotechnology, and alerted the world to its mind-boggling possibilities.

opposite Richard Feynman lecturing at CERN

Key Invention

First practical fuel cell created

Unlike a battery, which contains all its fuel internally and eventually runs out, a fuel cell requires a constant input of fuel from outside, normally in the form of hydrogen and oxygen. As long as the supply is kept up, the fuel cell will make electricity indefinitely.

William Grove created the first fuel cell in 1839. It was already known that a process called electrolysis could split water into hydrogen and oxygen by using electricity. Grove simply reversed the process, recombining the two elements to produce electricity and water.

British engineer Francis Bacon became the powerhouse behind the development of the technology. He spent twenty years refining the process and making better catalysts for the fuel cell reactions. Finally, in 1959, Bacon and his team demonstrated an array of forty linked fuel cells that could produce a staggering five kilowatts of power, and operate at sixty percent efficiency – enough electricity to power a welder.

In the 1960s, fuel cells offered an efficient and clean method of powering manned spacecraft. Hydrogen and oxygen tanks were already integral parts of spacecraft propulsion systems; fuel cells used this existing supply to create electricity. Conveniently, the waste product was pure water that the astronauts could drink. U.S. aircraft engine manufacturer, Pratt and Whitney, which licensed Bacon's patents, continue to provide NASA with its fuel cells for the Space Shuttle to this day.

David Hawksett

Date 1959

Scientist Francis Bacon

Nationality British

Why It's Key Bacon made the large-scale use of fuel cells possible for the first time.

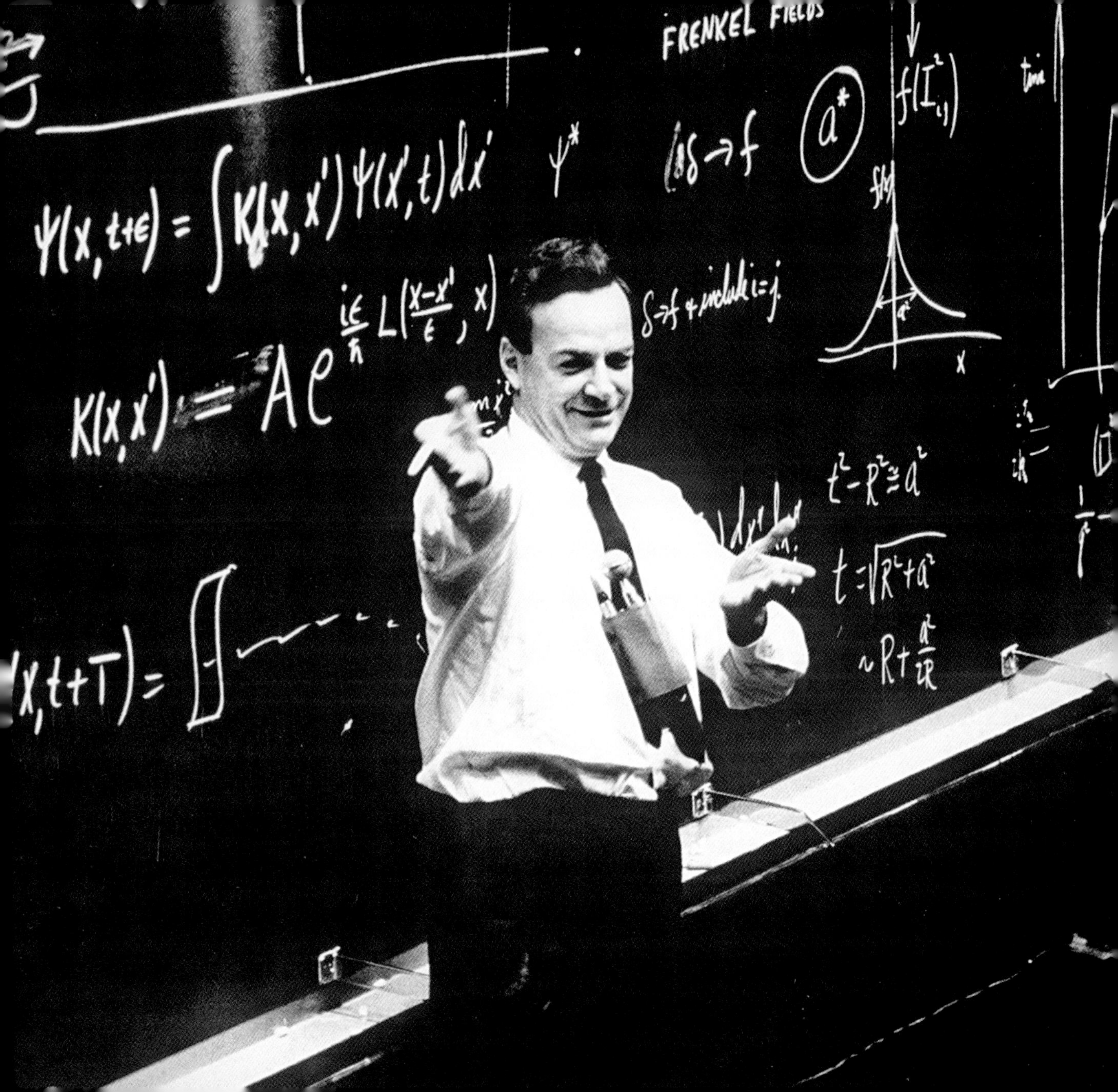
FRENKEL FIELDS
ψ(x, t+ε) = ∫ K(x, x') ψ(x', t) dx'
K(x, x') = Ae
t² − R² = a²
t = √(R² + a²)
~ R + a²/2R
x, t+T) =

Key Invention **The first fully automatic camera**

Smile please

Released in 1959, Agfa's Optima was the first camera with fully mechanical exposure. It needed no batteries and is still, therefore, one of the most environmentally friendly cameras ever to be produced. It sold a million in its first three years on the market. Due to its enduring popularity, several of the series are still in existence and early examples are relatively affordable.

Before the Optima, taking the simplest of snaps required the photographer to specify the width of the aperture (the opening that allows light to reach the film) and length of exposure. The Optima, and others that followed, automated this process, reinvigorating interest in amateur photography by making it a newly hassle-free hobby. Of course, "serious" photographers still wanted to have full control over their exposures, and preferred not to "cheat" by switching to automatic mode.

Recent advances in camera technology have seen automation taken to a whole new level – to the extent that there's almost no need for a photographer at all. In 2005, Canon created an automatic smile detection system, which could capture a "cheese" moment through the wonders of artificial intelligence. Fortunately for professionals, it's still going to be a while before any old Tom, Dick, or Harry can afford to pay for one of these new fangled snappers at his wedding.

Vicky West

Date 1959

Scientists Agfa

Nationality German

Why It's Key Made taking holiday snaps less of a chore.

Key Event

Russians reach the Moon

A single luminous object dominates our night sky: the Moon. It has been the source of a large number of myths and legends, has been attributed God status, and remains a source of fascination to many people the world over. In 1959, this astral body once again took center stage, as the first ever Moon landing took place.

The Lunik II (Luna 2) was launched on September 13, and took a mere 33.5 hours to reach its final destination. As it landed on the Moon, somewhere west of Mare Serenitatis, the capsule released a number of pentagonal Soviet pendants, detonating a small explosive to scatter them around the area. Lunik II then used an array of instrumentation to take detailed readings about conditions on the surface of the Moon, including radioactivity and magnetic fields.

Lunik II was closely followed by Lunik III, on October 4 that same year. Instead of landing on the Moon, however, Lunik III used a slingshot maneuver to fly around it, coming to within 6,000 kilometers of the lunar surface, before swinging round and heading back toward Earth, where it burned up in the atmosphere.

Lunik III's mission was to take pictures of the far side of the Moon. This it did, yielding seventeen poor-quality images of the "dark-side" that were used to create the first attempt at a lunar atlas.

Barney Grenfell

Date 1959

Country USSR (now Kazakhstan)

Why It's Key The Lunik II and III were just two of over twenty Soviet missions to the Moon, but because none of them were manned, their significance is often overlooked.

opposite Luna 1 passes by the Moon after failing to land, but Luna 2 hit the spot

Key Invention **The father of alkaline invents long lasting batteries**

We've all seen the toy bunnies in TV commercials that just keep going and going. But what do they owe their longevity to? It is of course the alkaline battery, invented by Lewis Urry.

Urry was working for the Eveready Battery Division when they tasked him with improving the carbon-zinc battery. These batteries had a short lifespan and were limiting the sale of toys that used them. He quickly deduced that carbon-zinc batteries were a dead-end and turned his attention to designing a new battery.

All batteries have positive and negative electrodes, the cathode and anode respectively, which are immersed in an electrolyte liquid. When connected, the electrons flow from the negative to the positive electrode through the electrolyte, generating electricity. As they do this, a chemical reaction occurs, and once all of the chemical reactants in the battery are used up, it dies.

Urry investigated the use of an alkaline electrolyte but had limited success; it wasn't until he changed the anode from solid zinc to powdered zinc that he hit the jackpot. The powdered zinc had increased the surface area of the anode, and had resulted in a long-lasting battery. He gathered the presidents of the company in the canteen and showed them that a toy car powered by his new battery kept going long after an identical car powered by a carbon-zinc battery. It is estimated that 80 per cent of the dry cell batteries in the world today are based on Urry's original design.

Riaz Bhunnoo

Date 1959

Scientist Lewis Urry

Nationality Canadian

Why It's Key Without the long-lasting alkaline and lithium batteries developed by Urry, many of the portable devices around today such as laptops, personal stereos, and mobile phones would simply not exist.

Key Discovery **The structure of hemoglobin**
Structurally sound

Hemoglobin is a protein in blood that carries oxygen around the body, and its structure was first determined by Max Perutz in 1959.

In the early twentieth century, the structures of large proteins such as hemoglobin were still a mystery. It wasn't until Perutz started analyzing crystallized horse hemoglobin, given to him by a colleague, that the story started to unravel. Perutz analyzed the blood using X-rays, but the images produced from X-ray diffraction of the large hemoglobin protein were complex and difficult to interpret.

He realized that if he re-analyzed hemoglobin with heavy metal atoms attached, he would gain more information about the molecule's structure. Perutz and his team carried out an extensive range of experiments using different heavy metals. Eventually, their work allowed the determination of the molecular make-up of this complex protein.

In 1960, after 20 years of careful experimentation, Perutz finally published the results of his research in the journal *Nature*. His studies on hemoglobin had allowed its precise structure to be determined, and provided an essential basis for the understanding of the biological mechanism of oxygen binding and gas exchange in the molecule itself. Perutz's work was recognized when he received the Nobel Prize in Chemistry in 1962 (jointly with colleague John Kendrew), and his technique of protein analysis is still widely used today.

Katie Giles

Date 1959

Scientist Max Perutz

Nationality Austrian

Why It's Key Revealed the structure of one of life's most important proteins, developing analytical techniques along the way.

Key Invention
Beta blockers

In the early part of the 20th century it was discovered that adrenaline could cause the constriction of blood vessels or their relaxation, depending on their location. A couple of decades on and James Black was forging new ground in clinical pharmacology by specifically searching for a drug that reduced the stress on the heart and arteries that adrenaline had been linked with. At Black's Glasgow Veterinary School in the late 1950s he had success when he developed beta blockers.

By binding to the beta receptors for adrenaline (beta-adrenoceptors) – found in the heart, arteries and other parts of the body – Black found that beta blockers inhibit the reuptake of this chemical and thus reduce the effects of activation of the sympathetic nervous system. They can thus cause arteries to widen, slow the heart rate, and decrease its force of contraction, resulting in a drop in blood pressure. Since widely prescribed for patients who have suffered heart attacks, and for reducing blood pressure, they also reduce the symptoms associated with hypothyroidism such as rapidly beating heart (palpitations), diarrhoea, shakiness and muscle aches. Put simply, they reduce the symptoms of the so-called "fight-or-flight" response and, known as the "musicians' drug", have been used by performers for their ability to reduce these physical effects of anxiety.

Their immediate impact on treatment and research in heart disease in the 60s was significant – acknowledged when Black was awarded Nobel Prize in 1988 – and they have become synonymous with the type of lock and key diagrams used to represent receptors and chemical messengers that act on them.

Fran Archway

Date 1959

Scientist James Black

Why It's Key The discovery of beta blockers was a significant milestone in understanding the molecular action of drugs and had a lasting effect on the world of clinical pharmacology.

Key Publication **Pheromones**
We've got chemistry!

What is it that attracts you to the object of your affections? Those piercing blue eyes? That razor sharp wit? Or is it – could it be – that lingering smell of sweat that rises like a mist from your loved one's armpits after a quick sprint round the block?

In truth, it's unlikely that anyone with a really pungent underarm odor is going to be beating off admirers with a stick. But there are a few sexy little chemicals in sweat that some claim can be a real turn on, although the exact effects and potency of these pheromones in humans in still up for debate.

"Pheromones" was a term coined by biologists Peter Karlson and Marin Lüscher in 1959, in a letter to the scientific journal *Nature*. Over the years that followed, researchers observed the physiological effects of pheromones in insects, mice, moths, and hamsters, to name but a few; these effects seemed to manifest themselves mainly as animals getting jiggy. It took until much later, however, for scientists to realize that it wasn't just other species that the chemicals were working their magic on.

It was in the 1980s that pheromone research started to hot up. In one experiment, women were asked to wear necklaces that would secrete the pheromone androstenol while they slept. In the morning, it became apparent that women wearing the pendant were far more likely to have had "intersexual contacts" with men in the study group. Large-scale studies are, however, still needed to confirm the mysterious – and racy – effects of pheromones.

Hayley Birch

Publication *Nature*

Date 1959

Author Peter Karlson, Martin Lüscher

Nationality Germany, Switzerland

Why It's Key The discovery of chemicals that could explain why we find people attractive – even if they look like they've hit every branch while falling out of the ugly tree.

Key Person **Rosalyn Yalow**
An unlikely medic

Born in New York in 1921, Rosalyn Yalow was a determined academic, excelling at a time when expectations of women were not high. She was brilliant in mathematics and chemistry, but it was physics that piqued her interest during the 1930s. Her future was decided when she sat through a lecture about nuclear fission.

She started teaching at the University of Illinois in 1941, the only woman in a four-hundred-strong faculty – and the first for twenty-four years. She worked throughout the war and obtained her PhD in 1945, an expert in the measurement of radioactive substances.

Back in New York, her husband introduced her to medical physics, an ideal platform for her knowledge. She first developed a major radioisotope service and then began to research the clinical applications of radioisotopes. While working on insulin, the idea of antibodies as a measuring tool occurred to her. It took several more years to develop the radioimmunoassay, but in 1959 the technology was finally launched.

Now found in laboratories worldwide, the radioimmunoassay measures with extraordinary sensitivity hundreds of chemical substances in body fluids, using antibodies and radioactive "labels." It has countless uses, from screening donated blood for contaminants to identifying drug abusers. Remarkably, Yalow had no formal medical training despite this massive contribution to medicine. She was awarded the Nobel Prize for Physiology or Medicine in 1977, for the technique she had devised with her colleague Solomon Berson nearly twenty years earlier.

S. Maria Hampshire

Date 1959

Nationality American

Why She's Key Yalow flourished in a male-dominated field, with no formal medical education, and gave the world of medicine a gold-standard tool for clinical analysis.

opposite Rosalyn Yalow working in the laboratory

Key Publication **The Dyson Sphere**
A modest proposal for Solar System engineering

In 1959, Giuseppe Cocconi and Phillip Morrison speculated that highly advanced civilizations around other stars would be sending radio signals toward our Solar System in order to welcome us to "the community of intelligence" as soon as we developed radio telescopes.

A year later, physicist Freeman Dyson proposed that Cocconi and Morrison might have been thinking too small. He wrote that a million-year-old technical civilization would have enormous energy needs, comparable to the total energy output of their star. This energy could be captured by dismantling Jupiter-like planets to manufacture "a loose collection or swarm of objects traveling on independent orbits around the star." A hypothetical shell of loose planetoids would absorb the star's visible light, while also emitting infrared radiation. Dyson was not the first to come up with the concept. John Desmond Bernal proposed it in 1929, and Olaf Stapledon featured the idea in his 1937 science fiction novel *The Star Maker*. Dyson believed that "a solid shell or ring surrounding a star is mechanically impossible," but that hasn't stopped science fiction authors from writing about solid Dyson Spheres and Ringworlds.

In 1964, Nikolai Kardashev suggested that Dyson himself may have been thinking on too small a scale. He proposed civilizations able to harness all the power available to an entire galaxy. Astronomers have performed numerous radio, optical, and infrared searches for extraterrestrial intelligence over the past fifty years, but so far have yet to come across any.

Eric Schulman

Publication "Search for Artificial Stellar Sources of Infrared Radiation"

Date 1960, *Science*

Author Freeman Dyson

Nationality British-American

Why It's Key Expanded our ideas about how to search for extraterrestrial intelligence.

Key Event **America strikes back in the space race**
The first communications satellite

Most of us are aware of Sputnik 1, the first ever manmade satellite to be launched into Earth orbit. To the USSR, the cheery beep of Sputnik symbolized all that was great about their mighty country, and confirmed that communism could achieve anything. To the Americans it was bad news. Not only did it create a sense of paranoia that the Russians were able to spy on them using their amazing satellite technology, but it also left them rather embarrassed.

America needed to fight back to prove their worth against their rival superpower. Enter Echo 1. Or to be more precise, Echo 1A, as Echo 1's launch vehicle failed. Echo 1 was the world's first communication satellite and, as with many of the early satellites, was incredibly simple. Essentially a big reflective, metallic ball in the sky, Echo was affectionately nicknamed a "satelloon" by those working on the project. As a passive communication satellite, it was used to bounce electromagnetic signals off, and so communicate around the curvature of the earth. Although this was a fairly ineffective method of communication – as the signal was greatly weakened – it did prove that signals could be communicated to and from space, thereby paving the way for more advanced communication satellites.

As well as being a forerunner in the field, and a political ego booster, Echo also assisted in a neat bit of science by helping calculate atmospheric density and solar pressure. Its main role, however, was to remove some egg from NASA faces in the emerging space race.

Jim Bell

Date 1960

Country USA

Why It's Key Most of us use communication satellites every day of our lives, and this was the first.

opposite The Echo 1 satellite undergoes an inflation test

Key Person **Harry Hammond Hess**
Giant of deep-sea geology

When Harry Hess died just a month after Apollo 11 landed the first men on the Moon – a project which he had helped to plan – humanity lost a man who had peered with curiosity at another equally mysterious frontier: the ocean deep.

Harry Hess, a professor of geology at Princeton University, specialized in studying arcs of active volcanic island chains. Even on active service during World War II, in which he commanded a submarine, neither tropical temptation nor the sound of gunfire could deter him from his illustrious research career. As he cruised from battle to battle, he often carried out echo-sounding surveys of Pacific Ocean basins. In 1945, Hess measured the ocean's depth to seven miles, a record at the time. Throughout his exploration, Hess began to notice a number of oddities in the ocean depths. He was especially puzzled by hundreds of flat-topped, deep-sea mountains that he dubbed "guyots."

After the discovery of a vast underwater Atlantic mountain range, two American physicists discovered that the range fell into a canyon, the Great Global Rift. In 1960, Hess used this and his own data to construct a theory of sea floor spreading, neatly explaining a number of geological problems, including the oddly eroded guyots, island arcs, and the surprising youth of the ocean floor. His paper, one of the pillars upon which the theory of plate tectonics was based, detailed how magma flows up from inside the Earth to create a new sea floor and how it moves, eventually returning to the Earth's interior.

Logan Wright

Date 1960

Nationality American

Why He's Key Hess was vital in bringing about the now accepted theory of plate tectonics, shedding light on the way our planet's surface moves.

N.A.S.A.

Key Event The contraceptive pill approved for use in the USA

After a hard-fought battle for birth control, the contraceptive pill's acceptance in America in 1960 was a heralded as a triumph. "The Pill," as it commonly known, is now used by an estimated 100 million women worldwide.

Gregory Pincus, an American physician and researcher, made a major breakthrough in 1934 in fertility research by successfully carrying out in vitro fertilization in rabbits. In 1953, he was approached by Margaret Sanger and Katherine McCormick, two of the catalysts in the development of the Pill, who offered him the financial backing to develop an oral contraceptive. By 1957, the Pill was available as a treatment for gynecological disorders and in 1960, was approved by the Food and Drug Administration.

The basis of the Pill is a combination of synthetic hormones, similar to those produced by women naturally during pregnancy, which effectively suppresses the release of eggs from the ovaries.

Prior to the introduction of the Pill, contraception was primarily the responsibility of the man. The Pill heralded the start of a sexual revolution; far from being simply a new medicine, it gave a new lease of life to women, who were effectively able to take charge of their own bodies. And not only were they liberated sexually – education and freedom to work became realities. The lifelong battles fought by Sanger and McCormick had won a victory for womankind.

Ceri Harrop

Date 1960

Country USA

Why It's Key A new lease of life for women, and the start of a sexual revolution as the sixties swung into action.

Key Invention First pacemaker fitted

When we speak of heart disease, or a heart attack, or myocardial infarction, most people are actually talking about a problem with ischemia. In other words, not enough blood is being channeled to the heart's own muscles. This represents only part of the heart's workings, however. It also has its own unique electrical system that tells it when to beat, without which it cannot function. When the heart develops an electrical problem, it can be just as deadly as ischemia, but must be treated entirely differently.

In the 1950s, the idea of using an external electrical device that could tell the heart when to beat started to become a reality. Initially, such devices were cumbersome and had to be attached to external electrical sources – they were literally plugged in to walls. In 1957, an electronics repairman named Earl Bakken dramatically improved the design by both shrinking the device to the size of a small book, and switching it to battery power. These innovations made pacemakers substantially more practical, but still less than perfect.

The first implantable pacemaker was designed quite by accident by Wilson Greatbatch. While trying to invent a device that would only record heart sounds, Greatbatch accidentally inserted the wrong size resistor into a device. The resulting circuit fired exactly like a heart beat. Greatbatch modified his device further to be completely impervious to conditions inside the body, and also to have long battery life. The result was an implantable, largely maintenance-free pacemaker that saved countless lives.

B. James McCallum

Date 1960

Scientist Wilson Greatbatch

Nationality American

Why It's Key Cardiac pacemakers assume the role of the heart's natural electrical system in people who have arrhythmia – staving off death in many instances.

Key Event **Plumbing the depths**
Diving the Marianas trench

Diving into the abyssal depths of Davy Jones' locker, where no human being had been before, must have taken a lot of gumption. Luckily, Jacques Piccard and Don Walsh were up to the task.

On January 23, 1960, they journeyed to the very bottom of "Challenger Deep," part of the Mariana trench – the deepest known point in the ocean. They descended in a bathyscaphe, a self-propelled free diver that requires no power to ascend, and relies on condensed air and its own buoyancy to reach the surface. The intrepid twosome's bathyscaphe, designed by Piccard and his father, was christened the *Trieste* and remains to this day the only manned submersible to have visited the Challenger Deep.

In a descent that took over five hours, Walsh and Piccard sunk down to the ocean floor, measuring the depth at 10,916 meters. To appreciate this depth, just consider that Mount Everest – at 8,840 meters, the world's tallest mountain – would nestle comfortably in the trench, hidden more than a mile under the waves.

Piccard reportedly saw signs of life in the depths, describing animals that appeared to be species of sole, flounder and sea cucumber. This marked the beginning of a new age of oceanic exploration and has encouraged the ongoing analysis of sea environments and deep-dwelling aquatic life.

Nicola Currie

Date 1960

Location Pacific Ocean

Why It's Key Man has never dived deeper beneath the waves.

Key Invention **CPR**
The girl from the River Seine

The birth of modern cardiopulmonary resuscitation (CPR) occurred during a two-day car trip in October 1956. Doctors Peter Safar and James Elam were returning home from a medical conference. Elam had just presented a paper showing that a rescuer could adequately oxygenate a non-breathing victim with exhaled breath passed through a tube. Safar, invigorated by the discussion, went straight off to find out whether you could simply oxygenate someone mouth-to-mouth. He confirmed it by oxygenating sedated volunteers who had been given curare, the paralyzing toxin derived from South American poison arrow frogs.

The two men, and many others, refined the technique further to include opening the airway and chest compressions. By 1960, CPR had become the preferred method of resuscitation.

Modern CPR is also due to a chance encounter. Safar attended a conference in the Netherlands when he met Bjorn Lind, who later introduced him to Asmund Laerdal, a Norwegian toymaker. Laerdal agreed to produce a mannequin and had a face in mind: "L'inconnue de la Seine." In the late nineteenth century, the body of an unknown young woman was retrieved from the River Seine, and a death mask was made of her features. There were no marks on her body, and it was assumed she had committed suicide. It was a copy of the death mask of this young woman that became the face of the mannequin – Resusci Annie – which has helped teach CPR to countless people around the globe.

Stuart M. Smith

Date 1960

Scientists Peter Safar, James Elam

Nationality American

Why It's Key Empowered non-medical professionals; saved countless lives.

Key Discovery **Out of Africa** Early hominid fossils discovered

It seems unthinkable now, but in the early twentieth century, fossil finds in Java and China pointed to man's origins having been in Asia. Kenyan-born Louis Leakey, in patriotic disagreement with this view, began his excavations in East Africa in the late 1920s.

In 1936, Louis married illustrator Mary Douglas Nicol and the couple traveled to East Africa to excavate several sites, including Olduvai Gorge in Tanzania. Although they made a number of discoveries of mammal fossils and stone tools, hominid fossils eluded them.

Their luck changed in 1959, when a fragment of bone caught Mary's eye. Upon further investigation, some suspiciously hominid-like teeth were revealed. The remains were classified as *Australopithecus boisei*; the skeleton characterized by a robust skull and a huge set of teeth. The Leakey's lucky streak continued and in 1960, their son Jonathan discovered a second type of hominid at the Olduvai site. This skeleton had a larger brain and less robust skull than *A. boisei*, and Louis proposed that it was the maker of the stone tools they had discovered. He named this species *Homo habilis* – man of skill.

Homo habilis was a controversial addition to the *Homo* genus as it lacked characteristics of the later hominids. Some considered the species to be more ape-like than human. At 1.75 million years old, Louis's *H. habilis* represents the earliest hominid to be placed into the genus, vindicating his conviction that humans started out in Africa.

Helen Potter

Date 1960

Scientists Louis Leakey, Mary Douglas Nicol, Jonathan Leakey

Nationality Kenyan, British

Why It's Key Traced man's origins to Africa and discovered a new link in the evolutionary chain.

opposite Louis and Mary Leakey studying fossils in Africa

Key Discovery **Laser beams** A solution seeking a problem

Despite the fact that lasers have been around for decades, they inevitably conjure up futuristic images that have little to do with their many, and often mundane, applications. They're at work everywhere, from the inside of your CD player to the barcode scanner at the local supermarket.

It was Albert Einstein who laid the foundations for laser beam technology, back in 1917. He figured out that when a packet of light called a photon passes an atom, the atom can emit its own photon – crucially, the new photon would have the same direction and the same frequency.

Development of Einstein's ideas continued for many years, and in 1953, a device that could produce microwaves was built. They called it a MASER, standing for Microwave Amplification by Stimulated Emission of Radiation. Microwaves are bigger than visible light, so these were easier to produce. Then, in 1957, Gordon Gould described the important components that you would need to build what he called a LASER. In 1959, Gould introduced his ideas to the public and tried to get a patent for it, but was denied.

Finally, on May 16, 1960, Theodore Maiman operated the first working laser, using a tubular flash lamp coiled around a manmade ruby. At the time, the laser was described as "a solution looking for a problem," but now we have all sorts of uses for them. Among other things, they've been used to measure the exact distance to the Moon, guide missiles, remove tattoos, catch highway speed junkies, and make pretty patterns at concerts.

Douglas Kitson

Date 1960

Scientists Gordon Gould, Theodore Maiman

Nationality American

Why It's Key Lasers have scientific, military, and medical applications. They make your CD player run, and almost cut James Bond in half.

Key Event **First weather and communication satellites go into space**

The first pictures of Earth from space were taken in 1947 by an unmanned rocket, carrying a camera; they demonstrated that weather observations could be made from space.

In 1960, the first weather satellite TIROS-1 was launched by the United States, carrying a video camera. It made regular observations of the atmosphere below, which were compared with ground-based measurements. The cylindrical satellite was powered by solar panels and was equipped with two cameras that took snapshots of the scene below every ten seconds, transmitting them back to the ground station. It was also equipped with tape for recording images when the satellite was out of range; these were played back to Earth the next time it made contact. The transmitted data was recorded onto 35-millimeter film at the ground station for analysis and, for the first time, 3D images of severe thunderstorms were a reality.

Later on in 1960, NASA launched the Echo project. This investigated the use of a metallic balloon to reflect microwave signals back to Earth as a means of communication. Microwaves are used in radio, television, and telephone transmissions. Prior to this, the U.S. Navy had looked at bouncing microwaves off the Moon but this wasn't ideal. The first balloon launched was Echo-1 which was thirty meters in diameter and had a smooth, reflective aluminum surface. Once in space, it was visible over most of the Earth, and was reportedly brighter than most stars. The balloon worked well, providing high quality verbal communication to and from different points on Earth.

Riaz Bhunnoo

Date 1960

Country USA

Why It's Key TIROS-1 enabled accurate weather prediction and gave early warning of severe thunder storms. Echo-1 provided a reliable method of satellite communication and tracking, via microwaves.

Key Experiment **Skinner's theory of operant conditioning**

Discovering what makes people tick has always fascinated psychologists, but one man arguably made a bigger contribution to understanding human and animal behavior than any other. That man was Burrhus Frederic Skinner.

Skinner is most famous for coining the term "operant conditioning" – to describe a theory of learning – after spending years studying the behavior of rats and pigeons. He observed rats in a special cage called the "Skinner box." The box had a pedal that, when pressed, released a small amount of food. The rats initially stepped onto the pedal unintentionally whilst exploring their cage, but when rewarded with a tasty treat, they soon learned to step on the pedal whenever they felt hungry. Skinner called this behavior "operant conditioning," because the rats had "operated" on their environment in a way that resulted in a reward, reinforcing that behavior and making them more likely to try it again. The rats had been "conditioned" to press the pedal for food. Skinner used his theory to teach simple animals to do complex tasks – he even taught pigeons to play table tennis.

Skinner's theories were so successful that they have been applied to human psychology in clinical and educational settings, and he has been heralded as one of the most influential psychologists of the twentieth century.

Hannah Isom

Date 1960

Scientist Burrhus Frederic Skinner

Nationality American

Why It's Key Skinner used the term "operant conditioning" to explain how individuals learn by the consequences of their behavior.

opposite A hungry rat learns to get food whilst being observed in a Skinner box

Key Invention **Quicksort algorithm**
A method for sorting stuff, quickly

An algorithm is a list of instructions, often represented as a flowchart, which tells a computer how to do a specific task. One of the most famous is the Quicksort algorithm, created by British computer scientist Charles Antony Richard Hoare in 1960.

Given any list of random numbers, Quicksort works to put them in order; as its name would suggest, it sorts them. Using a divide and conquer technique, Quicksort picks one of the numbers in the list – a "pivot" point – and rapidly rearranges the list so that all the numbers smaller than the pivot number are placed before it, while all the larger numbers are placed after it. This effectively leaves two lists of numbers, one of which contains all the big numbers and the other the small ones. Quicksort then does the same thing to each of the two smaller lists, and keeps going until all of the numbers from the original random list are now in a perfect sequence of size.

Hoare's algorithm was soon widely adopted as well as extensively analyzed by computer scientists around the world. It turned out to be around twice as fast as any other method for getting a computer to sort numbers, and is today the most widely used algorithm of its kind in the world. Charles Antony Richard Hoare was knighted for his services to computer science and education.

David Hawkset

Date 1960

Scientist Charles Antony Richard Hoare

Nationality British

Why It's Key Hoare's Quicksort algorithm became the most widely-used sorting algorithm in the world.

Key Event **The Great Chilean Earthquake**
The power of the Ring of Fire is unleashed

The "Ring of Fire" refers to the region surrounding the Pacific Ocean, home to about 90 per cent of the world's active volcanoes and the location of over 80 per cent of the world's largest earthquakes. Regions including South America, Indonesia, and the western United States have become used to regular seismic activity; the U.S. Geological Survey (USGS) estimates that since 1900 these areas have experienced nearly twenty earthquakes per year that have measured over 7.0 on the Richter scale.

The Richter scale measures the magnitude of the largest wave caused by the earthquake. Because the scale is logarithmic, every 1.0 increase on the scale represents a ten-fold increase in the severity of the quake. Small quakes, on the scale of 2.5–4.0 occur on a daily basis worldwide; above 6.0 fatalities are common.

Situated in the Ring of Fire and perilously close to the Nazca, South American, and Antarctic tectonic plates, Chile lives under constant threat of quakes. In 1960, the country was hit by the largest ever recorded quake, registering a huge 9.5 on the Richter scale. Although massively destructive, it was responsible for the deaths of only 1,655 people; the low death toll is thought to be because of an earlier foreshock that had alerted the people to the possibility of a large quake.

The earthquake itself was a "megathrust," meaning that one tectonic plate actually dropped below another. The massive power of this movement also triggered a tsunami responsible for the deaths of sixty-one people on Hawaii, over 10,000 kilometers from the epicenter.

Jim Bell

Date 1960

Country Chile

Why It's Key Although probably not the largest earthquake ever, it was the biggest ever recorded and is an important reminder of the massive destructive power our planet can unleash.

Key Event
WWF registers as a charity

On September 11, 1961, the WWF (formerly the World Wildlife Fund and now the World Wide Fund for Nature) was registered as a charity. The birth of this organization initiated an international fundraising mission for ecological conservation.

Public awareness of the need to protect the world's wildlife was raised after British biologist Sir Julian Huxley published a series of discoveries on the poor conservation of wildlife in East Africa. His articles stirred up much public concern, and led to the call for an international fundraising organization to promote the protection of wildlife. The organization was set up by a group of scientists, advertisers, and PR professionals, who set out to educate the world about the need for conservation. With offices in different countries, and working with other non-governmental organizations, the WWF set about raising funds to support vital research projects, education programs, species and habitat management programs, the establishment of protected areas, and so on. Grants were, and still are, issued based on the best scientific knowledge available, and the scale of projects varies.

The WWF now has five million supporters, and offices in 90 countries. In the last 22 years they have invested over US$1billion in more than 12,000 projects focusing on a wide range of environmental issues. The international extent of conservation organizations is immense. The IUCN (World Conservation Union) oversees the largest international conservation network having 1,056 members. This includes national and international NGOs and government agencies.

Emma Norman

Date 1961

Country Switzerland

Why It's Key The WWF raised international awareness of threats to the world's wildlife and natural environments, and began the global mission to raise money for nature conservation projects.

Key Event **The Drake Equation**
How many needles are there in our haystack?

When Frank Drake was organizing the first ever Search for Extraterrestrial Intelligence (SETI) conference in West Virginia, in 1961, he was looking for a way to structure the discussions. Unlike ET, Drake couldn't phone home for the answer, so he devised a new equation which broke down the search into individual fields, each to be covered at the event.

Drake multiplied a number of probabilities and rates together in order to yield an estimate for the number of communicating civilizations in our galaxy at any one time. The factors involved were the number of suitable stars formed each year; the portion of these that have planets; the number of habitable planets per such system; fractions concerning the development of life, intelligence, and technology; and the average number of years for which able civilizations communicate.The evolution of life, particularly into intelligent life, is considered by some to be miraculous and therefore, to them, the solution to the equation is one. To others, life is inevitable, so to them the solution to the equation is in the millions. However, the real benefit of the equation is not the numbers produced, but the discussions initiated.

While it may not role off the tongue like $E=mc^2$, Drake's Equation ($N=R^* \times f_p \times n_e \times f_l \times f_i \times f_c \times L$) certainly has its own charm, finding its way onto coffee mugs, T-shirts and mouse mats; with each new planet, microbe or potential signal we discover, fingers tap at calculators, while enthusiasts and scholars alike look up to the stars, contemplating.

Christopher Booroff

Date 1961

Country USA

Why It's Key This thought experiment opened up a cutting edge field of science to the masses, simplifying discussions and assessing the relevance of each astronomical discovery we make.

Key Event
Yuri Gagarin is the first human sent into orbit

Following in the footsteps of the many brave mice, monkeys, and houseflies that went before him, the Soviet cosmonaut Yuri Gagarin became the first human in space and the first to orbit the Earth, in the USSR's highly successful mission, Vostok 1, on April 12, 1961.

Taking off from the Baikonur Cosmodrome in Kazakhstan at 9:07 am, Gagarin was launched into space and completed one full orbit of the Earth, before returning back into the atmosphere at approximately 10:55 am. Gagarin ejected from the returning vessel and had his feet back on *terra firma* by 11:05 am.

To give you some sense of the length of this momentous space mission, imagine a Eurostar train making a journey direct from Paris to London. The train could leave Paris on the first chime of 9 am (giving it a generous seven-minute head start) and still Gagarin could have taken off, landed, boiled a kettle, and enjoyed the first few sips of a well-earned cup of tea before the Eurostar reaches its destination.

This was something of a blow for the United States in the ongoing space race. Though they followed right behind by sending their man Alan Bartlett Shepard Jr. up a few weeks later, at this point it seemed that the Soviets very much had the upper hand.

Chris Lochery

Date 1961

Country USSR (now Kazakhstan)

Why It's Key Successfully showed that human space travel and exploration was not pure fantasy.

opposite Yuri Gagarin onboard Vostok 1 on his way to space

Key Event **Valium**
Mother's little helper

Valium is the trade name given to the sedative drug diazepam, which is commonly used in the treatment of anxiety, insomnia, seizures, alcohol withdrawal symptoms, and muscle spasms. It has become one of the most frequently prescribed drugs of the last forty years, thanks to its effectiveness in treating a wide variety of conditions.

First synthesized by the Polish-Jewish chemist Leo Sternbach at the Hoffmann-La Roche Laboratories in the United States in the late 50s, the drug was approved for medical use just four years later in 1963. Being a much safer alternative to barbiturates, it gained rapid popularity, and by 1966 had been nicknamed Mother's Little Helper by none other than the Rolling Stones. Between 1969 and 1982, Diazepam was the top-selling prescription drug in the United States; generating over US$1 billion of revenue per year, it was the first of the "blockbuster drugs." In 1978, sales hit 2.3 billion pills.

By this time, however, evidence that Valium might not be such a wonder drug after all was mounting. Patients were becoming addicted, suffering severe withdrawal symptoms when they tried to stop taking the pills, and an underground drugs trade had sprung up. The public were alerted to the dangers, in no small part, by the 1979 book *I'm Dancing as Fast as I Can* by Barbara Gordon. Gordon, a high-flying television producer, had become so hooked on diazepam that withdrawal symptoms had left her hospitalized.

Diazepam is still in medical use today, but controls on its prescription are now much tighter.

David Hall

Date 1961

Country USA

Why It's Key The first blockbuster drug which revolutionized the relief of anxiety, tension, and seizures, but with sinister side effects. The drugs approval system is now much stricter.

Key Experiment
The nature of the genetic code is cracked

By the late 1950s, the central dogma of molecular biology – that DNA makes RNA makes protein – was established. However, little was known about the so-called genetic code, i.e. how the sequence of DNA relates to the sequence of amino acids making up a protein. Three key sets of experiments helped scientists crack this code, and by 1966, the sixty-four different genetic "code words" had been mapped to the twenty different amino acids in all living things.

U.S. scientist Marshall Nirenberg and German-born J. Heinrich Matthaei kicked off this process by creating a cell-free system that could produce protein. They added an artificial messenger RNA (the intermediate between DNA and protein synthesis), made only of a string of uracil molecules. This produced a protein made only of the amino acid phenylalanine, leading them to conclude that the "code" for phenylalanine was a string of uracil molecules: how many was unclear. Hearing of this experiment, Francis Crick and Sydney Brenner set to work building on these results. They showed that the genetic code was a triplet code, which means that each amino acid is coded for by a combination of three DNA bases, called a codon. They also showed that each codon followed in sequence without overlapping.

In 1964, Nirenberg and Philip Leder developed the "triplet binding assay." This elegant experiment, which involved radioactively labeled amino acids, allowed the scientists to identify which amino acids corresponded to which codons. Soon, the entire genetic code was cracked.

Christina Giles

Date 1961

Scientists Marshall Nirenberg, J. Heinrich Matthaei, Francis Crick, Sydney Brenner, Philip Leder

Nationality American, German, British

Why It's Key Produced concrete evidence of how DNA sequence relates to protein make-up.

Key Invention **Imperfect implants?**
The rise and fall of breast augmentation

Surgical breast augmentation has been around since 1890, when paraffin injections were introduced. These were replaced in the 1920s by fat transplants, in which fat from the abdomen and buttock areas was injected into the breast. Both processes led to abnormally shaped and hardened breast tissue, and were soon replaced with other techniques such as polyvinyl sponges and silicone injections.

In 1961, two plastic surgeons from Houston, Texas – Thomas Cronin and Frank Gerow – developed a silicone sack to replace injections. At the same time, the Dow Corning Corporation was developing silicone product lines for medical use. Dow Corning marketed the invention as the "Cronin breast prosthesis," and sales of the implants soared over the next two decades. The sixties also saw the introduction of the saline breast implant, developed in France in 1964. The initial decline of the silicone implant began in the 1980s, when claims of injuries and adverse reactions such as cancer and autoimmune diseases (thought to be due to leakage of the silicone gel) were brought against Dow Corning.

A lawsuit was settled in 1988 when the company paid out US$3.2 billion to 170,000 women in claims. The implants were taken off the market in the early 1990s. Recent scientific studies have shown that there may in fact be no link between silicone implants and cancer or connective tissue disease. The implants were approved again in 2006 by the Food and Drug Administration, who required manufacturers to follow up extensively with patients on any potential health risks.

Rebecca Hernandez

Date 1961

Scientists Thomas Cronin, Frank Gerow

Nationality American

Why It's Key Not only did silicone implants have a huge impact on plastic surgery, they made a comeback after being off the market for over a decade.

opposite A mammogram X-ray of a silicon breast implant

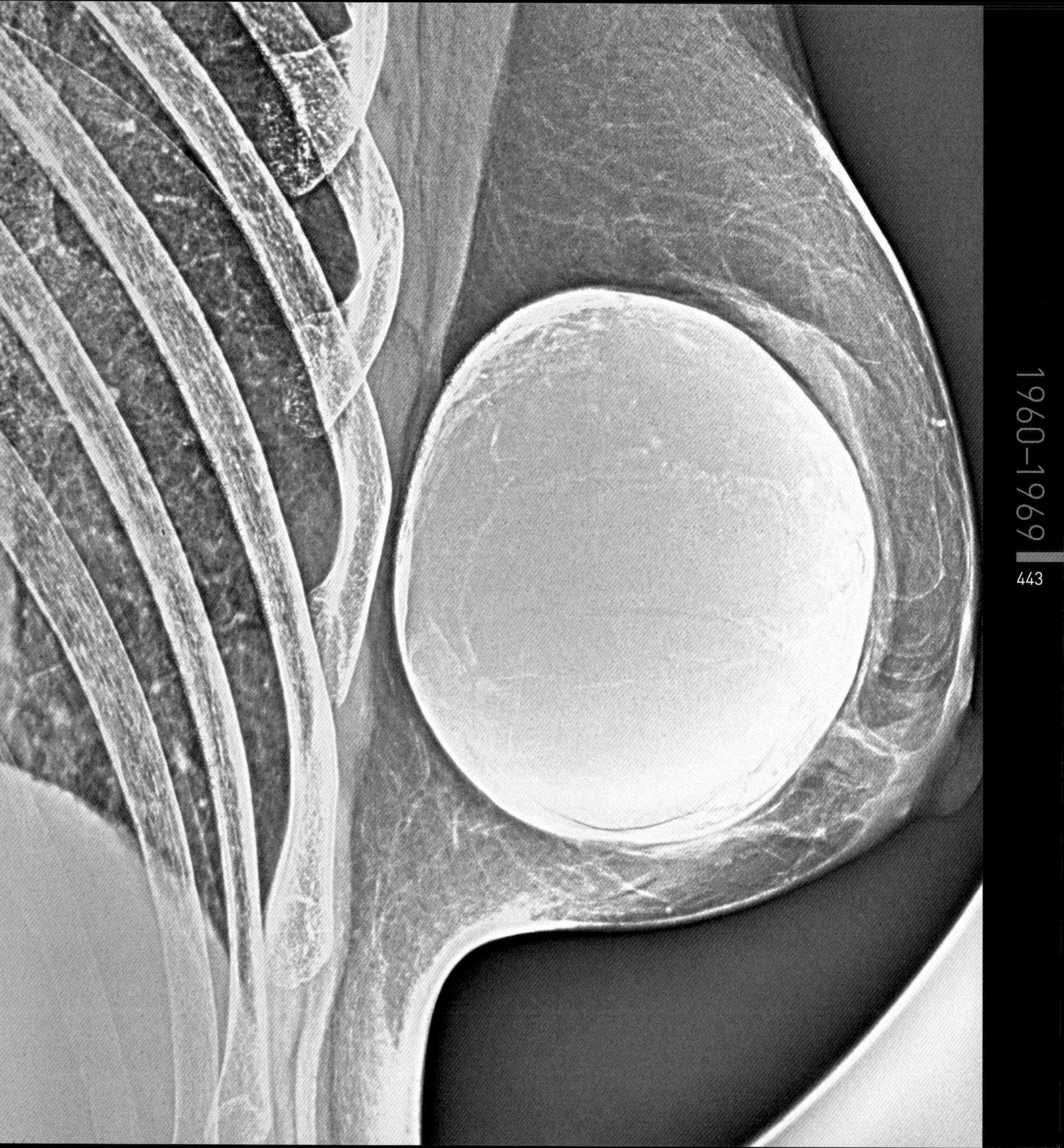

Key Event **Venera 1**
The first flyby of another planet

The 1960s saw the escalation of the space race between the United States and the USSR. In 1961, the Russians gained another small victory over the Americans by successfully performing the first flyby of another planet.

Not only was Venera 1 the first manmade object to pass another planet, it also was the first satellite to launch from space. The launch of Venera 1 took place in two parts. First, on February 12, 1961, the space station Sputnik 8 blasted off. Then, when it reached an orbit of 64,000 kilometers, the Venera 1 spacecraft was released toward Venus. On May 19 and 20, the spaceship passed the planet but entered into a heliocentric orbit – one with the sun at its center – and lost contact with Earth.

The first fully successful flyby mission was completed by Mariner 2 in 1962. This time it was a NASA spaceship which took the crown. The American craft managed to remotely survey Venus and relay the information back to Earth. Scientists at the time were given the first direct data from the planet and were able to confirm that it was extremely hot with no plate tectonic processes occurring. They were also able to refine the estimates of the planet's mass and the definition of an astronomical unit – roughly the distance from the Earth to the Sun.

After the mission was completed, the spacecraft was sent into a heliocentric orbit where it no doubt now hangs out with its Russian counterpart, Venera 1.
Josh Davies

Date 1961

Country USSR (now Kazakhstan)

Why It's Key The first manmade object to fly past Venus, or any other planet, marked a major achievement in space exploration.

Key Event **Thalidomide**
A silver lining? The FDA

These days the Food and Drug Administration (FDA) is seen by some as little more than a bureaucratic nightmare. However, as the body responsible for preventing another tragedy as horrifying as the thalidomide disaster, it is clear why FDA regulations are so tight.

Thalidomide was originally developed by a German company in the 1950s, and prescribed as a tranquilizer. During the late 1950s and early 1960s, thousands of pregnant women were prescribed thalidomide as an antidote to morning sickness. It was only afterwards that the terrible deforming effects of thalidomide on unborn babies were realized. Approximately 10,000 babies were born blind, deaf, and with missing or deformed limbs. If every cloud has a silver lining, then in this case it went to the FDA. In 1960, Dr Frances Kelsey, the FDA official in charge of overseeing the drug application for thalidomide, raised her concerns over the safety of the drug and refused to approve it. She later received a Distinguished Federal Citizen Service Award from US President John F. Kennedy for her work. The associated repercussions of Kelsey's decision kept thalidomide tied up long enough for the drug's effects on the unborn to become apparent in 1961.

By 1962, legislation had been tightened and the FDA was provided with extra resources to enhance regulation and increase public safety. As long as the public expect and demand only FDA-approved drugs, this will ensure that drugs are tested as safe, effective, properly manufactured, and accurately labeled.
Ceri Harrop

Date 1962

Country USA

Why It's Key The impacts of inadequate drug regulation became all too apparent and the consequent legislations enforced by the U.S. Congress meant much greater protection of public safety.

opposite The damaging effects of thalidomide on unborn babies is shown in X-ray

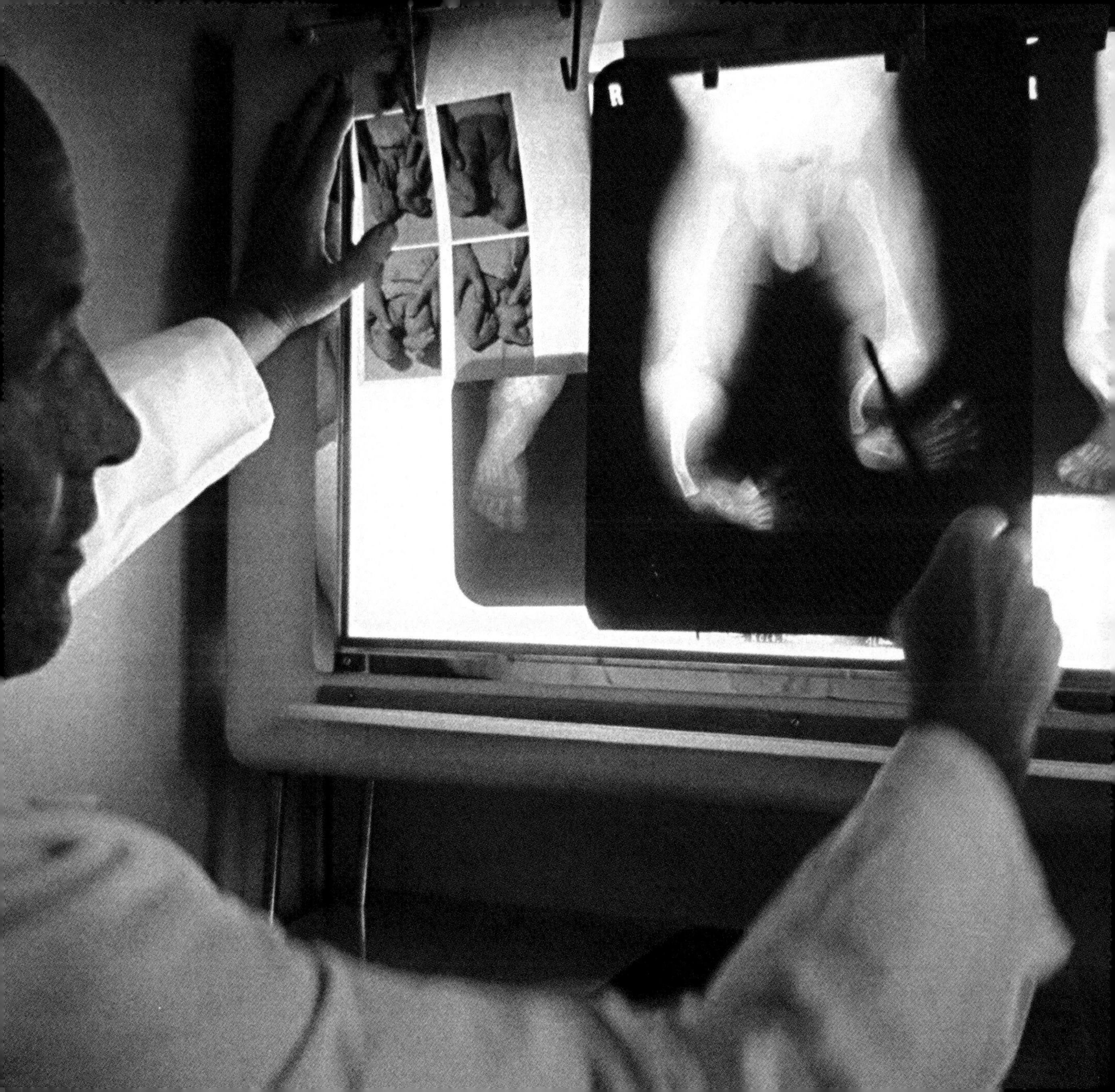
R

Key Event
Surtsey erupts

On November 8,1963 a fisherman reported seeing smoke belching out of the sea 80 kilometers off the coast of southern Iceland – something strange and decidedly volcanic was happening beneath the Atlantic waves. A week later the world's newest island, spitting fire and brimstone, poked its head above water for very first time.

The volcanic eruption that had formed the new island continued for another four years, eventually petering out as the island reached 174 metres above sea level, covering an area of 2.7 square kilometers. For geologists and biologists this was a dream come true – a blank canvas on which they could study the formation and colonization of new land.

Closed to all but a few invited scientists, Surtsey has remained virtually free from human disturbance; any organisms that have colonized have arrived by purely natural means. The first plants arrived even before the volcano had stopped erupting had finished. Sea rocket was found growing there as early as 1965. Nonetheless, colonisation in these early years was slow and difficult; during its first twenty years, just ten plant species would make Surtsey their home.

It took the establishment of a seagull colony on the island to speed things up a bit. Not only did the gulls inadvertently bring a plentiful supply of seeds, they also helped fertilise the land via the power of guano - bird droppings. By 2004 over sixty species of plant had been spotted. Today, scientific studies on Surtsey continue to provide interesting and valuable data about the mechanisms of colonisation.

Mark Steer

Date 1963

Country USA

Why It's Key The entire process of ecosystem colonization is being monitored, recorded and analysed. Nowhere else has this been possible.

Key Publication
The Structure of Scientific Revolutions

Way back at the beginning of the sixteenth century, in the harbor town of Frombork, Germany (now in Poland), a servant of the bishop began working on a theory that was to rock the astronomical world to its core. Nikolaus Copernicus painted a new picture of the Universe that plucked Ptolemy's Earth from its majestic role at the center and sent it spinning across the outer reaches of the cosmos.

More than five hundred years later, Thomas Kuhn would call this a "scientific revolution"; a fundamental change in the way we had come to view our world. As one of the twentieth century's most influential – and controversial – thinkers on the scientific method, Kuhn's writings relied heavily on historical examples. The Copernican Revolution, which recognized the Earth's correct position and orbit around the Sun, was one of his favorites.*The Structure of Scientific Revolutions* was Kuhn's take on the nature of scientific progress. At the heart of this was the idea that all real progress was based on the rise and fall of different "paradigms," or world views. He envisaged a long, slow buildup of information, eventually leading to a sudden revolution in human knowledge and scientific understanding – a paradigm shift.

Kuhn's work faced attack from various angles, as it took issue with the ideas of luminaries such as Karl Popper. Yet Revolutions has sold into the millions and continues to be a core text in the study of the history and philosophy of science.

Hayley Birch

Publication *The Structure of Scientific Revolutions*

Date 1962

Author Thomas Kuhn

Nationality American

Why It's Key A new philosophical perspective on science and scientific theories.

Key Experiment
John Gurdon clones an animal

Dolly the sheep was undisputedly the most famous cloned organism science has ever produced. But it was actually a much smaller animal, a tadpole, which laid much of the groundwork for the understanding of cloning and cell differentiation.

John Gurdon, then of Oxford University, spent most of the late 1950s and early 1960s trying to solve an important question that had been nagging at developmental biologists for years: whether certain genes disappear or are simply turned off and on as a cell develops into its specialized type.

Continuing Robert Briggs and Thomas King's important work on nuclear transfer into eggs, Gurdon transplanted nuclei from the intestinal cells of African clawed frogs into frog eggs that were missing their nucleus, the result being ten normal tadpole embryos. What made Gurdon's work stand out was that he used nuclei from cells that had already differentiated into their specialized type. Despite this, the DNA was reprogrammed to turn the cells into an embryo, not more intestinal cells.

Gurdon later proved that in addition to egg cells, embryonic stem cells, which are a bit further along in the developmental process, could also be manipulated in the same way. For developmental biologists, this knowledge opened the door to a whole new world of scientific manipulation, one that years later would produce that infamous cloned sheep.

Rebecca Hernandez

Date 1962

Scientist John Gurdon

Nationality British

Why It's Key One of the earliest cloning moments in history, which led to subsequent scientific and ethical discussion.

Key Person **Francis Crick**
Father of the double helix

Francis Harry Compton Crick was born in June 1916 in Northampton, England. He began his academic career as a physicist, but found his work disrupted by World War II. He left the British Admiralty in 1947 with a different focus – the application of physics to biology.

Crick set about getting to grips with biology, organic chemistry, and crystallography, and in 1949 he began working in the Cavendish Laboratory at Cambridge, UK, specializing in X-ray diffraction – a technique that allows scientists to determine the 3D structure of molecules.

In 1951, a young American biologist, James Watson, joined the lab. He was interested in trying to work out the structure of DNA, and he and Crick soon started working together. In 1953, in a paper published in the journal *Nature*, Watson and Crick revealed the iconic double helix structure of DNA to the world. This discovery won Crick a share of the 1962 Nobel Prize in Physiology or Medicine. He gained his PhD in 1954 and remained at Cambridge, teaming up with Sydney Brenner, with whom he helped unravel the genetic code.

In 1977, he moved permanently to the Salk Institute in California to pursue his interest in neuroscience, specifically the belief that there is a biochemical basis to consciousness. He continued to wrestle with this problem until his death from cancer in 2004, one year after the fiftieth anniversary of his – and possibly modern science's – most famous discovery.

Christina Giles

Date 1962

Nationality British

Why He's Key One of the architects of modern science's greatest discovery.

Key Person **Linus Pauling**
Nobel double-winner

"A tall thin man with a shock of grey, unruly hair" according to a sixties newspaper column; Linus Pauling (1901–1994) made a massive impact on science in the twentieth century. He made important contributions to chemistry, biology, medicine, and even world peace, and is the only person who has ever won two unshared Nobel prizes in different fields.

Born in Portland, Oregon in 1901, he entered university at the age of seventeen to study for a science degree, which he received in 1922. He was to remain in academic circles all his life, as his intellectual journey took him through the disciplines of physics, chemistry, biology, and medicine.

After fifteen months in Europe on a Guggenheim scholarship, he became a faculty member at the California Institute of Technology where he used quantum mechanics to revolutionize contemporary understanding of chemical bonding in his 1939 book, *The Nature of the Chemical Bond*.

Next he turned his attention to biology and founded a new discipline called "Molecular Biology." He determined the structure of various proteins, especially the secondary structure called the "alpha helix." He was awarded the Nobel Prize in Chemistry in 1954 for this work.

The rest of his life was defined by two crusades. First was his championing of the importance of Vitamin C in maintaining health; then came his anti-nuclear campaigning. It was for this second activity that he was awarded the Nobel Peace Prize in 1962.

Simon Davies

Date 1962

Nationality American

Why He's Key A leading scientist who contributed hugely to the fields of chemistry, biology, and medicine.

opposite Linus Pauling with a ball and socket model of the water molecule

Key Discovery **Quarks**
Taming the subatomic zoo

Ever since the first particle accelerators were created and physicists could make their own bizarre particles instead of, quite literally, waiting for them to fall from the sky, the number of known subatomic particles just kept growing. Not only sensible particles like protons, neutrons, and electrons appeared, but also whole families of other particles which seemed to have very little in common.

They were loosely grouped into leptons (lightweights, like the electron), mesotrons or mesons (middleweights), and hadrons (heavies, like the proton), but this was as good as the classification got. What governed when they came into existence, what they changed into, and how long they took to do it was a mystery. The plethora of particles was called the "subatomic zoo," and was in need of particle taxonomists.

Physicists Murray Gell-Mann and George Zweig independently proposed that the new particles were different combinations of further sub-particles, "quarks" according to Gell-Mann or "aces" as Zweig called them. These quarks came in different types or "flavors," some had a positive charge of two thirds; some had a negative charge of one third. Initially the theory seemed odd, especially since no particles with fractional charges had ever been seen, but it explained the behavior of the subatomic zoo so concisely that they were accepted, in Gell-Mann's own words, as a "convenient mathematical fiction." As time progressed the idea of quarks began to mesh with more and more discoveries in particle physics; and they were accepted as part of the true picture of the subatomic world.

Kate Oliver

Date 1962

Scientists Murray Gell-Mann, George Zweig

Nationality American, Russian

Why It's Key A simple(r) way of taming the subatomic zoo; organizing it by relating the properties of each particle to what it was made of.

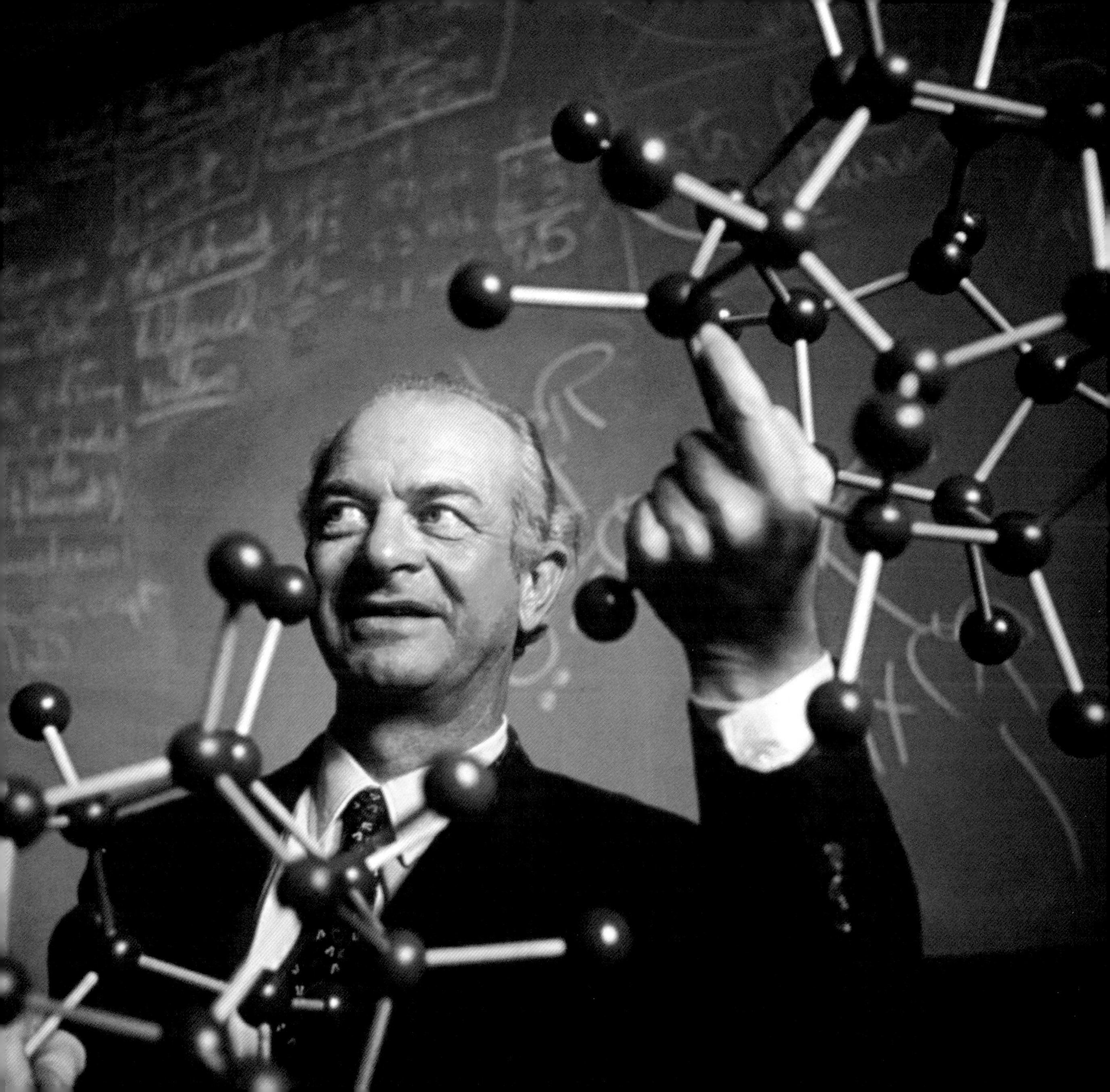

Key Person **J.C.R. Licklider**
The man who imagined the Internet

Dr J.C.R. Licklider – known as "Lick" to most – was a true visionary. Writing in 1960, he predicted handwriting recognition and touch-screen interfaces at a time when computers were still being fed with instructions punched into pieces of card. Perhaps the reason that Licklider's predictions came true so often was that he worked hard to make them happen, molding the world to the shape of his dreams.

Licklider's two years at the Department of Defense's Advanced Research Projects Agency (ARPA) were some of his most influential. Hired to manage research funding, he broadened ARPA's narrow horizons of defense contracts and computer war gaming. As well as contracting dozens of companies for all manner of computer research projects, he kick-started the first four PhD programs at American universities, and helped fund Stanford's Augmentation Research Center, which within a year would invent the first computer "mouse."

In a pre-networking age, Licklider nicknamed his collection of academics and private companies the "Intergalactic Computer Network," and set them off in the direction of his vision: that computers could be shared by many people, and that their true value was in letting them communicate with each other like never before.

Licklider left ARPA in 1964, but the teams he left behind followed through on his ideas. Within five years, the first message was sent over the ARPANET, the first true computer network, which would eventually become today's Internet.

Matt Gibson

Date 1962

Nationality American

Why He's Key He was the right person, in charge of a big stack of money, at exactly the right time to make a difference. If you can surf the Internet today, you probably have "Lick" to thank for it.

Key Person **James Watson**
DNA pioneer

James Dewey Watson will forever be known as one of the great scientists of the twentieth century. In helping uncover the double-helical structure of DNA, Watson unlocked an area of scientific enquiry that ushered in a new era of genetics, still gathering momentum today as we enter the era of personalized medicine.

Watson was born in Chicago on April 8, 1928, and gained his PhD in zoology from Indiana University in 1950. From there, he moved to Copenhagen and on to the Cavendish Laboratory in Cambridge. It was here that he met Francis Crick, the Englishman with whom he would later deduce the structure of DNA.

Watson's key input into the structure was to work out how four repeating structures in DNA, known as the bases A, C, T, and G, spatially pair up. He made the deduction after playing with simple models of the four bases. Further modeling work with Crick led to the correct overall structure, for which they won the 1962 Nobel Prize in Medicine or Physiology with Maurice Wilkins, who had supplied X-ray data hinting at the structure.

Watson went on to write *The Double Helix*, a bestselling account of the molecule and its structure. His later career has been tainted by controversial comments on race and gender, culminating in his resignation from Cold Spring Harbor in 2007 after a remark about race and intelligence. In 2007, he became one of the first individuals, along with J. Craig Venter, to have his genome sequenced.

Matt Brown

Date 1962

Nationality American

Why He's Key Whole disciplines of science are founded on the structural DNA work of Watson and his colleagues.

Key Event **Rise of the robots**
Unimate is the first industrial robot

Mention robots, and the image that springs to mind is of metallic men, often bent on destruction. The truth is rather more prosaic. The first industrial robot, Unimate, weighed 400 pounds and resembled a motorized arm.

Unimate was designed and built by inventors George Devol and James Engelberger. They set up the company Unimation in 1956 specifically to manufacture robots, but initially had a hard time convincing industry of the benefits of going robotic. It was only after the company was taken over by the Condec Corporation that things started to look up. Their first robot shipped in 1961; it could repeat a series of pre-programmed maneuvers.

The next year that robot was finally installed in a General Motors plant to lift and stack hot pieces from a die casting machine, but the robots really took off in the car manufacturing centers of Europe. The robots were used to weld together pieces of the automobile bodies, a task that was unpleasant for humans.

The great strength of Unimate was its ease of use and easy programmability. The robotic arm could be programmed in six axes of motion, and manipulate parts weighing up to 500 pounds, making it ideal for use on an assembly line.

As for the eventual rebellion of robots and their domination of humankind, George Devol has a simple philosophy: "If people are stupid enough to let that happen, then they deserve it."

Helen Potter

Date 1962

Country USA

Why It's Key Revolutionized industrial production and assembly lines, and created the science of robotics.

Key Publication ***Silent Spring***
Environmental call-to-arms over DDT

The year 1962 was a monumental year in the career of Rachel Carson, an American marine biologist and author of several books and articles on the natural world. It was in this year that her groundbreaking book, *Silent Spring*, was published. After four years of research and several rejections from magazines, who declined to publish her work on the subject, Carson was finally able to expose to the world the dangers of Dichloro-Diphenyl-Trichloroethane (DDT), a dangerous pesticide that had been in widespread use since the end of World War II.

Carson was one of only a few scientists who questioned the use of DDT when it first became available to the civilian population. Her concerns were highlighted when she received a letter from a friend living in Cape Cod, Massachusetts, describing the death of songbirds after the area was sprayed with the pesticide. The book describes in scientific detail the shocking effects DDT had on ecosystems, such as insects becoming resistant to it and therefore populations increasing, and its concentration in the water supply.

Silent Spring faced a number of foes before and after its release, including an attempt by the chemical industry to stop its publication. This did not prevent the impact it had on the public, however. One year after its release, President John F. Kennedy ordered a scientific advisory committee to be formed to look into the harmful effects of DDT. The chemical came under much scrutiny following this, and a federal ban was finally imposed in 1972.

Rebecca Hernandez

Publication *Silent Spring*

Date 1962

Author Rachel Carson

Nationality American

Why It's Key Carson's book exposed DDT's dangerous effects and increased environmental awareness of the general public.

Key Experiment **X-Ray astronomy**
Studying hot gas

On June 18, 1962, Riccardo Giacconi and his team at White Sands Missile Range launched an Aerobee sounding rocket that flew three Geiger counters above the atmosphere for almost six minutes. They detected one X-ray source, Scorpius X-1. It was the first time that anyone had proved that X-rays were emitted from celestial bodies, which came as quite a surprise to astronomers. We now know Scorpius X-1 is a neutron star, about 9,000 light years away, that is stealing material from a companion star of lower mass.

X-ray astronomy cannot be carried out from the ground because the Earth's atmosphere is an efficient absorber of X-rays. The first dedicated X-ray astronomy satellite, Uhuru, was launched in 1970. Uhuru found hundreds of X-ray sources like Scorpius X-1, but also discovered that supernova remnants, active galaxies, and the space between clusters of galaxies contain clouds of gas, which can reach 100 million degrees Celsius. X-rays have given astronomers another window on cosmic objects and have enabled them to study energetic environments that could not otherwise be observed. They have also allowed astronomers to carry out all sorts of complex calculations. For example, by measuring how X-rays from the Crab Nebula were blocked by Saturn's moon Titan, a team were able to calculate the width of Titan's atmosphere.

In 2002, a century after Willhelm Röntgen won the first ever Nobel Prize for discovering X-rays, Giacconi received the award for "pioneering contributions to astrophysics, which have led to the discovery of cosmic X-ray sources."

Eric Schulman

Date 1962

Scientist Riccardo Giacconi

Nationality Italian

Why It's Key Allows us to study neutron stars, black holes, supernovae and their remnants, and hot gas in interstellar and intergalactic media that cannot be observed from the ground.

Key Discovery
A new neutrino

By the early 1960s, scientists knew that some particles could spontaneously decay into something called a lepton; a particle which comes in one of three "flavors" – an electron, a muon, or a tau. At the same time as a lepton was produced, a neutrino – a particle with no detectable electric charge and virtually no mass, that moves almost at the speed of light – would appear. Quite what a neutrino was scientists hadn't yet decided, but they definitely seemed to partner the leptons. The question was, were there different types of neutrino in the same way as there were different types of lepton?

To answer this question, American physicists Leon Lederman, Melvin Schwartz, and Jack Steinberger took a supply of pions – particles which only break down into muons and neutrinos – and fired them at a 5, 000-ton steel wall. Made from the armor plating of old battleships, this wall was enough to block all particles except neutrinos from passing through.

Once on the other side, some of the millions of neutrinos reformed back into leptons. If the muon neutrino was the same as the electron neutrino, equal amounts of electrons and muons should be produced. Instead, all of the recorded particles were identified as muons. The conclusion: the neutrinos produced at the same time as muons are a distinct type of particle.

Understanding neutrinos is becoming increasingly important to physicists, as they have the potential to give us our clearest view yet into the heart of the Sun and the Milky Way.

Kate Oliver

Date 1962

Scientists Leon Lederman, Melvin Schwartz, Jack Steinberger

Nationality American, German-American

Why It's Key Not only was it an experimental triumph to detect a distinct species of the elusive neutrino, the "muon neutrino" fitted nicely into the family tree of subatomic particles.

opposite X-ray image of star cluster Trumpler 14

Key Event **Valentina Tereshkov-Nikolayeva is first woman in space**

On June 16, 1963, Russia's Valentina Tereshkova became the first woman to fly into space aboard the Vostok 6 mission, just two years after Yuri Gagarin had made man's maiden space flight.

Being an expert parachutist – she had made over 125 jumps before entering space-flight training school – helped in her selection and training as a cosmonaut in 1962. She was originally selected as the pilot for Vostok 6's twin, Vostok 5, with fellow female hopeful Valentina Ponomareva set to follow in Vostok 6, but a change of plan saw a male cosmonaut take over duties in the first of the two linked missions. Two days after Vostok 5 went into orbit, Vostok 6, piloted by Tereshkova, blasted off into space with the other female candidates looking on. During the mission, Tereshkova traveled round the Earth forty-eight times, and spent a total of nearly three days in space – still the longest time in space for any woman on a solo flight. On her successful return to Earth, she became a cosmonaut engineer and later went into politics, becoming a prominent member of the Soviet government.

She is much decorated, receiving the prestigious Hero of the Soviet Union – Russia's highest honorary – title in 1963. She was also voted the "greatest woman achiever of the century" award in London in 2000, by the Annual Women's Assembly. There is a moon crater named after her and Tereshkova now heads the Russian government's Center for International Scientific and Cultural Co-operation.

David Hall

Date 1963

Country USSR (now Kazhakhstan)

Why It's Key Voted the "greatest woman of the 20th century", the first woman in space reached new heights in terms of both exploration and equality.

Key Event **Staring into space** Arecibo Observatory opens

Arecibo Observatory is best known to film fans as the Bond villains' lair in *Goldeneye*, or the setting for the sci-fi thriller *Contact*, but to the scientific community it is the largest single-aperture telescope ever constructed.

Situated in a depression in Puerto Rico, the telescope has a collecting dish 305 meters in diameter. The moveable receiver weighs 75 tons and is suspended on a network of cables 137 meters over the dish. The telescope covers a 40 degree cone of the sky, enabling it to observe all the planets in the Solar System.

Opened in 1963, the observatory quickly made its mark as it collected data that showed the rotation rate of Mercury was fiftey-nine days, instead of the previously established eighty-eight. Arecibo also discovered the first evidence of neutron stars, the first binary pulsar, and the first extra-solar planets.

In 1974, the Arecibo message was beamed to the M13 cluster of stars 25, 000 light years away. The message was in binary, encoding an image that contained a picture of the telescope as well as a map of the Solar System and a picture of a human being.

The telescope is run by Cornell University and the U.S. National Astronomy and Ionosphere Center. Since 1999, Arecibo has been the base for the SETI@home project, collecting data that is analyzed by a network of home computers looking for evidence of extra-terrestrial communication.

Helen Potter

Date 1963

Country Puerto Rico

Why It's Key The opening of the largest single-aperture telescope on the planet allowed astronomers to search the galaxy for new phenomena and evidence of extra-terrestrial life.

opposite The Arecibo radio telescope

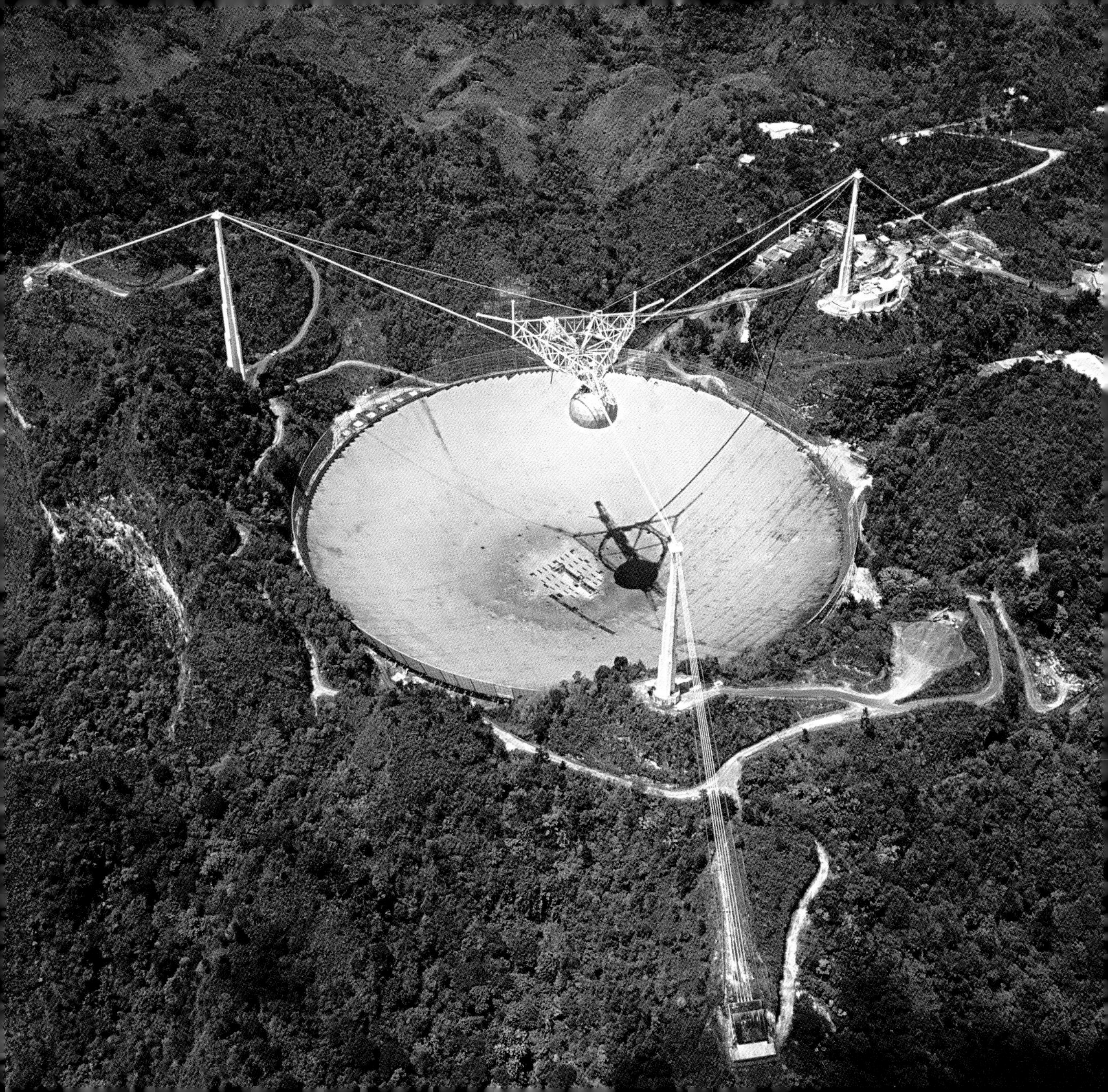

Key Discovery **Stem cells**
Be anything you want to be

During the height of the Cold War, two young men tried to figure out just how much radiation mice could take. What they found was one of the most important discoveries of the twentieth century.

Ernest McCulloch and James Till's research involved literally irradiating mice, and performing bone marrow transplants on them. The experiments were part of an effort to find ways of saving lives in the event of a nuclear war, but they ended up advancing medical science in a totally unforeseen way.

While studying some of their mouse bone marrow transplants, the researchers became aware of strange nodules within the animal's spleens, all of which appeared to have originated from a single cell. Moreover, the nodules appeared to harbor cells which were differentiating into each of the three major types of blood cells – red blood cells, white blood cells, and platelets. McCulloch and Till had found a type of blood cell that could differentiate into any other blood cell – the so called pluripotent stem cell.

While their findings revolutionized hematology, McCulloch and Till's lasting contribution to science is much broader. Although still effectively in its infancy, stem cell research has the potential to solve a great number of the diseases known to man. Already used in bone marrow transplants to treat leukemia, stem cells have been touted as potential cures for multiple sclerosis, Parkinson's disease, Alzheimer's disease, and even diabetes.

B. James McCallum

Date 1963

Scientists Ernest McCulloch, James Till

Nationality Canadian

Why It's Key The use of stem cells to treat disease has the very real possibility of eliminating many of the illnesses that plague society today.

Key Event **First vaccine for measles developed**
The journey to MMR begins

Famously described as being "as inevitable as death and taxes," measles was once the single most lethal infectious virus in the world. In the early 1960s, it was claiming 6 million lives every year. That was until 1963, when John Enders and Thomas Peebles developed the first effective vaccine for measles that would go on to save the lives of countless numbers of children worldwide.

In 1954, American-born Enders had already won the Nobel Prize for successfully growing polio virus in the laboratory. But it wasn't until he joined forces with Peebles at Harvard University that the pair managed to isolate the measles virus, in cells taken from human and monkey kidneys.

Soon afterwards, they managed to obtain a sample of the virus from the blood and throat washings of an eleven-year-old boy named David Edmonston, which they used to develop a measles vaccine. The Edmonston B strain, as it became known, was licensed in 1963 and was used widely until 1975.

As research into viral infections advanced, vaccines for two other devastating childhood diseases, mumps (1967), and rubella (1969) were licensed. In 1971, the decision was taken to combine measles, mumps, and rubella into a single vaccine – MMR. The MMR vaccine is now given routinely to all babies over twelve months old, and has reduced the reported cases of the three diseases by a staggering 99 per cent. Scientists are even hopeful that, like smallpox, measles could one day be banished to the realms of medical history.

Hannah Isom

Date 1963

Country USA

Why It's Key The measles vaccine has saved millions of lives across the world, and unlocked the door to the production of the combined measles, mumps, and rubella (MMR) vaccine.

Key Experiment **Producing proteins**
A chemical pep-talk

Robert Bruce Merrifield was a young biochemist at the Rockefeller University who, in the early 1960s, was attempting to manufacture chemicals called polypeptides using established methods. His problem was that the established methods took a long time and gave very low yields. In fact, the established methods were pretty rubbish. He began to think that there must be an easier way.

Polypeptides are made up of long chains in which lots of molecules called amino acids are joined by "peptide" bonds. These molecules are vitally important in nature, where they go by the pseudonym "proteins".

Merrifield eventually published a paper in 1963 in which he outlined his completely new method. It was simpler, easier, and, importantly, had the potential to be automated. By attaching the first amino acid in the chain to an insoluble base he could then add further links in the chain one by one until the required polypeptide had been produced.

By the mid 1960s Merrifield's lab had developed the methods to produce medically significant chemicals such as insulin, oxytocin, and bradykinin (which forms the basis of many anti-hypertension drugs). For the first time biochemists had the capability to produce important proteins on an industrial scale.

The technique wasn't just important for industry, however. When Merrifield and his colleague Berndt Gutte first synthesized the enzyme ribonuclease A, in 1969, they were able to show that the sequence of amino acids in the enzyme affected both the shape and biological activity of enzymes.

Simon Davies

Date 1963

Scientist Robert Merrifield

Nationality American

Why It's Key This method led to the automated mass production of proteins like insulin and also nucleic acids, and the synthesis of other complex organic molecules.

Key Invention
The mouse

What would computer users do without the mouse? It's a simple device, yet of such enormous use that it's barely given a second thought. Conceived in 1963, the potential of the computer mouse was not completely realized until the advent of graphical user interface-based computing two decades later.

The mouse began as the brainchild of Stanford electrical engineer Douglas Engelbart. Managed by his colleague Bill English, he designed equipment which could select objects on an interactive display workstation. Up until this point, light pens were mainly used to select objects. A variety of gadgets were tested, and out of the experiments came the mouse, a wooden shell composed of one button with a cord coming out of one end.

The real test of the mouse occurred in a demonstration performed in 1968 by Engelbart and his associates at the Fall Joint Computer Conference held in San Francisco. There, the group demonstrated their NLS (oNLine System), which consisted of a networked computer system including use of the mouse, object links, and videoconferencing. The system ran on a 192-kilobyte mainframe computer twenty-five miles away from the demonstration site. Engelbart's group received a standing ovation for their efforts, and the system was considered by many to be revolutionary.

The mouse received a patent in 1970 under the name "X-Y position indicator for a display system," and Engelbart is recognized today as one of the fathers of modern computing.

Rebecca Hernandez

Date 1963

Scientist Douglas Engelbart

Nationality American

Why It's Key The computer mouse is an indispensable element of the modern computing world.

Key Invention **Audio cassettes**
Music for the masses

The original tape recorders exploited the fact that a microphone could turn a sound into an electrical signal, which created a magnetic field. A magnetic tape passed through the field would record what the field was like, and running the process in reverse allowed playback. But early reel-to-reel tapes were bulky, temperamental, and the tape needed to be threaded in carefully by hand.

The cassette tape developed by Philips in the 1960s was half the size of previous magnetic recording tapes, but still used the same idea of tape coated in magnetic powder. Serious sound-lovers spurned it for its low fidelity, but for people who liked to be able to transport their music in a pocket, it was perfect.

Sales were initially slow, but thanks to Philips not putting royalties on the cassette design, and merely requiring a common standard, the market bloomed.

Many companies developed players and manufactured tapes. Philips were surprised by the demand for blank tapes by people wanting to record their own music, dictations, or even – shock horror – music from the radio. The British music industry was so worried by this turn of events that it launched a "home taping is killing music" campaign.

Luckily for us, music somehow managed to survive. Audio tapes have since been replaced by a succession of more compact and longer-lasting media. Although cassettes are rarely seen in Western countries nowadays, in countries like India and China, they are still the major music medium.

Kate Oliver

Date 1963

Scientist Philips

Nationality Dutch

Why It's Key Smaller cassettes made portable music systems possible, not least the Walkman and the mighty Boom Box. Home recording media put music recording in the hands of people rather than companies.

Key Discovery **Radio star**
The discovery of quasars

In the late 1950s, curious signals began to be picked up on Earth from space: highly focused beams of radio waves. Observers decided that the waves must originate from "QUAsi-stellAR radio sources," or quasars, but aside from the fact that they were like stars and emitted radio waves, they remained a mystery. That is until Maarten Schmidt, a Dutch astronomer, started analyzing the emissions, in 1963.

Schmidt noticed that the spectrum of light received from one of these quasars looked like that of a visible star, only shifted way back toward the red end of the light spectrum. A possible explanation for this was the Doppler effect, which causes light waves to change color if the source is traveling away from or toward you (in the same way that an ambulance siren changes pitch as it moves toward and away from you), but no-one had ever seen a light wave shifted this far before. It looked like the quasar was moving away from our observation point at an astonishing rate: 47,000 kilometers a second.

Given that we think everything in the Universe started off at the same place, Schmidt calculated that this quasar must be 1.5 billion light years away, meaning that what we see of the quasar is 1.5 billion years in the past. It takes an enormous amount of energy for emissions from so far away to reach us, and astronomers now think quasar radiation is produced by matter falling into black holes.

Kate Oliver

Date 1963

Scientist Maarten Schmidt

Nationality Dutch

Why It's Key Quasars mark the edge of our knowledge of the Universe, and provide us with a window onto what galaxies looked like over a billion years ago.

opposite A Hubble telescope image of quasar PG 0052+251 at the centre of a spiral galaxy

Key Discovery
Spreading stripes on the deep sea floor

The phenomenon of seafloor striping was first observed in the 1950s, when deep-sea explorers used World War II equipment to show that different strips of seafloor possessed different magnetic fields.

Neighbouring strips of sea floor appeared to have reversed magnetic -poles – while one strip would point to magnetic north, the field of its parallel neighbor would point to magnetic south. These reversals of polarity were explained by two British geophysicists, Frederick Vine and Drummond Matthews, as responses to switches in the Earth's magnetic field. They proposed that magnetite, a magnetic metal present in magma on the seafloor, aligns to the Earth's magnetic field. The magnetite then becomes fixed in this position as the magma cools to rock. As a reversal in the Earth's magnetic field occurs, so a matching reversal in magnetite alignment would be observed. But why was the rock forming in stripes?

The answer to this question had in fact already been provided by American geologist Harry Hess, in 1960. He noted that rocks of the same age were distributed at equal distances from both sides of the seafloor's deep ocean ridges. He then suggested that magma erupts periodically from these deep ocean ridges, spreading out on either side to form stripes of rock. Hess' theory was aptly named "seafloor spreading," and is now used alongside Vine and Matthews' idea to explain magnetic seafloor stripes.

Hannah Welham

Date 1963

Scientists Frederick Vine, Drummond Matthews

Nationality British

Why It's Key Confirmed Harry Hess' ideas about seafloor spreading.

Key Experiment **Obedience to authority**
This Might Shock You

"Sixty-five Percent in Test Blindly Obey Order to Inflict Pain." This was the *New York Times* headline after Stanley Milgram published his controversial work on obedience to authority in 1963.

Milgram, a psychology professor at Yale University, had just completed an ambitious project to test how humans respond to orders from authority. The contentious experiment involved scientists ordering research subjects to give high voltage shocks to a "learner" after they made mistakes on a word-matching test. Surprisingly, the majority of the participants continued to give shocks of up to 450 volts to their victim simply because they were told to do so by the experimenter. Later it was revealed that the "learner" was actually an actor who was simply pretending to be in pain, however the impact on the experimental subjects was powerful. Milgram's experimental methods were criticized by many, but others considered the work groundbreaking. Milgram's research showed that ordinary individuals, under only verbal command and no physical threat, could inflict pain on an innocent victim simply because they were told to do so.

Part of the drive behind Milgram's interest in obedience was his interest in the suffering of his fellow Jews during the Holocaust, and attempting to explain the behavior of the Nazis toward their victims. The results of his obedience experiments continue to make an impact on the field of psychology, and his research has since been dramatized in plays, movies, and music.

Rebecca Hernandez

Date 1963

Scientist Stanley Milgram

Nationality American

Why It's Key Showed that even ordinary citizens have the capacity to commit evil acts simply if they are instructed to do so.

Key Person **Maria Goeppert Mayer**
Coaxing the nucleus out of its shell

Maria Goeppert Mayer (1906–1972) was never going to be an under-acheiver. Born into a family of seven generations of university professors, her academic prowess would have come as no great surprise. But she surpassed all expectations, contributing greatly to the field of nuclear physics and bagging a Nobel Prize to boot.

Mayer began her university studies in physics at a time when quantum mechanics was new and exciting. Although her academic career was interrupted by the depression during the 1930s, which limited her research posts, she made many important contributions to the emerging field, despite often having to work for no money. In 1946, with only very limited knowledge of the area, she began working on a model of the atom. The Nobel Prize she received in 1963 was for her discovery and application of this model, which described how protons and neutrons are arranged into shells within the nucleus. They had previously been thought to be randomly colliding, but Mayer showed that discrete energy states existed. Her work also showed why isotopes of elements – forms of elements that have different numbers of neutrons in the nucleus – are less stable.

Goeppert Mayer's career is remembered yearly in the form of an award to a female scientist or engineer at an early stage in their career. Mayer often encouraged women to engage in scientific careers and follow in her distinguished scientific footsteps.

Nathan Dennison

Date 1963

Nationality German

Why She's Key She developed a new model of the atom and was one of the pioneers for women in scientific research.

Key Person **Jonas Salk**
Master medic

On October 28, 1914, Jonas Salk was born; forty years later, millions of people would have cause to celebrate that fact. The son of Russian-Jewish immigrant parents, Salk would be the first of his family to enter college. His innate inquisitiveness and intellect meant he would soon be asking questions; the answers to many of which he would ultimately challenge and redefine. The world would be grateful for this when, in the 1950s, he created an effective polio vaccine.

A brilliant scholar who graduated from high school at fifteen, Salk had initially intended to follow a legal career and enrolled as a pre-law student at college. After some soul searching, however, he changed his focus to medical science. His mother had hoped he would become a teacher, but such a path held little interest for a man whose default mode was to question the perceived wisdom of others, rather than reinforce it.

Something of a loner (attributed in part, by Salk himself, to an over-protective mother), he continued to enjoy his own company through his later years. His sense of curiosity and wonderment would never leave him, however. He wrote: "Hope lies in dreams and in the courage of those who dare to make dreams a reality."

In 1963, Salk founded the Jonas Salk Institute for Biological Studies, a center for medical and scientific research. His final years were spent searching for a vaccine against AIDS. He died on June 23, 1995, of heart failure, aged eighty.

Mike Davis

Date 1963

Nationality American

Why He's Key Salk's research and developments have directly and indirectly saved literally millions of lives. His altruistic, humanist philosophy lives on as an example to scientists and researchers everywhere.

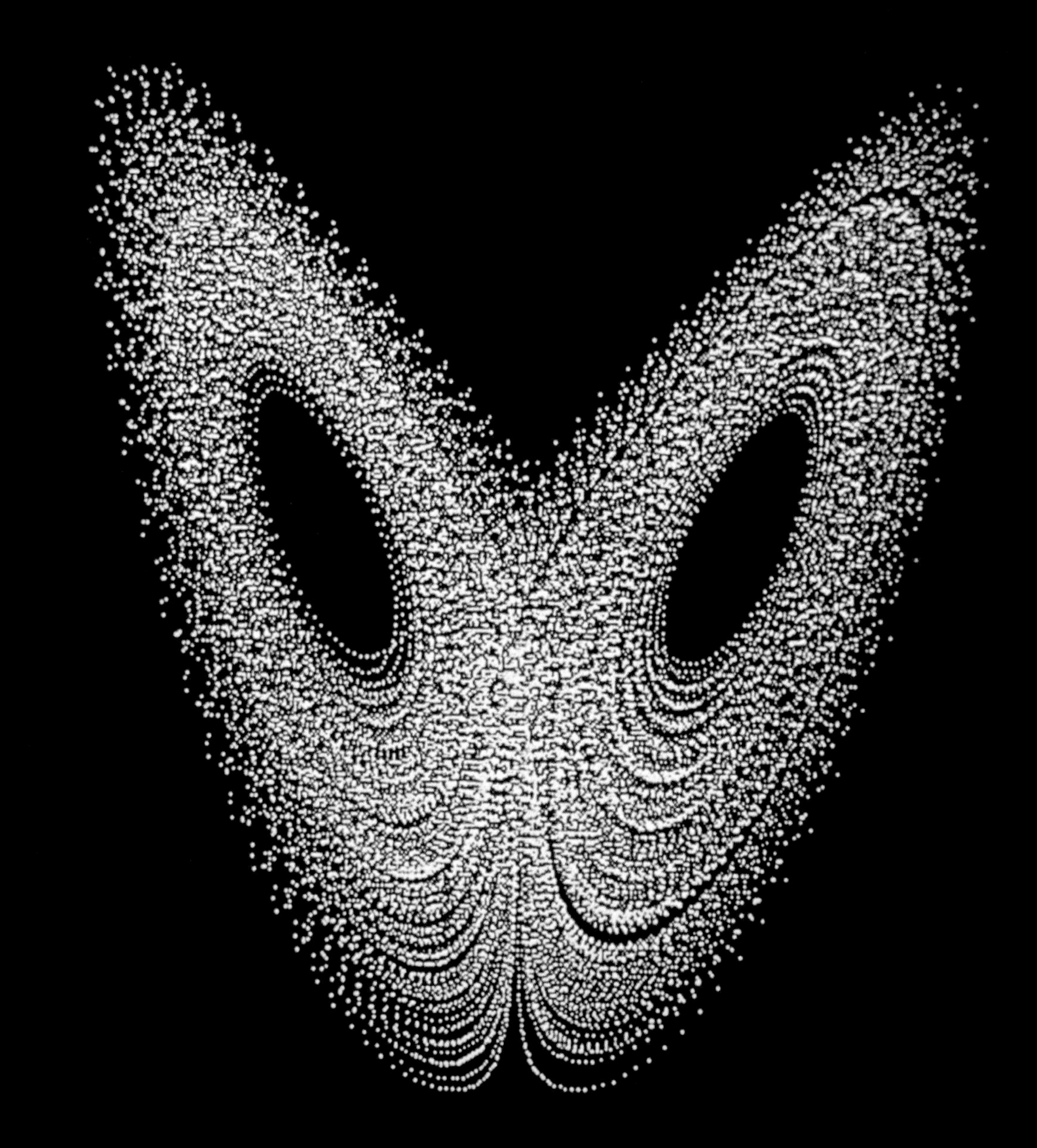

Key Discovery
The butterfly effect

It's virtually impossible to predict what the weather will be like in one month's time. But the weather is not completely random because it demonstrates recurrent behavior – average temperatures are always higher in summer than winter, for example. So why is long-term weather prediction so difficult? In the early 1960s Edward Lorenz was working on weather prediction when he stumbled across the answer – chaos. Lorenz was modeling weather patterns on a computer, using numbers to represent the contributing factors. The computer produced line graphs that enabled him to predict the weather accurately for the next day.

One day, he decided to run his computer experiment twice with the same values for each of the factors. Over short time periods the lines on the graphs were identical, but as the time scale of the predictions got longer they began to diverge, and ended up completely different. He soon worked out what had happened – the initial experiment had used data from the computer to six decimal places, but when he typed in the data from the printout, it was only to three.

It was strange that such a tiny difference in the initial values could lead to such a big difference in outcome. Lorenz famously suggested that something as small as a butterfly flapping its wings in Beijing could affect the weather days later and thousands of miles away. He termed this the butterfly effect and it explains why it is impossible to predict the weather long into the future.

Riaz Bhunnoo

Date 1963

Scientist Edward Lorenz

Nationality American

Why It's Key The butterfly effect neatly describes chaos theory. The fact that an outcome can be so sensitive to small differences in initial values explains why long-term weather prediction is impossible.

opposite The Lorenz Attractor, a three dimensional graph produced by chaos mathematics

Key Invention
Home dialysis machine

For patients with kidney failure, hemodialysis (dialysis for short) is literally a lifesaver. The kidneys play a vital role in keeping the balance of water and minerals in our blood at the right levels. When these organs fail, people rely on dialysis or a kidney transplant to keep them alive. During dialysis, the patient's blood is passed through a membrane surrounded by a tank full of dialyzing solution, allowing toxins to move out of the blood and into the solution by the process of osmosis.

Patients normally need to dialyze three times a week – it's a severely disruptive treatment. To try and minimize the inconvenience and stress of dialysis, home systems were developed soon after the first hospital-based machine came into use. But who developed home haemodialysis? Claims that a Japanese researcher was the first to describe the use of hemodialysis outside of the hospital setting in 1961 have been strongly contested by some. What is more certain though is that three groups from Boston, Seattle, and London worked independently on the system, reporting their experiences at a meeting in Seattle in late 1964.

Despite the cost-effectiveness and convenience of home hemodialysis, its use has been falling in many countries over recent years. The reasons for this decline are thought to include the introduction of an alternative form of dialysis – continuous ambulatory peritoneal dialysis – that can be done at home without a machine, as well as an increase in live-donor kidney transplants.

Christina Giles

Date 1964

Scientists Constantine Hampers, F. K. Curtis, Stanley Shaldon

Nationality American, British

Why It's Key Brought life-saving treatment into patients' homes, giving them greater independence and freedom.

Key Publication **Hamilton's genetic evolution of social behavior**

Royal Society Fellow William Donald Hamilton – Bill to the people who knew him – was one of the greatest thinkers in evolutionary biology since Darwin himself. His two papers on the genetic evolution of social behavior, published in *The Journal of Theoretical Biology* in 1964, began what is sometimes referred to as the "second Darwinian Revolution."

These two publications are now seen as ground-breaking, but at the time, and for years afterwards, they were given little attention. Through reference to his work by countless other articles, Hamilton's idea for a genetic basis to social behavior has grown in acceptance and is now a standard topic of study for undergraduate biology students. His contribution to evolutionary biology has been recognized with many awards, most notably the 1993 Crafoord Prize, which is comparable to a Nobel, as Nobel Prizes are only awarded for medical biology.

The theory, as proposed in these two papers, shows how altruistic, unselfish, behavior can be beneficial in social groups in which individuals are related to one another. Unselfish behavior can have a genetic origin, as altruistic acts can help the survival of the altruistic individual's genetic material. If, for example, an animal shares food with a starving sibling, the cost to the individual is to lose some food, but if this stops his sibling – who shares 50 per cent of his genetic material – starving to death, the act will improve the chances of survival of the altruistic gene.

Jim Bell

Publication *The Journal of Theoretical Biology*

Date 1964

Author William Donald Hamilton

Nationality British

Why It's Key The basis for massive steps forward in our understanding of evolutionary genetics, and a realization of how much of our lives are controlled by our genetic makeup.

Key Discovery **"Terrible Claw" links dinosaurs to birds**

Jurassic Park popularized a branch of dinosaurs known as Dromaeosaurs in the form of Velociraptors. In the film, these terrifying lizards introduce the idea of a previously unknown threat, every bit as frightening as Tyrannosaurus Rex, because of the fact that these "raptors" were intelligent.

Deinonychus antirrhopus, the original "Terrible Claw," is one of these same Dromaeosaurs – the first to be discovered in fact. Measuring three meters long and weighing as much as a full-grown man, this dinosaur combined speed, intelligence, and killing claws, to make it a deadly efficient killer. Paleontologist John Ostrom's discovery of *Deinonychus*, in 1964, represented a very significant find in modern paleontology, radically changing our perception of dinosaurs. Previously these had been considered to be slow, lumbering beasts, with physiology similar to lizards, requiring heat from the sun to become active.

Ostrom suggested an alternative. Based on his observations of the specialized adaptations in *Deinonychus*, Ostrom proposed that dinosaurs were actually much closer in design to birds: fast, active, agile predators. And, he speculated, they would probably have been at least partially warm blooded. His revelations reinvigorated long-running debates about dinosaurs, seemingly confirming an idea originally suggested a century earlier – that birds are direct descendents of the dinosaurs. The debate raged on until the early 1990s, when the first example of a non-flying, feathered dinosaur fossil was found and Ostrom's theory was fully vindicated.

Barney Grenfell

Date 1964

Scientist John Ostrom

Nationality American

Why It's Key Ostrom's discovery of *Deinonychus*, arguably the most important dinosaur fossil ever, changed people's perception of what dinosaurs were and heralded a new "dinosaur renaissance" in the 1970s.

opposite Model of a *Deinonychus antirrhopus*, being checked before an exhibition in Tokyo

Key Discovery
The Higgs Boson

In 1964, British physicist Peter Higgs went out for a walk in the Cairngorms, a rocky, mountainous plateau in the east Scottish Highlands. Higgs returned from the wilderness with a truly spectacular idea – one that proposed an explanation for why subatomic particles, and in fact all matter, have mass.

Theoretical physicists have been hard at work since the early part of the twentieth century constructing what is called the "Standard Model" of fundamental particles. The electron is the first known particle that the Standard Model states is non-divisible; it is not comprised of smaller components, and hence is fundamental. While the Standard Model has been successful in describing and observing all the particles of which matter is made – as well as the particles which carry the forces we observe in nature – it does not explain why these particles should have mass.

Higgs proposed that just after the Big Bang, particles had no mass. As the temperature dropped and the Universe cooled, an energy field – "The Higgs Field" – formed, as well as the Higgs boson, the so-called "God Particle" which gives the Universe its mass. The field is theorized to permeate the entire Universe, and particles that interact with it acquire mass.

At the time of writing, there have been no confirmed observations of the Higgs boson. However, the Large Hadron Collider (LHC) – a 26-kilometer ring shaped particle accelerator, being built at CERN in Geneva and scheduled to be completed in July 2008, is thought to have the capability to reveal the elusive particle.

Andrey Kobilnyk

Date 1964

Scientist Peter Higgs

Nationality British

Why It's Key The Higgs boson could not only explain the nature of mass, but also be a valuable contribution towards the validation of the Standard Model of fundamental particles and theories of the future.

Key Event **Faster than a speeding bullet train**
High-speed rail transportation appears

Japan Railways' Tokaido "Shinkansen" service, connecting Tokyo with Nagoya, Kyoto, and Osaka, was inaugurated in 1964 as the world's first truly high-speed train line. At that time, the trains operated at speeds of around 200 kilometers per hour; however, with improved technologies, modern speeds have risen to over 300 kilometers per hour, with some test trains even having exceeded the 400 kilometers per hour mark.

The railway line ran parallel to the renowned Tokaido Road that had linked Tokyo and Kyoto in Samurai times. It would subsequently be extended beyond Osaka to reach Hiroshima and Fukuoka. An incredible example of technological development, and popularly known outside Japan as the "Bullet Train" service, the name "Shinkansen" actually means nothing more glamorous than "new trunk line." It would be 1981 before Europe saw its own high-speed rail service, in the form of the French "TGV" (Train a Grande Vitesse). Also running on specially designed and constructed high-speed rail tracks, the idea for these very efficient and environmentally preferable transport systems originated as an alternative to local flights. When the first TGV line from Paris to Lyon opened in 1981, it was an immediate commercial success. The TGV service now operates across most of France and other parts of Europe.

Persistent breakthroughs in related technology continue to further the development of high-speed trains; the future can hold only faster, more efficient, and more comfortable examples.

Mike Davis

Date 1964

Country Japan

Why It's Key Bullet and TGV services have revolutionized rail travel on their respective continents; such technology offers a viable, and environmentally friendlier, alternative to flying between many locations.

opposite The Bullet train running in Kyoto

12
K5

Key Discovery **CP violation**
An imbalance of matter and antimatter

The Big Bang should have created equal amounts of so-called "normal" matter and antimatter – matter that seems to be made of normal particles but has the opposite charge. But when matter and antimatter collide, they destroy each other, so how come the Universe exists at all? An obscure effect called CP violation seems to hold the answer. Now hold on to your hats, the next paragraph is going to get intense.

In 1964, the American nuclear physicists James Cronin and Val Fitch discovered that certain particles, called neutral kaons, did not obey CP conservation – whereby equal numbers of particles and antiparticles are formed. It turned out that any particles which interact via the weak nuclear force (one of the four fundamental forces in nature) could violate CP conservation. Since these interactions were present at the Big Bang, an unequal mix of matter and antimatter was created. Most scientists now agree that CP violation is the effect that caused the Universe as we see it today to be composed mainly of matter, rather than equal parts of matter and antimatter. The antimatter that was created initially was annihilated by some of the abundant matter.

Particle physics often seems somewhat removed from reality – physicists with huge particle accelerators crashing particles into one another to create others that exist for miniscule lengths of time. The discovery of CP violation, however, is a good example of particle physics giving a deeper understanding of one of the fundamental questions in science – how was the Universe created?

Leila Slattary

Date 1964

Scientists James Cronin, Val Fitch

Nationality American

Why It's Key The effect helps understand why our universe is made up of matter and what happened at the Big Bang.

Key Invention **BASIC principles of computer programming**

When John Kemeny and Eugene Kurtz invented the BASIC computer language they opened computing up to a vast number of people. In the early 1960s students learning about computing had to write their programs on punched cards and then take them to the nearest laboratory that actually had a computer. For Kemeny and Kurtz this was a 135-mile train ride, and the results of their programs would sometimes take days to arrive back at Dartmouth University.

Between 1963 and 1964 they designed and created the Beginner's All-Purpose Symbolic Instruction Code – BASIC – to allow students to learn about computing without the need for train journeys. At the time, most computers were outrageously expensive set ups designed for special tasks, with their programs written by mathematicians and scientists. BASIC was designed with eight principles: easy for beginners; a general-purpose language; adaptable for the addition of advanced features; interactive; able to give clear error messages in English; able to respond quickly for small programs; usable without an understanding of computer hardware; and shield the user from the operating system.

These features meant any computer could be used for teaching students without any risk to the fundamental operating system. By the 1970s, BASIC and its variations had become universal. Bill Gates and Paul Allen, co-founders of Microsoft, released Altair BASIC in 1975 and a myriad of companies began making and selling machines that could run BASIC. The home computer industry was born.

David Hawksett

Date 1964

Scientists John Kemeny, Thomas Eugene Kurtz

Nationality Hungarian-American, American

Why It's Key BASIC became the language that gave most users their first taste of computer programming.

Key Event **Surgeon General reports that smoking is a health risk**

The Scottish philosopher Thomas Carlyle is generally considered the architect of the "Great Man Theory" of history, which states that great individuals shape the history of the world through their vision. If that is the case, then Luther Leonidas Terry may end up being lung cancer's worst enemy. The former Surgeon General of the United States, born in Red Level, Alabama in 1911, made his most famous contribution to society in the mid 1960s.

As evidence that tobacco was hazardous to health began to mount, Terry took the important step of appointing a panel to investigate the phenomenon. The result was a 1964 report that concluded that both lung cancer and chronic bronchitis were caused by smoking, and furthermore, that emphysema and heart disease, among others, may be related as well. The report generated a dramatic amount of publicity about the hazards of smoking. It ultimately resulted in health warnings being printed on packets of cigarettes, and prevented countless cases of disease and death.

To this day the American Cancer Society still presents the Luther L. Terry Award for exemplary leadership in tobacco control. Past winners have included a wide range of individuals and groups including The Ministry of Health and Family Welfare, Government of India; the Director of the WHO, Gro Harlem Brundtland; and the Non-smokers Rights Association of Canada.

B. James McCallum

Date 1964

Country USA

Why It's Key The first report with clear evidence that tobacco is a health risk. It changed consumer behavior overnight and, in the process, saved thousands of lives.

Key Publication **"Bell's theorem"** Sorting spooky actions

"God does not play dice with the Universe" was how Einstein summed up his objections to the seeming randomness of nature predicted by quantum mechanics. He was unhappy with the notion that if the position of a particle was known, its momentum was said to be completely random and unknowable.

Others sided with Einstein and together they outlined a thought experiment, described by imagining a subatomic particle, called a pion, as a football with no spin. The football decays into two photons, or golfballs. The golfballs are also spinning but, as they came from the same football they are "entangled" and their spins must add up to zero and cancel each other out. You can measure the spin of one golfball, but not its momentum, because quantum mechanics forbids you from knowing both. But if the spins total zero, surely you could then work out the spin of the other ball, while also being able to measure its momentum? In such a case you would know both values and, if this were possible, then quantum mechanics was missing some key components.

This paradox remained for twenty-nine years until physicist John Bell published his famous paper, "On the Einstein, Podolsky, and Rosen Paradox" in 1964. In it he devised a test for these missing values in quantum mechanics. He proved that that if quantum mechanics relied upon hidden properties of particles, then their effects would be non-local, and that entangled sets of particles could exchange information instantly in what is known as "spooky action at a distance."

David Hawksett

Publication "On the Einstein, Podolsky, and Rosen Paradox"

Date 1964, *Physics*

Author John Bell

Nationality British

Why It's Key Bell's theorem of inequality has been called "the most profound in science" and removed what had been a spanner in the works for quantum physics.

Key Experiment
Scientists prove DNA-protein link

In 1941, two scientists proposed a radical idea to explain the production of proteins from genes. DNA had long been dismissed as being incapable of encoding the many proteins within the body. Nevertheless, George Beadle and Edward Tatum provided an exciting new hypothesis – that one gene encoded one protein. However, at the time there was no clear evidence to support this hypothesis, and knowledge of protein structure was too limited.

By the 1950s, however, technology and knowledge had advanced greatly, culminating in the famous discovery of the double-helix structure of DNA by Watson and Crick, in 1953. Scientists realized that DNA, with nucleotide units of four different varieties, could create the massive array of combinations required to code for many complex proteins. But how?

A decade later, biologists Charles Yanofsky and Sydney Brenner were experimenting with the bacterium *E.coli* and found that particular gene mutation caused a defect in the protein it encoded. By altering specific regions of this gene, they discovered specific defects in the sequence of amino acids which make up the protein. They showed that the linear sequence of DNA chemicals in the gene corresponded to the linear sequence of amino acids in the protein – a phenomenon known as known as co-linearity. Their paper was published in 1964 and heralded a breakthrough in genetics – the elusive link between DNA and proteins had finally been found.

Steve Robinson

Date 1964

Scientists Charles Yanofsky, Sydney Brenner

Nationality American, South African

Why It's Key Yanofsky and Brenner's experiment provided a massive step toward fully understanding the genetic code and how it works to produce proteins – a cornerstone of life.

opposite Sydney Bremmer lecturing on DNA structure at the University of Cambridge

Key Person **Charles Hard Townes**
Making lasers and masers

Charles Hard Townes (b. 1915) was attracted to physics by what he called its "beautifully logical structure." Born in South Carolina, he achieved a bachelor's degree in physics by the age of nineteen, as well as one in modern languages. A year later, he received his master's degree before studying nuclear physics at the California Institute of Technology.

The year World War II broke out, Townes joined Bell Labs, where he helped develop a new radar system for U.S. military aircraft. He returned to academia in 1948 with a post at Columbia University, where he was partially funded by the military. Their desire for better radars, operating at smaller wavelengths, led him to conceive the "maser."

Microwave Amplification by Stimulated Emission of Radiation was a technique to produce coherent radiation – where photons correlate with one another, as opposed to being randomly distributed. He had wanted to obtain strong radiation with very short wavelengths using a source other than a vacuum tube, and followed work done by Einstein in 1916 which showed that atoms could emit radiation when stimulated by radiation. The device Townes built contained ammonia molecules, which he illuminated with weak microwave radiation. Masers were soon put into practical use, for example in atomic clocks, due to their extremely precise nature of frequency. Townes then went on to show that it would be possible to achieve similar radiation "coherence" using visible light instead of microwaves, and this eventually led to the invention of the laser.

David Hawksett

Date 1964

Nationality American

Why He's Key For his work on masers and lasers, Townes was awarded the Nobel Prize for Physics in 1964.

TCC
AGG

СССР

Key Discovery **Taxol**
A cure for cancer

Outwardly, the Pacific yew tree is a spectacularly unremarkable tree. Hiding out in the understory of the north-western forests of the United States, it's not very tall, it doesn't provide any edible fruits, and the wood doesn't even burn very well. Is there anything, you might ask, interesting about it at all? Well of course there is.

In the early 1960s a chemist, Jonathan Hartwell, working at the U.S. National Cancer Institute organized for thousands of plant samples from the United States to be tested for potential anti-cancer drugs. In 1964, amongst the myriad botanical non-starters, one sample stood out; something in the bark of the Pacific yew was killing cancer cells very effectively. The compound was eventually isolated and named Taxol.

The drug soon went into large scale production, being used to treat various forms of lung, ovarian, breast, head, and neck cancer. There was only one problem. The bark from a single tree only yielded enough Taxol for about one dose of the drug, and harvesting it killed the tree. The Pacific yew had become a victim of its own success, especially since initial attempts to artificially synthesise Taxol proved fruitless.

It wasn't until the late 1980s that biochemist Robert Houlton developed a method to turn a similar, but much more abundant, chemical from the common yew into Taxol. For environmentalists worried about the effect the drug collection was having on the yew, the discovery came not a moment too soon.

Mark Steer

Date 1964

Scientists Jonathan Hartwell, Monroe Wall, Mansukh Wani and others

Nationality American, Indian

Why It's Key In the battle against cancer, Taxol is an important drug, and big business; in 2000, sales in the United States topped US$1.6 billion.

Key Event
Leonov walks in space

Alexei Leonov was selected to become one of the Soviet Union's first group of cosmonauts in 1959, along with Yuri Gagarin. After Gagarin's historic first space flight in 1961, the space race between the Soviets and Americans was truly underway, and each superpower began flying more and more elaborate and technically complex flights.

Voskhod 2 was to be no exception. It was the Soviet's eighth manned flight and Leonov was to attempt to leave the capsule on a tether and float in space, just in his space suit, while his fellow cosmonaut Pavel Belyayev remained inside. Ninety minutes into the mission, Leonov extended the inflatable airlock and ventured outside. Ten minutes into his space walk, Leonov encountered a problem. His space suit had ballooned and stiffened in the vacuum of space, and he was unable to fit back inside the airlock. There was no contingency for rescue and, if Leonov could not overcome this problem, his fellow cosmonaut would have to cut him loose and return to Earth without him.

Leonov decided on a risky move. He partially deflated his space suit by allowing air to bleed into space, and was eventually able to squeeze back inside the capsule safely. This was not the end of the mission's problems, however. Their re-entry systems failed and Leonov and Belyayev had to fire their retrorocket manually. They landed 1,000 kilometers away from the planned site – in forest shrouded in deep snow – and had to spend two nights surrounded by wolves before being rescued.

David Hawksett

Date 1965

Country USSR (now Kazakhstan)

Why It's Key Leonov's space walk beat the Americans' first space walk by three months.

opposite Alexei Leonov performing the historic first space walk

Key Invention **The portable defibrillator**
Heart thrills on wheels

Defibrillation or "shocking" the heart with a therapeutic dose of electrical energy is sometimes the only successful treatment for life-threatening changes in heart rhythm. In 1965, Irish cardiologist James Francis "Frank" Pantridge set pulses racing when he created the first portable defibrillator.

Pantridge had previously been key to establishing the principles of cardiopulmonary resuscitation (CPR) for cardiac arrest in the 1950s. He realized that unless patients were in hospital when their heart stopped beating, they had a very poor chance of survival.

His original 70-kilogram prototype could barely be described as portable, but his defibrillator-equipped ambulances became a welcome addition to Belfast's streets in 1965. Pantridge's invention proved invaluable in the treatment of out-of-hospital cardiac arrests, and his treatment protocols – the "Pantridge Plan" – revolutionized emergency medical treatment worldwide.

Pantridge continued to refine his design, and today's compact automated external defibrillators (AEDs) are a far cry from the hefty car-battery-powered prototypes of the 1960s. These compact, sophisticated machines can both analyze and shock dangerous heart rhythms with minimal human assistance, and are commonplace in many public areas. Increasing accessibility to defibrillators has made a real difference to survival rates. Research has shown that in cardiac arrest the chance of survival drops by 10 per cent for every single minute without treatment, so time really is of the essence.

Katie Giles

Date 1965

Scientist Frank Pantridge

Nationality British

Why It's Key Changed management of out-of-hospital cardiac arrests and saved thousands of lives.

Key Discovery **Background radiation**
The birth of the Universe, or pigeon droppings?

In 1963, astrophysicists Arno Penzias and Robert Wilson made use of a six-meter radio antenna that was no longer needed for satellite tracking. The disused antenna housed a flock of pigeons. By mid-1964 the two physicists' experiments were being plagued by an annoying hiss that seemed to have no particular source. They blamed the pigeons, and decided to have them trapped and relocated. When that didn't work, they had them shot. But the noise remained. It was later found to be long-wavelength, low-temperature electromagnetic radiation from the beginning of the Universe.

The Big Bang Theory was first proposed in 1927 by the Belgian Catholic priest Georges Lemaître, and later championed by cosmologists George Gamow, Ralph Alpher, and Robert Herman, who predicted, in the late 1940s, that there should be radiation left over from the Big Bang itself. It was just unfortunate that Penzias and Wilson didn't come across the theory until after the deaths of those ill-fated pigeons.

Still confused about what was causing the noise, the two physicists rang a colleague, Robert Dicke, to see if he had any suggestions as to its source. Dicke knew immediately what it was they were describing – after all, he was actively searching for Gamow's theoretical radiation. "Well boys," he said as he turned to his own co-workers, "we've been scooped." Without realizing it, Penzias and Wilson had found the oldest echoes in the Universe.

Stuart M. Smith

Date 1965

Scientists Arno Penzias, Robert Wilson

Nationality American

Why It's Key By studying background radiation we can peer 13 billion years into the past at events that occurred at the very beginning of the Universe.

opposite Penzias and Wilson with the previously disused radio antenna

Key Person **Timothy Leary**
Turn on, tune in, drop out

Timothy Leary (1920–1996) was an innovative professor of psychology at Harvard University. In encouraging people to think for themselves and question authority, he pioneered group therapy to empower individuals to treat "hir" (Leary's gender-neutral term) own condition.

In the 1950's, Leary read an article in *Life* magazine on the hallucinogenic experience of a vice-president of J.P. Morgan. He was excited at the possibilities for exploration of human consciousness, and rushed to Mexico to try hallucinogenic mushrooms. He continued researching psychedelic substances to probe consciousness on himself and others (before they became illegal in the United States). Leary believed that, if used correctly, they could combat self-destructive behaviors, thoughts, and personality disorders. Unsurprisingly, he was dismissed from Harvard. He argued against criminalization of psychedelics and called for the abandonment of monotheism. President Nixon called him "the most dangerous man in America." He was arrested for marijuana possession in 1965, but escaped from jail, and the country, helped by the radical group Weather Underground. However, he was later recaptured.

In his exploration of the mind, Leary proposed the "eightfold model of human consciousness," which claimed new environmental signals such as cyberspace have caused an evolution of the human brain. Timothy Leary died from prostate cancer in 1996 at the age of seventy-six; his final project was to chronicle his slow death via the Internet.

Umia Kukathasan

Date 1965

Nationality American

Why He's Key Leary was a part of the beat generation, in a time when many questioned authority. Although most (in)famous for advocating drugs, Leary also pioneered group therapy.

Key Invention **Kevlar**
For super-safe soldiers

Some say the best form of attack is defense. With that in mind, Kevlar is crucial to the protection of our modern-day heroes, from riot squads to front-line soldiers. Like the plate metal armor of medieval knights, Kevlar can really take a hit. Luckily, unlike Sir Lancelot's heavy armor, our warriors' protection is a lot less clunky and cumbersome.

Since its creation in 1965, by American chemist Stephanie Kwolek, Kevlar has been put to all manner of different uses. It is a high strength, low weight material, made from a chemical known as a polyamide, which is five times stronger than steel. It is also non-flammable, chemically resistant, and highly durable.

As a fiber, it functions as a lightweight reinforcement in ropes, cables, protective gloves, and is used to strengthen tires and hoses. Kevlar threads can also be used to increase the effectiveness of fire-resistant items such as mattresses, and are an integral part of fire-fighter uniforms owing to their ability to withstand scorching temperatures. Talk about hot stuff!

More poignantly, it is used in the armored vehicles, bulletproof vests, and helmets that protect our world's policeman, security, armies – you name it. This is possible because of the revolutionary high strength and lightweight properties of Kevlar.

Nicola Currie

Date 1965

Scientist Stephanie Kwolek

Nationality American

Why It's Key Not only is Kevlar a multi-purpose material used to fortify a variety of products, this champion of the chemical world also continues to save lives through its use in protective clothing and headgear.

opposite Kevlar is hit by a bullet traveling at 220 meters per second

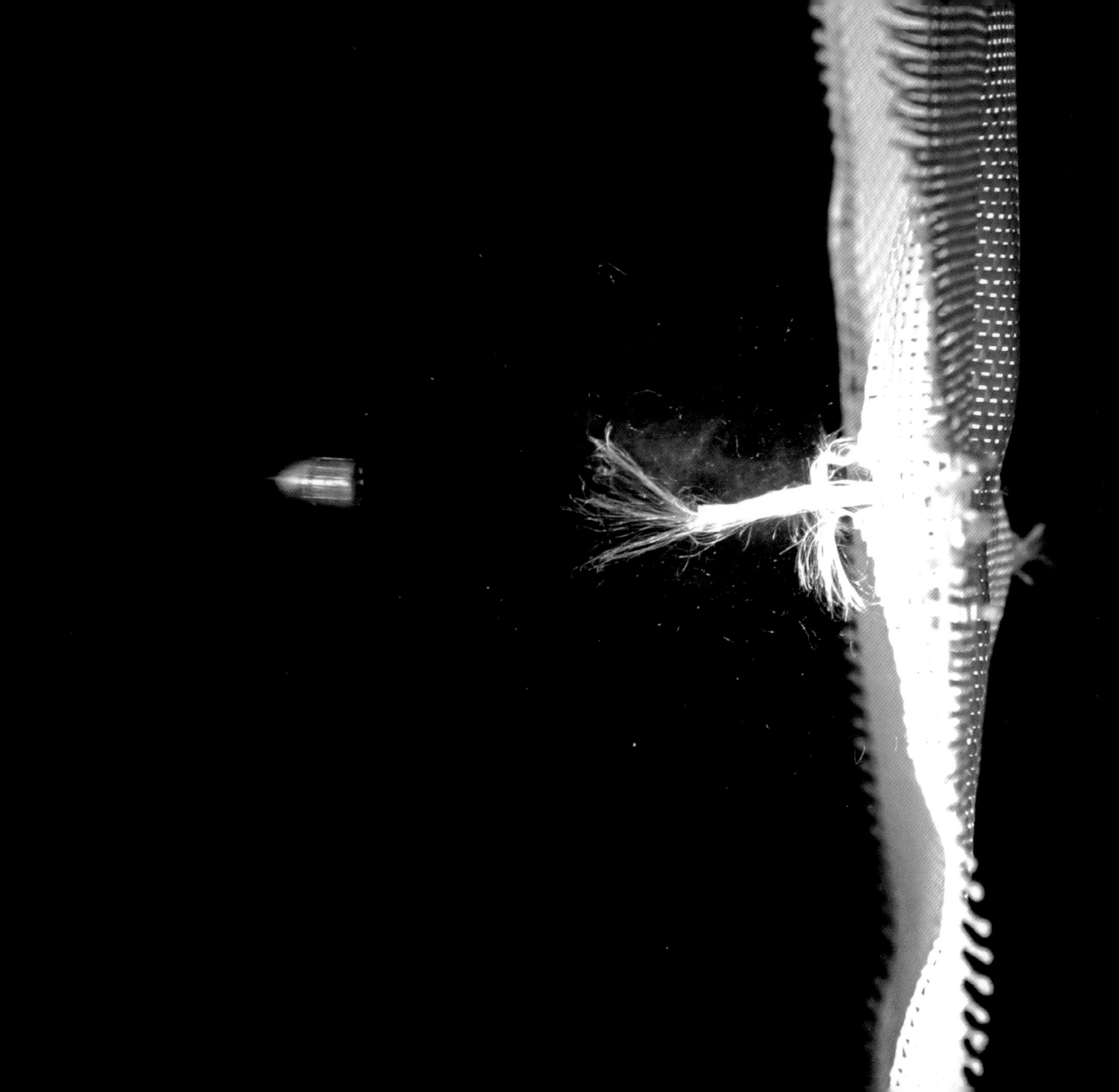

Key Invention **Easier on the eyes** Soft contact lenses

Leonardo da Vinci sketched out his ideas for lenses to be worn directly on the eyes way back in 1508, but it took until the early 1800s before the first lenses were in use, and these were far from comfortable. Made from glass, these early contact lenses put brave wearers at risk of splinters, and caused discomfort on a daily basis. It took until the early 1900s for plastic versions to take their place. However, there was still room for improvement, as they caused a certain amount of irritation to the wearer over relatively short periods of use.

Enter Czechoslovakian chemist Otto Wichterle and his assistant Drahoslav Lim, who spent the 1950s experimenting with a plastic invented by Lim in 1945. Hydroxyethyl methacrylate was soft and water-absorbent, making it an ideal material from which to produce contact lenses. Using only a few spare bicycle parts, Wichterle created a unique spin-casting machine which he used to produce the world's first soft contact lenses. Seeing huge commercial potential for his work, Wichterle sold his product to Bausch & Lomb, who cashed in, producing the first lenses for the world market in 1971.

Wichterle and Lim revolutionized the world of contact lenses, making spectacle-free life a reality for thousands who were previously unable to wear hard lenses, and increasing the comfort vastly for those who could. In America today, over 90 per cent of contact lens wearers use soft lenses.

Katherine Ball

Date 1965

Scientist Otto Wichterle, Drahoslav Lim

Nationality Czech

Why It's Key Enhanced the lives of lens wearers through increased safety and greatly improved comfort.

Key Event **First photos of Mars taken by Mariner 4 probe**

The first ever pictures of another planet in our Solar System were taken by Mariner 4, a space probe designed to fly close to Mars. It passed over the planet at an altitude of 9,846 kilometers on July 15, 1965, and subsequently sent twenty-one pictures to Earth, which took more than a week to transmit.

Mariner 4 was identical to Mariner 3, which never reached its destination because of a fault after launching. Three weeks after Mariner 3 failed, however, Mariner 4 was successfully launched. Its useful life exceeded expectations; it was anticipated that the craft would last eight months, but it continued working for nearly three years. In July 1967, it swung back close enough to Earth to allow engineers to conduct experiments with it that had not been planned originally because of the unexpected longevity of the probe. It stopped operating in December of that year.

The images taken finally disproved the existence of canals, the construction of which was thought to have been carried out by intelligent beings; showing instead that they were simply random geological configurations. But the images did confirm that some of the surface photographed was covered in a red dust, and cavities and peaks similar to those on the Moon, and assumed to have been made by the impact of meteors. Later images showed that these were not typical of the whole surface.

Other experiments carried out by Mariner 4 enabled scientists to build up a picture of the planet's composition.

Fabian Acker

Date 1965

Country USA

Why It's Key The data obtained from this probe shaped the nature of future programs.

opposite The Mariner 4 launch

Key Person **Ronald David Laing**
Alternative therapy

Ronald David Laing was born in Glasgow in 1927. On completing his degree in medicine, he joined the British Army as a psychiatrist, where he discovered a talent for communicating with the mentally distressed. On leaving the army, he returned to Glasgow to complete his psychiatric training and to formulate the ideas that would shape his career.

Laing felt that mental illness was a reaction to an individual's inability to cope with the pressures of society and other people, and that psychotic episodes were an expression of this distress. He shunned the psychiatric dogma that mental illness was caused by a biological trigger, instead seeing madness as part of a transformative journey for the patient. He used this theory to steer the treatment of schizophrenia away from traditional remedies such as electroconvulsive therapy and toward a more interpersonal approach, trying to break down the boundary between doctor and patient.

In 1965, he co-founded the Philadelphia Association, a charity dedicated to helping those in mental distress. He is often cited as an icon of the anti-psychiatry movement, but this is incorrect as he never advocated not treating mental illness, simply an alternative way of looking at it.

Laing had a colorful personal life, marrying twice and fathering ten children. He published in the fields of psychiatry, philosophy, and poetry, and inspired a play entitled *Did you used to be R. D. Laing?* He died while playing tennis in St Tropez, at the age of sixty-one.

Helen Potter

Date 1965

Nationality British

Why He's Key Championed an alternative view of madness and treatments for schizophrenia.

Key Person **Richard Feynman**
A very distinguished history

A glass of icy-cold water, a New York meal, and an inevitable Nobel Prize defined the career of Richard Feynman (1918–1988). Known for his fresh thinking and relentless curiosity, he taught himself to decipher hieroglyphics and pick locks, putting the latter to use while working on the Manhattan Project, opening safes and leaving mischievous notes behind.

Feynman coped well with the burden of working on the bomb during World War II, initially appreciating the severity of the threat from Germany. While enjoying a meal in a New York restaurant, however, his wandering mind began to calculate the destruction that the bomb could have caused to Manhattan; he began to see society as doomed, and construction projects as futile. In 1965, Feynman shared the Nobel Prize with two other scientists for his work in quantum electro-dynamics, his methods typically differing from those of his counterparts. He challenged the assumption that particles had singular histories, and stated that particles had multiple histories, but that the sum of these would cancel to provide individual observations.

Having already written a bestselling book the year before, in 1986 Feynman's fame grew while investigating the Challenger Space Shuttle disaster. Live on television, he dropped a vital ring made of rubber into ice-cold water, in order to demonstrate that it became brittle at low temperatures. At the age of sixty-nine, Feynman's own brittleness was exposed by the abdominal cancer that ended his life.

Christopher Booroff

Date 1965

Nationality American

Why He's Key Feynman is recognized as one of the greatest communicators of scientific ideas to have lived, winning a Nobel Prize and writing a bestselling book along the way.

Key Experiment **Mutant flies make genetic modification a reality**

Once purely the realm of science fiction, genetic engineering became a real life possibility in 1966, thanks to Allen Fox and Sei Byung Yoon of the University of Wisconsin. Whether you embrace it or fear it, genetic engineering technology has a potentially huge impact on our lives, from GM foods to gene therapy for life-threatening diseases.

Fox and Yoon's experiment would be considered somewhat crude by today's standards. They simply took some fruit fly embryos and dunked them in a solution of DNA made from flies with certain identifiable genetic faults (mutations). Some of the embryonic cells took up the DNA, and incorporated it into their genomes. When the embryos hatched into adult flies, the researchers found that they now had the same mutations as the DNA donors.

But how did they know this was true genetic modification? The key was in the details. If the donor flies had darker colored eyes, Fox and Yong found that patches of eye cells in the recipients were also that color. The patches had grown from the original cell that first took up the DNA, while the cells around it were unmodified. Not only that, but the fly's DNA had been permanently changed, and its new look was passed on to the next generation.

This was the first time that an organism other than a simple bacterium had been successfully genetically altered in this way, allowing scientists to begin wielding the incredible – and controversial – power to manipulate the genes of animals, plants, and even humans.

Kat Arney

Date 1966

Scientists Allen Fox, Sei Byung Yoon

Nationality American

Why It's Key A simple experiment with fruit flies made the possibility of genetic modification a reality.

Key Discovery **High-yield rice** Bumper crops for all

In the mid twentieth century, it became apparent that rapidly growing populations – particularly in Asia – would require significant increases in rice production, however, insufficient areas of suitable land were available. To combat the ensuing humanitarian disaster, the International Rice Research Institute (IRRI) developed a new strain of rice, capable of producing much higher yields than existing strains. Its development was officially announced on November 28, 1966.

The IRRI had been set up in 1960 specifically to search for a way to reduce the threat of increased famine and poverty due to insufficient rice crops. Based in the Philippines, it still works with numerous organizations worldwide. The first new rice hybrid developed by IRRI was called IR8, and was a cross-breed of Indonesian and Taiwanese varieties. It was shorter than most species, preventing it from matting and then rotting, and making harvesting easier. It could be planted any time of year and had a shorter growing period, which meant it could be cropped multiple times in a year.

The hybrid also had its disadvantages – it wasn't as tasty, it had a chalky consistency, and it required large amounts of fertilizer. The hybrids did not fertilize themselves, and so seeds had to be purchased for each crop. This, however, didn't affect its success. The new hybrid significantly increased yields and reduced the cost of rice production, enabling an increased income per weight of rice.

Emma Norman

Date 1966

Scientists Peter Jennings, Henry Beachell, Surajit De Datta, Te Tzu Chang

Country Philippines

Why It's Key Reduced widespread famine in countries heavily dependent on rice, and improved their economic growth.

Key Event *Star Trek*

To boldly go where no TV show has gone before

Gene Roddenberry was a struggling script writer and producer in Los Angeles when he came up with an idea that would change the world of television and create a global following of fans who would become almost as famous as the program itself.

Star Trek was a slow burner. The original series, first aired on November 8, 1966, received low ratings and was canceled after just three series. After NBC syndicated the show, however, allowing it to be shown on other channels and at different times, it became much more popular, enough so for a *Star Trek* movie to be created in the mid-1970s.

One of the notable things about *Star Trek* is its attention to "scientific" detail; the technology demonstrated in the show's numerous incarnations has inspired many people to speculate about the feasibility of things like warp drive, replicators, and phasers.

Is this so far fetched? Scientists at MIT are looking at ways of sequencing proteins together to create new materials from scratch; in the not too distant future we might be able to punch in a code to create whatever we want, rather than shopping for it. Mexican physicist Miguel Alcubierre speculated about the possibility of faster than light travel with a form of "warp-drive." The phaser too may be a not-too-distant reality – the U.S. military is currently developing a "pulsed energy projectile" weapon, which can stun or kill.

Barney Grenfell

Date 1966

Country USA

Why It's Key Although much Star Trek technology is not with us yet, the series, and all its various spin-offs, remain a source of inspiration for "trekkies" and scientists alike.

opposite Captain Kirk, Dr Bones McCoy and Mr Spock in Star Trek

Key Discovery **Dynamic earth**

The theory of plate tectonics

A relatively new theory in the history of geological understanding, the phenomenon of plate tectonics was first identified in 1967 by three scientists: Dan McKenzie and Robert Parker of the U.K., and W. Jason Morgan of the United States.

Plate tectonic theory promotes an image of the Earth's surface as a series of large plates, moving over hot magma below. It goes on to show that these plates are able to move relative to one another. This movement may take several forms, whether it be plates drifting apart, scraping past one another, or even colliding head-on.

Since McKenzie, Parker, and Morgan first published their thought-provoking idea in 1967, plate tectonic theory has been universally accepted among the Earth science and geology fraternities. It has been used to explain such phenomena as earthquakes, thought of now as the direct result of rocky plates scraping past one another. It also shines light on the formation of certain features of the Earth's landscape, for example mountain ranges. A direct collision of plates forces up a region of rock, which then forms a mountain range along the length of the collision area. The Himalayas were formed when the Indian subcontinent smashed into Asia.

This theory of "continental drift" is also used to explain the apparent breaking apart of an ancient supercontinent, Pangaea, into the individual continents we know today. Without this revolutionary theory, we would be severely limited in our knowledge and understanding of the Earth's surface.

Hannah Welham

Date 1967

Scientists Dan McKenzie, Robert Parker, W. Jason Morgan

Nationality British, American

Why It's Key Cleared the way for a new understanding of earthquakes, continental drift, and volcanic activity.

Key Invention **The fertility drug**
That's my boy, and so is that, and that…

One of the unfortunate realities of life is that for some couples desperate to have children, it is virtually or completely impossible for them to conceive naturally. For the vast majority of recorded history, such couples have been relegated to the roles of doting aunts and uncles, using offspring of family members or close friends as surrogate progeny. That all changed, however, in 1967, with the invention of fertility drugs.

The greatest of these drugs is undoubtedly clomiphene. The drug increases the probability of pregnancy by stimulating the increased release of two hormones, luteinizing hormone and follicle stimulating hormone, from a gland at the base of brain called the pituitary. It does this by inhibiting the action of the female hormone estrogen. These hormones subsequently support the formation of the ovarian follicle and the corpus luteum – events which are vital for a successful pregnancy.

While the drug has helped many formerly infertile couples conceive, it has not done so without some side effects. Approximately five per cent of patients taking the drug end up having twins, and it is not uncommon for even larger numbers of children to pop out.

Unfortunately, multiple births can have graver consequences than simply having more college fees to pay – it often complicates the birth itself. Multiple birth babies are more likely to be premature and suffer from fetal crowding in the womb.

B. James McCallum

Date 1967

Scientist Frank Palopoli

Nationality American

Why It's Key Many couples who were previously unable to know the joys and pains of childbirth can now do so, although sometimes they get more than they bargained for.

Key Invention
A vaccine for rabies

Dying is never pleasant. But if you can pick one way to avoid going, then avoiding rabies might be an expedient choice. Were you unlucky enough to be bitten by an infected animal, the rest of your life might go something like this. After a couple of months of feeling fine whilst the rabies virus is working its way steadily towards your central nervous system you get flu-like symptom which quickly give way to slight paralysis, paranoia, and hallucinations. You're now constantly salivating and, being unable to swallow, froth at the mouth whilst becoming hydrophobic. The symptoms worsen until you finally slip into a coma and die. There is no cure. Of the six humans known to have survived full-blown rabies, only one has been left without serious brain damage. Luckily, however, there is now a vaccine.

Louis Pasteur created the first rabies vaccine in the 1880s, but it lacked efficacy in many cases and was extremely painful. It wasn't until 1967 that a truly effective vaccine was first produced. A new technique was developed by immunologist Hilary Koprowski – who had previously been instrumental in developing vaccines for polio – and his colleagues to grow the rabies virus in human cells. The viruses could subsequently be inactivated and used to confer immunity to patients. The vaccine was so effective that it is now held as the gold standard by which all other vaccines are measured.

Mark Steer

Date 1967

Scientist Hilary Koprowski and colleagues

Nationality Polish

Why It's Key A revolutionary technique for producing vaccines and a weapon to combat one of the world's worst diseases.

opposite False-colour transmission electron micrograph of the rabies virions

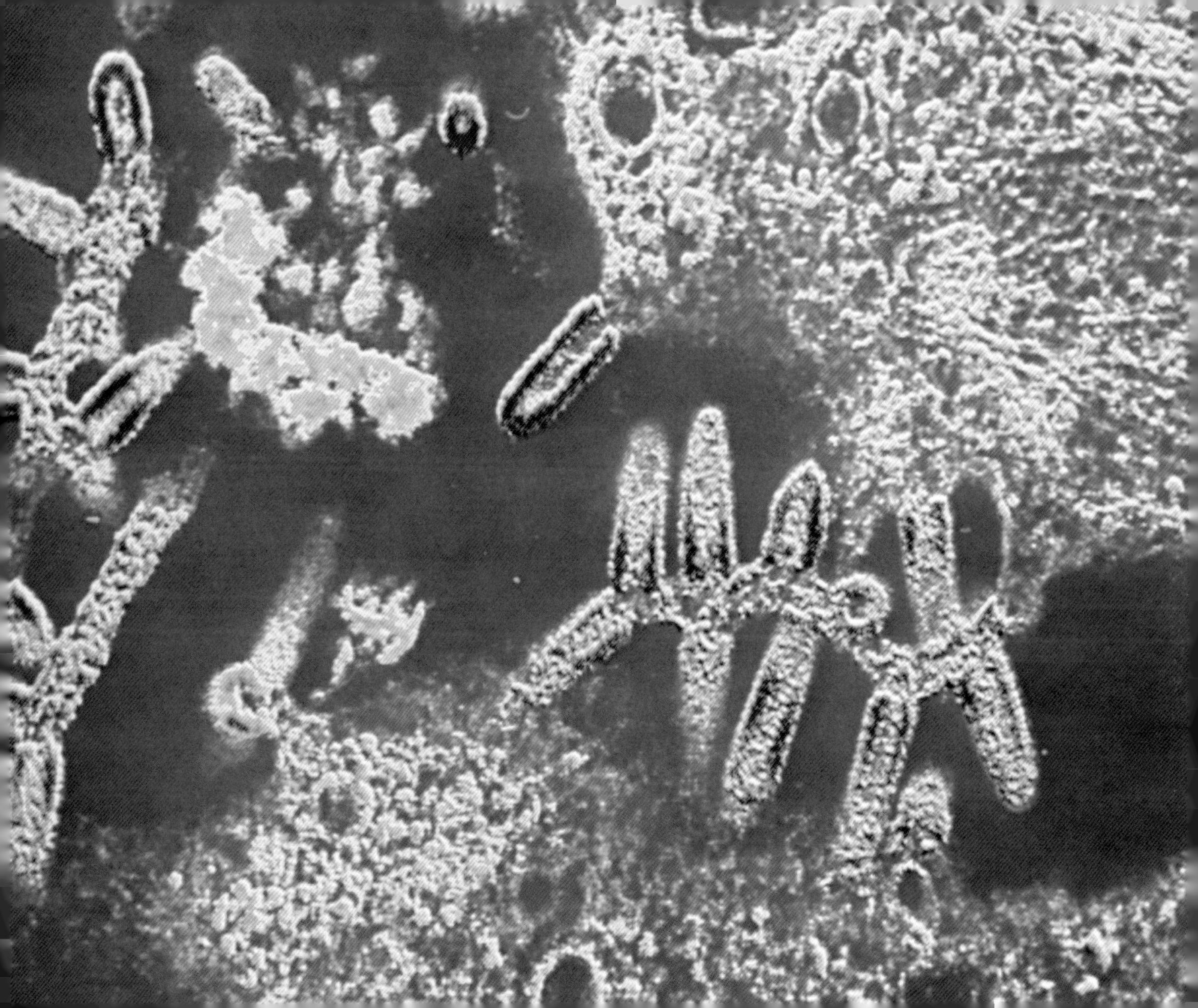

Key Event
First human heart transplant

The first human-to-human heart transplant occurred in December 1967, at the Groote Schuur Hospital in Cape Town, South Africa. A team, led by Dr Christiaan Barnard transplanted twenty-five-year-old Denise Darvall's heart into a dying patient. Darvall had been fatally injured in a road accident, and shared the same blood type as the recipient. Louis Washkansky, fifty-five, lived for eighteen days after the procedure before dying of pneumonia, probably as a result of his suppressed immune system. Barnard went on to enjoy international fame and fortune.

Credited as the father of transplant surgery, and an instructor of Barnard, Dr Norman Shumway transplanted the first human-to-human heart in the United States a few months later. Early transplant recipients typically died after only a few days or weeks; victims of organ rejection or infection. Within three years of the first transplant, 146 of the first 170 heart transplant recipients were dead.

Shumway focused on the dilemma of organ rejection. The solution came from a most unlikely source; a soil fungus discovered in Norway contained a chemical that would help prevent organ rejection but spare some of the body's immune system. Ciclosporin was first used to prevent rejection in a transplant patient in 1980.

Building on Shumway's research, Dr Bruce Reitz performed the first successful heart-lung transplant in 1981. The recipient, Mary Gohlke, a forty-five year-old advertising executive, lived five more years, and wrote a book about her experiences.

Stuart M. Smith

Date 1967

Country South Africa

Why It's Key Revolutionized transplant medicine.

Key Discovery **Identical amino acids found across species** Singing from the same hymn sheet

In 1967, geneticists Charles Caskey, Richard Marshall, and Marshall Nirenberg decided to study the response of guinea pig livers, *E. coli* bacteria, and African clawed frogs to certain sequences of transfer RNA – the sections of genetic material which actually translate the DNA code into protein. What they found was that the RNA sequences were translated into almost identical sequences of amino acids.

Though they subsequently found a small degree of variation, their results suggested very strongly that most forms of life on this planet not only use the same genetic code, but respond in nearly the same fashion. Why this occurs is anybody's guess, but it has been suggested that the code was effectively frozen after life had advanced to the bacterial level because any further variance would likely have proved lethal. That said, it must be pointed out that the genetic code is only "mostly" universal – some codes are not recognized by some species and others are more efficient in some species than others.

What this means is that the same basic genetic code can be combined to create things as diverse as the larch tree, the hammerhead shark, and the penguin. Despite the fact that there are a few changes in the genetic code – for example a sequence that would normally code for the amino acid arginine in most organisms halts the process of protein manufacture in mammals – it implies that all of us have a common genetic ancestor.

B. James McCallum

Date 1967

Scientists Charles Caskey, Richard Marshall, Marshall Nirenberg

Nationality American

Why It's Key Basically all life on Earth, from mycobacteria to Michael Jackson, is made up of the same genetic material.

Key Publication
The climate change debate begins

Climate change dominates the popular science arena in the first part of the twenty-first century, but it was a study published back in 1967 that really kick-started the debate.

The concept of the "greenhouse effect" – in which gases like carbon dioxide (CO_2) in the atmosphere absorb and trap heat radiated from the Earth, causing a rise in temperature – has been around since the middle of the nineteenth century. In 1896, a Swedish chemist called Svante Arrhenius first suggested a link between increases in CO_2 and an increase in global temperatures, and his findings predicted a 5 degrees Celsius warming from a doubling of atmospheric CO_2.

It wasn't until more than half a century later that the Japanese climate scientist Syukuro Manabe made the first modern estimate of the effect of a doubling in CO_2 levels on global temperatures, this time using a computer model. His prediction was a 2 degrees Celsius rise from a CO_2 doubling, strikingly close to that made by Arrhenuis without the aid of modern computation. Manabe went on to pioneer the first combined atmosphere-ocean computer model, and used this to reveal the capacity of the ocean to act as a "giant heat sink" in slowing the pace of global warming.

Following Manabe's contributions, the debate on climate change spilled into the public arena, and has now become one of the biggest challenges to face human existence. The latest models predict a rise in global surface temperatures of between 1.1 and 6.4 degrees Celsius before the end of this century.

Hannah Isom

Publication *Philosophical Magazine*

Date 1967,

Author Syukuro Manabe

Nationality Japanese

Why It's Key The first modern study to suggest a causal link between man-made CO_2 emissions and an increase in global temperatures.

Key Discovery **Pulsars**
Little green men or spinning neutron stars?

In 1932, James Chadwick discovered the neutron. Two years later, Walter Baade and Fritz Zwicky proposed that there could be stars composed almost entirely of neutrons and that "the supernova process represents the transition of an ordinary star into a neutron star." Thirty-three years after that, a Cambridge graduate student searching for quasars found weird signatures in her data. They turned out to be the first four neutron stars ever discovered.

In July of 1967, Jocelyn Bell had sole responsibility for using the radio telescope, designed by Professor Antony Hewish, to search for quasi-stellar radio sources. Two months into the survey, she noticed peculiar features repeating in the data at the same location in the sky every day. The pulses were regular and were coming from beyond the Solar System. Could they be signals from alien intelligence?

While the "'Little Green Man Hypothesis," as it was known, was given serious consideration for a time, even at Cambridge University, several features in Bell's data made an alien-based solution unlikely. So Bell and Hewish came up with a less exciting but more compelling explanation: they had discovered rapidly spinning neutron stars, soon to be known as pulsars.

Antony Hewish was awarded the Nobel Prize in Physics in 1974 for his "decisive role in the discovery of pulsars." Jocelyn Bell's rather crucial role, however, was overlooked.

Eric Schulman

Date 1967

Scientists Jocelyn Bell, Antony Hewish

Nationality British

Why It's Key Increased our understanding of neutron stars, stellar evolution, and general relativity.

Key Publication *The Naked Ape*
Desmond Morris unwraps humanity

Written by Desmond Morris, zoologist and long-time curator of mammals at London Zoo, *The Naked Ape* stands out, like the sore thumb of publishing history, for providing a controversial insight into human behavior in the 1960s.

The book was one of the first to describe humans, the so-called "naked apes," as animals. The infamous work includes chapters on fighting, rearing young, and feeding; terms typically used to describe strictly non-human behaviors. Not surprisingly, this departure from the anthropomorphic turns of phrase usually seen in writings on humanity caused a bit of a stir among the scientific community. Morris also managed to raise public eyebrows with several suggestive and controversial written statements, in which he described man as "the sexiest primate alive," with "erogenous earlobes and rounded breasts to be used as sexual signals." Despite the controversy surrounding its publication, *The Naked Ape* remains a bestseller to this day. It has been translated into twenty-three languages worldwide, and has sparked the release of a successful television series and film relating to the book. Morris himself has become quite the celebrity in the scientific community since the book's publication, and has been probed for his thought-provoking opinions on several radio and television productions.

As well as being a fine scientist, Morris is also a surrealist painter of some repute. In 1957, he even organized an exhibition of paintings by chimpanzees at the Institute of Contemporary Arts, London.

Hannah Welham

Publication *The Naked Ape*

Date 1967

Author Desmond Morris

Nationality British

Why It's Key A brave departure from the tired anthropomorphic writing style of the 1960s, it sparked controversy within the scientific world.

opposite Desmond Morris and one of his subjects

Key Publication
Dancing bees' mysteries unmasked

You wouldn't think you could get a Nobel Prize for studies of dancing bees, but that's exactly what Karl von Frisch got after publishing a book called *The Dance Language and Orientation of Bees*. This brought together forty years of his work on decoding the "waggle dance" which returning honeybees use to communicate the location of flowers to their hive-mates, a phenomenon which had puzzled even Aristotle.

Von Frisch's study illuminated the complexity of bee communication, and showed that they are able to learn from each other. He correlated the runs and turns of the returning bees doing the waggle dance to the distance and direction of a food source from the hive. The orientation of the dance relates to the relative position of the Sun, and the length of the dance is correlated to the distance from the hive – a longer dance means food is further away. Foragers communicate their floral findings to persuade other worker bees to forage in the same area.

The discovery turned accepted theory on its head. Before this, it was thought that bees were not capable of anything as complicated as communication and learning. Von Frisch's experiments helped to re-invigorate the study of animal behavior and turn it into a proper science. He was rewarded for his efforts, in 1973, with the Nobel Prize for Physiology or Medicine.

Catherine Charter

Publication *The Dance Language and Orientation of Bees*

Date 1967

Author Karl von Frisch

Nationality Austrian

Why It's Key Transformed studies of animal behavior and communication, and proved that bees were more than just little buzzers.

Key Discovery **Electroweak theory**
Two fundamental forces of the world united

The unification of fundamental forces (strong nuclear, weak, electromagnetic, and gravitational) has been a goal of scientists since Einstein in the early twentieth century, and remains so today. The first step toward achieving this goal came in 1967 in the form of the electroweak theory – the unification of the electromagnetic and weak forces, which were discovered to be different aspects of the same force.

Under certain conditions, it emerged, the two forces acted in a similar way, even though they were known to be completely separate. It took temperatures in excess of 100,000 degrees Celsius to establish these similarities. Since electromagnetic and weak forces are carried by two different types of particles, the discovery came as something of a surprise to physicists. The weak force, true to its name, has a very small range and affects only the electrons within an atom. The electromagnetic force, on the other hand, is the force between electrically charged particles that defines the behavior of atoms. It is thought that during the Big Bang explosion, all four forces were part of a "superforce" and have since divided to play their individual roles.

In 1979, the Nobel Prize in Physics was awarded to Abdus Salam, Sheldon Glashow, and Steven Weinberg for their "electroweak theory." The theory later predicted the presence of so-called W and Z particles for the weak force; their existence was confirmed in the 1980s, as were the characteristics of the strong force.

Nathan Dennison

Date 1967

Scientists Abdus Salam, Sheldon Glashow, Steven Weinberg

Nationality Pakistani, British, American

Why It's Key The electroweak theory marked the first step on the road to a grand unified theory.

Key Publication **Island hopping**
New ideas about biodiversity

In 1967, the biologists Robert MacArthur and Edward Wilson published a fundamental text that would redefine scientists' ideas about species diversity and how they should be creating nature reserves. Their *Theory of Island Biogeography* explained how the species diversity in island communities was affected by the island's size and its proximity to the mainland.

One of the main conclusions was hardly surprising: Small, isolated islands are home to fewer species than large islands near the coast. MacArthur and Wilson even produced a formula that would predict the number of species on an island by estimating immigration and extinction rates.

It wasn't long, however, before the importance of the relationships were realized. Islands don't just have to be lumps of land dotted amid the waves – nature reserves can also be seen as islands of natural habitat, essential for the conservation of some species, in a sea of human-altered environments. What's more, as islands of wild habitat diminished in size and became fragmented from each other, the theory could predict how many species would go extinct. Conservation had become a science.

The theory of island diversity has been hugely influential in the design of nature reserves, and continues to help guide ecological decision-making to this day. It gave conservationists the tools to start predicting future changes in biodiversity with a degree of confidence and scientific rigor.

Katie Giles

Date 1967

Authors Robert MacArthur, Edward Wilson

Nationality American

Why It's Key Prompted discussion about biodiversity and shaped future conservation efforts.

Key Publication **The origin of cells' machinery**
The Endosymbiotic Theory

The pioneering work and dogged persistence of one biologist saw a once-unorthodox theory turn into one of the great contributions to evolutionary biology. In 1967, after fifteen rejections, Professor Lynn Margulis finally published her work "The Origin of Mitosing Eukaryotic Cells" explaining how complex cellular organisms came into being.

As life on Earth evolved, two broad categories of organisms appeared: the prokaryotes (mainly bacteria), and eukaryotes, like humans (and fungi, plants, dogs, cats; in fact, pretty much everything that isn't bacteria). So what's the difference? Why are humans in the same category as a mushroom whilst bacteria get a category virtually to themselves? Prokaryotes have a simple cellular structure, containing very little internal organization of their component parts. The eukaryotes, on the other hand, are made up of cells which contain small internal structures called organelles, for example the nucleus, where DNA is stored, and mitochondria, which act as cellular power stations, supplying the cell with energy.

Margulis' Endosymbiotic Theory explained how the more complex organization of eukaryotic cells arose. During early evolutionary history, as conditions on Earth changed, prokaryotic organisms began to rely heavily on each other for survival. Some moved within other cells to find protection, in return providing their host with benefits – in the case of mitochondria, energy production. These were the prototype organelles. Over billions of years, these evolved into compartmentalized and organized cells that make up you and me.

Ceri Harrop

Publication *"The Origin of Mitosing Eukaryotic Cells"*

Date 1967

Author Lynn Margulis

Nationality American

Why It's Key Described how complex cells, containing structures such as mitochondria and chloroplasts, evolved.

Key Event **The 36-foot Radio Telescope**
Peering into the interstellar dust clouds

Traditionally, astronomers turned their eyes and telescopes to the hot things in the Universe – the stars and galaxies that throw out visible light. But as knowledge of the electromagnetic spectrum improved, it became clear that colder objects could also be detected, even those that give off no visible light.

This included dust clouds and gaseous expanses of molecules where new stars and galaxies are born. These are tenuous and cold, and invisible to the naked eye. But as they vibrate, the small molecules in the clouds emit radio waves, and with the right detectors, these can be collected and analyzed.

Enter the 36-foot Radio Telescope, so named for the size of its dish. The telescope was opened by the U.S. National Radio Astronomy Observatory (NRAO) in 1967 in Kitt Peak, Arizona, and it began full operation in January 1968. It observed radiation in the millimeter wavelength – that is, light with wavelengths between infrared and radio waves. Different molecules emit this radiation at slightly different wavelengths, allowing the composition of the dust clouds to be deduced.

The Kitt Peak telescope has discovered many of the molecules in the interstellar medium, such as carbon monoxide, and even the first sugar found in space. It was refurbished, extended, and went metric in 1984, to become the 12-Meter Radio Telescope, and now falls under the auspices of the Arizona Radio Observatory, having been closed down by the NRAO in July 2000.

Matt Brown

Date 1967

Country USA

Why It's Key The 36-foot Radio Telescope pioneered millimeter wavelength astronomy, finding many of the molecules in interstellar space.

Key Invention **QWERTY**
Still going strong

Try to imagine how these words would have reached the page without the benefit of a modern computer keyboard. It would have been a tedious affair. First they would have been typed into a machine that transferred them onto a punch card, which would then have had its data verified by someone inserting it into another machine and typing it in again. The data could only have been read by feeding it into a reader that would then print out a copy. This combination of typewriter and punched card system, called a keypunch, was the only method available in the 1950s.

In the 1960s, a new form of output technology was proposed – the visual display unit (VDU). This allowed users to see their text as they went along, enabling them to create, edit, and delete it as necessary. The invention of the electric keyboard removed the many electromechanical steps – between hitting a key and the letter appearing on the screen – which had caused the process to be so slow. Instead, electrical impulses were transmitted directly from the keyboard to the computer, drastically increasing the speed of input.

The QWERTY keyboard layout, invented more than sixty years earlier by an American, Christopher Scholes, and originally designed to slow typists down and so prevent typewriter jams, was retained. The validity of this decision has been questioned, as computer keyboards contained no moving parts that needed to be protected. More efficient typing systems such as DVORAK have been proposed but none have caught on and the QWERTY arrangement is still the standard for English-language keyboards.

Helen Potter

Date 1967

Scientist MIT, Bell Labs and General Electric

Nationality American

Why It's Key Increased the speed of computer programming and data entry, while making it more accessible to the home user.

Key Event
First coronary artery bypass

Coronary artery bypass grafting is a special kind of open-heart surgery. It creates a route for blood to reach the oxygen-starved muscles of the heart when the normal route becomes blocked. It is better known by its abbreviated name, CABG, pronounced "cabbage."

During the 1960s, surgeons at the Cleveland Clinic in America were trying out different ways to repair blocked blood vessels. They monitored their results with a special imaging technique that uses injected dye to reveal the flow of blood through the heart's arteries.

René Favaloro, originally from Argentina, studied thousands and thousands of these images at the clinic. He devised a new technique using a vein that runs the entire length of the leg, called the saphenous vein. It had been used already for making "patch" repairs, but with limited success. His idea was to take a section of this healthy vein and fix (or graft) its open ends onto healthy parts of a diseased artery, on either side of the blocked part. The blockage would be bypassed and blood would flow easily to the heart again.

In May 1967 he performed the first CABG procedure on a fifty-one-year-old woman. It worked perfectly, giving her a new lease of life, and gaining him international recognition. Thus a new era of cardiovascular treatment began. Now we commonly hear of single, double, triple, and even quadruple bypasses. Better graft materials are now available, and the techniques are more sophisticated, but CABG is still the mainstay of treatment for coronary artery disease.

S. Maria Hampshire

Date 1967

Country USA

Why It's Key This first true bypass surgery revolutionized the treatment of coronary artery disease, prolonging the lives of millions of people every year around the world.

opposite Colored CT scan of the chest of a patient after heart bypass surgery

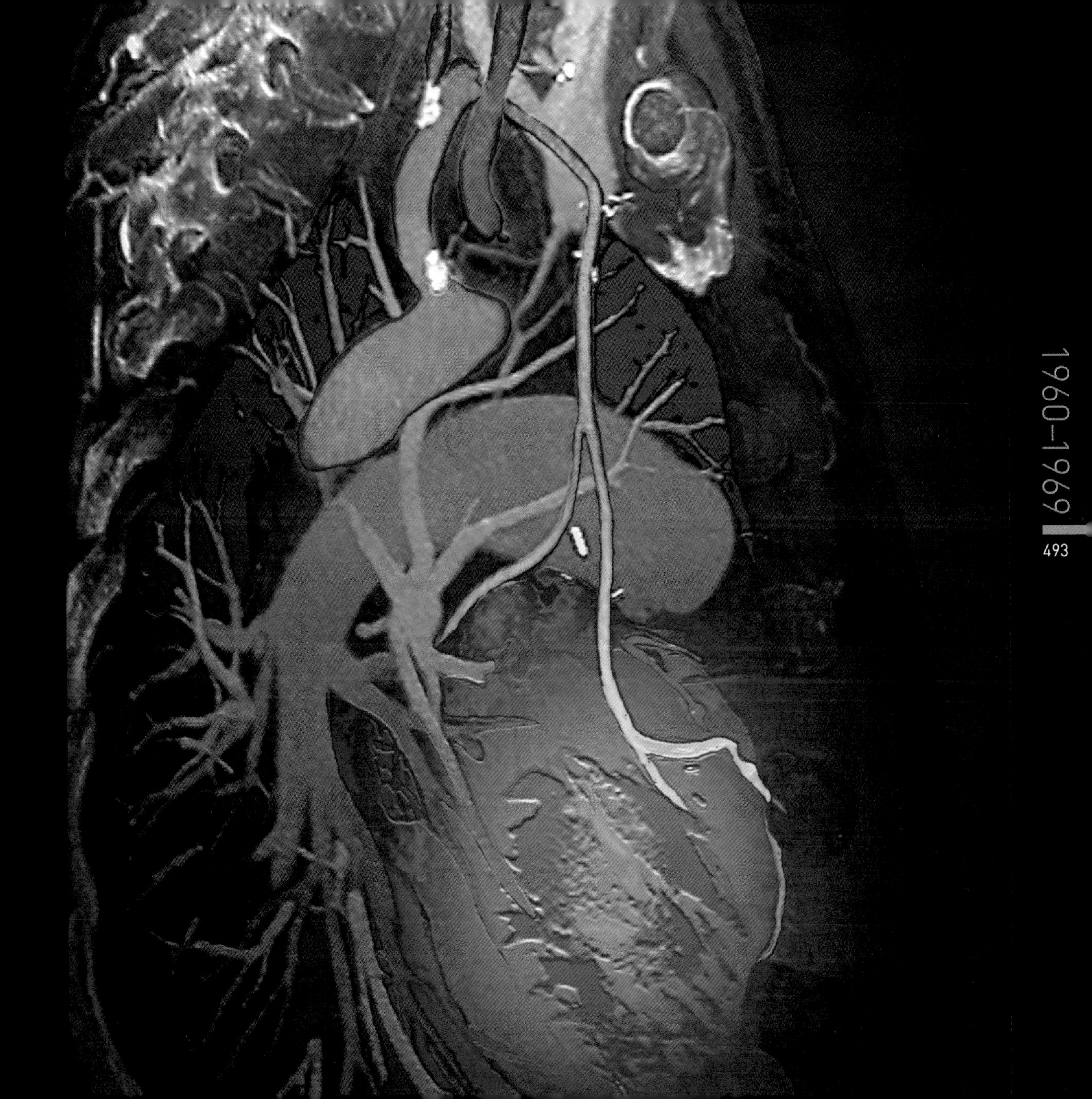

Amour
Amore
Amor
Love
Liebe

Key Event **Welcome to Our World**
The Beatles star in the first satellite broadcast

"Our World" was the name of the first ever international satellite broadcast, transmitting to over 400 million people. From a technological perspective, it was no mean feat; from a social perspective, it changed the views, quite literally, of millions. On June 25, 1967, nineteen countries participated in the event, with the intention of introducing the globe to a variety of different cultures. Politicians were banned as Our World showcased the best artists from each participating country in a bid to promote worldwide acceptance.

Most famously, perhaps, the broadcast kick-started a global phenomenon we now know as "Beatlemania." The act that everyone remembered was The Beatles' performance of *All You Need Is Love*, a song written especially for the event in protest against the Vietnam War. The international scope of the broadcast allowed the voice of protest to reach every corner of the world.

As well as live music, Our World featured scientific and cultural reports, offering new ideas and views of previously unknown cityscapes to gripped viewers. For the first time, many were given the opportunity to step into lives and countries thousands of miles from their own homes; lives and countries in some ways similar to and in other ways markedly different from their own.

Satellite broadcasting is now part and parcel of modern telecommunications; sounds and images can be beamed, almost instantaneously, around the world to eager audiences in far-flung places.

Nicola Currie

Date 1967

Country International

Why It's Key Helped raise the bar for worldwide communications.

opposite The Beatles perform in front of the world via satellite

Key Experiment
Solar neutrinos tracked down to source

Filling a million-gallon tank full of dry cleaning fluid and sticking it under a mountain seems a bit of an indirect way to detect cosmic particles, but neutrinos like to play hard to get. It's not that there aren't enough of them – 60 million million solar neutrinos go through your head every second – but because neutrinos hardly interact with matter at all, no-one suspected until recently that they even existed.

To look for the least-interacting particle we know of you have to conduct experiments deep underground. Whilst all the other particles are blocked by the ground above, neutrinos go sailing gaily through the rocks and reach the detector – that's your million gallons of fluid. The results of two experiments starting in 1968, but not fully confirmed until 1986, finally revealed the presence of solar neutrinos. Physicists breathed a big sigh of relief – the neutrinos that were supposed to exist, did exist, even fewer of them were detected than the scientists expected.

Since neutrinos travel at the speed of light, we can study them to learn about the fusion reactions that were happening at the very centre of the Sun just 8 minutes ago. And not just the Sun – all astral bodies that undergo nuclear reactions emit neutrinos. The neutrino's lack of interaction means it can reach us over astronomical distances, bringing with it information about stars light-years away. Neutrino astronomy had been born.

Kate Oliver

Date 1968

Scientist Raymond Davis Jr. and colleagues

Nationality International

Why It's Key It was confirmation that these inconceivably tiny and plentiful particles are produced in solar fusion and can be used as a new tool in astronomy.

Key Person **Hans Bethe**
The inventor becomes the objector

Hans Albrecht Bethe (1906–2005) was a German-born American physicist who was called the "supreme problem solver of the twentieth century" by fellow physicist Freeman Dyson. He obtained his PhD in physics at the University of Munich before eventually leaving the country in 1933 after the Nazis gained power. He ended up in the United States, where he joined Cornell University. Like many physicists who became involved in the U.S. war effort, Bethe was originally an astrophysicist – his primary interest was in the nuclear reactions that take place in the hearts of stars and make them shine.

Bethe became a naturalized American citizen in 1941; this allowed Robert Oppenheimer to invite him to a special meeting at the University of California in 1942, to discuss the early first designs for an atom bomb.

The following year, Oppenheimer was given the task of creating a new top secret lab for designing the bomb and he appointed Bethe as his Director of the Theoretical Division. The operation was known as the Manhattan Project and it was Bethe who calculated the critical mass of uranium-234 needed for nuclear detonation, as well as working out the explosive yields of the test device used in the Trinity test explosion and the "Fat Man" bomb dropped on the Japanese city of Nagasaki. After the war, he went on to work on the hydrogen bomb, before campaigning against nuclear testing along with other scientists including Albert Einstein. In 1967, he was awarded the Nobel Prize for Physics.

David Hawksett

Date 1967

Nationality German-American

Why He's Key As well as making many important advances in the field of astronomy, Hans Albrecht Bethe was instrumental in the design of the atom bomb, despite his later objections to nuclear testing.

Key Event **Apollo 8 Orbits the Moon**
First people to see the far side of the Moon

In 1961, President John F. Kennedy proclaimed that America "should commit itself to achieving the goal, before this decade is out, of landing a man on the Moon and returning him safely to the Earth." By the summer of 1968, however, problems with the lunar module had delayed its first flight and threatened to postpone the first lunar landing beyond 1969.

So, in August 1968, NASA announced a plan to accelerate the flight schedule and make the first manned launch of a Saturn V rocket – previously planned for December – the first manned mission to the Moon. Apollo 8 took off that year on December 21. Frank Borman, James Lovell, and William Anders made ten orbits of the Moon, coming within 112 kilometers of the lunar surface. On December 24, Anders became the first person to see the Earth rise from behind the Moon. He took a number of pictures, one of which became an iconic image for environmental groups around the world. As Lovell said during a live television broadcast: "The vast loneliness up here of the Moon is awe inspiring, and it makes you realize just what you have back there on Earth. The Earth from here is a grand oasis in the big vastness of space."

Apollo 8 was the last spaceflight for Borman and Anders. Lovell was to have landed on the Moon as the commander of Apollo 13, but an in-flight explosion wrecked the mission and the Apollo-13 crew barely made it back to Earth alive.

Eric Schulman

Date 1968

Location The Moon

Why It's Key The first time humans traveled in a region where the gravitational pull of another celestial body was stronger than that of the Earth.

opposite Apollo 8 image of the Earth as seen from 112 kilometers above the Moon

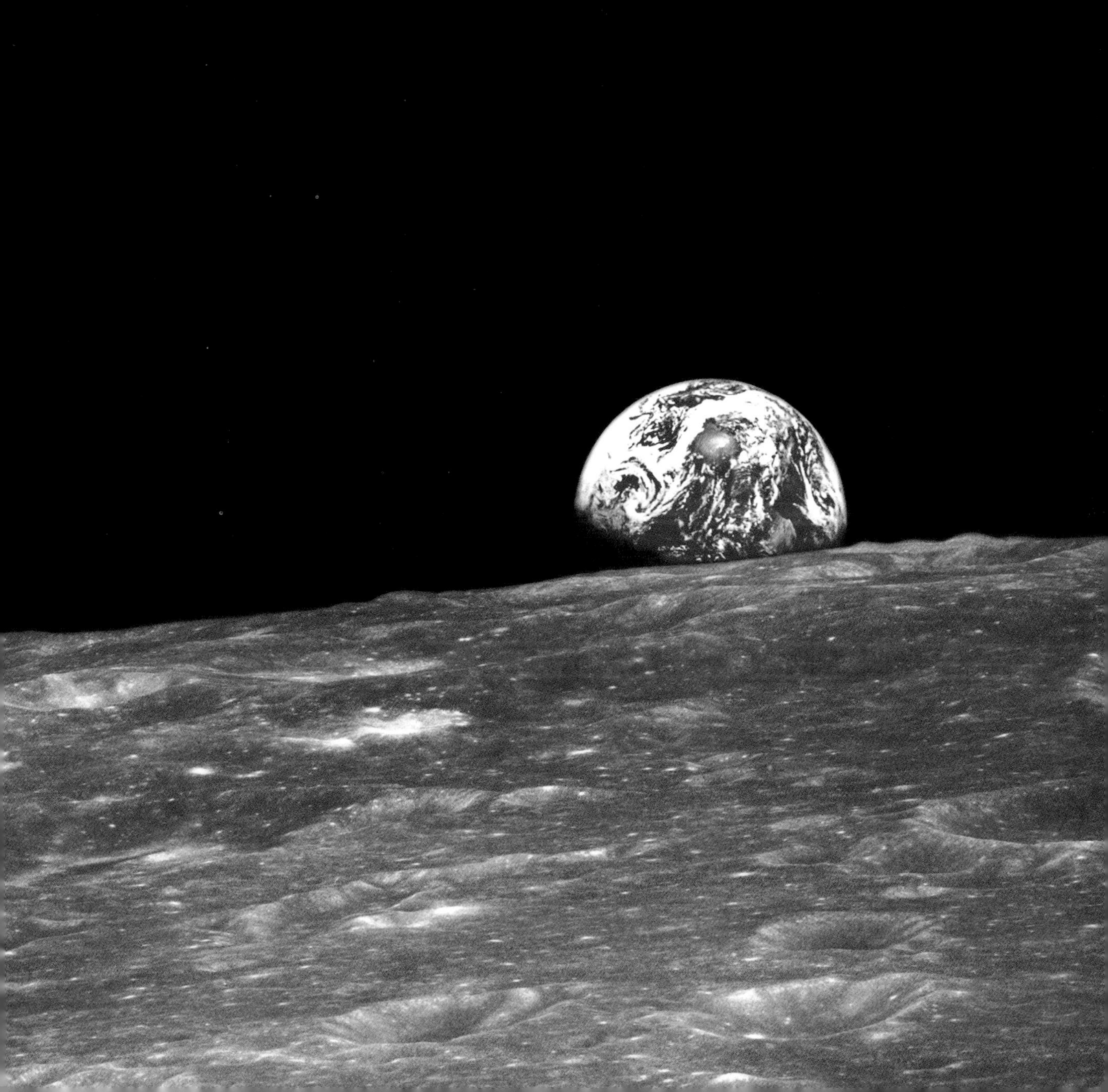

Key Invention **DRAM patented**
One giant leap for computer kind...

There is little doubt that one of the key inventions of the twentieth century is the electronic computer. When we look at the sleek and elegant hand-held devices of today it is hard to believe that as recently as the middle of the last century, an average computer would have filled a small room. At this time, computers cost tens of thousands of dollars, and the ownership of one was limited to large companies. The idea of people having a computer in their home, let alone their pocket, was unheard of.

Prior to the invention of DRAM (Dynamic Random Access Memory), a computer's memory was comprised of vacuum tubes, which took up an enormous amount of space. It was into this world – early "first generation" computing – that Carnegie Tech graduate Robert Dennard stepped.

Dennard's key innovation was to recognize that the 0s and 1s of binary code – the "language" spoken by computers – could be stored on a capacitor as a positive or negative charge; a single transistor attached to this would act as the "switch." In his "One-transistor Random Access Memory," Dennard had created a simple "cell" that could be arranged in units numbering in the billions, allowing a large amount of data to be stored in a very small space.

The rest, as they say, is history. By the mid 1970s RAM memory of Dennard's design was standard, enabling the production of increasingly smaller and more powerful computers at affordable prices.

Barney Grenfell

Date 1968

Scientist Robert Dennard

Nationality American

Why It's Key The invention of Random Access Memory allowed the development of smaller, more powerful computers, accessible by the general public as well as large companies.

Key Person **Luis Alvarez**
Some people just don't know when to stop

Luis Alvarez (1911–1988) wasn't going to let himself be pinned down to one branch of science. Starting out in experimental nuclear physics, he made a number of significant discoveries – including the radioactivity of tritium – as well as inventing several machines.

His career suffered a minor interruption due to World War II, during which he designed three types of radar. Having moved to Los Alamos to join the team working on the atom bomb, he was responsible for producing the detonator mechanism and was the scientific observer who detailed the bomb drop on Hiroshima. After the war was over, Alvarez returned to non-military research, working in high-energy particle physics. He was awarded his Nobel Prize in 1968, for pioneering work in developing the bubble chamber into a scientific research tool. Used for detecting exotic particles, a bubble chamber is filled with a liquid just below its boiling point and responds to the presence of ionizing particles by forming bubbles around the path they have taken.

For some variety, one assumes, Alvarez started working with his geologist son in 1965, beginning with an expedition to hunt for hidden chambers in the Egyptian pyramids, using subatomic particles. Later, they were the first scientists to develop the theory that a meteorite was the cause of the extinction of the dinosaurs. When not inventing things, discovering things, or investigating the assassination of President Kennedy, Alvarez found time to design an indoor golf-practicing machine for President Eisenhower.

Kate Oliver

Date 1968

Nationality American

Why He's Key Whether in geology or archaeology, developing bombs or conducting experimental physics, Alvarez was an interdisciplinary success.

Key Person **Andrei Sakharov**
Nuclear pioneer who called for peace

Andrei Sakharov (1921–1989) was a famous Soviet nuclear-physicist-turned-activist. He initially studied cosmic rays, before becoming a key player in the development of the Soviet atomic bomb, tested in August 1949. Sakharov then moved his research on to the next stage: the even more powerful hydrogen bomb. The H-bomb was tested in 1953, the same year that he received the first of his three Hero of Socialist Labor awards.

By the time his fifty-megaton super hydrogen bomb was tested in October 1961, Sakharov had begun to express moral concern over his work, and had become politically active. He was involved in the 1963 Partial Test Ban Treaty forbidding nuclear weapons tests in the atmosphere, space, and underwater, and signed by the Soviet, U.S., and UK governments.

In 1967, he urged the Soviet government to stop development of antiballistic missile technology, fearing it would lead to a new arms race, but his requests fell on deaf ears. Under increasing scrutiny by the KGB, Sakharov managed to circumvent government censors and circulate an essay outside the USSR in 1968, in which he described antiballistic missile defense as a great threat to the world.

It was efforts such as these that won him the Nobel Peace Prize in 1975; the Soviet government, however, would not allow him to leave the country to receive it. After publicly protesting against the Soviet invasion of Afghanistan in 1980, he was arrested and exiled to the closed city of Gorky, where he remained until 1986 when Mikhail Gorbachev released him.

David Hawksett

Date 1968

Nationality Soviet

Why He's Key Sakharov was described by the Nobel Peace Committee as a "spokesman for the conscience of mankind."

Key Person **Harry Harlow**
Love makes the world go round

There was a time when ethical issues in science didn't occupy everyone's thoughts as they do today. Notorious psychologist Harry Harlow (1905-1981) exploited these times with his investigations concerning, of all things, love. Born on Halloween in 1906, Harlow earned both undergraduate and doctoral degrees at Stanford University before joining the psychology department at the University of Wisconsin, where he performed his most famous experiments.

Fascinated by the bonds that form between individuals, Harlow staged a series of investigations, between 1963 and 1968, known as the surrogate mother experiments, during which he observed the behavior and development of baby rhesus macaques. He followed these with altogether darker studies. The Iron Maiden was an artificial monkey covered in spikes that blasted out cold air; could baby macaques love this "evil mother," Harlow wondered. He found that, no matter how bad the torture, the babies would not let go. Other experiments employed the "well of despair," within which monkeys would be isolated from all outside contact for up to two years, becoming severely and irreversibly traumatized. If this was the case for monkeys, he said, the same would be true for people.

Whatever you might think of his methods, Harlow highlighted how important "love" is in our development. Since the 1930s, medics and psychologists had been advocating that children should be brought up with the minimum of affection. Harlow's work made it obvious how damaging this approach could be.

B. James McCallum

Date 1968

Nationality American

Why He's Key Showed that love is just as important in development as food, shelter, and water. Many of society's modern trends for greater ethics and compassion can be traced back to Harlow's ground-breaking experiments.

Key Event
2001: A Space Odyssey

The year 2001 has come and gone, but humanity is not quite the space-savvy, robot-commanding race that appears in the famous book by Arthur C. Clarke, and film by Stanley Kubrick. The work's predictions are off, but its technological glory, robotic conflicts, and mind-bending finale have impacted scientists and science fiction fans alike.

The novel and the film were released in 1968 as a collaborative project between Clarke and Kubrick. Initially critics were not thrilled, but the film has gained steam and both works are considered classics. The work is profoundly mystic, dominated by silence, wonder and scenes that appear to have been taken from an LSD episode. The audience receives little in the way of verbal narration, but is fed load after load of philosophical questions about technology and science.

The potential pitfalls of artificial intelligence are shown with HAL-9000, the cunning computer that controls the main characters' spaceship, the Discovery. Thanks to his incredible intelligence and learning capabilities, it takes HAL little time before he realizes that he doesn't have to take orders.

2001: A Space Odyssey also ponders the origin of life, extraterrestrials, and postulates on the potential of technology. Its insights, still relevant today, warn us of being careless with the advance of technology, but ultimately it is an optimistic work that has inspired innovation. There are at least fourteen different devices imagined in the film that have since become reality, including flat-screen televisions, biometric identification, and voice-recognition computing.

Logan Wright

Date 1968

Country USA

Why It's Key Many works of film and literature have inspired debate about the future use of technology, and influenced the course of scientific research. 2001 is arguably the most important of these.

opposite Keir Dullea as David Bowman in *2001: A Space Odyssey*

Key Publication
Evolutionary rate at the molecular level

The neutral theory of molecular evolution was first proposed by Motoo Kimura in 1968. It was, and to some extent still is, a highly controversial area of genetics; the theory has been seen by some as an alternative to the theory of evolution through means of natural selection.

Essentially what was proposed by Kimura was that the observed rate of mutation on a molecular level did not match with the observed emergence of new mutations in a population. Kimura proposed therefore that the majority of mutations that occur have no effect on fitness whatsoever – most mutations are neutral. According to previous thinking, when a mistake is made in the genetic code, this mistake causes variation in the population; variation being the key component that drives evolution by natural selection.

Kimura's neutral theory enraged Darwin's followers because it appeared to suggest that two different populations could become genetically different thanks to random genetic drift. It seemed contrary to Darwinian theory to suggest genes could change and new forms could be created by random chance, instead of being driven by selection pressure.

Kimura was well prepared for the ensuing criticism and had backed up his theory extensively and elegantly by the time he published *The Neutral Theory of Molecular Evolution* in 1983. It is now broadly accepted that Kimura's theory is correct and genetic drift does occur. Most modern biologists do not think that the neutral theory replaces natural selection, but rather sits alongside it.

Jim Bell

Publication 1968, *Nature*

Date *The Neutral Theory of Molecular Evolution*

Scientist Motoo Kimura

Nationality Japanese

Why It's Key The neutral theory is not a rival to natural selection, but rather an additional consideration that helps us to understand how our intricate system of inheritance works.

Key Event **Man on the Moon** The lunar legacy of Apollo 11

The purpose of the mission was simple: Perform a manned lunar landing and return. The Apollo lunar landing program began in the 1950s as a NASA initiative under the Eisenhower administration, but it wasn't until the 1960s that it really took off, literally and figuratively.

After Soviet astronaut Yuri Gagarin became the first human in space, in April 1961, President Kennedy realized that the U.S. space program needed to be more aggressive if it was to keep up, and he approved a multi-billion dollar plan to rev up the program, with particular emphasis on a lunar landing.

The Apollo 11 mission touched down eight years later, led by astronauts Neil Armstrong, Michael Collins, and Edwin "Buzz"Aldrin. Seventy-six hours after launching on July 16, 1969, the spacecraft reached lunar orbit. Armstrong and Aldrin, contained in the lunar module "Eagle", landed on the Sea of Tranquility. Taking his first step on the Moon at 10:56 pm EDT, Armstrong famously declared to a television audience of approximately 700 million people: That's one small step for man, one giant leap for mankind. Armstrong was followed by Aldrin, and they both spent about two-and-a-half hours on the Moon, collecting samples, taking photos, planting flags, and performing scientific experiments.

A simple mission, completed quite flawlessly. It not only put the Americans ahead in the space race, but is deservedly one of the most important moments in human history.

Rebecca Hernandez

Date 1969

Country USA

Why It's Key A defining moment in human history as we reach out towards the stars.

opposite Buzz Aldrin walking on the Moon

Key Discovery **Up, down, charm, strange, top, and bottom** Quirky quarks

In the early part of the twentieth century, physicists were able to begin investigating the composition of the atom. By the 1930s it was known that atoms were comprised of an inner core, or nucleus, around which electrons "orbited." The nucleus itself was made of two types of particles – neutrons and protons.

The concept of quarks arose from an attempt in the 1960s to understand hadrons – particles which were observed to be influenced by a force known as the "strong interaction" – which binds protons and neutrons together in the nucleus of an atom. In 1969, a new experimental technique involving the firing of electron "bullets" at hadrons revealed that they had internal structure – quarks were no longer mathematical concepts, but detectable particles.

The name "quark" was coined by physicist Murray Gell-Mann, who lifted the phrase from James Joyce's *Finnegans Wake*, in which Joyce writes "Three quarks for Muster Mark"; meaning three "cheers." The number three, as well, was meaningful, as protons and neutrons have combinations of three quarks each; the proton, two up quarks and one down; the neutron, one up quark and two down. And the fun doesn't stop there – the six different types of quarks (which, incidentally are known not as types, but as "flavors") are called "up," "down," "charm," "strange," "top," and "bottom."

In a testament to the long reach of the Standard Model of Elementary Particles, the last quarks to be predicted and discovered were the charm quark in 1974, the bottom in 1977, and the top quark in 1996.

Andrey Kobilnyk

Date 1969

Scientists Jerome I. Friedman, Henry W. Kendall, Richard E. Taylor

Nationality American, Canadian-American

Why It's Key Quarks were a necessary component to complete what is known as the Standard Model of Elementary Particles – the most complete description of matter and forces known to physics.

Key Person **Jocelyn Bell**
Discoverer of "Little Green Men"

Joceyln Bell (b. 1943) was studying at Cambridge University in 1967, under the guidance of Antony Hewish when she discovered a regular, pulsing radio emission emanating from outer space; jokingly she labeled it "Little Green Men 1" on charts. It was later discovered to be a radio pulsar – a fast-spinning collapsed star – and Hewish's team published their findings in a scientific paper. Discovery turned to controversy, however, as Hewish and co-author Martin Ryle later won the Nobel Prize to the exclusion of Bell, a decision debated ever since.

Bell joined Hewish's team after completing a physics degree at Glasgow University in 1965, helping to build a massive radio telescope to study the phenomenon of quasars while studying for her PhD in 1967. She soon noticed a series of repeating radio emissions on the printouts, occurring just over once every second – too frequent to be from a quasar. The team realized the emissions must be coming from rapidly spinning celestial corpses called neutron stars – the collapsed remnants of a dead star.

The discovery prompted a flurry of media attention and the discovery of more pulsars, and Bell earned her PhD in 1968. In 1974, the Nobel committee rewarded Hewish and Ryle, leaving out Bell, a move that angered cosmologist Fred Hoyle. Bell remained unconcerned, however, and went on to study astronomy at various UK institutions. Her work is remembered as an important discovery in the modern description of our Universe, later acting as further evidence for the existence of black holes.

Steve Robinson

Date 1969

Nationality British

Why She's Key Bell's discovery added further insight into the phenomenon of collapsed stars, and spawned a whole field of study around neutron pulsars.

opposite Dr Jocelyn Bell-Burnell

Key Discovery **Acid rain**
An environmental success story?

Acid rain is caused by human activity releasing acidic chemicals, particularly sulphur dioxide, into the atmosphere. These chemicals react with moisture in the atmosphere, so increasing the acidity of any rain or other forms of precipitation, such as snow. Rain is naturally slightly acidic, with a pH of around 5.6 (pH neutral is 7), but human activity can produce rain as acidic as pH 2.5, similar to that of vinegar.

During the rapid industrialization of England during the nineteenth century, acid rain was reported by Robert Angus Smith in Manchester, who became the first Royal Alkali Inspector. But although we were aware of acid rain, it was not until the 1960s that the problem came to prominence when a Canadian scientist, Harold Harvey, was investigating fish populations in freshwater lakes in Canada.

Harvey discovered surprising crashes in the populations of fish in the so called "dead" lakes due to highly acidic water. He realized that the origin of the acid was coal-burning power stations across the border in the U.S.A. The UK faced a similar problem in the 1970s, when emissions from coal-fired power stations caused environmental problems both in the UK and abroad, particularly in Scandinavia. Since then, governmental action, technological innovation, and moves toward gas fuelled power, have dramatically decreased the effects of acid rain. It may read like an environmental success story, but acid rain is still a major ecological concern in many countries of the world, particularly in rapidly developing China, and areas of Russia.

Jim Bell

Date 1969

Scientist Harold Harvey

Nationality Canadian

Why It's Key Although it's encouraging that governments in North America and Europe were able to take effective steps to confront acid rain, it still poses a problem in many areas of the world.

Key Person **Max Ludwig Henning Delbrück**
Random mutation fights virus infection

As a boy, Max Ludwig Henning Delbrück (1906–1981) was fascinated by astronomy, but during his graduate studies he shifted to theoretical physics after there had been some major breakthroughs in quantum mechanics. But life in Nazi Germany was tough for scientists. Living in Berlin at the time, Delbrück believed politics would stop scientific advancements in the country.

To become a university lecturer, Delbrück would have to pass a political maturity test, which he wasn't comfortable with. So, a small group of biologists and physicists – including Delbrück – began to meet privately in 1934. Out of these meetings came a paper on mutagenesis (the process of creating genetic mutations) which would have a strong influence on the development of molecular biology.

In 1937, a fellowship from the Rockerfeller Foundation meant that Delbrück could move to the United States. Soon after, he co-authored a revolutionary paper with E.L. Ellis on "The Growth of Bacteriophage," demonstrating that viruses reproduce in one step, and not exponentially like bacteria.

The fellowship ran out in 1939 but, due to the war back home, he decided to stay in America. In 1942, with Salvidor Luria, he published his most notable work on bacterial resistance. They demonstrated that resistance to virus infection is caused by random mutation, not adaptive change. Called the Luria-Delbrück experiment, it won the Nobel Prize in Physiology or Medicine that year.

Faith Smith

Date 1969

Nationality German

Why He's Key Made key discoveries in the reproduction of viruses and our resistance to their infections.

opposite Max Delbrück

Key Person **Murray Gell-Mann**
Universal genius

Murray Gell-Mann (b. 1929) is unarguably a genius. He is also, debatably, a polymath, a curmudgeon, and incapable of knowing when to stop. Graduating from high-school at age fourteen, he was driven into physics by his father's insistence he study something practical, instead of archaeology. He had a PhD by the age of twenty-two, and was awarded the Nobel Prize in Physics in 1969.

He is most famous for his work on classifying subatomic particles, which he first arranged in a pattern known as the Eight-fold Way – a reference to the eight components of the pattern, and of the Buddha's path – and then reduced to three subunits. These subunits were a whole new type of particle, which Gell-Mann proceeded to name with gay abandon. The particles as a family he named "quarks" – pronounced quorks – a combination of a sound which pleased him and a spelling taken from James Joyce's *Ulysses*. Quarks came in different types, initially "up," "down," and "strange"; and colors: red, blue, and green. "Color" here simply describes the particles' different properties, and is not meant in the common, visual, sense. Gell-Mann's subsequent work defined how quarks of different colors interact, via the exchange of particles which stick them together, called "gluons." As may be inferred from his contributions to the language of twentieth-century physics, Gell-Mann is also a keen linguist. He has also recently invested his energies in environmental concerns, following a lifelong love of natural history. Generally, however, there seems to be little he would not consider worthy of his interest.

Kate Oliver

Date 1969

Nationality American

Why He's Key Diverted momentarily into categorizing the underlying structure of the Universe, Gell-Mann continues to dive into everything that interests him.

Key Invention **The smoke detector**
A nuclear seatbelt

It's a little invention that relies on radioactivity to save lives: In 1969, Kenneth House and Randolph Smith patented the neat – and nuclear battery-powered – smoke detector. It's a heroic device, which can stand in as a last-minute kitchen timer, but it was designed to alert a building's occupants to a fire. Its buzzing call is really just code for "Save yourselves!"

Within the smoke detector's ionization chamber, there is a minute quantity of the highly radioactive element americium-241. The americium emits a steady flow of alpha particles – large radiation particles that ionize atoms in air. Ionization is the process of removing an electron from an atom. It leaves behind a positively charged atom, while the removed electron can go off and have a negatively charged whale-of-a-time all on its own.

The negative and positive terminals of the detector's battery are connected to a pair of conducting plates, giving one plate a positive charge and the other a negative. Opposite charges attract, so the positively charged atoms in the air move toward the negative plate and the electrons move towards the positively charged plate. This completes an electrical circuit.

When smoke passes between the plates, it disrupts the circuit. The electronics in the detector sense the disruption and the alarm goes off to alert us of a fire. Or the neglected turkey in the oven.

Logan Wright

Date 1969

Scientists Kenneth House, Randolph Smith

Nationality American

Why It's Key Think of it as a nuclear seatbelt. The smoke detector is a superhero of an invention: it's powered by radioactivity and it's brutally effective.

Key Invention **Quartz watches**
Just in time for Christmas

The invention of the quartz clock in 1927 was a big achievement. Forty-two years later, in Japan, Tsuneya Nakamura was hoping for a much smaller one: a quartz clock that he could wear on his wrist. But fitting the quartz "tuning fork" and its circuitry into a wristwatch-sized package was going to mean some serious miniaturization; at the start of the 1960s, a typical quartz clock was the size of a filing cabinet.

Nakamura was an engineer at Suwa Seikosha – part of the Seiko group. Throughout the 1960s, they worked tirelessly on making quartz clocks smaller, developing the chronometers used for the 1964 Tokyo Olympics, and fitting their clocks in the cockpits of the Shinkansen "Bullet Trains."

During the late 1960s, Nakamura's team built on developments in integrated circuits and battery technology to produce prototype watches. In 1969, working with colleagues at the Suwa Seikosha factory, they finalized the design for a quartz wristwatch that could be mass-produced. Using electronics packaged into a single hybrid module, and with a miniature stepper-motor driving its hands, it could survive the rigors of everyday use, was accurate to five seconds a month, and would run for a year on its tiny battery.

The Seiko Quartz Astron 35SQ was launched on Christmas Day, 1969. In its 18-carat gold case, it proved so attractive to consumers that a hundred were sold before the New Year arrived, even though, at ¥450,000, the watch cost as much as a family car.

Matt Gibson

Date 1969

Scientist Tsuneya Nakamura

Nationality Japanese

Why It's Key Today, 90 per cent of the billion watches produced each year are quartz watches, bringing low-cost, high-accuracy timekeeping within the reach of nearly everyone on the planet.

Key Person **Dorothy Mary Crowfoot Hodgkin**
A Dame with attitude

Dorothy Hodgkin (1910–1994), one of the most famous British women scientists, was actually born in Egypt. For thirty-five years of her illustrious career she worked on the structure of insulin, and improving the technique of X-ray crystallography to enable it to work out the structure of large and complex biological molecules. The structure of insulin was finally resolved in 1969, but, always conscious of the importance of applying her research to improve humanity, Hodgkin continued to work on insulin and its role in understanding and treating diabetes.

Alongside her work on insulin, she used X-ray crystallography to determine the structure of other biologically important molecules such as cholesterol, penicillin, and vitamin B12, which led to the award of the Nobel Prize for Chemistry in 1964.

In 1965, Dame Dorothy followed in the footsteps of Florence Nightingale as only the second woman ever to become a member of the Order of Merit.

Hodgkin was heavily influenced by the eminent chemist and social historian, John Desmond Bernal – not just in her research, but also in developing a strong social conscience. She was an active campaigner for social justice and peace, and was President of the Pugwash Conferences on Science and World Affairs from 1976 to 1988. During this period, she asked all living Nobel scientists to sign a Declaration against nuclear weapons; one hundred and eleven of them did so.

Richard Bond

Date 1969

Nationality British

Why She's Key
Hodgkin's contributions to understanding the structure of natural molecules have led to the development of crucial drugs and saved many lives.

Key Invention **Scanning electron microscopes**
3-D glasses for small stuff

Ordinary light microscopes have been around for over four hundred years, but reached their limits in resolution and magnification a long time ago. Limited by the wavelength of visible light itself, their maximum useful magnification is 1000x.

If we use something with a smaller wavelength than light, however – like electrons – then we can "see" smaller objects. The first electron microscopes were similar to light microscopes except that instead of visible light they use beams of electrons focused through magnetic lenses. But there was a problem. The object under observation would generate secondary electrons or "backscatter." This was a nuisance at first as it could create "noise" and degrade the desired image. But Manfred von Ardenne was a man who saw every problem as a potential solution; he realized this backscatter could be used to generate an image. This image could also be "scanned" so larger areas could be viewed. Unfortunately, his prototype was destroyed during an Allied air raid, and he was forced to abandon his work. It fell to British engineer, Charles Oatley, to carry on Von Ardenne's good work and make a functional system. Over the course of fifteen years, a succession of his research students built five microscopes of increasingly improved performance, culminating in the production of Stereoscan 1 in 1965.

Modern scanning electron microscopes reach magnifications of 1,000,000x and have been used to produce amazing three dimensional images, revealing the intimate details of subjects as diverse as pollen grains and semiconductors.

Stuart M. Smith

Date 1969

Scientist Charles Oatley

Nationality British

Why It's Key Enabled scientists to view the world at a whole new resolution.

Key Invention
Touch screen technology

Rapidly becoming more commonplace in our modern lives, touch screen technology has its roots in the early 1970s and certain patents concerning touch systems date to 1969. The computer keyboard and mouse had recently revolutionized methods of inputting data into computers, but that didn't prove to be enough for one man.

Physicist Sam Hurst had taken leave from Oak Ridge National Laboratory for two years to teach at the University of Kentucky when he was faced with the task of reading a huge amount of data from graphs. It was a job that would have taken months, and Hurst just wasn't having it – there must be a better way, he thought. A few months later he had invented the Elograph coordinate measuring system – the world's first touch screen input device.

Realizing he was on to a good thing, Hurst promptly formed the Elographics company and six years later, in 1977, the company unveiled what has become the most widespread technology for touch screens to date. Five-wire resistive touch screens are made up of transparent layers which can be pushed together. Pressure can then easily be translated into electrical data.

It took a long time for touch screens to become truly mainstream, but these days ATMs and info-centres, juke boxes and quiz machines all make use of the technology. And with the recent release of the Microsoft Surface – a solely touch screen computer – they are set to become a household item.

William Scribe

Date 1969

Scientist Samuel Hurst

Nationality American

Why It's Key Touch screen technology looks set to become an increasingly common part of everyday life in the 21st century.

Key Experiment **From test tube to baby**
First in vitro fertilization of a human egg

In vitro fertilization (IVF), where eggs are fertilized in a test tube, is now a standard treatment for couples who are unable to conceive naturally, but the science behind it is relatively recent. The 1950s and 60s were known as the golden age of IVF, with many key successes leading to the first in vitro fertilization of a human egg in 1969.

Early attempts at IVF focused on mammals' eggs such as rabbits and rats, but results obtained from these were misleading, as both could be activated to form embryos by means other than fertilization with sperm. A significant breakthrough was the discovery of capacitation – the need for sperm to undergo changes in the reproductive tract of the female before fertilization is possible. Another vital experiment was the implantation of fertilized rabbit eggs into a mother and the production of normal young, paving the way for IVF to treat infertility.

The most important experiments leading up to the first successful laboratory fertilization of a human egg were, surprisingly, done on hamsters. The work on hamster IVF showed that the acidity of the culture medium was vitally important for the success of the overall fertilization.

Although the fertilization process was a success, it took nearly ten years before the first IVF baby was born in 1978. The technique has been massively successful; it is estimated that more than a million IVF babies have been born worldwide.

Helen Potter

Date 1969

Scientist Robert Edwards

Nationality British

Why It's Key Helped millions of couples to conceive and fulfill their dreams of starting a family.

opposite Colored scanning electron micrograph of a 3-day-old, 16-cell human embryo on the head of a pin

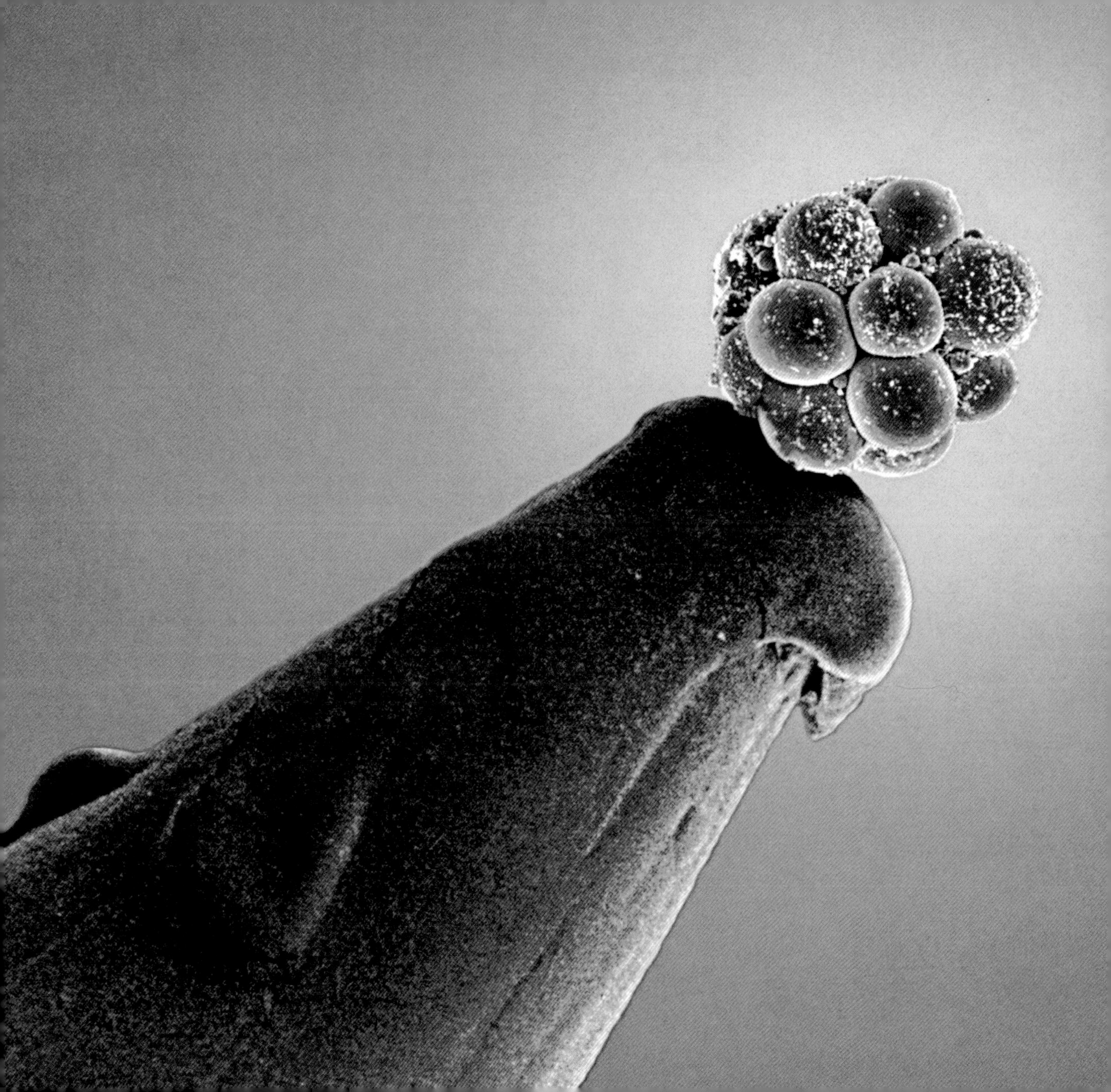

Key Invention
CDs are first produced

James Russell was born in Bremerton, Washington in 1931. Aged just six, he invented a remote-control battleship with a storage chamber for his packed lunch. He eventually went on to earn a BA in Physics from Reed College in Portland.

In 1965, Russell joined the Battelle Memorial Institute as a senior scientist, yet it was his interest in music which was to influence his research. Like many music fans of the era, he was frustrated with the short shelf life and poor quality of his vinyl records. He visualized an audio storage system that would have no contact between player and storage medium, and set about creating it, using a beam of light as the reading device. Russell's previous research had involved the use of digital storage using film and tape; expanding on this, he created the equivalents of binary zeros and ones using light and dark. Battelle supported Russell's research and eventually he developed an early prototype of the digital compact disc (CD), patented in 1970. Light and dark pits, one micron (one millionth of a meter) wide were printed onto a photosensitive surface which could then be read by a laser and converted into a digital signal.

This signal was then fed through a processor, and recreated as sound. Russell continued to develop the CD, expanding further into the digital storage market and creating the CD-Rom. Sony, Philips, and JVC have all co-developed compact disc technologies via strict licensing agreements, but Russell still claims to have "hundreds of ideas stacked up, many worth more than the compact disc."

Vicky West

Date 1970

Scientist James Russell

Nationality American

Why It's Key Changed the way we listen to music.

Key Event
Apollo 13 mission abandoned

Although the first Moon landing of Apollo 11 undoubtedly takes the plaudits as the most important and most famous of the Apollo missions, Apollo 13 certainly comes a close second in terms of notoriety.

Moon bound and 200,000 miles from Earth, a catastrophic explosion in one of Apollo 13's oxygen tanks left the ship venting oxygen into space from its one remaining tank. With the power and life support source disappearing, the Moon landing mission was abandoned and all effort was put into getting the crew home alive.

Thanks to the amazing efforts and ingenuity of ground control, and incredible endurance of the astronauts, the crew were returned safely. They had spent over 140 hours in freezing conditions with little to eat and surviving on about a fifth of their normal drinking water supply.

While 13 is traditionally thought of as an unlucky number, Apollo 13 was less bad luck and more an accident waiting to happen. The failure of the oxygen tank was caused by 28 volt fuel tank heaters being hooked up to a 65 volt supply. All the Apollo missions up to number 13 had flown with this set-up, so it was almost inevitable that at some point this would happen.

Jim Bell

Date 1970

Country USA

Why It's Key Reignited American enthusiasm for lunar exploration and re-wrote the manual in terms of safety and emergency procedure for spaceflight.

opposite The command module of Apollo 13 is recovered after landing

NT
NAVY
66
HS-4

Key Discovery **Reverse transcriptase**
Reversing molecular biology's central dogma

Reverse transcriptase is not a familiar phrase for most non-scientists, but its discovery explained a riddle that had been bothering scientists for years. Tumor viruses, which infect cells and can cause cancer, contain either DNA or RNA in their genome. Before the discovery of reverse transcriptase, it was known that DNA was made into RNA, and how it happened. It was also known that somehow RNA from tumor viruses had the ability to give rise to DNA copies inside their infected cells. But the mechanism of this reverse RNA to DNA reaction remained a mystery.

In 1970, two groups of American scientists, one led by Howard Temin, the other by David Baltimore and Satoshi Mizutani, appeared to have tackled this problem by identifying the enzyme responsible for the reaction. It was Temin in the early 1960s who originally postulated that an RNA-containing virus could give rise to a DNA copy in its infected cell. The discovery of this enzyme, by purification of the tumor virus particles, added one more level of complexity to the DNA-to-RNA-to-protein concept of molecular biology, and provided scientists with a useful new tool in their manipulations of genes and understanding of cancer.

Baltimore and Temin, along with Italian scientist Renato Dulbecco, were awarded the Nobel Prize for medicine in 1975 for this discovery. A blatant omission from this prize-winning group was Satoshi Mizutani, Temin's postdoctoral fellow who, it has been said, was just as responsible for the discovery and design of the reverse transcriptase experiments as Temin himself.

Rebecca Hernandez

Date 1970

Scientists Howard Temin, David Baltimore, Satoshi Mizutani

Nationality American, Japanese

Why It's Key Explained how RNA-containing viruses infect their hosts, and revealed an important missing link in the study of molecular biology.

Key Event **Huge chunks of flying steel**
Jumbo jets take to the skies

In the 1960s, airlines realized that they were going to need something big to accommodate the increasing demand for air travel. The president of Pan American Airways, Juan Trippe, felt that big would not suffice and looked instead for something else; something jumbo.

He talked to Boeing, who concluded that the solution would be a plane that could comfortably house over three hundred passengers. After receiving financial support from Pan American, Boeing went to work. It was a monumental project, involving over 50,000 people and 75,000 engineers' drawings. The jet's engines were a project all on their own, and proved to be the most efficient, quiet, and powerful airplane engine yet.

When the smoke cleared, 4.5 million parts were connected together, in a plane to be admired by all. The jumbo jet is 70.4 meters long, and carries more than four hundred passengers and a ton (literally) of air. Indeed, just the economy section is longer than the Wright Brother's first flight.

Although its maiden flight was a little shaky (engine trouble slowed its journey), the 747 became a hit. In 1970, Pan American began to use the plane to connect New York to London. Almost forty years later, the jet has an admirable safety record and, as of 2005, had served 3.5 billion passengers. It is popular with customers, for its economy, and with the president of the United States, who relies on a pair of 747s known as Air Force One for personal transport.

Logan Wright

Date 1970

Country USA

Why It's Key The jumbo jet, or "Everyman Plane," has helped to make air travel affordable and bridged the once-daunting pond known as the Atlantic Ocean.

opposite The first Boeing 747 takes off

Key Experiment **Chimp talks back**
Washoe learns American Sign Language

Our closest relative, biologically speaking, is the chimpanzee. How similar chimpanzees and humans actually are is a subject of ongoing debate; a debate that intensified in 1970 when Washoe the chimpanzee communicated in American Sign Language.

Allan and Beatrice Gardner acquired "Kathy," as she was then known, from the United States Air Force, for their research into teaching chimps human language. They renamed her Washoe after the county in Nevada where she was to grow up. The Gardners quickly concluded that attempting to teach Washoe vocal communication would be a waste of time, since chimps don't have the necessary vocal physiology, so they focused instead on sign-language.

The idea that apes might learn sign language was not a new one. In 1925, primatologist Robert Yerkes had theorized that apes might be capable of this, as had a number of other scientists even earlier.

The first sign Washoe learned, initially through a technique called "operant conditioning," was the signal for "more." This was quickly followed by others, building up to Washoe's current vocabulary of around two hundred different signs. In addition to knowing all these signs, Washoe has demonstrated the ability to learn new ones without operant conditioning (through simple observation) and to combine signs into novel and meaningful sentences; no-one has taught her to do this.

Barney Grenfell

Date 1970

Scientists Allan Gardner, Beatrice Gardner

Nationality American

Why It's Key The Washoe project is the first to offer real evidence that non-humans can be taught a human language and communicate successfully with it, an important milestone in our understanding of primate intelligence.

Key Event **The Aswan Dam completed**
Engineering feat, ecological defeat?

The Aswan High Dam was fully completed in 1970. Spanning the River Nile in Egypt, just north of the Sudanese border, the dam was built to provide flood protection, hydroelectric power, and irrigation for the growing population downstream in the Nile floodplain.

Although deemed an impressive engineering achievement, the construction of the dam has had many negative impacts, both social and ecological. Stopping the annual floods actually led to agricultural problems downstream, as the nutrient-rich silts normally deposited on the soil are now held behind the dam. An average flood contained around 5,500 tons of phosphate and 280,000 tons of silicate. These nutrients were also crucial to the ecology of the river and the Mediterranean Sea; their absence has led to declines in river and marine species, including commercially important shrimp. The lack of sediment transported downstream has also meant that the Nile delta is decreasing in size since the sands are no longer being replenished once they are washed away.

Before the reservoir was flooded, 90,000 Nubian people were displaced, forcing them to leave their traditional farming lifestyles along the Nile. The construction of the reservoir also threatened ancient sites such as the Abu Simbel temples, prompting the international body UNESCO to launch a campaign to dismantle and re-site the ancient monuments. It was this campaign that initiated the development of the UNESCO World Heritage List, protecting sites of world class cultural and natural value.

Emma Norman

Date 1970

Country Egypt

Why It's Key The social and environmental impacts of the dam led to international controversy over the construction of large dams, and contributed to the designation of the UNESCO World Heritage List.

opposite Satellite image of the Aswan Dam

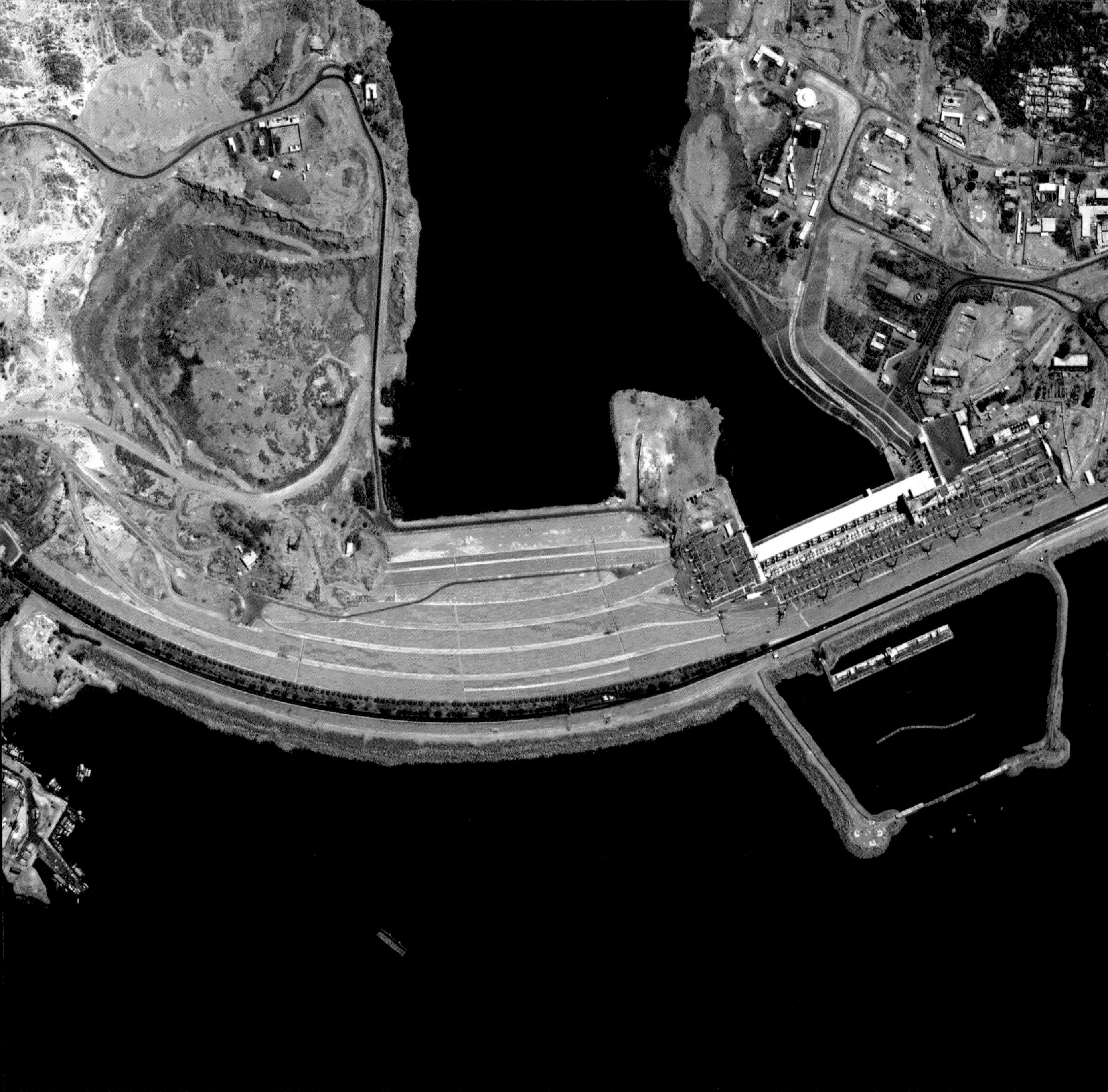

Key Invention **Caught in the light** Optical tweezers

Optical tweezers use tightly focused laser beams to manipulate microscopic objects such as cells or even atoms. They were invented by physicists Art Ashkin, Steve Chu, and their colleagues at Bell Labs in Holmdel, New Jersey, and are similar to the tractor beams in sci-fi films, just on a much smaller scale.

Ashkin had considered the potential applications of using radiation pressure from light for some time and, in 1970, decided to try and prove his theories. He began by focusing a laser beam down onto tiny latex spheres in water, and watched as they were pushed along. He then tried using another, opposing, beam to hold the particles in place. It worked – the spheres were being held with just light; this is now commonly referred to as an "optical trap." Despite a clear understanding of the process, it wasn't until 1986 that Ashkin and Chu first demonstrated the process in detail. In practical applications they have been used to organize cells, track the movement of bacteria, and alter cell membranes. Most recently, they have been crucial pieces of equipment in the development of molecular motors, in cloning experiments, and during in vitro fertilization (IVF).

In addition to their accuracy, there are two more big advantages of optical tweezers – they never get dirty, and storage is just a matter of turning out the light.

Mike Davis

Date 1970

Scientists Art Ashkin, Steve Chu

Nationality American

Why It's Key Enabled the manipulation of objects as small as cells, or even individual strands of DNA, without disruption.

Key Invention **Floppy disks**

Floppy disks - remember them? Today, in the age of USB sticks and their flash memory, the floppy disk is more like the backward yokel cousin of his shinier, brainier relatives. Now left disowned and forgotten in desk drawers and offices the world over, there was a time when floppy disks were considered both costly and hi-tech.

The first floppy disk to be widely used was invented by IBM engineers Alan Shugart and David Noble in 1970, but the innovation had been patented in Japan some twenty-eight years earlier by inventor Yoshiro Nakamatsu.

The disk was an eight-inch circle of plastic coated with iron oxide, which would be "written" to, by using a magnetic head to align iron particles in specific directions, meaning it could be read rather like a cassette tape. The first disks could hold 80 kilobytes of information, as well as being lightweight, compact, and therefore portable. It's no wonder the floppy swiftly became a techie's best friend in a time when hard drives weren't big enough to store operating systems.

Floppies enjoyed their heyday in the late 1990s but have since fallen almost entirely from use, with drives no longer being installed as standard on the vast majority of computers. Innovations in flash memory have led to smaller USB devices with a much greater memory capacity than the 1.44 megabytes the final floppies could manage.

Nicola Currie

Date 1970

Scientists Alan Shugart, David Noble

Nationality American

Why It's Key The advent of the floppy disk allowed greater, quicker file storage, and became so popular that its global use continues today.

Key Publication
Everything is made of string

When we consider the matter that makes up everything around us, we assume it to have three dimensions. But when describing tiny particles of matter, it is common for scientists to assume that these 3-D particles become so small that they are actually zero-dimensional, point-like particles. While this description serves the majority of science well, one aspect it cannot account for is gravity.

In nature, there exist four fundamental forces – the strong nuclear force, the weak nuclear force, the electromagnetic force and gravity – all of which, except for gravity, can be explained using quantum mechanics. String theory tries to unite these forces in a grand "Theory of Everything". In string theory, properties of forces, and mass, are explained by the way in which strings vibrate.

The theory, published in 1970, assumes that fundamental particles are not made up of tiny point-like pieces, but rather tiny one-dimensional strings folded into ten or more dimensions – something which is extremely difficult to comprehend since we live in a world of three spatial dimensions (length, breadth and height) and one time dimension.

Yet if proved to be true, string theory would be the first description of all four fundamental forces using quantum mechanics; a result which even eluded Einstein throughout his career. The predictions, however, of the original theory didn't hold up to experimental testing and it fell from general favour in the mid-1970s. But in 1984 string theory came back with a bang in the first "superstring revolution".

Gavin Hammond

Publication "Dual-Symmetric Theory of Hadrons"

Date 1970

Author Larry Suskind

Nationality American

Why It's Key String theory could unify all four fundamental forces, leading to a "Theory of Everything" – one of the most elusive goals in physics.

Key Event
Wow! It's a search for little green men

Evolution has rendered humans fascinated by unfamiliar faces and threateningly intellectual rivals. Our enthrallment with the prospect of making contact with extra-terrestrial intelligence, therefore, is of little surprise. In 1971, NASA decided to analyze the scientific and technological requirements for an effective search; the Search for Extra-Terrestrial Intelligence (SETI) had begun.

Although SETI has now branched out into searching for optical signals, the core principles have remained unchanged since Frank Drake first pointed an antenna at two nearby Sun-like stars, in early 1960. Drake monitored the frequency (1,420 megahertz) associated with the Universe's most abundant element, hydrogen. Sun-like stars were chosen as they were more likely to harbor Earth-like planets, and therefore forms of life which would be easiest for egocentric humans to spot. Due to the sensitivity of the telescopes required, there have been plenty of false alarms; however the imagination of the astronomer present tends to determine the notoriety of these incidences. On August 17, 1977, Jerry Ehman scribbled "Wow!" on a print-out from the "Big Ear" radio telescope in Ohio. He was amazed by how closely the signal matched the expected signature of an interstellar communication.

The "Wow!" signal remains a great mystery, as it has never been explained, nor heard again. Despite this, the interest that is generated from a few seconds of noise from a distant star or galaxy is a clear indication of the importance that this search holds for the human race.

Christopher Booroff

Date 1971

Country USA

Why It's Key Understanding whether intelligent life is miraculous or inevitable is of great importance to us.

Key Publication
The Saffir-Simpson Scale

Hurricanes can be one of nature's most devastating creations. They can shred roofs, destroy walls, flood towns, and ravage entire states in certain circumstances. It's strange then that, up until the late 1960s, nobody thought of a good way to compare hurricanes to one another.

Herbert Saffir was a structural engineer working for the United Nations when he decided to codify the scale that now bears his name. Saffir based the code largely on the prevailing wind speed and what that would do to extant structures. Bob Simpson, then the director of the National Hurricane Center in the United States, added an extra dimension to the scale, that of the devastating torrent of water that precedes the hurricane: the storm surge. The two together are able to predict fairly accurately the amount of damage a storm can do to a given area, and allow comparisons between storms.

The Saffir-Simpson scale is familiar to most people who live on the east coast of the United States. The scale ranges from category one to category five. A relatively benign category one storm caps out with 150 kilometers-per-hour winds and causes little structural damage. In contrast, a huge category five storm doesn't even start until wind speeds reach 250 kilometers an hour, and has no top wind speed. And the damage level? As per the scale itself: catastrophic.

B. James McCallum

Date 1971

Authors Herbert Saffir, Bob Simpson

Nationality American

Why It's Key The scale allows hurricane damage to be predicted and pre-empted, and allows comparisons among storms.

opposite View of Hurricane Bonnie a category 2 storm as seen from Shuttle Mission STS-47

Key Experiment
The Stanford Prison experiment

It is one of the most infamous, and terrifying, experiments of all time. Twenty-four normal, middle class, law-abiding men were randomly assigned roles as either inmates or guards in a "prison." So rapid was the dehumanization of the prisoners and rise of brutality among the guards, that the two-week experiment was cut short after just six days, but not before even the scientists themselves had felt the lines between experiment and reality blur.

The experiment was carried out by psychologist Philip Zimbardo and his colleagues, who constructed their mock prison in a basement in Stanford University. The prisoners were dressed in ill-fitting smocks, known only by number, and locked in cramped cells. The guards were issued with wooden batons, military-style uniforms, and mirrored sunglasses. The situation was soon virtually out of control. Prisoners experienced, and accepted, increasingly sadistic treatment at the hands of the guards; two were so traumatized that they had to be released early. Even Zimbardo, who assumed the role of prison superintendent,felt drawn into the role – when rumors of a planned rescue attempt circulated, he contacted the police to ask if he could transfer his prisoners to their cells.

The Stanford Experiment showed how easily good people can be manipulated by their environment to perpetrate or accept heinous acts. Zimbardo has since campaigned for better prison conditions and has drawn parallels between his experiment and the atrocities committed at, for example, Abu Ghraib prison in Iraq.

Mark Steer

Date 1971

Scientist Philip Zimbardo

Nationality American

Why It's Key Showed how easily people can be affected by their environment to carry out inhuman acts.

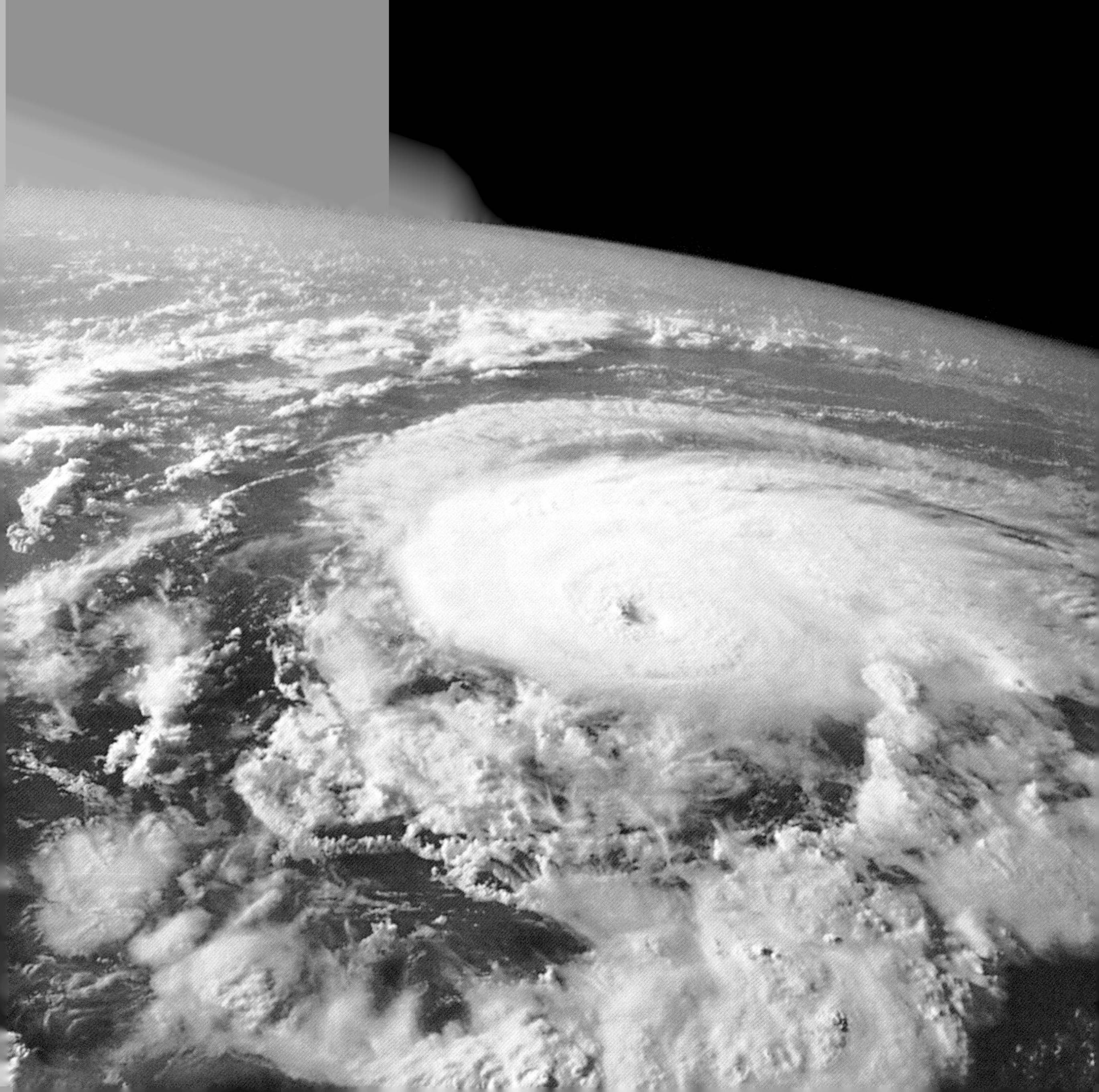

Key Event
Three Soviet cosmonauts dock with Salyut-1

Despite the occasional tragedy plaguing its otherwise successful story, Salyut-1 is one of the most significant objects that has ever launched at high speed off the face of the Earth. The first manned space station to be put into orbit, Salyut-1 had a wide range of objectives, including photographing the Earth, the study of plant life in space, and experiments with intense gamma rays, as well as a number of military experiments. Getting people on board to carry out these tasks, however, was another matter.

The intention was to send the first crew of cosmonauts aboard Soyuz 10 to dock with the station, and for them to carry out a number of these objectives. Unfortunately, the mission did not go entirely to plan. Although the craft did manage to make prolonged contact with the space station, the hatch of Salyut-1 refused to open, preventing the cosmonauts from gaining access. The mission was aborted and the crew returned to Earth.

The following mission, Soyuz 11, was even more of a mixed bag. The three cosmonauts docked successfully with Salyut-1 and spent twenty-two days aboard, setting new space endurance records. Sadly, on their journey back to Earth, one of the return capsule's hatch valves loosened, causing a fatal loss of pressure. All three members of the mission crew were dead when they landed.

But, however sad the situation and however huge the sacrifice, this was still a great advance in the Soviet Union's ongoing space exploration program.

Chris Lochery

Date 1971

Country USSR (now Russia)

Why It's Key The first manned space station in orbit, proving (albeit across two separate missions) that it was entirely possible to conduct human experiments over an extended duration in space.

Key Discovery
Restriction enzymes added to the genetic engineering tool box

In every biological laboratory across the world, there is most likely a freezer stocked full of restriction enzymes. Scientists working in all manner of biological fields use them day in and day out as tools, as a picture-framer would a craft knife. A natural phenomenon, restriction enzymes have been utilized to progress our understanding of genetics and advance the possibilities of genetic engineering. Their initial discovery by Werner Arber has changed the face of genetic research.

In 1968, Arber observed viruses' DNA being destroyed upon entry into bacterial cells. It later emerged that the DNA was being cut up into little pieces by proteins called restriction enzymes, made by the bacteria, which protected them from the virus. These results were reported in 1971.

Daniel Nathans and Hamilton Smith went on to discover that these enzymes were able to scan DNA and then cut it at specific places; over 900 such enzymes – each with their own cutting point – have since been isolated from 230 species of bacteria. Nathans and Smith were awarded the Nobel Prize in 1978 for their discovery.

This revelation meant that genes could now literally be cut and pasted, heralding the dawn of the genetic engineering era. Using different restriction enzymes, genes could be removed, inserted, or swapped around. These natural chemical knives are used today to show the involvement of genes in disease and development.

Nathan Dennison

Date 1971

Scientist Werner Arber

Nationality Swiss

Why It's Key Enabled scientists to harness the ability of bacteria to cut up DNA, opening up new possibilities for genetic engineering.

Key Invention
First commercial microprocessor

Integrated circuits had developed well through the 1960s, miniaturizing complex electronic circuitry onto slivers of silicon. Each circuit had to be specially tailored to its job, however, and that cost time and money. At the same time, programmable computers had been around for many years, and could do lots of different things, depending on the instructions they were given. They were still big devices, though, often the size of a filing cabinet.

Combining these two technologies to produce a miniaturized computer was an inevitable idea. At the end of the 1960s, Ted Hoff, an engineer at Intel, was given the job of making a few different calculators for a Japanese company, and decided that instead of making several different integrated circuits, he'd be better off producing one single chip capable of doing all the jobs, depending on the instructions it was given. During the development of this chip, Intel realized what they had, quite literally, in the palm of their hands: a programmable, general-purpose computer with the power of machines that, a decade before, had taken up entire rooms. Making a deal with the calculator manufacturer, Intel kept the rights to sell the chip to other people, and released the Intel 4004 processor in 1971.

Packing 2,250 transistor "switches" into a chip the size of a fingernail, the 4004 began Intel's transition into a processor-manufacturing giant. Physically, the latest Intel microprocessors aren't much bigger than the 4004, but they contain around 300 million transistors.

Matt Gibson

Date 1971

Scientists Ted Holt and colleagues

Nationality American

Why It's Key The arrival of the Intel 4004 was the advent of the microcomputer, the revolution that would free computers from their huge, air-conditioned rooms; bringing them into our offices and homes, and even into our cars, televisions, and toasters.

Key Invention **Liquid Crystal Displays**
Fergason's flat screen revolution

The first suggestion of being able to use liquid crystals in a visual display came in the 1960s, but it was still a relatively new field in modern science. During his research at the University of Pennsylvania, American physicist James Fergason became interested in liquid crystal research, and noted that liquid crystals, unlike their solid cousins, reflect light when they are hit with an electric current.

To construct an effective liquid crystal display (now commonly known as LCD), a layer of liquid crystals is sandwiched in a thin space between two sheets of glass fitted with tiny electrode bars. Fergason found that when he passed an electric current through very specific combinations of these electrodes, he could cause the glass to light up at certain points. This created contrasting areas of light and dark across the glass, which appeared as basic images. Early LCDs used a charge to scatter the light, but they suffered from high power consumption, limited lifespan and poor contrast. However, by 1971, Fergason had patented a new technique, without the need for a current flow and therefore allowing for more portable devices.

LCDs are now used in calculator, computer, and digital watch screens the world over. The industry is estimated to be worth an astonishing US$10 billion dollars, and Fergason now holds over five hundred foreign patents.

Hannah Welham

Date 1971

Scientist James Fergason

Nationality American

Why It's Key Used simple electrochemistry to provide an innovative and popular new screen system.

Key Person **Carl Sagan**
A stellar performer

The Earth is a very small stage in a vast cosmic arena, and Carl Sagan (1934–1996) was a brilliant scientist with an extraordinary talent for communicating scientific ideas to the non-scientific community. The opening twelve words of this piece were Sagan's own, inspired by a distant image of the Earth in which it appeared to be nothing more than a pale blue dot. Had Sagan been an actor on stage himself, his greatest skill would have been the ability to connect with almost any audience.

Born in 1934, Carl Edward Sagan spent most of his life looking up to the stars, fascinated by the prospect of finding extra-terrestrial intelligence. He devoted much of his life to understanding the evolution of intelligence on Earth; the conditions under which life could prosper; and the technologies required to search for life elsewhere in our Solar System and beyond. Sagan made huge contributions to the American space program, designing missions to the Moon, Venus, and Mars, and interpreting the data returned. But if his contributions to fellow scientists were great, his contributions to the general public were gargantuan; over a billion people watched his 1980s television series *Cosmos* and he won the Pulitzer Prize for his book *The Dragons of Eden*. These award-winning projects helped educate people about the cosmic arena in which we exist.

On December 20, 1996, having previously required a bone marrow transplant to prolong his existence, he was finally defeated by pneumonia, aged sixty-two. Sagan had performed his final encore.

Christopher Booroff

Date 1971

Nationality American

Why He's Key What Sagan offered, that perhaps nobody in any discipline has ever matched, was a genuine link between science and the rest of society.

opposite Carl Sagan

Key Invention **Email**
A hotline to God?

"What hath God wrought?" was the haunting first message sent by telegram, signaling the dawn of the age of electronic communication. The first email, however, was considerably less poignant than the telegram. According to its sender, Ray Tomlinson, it was just some random nonsense used to test the system. Some might think this is fitting, however, as the vast majority of the billions and billions of emails that are sent daily are just spam and viruses.

Electronic mail had been sent through the Massachusetts Institute of Technology's "Compatible Time-Sharing System" since around 1965. Tomlinson's achievement was to enable users to send mail between computers that were not connected to a common network. By his own admission, it wasn't a very hard job, taking just five or six snatched hours of spare time when he should have been working on other things. When he showed it to a colleague, he is reported to have said: "Don't tell anyone! This isn't what we're supposed to be working on."

Scams aside, email allows anyone in the world to remain in contact with anyone else, as long as they have internet access. You can now email the president of the United States (president@whitehouse.gov), or even the Pope (benedictxvi@vatican.va) if you so wish. If, for example, the president of the USA was to email the Pope, he (or she) would be sending a message to the server their computer is connected to (in this case whitehouse.gov). This server then locates the Vatican server (vatican.va), which finally finds the Pope among its users and puts the message in his inbox.

Jim Bell

Date 1971

Scientist Ray Tomlinson

Nationality American

Why It's Key The dawning of an era of rapid communication, email-related procrastination, and of course, spam.

Key Event **CANDU can do**
Heavy-water nuclear reactor produces electricity

The marvellously affirming acronym "CANDU" stands for Canadian Deuterium Uranium and, in the context of CANDU reactor, refers to the model of the first heavy-water nuclear reactor to go into operation. Tested in power plants throughout the 1960s, the first large scale CANDU power plant, Pickering A 1, came to life in 1971.

The Canadian part of CANDU is self-explanatory. Developed and built in Canada by a number of Canadian organizations and individuals, the reactor's design currently accounts for 100 per cent of the country's active power reactors – making the CANDU more Canadian than Mounties, moose, and maple leaves combined.

Deuterium refers to the chemical compound deuterium oxide. Deuterium is a stable isotope of hydrogen and when it is oxidized, to form D_2O, it is known more commonly as "heavy water." The D_2O acts in the reactor as a neutron moderator, encouraging neutrons expelled from a fission event (where an atomic nucleus is split to release energy) to induce further fissions. This is essential to the operation of a nuclear reactor, as it leads toward the chemical chain reaction that releases energy.

Finally, Uranium (chemical symbol U and atomic number 92, fact fans) is the fuel of choice for your modern nuclear reactor – particularly the CANDU, as it can make use of natural, unenriched uranium – so no chemical power plant would be complete without some.

Chris Lochery

Date 1971

Country Canada

Why It's Key Deuterium oxide allows the reactor to make use of a wide range of fissile fuel types, which leads to significant savings in the generation of energy.

Key Person **Stephen Jay Gould**
America's evolutionist laureate

After graduating from Columbia in 1967, with a degree in geology, Stephen Jay Gould began his scientific career at Harvard University, where he was a teacher and spent his entire professional life. He has since become one of the most publicly noted biological figures in the world, and was even immortalized in yellow in the hit cartoon *The Simpsons*.

Gould is most famous for challenging Darwin's theory of gradualism – the idea that evolutionary change occurs slowly. In 1972, with his colleague Niles Eldridge, he argued that instead, species experience long periods of stasis (evolutionary inactivity) interspersed with bursts of evolutionary change. For his theory, known as "punctuated equilibrium," Gould won the Scientist of the Year accolade from *Discovery* magazine.

Often referred to as "America's evolutionist laureate," Gould spent much of his spare time fighting creationism. In 1981, he was a courtroom witness in the famous Mclean vs. Arkansas court case regarding the teaching of evolution in schools.

In 1982, Gould was diagnosed with peritoneal mesothelioma, a form of abdominal cancer. Despite the statistics stacked against him, he survived and went on to write a column, "The Median isn't the Message," in *Discovery*. His contribution brought peace of mind to cancer sufferers worldwide, explaining the statistics of survival rates. He lived for another twenty years before passing away in 2002, but he will always be known as one of the most influential evolutionary scientists who ever lived.

Faith Smith

Date 1971

Nationality American

Why He's Key Revolutionized evolutionary biology with his theory of punctuated equilibrium and well-written science books.

opposite Stephen Jay Gould

Key Event **Computer Space launches**
The video game that didn't quite take off

Billed as "a simulated space battle that pits computer-guided saucers against a rocketship that you control," Computer Space brought video games out from the science labs and into the big bad world.

Inspired by the interactive exhibits that were created and showcased in some of the world's most pioneering laboratories and universities, but restricted by the prohibitive costs of expensive embryonic computer processors, electrical engineer Nolan Bushnell hit upon the idea of using a television screen and a series of transistors and diodes to make a cost-efficient video game system. To make it look a little more impressive, the circuit was housed in a fancy glittering cabinet and, though it looked a little like the type of cash machine you'd expect the Jetsons to use, this finished product marked the beginning of arcade machines as we know them today. Despite the novelty, Computer Space was not a commercial success. Accustomed to live-action pinball machines, the average arcade player found the gameplay difficult, the controls counterintuitive, and the general concept a bit baffling. Popping a coin into a slot, only to be confused by a bunch of dazzling lights and glowing buttons was not an appealing way of burning a bit of disposable income, and so players preferred to spend their money on drinks instead.

Undeterred by the relative indifference shown towards the machine by the gaming community Bushnell and his business partner Ted Dabney set to work designing and manufacturing new games for their own company, Atari.

Chris Lochery

Date 1971

Country USA

Why It's Key The first coin operated video game. It kick-started a whole new form of entertainment that has since become a hugely successful multi-billion dollar industry, with revenue to challenge Hollywood's.

Key Event **DDT banned**

DDT (dichloro-diphenyl-trichloroethane) has become a notorious example of man failing to understand the side effects of his powerful chemical agents on the environment. The story begins in 1939, when Swiss chemist Paul Hermann Muller, uncovered DDT's potential as a powerful insecticide. It was used effectively during World War II to battle malaria-carrying mosquitoes, and earnt its inventor a 1948 Nobel Prize. In the 1950s, even the World Health Organization declared it would try to eradicate malaria using DDT.

Doubts over the chemical's safety didn't emerge until the 1960s when reports of fish poisoning caused by DDT led to local bans in the USA. Pressure mounted to ban the substance, with organizations such as the Environmental Defense Force joining the campaign to highlight its environmental destructiveness, including causing the death of entire populations of birds of prey. By 1970, DDT was banned in Norway and Sweden. The United States followed their example in 1972, as did the UK, but not until over a decade later.

DDT is a persistent organic polluter, lingering active in water for up to two months, and accumulating in the food chain. Although there are few acute human symptoms associated with its use, it has been linked to a worrying number of conditions that take their effect more gradually. Neurological problems, liver damage, and, more recently, cancer, have all been suggested as side effects of its ingestion.

Today DDT is still banned in many countries, but there have been calls for it to be used again in the renewed fight against malaria.

Steve Robinson

Date 1972

Country USA

Why It's Key Banned because of its threat to the environment, DDT's effects on humans are still not fully understood.

Key Publication **Evolutionary change**
Gould and Eldredge posit punctuated equilibria

Before 1972, the general view of Darwinian evolution was that it involved a process of gradual change, that species only turned into new species over very long periods and only very slowly.

"Punctuated equilibria: an alternative to phyletic gradualism," Niles Eldredge and Stephen Jay Gould's landmark paper, proposed that species actually evolved in rapid bursts. When we say rapid this is of course relatively speaking, it still took tens if not hundreds of thousands – of years. But between these bursts of change were long periods when little, if any, change took place.

Building on the work of Ernst Mayr in the 1950s, Eldredge and Gould noted that there were relatively few fossils of any particular species that showed much variation in between the times when significant changes occurred. They surmised that the absence of intermediary forms, compared to the number of stable forms, meant that when change happened, it did so rapidly. Although he gave Eldredge the credit for the idea, "punctuated equilibrium" was a term coined by Gould to describe this process. Gould himself, of course, went on to be a hugely successful author of popular science books.

The idea of punctuated equilibrium, although still not widely accepted, has prompted much debate and has been influential in social and political thinking. The principle that change tends to happen suddenly, interspersed with long periods of stability, has been applied to organizations, political systems, social groups, and even to the history of science itself.

Richard Bond

Publication "Punctuated equilibria: an alternative to phyletic gradualism"

Date 1972

Authors Niles Eldredge, Stephen Jay Gould

Nationality American

Why It's Key Evolution isn't necessarily gradual, but might happen rapidly in short bursts.

Key Experiment **Cops don't feel the beat**
Preventative patrol

In 1972, the Kansas City Police Department performed a groundbreaking social experiment which attempted to discover whether routine police patrols actually achieved anything. At the time, a great deal of money was being pumped into routine marked-car patrolling, on the basis that it kept crime rates down and made people feel safer. The preventive patrol experiment aimed to answer four key questions: Would citizens notice a change in police presence? Would a change in police presence affect crime rates? Would people fear crime more? And would people still think the police were doing a good job?

Three areas in Kansas City were targeted for participation in the experiment, which lasted into 1973. Police presence within these areas varied between no routine patrols at all, normal patrols, and very heavy patrols. The results were surprising. The degree of police presence seemed to make absolutely no difference to the amount of crime actually perpetrated, or to the rates at which crime was reported, in fact the study seemed to suggest that the change in patrol density was barely even registered by the general populace. From a sociological point of view, it was one of the most important non-results ever.

The results of the study were used to call for a change in police procedures in many countries, with regular patrols being decreased in favor of other methods of preventing crime.

Barney Grenfell

Date 1972

Scientists George Kelling and colleagues

Nationality American

Why It's Key A sociological study that yielded practically applicable results, and the first time that the tradition of the police patrol had been called into question.

Key Event
First video games console goes on sale

The history of video gaming is surprisingly long and unsurprisingly geeky. Although there are many claimants to the title "inventor of video games," Ralph H. Baer was indisputably the driving force behind the first commercially available games console.

The bizarrely named Magnavox Odyssey was released in 1972 to reasonable commercial success. Around 100,000 units were sold in its first year of business, despite a rather steep retail price of US$99 and an awful marketing campaign which suggested that the console would only work on Magnavox television sets. There are some suspicions that this was a ploy intended to boost Magnavox's sales.

Rather than create graphical backgrounds on the television screen, the console used plastic overlays to play simple games such as Wipeout, Fun Zoo, and the highly popular Table Tennis. Because of the extremely limited processing power of the machine, the games were all fairly similar, but the overlays gave gamers some much needed, if low-tech, variety. The console had no ability to produce sound, and many games also required the use of dice, poker chips, or score cards.

Light guns (controllers) and Magnovoz's Shooting Gallery game were more innovative additions to home gaming. Costing US$25, the light gun had previously been available in arcades, but didn't make a big impact on the home gaming industry until the mid-1980s, with the success of the Nintendo Entertainment System and Sega Master System consoles.

The Odyssey was available until 1975, by which time it had sold around 350,000 units.

Jim Bell

Date 1972

Country USA

Why It's Key Although it might have been hard to foresee back in 1972, the Magnavox Odyssey was the start of a multi-billion dollar entertainment industry.

Key Invention **CT scans and MRI**
Seeing is believing

Inspiration can strike in the strangest of places, and little did Godfrey Hounsfield realize that an idea that came to him whilst he was out rambling would forever change medical diagnostics.

An electrical engineer by trade, Hounsfield developed X-ray computed tomography, better known as CT scans. These work by taking X-rays of the same area at numerous angles; a computer then combines the images to give a cross-section of that area. CT scans were first used in the early 1970s for diagnosing cysts and lesions, especially in the brain, and continue to be used today.

A similar brainwave struck chemist Paul Lauterbur whilst eating a burger one summer's afternoon in 1971. He realized that, by applying the principles of nuclear magnetic resonance (NMR), a technique used to study the chemical structures of substances, he could produce images of the body in a similar way; magnetic resonance imaging, one of the most sensitive diagnostic tools in a medic's armory, was born.

Lauterbur, viewing the human body as a collection of different atoms, discovered that by applying what was essentially a huge magnet to it, the nuclei of hydrogen atoms inside the body could be made to line up with the magnetic field. Applying a radio pulse to these atoms caused them to resonate at a certain frequency, which could be picked up by a computer, and used to generate an internal image of the body. MRI is particularly useful for imaging soft tissues, as these contain more water, and therefore more hydrogen, than bone.

Ceri Harrop

Date 1972

Scientists Godfrey Hounsfield, Paul Lauterbur

Nationality British, American

Why It's Key Provided new ways for doctors to see inside the body, in great detail, without the need for invasive surgery.

opposite A colored CT scan of a healthy, 23-year-old heart

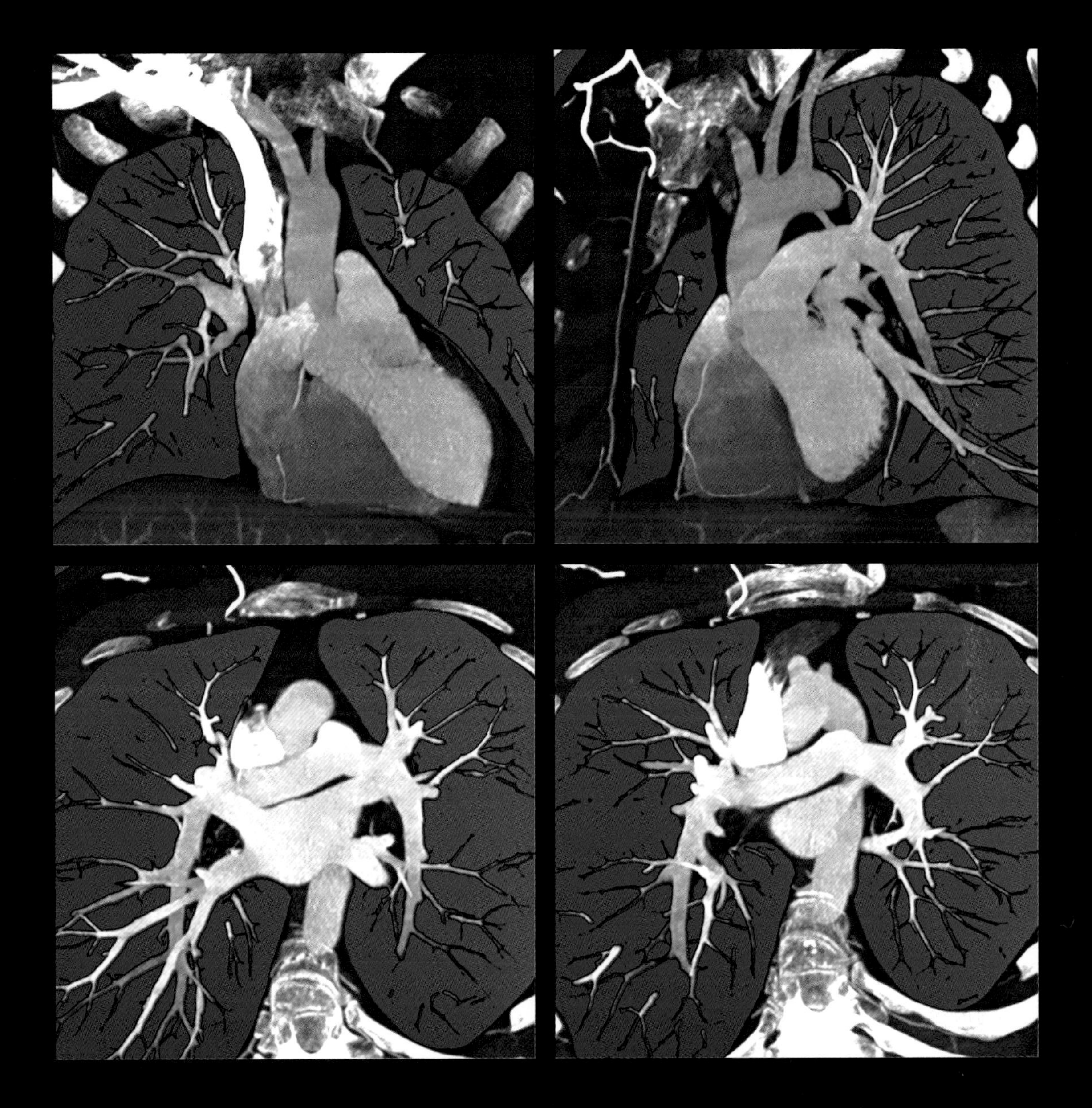

Key Experiment **Rosenhan Experiment**
Being sane in insane places

David Rosenhan's experiments entitled "On being sane in insane places" caused great consternation in the psychiatric community but have brought about important changes in the thinking on mental health.

His first and best known experiment involved gathering eight perfectly sane individuals who were to request to be committed to psychiatric hospitals. The "pseudo-patients," under Rosenhan's direction, complained of hearing unfamiliar voices in their heads that said the words "empty," "hollow," or "thud." These words were chosen because they relate to an existential fear of how meaningless life is. If committed, the volunteers were instructed to behave completely normally and record their experiences. Worryingly, all but one of the twelve hospitals used in the study admitted these perfectly sane individuals under a diagnosis of schizophrenia or bipolar disorder, and took on average about three weeks to diagnose them as in remission. As "sane" observers of the other patients, the pseudo-patients reported a depersonalization of real psychiatric patients. They had little contact with, and were frequently ignored by, medical staff. Also normal behavior, such as the pseudo-patients' writing of notes, was interpreted as pathological.

Due to the outrage from hospitals, Rosenhan conducted a second experiment in which he pre-warned the hospitals of more pseudo-patients, and then cunningly neglected to send any out. In subsequent weeks, nineteen patients – all of whom may have been completely barking – were wrongly identified as sane pseudo-patients.

Jim Bell

Date 1972

Scientist David Rosenhan

Nationality American

Why It's Key Made people think about the usefulness of labels as a diagnostic tool in mental health, and brought about improvements in the care of psychiatric patients.

Key Event **Pioneers 10 and 11 launched**
Exploring beyond the Solar System

Launched from Cape Canaveral in 1972, Pioneer 10 was the first spacecraft to send back close-up images of Jupiter, and has since become the first human-made object to leave our Solar System. The main aims of Pioneer 10's mission were to study interplanetary and planetary magnetic fields and to photograph Jupiter and its satellites. Pioneer 10 and its sister spacecraft Pioneer 11, launched the following year, also provided crucial information to help plan future long-distance missions.

Pioneer 10 sent back valuable data throughout its mission and became the first spacecraft to investigate Jupiter, in 1973. As it drifted even further away from Earth, it continued to send back information until its mission officially ended in 1997. Its last weak signal was received in 2003 when it was some 7.5 billion miles from Earth. It is thought to be heading towards the star Aldebaran in the Taurus constellation, but is likely to take around two million years to reach it.

Pioneer 11 also investigated Jupiter, but went on to become the first spacecraft to explore Saturn and its rings close up, before following its sister ship out of the Solar System.

Pioneers 10 and 11 were both fitted with a gold plaque with a message for any extraterrestrial beings that might happen upon them. The plaque features human male and female nude figures and a diagrammatic map showing where the spacecraft had come from. So far no one has got in touch.

Richard Bond

Date 1972

Country USA

Why It's Key Provided first close-up views of Jupiter and Saturn, and was crucial in planning explorations of the outer reaches of the solar system and beyond.

Key Invention
Pocket calculators

On the big screen in 1972 *The Godfather* enthralled cinema-goers the world over; the Olympic Games were held in Munich; and Paul McCartney appeared live on stage for the first time with his new band "Wings." But all of that pales into insignificance against the invention and release of the world's first "pocket" calculator, the HP-35.

Prior to 1972, calculators had been large cumbersome affairs, available only as desktop versions. They were limited to just the four basic function keys: addition, subtraction, multiplication, and division.

Hewlett-Packard co-founder, Bill Hewlitt, was determined to produce a calculator that would fit in the average shirt pocket and have a greater degree of functionality – a scientific calculator capable of performing trigonometric and exponential functions. Despite studies that showed there was no market for such an expensive gadget – it retailed at US$395, or US$1,800 in today's money – Hewlett drove the development of the "HP" calculator forwards.

The HP-35 (35 was actually added later when the HP-45 was released, earlier models were called simply HP) defied market predictions, selling 100,000 units in its first year, and going on to sell a further 300,000 in the three and a half years it was on sale. Before the HP-35, and other similar scientific calculators, trigonometric and exponential calculations had still been done on a slide rule, a device created in the 1600s. The HP-35 blasted scientific calculation into the twentieth century.

Barney Grenfell

Date 1972

Scientist Bill Hewlett

Nationality American

Why It's Key Creation of compact hand-held devices capable of performing complex scientific calculations with great accuracy. And no more long division!

Key Event **Evolution and creationism clash in Californian schools**

The teaching of evolution in schools has been controversial, especially in the United States, ever since Charles Darwin published his *Origin of Species* in 1859.

The U.S. Constitution does not allow science, or any other subject, to be taught in schools according to any particular religion or doctrine. As a result, in a test case in 1925, John Scopes, a high school biology teacher in Tennessee, was charged and found guilty of teaching the theory of evolution. He was fined US$100.

Although the ruling was widely ignored, it wasn't until 1968 that the U.S. Supreme Court finally decided that any state laws prohibiting the teaching of evolution in schools were invalid. But the powerful creationist lobby continued to challenge school curricula and, in something of a backlash four years later, the Californian State Board of Education ruled that all theories of human origin should be taught as speculation not fact, including evolution. In California at least, biblical accounts of the creation were therefore given equal footing with Darwinian theories of evolution by natural selection. The ruling was resisted but not overturned.

Other U.S. States ruled differently. In 1982, a court in Arkansas decided that such balanced treatment of evolution and creationism actually violated the U.S. Constitution, on the basis that while evolution was a scientific idea, creationism was not a form of science, but a religious doctrine, and should therefore not feature in science lessons. The controversy continues.

Richard Bond

Date 1972

Country USA

Why It's Key A key moment in the long-running debate between science and religion, and their conflicting accounts of the origin of humankind.

Key Event **Dawn of the Web** ARPANET and packet switching demonstrated at ICCC

If not for the threat of nuclear war, the Internet might not exist at all. It began in the late 1960s, developed by the U.S. Department of Defense, as a Cold War project to create a communications network that was immune to nuclear attack.

The Advanced Research Projects Agency Network (ARPANET) was assembled with the belief that "the promise offered by the computer as a communication tool between people, dwarfs into relative insignificance the historical beginnings of the computer as an arithmetic engine." Having been successfully established between the University of California, Los Angeles and the Stanford Research Institute on November 21, 1969, ARPANET was publicly demonstrated in October 1972 at the first International Conference on Computer Communication (ICCC), in Washington, DC. ARPA prime movers, such as J.C.R. Licklider, Robert Taylor, and Leonard Kleinrock, had worked with others to develop a communications network that would remain functional, even if some sites were destroyed. Fundamental to this would be Kleinrock's idea of using "packets," rather than hard-wired circuitry to "switch" or route the message.

In packet-switching protocols, each "packet" or block of data contains the information it needs to reach its destination – the sender's computer (IP) address, intended recipient's IP address, and so on. Each packet is transmitted individually and can follow different routes, depending on network availability, to its destination. Once all arrive, they are recompiled into the original message.

Mike Davis

Date 1972

Country USA

Why It's Key Heralded the arrival of the Internet age and packet-based, digital communications systems.

Key Discovery **The beast is stirring** Mapping of the Yellowstone caldera

The Yellowstone National Park attracts more than two million visitors every year, but most tourists are unaware that the park sits atop a sleeping supervolcano. If the volcano were to erupt, its effects would be worldwide and catastrophic.

The geographical region of Yellowstone was first mapped in the late 1960s by a team of scientists headed by Robert Christiansen, working for the U.S. Geological Survey. They discovered that part of Yellowstone rests in a caldera – a roughly circular depression in the landscape that is the crater from a volcanic eruption – measuring a whopping 48 kilometers by 72 kilometers.

It is believed that, over the past two million years, the supervolcano has erupted several times, leaving behind this huge volcanic caldera. The last time the volcano is known to have erupted is 640,000 years ago, and some scientists now fear that this dormant beast is long overdue blowing off some steam.

The only current signs of volcanic activity at the park are the hissing geysers and gurgling mud pools. This activity is the result of a vast pool of magma, or molten rock, bubbling beneath the Earth's surface and causing the land mass to rise and subside. Scientists are worried that if this molten mass breaks through the surface, it would cause an eruption so big, it would make the 1980 Mount St. Helens blow-out look like a sneeze. Luckily, most geologists seem to think that the probability of this happening over the next few thousand years is very low.

Hannah Isom

Date 1972

Scientist Robert Christiansen

Nationality American

Why It's Key Led to the discovery of one of the world's largest supervolcanoes.

opposite Yellowstone Lake lies inside the vast caldera, photographed from the International Space Station

Key Event **CGI heralds the dawn of a new cinematic era**

Cinema-goers today are used to being dazzled by extreme effects created by computers. Computer Generated Imagery (CGI) has transformed film-making beyond recognition. Ever more complex techniques are used to transport us to distant worlds, where we meet frighteningly realistic monsters and revel in a world of animated heroes. CGI has come a long way in the last thirty-five years.

The first time audiences caught a glimpse of this technology was in cult sci-fi hit *Westworld*, released to great acclaim in 1973. *Westworld* was written and directed by Michael Crichton, better known as the author of *Jurassic Park*.

Westworld used two dimensional graphics, employing computer software to pixelate film images, producing the viewpoint of evil robot *The Gunslinger*, played by Yul Brynner. It would be another three years until its successor *Futureworld* utilized 3D computer graphics for the first time to produce the hands and faces of the robots. A further advancement of the use of CGI came a year later when George Lucas used 3D vector graphics in his seminal film *Star Wars IV: A New Hope*.

As if paying tribute to the first use of CGI in film, the 2003 adaptation of Crichton's *Jurassic Park* later became a shining example of what CGI could produce. It proved to audiences the world over that CGI could be used alongside live action to create a thoroughly believable piece of film.

Harriet Ward

Date 1973

Country USA.

Why It's Key CGI has taken movie-going pleasure to a whole new level, and in many cases, whole new worlds.

Key Invention **The Ethernet**

In 1973 at the Xerox PARC Labs, Silicon Valley, electrical engineer Robert Metcalf was looking at ways of connecting Xerox's new computers to a new laser printer and he began to consider possibilities of the ethernet's potential. The following year Metcalf and colleague David Boggs presented their "Draft Ethernet Overview" to the company, explaining how this new technology could be used to connect computers and other devices into a network.

An Ethernet network involves special cabling in which information can be carried between machines, a unique address for each connected device, and a standard protocol (like a language or code) for all devices to use. Once connected, data packets can be sent between any device (called a node). These data packets, called frames, must be written in the same protocol and include the address of the sender and the recipient. An advantage of this system is that nodes can be added to and removed from the network without having to modify the existing devices.

The name "Ethernet," coined by Metcalf and Boggs, was based on the likening of the single communication cable to the concept of the ether, an imaginary medium filling space through which waves may travel.

The Ethernet system was successfully deployed at PARC; in 1979 Metcalf left Xerox to promote the use of personal computers and local area networks (LAN) using Ethernet as a standard. Today, Ethernet is one of the most common LAN protocols used in local networks, particularly in businesses, across the world.

Shamini Bundell

Date 1973

Scientist Robert Metcalf and colleagues

Nationality American

Why It's Key It forms the backbone of virtually all computer network systems.

Key Publication **"A New Evolutionary Law"** The Red Queen Hypothesis

Evolution is hard. The simple fact that environments keep changing means that species have to keep adapting to ensure their continued existence.

But it's not just the non-living parts of the environment – the climate, for example – that might change; other organisms are evolving, gradually changing to become better and better at exploiting the natural resources around them. Just to maintain a level of fitness, and be able to compete with all their neighbours, species must continuously adapt.

This constant need to keep up with your evolutionary neighbours is what Leigh Van Valen highlighted in 1973 when he introduced the world to the Red Queen Hypothesis. The hypothesis is named for a living chess piece Alice meets on her adventures through the looking glass. The piece must keep running in order to stay in the same place just as species must keeping evolving in order to stay competitive.

The outcome of this scenario is the emergence of "evolutionary arms races" between predator and prey, host and parasite. As predators adapt to become faster, sharper, meaner, better at catching prey, so prey have to evolve better and better ways of evading attackers.

The hypothesis was later used to provide a neat explanation for the curious persistence of sexual reproduction which is, after all, a most inefficient way to go about passing on genes. If parasites are constantly evolving to get through a species' resistances, then individuals with novel combinations of genes may stand a better chance of fending them off.

Jamie F. Lawson

Publication "A New Evolutionary Law"

Date 1973

Author Leigh Van Valen

Nationality American

Why It's Key Formalized the idea of evolutionary arms races and provided evolutionary explanations for predator-prey and parasite-host relations as well as for the persistence of sexual reproduction.

Key Discovery **Bugs and biotechnology** The birth of genetic engineering

As Herbert Boyer and Stanley Cohen discussed their work over corned beef sandwiches in a Waikiki deli, neither could have envisaged the multi-billion dollar industry that would stem from their chance encounter.

Herbert Boyer, then working at the University of California, San Francisco, was involved in the isolation of a restriction enzyme. Essentially a molecular pair of scissors, this enzyme cuts DNA into chunks, leaving behind "sticky" ends. With the help of another enzyme (ligase), any two pieces of DNA with the same sticky ends can be glued together. At the same time, Stanley Cohen, professor of medicine at Stanford University, was studying bacterial plasmids. Plasmids are small circular pieces of DNA that often carry the genes involved in antibiotic resistance. Cohen discovered that plasmids could be transferred from one bacterium to another, and with them antibiotic resistance.

In November 1972, at a conference in Hawaii, the paths of Boyer and Cohen crossed. They realized that by combining their expertise they could, in theory, "cut-out" genes from any organism, stick them into plasmids cut open with the same enzyme, and put them into the bacteria *E. coli*. The bacteria should then act as factories, using the information encoded in the inserted genes to make biological chemicals. By early 1973, their idea was realized. For the first time, the commercial potential of such a discovery was recognized, with bacteria engineered to produce medicines such as insulin and human growth factor, heralding today's biotech industry.

Becky Poole

Date 1973

Scientists Stanley Cohen, Herbert Boyer

Nationality American

Why It's Key Proving that genes could be transferred from one species to another was the birth of genetic engineering and the multi-billion dollar biotech industry.

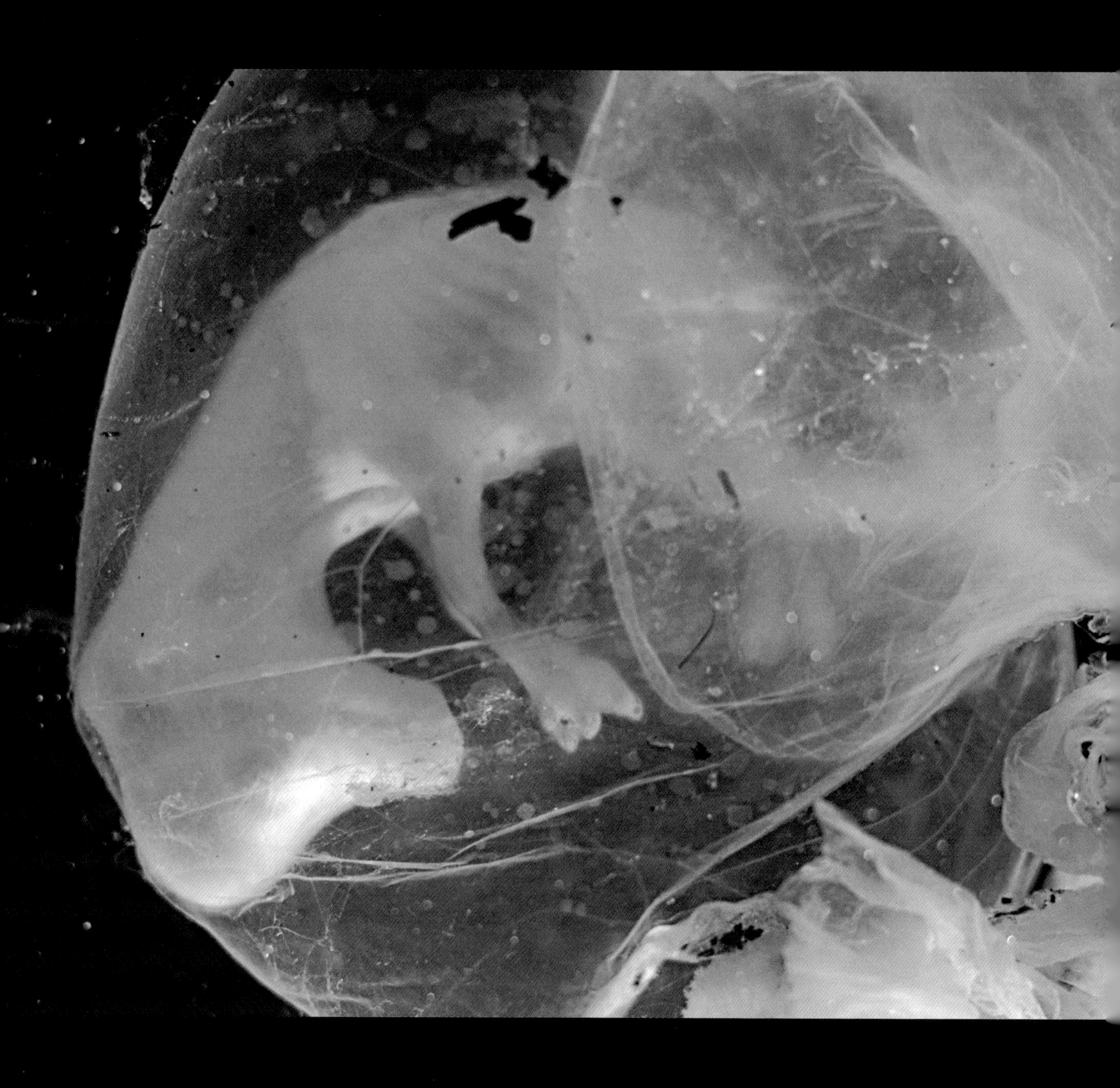

Key Person **Percy Julian**
Breaking the mold to make hormones

An African American born in 1890s Alabama and descended from a slave, Percy Lavon Julian (1899-1975) broke the traditions of the Deep South to pursue his scientific education. His pioneering work in the field of chemistry led to the production of human sex hormones, whose uses range from contraceptive pills to illegal body-building steroids.

Julian's first major achievement was creating physostigmine – a chemical found in the West African calabar bean – in the lab. He then went on to show that another chemical from the bean, stigmasterol, could be converted into the human sex hormones oestrogen, progesterone, and testosterone. This discovery led to the large-scale manufacture of these so-called steroid hormones, for use in a number of medical applications.

Julian next turned his attention to cortisone, another chemical steroid in the group. At the time, this was being produced by the company Merck at great expense, and used for the treatment of rheumatoid arthritis. Using his chemistry skills, Julian successfully managed to produce cortisone from acids found in bile at a much lower cost. Over the years, Julian turned his hand to creating many other valuable chemicals from humble plant or animal origins, eventually setting up the firm Julian Laboratories, Inc., and encouraged other African American chemists to come and join him.

In recognition of his important achievements, Julian was elected to the prestigious U.S. National Academy of Science in 1973. He was the second black man ever to be inducted in the Academy's history.

Kat Arney

Date 1973

Nationality American

Why He's Key Showed that plants could provide valuable drugs, and blazed a trail for black scientists.

Key Event
Calf produced from a frozen embryo

Wouldn't it be great if all of a farmer's cows could give birth to the calves that have the best tasting beef or produce the most milk? Well, that's largely the case these days due to the advancements made in the field of embryology by Ian Wilmut. Long before he facilitated the birth of Dolly, the worlds first cloned sheep, he pioneered the technique of freezing embryos.

Wilmut's idea was always to genetically enhance the barnyard. Cows produce precious few offspring in a lifetime and Wilmut realized that he could substantially improve herds of livestock if he could figure out a way to place embryos from the best cows into inferior livestock. Fresh from conducting research with frozen boar sperm and embryos for his doctorate, he postulated that he could do the same thing with cows. In 1973, he succeeded in producing the first animal grown from a frozen embryo. The animal's name was Frosty.

The implanting of embryos from prized cattle into other livestock was already being done at the time, but had to be done almost immediately. But while it was a relatively simple process to quickly implant an embryo into a single cow, what about a whole herd? Wilmut took away the time factor by freezing the embryos, allowing farmers to implant them in other cows at their leisure. The same basic technique is now being used to help humans with fertility problems.

B. James McCallum

Date 1973

Country UK

Why It's Key Wilmut revolutionized the field of farmyard genetics – giving us higher quality beef and milk, as well as laying a path for human fertility treatments.

opposite A cow embryo at six weeks old

Key Experiment **Creating vitamin B-12**
Chemistry in stereo

The synthesis of vitamin B12 was one of the most complicated chemical constructions ever undertaken. But in 1973, after ten years of painstaking work, Robert Burns Woodward and Albert Eschenmoser finally announced that they had managed it. In fact, most of the lab work was carried out by an army of over 100 students, who did all the bench chemistry, but Woodward and Eschenmoser directed proceedings and planned exactly which reactions would take place.

The molecule in question is undoubtedly a tricky one, due to a property called stereoisomerism. Organic molecules often come in "left-handed" and "right-handed" versions – spiraling clockwise or anticlockwise. Just as your hands can't be superimposed on one another – they mirror each other – neither can a stereoisomer, even though it contains all the same component parts as its opposite number.

Vitamin B12 has nine places where the stereochemistry is important and getting each of these just right took time – unless each was perfect, it wouldn't be the same molecule at all. Whilst battling with the problem Woodward came up with a set of rules to explain how stereoisomers would behave in reactions. These rules were produced in collaboration with Roald Hoffman and became known as the Woodward-Hoffmann rules. In 1965, Woodward had been awarded a Nobel Prize for his work." Nearly two decades later, Hoffmann gained another for the Woodward-Hoffmann rules, a prize that Woodward would have shared had he been alive.

Anne-Claire Pawsey

Date 1973

Scientist Robert Burns Woodward, Albert Eschenmoser

Nationality American, Swiss

Why It's Key Not only could scientists now create an important vitamin in the lab, but methods were in place to produce other complex molecules.

Key Person **Stephen Hawking**
Singularly brilliant

Born three hundred years to the day after the death of Galileo, Stephen Hawking is arguably the most renowned living physicist. He's certainly the only scientist to have appeared in a Pink Floyd album, three episodes of *The Simpsons* and even an episode of *Star Trek: The Next Generation*.

But why so famous? Well, not only has Hawking made some incredible advances in the field of astrophysics, and authored one of the most popular science books of all time, he has done so while suffering a disease that would have put paid to mere mortals such as you and I. At the age of 21, having recently begun his doctorate studies at the University of Cambridge, Hawking was diagnosed with motor neurone disease and given just two years to live. That was in 1962.

Hawking's first major contribution to science was the concept of singularities: points of infinite gravity, where space and time are infinitely warped, the like of which are found at the centres of black holes. His continuing fascination with black holes has led to further revelations about how they must emit radiation and that there might be millions of tiny black holes throughout the universe. Not one to shirk a challenge, Hawking said in 1995 "My goal is a complete understanding of the universe, why it is as it is and why it exists at all." According to another interview, he also hopes to design a warp drive. And who would bet against him?

William Scribe

Date 1973

Nationality British

Why It's Key An inspirational figure in countless ways, Hawking opened the eyes of scientists and the public alike to the wonders and mysteries of modern physics.

Key Publication
What happens when animals play games

Game theory was developed in the 1950s by economists and mathematicians trying to explain why people interact with each other. It wasn't too long before biologists started to apply the same principles to other animals.

John Maynard Smith was the first person to put animal decision-making into a mathematical framework. His 1973 paper, "The Logic of Animal Conflict," co-authored by George Price, set these ideas out in full for the first time. Game theory is useful for understanding situations where an animal's fitness depends in part on the behaviors of the other animals in the population.

One of the most important of Maynard Smith's concepts was the Evolutionary Stable Strategy, which explained all sorts of phenomena which seemed to be at odds with the theories of evolution. For example, most male deer die never having fathered any young, since the dominant individuals in any herd monopolize the females. Why then do female deer give birth to just as many males as females, if males are so unlikely to pass their genes on to the next generation?

What game theory showed was that, if you were to have a population of deer where the female animals give birth to lots of daughters, a mother who gives birth to many sons biases the probably that one of her sons will become one of the few dominant individuals. Therefore the only strategy from an evolutionary perspective is to give birth to equal numbers of sons and daughters.

Catherine Charter

Publication "The Logic of Animal Conflict"

Date 1973

Scientists John Maynard-Smith, George Price

Nationality British, American

Why It's Key Game theory underpins many modern theories of animal behavior.

Key Discovery **Mysteries of aurora borealis finally uncovered** Skylab sent into orbit

America's Apollo Program of lunar exploration was called off early in 1971, with two further missions scrapped due to budget cuts. But scientists were still crying out for access to space, so NASA embarked on a new project, utilizing parts and expertise from the Moon program.

Project Skylab used a Saturn V rocket with a "dry" third rocket stage that was outfitted as a laboratory. America's first space station was sent into orbit unmanned on May 14, 1973, and the first manned expedition to this new outpost launched the following month.

The 75-ton Skylab space station was to provide astronauts with the opportunity to study the Sun and its effects on Earth. The lab included the Apollo Telescope Mount – a solar observatory for watching solar radiation at a variety of wavelengths, including X-rays. In total, some 941 hours of solar observations were performed from orbit during Skylab's 171 days of crewed activity.

One of the major highlights of the project was the discovery of the nature of the Sun's coronal holes. Previous unmanned missions to other planets had encountered regions of the solar wind – the flow of charged particles from the Sun – that were significantly faster than predicted. Skylab data showed that holes in the Sun's outermost layer were responsible for this variable wind, and enabled scientists to link these phenomena to events on Earth such as geomagnetic storms and auroral displays. Skylab burned up in Earth's atmosphere in 1979.

David Hawksett

Date 1973

Scientists NASA

Nationality American

Why It's Key The Skylab intensive solar observation program paved the way for modern space weather prediction.

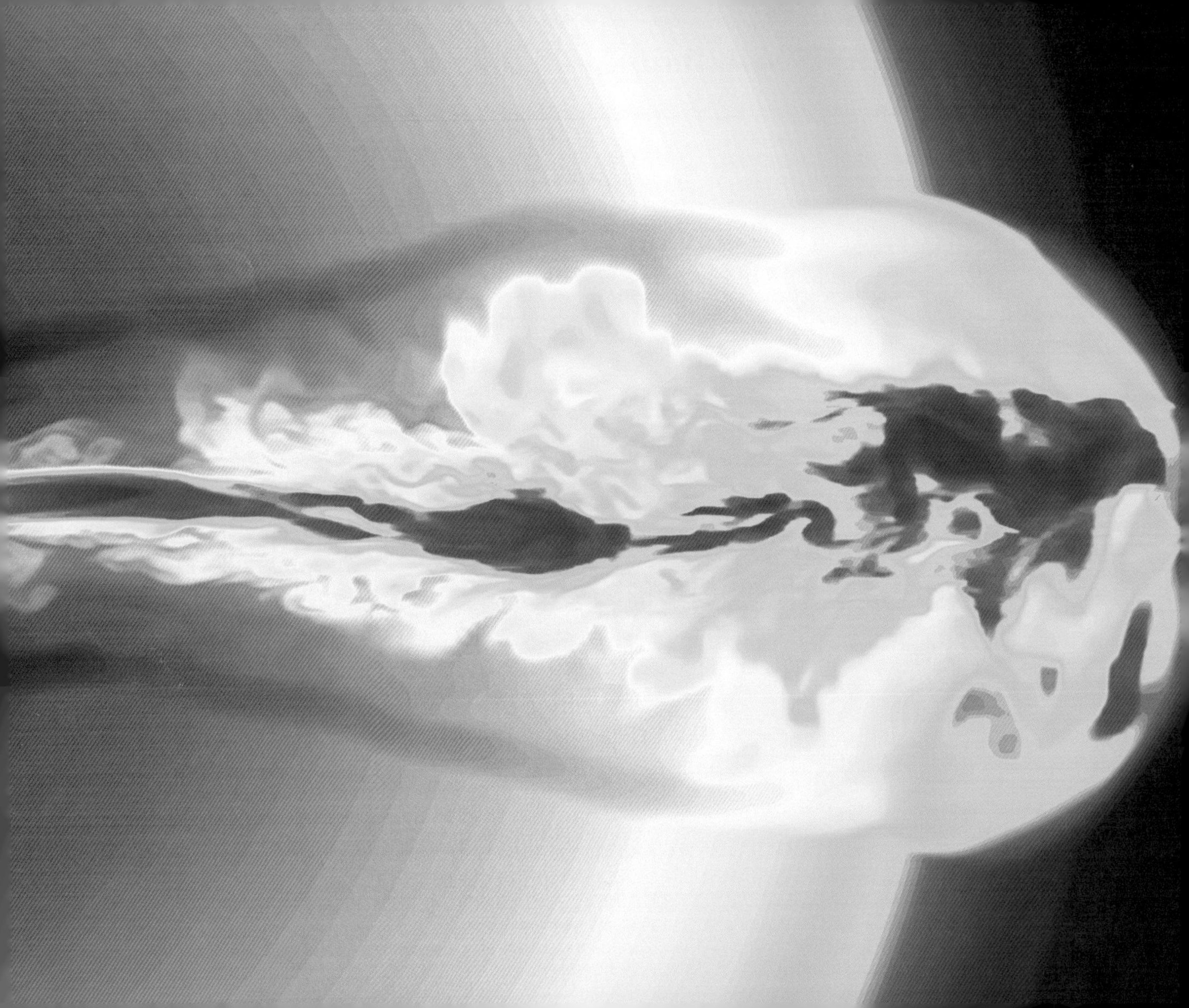

Key Discovery **Gamma-ray bursts**
The largest explosions since the Big Bang

The Vela satellites were designed to look for nuclear explosions in the atmosphere and outer space. Instead they discovered the most luminous events in the Universe.

The United States launched the first pair of Vela satellites fearing that the Soviet Union might secretly test nuclear weapons, despite having signed the Nuclear Test Ban Treaty in 1963. Each satellite had six gamma-ray counters designed to detect nuclear explosions in space. The satellites never picked up any signs of nuclear tests, but did detect unusual bursts of energy coming from deep space. When the sixteen extraterrestrial bursts of gamma-rays were first detected astronomers thought they might be associated with supernovae in distant galaxies. It is now thought that most gamma-ray bursts (GRBs) are caused when enormous stars collapse in on themselves to form black holes. However, different bursts vary greatly from each other, suggesting there may be a number of other reasons for their formation.

GRBs gained notoriety in the 1990s when it was suggested that an explosion from within the Milky Way could cause mass extinctions on Earth. Recent studies, however, have indicated that the probability of Earth being in the way of such a burst is extremely low.

Ancient gamma-ray bursts can reveal young galaxies and provide information on conditions in the early Universe. In 2005, the "Swift" satellite detected the most distant gamma-ray burst yet. The explosion took place almost 13 billion years ago when the Universe was only 900 million years old.

Eric Schulman

Date 1973

Scientist US Government

Nationality American

Why It's Key Gives us information on the deaths of massive stars, neutron stars, and black holes, and can act as a probe of the intergalactic medium.

opposite A gamma ray burst formation

Key Event
The first mobile phone call

If you were going to make the first mobile phone call in history, who would you call? This was the problem that Dr Martin Cooper, Corporate Director of Research and Development at Motorola, faced as he paced the streets of New York in 1973. In the end, his wicked sense of humor caught up with him and he called his rival Joel Engel, head of research at Bell Labs.

"Joel, I'm calling you from a real cellular telephone. A portable handheld telephone." This simple gloating call caused a communications revolution, not to mention much consternation among passers-by.

The potential impact of the mobile phone was summed up by Motorola Vice President John F. Mitchell at the time. In a press release he said, "it will be possible to make calls while riding in a taxi, walking down the city streets, sitting in a restaurant, or anywhere else a radio signal can reach." He predicted that mobile phones would be used by "a widely diverse group of people – businessmen, journalists, doctors, housewives, virtually anyone who needs or wants telephone communications in areas where conventional telephones are unavailable."

The first handset was described as weighing under three pounds (1.36 kg) and being as simple to operate as a conventional touch-tone telephone. It took nearly ten years (and considerably smaller handsets) for the first mobile phones to become commercially available, but in 2007 the number of worldwide mobile phone subscribers passed the three billion mark.

Helen Potter

Date 1973

Country USA

Why It's Key Revolutionized communication by removing the need for wires.

Key Publication **Predicting the future**

Stability and Complexity in Model Ecosystems

By the early 1970s, it was becoming increasingly apparent that human activity was degrading the environment. Acid rain produced from power stations was killing forests and lakes; due to the action of CFCs a hole had appeared in the ozone layer; and over-fishing was leading to the collapse of economically important fish stocks. All of a sudden the future for life on Earth wasn't looking so secure, and scientists found themselves in need of predictions and recommendations that would help avert disaster.

The answer came from theoretical physicist-turned-zoologist Bob May: mathematics. His landmark book, *Stability and Complexity in Model Ecosystems*, developed techniques which are now used across a wide range of sciences. Medics and veterinarians, for example, model how disease spreads through human or livestock populations; knowledge we can use to protect ourselves against bird flu or foot and mouth disease. We can calculate how much of a natural resource can be harvested sustainably, and avoid chopping, hunting, or fishing livelihoods into extinction.

May, now Lord May of Oxford, is now one of the world's most influential scientists and acted as both President of the Royal Society and Chief Scientific Advisor to the UK government. One of the problems that has beset the practical use of ecological forecasts, however, has been the difficulty of persuading politicy-makers to use their results. While May's work gave us the tools to predict an uncertain future, getting government and industry to act on their recommendations is a totally different problem.

Mark Steer

Publication *Stability and Complexity in Model Ecosystems*

Date 1973

Author Bob May

Nationality Australian

Why It's Key Oversaw the transition of ecology from an essentially descriptive discipline to a solid science that could make useful predictions about the future.

Key Person **Niko Tinbergen**

By his own admission, Nikolaas Tinbergen (1907–1988) was an unremarkable student and preferred to spend his time among Holland's "unparalleled natural riches". He had to be persuaded to study biology at Leiden University, but there he developed the interest in animal behavior that would define his career.

In 1936, Tinbergen met Konrad Lorenz, who became his lifelong friend. Together, they laid the foundations for ethology, the new science of animal behaviour. Tinbergen realized many scientists were too focused on what animals were doing. Understanding a behavior involved working out what it was for, how it worked, how it evolved and how it developed. These Four Questions provided a framework that biologists have used to rigorously examine animal behaviour for decades. Tinbergen himself was renowned for devising cunning experiments to answer his four questions in species as diverse as wasps, falcons and humans. In *The Study of Instinct* (1951), he showed instinctive behaviour in sticklebacks, like aggression and courtship dances, could be triggered by simple visual signals. In *The Herring Gull's World* (1953), he worked out which aspects of a parent's bill get chicks fired up by creating a series of puppets with bills of varying shapes and colours.

In 1973, he was awarded the Nobel Prize for his work on animal behavior, with Konrad Lorenz and Karl von Frisch. He died in 1988 but not before influencing a new generation of scientists at Oxford University, including one Richard Dawkins.

Ed Yong

Date 1973

Nationality Dutch

Why It's Key Laid the foundations for an entirely new way of studying animal behavior

Key Person **Karl von Frisch**
The grandfather of animal behavior

As a zoologist, it is unlikely that Karl von Frisch ever contemplated winning a Nobel Prize. There are only five areas that a person can win a Nobel Prize in, and zoology is not one of them. What's more, von Frisch's parents, particularly his father who was a physician, thought the subject was unlikely to take their son very far as a scientist, and so enrolled him at medical school. Although he lasted until the third year of the degree, young Karl was increasingly frustrated by the "medical character" of the course. He dropped out of medical school and, in spite of his family's misgivings, enrolled on a course in zoology, quickly making up for lost time.

Von Frisch's first major finding came as part of the thesis for his degree. It had long been assumed that most animals were color-blind, but he was able to prove that minnows reacted to different colors. He then went on to show that many other animals, including bees, were able to perceive colors just as well as, if not better than humans could.

It must have been with a unique sense of accomplishment that Karl von Frisch accepted the 1973 Nobel Prize in Physiology or Medicine, for his work in animal behavior, most notably communication in bees. His work in connecting animal and human behavior as one and the same was considered too important to overlook.

Jim Bell

Date 1973

Nationality Austrian

Why He's Key Karl von Frisch's work was crucial in the advancement of our understanding not only of how animals behave, but also of our own place in the natural world.

Key Event **The Standard Model of particles**
Elementary, my dear

Willis Lamb won the Nobel Prize in 1955 for his work on the fine structure of the hydrogen atom. In his Nobel Lecture he told how he had heard it said that, "the finder of a new elementary particle used to be rewarded by a Nobel Prize, but such a discovery now ought to be punished by a US$10,000 fine."

Perhaps whoever uttered those words thought the "sea" of subatomic particles that was emerging would leave scientists with too much to do. From the 1930s onwards, regular discoveries were revealing ever finer levels of detail in the subatomic world. The electron is the oldest known elementary particle – one that isn't made of even smaller particles – while protons and neutrons are comprised of particles called quarks.

At a conference in 1974, physicist John Iliopoulos tried, for the first time, to put all the particles into a framework now known as the Standard Model. This construct explained the interaction of particles using three of the four fundamental forces in nature, these being electromagnetism, the weak interaction, and the strong interaction – forces responsible respectively for light and electricity, nuclear fusion, and the cohesion of the atomic nucleus. Gravity, the weakest force, is a shortcoming of the Standard Model as it is not included.

The Standard Model proposes that our observable Universe was composed of particles of matter such as electrons and quarks, as well as bosons – particles which carry the fundamental forces. These elementary components interact with each other to produce the physical Universe.

Andrey Kobilnyk

Date 1974

Country Greece

Why It's Key The Standard Model of Elementary Particles is the most complete theory we have describing the composition and behavior of the smallest particles of matter and forces of nature.

Key Discovery **Lucy**
Oldest human ancestor's remains found in Africa

On November 24, 1974, two paleontologists were returning from a hard day's digging under the baking Ethiopian sun, when one spotted a tiny fragment of bone lying in a ditch. Picking it up, Don Johanson quickly identified it as being from the forearm of a hominid, an early human ancestor. Nearby, Johanson and his colleague, Maurice Taieb, also uncovered part of a skull, as well as ribs, a pelvis, a lower jaw, and a thigh bone. Their remarkable discoveries made up part of the oldest hominid skeleton ever found.

Over the following two weeks, the team, working on the Hadar Formation in northern Ethiopia, unearthed several hundred fragments of bones, all apparently from the same individual. By the time they finished, they had uncovered 40 per cent of the skeleton. That night, it is said, the team celebrated their discovery back at camp by playing the Beatles song "Lucy in the Sky with Diamonds," which is how the skeleton came to be named "Lucy."

From analysis of the skeleton, they were able to deduce that it belonged to the female of a new hominid species, dubbed *Australopithecus afarensis*. Detailed study has revealed that Lucy was only 3ft 8 inches (112 centimeters) tall, and walked upright like a human, yet retained many ape characteristics. The skeleton, at over three million years old, provides us with a potential common ancestor of man and chimpanzees; its importance therefore cannot be overstated.

Steve Robinson

Date 1974

Scientists Don Johanson, Maurice Taieb

Country Ethiopia

Why It's Key At the time they were the oldest hominid remains ever found.

opposite Lucy's remains

Key Event **Is there anyone out there?**
The first attempt to contact extraterrestrial life

In what must be one of the most beautifully optimistic experiments ever conducted, a radio communication was sent from the Arecibo observatory in Puerto Rico out into space. It was directed toward Messier 13, a globular cluster in the Hercules constellation some 25,000 light years away. The message – dubbed the Arecibo message – is the strongest manmade signal ever sent to communicate with extraterrestrials.

The message was a binary code designed to be decoded as a graphic. The information contained therein included the atomic numbers of the elements which make up DNA; numerical data about biochemicals; and crude depictions of a human (essentially a stick man); our Solar System (a bunch of dots); and the Arecibo telescope itself (an "M" with a hat on).

It goes without saying that this was more of a demonstrative experiment than an actual attempt to initiate a bit of chit-chat with aliens since, provided that there is some communicative life form at Messier 13, and that they have no urgent errands keeping them from replying immediately when they finally get the message, it'll be 50,000 years before we get the reply.

Chris Lochery

Date 1974

Country Puerto Rico

Why It's Key Even though it was only a token attempt, the Arecibo message prompted a great deal of thought about how best to communicate with other life forms across space and time.

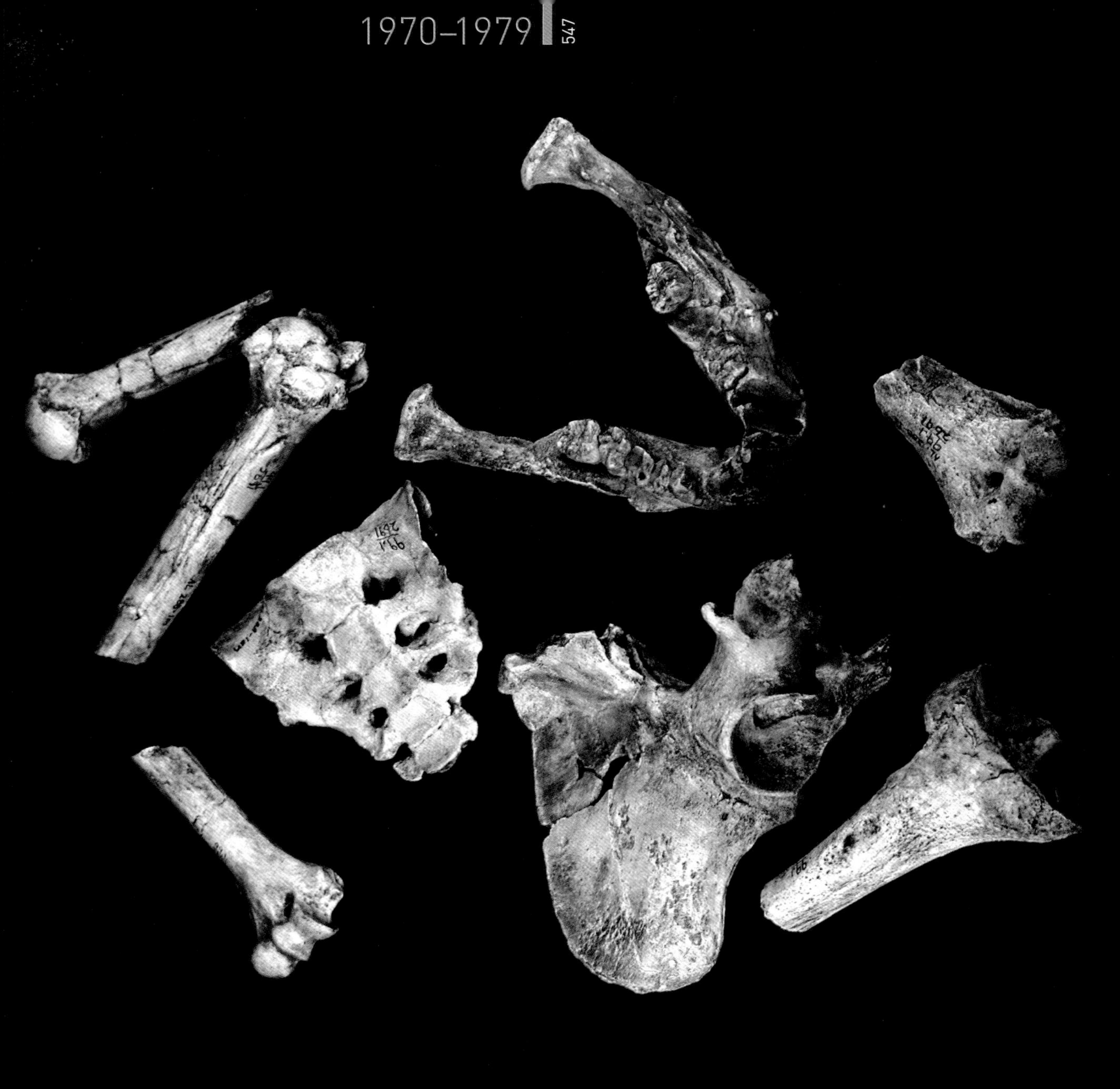

Key Person
James Lovelock

James Lovelock (b. 1919) is the sort of intriguing and increasingly rare type of scientist who is virtually impossible to pigeonhole. He works as independently as is possible in an age of commercial science, and, unusually for a chemistry graduate holding a PhD in medicine, has environmental studies as a focus. Having published over two hundred papers on a range of scientific subjects and filed more than fifty patents, Lovelock is a unique and eclectic figure.

He was elected a Fellow of the Royal Society in 1974 for his diverse contributions to science, which included cryogenically frozen hamsters, geophysics on Mars, and the invention of an electron capture detector. Despite appearing to be the most disparate areas of study imaginable, a theme was arising in Lovelock's thinking.

The Gaia theory, which proposes that our planet is a living superorganism, is not accepted by all of the scientific community, and yet Lovelock remains a popular, if unconventional figure. He is credited with the beginning of modern-day environmental awareness,though he has attracted criticism for his support of nuclear power as means to cut emissions.

Lovelock more recently proposed a system of deep sea pipes to draw nutrient-rich water to the ocean's surface, promoting algal growth and, consequently, carbon absorption. The proposal attracted much media attention whilst being dismissed by many as unfeasible. It might take some unconventional thinking, however, to tackle a problem on the scale of climate change.

Jim Bell

Date 1974

Nationality British

Why He's Key He has changed the way we think about our planet, and greatly influenced our understanding of the problems posed by climate change.

opposite Professor James Lovelock, with his electron capture detector

Key Publication **CFC warning**
There's an ozone no zone

On June 28, 1974, the journal *Nature* published a revolutionary paper by chemists Sherwood Rowland and Mario Molina.

Throughout the 1960s and 1970s, CFCs, or chlorofluorocarbons, were being pumped into the atmosphere at an astonishing rate. Nobody realized that the compounds, now much more limited in their use, could be having negative effects on the Earth's atmosphere. They were simply thought of as efficient molecules for use in refrigerator coolants and aerosol pressure cans that supported a technological revolution.

Luckily, Rowland and Molina realized, before it was too late, that there might be a problem associated with CFCs. Typically, compounds that are released into the atmosphere are broken down in several ways. They may be broken apart by sunlight, washed out by rainwater, or zapped in a reaction with ozone. Rowland and Molina found that none of these processes can happen when CFCs are in the equation, because the compounds are transparent, insoluble, and unreactive in the presence of ozone.

The two pioneers went on to suggest that CFCs must therefore be building up in our atmosphere. We now know that these CFCs will eventually be broken down high in the atmosphere, but only to release chlorine atoms. These highly reactive atoms then go on to destroy the ozone layer, leaving the Earth exposed to deadly ultraviolet rays.

Hannah Welham

Date 1974, *Nature*

Scientists Sherwood Rowland, Mario Molina

Nationality American, British

Why It's Key Brought about the large-scale elimination of CFCs, preventing further damage to the ozone layer.

Key Event
Greenpeace start anti-whaling campaign

Since the 1700s, whales the world over have been hunted for their expensive meat, oils, and by-products. After nearly three centuries of hunting, the effects of the whaling trade had been truly devastating on wild whale species.

In 1946, an International Whaling Commission was set up to control the whaling industry, but to little effect. Despite persistent activity within the organization, the price of whale products continued to rise, and greedy hunters simply couldn't get enough. The trade had entered a vicious cycle.

In 1964, a Canadian organization similar to Friends of the Earth was founded, with the interests of the environment and its species at heart. It was called Greenpeace. Whaling was one of the hottest topics on the organization's agenda, and in 1975 they started a very public anti-whaling campaign. High profile stunts, including members jumping into the sea in front of the massive whaling ships, coincided with legal actions against some of the biggest cheeses in the whaling industry. As a result, Soviet, Australian, Peruvian, Spanish, and Brazilian fleets stopped their hunts completely, allowing for some recovery of natural whale populations. Now, only Japan, Iceland, and Norway remain in opposition to the ban.

Greenpeace continue their anti-whaling work to this day, stirring up public opinion on the subject, in the face of whaling nations. There is increasing evidence that whales have managed to take advantage of their fortune – numbers are on the up. For now.

Hannah Welham

Date 1975

Country Canada

Why It's Key Massively effective campaign swung general opinion against the whaling trade. Greenpeace had become a force to be reckoned with.

Key Publication **Radiation from black holes**
How can anything escape?

Black holes are not supposed to emit anything. Full stop. A black hole has such a strong gravitational field that even the fastest and lightest thing going, light, can't get out. Nothing else stands a chance. It's funny then that one of the world's most pre-eminent physicists, Stephen Hawking, should believe that black holes give off a type of radiation.

One of the fundamental facts of quantum mechanics is that you cannot know the energy of a particular system, and how it is changing with time. Given a vacuum with nothing in it, quantum mechanics would have you believe that instead of the energy levels being zero, they fluctuate up and down a little bit. And in this vacuum, the energy fluctuations cause the spontaneous creation of pairs of particles: antimatter and matter. But, having opposite charges, matter and antimatter attract each other and annihilate, giving off a load of energy.

If there's a black hole in the way, however, one particle may get sucked down into it while the other escapes. Since nothing can get out of the black hole, then there is no way the pair of particles can annihilate, and the other particle has got away scot-free. An outside observer would see a trail of lonely particles being emitted from the black hole. And this is the black hole radiation that Stephen Hawking, predicted, in his 1974 paper "Black Hole Explosions?". Hence the name: Hawking radiation.

Kate Oliver

Publication *A Brief History of Time*

Date 1974

Author Stephen Hawking

Nationality British

Why It's Key A way of detecting black holes, and some interesting thoughts about how the ideas of quanta work near intense gravitational fields.

Key Event **A halt in the development of genetic engineering** What limits on biotechnology?

The discovery of restriction enzymes in the late 1960s opened up a brand new field of biology – genetic engineering. Use of these enzymes, which cut DNA in a sequence-specific way, soon led to the production of the first recombinant DNA molecules, which contained DNA from two different species.

The potentially huge implications of this technology triggered a group of American scientists to write to the journal *Science* expressing their concerns. They called for a voluntary halt to all experiments involving recombinant DNA and urged the National Institutes of Health to come up with guidelines regulating the use of this technology.

One of the aims of the moratorium was to provide time for a conference for people to consider the technology and its potential risks. In 1975, a group of some 140 people, mostly biologists, met at the Asilomar Conference Center, near Monterey, USA. "Recombinant DNA was the most monumental power ever handed to us," said David Baltimore, a past president of the California Institute of Technology and one of the organizers of the conference. "The moment you heard you could do this, the imagination just went wild."

At the time, there was little scientific evidence to guide the conference attendees, who had decided to tackle the safety concerns of the technology but not the ethical ones. After some debate, an agreement was made to allow recombinant DNA research to proceed, but in accordance with guidelines produced by the National Institutes of Health.

Christina Giles

Date 1974

Country USA

Why It's Key Scientists worked together to set limits on the use of a new, unknown, and potentially dangerous technology.

Key Experiment **The car crash experiment** Led astray by leading questions

In the early 1970s, Elizabeth Loftus and John Palmer led a pioneering investigation into the validity of eyewitness statements. They showed 150 people a film, lasting less then one minute, of a multiple car accident (there was probably no popcorn). The accident itself lasted just four seconds. The film was followed by a questionnaire and the key question was on the speed at the point of impact.

A control group of fifty subjects weren't asked about the speed. Fifty subjects were asked "how fast were the cars going when they smashed into each other?" Their average estimate was 10.46 miles per hour. Another fifty subjects were asked about the speed when they "hit into each other." This sample estimated an average of 8 miles per hour.

A week later, the volunteers were recalled for more questioning. The key question this time was "Did you see any broken glass?" In truth, there was no broken glass, but between 10–15 per cent of those in the control and "hit" group thought they did see the phantom broken glass. People presumed there had been broken glass simply because they were asked; a week after witnessing, their memories had been distorted.

In the "smashed" group, however, the number was even more astounding. A third of them claimed they saw broken glass. With just one word different between the groups, the researchers had proved the risk of leading questions, and the impact: false memory creation.

Umia Kukathasan

Date 1974

Scientists Elizabeth Loftus, John Palmer

Nationality American

Why It's Key A cautionary tale, leading questions risk leading eyewitnesses to false memories. This evidence on memory has been used in numerous criminal trials.

Key Publication **The first big GUT**
Physicists publish a Grand Unified Theory

Throughout the modern evolution of physics, its major discoveries have gradually made the Universe seem simultaneously more complex and more straightforward. Over time, complicated theories of how the Universe functions have begun to merge, giving a tantalizing glimpse of an overall theory of everything.

When American physicists Howard Georgi and Sheldon Glashow published their Grand Unified Theory in 1974, it contained many features attractive to physics. Among these was an important notion relating the magnitude of the electric charge on leptons (the family of particles to which the electron belongs) to the charge on quarks (the separate particle group that are the building blocks of protons and neutrons). By attempting to unite the contrasting results of quantum mechanics and special relativity, they had taken a step toward boiling the laws of physics down to their absolute simplest forms.

Critics have described Grand Unified Theories (GUTs) as neither grand nor unified, as they do not deal with the force of gravity. However, as gravity is so weak compared to the other three fundamental forces – strong nuclear, weak nuclear, and electromagnetic – it is usually discounted when dealing on the atomic and subatomic level.

The search for a true Grand Unified Theory, one which also describes how gravity works, could be announced tomorrow, next year, or never. If it happens at all, it will almost certainly be the greatest advancement in science of all time.

David Hawksett

Date 1974

Scientists Howard Georgi, Sheldon Glashow

Nationality American

Why It's Key Georgi and Glashow's work represents a true step toward a Grand Unified Theory of everything.

opposite Professor Sheldon Glashow

Key Invention **Catalytic converters first produced**
On the road to lead free petrol

The 1970s were a time of increasing environmental awareness, especially in the United States. In 1970, in the wake of the first Earth Day, Congress passed the Clean Air Act, which planned to reduce pollution from automobiles by 90 per cent within five years.

Scientists rushed to find ways to reduce emissions. One avenue looked particularly promising: catalysts. A catalyst is a material that encourages a chemical reaction to happen, without necessarily taking part in it – and therefore without being "used up" or wearing out. Catalysts were already being used to break crude oil down into gasoline in refineries.

A catalytic converter works by providing a surface which can "grab hold" of a bit of a toxic molecule, weakening the bonds between its constituent parts. The surface – made of precious metals like platinum, rhodium, and palladium – can bring molecules closer together, helping them react with each other. In the converter, toxic carbon monoxide molecules from the exhaust fumes react with oxygen to make carbon dioxide, which is far less toxic. A modern "cat" can also break down nitrogen oxides, which cause smog, back into nitrogen and oxygen, to release into the air.

By 1974, mass-production of catalytic converters had begun, just in time for them to be fitted to production automobiles by the 1975 Clean Air Act deadline. Cats soon hastened the introduction of unleaded gasoline, as lead was found to "clog up" the surface of the converter.

Matt Gibson

Date 1974

Scientists Irwin Lachman and colleagues

Nationality American

Why It's Key Without them we'd have a lot more toxins in our air, lead in our lungs, and smog in our skies.

SU(2)

Key Invention **Stereolithography**
Printing in 3D

Rapid prototyping may be unique in being the only multi-billion-dollar industry founded on a joke. In 1974, chemist David Jones was writing spoof inventions for *New Scientist*, under his mad-scientist *nom-de-plume* "Daedalus." In the October 3 edition, he suggested that a laser beam could be steered by computer-controlled mirrors to "scribble" in a vat of liquid plastic monomers (a type of hydrocarbon); where the beam focused, the monomers would photopolymerize (join together to form long chain molecules) to build up a solid.

That is exactly how stereolithography works, and Jones' design was subsequently published as a patent in 1977 by Wyn Kelly Swainson, who'd had the same idea some years before. Stereolithography was the first of many systems that now exist for printing 3-D objects directly under the control of a computer, and it is still a significant one. Though the materials it requires can be messy and inconvenient, stereolithography produces mechanically strong parts at a high resolution. Satoshi Kawata and his colleagues at Osaka University, Japan, have used the technique to make a micro-sculpture of a bull that is only eight microns long (about one hundred-thousandth of a meter) or about the size of a red blood cell. It was listed as the world's smallest sculpture in *Guinness World Records 2004*.

Stereolithography has applications in everything from manufacturing to art, and as the technology continues to improve, and costs decrease, who knows how important it could become in our daily lives. You just never know when you might need a miniature bull.

Adrian Bowyer

Date 1974

Scientist David Jones

Nationality British

Why It's Key
Stereolithography is one of the techniques that manufacturers use to take designs from computer programs and turn them into real, solid objects.

Key Person **Roger Penrose**
The polymath mathematician

Sir Roger Penrose, KBE (b.1931), is a mathematician who has worked in and influenced many diverse fields, including cosmology, mathematics, quantum mechanics, neuroscience, and even art and design.

As a teenager, Penrose, along with his father the geneticist Lionel Penrose, designed the Penrose Staircase, an impossible two-dimensional, self-perpetuating staircase made famous by the art of M.C. Esher.

Penrose's work in mathematical physics led to his proving that a collapsed star can form a singularity – a place where huge quantities of matter can be confined in an infinitely small space. His other contributions to physics have included his being at the vanguard of those trying to develop a Grand Unifying Theory of Everything, notably with his 1967 Twistor Theory, which is concerned with mapping coordinates in four-dimensional space-time. He has also contributed to more earthy matters, such as recreational mathematics, with the discovery of a form of non-periodic tiling – better described as "a peculiar sort of pattern" whereby a shifted copy will never match the original exactly – in 1974, which now bears his name.

Recently Penrose's work has been brought to wider attention through his theories on the nature of consciousness. According to Penrose, consciousness may be a by-product of quantum mechanical action within the brain. He also believes that these quantum processes may be impossible in an artificial computer, and as such we may never see conscious machines.

Neal Anthwal

Date 1974

Nationality British

Why He's Key The work of Sir Roger Penrose is at the forefront of bodies of research in mathematics, physics, and even the nature of consciousness.

Key Event **Production of Altair 8800 begins**
Computers get personal

In 1975, the January edition of *Popular Electronics* magazine featured the MITS Altair 8800, a build-it-yourself "micro-computer" kit that could be purchased by mail order. Within two months, Ed Roberts, the Altair's creator, was struggling with thousands of orders, marking the launch of the personal computer industry that has revolutionized the modern world.

With only 256 bytes of memory, and no keyboard, monitor, or any software, it couldn't exactly achieve much. In fact, all that users could do after their toil to construct it was create programs that made LED lights on the front panel blink on and off. The first computer game might well have been "Kill the Bit," which involved guessing which light would come on and flicking a switch before the light went out.

But the Altair's popularity grew, especially among hobbyists. In particular, it grabbed the attention of two young men, Bill Gates and Paul Allen, who saw potential for developing a user-friendly programming language for the memory-deficient machine. Soon, they had licensed their BASIC programming software to Roberts, and had founded Microsoft, one of today's software giants.

Old companies like IBM and new ones such as Apple, soon tried to make their mark in the wake of the Altair's commercial success, creating both fierce competition and fruitful collaboration in an industry that continues to make ever more powerful, compact, and capable personal computers.

James Urquhart

Date 1975

Country USA

Why It's Key A hobbyist's toy and one of the earliest PCs, later transformed into an indispensable tool that has revolutionized the way we communicate, deal with information, work, play, and learn.

Key Event **Bill Gates forms Microsoft**
BASIC stuff

With revenues for the year of less than US$17,000, three employees including the CEO, and a product that most of the world had never heard of, would you invest? With hindsight, you probably wish that you'd had the opportunity. The company was Microsoft; the year 1975.

Ex-Lakeside Prep School buddies Bill Gates and Paul Allen knew they were in the right place at just the right time when, the previous December, they had sat reading a copy of *Popular Electronics* magazine which carried a review of "The World's First Microcomputer Kit to Rival Commercial Models"; the Altair 8800 Programmer. One-time hacker and entrepreneur Gates wasted little time in contacting Micro Instrumentation and Telemetry Systems, (MITS), the makers of the Altair, to tell them how he had written a computer language that was perfect for their machine. He was lying.

BASIC (Beginner's All Purpose Symbolic Instruction Code), had been developed to assist computer programming in an era when input commands, and the resulting output, were typically realized via punched cards or paper tape. Gates and Allen were confident they could create their own home-use version ready for an arranged demonstration at MITS some two months later, but at the time of the sales pitch, not a single line had been written.

The demonstration worked faultlessly. Allen joined MITS as director of Software Development; Gates followed that year and created an informal partnership known as "Micro-Soft." An empire had been born.

Mike Davis

Date 1975

Country USA

Why It's Key Facilitated computer use among non-techies, and led to the exponential growth of personal computing.

Key Publication **Beautiful mathematics**
Les objets fractals: forme, hazard et dimension

Fractal geometry is often pictorially represented by swirls of complex patterns of color, which keep their structure no matter how close you look. The Mandelbrot set, named after its French/American founder, is the most famous example of a fractal and its image has become somewhat of an icon of science.

Fractals are complex shapes that even when they are magnified still show the overall structure and can be found in nature as well as mathematics. Normal objects do not continue to show a reduced copy as you look infinitely close; nature only shows fractal-like structures. If you compare a fir tree to a singular fir cone and then to a pine on the fir you will find that they all show the same structure. Similar patterns appear in a variety of forms like in snowflakes, lightning bolts, and blood vessels.

Benoît Mandelbrot first published work in this area and coined the term "fractal" in his book which translates from the French as *Fractals: Form, chance and dimension*. He showed that a single formula using fractal geometry can describe lots of data, which of course made mathematicians happy.

Fractals have been used to model and describe, though not predict, a number of complex phenomena such as water and air turbulence, coastlines, the internet, galaxy clusters, and the fluctuations of stock markets.

Leila Sattary

Publication *Fractals: Form, Chance and Dimension*

Date 1975

Author Benoît Mandelbrot

Nationality French-American

Why It's Key Mathematical way of describing complex shapes that when magnified, still show the same structure.

opposite Benoît Mandelbrot

Key Person **Edward Osborne Wilson**
The world's first sociobiologist

It might seem unlikely that an encounter between a seven-year-old boy and a fish spine would be the catalyst for the creation of a new scientific discipline. But Edward Wilson (b. 1929) later realized the importance of his unfortunate and, as it turned out, partially blinding accident. As Wilson said himself, "The attention of my surviving eye turned to the ground. I would thereafter celebrate the little things of the world." It was his fascination with these little things – insects, and ants in particular – that led to the creation of the controversial field of sociobiology.

From his studies of ant behavior, Wilson began to develop an understanding of social behavior in evolutionary terms. He has met fierce opposition for suggesting a genetic component to human behaviors such as aggression, altruism, and sexuality. This led naturally to the ongoing nature-versus-nurture debate in which Wilson described our behavioral free will as being on a "genetic leash."

Despite the controversy surrounding his work, including having a jug of water tipped over him at the 1978 meeting for the American Association for the Advancement of Science, Wilson has since been presented with many awards, among these the prestigious Crafoord Prize of the Royal Swedish Academy of Sciences, in 1990. He continues to study psychological, sociological, and anthropological ideas, and is an influential figure in ecology and environmentalism, being particularly concerned with maintaining biodiversity during the current ecological crisis.

Jim Bell

Date 1975

Nationality American

Why He's Key Responsible for the creation of a new field of biology, and one of the foremost champions of environmentalism.

Key Discovery **Endorphins**

Blocking pain in the brain

By the age of seventy-two, most of us probably plan to be happily retired, with the grind of a nine-to-five job far behind us. But not Dr Hans Kosterlitz. Despite being officially retired from his post at the University of Aberdeen, Kosterlitz was active in the laboratory up to the age of ninety-eight.

In 1975, seventy-two-year-old Kosterlitz and his colleague John Hughes isolated two proteins from pigs, which bound to opiate receptors in the brain. They named the chemicals "enkephalins" from the Greek for "in the brain."

Their discovery was part of the global attempt to answer a question that had been occupying researchers since 1973, when three independent research groups – two in the United States, one in Sweden – discovered that opiate receptors were found in our brains. These receptors were the correct fit for drugs such as morphine, but why would the body evolve receptors just for "foreign" chemicals? The answer was that it hadn't; Hughes and Kosterlitz had discovered the first endogenous opiates – those made by the body itself.

According to the International Narcotics Research Conference, the name endorphin came about after a "nomenclature brouhaha." The U.S. pharmacologist Eric Simon, who had been part of one of the groups that identified opiate receptors in 1973, eventually came to the rescue by suggesting the name "endorphine," a contraction of "endogenous morphine." A colleague recommended that the final "e" be dropped, and endorphins were born.

Christina Giles

Date 1975

Scientists Hans Kosterlitz, John Hughes

Nationality British

Why It's Key Marked the discovery of the first of the body's own painkillers.

opposite Polarized light micrograph of beta-endorphin crystals

Key Invention **VHS vs Betamax**

Video system wars

Video systems were developed in the 1960s, but never caught on domestically as they weren't particularly user-friendly. They were designed with the simple aim of playing back pre-recorded videos, however, Japanese companies sought to produce a system that was able to record as well as play, giving consumers more viewing freedom.

In 1974, Sony presented a prototype to other manufacturers in the expectation that they would pursue the same format. Sony introduced its video recording system, Betamax, in 1975. But a year later, JVC produced its own version using a different format – VHS. VHS originally provided the longer recording time of two or three hours – long enough to record a movie – whereas Betamax only lasted sixty minutes. Picture and sound quality, on the other hand, were better on Betamax. Competition soon led the manufacturers to produce improved versions of each format, and the differences between the two diminished.

For around ten years VHS and Betamax battled it out, but ultimately consumer choice came down to price; VHS machines were simpler to operate, cheaper to produce, and provided more recording time for the cost.

By 1988, 170 million home video systems had been built, but only 20 million of them used the Beta format. Eventually Sony admitted defeat and turned to making VHS. The availability of movies to rent became more extensive on VHS than Betamax. Finally movie studios abandoned the failing Beta market and released their movies solely on VHS.

Emma Norman

Date 1975

Scientists JVC

Nationality Japanese

Why It's Key The start of a major thrust in the home entertainment industry, and a fascinating case study on the effect of markets on technologies.

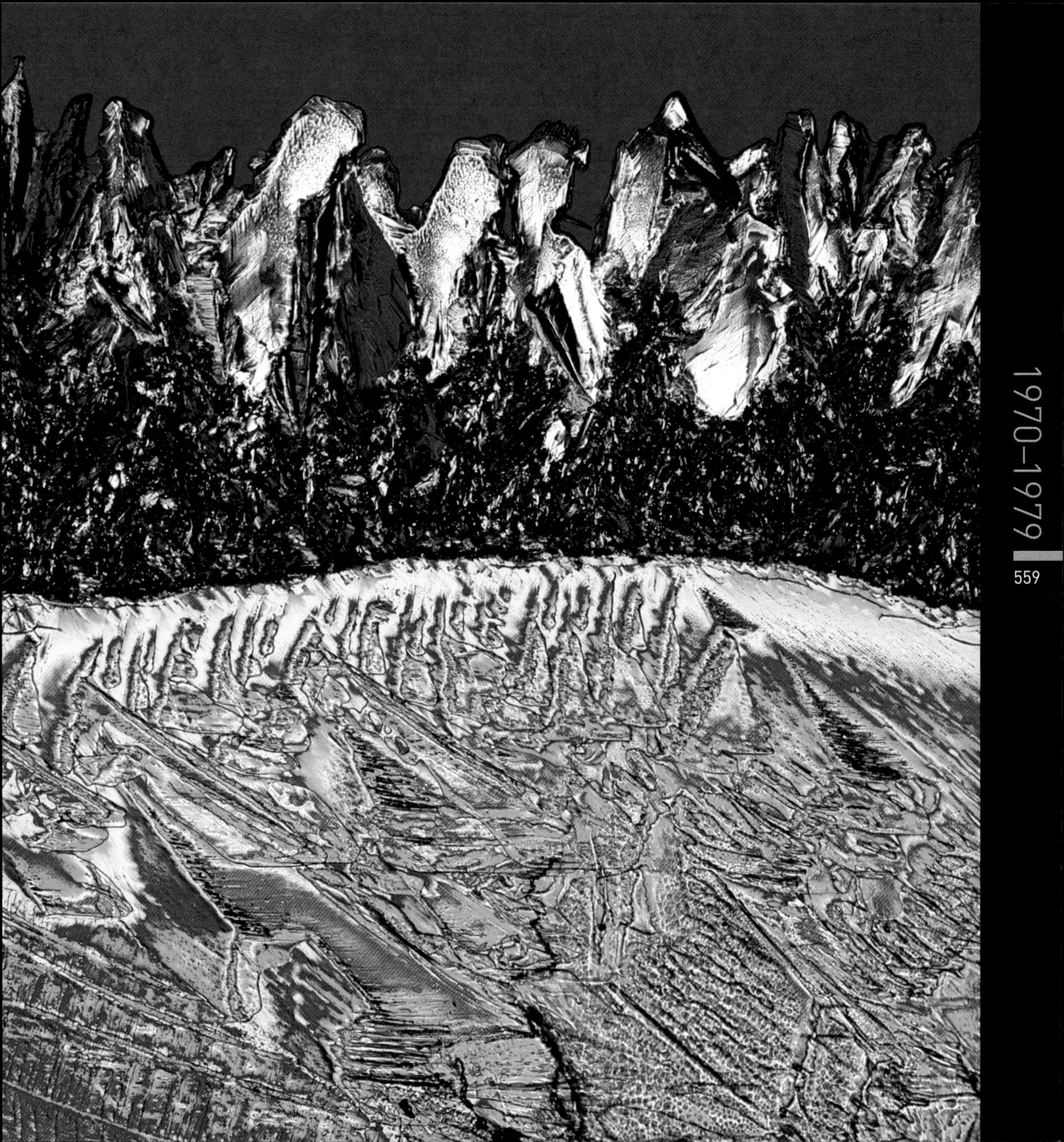

Key Publication
Isolating antibodies

Your immune system has at its disposal a billion different types of antibody. Each is able to latch on to a different invader, such as a cold virus or rogue bacterium, sometimes rendering it useless but more often attracting the attention of a white blood cell, which then destroys the miscreant.

Antibodies' capability for distinguishing between millions of molecular forms makes them valuable tools for the bench scientist. Why? Imagine you are presented with a bag full of assorted screws and asked to identify the iron ones by looking at them. It seems impossible. But you know that the simple application of a magnet could easily pull out all the screws you are searching for. In the same way, the right antibody can pull out all the molecules or cells a biologist is interested in when carrying out a particular experiment, diagnosis, or treatment.But with billions of different antibodies floating around, how does a scientist obtain enough of the one they're looking for? High levels of particular antibodies are present in serum from lab animals injected with the corresponding antigens – the proverbial "invaders" – but what is really needed is a pure form.

Georges Köhler and Cesar Milstein solved the problem in 1975. They described a method for fusing together normal, antibody-producing B cells and immortalized tumor cells. The resulting "hybridomas" manufactured bulk quantities of identical antibodies which we now call monoclonal antibodies. These can effectively sort through nature's bag full of screws to find a relevant protein, disease indicator, or cell type.

Hayley Birch

Publication *Nature*

Authors Georges Köhler, Cesar Milstein

Nationality German, Argentine

Why It's Key A pure source of antibodies is useful for a broad spectrum of biological applications, including diagnosing disease, therapeutic interventions, and imaging biological systems.

opposite Stained fluorescent human monoclonal antibodies cling to brain cancer cells

Key Event **Apple Computer Company formed**
A big bite of the apple

In computing and new media technologies, one name in particular is hard to escape: Apple. Innovations such as the iPhone, iPod, Apple TV, and a range of PCs, which appeal to the masses through a combination of usability and sexy designer looks, have secured Apple's place as a leader in the field of consumer technology.

Steve Jobs, Steve Wozniak, and Ronald Wayne founded Apple in 1976. Wayne sold his shares back to Jobs and Wozniak just a few weeks later for just US$800, a decision he claims he does not regret.

Apple's first PC came in kit form; a loose assemblage of CPU and RAM, which purchasers had to wire up themselves. The imaginatively titled Apple I found buyers and, by the end of the 1970s, Apple had a team of staff and a production line.

The 1980s saw the production of two key Apple products. Lisa was the first computer to have a graphical interface, something now ubiquitous in computing, but did not sell well due to a hefty price tag. The Macintosh was better received, and became the main competitor for the Microsoft and IBM products that were dominating the home PC market at the time.

Legal battles with Microsoft, internal wrangling, and poor products all took their toll in the 1990s. But Apple's ability to diversify and innovate, combined with a loyal consumer base in desktop publishing and education, saw the company through difficult times and has elevated it to the position of a market leader today.

Barney Grenfell

Date 1976

Country USA

Why It's Key Apple computers have been a constant source of innovation, helping to shape computing as we know it today. Graphical interface, Personal Data Assistants, and the laptop have all been popularized through Apple products.

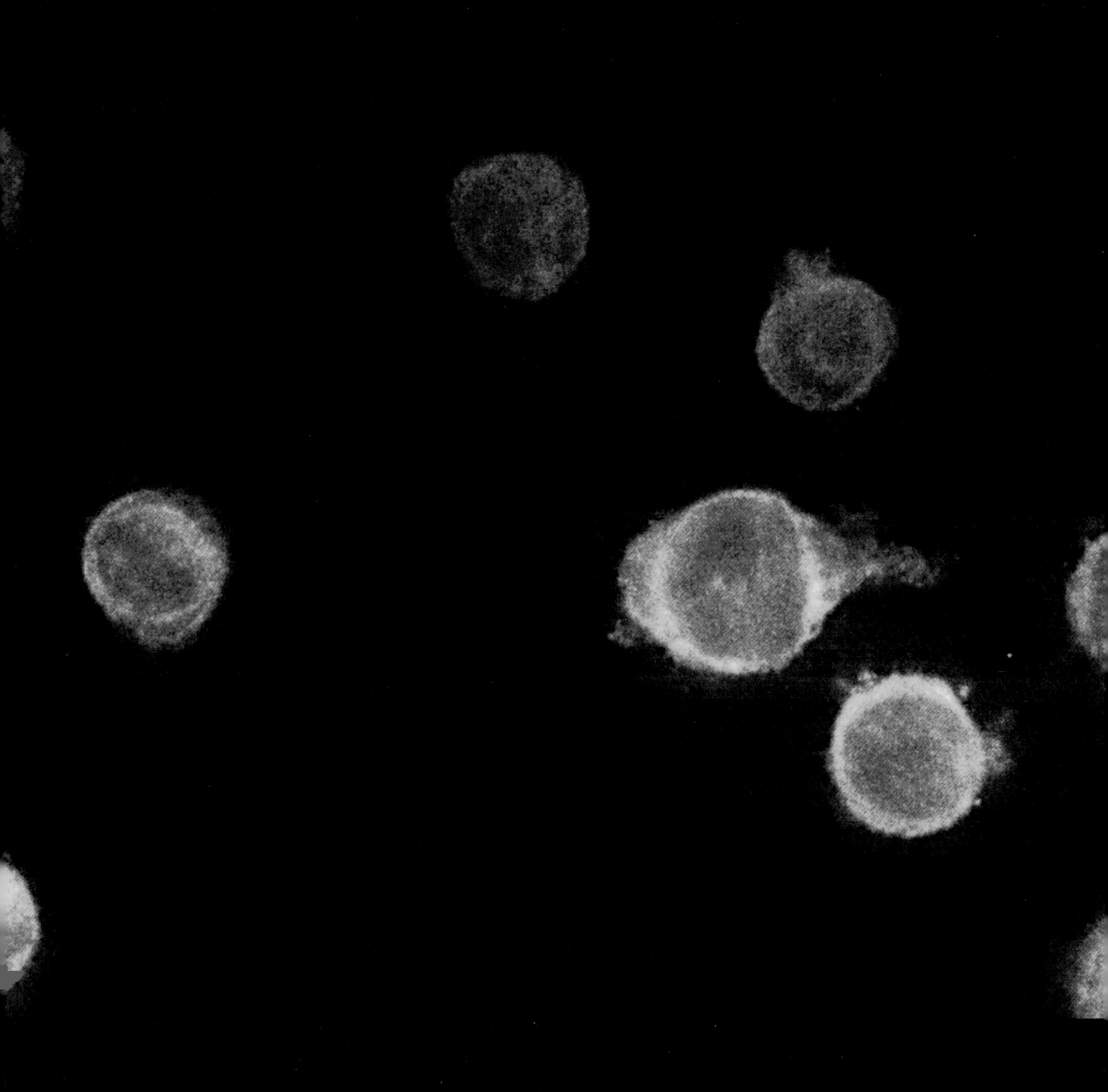

Key Event
Vikings' red planet quest

The Viking craft circled the red rock for thirty-one days, searching for a suitable landing site, hunting for signs of life, and surveying this curious land. For any beings living on the surface of Mars during the summer of 1976, it would have been an anxious time. Viking 1 and Viking 2 were NASA spacecraft sent to Mars in the hope of finding life.

Each craft consisted of an orbiter and a lander, each lander being equipped with water-filled test tubes in which Martian soil was to be analyzed. The theory and hope behind this was that there would be some dormant life that was waiting for Mars to turn wet once more, and would spring back into action in the presence of water.

Carl Sagan compared the experiment to waking Sleeping Beauty and, in truth, success would have been quite a fairytale. Indeed the princely robots initially reported that signs of life had been found in the soil. Hopes of a fairytale ending were dashed however, when scientists later attributed the results to chemical reactions rather than biological processes.

Despite this official conclusion, thirty years on there is still uncertainty over the findings; geological tests conducted on Earth soil have suggested that the thresholds used in the Viking experiments were too high, and that life may have been detected in 1976 after all!

Christopher Booroff

Date 1976

Location Mars

Why It's Key The implications of finding life would be huge. However, being able to analyze extraterrestrial soil is a great achievement in itself.

opposite A Martian sunset, captured by Viking 1 lander

Key Person **Steve Jobs**
Apple, and the cool computer

When Steve Jobs (b. 1955) founded the Apple Computer Company, in 1976 with Steve Wozniak, he could hardly have expected Apple and the Apple Macintosh in particular would play such a significant role in the popularization of personal computers.

Jobs and Wozniak's partnership began in 1974, producing high-tech gadgets that enabled people to make illicit, free, long distance telephone calls. Moving into more legitimate work, they soon saw the potential for a personal computer that was functional, but also appealed to the eye. They set up Apple to market their new product. After success with early Apple PCs, they introduced the Macintosh in 1984, the first commercial computer with a user-friendly graphics-based interface. The product revolutionized the design of PCs, which began to focus on aesthetics as much as functionality.

Always looking to move things forward, Jobs left Apple in 1985 to set up another company, NeXT Computer. In doing so he played a significant role in the origin of the World Wide Web, developed by Tim Berners-Lee using a NeXT workstation. Around the same time, Jobs bought The Graphics Group from George Lucas (of *Star Wars* fame), which became Pixar and pioneered computer-based animation in films like *Toy Story* and *Finding Nemo*.

Steve Jobs returned to Apple in 1997, following its take over of NeXT, and helped reinvigorate the company, leading to the development of a range of aesthetically and commercially successful products such as the iMac computer, the ipod, itunes, and the iphone.

Richard Bond

Date 1976

Nationality American

Why He's Key Recognized the importance of design and aesthetics, as well as functionality, in the development and marketing of high tech products.

Key Event **Space Shuttle Enterprise**
NASA bows to trekkies

The first Space Shuttle was meant to have been called "Constitution," to honor the U.S. Constitution's Bicentennial in 1976. However, a huge letter campaign organized by artist, writer, and massive *Star Trek* fan Bjo Trimble caused it to be renamed "Enterprise" after the TV show's starship. Over 200,000 letters helped in the petition.

The Space Shuttle itself was first built in 1974, and two years later was rolled out of Rockwell's plant at Palmdale, California. To the sound of the *Star Trek* theme tune, the shuttle, which represented a new era in space travel, was christened at a dedication ceremony attended by the show's creator Gene Roddenberry and most of the original *Star Trek* cast, but not William Shatner. Unlike its fictional namesake, however, Enterprise was never designed to travel into space. It was built and used to check the performance of the Space Shuttle Orbiter's design on the final approach and landing. It was retired in 1985, and now resides in the Smithsonian's National Air and Space Museum in Washington DC, where it can be seen today.

As a fitting tribute, the Space Shuttle Enterprise made cameo roles in the TV show on several occasions, including the wall of Jonathan Archer's ready room, and in Benjamin Sisko's office on Deep Space 9. It also appeared on USS Enterprise's recreation deck in *Star Trek: The Motion Picture*.
David Hall

Date 1976

Country USA

Why It's Key Enterprise heralded a new era in space flight, while its naming encapsulated popular enthusiasm for space exploration.

Key Publication ***The Selfish Gene***
Dawkins' Magnum Opus

Few books in recent times have been as influential as Richard Dawkins' insight into evolution, *The Selfish Gene*. Building on the work of evolutionary biologists such as George Williams, Bill Hamilton, and Robert Trivers, Dawkins put forward the idea that evolution worked not on groups or individuals, but on genes themselves. And not only did he do this in an original, comprehensive, and erudite way, he also did it in a way that everyone could understand. His book has since become internationally renowned, selling well over a million copies worldwide, and has been translated into at least twenty-five languages.

The public imagination was caught by the idea that our bodies are merely vehicles for genes to carry on a blind quest for immortality. Everything that makes us who we are – our minds, personalities, and actions – are just part of a complicated solution to pass genes on from generation to generation, replicating themselves through the ages.

One of the most popular, and controversial, of Dawkins' ideas only appeared in the last of the book's thirteen chapters. Here he wondered if, like genes, ideas themselves could be classed as replicators that are passed down the generations, strengthened and modified by forces similar to natural selection. Our cultural heritage – the very ways in which we are taught to think – could be shaped by evolutionary mechanisms in much the same way that they have molded our genes to stand the trials of time.
Mark Steer

Date 1976 (Oxford University Press, Oxford)

Author Richard Dawkins

Nationality British

Why It's Key Eye-opening and jaw-dropping for scientists and the general public alike, Dawkins changed the way we think about evolution and our very existence.

Key Discovery **The four colour theorem** A computer versus the written "poetry" of mathematical proof

The four color theorem is often described in terms of maps. The reason for this is perhaps because it was first thought up by the South African mathematician Francis Guthrie, in 1852, as he colored in a map of the UK. It is now most remembered because it was the first theorem to be proved by a computer.

Guthrie proposed that, for any map, just four colors would be required to shade in the regions in such a way that assured no two adjoining regions were the same. Thus the theorem was born, but would it always hold true? It was not until 124 years later that Guthrie's hypothesis was finally proved by fellow mathematicians Kenneth Appel and Wolfgang Haken at the University of Illinois. To complete the proof by hand would have been impossible, because of the vast number of calculations required to do it. Appel and Haken's work is accepted as a valid proof of the theorem but it was and still is a contentious issue among the mathematical community because it relies so heavily on whether the computer programs are exactly right. A small bug in a program can give a wildly different result.

In fact, since first being proved in 1976, other unsatisfied mathematicians have continued to search for alternative proofs that can be written on a piece of paper and not on a hard drive.

Helen MacBain

Date 1976

Scientists Kenneth Appel, Wolfgang Haken

Nationality American, German-American

Why It's Key The first mathematical theory proved using a computer.

Key Person **Superphysicist Jim Gates believes in supersymmetry**

James Gates (b. 1950) has been referred to as the "Tiger Woods of Physics," an enthusiastic guide to superstring theory, or, more simply, a professor of Particle Physics. From the day his father bought him a copy of the *Encyclopedia Britannica*, Gates has had a fascination with physics and our Universe. Mesmerized by Schrödinger's equation, which he read in that very book, he has been intent on understanding the most fundamental issues of our Universe ever since.

During the 1970s, work on particle physics gathered pace and Gates wrote the first ever paper upon the subject of supersymmetry – a theory in which all particles have super-sized "superpartners". When he began his work on supersymmetry as a young graduate student at the Massachusetts Institute of Technology, nobody had heard of it. Now, due at least in part to Gates' knack of communicating mind boggling physics in a way that people can understand it, supersymmetry forms an essential part of string theorists' version of the Universe. And given a few years to bash particles about with a supersized particle accelerator, otherwise known as the Large Hadron Collider, scientists might even be able to prove it exists.

In 2006, Gates picked up an award from the American Association for the Advancement of Science for "sustained and career-long contributions to public understanding of physics." Whatever the use of these super theories, Gates has continually communicated these ideas eloquently over a distinguished scientific career.

Nathan Dennison

Date 1976

Nationality American

Why It's Key A super-scientist in every sense of the word, Jim Gates has made outstanding contributions to physics, as well as our understanding of it.

Key Event **First commercially developed supercomputer sold**

Seymour Cray was a dreamer; an eccentric who once claimed that elves would come to him with solutions to technical problems. And unlike most computer designers, he didn't worry about the price tag. In a speech in 1974, he explained: "In all of the machines that I've designed, cost has been very much a secondary consideration. Figure out how to build it as fast as possible, completely disregarding the cost of construction."

Based on this philosophy, he built the Cray 1, which was completed in 1976. It could perform over a hundred million arithmetic operations every second and had an 8 megabyte (one million word) memory. The unique "C" shape of the machine allowed integrated circuits to be physically closer together, and no single wire in the computer was more than four feet long. Freon liquid was used to keep the system from overheating.

Such was the excitement about this revolutionary computer that a bidding war ensued between Lawrence Livermoore National Laboratory and Los Alamos National Laboratory in the United States for the right to own the very first Cray 1. Los Alamos won and became the first customer – paying US$8.8 million.

Cray expected to sell around a dozen machines but, eventually, eighty-five were built and sold to customers paying US$5 to 8 million dollars. The Cray 2 was introduced in 1985, with a tenfold increase in performance, by which time Cray himself had left his company to become an independent contractor focusing on designing even better machines.

David Hawksett

Date 1976

Country USA

Why It's Key The Cray 1 became one of the most famous and successful supercomputers in history.

opposite Cray 1 supercomputer installed at Lawrence Livermore National Laboratory, California

Key Publication **"Topology of Cosmic Domains and Strings"** Cosmic string theory proposed

Just when we thought we'd heard it all with string theory, along comes another baffling, cosmically proportioned idea for us to get our heads around – cosmic string theory. In fact, although these two theories might seem equally absurd, they are usually considered to be independent of each other. String theory proposes that all matter is made up of tiny vibrating strings. Cosmic string theory is at the other end of the scale – it is the notion of a Universe-sized string vibrating at near light speed; a one-dimensional line so concentrated with mass-energy, that one meter of its length, it has been said, might have the mass of a mountain.

Implausible as it might seem, there are in fact a number of very good arguments for cosmic string theory. In 1976, Thomas Kibble made the first stab at pinning down one of these giant strings – albeit in mathematical form. While developing models of the Big Bang, he suggested that the force of such a massive explosion, followed by rapid cooling, could have created huge, string-like faults in the Universe. Some scientists now think that these strings could play a role in holding the whole Universe together.

We still can't tell whether cosmic strings exist at all. Maps of cosmic radiation – sometimes described as echoes from the Big Bang – can be used to show the presence of distortions that indicate their presence. But it all hinges on the accuracy of the data – as well as the computing power you throw at it.

Hayley Birch

Publication *Topology of Cosmic Domains and Strings*

Date 1976

Author Thomas Kibble

Nationality British

Why It's Key If it exists, cosmic string might span the entire Universe, and even be what's holding it together

Key Event
Voyagers 1 and 2 launched

No manmade objects have traveled further from Earth than the sister spacecraft Voyagers 1 and 2, launched from Cape Canaveral in 1977.

Originally conceived as part of the Mariner program, these two probes were reallocated when NASA project planners realized, no doubt with an excited glint in the eye, that the Solar System's four largest planets, Jupiter, Saturn, Uranus, and Neptune were lining up in a way that only occurs every 176 years. It meant that a space probe could visit them all in a relatively short time frame (twelve years), using minimal amounts of fuel.

Confusingly it was Voyager 2 which was first sent spacewards, on August 20. Voyager 1 followed on September 5, but was set on a faster course, and so reached Jupiter first.

Both probes made flybys of Jupiter and Saturn before going their separate ways. In January 1986, Voyager 2 became the first probe to visit Uranus, the coldest planet in the Solar System. It then chalked up another first by flying past Neptune in 1989, photographing both the planet itself and Neptune's nitrogen-frost covered moon Triton.

Since leaving Neptune behind, Voyager 2, like its twin, has been striking out for the edge of the solar system. On February 5, 1998, the quicker Voyager 1 overtook NASA's previous interstellar craft, Pioneer 10, to become humanity's furthest outpost and in 2004, Voyager 1 entered the heliosheath – the zone at the very edge of the Sun's influence.

Mark Steer

Date 1977

Country USA

Why It's Key Made the first flybys of Uranus and Neptune before heading out toward the outer limits of the Solar System. Who knows what discoveries are still to come.

opposite Shot of planet Neptune produced from images taken by Voyager II's wide-angle camera lens

Key Discovery
The rings of Uranus

Uranus is the seventh planet from the sun and the third largest in diameter; it was identified as a planet by William Herschel in 1781. Thirteen rings encircle it, the most prominent of which, with low magnification telescopes, look like thin solid discs. In fact they are composed of fine particles, rocks, or lumps of ice, some of which extend to 10 metres in diameter. On average, the rings are about 100 meters thick and vary in width from 1 kilometer to 2,500 kilometers.

The existence of nine of the rings was confirmed in 1977 by astronomers on board the Kuiper Airborne Observatory; two more were photographed by the Voyager 2 probe in 1986, bringing the then total to eleven. In 2005, the Hubble Space Telescope detected a pair of previously unknown rings, now called the "outer ring system," which brought the total to thirteen. There is some evidence to suggest that Herschel saw at least one of the rings, despite the limitations of his telescope, which he had made himself. Drawings, which are apparently based on his observations, include a sketch of a ring showing it in the right place and at the right inclination.

Uranus also has twenty-seven moons circling it, all of which have been identified and named (unusually they are named after characters from the works of Shakespeare and Alexander Pope), but it is suspected that there might be more within or between the rings.

Fabian Acker

Date 1977

Scientist James Elliot

Nationality American

Why It's Key Although the rings of Uranus were the second set to be discovered in our planetary system, knowing their composition and behavior have improved our understanding of the formation of the planets.

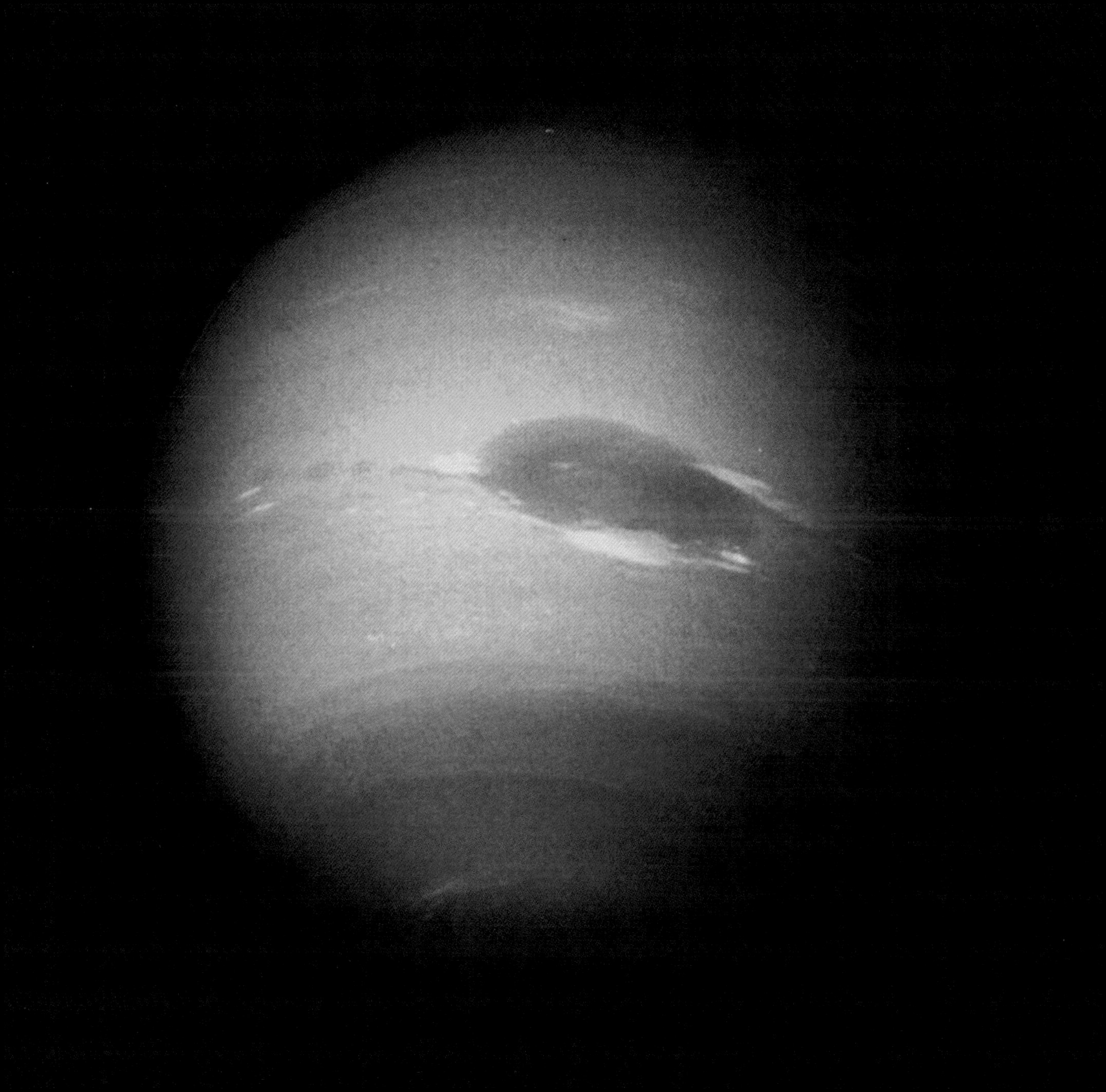

Key Person **Jane Goodall**
Queen of the chimps

Jane Goodall (b. 1934) did far more than simply monkey around in the jungle. As well as being one of only eight people in the world to have been awarded a doctorate without a bachelor's degree, in 1977 she set up the Jane Goodall Institute, now a powerful conservation body.

In 1960, Goodall started studying chimps in the Kakombe Valley, Tanzania, a feat in itself at a time when it was almost unheard of for women to head off into the wilds of the African forest. Her first task was simply to acclimatize the chimps to her presence, so that she could start to study their behavior.

She defied scientific convention by giving the chimps names instead of numbers, and revolutionized both the way they were studied, and the accepted knowledge about their behavior. Goodall unmasked complex, emotional behaviors such as adoption, tool making, and even warfare, and insisted on the validity of her observations that animals have distinct personalities, minds, and emotions.

The Jane Goodall Institute was set up to provide continuing support for chimpanzee research in Africa. Today, its mission is to advance the power of individuals to take informed and compassionate action to improve the environment for all living things. With nineteen centers around the world, the Institute is a leader in the effort to protect chimpanzees and their habitats, and is widely recognized for establishing innovative community-centered conservation and development programs across Africa.

Catherine Charter

Date 1977

Nationality British

Why She's Key Single-handedly revolutionized the branch of zoology that studies the behavior of animals in their natural habitats. The discoveries she has made and the way she has changed public and scientific perceptions of chimps are huge.

opposite Jane Goodall with one of her research subjects in the Gombe National Park

Key Discovery **Woolly mammoth, perfectly preserved in permafrost**

The last woolly mammoths to walk the Earth died out around 10,000 years ago, at the end of the last ice age. Amazingly, some ten millennia later, in 1977, the intact remains of a baby mammoth were uncovered at a construction site in Siberia.

The baby mammoth had been preserved in the permafrost and was unearthed by a bulldozer near the tributary of the Kolyma River. The little mite was named "Dima" after a nearby stream and measured 115 centimeters long and 104 centimeters high. From his size, he was thought to have been around seven or eight months old when he died, but at 100 kilograms, he still weighed more than an adult man.

Carbon dating showed the baby to be 40,000 years old. It is not known how he died, but scientists speculate that he may have fallen into a ditch, starved to death, and become immersed in mud which then froze. The remains of around forty mammoths have been uncovered worldwide, but few have been found completely intact like little Dima. He was so well preserved in fact, that he even had tufts of chestnut colored hair on his legs.

In 2007, the most perfectly preserved mammoth ever discovered was unearthed, again in the frozen soils of Siberia. Scientists now hope to use new technology to extract DNA from this specimen, which they could, controversially, use to create the world's first woolly mammoth clone.

Hannah Isom

Date 1977

Country USSR (now Russia)

Why It's Key Until 2007, Dima was the best-preserved woolly mammoth ever unearthed.

Key Discovery
Life is found near deep ocean vents

In 1977, the famous deepwater submersible "Alvin," built in 1964 and named after its inventor Allyn Vine, journeyed to the Galapagos Rift via the Panama Canal. It was here that scientists on a geological expedition observed for the first time the existence of deep-sea hydrothermal vents, supporting exotic ecosystems in complete isolation from sunlight. In the impenetrable dark and under intense pressure at over 2,500 meters beneath the waves, enormous tube worms, mussels and giant clams were found to be thriving off energy created by bacteria, which use hydrogen sulfide purged from the vents.

Smokers, as they are known, are chimneys emitting mineral-rich waters, sometimes black, gray, or white, at temperatures up to 380 degrees Celsius. The otherwise toxic metal ions cause no concern for the abundant oases of life surrounding them. Hydrogen sulfide-processing bacteria were seen to cloak surrounding rocks, as well as being found actually in the cells of some worms and mollusks.

At about the same time, the bacteria-like archaea were distinguished as an entirely different life-form encompassing single-celled organisms capable of living in extreme environments, such as those able to live around the scorching waters of the deep-sea vents.

Now recognized as the most primitive organisms on Earth, the study of vent communities has provided new ideas about the origins of life, and the possibilities for its evolution in many locations unrelated but for the presence of water and geothermal energy.

Mel Wilson

Date 1977

Scientists National Oceanic and Atmospheric Administration

Location Pacific Ocean

Why It's Key The discovery has led to a new understanding of the origins of life on Earth, and raised the possibility that life could exist elsewhere in the Solar System – Jupiter's moon Europa for instance.

Key Discovery **The discovery of archaea**
Redrawing the tree of life

Life is divided up into five kingdoms, right? Animals, plants, fungi, protists, and bacteria. Wrong. The classification system which most of us were taught at school is no longer accepted; instead we now tend to group life into three domains: bacteria, eukaryotes (which include all the animals, plants, fungi, and protists you could possibly think of), and a group only discovered in the late 1970s – archaea.

Like bacteria, archaea are single-celled organisms which lack structures within their cells such as nuclei and mitochondria. But don't be fooled by appearances, evolutionarily speaking, the archaea are as different to bacteria as bacteria are to you and me.

They are also capable of living in the most extraordinary of places. Some live near deep sea volcanic vents where the water is over 100 degrees Celsius, others live only in super-salty pools of water, whilst another group while away their time in petroleum deposits deep underground. Not all archaea live in such extreme places, however; there are millions of them inside you right now, living in the oxygen-less tracts of your digestive system, merrily producing methane.

It was a team led by Carl Woese at the University of Illinois who first realized that these microscopic hard-nuts were so different to the bacteria they resembled, and deserved their own category in the scheme of life. It took a long time for their results to be accepted but these days most scientists recognize the three domains as a reasonable classification of life on Earth.

Mark Steer

Date 1977

Scientist Carl Woese

Nationality American

Why It's Key Archaea are probably similar to the first life-forms ever to evolve. They also have great potential for industry in, for example, producing biogas.

Key Event
Proposition of the theory of panspermia

The truth is out there; but is it just emptiness, rock, and dust? Or could space be teeming with life? In 1974, one scientist claimed that infrared absorption patterns in interstellar dust could be explained by the presence of organic polymers. It might not have been evidence of aliens but it was almost as exciting.

That year Sir Fred Hoyle and author N. Chandra Wickramasinghe proposed what has become the modern theory of panspermia. They integrated the finding of organic and potentially bacterial matter in interstellar space with the long-standing idea that life on Earth has an extra-terrestrial origin. The resulting theory suggested that bacteria are relatively widespread in space, traveling between planets in comets and asteroids. Some of these arrived to "seed" Earth soon after its creation, and they're still arriving today. Living material from Earth could even be transported to other planets by the same means.

Various pieces of evidence have been put forward to support the theory of panspermia, all of which are controversial. Initially, there was the evidence from the infrared absorption spectrum and the fantastically early appearance of life on Earth at a time when it was still being bombarded by meteors. More recently, there has been the discovery of extremophilic bacteria which could potentially survive in space, perhaps within an asteroid.

Aliens seeding the galaxy with humanoid life-forms may still lie in the realms of science fiction, but the controversial idea of panspermia has yet to be disproved.

Shamini Bundell

Date 1977

Scientists Chandra Wickramasinghe, Fred Hoyle

Nationality British

Why It's Key A potential – though controversial – explanation for the origin of life on Earth.

Key Discovery Bacteria rival scientists when it comes to genetic engineering

Few scientific topics have been as hotly debated in recent years as genetic engineering. But what amounts to the same process was discovered as early as 1977, occurring naturally in plants.

Mary Dell Chilton showed that the bacterium *Agrobacterium tumefaciens* was able to insert some of its own genetic material into plants. The newly introduced DNA caused the plants to form what was called a "crown gall tumor," a familiar site on trees the world over. Chilton also found that removing the bacterium from the plant and putting it on a healthy plant would cause a tumor to form.

Agrobacterium, like many bacteria, house their DNA in two distinct places – in a chromosome, and in something called a plasmid. This is an extra piece of DNA, containing special genes that are beneficial to the bacteria. When inserted into a plant, the plasmid is replicated to produce the "foreign" protein coded for by the bacteria, which causes a tumor.

Geneticists were able exploit the mechanism to insert non-plant DNA into plants. The plasmid is first disarmed to remove its cancer-causing gene, leaving only the genes responsible for its movement. Genes can then be added for increased growth or insecticide resistance, thus improving the plant. Hostile genetic takeover by the *Agrobacterium* plasmid showed that DNA was universal and that it could be fairly easily swapped around. The resulting technology has since been used to create genetically modified plants, now grown all over world.

Nathan Dennison

Date 1977

Scientist Mary Dell Chilton

Nationality American

Why It's Key The discovery that DNA could naturally cross biological kingdoms led scientists to develop a method for genetic engineering.

Key Invention **Balloon angioplasty**
Healing a broken heart

Werner Forssmann performed the first cardiac catheterization on a human in 1929 – on himself. With the tube still running into his heart from an incision in his arm, Forssmann walked to the X-ray department at the August Victoria Home, in Berlin, where he was studying surgery, and took a photograph.

Almost half a century later, Andreas Gruentzig, at the University Hospital in Zurich, wondered if a catheter could be used to clear blockages in the coronary arteries – the blood vessels that feed the heart. Blockages comprised of fatty plaques can narrow the vessels and deprive the heart of oxygen, leading to a condition called angina, first described in 1772 by the English physician William Heberden. He had observed, in patients complaining of chest pains, "a painful and most disagreeable sensation in the breast, which seems as if it would extinguish life, if it were to increase or to continue."

Gruentzig began experimenting with balloons in his kitchen, testing his ideas initially on dogs. He planned to attach a balloon to the end of a catheter and inflate it inside the body, where it would push fatty plaques back against the walls of an artery, allowing blood to flow more freely. The pathologist who looked at one dog's arteries after the procedure admonished Gruentzig, warning that it should never be attempted on a human.

On September 16, 1977, Gruentzig performed the first balloon angioplasty on a human. It is now a procedure that is carried out over two million times a year worldwide.

Stuart M. Smith

Date 1977

Scientist Andreas Gruentzig

Nationality German

Why It's Key A safe and cheap way to clear blocked coronary arteries.

opposite Coloured x-ray of a cardiac angioplasty in progress

Key Event
The DNA library receives its first book

An organism's genome can be viewed as a book, a collection of chapters, pages, words, and letters that define everything from the way it looks, to when it would like to eat lunch. The sequencing of the first ever genome of an organism, published by Frederick Sanger and colleagues in 1977, was the first volume in a library of all natural life.

This library owes its origins to a humble virus, the imaginatively named phage phi X 174 that lives in bacteria. With only 5368 base pairs of DNA, phi X has a small genome, when compared to almost any other organism; some of its genes are squeezed on its single strand of DNA in overlapping sequences. But, back in the 1970s it was a painstaking process that had to be carried out by hand, allowing only a few hundred base pairs to be deciphered every day. Phi X became a first step towards much larger, faster progressing projects with higher levels of automation.

In 2003, Craig Venter, by then famous for his human genome sequencing triumph at the reins of Celera Genomics, produced a synthetic genome based on the circular chromosome of phi X 174. More than two decades of research on the virus since the reading of its DNA sequence had made it an easy choice. And as the first organism to have had its genome sequenced, Venter saw it as a symbolic choice.

Nathan Dennison

Date 1977

Country UK

Why It's Key Sequencing an organism's genome is now a regular occurrence and advances our understanding of an organism. A tiny virus became the first organism ever to have its genome sequenced.

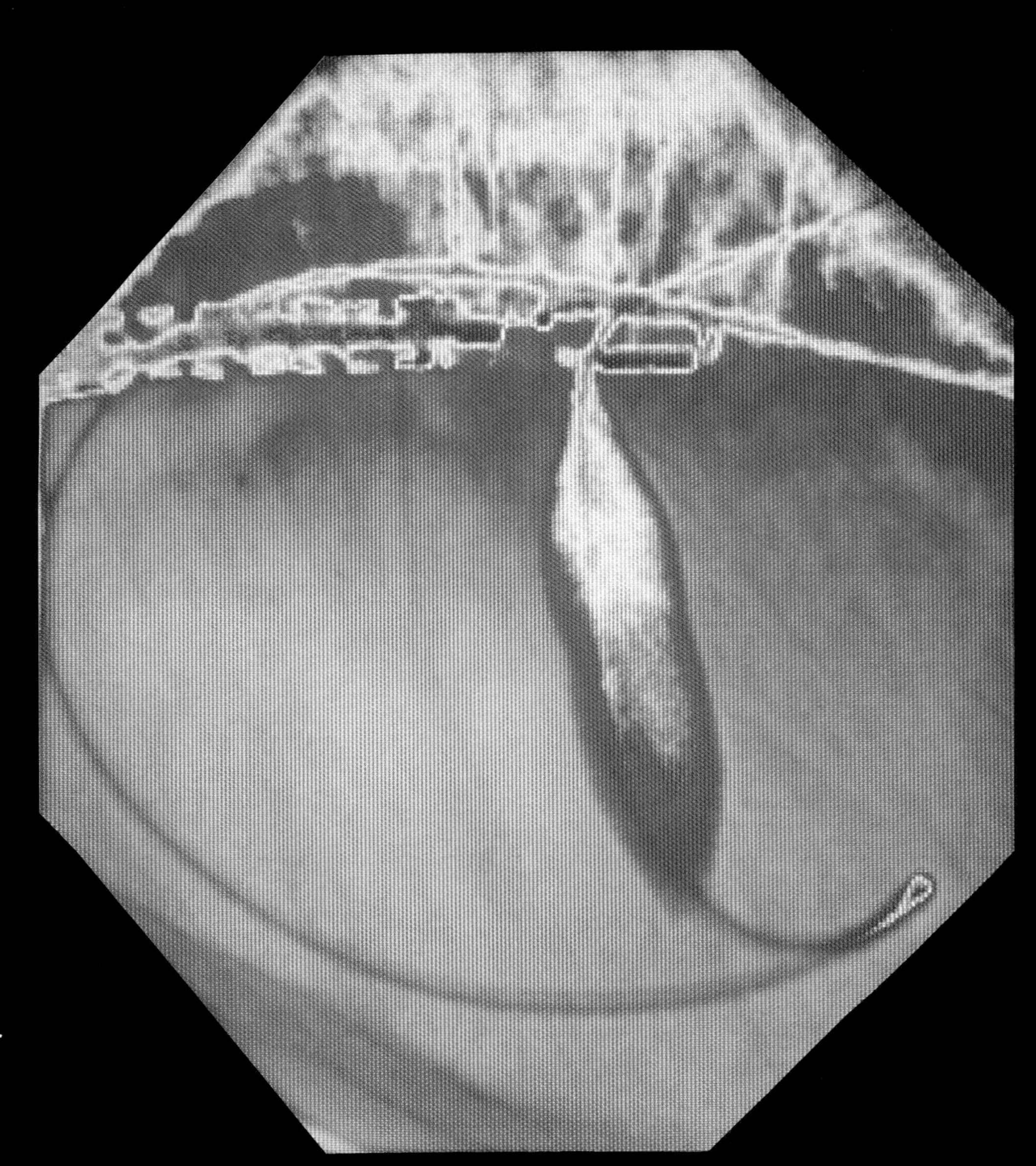

Key Invention Driverless car

In 1977, Japanese mechanical engineering firm Tsukuba created the first driverless car. It achieved a speed of 32 kilometers per hour on a marked-out course and ever since, mechanics and engineers have been striving to perfect the autonomous car. In the early 1980s Mercedes improved on Tsukuba with a car that hit 100 kilometers per hour on empty roads. In the Darpa 07 challenge, one of the entries, Junior, a Volkswagen Passat, used lasers to sense what was around it and "see" where it was going as well as probabilistic algorithms that processed the surrounding environments to aid its decision-making.

Given that cars are available with sensors, radar, cruise control and onboard computers, experts are predicting driverless cars could be a reality by 2020. Even earlier could be automated public transport systems. CityMobil is a EU-wide project to improve public transport across the continent. In conjunction with BAA, they have been testing PRT (personal rapid transport) vehicles at Heathrow airport. Electrically powered with carbon emissions 50 per cent lower than buses, the plan is to have an environmentally friendly, efficient service that could be used more widely than at airports, with testing sites at Rome and Castellón.

Whether people want driverless cars or not (issues like privacy concerning recorded data on car computers, the enjoyment of driving, what happens if the systems fail) a car that could remove the element of human error in judgement could be a good thing, especially as so often, driving is the last thing on people's minds when actually behind the wheel.

Fiona Kellagher

Date 1977

Scientist Tsukuba

Country Japan

Why It's Key The wheels start turning on an idea that could revolutionize personal and public transport.

Key Discovery **Hominids walked upright**
Fossilized *Australopithecus* footprints lead the way

In 1978, Mary Leakey and her team came across what was to become one of the most important set of footprints ever found. They belonged to two individuals, who walked over the wet volcanic ash at a site in Laetoli, Tanzania, some 3.5 million years before.

For the first time, the footprints showed that human ancestors known as *Australopithecus* had walked upright using a perfect two-footed stride, with no knuckle imprints that would have suggested the use of all fours. Before this discovery, we had no time scale for when human ancestors might have gone from getting around on four legs to walking on two. The prints showed the characteristically human arch in the foot and didn't have the same large big toe that other apes have. The footprints were not damaged because they had been covered in powdery ash. This was then cemented by soft rain, making a perfect imprint that would not be destroyed over millions of years. Information from the footprint size and the length of pace suggested that the two hominids would have been about 1.34–1.56 meters tall and 1.15–1.34 meters tall respectively.

The Laetoli site is located just 45 kilometers from another famous archaeological site, Olduvai Gorge, where many human tools and skeletons have been found over the years including those of *Homo habilis*. Mary Leakey, along with her husband Louis and son Richard, carried out significant research in the field of archaeology.

David Hall

Date 1978

Scientist Mary Leakey

Nationality British

Why It's Key This was the first time that it had been shown that our ancestors had walked upright. Even more importantly, it gave scientists a first glimpse of the timescales for which we might have been doing so.

opposite Mary Leakey kneels next to one of the Laetoli footprints, just uncovered

Key Experiment
Laser used to initiate a fusion reaction

Physicists have been trying to start a sustainable nuclear fusion reaction since the 1940s, as it would provide clean, almost limitless energy. Fusion requires small atoms to stay in contact, at a very high temperature, until they fuse together to form one heavier atom with less mass than the two previous ones added together. The spare mass is given off as energy. The problem is, heating the atoms up gives them a lot of energy, causing them to fly apart. Laser fusion is one solution to this.

The best fuels for nuclear fusion are deuterium and tritium (both forms of heavy hydrogen). For laser-initiated fusion, they are contained in tiny glass pellets. These pellets, less than one milimeter across, are irradiated for one billionth of a second by a powerful laser. When the powerful burst of light strikes the pellet, the outer layers are converted into a plasma, compressing the inner layers in a shell of hot gas and heating them to fusion temperature. The deuterium and tritium fuse together to form helium, giving off heat that, theoretically, can be converted into electricity in just the same way as in any power station.

In 1977, scientists proved initiating fusion burning this way was possible. Laser fusion is still being researched, but it has not yet progressed to the point of giving off more power from the fusion process than it uses to power the laser, and to process the pellets. It's not a solution to our energy needs yet, but it has promise.

Kate Oliver

Date 1977

Scientist Eugene Stark

Nationality American

Why It's Key A different approach to nuclear power, that is clean, controllable, can be moderated to meet demand, and gives a high energy yield. Now if only it worked...

Key Person **Robert Winston**
Reproductive king

Robert Winston, born in 1940, must be the only practicing scientist to have won a peerage, a BAFTA, and a Royal Society medal. Most people know him as the avuncular, soft-spoken presenter of television series such as *The Human Body* and *Superhuman*. With his trademark moustache and a talent for making complex science seem easy, Winston has explored our bodies, minds, childhoods, and our religious beliefs, on television and over fourteen books.

Winston has championed the cause of science in parliamentary halls and London hospitals as well as on TV screens. As a Life Peer, he regularly speaks in Parliament on scientific issues from stem cell research to pseudoscience. His research in reproductive medicine has allowed thousands of infertile women the world over to become mothers. From humble beginnings at the Hammersmith Hospital, London in 1970, Winston made several breakthroughs that led to international improvements in the field, including ways of precisely operating on Fallopian tubes under the microscope and reversing sterility. He organized the first team to offer in vitro fertilization (IVF) for free on the UK National Health Service. In 1978, he pioneered a technique for genetically screening embryos, to allow families with a history of genetic diseases to have healthy children, free of fatal conditions.

Still active in research, Winston hopes that by genetically manipulating stem cells that could develop into eggs or sperm, ways of growing eggs outside the body can be developed, in order to make IVF treatments cheaper and less intrusive.

Ed Young

Date 1978

Nationality British

Why He's Key His scientific advances have allowed thousands of women to become mothers, and his programs have educated millions of people about the human body.

opposite Robert Winston

Key Person **Donald Hopkins**
The medicine man

Working in Egypt as a young undergraduate, Donald Hopkins (b. 1941) was intrigued by the epidemic eye infection, trachoma, that was rapidly sweeping the country at the time. The encounter sparked a lifelong interest in pernicious tropical diseases and their treatment.

After completing a medical degree at the University of Chicago, and a Masters at the Harvard School of Public Health, Hopkins worked in the tropical region of Sierra Leone. Here he ran medical and research programs to tackle the smallpox and measles viruses, both prevalent and lethal in tropical areas. By 1978, thanks in no small measure to Hopkins' efforts, the smallpox virus had been completely eradicated from the world. The wiping out of this virus remains a completely unique case in modern medicine, where diseases are usually controlled rather than eliminated.

As a much respected figure within the United States Center for Disease Control and the World Health Organization, Hopkins continued his work on infectious disease. He still works closely with the organizations and he can currently be found at the Carter Center, USA. There, he is working on the elimination of the devastating Guinea worm parasite. Hopkins has become known as a true pioneer in the field of pernicious disease and, since his initial success, he has been irreplaceable as a vocal expert in the field, providing advice for major health programs around the world.

Hannah Welham

Date 1978

Scientist Donald Hopkins

Nationality American

Why He's Key One of the most influential players in the world of modern medicine, responsible for the eradication of the smallpox virus.

opposite Donald Hopkins

Key Person
John Maynard Smith

Rejected from the army for having poor eyesight, John Maynard Smith spent most of the war as an aeronautical engineer designing aircraft. Nothing special about that you might think, but this scholar would go on to become one of the world's greatest evolutionary biologists.

After a rather unfulfilling time at Eton College, Maynard Smith rather stumbled in engineering at university where he also nurtured a growing interest in the communist party. His first real job was working on the stress points in aircraft wings but after the war he decided he didn't even like planes, to him they were noisy and old fashioned. Instead he chose to follow his childhood interest and retrain as a zoologist in London.

In 1962, he became one of the founding members of the University of Sussex and by introducing mathematical models to biology, he revolutionised the concept of behavioural evolution. Key to this was his theory that the success of an individual in a population is intrinsically linked to the actions of others.

In addition to his research and his teaching, Maynard Smith is often considered one of the pioneers of science communication. He was passionate about breaking down the barriers of terminology and academic pretence and making ideas about evolution accessible to a wider audience, particularly through his books The Evolution of Sex (1978) and The Origins of Life (1999).

It's less well known, however, that he sighted science fiction literature as a factor in his fascination with genetics and evolution.

Andrew Impney

Date 1978

Nationality British

Why He's Key One of the most influential biologists of the twentieth century, Maynard Smith instigated many new fields of behavioral research, by combining theories from mathematics and biology.

Key Event
First test-tube baby is born

Every second of every day, somewhere in the world, about four babies are born. In 1978, there was one such bundle of joy who would take the world by storm. This little girl made the headlines, not because of who she would become, but because of how she was conceived.

Louise Joy Brown was born in Oldham, England on July 25, 1978 and became the world's first baby to be successfully conceived through *in vitro* fertilization (IVF). Her parents had been trying to conceive for over nine years and, in 1977, they agreed to undergo an experimental procedure as a last resort.

During the procedure, eggs were taken from the mother's ovaries and fertilized outside of her womb. Early tissue culture experiments were generally carried out in test tubes, and the press were quick to coin the phrase "test tube baby." In reality, the procedures are generally carried out in petri dishes, but somehow it doesn't quite have the same ring to it.

The fertilized egg, or zygote, was then transferred to the mother's uterus after just two and a half days. Previously, the doctors had waited five days until the egg had divided into sixty-four cells. This was considered the key to the success where previous attempts with other patients had failed.

In 2006, Louise Brown gave birth to a baby of her own after trying to conceive for around six months. The child was conceived naturally and the process had seemingly gone full circle.

Andrew Impey

Date 1978

Country UK

Why It's Key A triumph for medicine and a relatively simple technique that now helps millions of couples throughout the world to overcome infertility.

Key Publication **All backed up**
Why fiber keeps us regular

We all know fiber is good for us. Aside from keeping you regular, studies have shown that consuming more fiber reduces the likelihood of heart disease, Type II diabetes, and diverticulitis – a painful, age-related, gastric condition. Current recommended daily allowances for fiber are 30-38 grams for men and 21–25 grams for women, but most Americans consume only half that amount.

Dietary fiber is obtained in both soluble and insoluble forms. Soluble fiber is found in all plant-based foods and undergoes a fermentation process in the gut producing health-enhancing chemicals. Insoluble fiber, from sources such as whole wheat and bran, absorbs water and is used to bulk out waste and speed up the intestinal transit of waste material. In 1969, David Southgate developed a method for measuring the amount of fiber in foods. His work came into its own when, with Alison Paul, he began measuring the amount of fiber in common foods. Their results were published in 1978 in the influential book *The Composition of Foods* – the definitive guide to the nutritional content of food that was first published in 1940, and the basis for modern nutritional thinking.

Southgate and Paul's quantification of the fiber levels in food paved the way for studies on how dietary fiber affects general health, and for the nutritional advice we receive today.

Helen Potter

Publication *The Composition of Foods*

Date 1978

Authors Alison Paul, D. A. Southgate

Nationality British

Why It's Key Enabled dietary fiber to be quantified, helping establish a link between low fiber and cancer of the colon.

Key Discovery **Dark Matter**
More than meets the eye, or the telescope

In 1933, the irrepressible Fritz Zwicky demonstrated that clusters of galaxies contained "dark matter," when he found that galaxies in the Coma cluster were moving much faster than could be explained by their mass. No one paid much attention: it took decades for another astronomer to reference his discovery paper.

In the 1960s, however, astronomers began to find evidence for dark matter in individual galaxies. Morton Roberts made detailed observations of the Andromeda Galaxy with a radio telescope, while Vera Rubin and Kent Ford observed it with an optical telescope. They found that the gas and stars in the outer parts of Andromeda were moving faster than could be explained by visible matter alone, leading them to suspect that something was amiss. Zwicky's claims were starting to look less absurd.

Rubin and her team began feverish studies of other galaxies and, by 1980, observations of numerous galaxies and clusters had convinced most astronomers that dark matter was real. They didn't know what it was; they still don't. But that hasn't stopped particle physicists from proposing all manner of candidates. Examples include such hypothetical particles as the "axion" and "neutralino," as well as sterile neutrinos, whose existence some recent studies have found support for. As yet, however, none of these particles has been able to account for the elusive dark matter.

Recent observations of supernovae and the cosmic microwave background suggest that as much as 25 per cent of the Universe's mass could be made up of dark matter.

Eric Schulman

Date 1978

Scientists Fritz Zwicky, Vera Rubin

Nationality Swiss, American

Why It's Key Scientists now agree that a large amount of unseen matter is floating around the cosmos; what it is, nobody knows.

Key Event
An X-ray vision of the Universe

We're all familiar with the ability of X-rays to pass through skin and visualize the inner structures of the body. X-rays are also emitted by many high-energy sources throughout the Universe but, strangely, are unable to pass through the atmosphere.

Cosmic X-rays therefore remained completely unknown until 1962, when a crude detector was flown on a suborbital rocket. Just sixteen years after their discovery, a dedicated telescope was placed in orbit to study X-ray sources.

The High-Energy Astrophysical Observatory-2, or Einstein as it was more memorably dubbed post-launch, was lofted into a low-Earth orbit by NASA on November 13, 1978. It improved resolution of X-ray sources, such as supernova remnants and hot gas in galaxies, by a factor of more than 100 over previous efforts, and made more than 5,000 targeted observations in its three-and-a-half year lifetime.

As well as pioneering technology, Einstein was also a trailblazer for cooperative science. For the first time, the instrument was made available to the wider astronomical community, rather than being restricted to the partner universities and institutions who created it. Einstein brought X-ray astronomy into the mainstream, and paved the way for even more powerful missions, such as Nasa's Chandra observatory, launched in 1999. Its legacy has given modern astronomers the ability to detect and study neutron stars, black holes, and dark matter.

Matt Brown

Date 1978

Country USA

Why It's Key Our first detailed glimpse at the X-ray sky.

Key Person

Steven Weinberg can feel the force(s)

Since discovering the world of theoretical physics as a sixteen year-old living in New York City, Steven Weinberg has led a distinguished career in search of a model of nature that will unify all its fundamental forces.

Weinberg's key work surrounds the unity of two of the fundamental forces, the weak and electromagnetic forces. Work on the electroweak theory inspired the confirmation that a strong force exists to keep the nuclei of atoms together. It also inspired the Standard Model of all known forces except gravity; this has been called the bible of particle physicists.

He was awarded the Nobel Prize in Physics in 1979, along with Sheldon Glashow and Abdus Salam, for the implications of his work in unifying the two forces. Having developed the theory in 1967, his paper summarizing the electroweak theory is the most often cited paper in particle physics. He has also written numerous other publications including *The First Three Minutes* in 1977, which explained the Big Bang theory to a wider audience.

Weinberg is now a string theory enthusiast, although he is more of a spectator than a player, and has been elected to the US National Academy of Sciences and Britain's Royal Society. Weinberg's pioneering work may one day help to explain all of nature in the form of a handful of elementary particles.

Nathan Dennison

Date 1977

Nationality American

Why He's Key Steven Weinberg played a key role in attempts to unify the fundamental forces of nature.

opposite Sheldon Glashow and Steven Weinberg (right)

Key Event

Three Mile nuclear reactor meltdown

It is hard to know whether to celebrate or condemn the meltdown at Three Mile Island, but the sequence of malfunctions that occurred probably could and definitely *should* have been avoided. Ultimately though there was no loss of life or even injury, and various studies have determined that the environmental impact from the event was negligible. How much worse it could have been.

The business end of a power station is the reactor or core. Here, a controlled nuclear reaction takes place, regulated by control rods, which absorb neutrons and slow the reaction down. Heat generated in the core is taken away by water in the primary loop; this water is radioactive. Heat from the primary loop is transferred to water in a secondary loop, which converts to steam and passes through a turbine to generate electricity.

It was in this secondary loop that the accident started. A small malfunction caused pumps to shut down and this in turn caused pressure and temperature to rise in the primary loop. Automatic fail-safes vented the excess, but didn't close afterwards, as they should have. This caused a rise in temperature and pressure in the core, which became overheated and suffered a partial meltdown.

Luckily all radioactive leakage was contained within the site, probably due to the thick concrete walls that house the core. Damage was restricted to a minimum and there were no injuries. The radiation that local residents received was only one sixth of what you would be exposed to during routine a chest X-ray.

Barney Grenfell

Date 1979

Country USA

Why It's Key Three Mile Island should probably go down in history as one of the centuries "luckiest escapes." A potentially devastating meltdown, it had relatively little environmental impact, but taught us some important lessons.

Key Discovery **Making markers for disease**
Hide and seek, genetic style

Though DNA had been "uncoded," practical uses for its discovery were not immediately forthcoming. The fact that some diseases were genetic, and therefore had to be represented in the genetic code of sufferers, was clear enough, but how to translate that concept into a mechanism that could locate a gene was somewhat elusive.

Then, in 1979, the German-born British geneticist Sir Walter Bodmer suggested that gene markers could be used to find the strands of DNA coding for a particular disease in a given individual. Bodmer's idea was simple. He noted that, while an individual gene may be very difficult to find on a chromosome, there were often distinctive patterns of nucleotides on any given gene. He suggested that perhaps if one found a characteristic pattern of DNA located close to the actual gene coding for a certain disease process, then it was quite likely that that characteristic DNA sequence would be inherited with the DNA sequence that codes for the disease. One can then use the other, more recognizable series of DNA as a flashing light pointing to the actual gene that causes the genetic disorder.

Working almost like a table of contents, these markers can dramatically speed up the evaluation of the human genetic code. DNA markers have already been found near the genes for Alzheimer's disease, cystic fibrosis, and muscular dystrophy. It is hoped that one day, knowing the location of the marker for a disease may lead to the finding of the actual gene.

B. James McCallum

Date 1979

Scientist Walter Bodmer

Nationality British

Why It's Key Walter Bodmer's gene marker idea has led to the more rapid evaluation of the human genetic code and the location of several disease-causing regions of chromosomes.

Key Person **Arthur C. Clarke**
Visionary author and scientist

Arthur C. Clarke (1917–2008) was a believer. In his 1962 book *Profiles of the Future*, he looked with hope toward times ahead and stated his three laws: that virtually everything is possible; that the only way to discover our limits is to reach past them; and that technology will one day be so advanced that it will be "indistinguishable from magic." Clarke is most famous for his science fiction, notably his novel, *2001: A Space Odyssey*, but he also was a formidable scientist in his own right.

A member of the British Interplanetary Society since 1936, Clarke dabbled in many scientific areas. Before he published even one professional story, he had already published the technical paper "Extraterrestrial Relays," in 1945. This document outlined his idea for using satellites in geostationary orbits for global communication. A geostationary orbit is one whose orbit around the equator takes almost exactly a day, allowing it to be in position above a certain point on the planet all the time. By 2005, there were over 330 satellites in the Clarke Orbit, each invaluable for modern-day communications and weather observations.

According to Clarke himself, however, geostationary satellites will not be the idea for which he is most remembered in years to come. In the late 1970s he popularized the idea of the "space elevator" – a huge structure linking the Earth to orbiting bodies – in time, he says, these will replace space shuttles. This, he thinks, will be his lasting legacy.

Logan Wright

Date 1979

Nationality British

Why He's Key As both scientist and author, Clarke provides an imaginative spark that helps drive the wheel of technology forward.

opposite Arthur C Clarke, at home in Sri Lanka, by his satellite receiving dish

Key Event
Pioneer 11 passes Saturn

Just sixteen years after Sputnik became the first artificial object to enter Earth's orbit, mankind extended its reach to the gas giants of Jupiter and Saturn.

Pioneer 11 was launched by the United States on April 6, 1973. After a gravitational assist from Jupiter (previously visited by sister craft Pioneer 10), the probe headed on to Saturn. It swung past at a distance of 21,000 kilometers above the cloud tops on September 1, 1979, obtaining the first close-up images of the ringed planet. As well as sending back plenty of eye candy, Pioneer 11 also laid the path for two other probes, Voyagers 1 and 2, which were hot on its heels. Pioneer 11 was commanded to fly through the plane of Saturn's rings to see if tiny particles would pose a risk to spacecraft. This trajectory was necessary if the Voyagers were to fly on to Uranus and Neptune. Pioneer 11, living up to its name, proved that the maneuver was safe for the more ambitious probes to copy.

Pioneer's twelve main instruments also sent back a wealth of data on atmospheric conditions, the solar wind, cosmic rays, and much more. The spacecraft continued toward the outer Solar System, but encountered no further planets. The last communication with the probe came in September 1995. Pioneer 11 will reach the constellation of Aquila in about four million years, and carries a plaque showing an image of its planet of origin. Should the probe ever be picked up by an alien civilization, they'll know where to find us.

Matt Brown

Date 1979

Location Saturn

Why It's Key Humankind's first close peek at the Saturnian system pioneered the way for future probes, including the ongoing Cassini mission.

Key Discovery **Gravitational lenses**
Using galaxies as giant telescopes

Albert Einstein believed the gravitational lenses he predicted should exist in 1912 could never be observed. It took sixty-seven years to prove him wrong.

The theory of general relativity says that mass bends light. According to general relativity, then, light from a background star passing a nearer star could be bent around that star such that we would see two images of the background star, one either side of the closer one. Since this gravitational lensing effect requires a precise alignment of the foreground and background stars, Einstein concluded it was unobservable. He was persuaded to publish a paper on the subject, in which he concluded that "there is no hope of observing this phenomenon directly."

A year later, Fritz Zwicky suggested that galaxies offered a much better chance than stars for the observation of gravitational lens effects. He pointed out that such lenses would test the theory of general relativity, enable us to see galaxies at great distances, and provide us with the opportunity to measure the total mass of galaxies, in order to determine how much dark matter exists in galaxies and clusters of galaxies.

In 1979, astronomers Dennis Walsh, Robert Carswell, and Ray Weymann made the first discovery of a gravitational lens by accident. The object they saw became known as the Twin Quasar because it appeared as two identical objects. Since then, astronomers have used gravitational lenses to observe the most distant objects in the Universe, and to calculate that clusters of galaxies contain large amounts of unseen dark matter.

Eric Schulman

Date 1979

Scientists Dennis Walsh, Robert Carswell, Ray Weymann

Nationality British, American

Why It's Key Provides astronomers with a powerful tool for studying galaxies, clusters of galaxies, and the Universe.

Key Person **Sir David Attenborough**
Let me tell you about life on Earth

Every science discipline has its heroes, and over the last century, the field of natural history has had none greater than Sir David Attenborough (b. 1926). He is to wildlife documentaries what Darwin was to evolution, and he has probably done more for bringing the natural world to the attention of the general public than anyone in history.

Attenborough studied geology and zoology at Cambridge and, following two years in London studying anthropology, he was called up into the Royal Navy for National Service. By the 1950s, his career was somewhat drifting and he applied for a job with BBC radio – he was turned down. Luckily for us, someone saw enough in his CV to give him a chance, and instead he was offered a job in television. This sounds impressive now, but at the time hardly anyone even owned a television, let alone watched it.

In 1954, he appeared in front of the camera on *Zoo Quest*, and under his control, BBC2 became the first UK television channel to broadcast in color. Despite being offered the position of Director General of the BBC, Attenborough was clear where his passion lay, and that was in program making.

What followed was fifty years of the some of the finest wildlife documentaries seen anywhere in the world, centered around the "Life" series – the first being *Life on Earth* in 1979. These multi-award-winning educational programs have left millions of viewers spellbound with the wonders of the natural world.

Andrew Impey

Date 1979

Nationality British

Why He's Key It's rare to find an individual almost universally loved and respected, but David Attenborough has amazed and inspired generations for over fifty years.

Key Discovery **First human oncogene identified**
Tracking down the key to cancer

Scientists around the world are studying the genes and molecules that power cancer cells, in the hope of finding more effective treatments. And Robert Weinberg could be viewed as the grandfather of the field.

Weinberg and his colleagues discovered the first oncogene – a gene that, when faulty, causes cancer. The so-called Ras gene is found in all our cells, and normally controls cell division, switching the process on when and where it's needed. But if Ras is faulty, cell division can run out of control, leading to cancer.

Originally discovered by work with cancer-causing viruses, Ras' power was revealed by transferring small fragments of DNA, containing genes, into cells, and seeing if this caused them to multiply unchecked. The resulting discovery showed that the normal genes that control our cells are just a few small faults away from disaster. Weinberg became caught up in a race to find Ras with Michael Wigler, and the tense competition between the two rivals is dramatically described in the book *Natural Obsessions*.

Six years later, Weinberg made another impressive discovery. This was the retinoblastoma gene, known as Rb – the first tumor suppressor. Rb works as the "policeman" of our cells, acting to protect us by shutting off cell division when it goes awry.

Since Weinberg's impressive breakthroughs, many more oncogenes and tumor suppressors have been found. These have paved the way for the development of potential treatments for cancer, which may save many lives in the future.

Kat Arney

Date 1980

Scientist Robert Weinberg

Nationality American

Why It's Key Tracking down the first cancer gene revolutionized the field of cancer biology and opened the door to potential new treatments.

Key Event **WHO declares smallpox eradicated**
The world says goodbye to a killer

For over 3,000 years, smallpox wreaked havoc throughout the world, claiming millions of people's lives and leaving millions of others scarred or blinded. Thanks to Edward Jenner's pioneering work in the nineteenth century, the smallpox vaccine was the first to be developed. The challenge then was to distribute it to vulnerable populations around the world.

In 1958, the World Health Organization (WHO) called for the global eradication of the disease, prompting a number of governments across the developing world to establish programs to achieve a hundred per cent vaccine coverage within 3 to 5 years.

The success of the campaign was mixed. While the number of global cases dropped from an estimated 50 million in the early 1950s to 10–15 million in 1967, some countries did not fare so well. For example, India's program faltered throughout the 1960s, partly because of the geographical, political, and social diversity of this huge country. After a concerted effort, however, India achieved "Smallpox Zero" status in 1975.

In 1977, the last natural case of smallpox in the world was recorded, in a 23-year-old cook in Somalia, who made a full recovery. After extensive verification a commission of scientists declared smallpox had been eradicated across the globe, in December 1979. In May 1980, WHO confirmed the eradication.

Controversially, stocks of smallpox are still held by two secure laboratories, one in Atlanta, USA and one in Novosibirsk, Russia. Considerable research continues into smallpox vaccines and treatments today.

Christina Giles

Date 1980

Country International

Why It's Key A global effort finally confines a killer of millions to the history books.

opposite Coloured transmission electron micrograph of the smallpox virus

Key Event **Hepatitis B vaccine**
It's all in the blood

Hepatitis B, a debilitating and deadly disease that can cause liver cancer, has plagued humanity throughout history. But when scientists pioneered the first hepatitis B vaccine in 1980, this killer virus was delivered a hefty blow. Viral hepatitis is an inflammation of the liver caused by several different viruses (A, B, C, D, and E). Hepatitis B is caused by a blood-borne virus, transmitted when blood or bodily fluids of an infected person mix with someone else's, making it highly infectious – more so even than HIV. While it can be fatal, many people are oblivious to the fact that they have contracted it because symptoms do not always develop immediately, or at all.

Little was known about the cause of the disease until American Baruch Blumberg fortuitously spotted the virus while analyzing blood samples in 1965. It was a Nobel Prize-winning discovery, and one that led to a revolutionary blood test called the radioimmunoassay. The new test dramatically decreased the number of uninfected patients receiving contaminated blood through transfusions. In 1980, American scientists developed a Hep B vaccine using parts, or "subunits," of human blood-derived virus to induce immunity – the first vaccine ever to be created in this way. A few years later, researchers developed the first genetically engineered vaccine, which came into use in 1986.

But vaccines are only one step toward controlling the disease. Hepatitis B still kills around 1.5 million people worldwide every year, and much depends on effective health education to inform high risk groups of the need for vaccination.

James Urquhart

Date 1980

Country USA

Why It's Key Marked a giant leap forward, not only for the prevention of one of the world's most common and infectious diseases, but also for the disease associated with it - liver cancer.

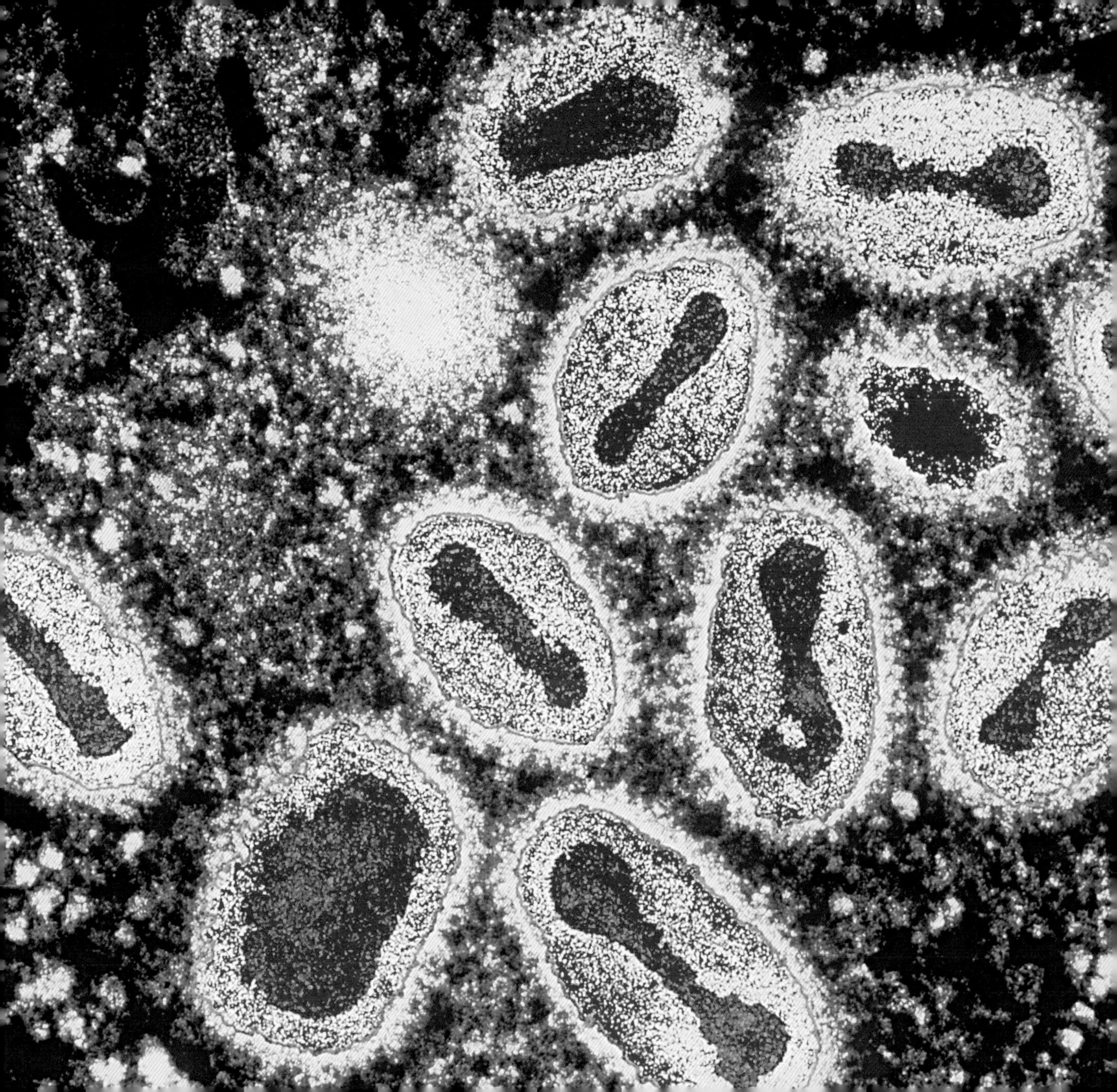

Key Discovery **The demise of the dinosaurs**
Meet the meteor

The fossil record shows that around 65.5 million years ago, 85 per cent of all of the species on Earth, including the dinosaurs, went extinct. There have been many explanations for why this occurred, but the theory we currently believe to be true is known as the Impact Hypothesis.

This theory holds that the mass extinction was caused by a meteorite slamming into the Earth. It was prompted by the discovery of high concentrations of the element iridium in layers of sedimentary rock dating from the same time period. Iridium is extremely rare in the Earth's crust but found in abundance in meteorites. By measuring the level of iridium in the rocks, and using the known percentages of iridium in meteorites, scientists estimated that a space rock ten kilometers in diameter smashed into the Earth. The impact is thought to have caused a global cloud of debris which blocked out sunlight, preventing photosynthesis and causing general death and destruction.

The theory received a significant boost in the 1990s with the discovery of a Mexican crater 180 kilometers wide; the impact site had been found. The composition of rocks in the region may have caused large amounts of sulphur dioxide and water vapor to be created by the impact, producing a global aerosol cloud. While a debris cloud may have lasted around a year – which would have been bad enough – an aerosol cloud could have lasted twelve years, blocking out the sun, cooling the Earth, causing acid rain, and ultimately sealing the dinosaurs' fate.

Emma Norman

Date 1980

Scientist Luis Alvarez, Walter Alvarez

Nationality American

Why It's key Widely supported theory explaining the extinction of the dinosaurs.

Key Person **Frederick Sanger**
Wins his second Nobel Prize

In 1980, Frederick Sanger (b. 1918) secured his place in history by becoming one of only a handful of people ever to receive more than one Nobel Prize. He was born in Rendcombe, England in 1918 and, instead of following his father into medicine, decided to study biochemistry because he thought it would provide more of a chance to solve problems.

And solve problems he did. In 1958, Sanger was awarded the Nobel Prize for being the first person to map the order of amino acids (the small building blocks of protein molecules which make up every living thing) in the hormone insulin. Spurred on by his achievement, he then focused his attentions on a more baffling genetic conundrum – the sequence of DNA in genes.

By the 1960s, it was known that DNA was made up of a linear sequence of compounds called "bases," but nobody had figured out how to find the order of these bases in a given genome. Over the next fifteen years, Sanger and his team managed to pioneer a way of "reading" the sequence of bases in a genome, which they called the "dideoxy" technique.

It was this work that earned Sanger his second Nobel Prize for Chemistry, in 1980. His work was so important that his techniques are still used to sequence DNA to this day, and have been instrumental in the Human Genome Project.

Hannah Isom

Date 1980

Nationality British

Why He's Key Decades of work studying proteins and genomes culminated in a method for sequencing DNA.

Key Invention **Single atoms come into focus**
The scanning tunneling microscope

Ever since scientists first peered through a microscope in the early 1600s, they have constantly worked to improve the magnification and see smaller and smaller objects. Eventually they hit a limit: the wavelength of light itself, around 2,000 times the width of an atom. Electron microscopes overcame this by relying on the much shorter wavelength of electrons to make images, but even these have their limits.

The most recent revolution in imaging the very small came when German and Swiss scientists Gerd Binnig and Heinrich Rohrer invented the scanning tunneling microscope, in 1981. It works using electricity and quantum mechanics. Instead of a lens, an extremely sharp stylus, with a tip just one atom across, is held the distance of one atom away from the object being magnified. Quantum mechanics allows electrons from the stylus to "tunnel" through the vacuum between the stylus and the sample, and produce an electrical signal that can be measured. The stylus can then slowly be moved across the sample, while being raised and lowered to keep its distance constant.

The process allows immensely fine studies of an object's surface to be made, atom by atom, that can be converted into an atomic contour map using a computer. Binnig and Rohrer were awarded a Nobel Prize for Physics just five years after their invention and their technique is widely used today in a variety of sciences including metallurgy, semiconductor research, and even genetics.

David Hawksett

Date 1981

Scientists Gerd Binnig, Heinrich Rohrer

Nationality German, Swiss

Why It's Key The breakthrough technique which finally enabled the study of materials at the atomic level.

Key Invention **Cell to cell**
Bacteria can talk

You might very well wonder how an organism smaller than a grain of salt can do something it takes the average person years to learn, but think about it this way: bacteria can do plenty of things that you can't do at all. They can spend thousands of years in a tomb and still come out alive. They can thrive around deep sea vents and in environments hot enough to boil water. And, conveniently for us, they can make yoghurt out of milk simply by sitting in it.

But how do they do it, and what kind of conversations go on in a troop of unicells? It's all down to a nifty little system called "quorum sensing" – essentially, a way of communicating using chemicals instead of words. In 1981, microbiologists pinned down the first quorum sensing molecule in a marine bacterium going under the name *Vibrio fischeri*. These rod-shaped bacteria were known to emit an eerie glow, but only in the company of other bacteria – never alone. This was due, as it turned out, to the presence of an "activator" molecule produced by the bacteria, which triggers luminescence when enough of it accumulates. The system helps the bacteria to tell when their population is becoming too dense.

Scientists are now aware of hundreds of species of bacteria that use quorum sensing, for everything from coordinating group movements to the release of molecules that gain them access to the tissues of multicellular organisms.

Hayley Birch

Date 1981

Scientists Anatol Eberhard and colleagues

Nationality American

Why Its Key Scientists can learn enough about bacterial conversations, they might be able to exploit them to combat serious infections, including superbugs.

Key Invention **Aspartame**
The sweetest thing

The diet colas and low-sugar ready meals we enjoy today have their genesis in a chemistry laboratory. In 1965, James M. Schlatter, working for G. D. Searle and company, happened to lick his finger while working on a drug synthesis. Schlatter had accidentally combined the amino acids aspartic acid and phenylalanine to create a sweet-tasting compound now known as aspartame. The white, crystalline additive, marketed as NutraSweet, is 180 times sweeter than sugar with virtually no calories.

It was set to revolutionize the food and drinks industry, but not before fifteen years of testing to prove it was safe for human consumption. Finally, the U.S. Food and Drug Administration approved the compound for use in dry foods in 1981. Further market encroachment occurred in 1983 when the sweetener was given approval for use in sodas. In 1996, all further restrictions were lifted. Today, sweeteners such as aspartame are widely used in the food industry as a low calorie alternative to sugar. It is also friendlier to teeth and an easily regulated source of sweetness for diabetics. Whenever you see the words "diet" or "light" on a food product, chances are it contains aspartame or one of its derivatives.

The compound has been dogged with claims that it causes adverse health effects, including an increased risk of cancer. However, little credible evidence supports these claims, despite over two hundred studies into aspartame's effects on health. Which is just as well; the increasingly corpulent population needs all the help it can get.

Matt Brown

Date 1981

Scientist James M. Schlatter

Nationality American

Why It's Key Few chemicals have had such a big impact on our shopping and dietary habits.

Key Event **Solar One**
The world's largest solar power station goes online

In much the same way that you'd choose to put a wind turbine somewhere fairly gusty if you wanted to maximize its energy yield, it stands to reason that if you want to make the most of a solar panel then you'd be wise to stick it somewhere sunny. Since places scarcely get sunnier than the Mojave Desert, it's no surprise that, when plans were drawn up for the large-scale solar power station, Solar One, this was the location they chose.

The Solar One system was comprised of 1,818 large sun-seeking mirror panels – known in the business as "heliostats" – all of which pointed toward a centrally positioned receiver that sat at the top of a 91 meter tower.

Each of these heliostats had a mirrored area of 40 square meters, and channeled sunlight up toward the receiver panel. Generally, somewhere between 80 and 95 per cent of the resulting reflected energy was absorbed by a heat transfer fluid and carried down to the base of the tower. There it was converted to electricity, through the use of steam-powered turbines, or used to thermally charge an energy-holding tank.

The system was capable of generating up to 10 megawatts of power, and clocked up an impressive 38,000 megawatt-hours over the course of its operational life span. Though technically Solar One still exists, the station was redesigned in 1995 to incorporate, amongst other things, another 108 much larger heliostats. The system was then renamed, rather logically, Solar Two.

Chris Lochery

Date 1981

Country USA

Why It's key Proved that "power towers" could operate successfully and provide electricity reliably – even when the sun was covered by cloud, or at night.

opposite The 1,818 panels at Solar One

Key Publication **Kaposi's Sarcoma in Homosexual Men** The first signs of AIDS

Human immunodeficiency virus (HIV) was undoubtedly mounting a silent attack throughout the 1970s. It wasn't until 1981, however, when U.S. clinicians spotted a small group of homosexual men suffering from an unusual disease, that scientists gained the first inklings of its existence.

The report, in *The Lancet*, centered on eight men suffering from a rare form of skin cancer called Kaposi's sarcoma, now widely recognized as the manifestation of full-blown AIDS (acquired immunodeficiency syndrome). That same year, a second publication in *The Lancet*, described a number of similar instances in Europe. Some of these cases were associated with pneumonia caused by *Pneumocystis carinii*, a fungus-like microorganism that resembles yeast. *Pneumocystis* normally poses no great threat to the body, as the immune system is capable of fighting it. HIV and AIDS sufferers, however, find their immune system severely compromised and struggle to fend off the invading microbe.

In 1981, the cause of these infections was not known and certainly wasn't attributed to a virus. The flurry of controlled studies that followed initially focused on small groups of gay men and identified links between a syndrome – a collection of different symptoms – and high levels of promiscuity. It later became apparent that the disease was not exclusive to homosexuals and, as cases began to notch up, pinning down the agent that carried it became the number one priority. The virus was finally isolated in 1985, as diagnosed cases globally raced into the thousands.

Hayley Birch

Publication *The Lancet*

Date 1981

Authors Kenneth Hymes and colleagues

Nationality American

Why It's Key Marked the recognition of a growing threat to global health, and led to research that identified the causes of AIDS.

Key Discovery **Embryonic stem cells first isolated**

In the early stages of embryonic development, there is a time when cells have an undefined future. They can become neural tissue, or muscle, or bone, or blood cells, or anything else they want to be. At this stage, they are called stem cells. Their existence was postulated by German pathologist Max Askanazy over a hundred years ago, but they weren't really considered practical for study until the early in 1980s.

By the 1950s, scientists had determined that there were cells in a rare type of tumor called a teratoma that could differentiate into other types of cells. In order to study these, however, they had to wait until a teratoma – a fairly rare tumor – occurred in a mouse and then isolate the cells.

This type of study was inefficient at best, but all that was about to change. In 1981, Nobel Laureate Martin Evans and embryologist Matt Kaufman isolated a group of cells from an early-stage mouse embryo called a blastocyst. The cells grew into teratoma when injected into other mice, and differentiated into other cells when left to their own devices in a petri dish – in other words, they passed every test needed to establish the fact that they were stem cells. What's more, the cells could be grown on culture, like bacteria or mold. The new, easy-to-obtain-and-propagate cell line could be used to study genetics as well as potentially facilitate gene therapy.

B. James McCallum

Date 1981

Scientists Martin Evans, Matt Kaufman

Nationality British

Why It's Key The isolation of stem cells from embryos allowed for a huge increase in the ease of their study. Stem cell technologies could be the defining medical treatments of the next century.

opposite An embryo of a mouse being used in stem cell research. The central organ is the heart

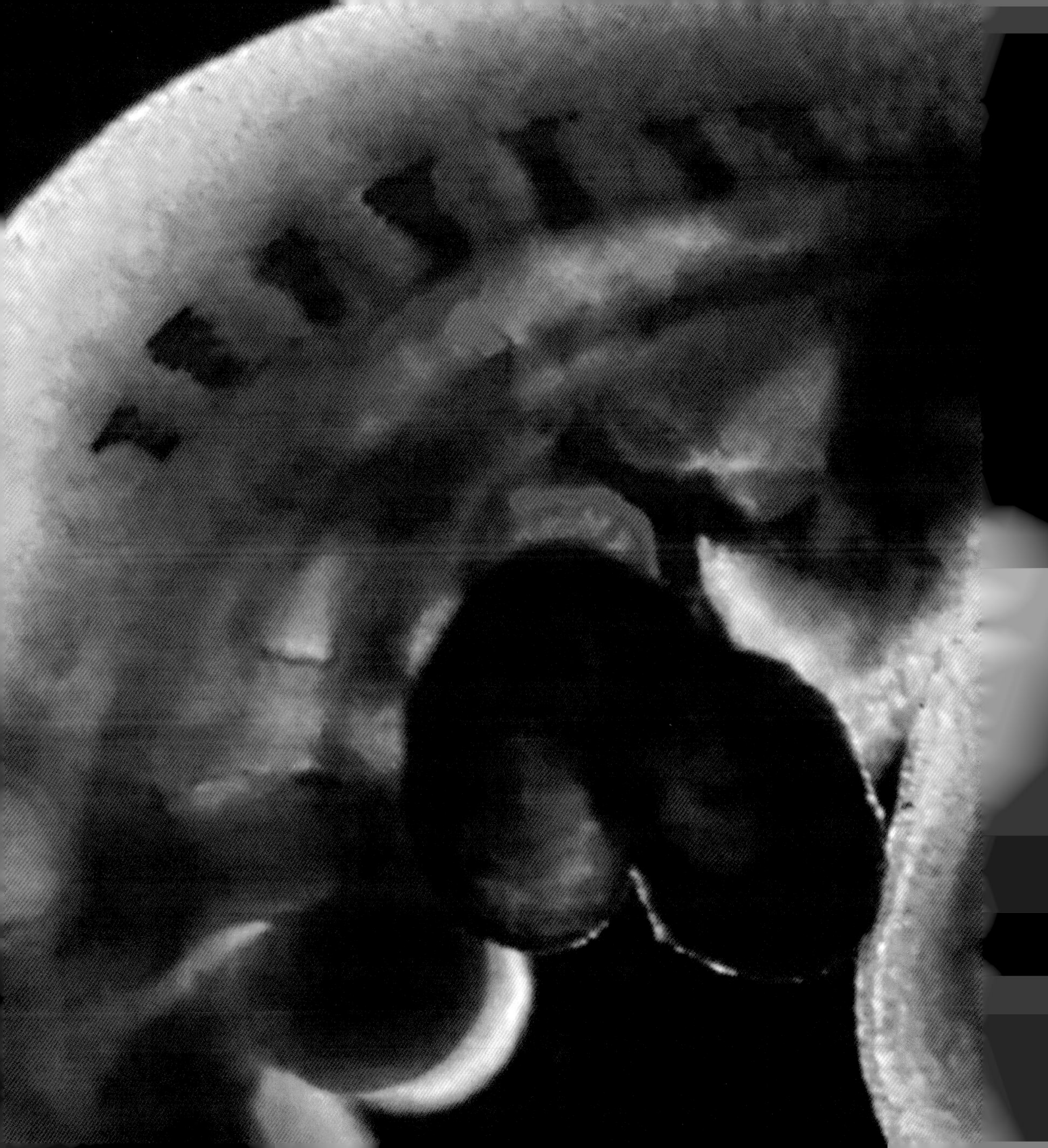

Key Invention **The backbone of the modern operating system** DOS-ing about

During the summer of 1980, IBM began working on a top secret project, known as Project Chess, to develop personal computers. They contacted Microsoft and, after getting them to sign a strict confidentiality agreement stopping them blabbing about the project, asked them to develop software for it.

Having never made an operating system before, Microsoft suggested IBM contact Gary Kildall of Digital Research and ask to use the system he had written called Control Program for Microcomputers (CP/M), which was the system most computers currently used. Digital Research, however, refused to sign a confidentiality agreement, so IBM returned empty-handed.

Running out of time, and in desperate need of an operating system, they turned to Seattle Computer Products (SCP). Tim Patterson who was working there had become tired of waiting for Digital Research to put CP/M onto the computers he was using, so he had written his own version, called the Quick and Dirty Operating System.

Microsoft called SCP and paid US$25,000 to use QDOS, and then charged IBM US$80,000 to use it. Microsoft later paid SCP another US$50,000 to buy the program outright, and many years later they paid US$975,000 for the licenses that go with it. As IBM got the rights at a low cost, it meant they could make the system available cheaply to its customers. On August 12, 1981, IBM officially launched the PC with DOS. DOS would go on to form the backbone of all Microsoft operating systems, right the way up to Windows 95.

Douglas Kitson

Date 1981

Scientists IBM

Nationality American

Why It's Key DOS was one of the most important pieces of computer software for fourteen years, by being the cheap industry standard, and forming a basis for Operating Systems to come.

Key Event **US space shuttle Columbia is launched** NASA reignites race with reusable rockets

Huge plumes of smoke engulfed the ground below as glowing beams of glorious light slowly eased the monument up toward the heavens. Tiles buckled under the vibrations and fell like leaves from this triumph of mankind, as it majestically overpowered the awesome force of gravity. Space Shuttle Columbia and her crew were on their way into orbit.

On April 12, 1981, Commander John Young and pilot Robert Crippen made history by guiding a space shuttle into space for the first time. NASA had not put a human into space for six years, but now they had a means to do so, with the added bonus of reusing most of the components. A space shuttle consists of an orbiter (the bit that looks like a plane), an external tank (the huge orange belly), and two Solid Rocket Boosters (SRBs). Only the external fuel tank is not reused.

Despite the near-perfect launch that Columbia demonstrated, the performance of the SRBs was not as expected. The boosters had been more powerful than anticipated and, at the point of separation, they had elevated the shuttle three kilometers higher than predicted.

During their flight, Young and Crippen tested over 130 features of the shuttle, from opening bay doors to landing the orbiter like a plane, just fifty-five hours after blasting-off like a rocket. The real achievement, however, came seven months later when Columbia and its SRBs were launched for a second time; a reusable Space Transportation System had been born.

Christopher Booroff

Date 1981

Country USA

Why It's Key The Space Shuttle has been instrumental in putting telescopes, space stations, and satellites into orbit, and has become an icon of the 1980s and 1990s.

opposite Columbia takes off on its first orbital flight from the Kennedy Space Centre

Key Publication **The cosmic inflation theory**

Big Bang not quite big enough

The evidence for the Big Bang is temptingly simple: everything is moving away from everything else, so extrapolating backwards, everything must have been in the same place at some point. This leaves some problems. The first is that the Universe looks pretty much the same in all directions. One would expect different regions of the Universe to evolve differently based on slight initial inconsistencies in the fabric of space-time, leaving a Universe that varies amazingly. The second is that, despite the Universe looking pretty much the same throughout, it was once lumpy enough to produce places with high concentrations of stellar dust, allowing stars, planets, and, ultimately, us to form.

It is possible to imagine a Big Bang that would produce a world like ours with no extra fiddling, but it would be extremely unlikely and highly dependent upon the initial conditions being just right.

A more plausible explanation was suggested by the theory of cosmic inflation, put forward by Alan Guth in his 1981 paper "The Inflationary Universe: A Possible Solution to the Horizon and Flatness Problems." In this theory, the Universe began from a tiny point where any inconsistencies affected the whole region. This point expanded so rapidly that it maintained its overall uniformity, apart from miniscule fluctuations. Once the Universe had reached about the size of a grapefruit, the period of rapid expansion stopped and a steady, slower expansion took over, leaving a Universe that is uniform on a large scale, but with small-scale blips from the magnified quantum fluctuations – matching what we observe.

Kate Oliver

Publication *"The Inflationary Universe: A Possible Solution to the Horizon and Flatness Problems"*

Date 1981

Author Alan Guth

Nationality American

Why It's Key Inserting the inflationary phase modifies the basic Big Bang theory to allow for a plausible cosmic evolution that, importantly, predicts what we see.

Key Event

Welcome home oryx

The twentieth century was pretty bleak for the Arabian oryx, but not all bad. A favourite target for hunters on the Arabian Peninsula, populations of this desert-dwelling antelope had been suffering for decades, but it was the widespread appearance of trucks and high-powered rifles following World War II that really spelt trouble, as herds were decimated. The last wild oryx was killed in Oman in 1972. Fortunately that didn't signal the end for the animal, which is thought to have spawned the legend of unicorns – in profile oryx often appear to have just one horn.

In 1962, the conservation charity Fauna and Flora International had sent nine individuals to Phoenix Zoo, USA, to set up a captive breeding program. The scheme was a success and, with the help of other zoos, the captive population grew large enough to enable head ranger Said bin Dooda Al Harsusi to reintroduce ten oryx back into the wild in Oman.

The Omani population enjoyed initial success and had risen to 450 individuals by 1996, before again dwindling due to poaching and habitat degradation. In 2007, Oman's Arabian Oryx Sanctuary became the first UNESCO World Heritage Site to be stripped of its status due to the Omani government's decision to open the area up to oil prospecting, decreasing available habitat by 90 per cent.

While the future is again uncertain for oryx in Oman, happily further reintroductions across the Arabian Peninsula mean currently there are 886 wild-living oryx, complimenting a captive population of over two thousand.

Mark Steer

Date 1982

Country Oman

Why It's Key A small but significant victory in the battle to safeguard the Earth's biodiversity.

Key Experiment **Eradicated by sex**
The sterile insect technique

If you were put in charge of eradicating a fly whose flesh-eating maggots were causing havoc, what would you do? Rearing millions and millions of the little blighters and releasing them into the environment probably wouldn't be top of your list. But then you're not Edward F. Knipling.

The New World screw-worm fly lays its eggs under the skin of livestock (and occasionally humans) leaving its maggots to feast on the fresh meat below. Aside from its general nastiness, the fly is a serious pest; in 1960 it was estimated to be costing the U.S. economy US$80 million every year.

Entomologist Edward Knipling and his colleague Raymond Bushland decided something had to be done. They knew that fruitflies could be sterilized by exposing them to X-rays. They also knew that female screw-worm flies only mate once in their lives. How about, Knipling and Bushland thought, if we made sure that females mated with sterile males. The eggs wouldn't be fertile, the maggots wouldn't hatch, and the fly would be eradicated.

Starting in the late 1950s, they reared and released truly astonishing numbers of sterile male flies, swamping local fly populations. By 1982, screw-worm had been eradicated from the USA and 500 million flies were being produced every week to carry on the assault through Mexico and on towards Panama.

The Sterile Insect Technique proved to be one of the most effective methods of pest control ever devised and is now being used to control a number of other pest species.

Mark Steer

Date 1982

Scientists Edward Knipling, Raymond Bushland

Nationality American

Why It's Key The Sterile Insect Technique showed that, with a little ingenuity, there are ways of controlling and even eradicating pests without having to resort to chemical insecticides.

Key Event
First computer virus infection

Richard Skrenta was a classroom joker, and the computer was his weapon of choice. He received an Apple II computer for Christmas while in the seventh grade, and this would prove to be a pivotal moment in his life.

He started off by simply annoying his friends by tampering with copies of their pirated computer games. Either he'd alter their floppy disks by adding a "booby trap" which would flash up a taunting on-screen message, or he'd sabotage the game completely so that it would self-destruct.

His friends started to get wise to his ideas so he was forced to evolve his reign of terror – albeit it, as a very scary ninth grader. In a moment of inspiration Skrenta realized that he could hide a virus on the RAM of a PC and then, every time a new disk was entered, a copy of the virus would be made to the disk. This in turn would spread it to a new computer using a method that has subsequently become known as the boot sector virus. The lack of any network meant that computers relied heavily on disks, and Skrenta was laughing all the way to the library.

Nearly a decade later his Elk Cloner virus was still popping up in the most unlikely places – even a sailor in the Gulf War fell victim to the prank. Skrenta was just a kid looking for kicks, but modern viruses have evolved to be much more malicious, corrupting documents and stealing privileged information.

Andrew Impey

Date 1982

Country USA

Why It's Key The first computer virus was merely a taste of what was to come and it was the dawn of a new era in computer technology and data protection.

Key Experiment **Making massive mice**
How growth hormones work

In 1982, the biochemist R.D. Palmiter created his first giant mouse. He had inserted the gene for rat growth hormone into the animals, making them grow to twice the size of their littermates. The experiment was big news; not only were these some of the first transgenic mice, but they carried a gene with a clear-cut function.

To create his giant rodents, Palmiter injected a modified version of the growth hormone gene into fertilized mouse eggs, where it was taken up by the mouse genome. The gene functioned well, producing copious amounts of hormone.

The experiment had many implications, including for the study of the biological effects of growth hormone, for accelerating the growth of livestock, and for farming valuable gene products. Most importantly, however, this seminal experiment paved the way for the first attempt to cure a genetic disorder.

Three years later another key breakthrough was made – human growth hormone (HGH) was produced by genetically engineered bacteria. HGH is usually secreted by the pituitary gland in the brain, but in affected children, insufficient hormone is made, stunting their growth. Since added growth hormone can make giant mice, it was thought that extra human growth hormone could help children with growth problems grow bigger. This turned out to be the case, and revolutionized the treatment of children with such diseases. By harvesting the growth hormone made by bacteria, scientists could produce enough HGH replace the missing hormone, enabling affected children to grow normally.

Catherine Charter

Date 1982

Scientist Richard D. Palmiter

Nationality American

Why It's Key The study of giant mice led directly to the ability to produce human growth hormone – a massively important medical advance.

Key Publication
Prusiner's prion hypothesis

There's nothing quite like a deadly, brain eating disease to strike fear into the hearts of a nation. And CJD is one that's struck a lot of fear into a lot of hearts. When the news hit in 1997 that scientists had found a strong link between a new form of CJD, or Creutzfeldt-Jakob disease, and eating cattle infected with a degenerative disorder known as "mad cow disease," all hell broke loose in the UK.

Roast beef was temporarily off the menu as parents panicked. It seemed the international community's decision to stop importing British beef after the previous year's mad cow disease outbreak had been vindicated. In retrospect, the crisis that ensued may have been unwarranted; only a handful of infections eventually emerged. But what was the agent in question?

The answer is debated to this day. Although Stanley Prusiner was awarded a Nobel Prize – some thought prematurely – in 1997 for his pioneering work on prions as infective particles, there are scientists who still deny their central role in what are often referred to as "prion diseases."

In a heavily criticized paper published in 1982, Prusiner had identified a completely new type of infective agent – essentially, a form of protein. His hypothesis went against the grain because it proposed that agents of disease didn't need to have DNA – the replicating material – to cause infection. Today the exact mechanism of the disease is still in question, although the importance of prions is now widely accepted.

Hayley Birch

Publication "Novel Proteinaceous Infectious Particles Cause Scrapie"

Date 1982

Author Stanley Prusiner

Nationality American

Why It's Key Paved the way for decades of research into prions and their importance in mad cow disease and CJD.

opposite Colored transmission electron micrograph of a prion protein, thought to transmit some brain degenerative diseases such as BSE and CJD

Key Event
Worst El Niño on record

Although the weather effect known as El Niño has only been well understood since the 1980s, the name itself relates to warm winter seas noticed by South American fishermen back in the nineteenth century. "El Niño" means "the boy child," a reference to the baby Jesus, as the change in water temperatures often begins around Christmas.

It is still unclear what triggers an El Niño event, but it occurs every four or five years, warming the seas on the western American coastline and spreading across the Atlantic, disrupting fish populations and weather patterns. The work of Jacob Bjerknes in the late 1960s had linked the El Niño event to the collapse of the Southern Oscillation wind pattern. This triggers dramatic global changes to weather patterns, including droughts in Australasia; warm, mild winters in eastern North America; and extensive rainfall in South America.

The 1982 El Niño event was important not only because it was particularly severe – being blamed for nearly two thousand deaths and more than US$13 billion in damage to property and livelihoods – but also because it caught scientists by surprise. An upsurge in interest in science, due to the launch of Sputnik, coinciding with a major El Niño in 1957 had paved the way for a network of observation devices but these were insufficient to make an accurate prediction of the 1982 event's severity.

Since then, further investment has provided a global network of satellites and ocean-going monitors that successfully predicted the severity of the 1996–1997 El Niño event.

Jim Bell

Date 1982

Country Global

Why It's Key Increases in the understanding of El Niño have dramatically improved our understanding of how the global climate operates.

Key Publication **How many species?**
Big thinking on biodiversity

In March 1982, a short scientific paper appeared in an obscure journal for beetle researchers. Standing at a little over 650 words long, it contained one firm piece of data, a few educated guesses, and a source of wild speculation. It was hardly the type of publication you'd think would change the world, but Terry Erwin's estimate of the number of species on Earth was revolutionary. Not only was it the first time anyone had attempted to estimate a scientific estimate, his result was an order of magnitude higher than the 1.5 million that was currently favored.

Erwin, an American entomologist, based his global estimates on the numbers of beetle species he collected from just nineteen trees in the forests of Panama. Over three seasons, he painstakingly identified 955 species, using this number to estimate that there were probably 12,448 species of beetle, and 31,120 species of arthropod – the phylum of creatures which have segmented bodies – in a single hectare of tropical forest canopy. If you extrapolate these numbers up to a global scale, you get 30 million species of tropical arthropods alone. Noah's biblical task had just got an awful lot bigger.

The results were controversial, but pounced upon by conservationists wanting to highlight the effect humans have when they destroy tropical rainforest.

We still don't know how many species there are on Earth. So far around 1.5 million different organisms have been described, but estimates for the total number still vary between 10 and 100 million.

Mark Steer

Date 1982

Author Terry Erwin

Nationality American

Why It's Key Showed that life on Earth was probably more diverse than we had previously imagined. There's more to save, and more reason to save it.

Key Event

The dawn of the quantum computer

Richard Feynman was one of the most renowned physicists of the twentieth century, perhaps most famous for his involvement in both the Manhattan Project and the investigation into the loss of the Challenger space shuttle. He also introduced the concept of nanotechnology, and was one of the first to realize the potential of using quantum mechanics to create a completely different system for computing.

A PC works on the same principles as the earliest computers. Both work with bits, the smallest unit of information, which are represented by a zero or a one. The computer manipulates and interprets the bits in order to perform any useful task.

Feynman's abstract model of a quantum computer in 1982 showed that once you get down to the atomic scale, the laws of nature are dominated by quantum mechanics. He suggested that, as computer components got smaller and smaller over time, eventually they would reach this scale, allowing a new kind of fundamental data piece – the "qubit." While a bit can be a zero or a one, a qubit could also be either of these but, crucially, could also be a blend of the two. The extra dimension of processing power that these atom-sized qubits would have would allow a quantum computer to tackle many mathematical problems at once.

Not only did Feynman's model suggest that a quantum computer could work, it also showed that it could be used as a simulator, allowing scientists to perform quantum physics experiments inside it.

David Hawksett

Date 1982

Country USA

Why It's Key Quantum computing represents the first departure from classical computing since the very first computers were built.

Key Event

Genetically engineered insulin goes on sale

The first practical commercial use of genetic engineering was not to make super humans, nor to make cows produce more milk, or even chickens lay more eggs, but to trick the lowly bacteria *E. coli* into making a substance that humans need to live.

Insulin, first isolated by Frederick Banting and Charles Best in the 1920s, is the hormone that prompts our cells to take up sugar from our blood, thus allowing it to be used for energy. Diabetics either do not produce this substance (Type I diabetes) or are resistant to it (Type II diabetes).

Prior to the early 1980s, diabetes patients were treated with insulin extracted from either pigs or cows. That all changed when scientists were able to place a small segment of human DNA into a ring-shaped strand of bacterial DNA known as a plasmid. *E. coli* bacteria were then made to absorb the plasmid and incorporate the enclosed DNA into their genetic structure. The resulting bacteria not only produced human insulin, but also passed this trait on to their offspring. The first bacteria were successfully produced in 1978, and the insulin was approved for medical use four years later.

Initially, the use of recombinant DNA – genetic material containing DNA sections from different sources – to produce proteins was solely the realm of bacteria, but now yeast have got in on the act, and play an important role in the production of insulin today. As a result, very few diabetics now take anything but recombinant insulin.

James McCallum

Date 1982

Country USA

Why It's Key Insulin was the first practical drug to be produced by genetic engineering.

Key Publication **Cycles of death**
"Mass Extinctions in the Marine Fossil Record"

We are one lucky species. Our ancestors, from fish-like sea creatures to scurrying shrew-like mammals, survived through some of the biggest catastrophes the Earth has ever known. At several points through its history, mass extinctions have killed off a huge chunk of the planet's species. Some species did survive and, in 1982, two of their descendants – a pair of paleontologists named David Raup and Jack Sepkoski – uncovered the events that nearly killed off their forefathers.

Two years earlier, Luis and Walter Alvarez had published their theory that an asteroid impact killed the dinosaurs, turning mass extinctions into a hot topic. At the time, Sepkoski had been collecting a huge reservoir of data on marine animals, including a quarter of a million species, both living and extinct. He noted when each appeared and disappeared in the fossil record, but it was his colleague Raup who suggested that he look for patterns in the disappearances.

When they plotted the number of extinctions over time, they saw five clear peaks, corresponding to five mass extinctions: at the end of the Ordovician, Devonian, Permian, Triassic, and Cretaceous periods. In 1986, the duo noted a periodic cycle in the peaks and troughs of their graphs that suggested mass extinctions happen in a 26-million-year cycle. The concept of these "Raup-Sepkoski" cycles is still heavily debated, but there is no question that their work told us much about the fate of extinct species, and makes us grateful that we're not amongst them.

Ed Yong

Publication "Mass Extinctions in the Fossil Record"

Date 1982

Scientists David Raup, Jack Sepkoski

Nationality American

Why It's Key Revealed the cataclysmic events that killed off most of the species that ever lived on Earth.

Key Person **Richard Dawkins**
Darwin's Rottweiler

Few scientists today are more capable of voicing or stirring up strong opinion than Richard Dawkins. Influential and controversial in equal measure, he is a proponent of evolution and atheism, and an antagonist of pseudoscience and religion. His forthright views and scientific contributions are matched only by his masterful writing ability, which has done much to push evolutionary theory into mainstream culture.

Born in Kenya in 1941, his interest in science was fuelled by his parents and later honed by the great biologist Niko Tinbergen at Oxford University, where Dawkins teaches. He rose to prominence through aggressively championing the "gene's eye view" of evolution and became its figurehead. He argued that evolution is really the story of genes; they are the true units of evolution, and animals and plants are merely their vehicles. This was the thrust of his bestselling book, *The Selfish Gene*, which also coined the term "meme" to describe units of cultural inheritance. In *The Extended Phenotype* (1982), Dawkins stretched the reach of genes, claiming that their influence extends to an animal's environment. Beaver genes, for example, control dams as well as beavers. Recently, Dawkins has courted controversy with his strong views on religion and his outspoken atheism. Once described as "Darwin's Rottweiler," he has encouraged atheists to make their voices heard, while launching an attack on organized religion. His latest book, *The God Delusion*, labels religion as "lethally dangerous nonsense." While his stance has been both criticized and lauded, he has, as ever, stirred up important debate.

Ed Yong

Date 1982

Nationality British

Why He's Key Championed the gene-centered view of evolution that has dominated modern biology, and acts as a figurehead for atheism.

opposite Richard Dawkins

Key Publication **Why Sex?** Explanations for a "silly" event

If the ultimate goal of life is to pass our genes on to the next generation, why on Earth do we do it by "doing it"? Sex just doesn't seem to make sense. Asexual organisms – ones which don't indulge in any reproductive hanky panky – pass on 100 per cent of their genes to each of their progeny. We, on the other hand, only pass on half, diluting our DNA with our partner's. So, in the words of the great evolutionary biologist Bill Hamilton, "why all this silly rigmarole of sex?" Beginning with a landmark paper in 1980, Hamilton argued that we put up with the genetic disadvantages of sex in order to keep parasites on their toes.

Our ability to resist parasites and disease is governed by our genetic make-up. If we reproduced asexually, all our children would have identical immune systems, giving would-be parasites just one problem to overcome – once they'd evolved the key to avoiding one set of defences, the whole family would be open to attack. If, on the other hand, we mix our genes up with others, each child will have a different set of defences for the parasites to conquer.

By 1982 Hamilton had further refined his ideas and argued that parasites were also at the heart of our mating strategies. The reason males tend to be such show offs is because they need to prove how good their immune systems are – a diseased peacock, for instance, simply couldn't have a resplendent tail.

Mark Steer

Date 1982

Scientists William Hamilton, Marlene Zuk

Nationality Irish, American

Why It's Key A convincing explanation not just for why we have sex, but also for why we make such a song and dance about it.

Key Experiment **The polymerase chain reaction** How to photocopy DNA

We've all seen images of DNA – it looks like a spiral staircase – but let's look a little closer. Imagine unwinding the spiral, and then flattening it. Each "step" on the staircase is made of two distinct halves – the chemicals of the DNA code – usually referred to by the first letter in their name, adenine (A), thymine (T), guanine (G), and cytosine (C). This is important because these chemicals always bond together with each other in predictable pairs – A bonds with T, and G bonds with C. When DNA is heated the bonds in the middle of each "step" break – the two strands float free.

In 1983, biochemist Kary Mullis, working in California, revolutionized the science of genetics by discovering a method to make many copies of DNA. He placed synthetic chemical markers called "primers" on each half of the separated DNA. Enzymes called "DNA polymerases" are attracted to these chemical markers. They move along step by step, reproducing the other half of the DNA, making an A when they find a T, and a C when they find a G.

After the third repetition of this process, two copies of a DNA sequence are produced. From this point, each cycle results in twice as many copies of the DNA: sixteen cycles produces 65,536 copies; twenty-four cycles produces 16.7 million copies. This means that scientists can create many copies of one strand of DNA – a vitally important process for geneticists, forensic scientists, and Crime Scene Investigation (CSI) teams.

Andrey Kobilnyk

Date 1983

Scientist Kary Mullis

Nationality American

Why It's Key Many genetics research projects require a large number of copies of the same DNA – the human genome project, for instance, would not have been possible without the polymerase chain reaction.

opposite A PCR amplification cell of DNA fragments

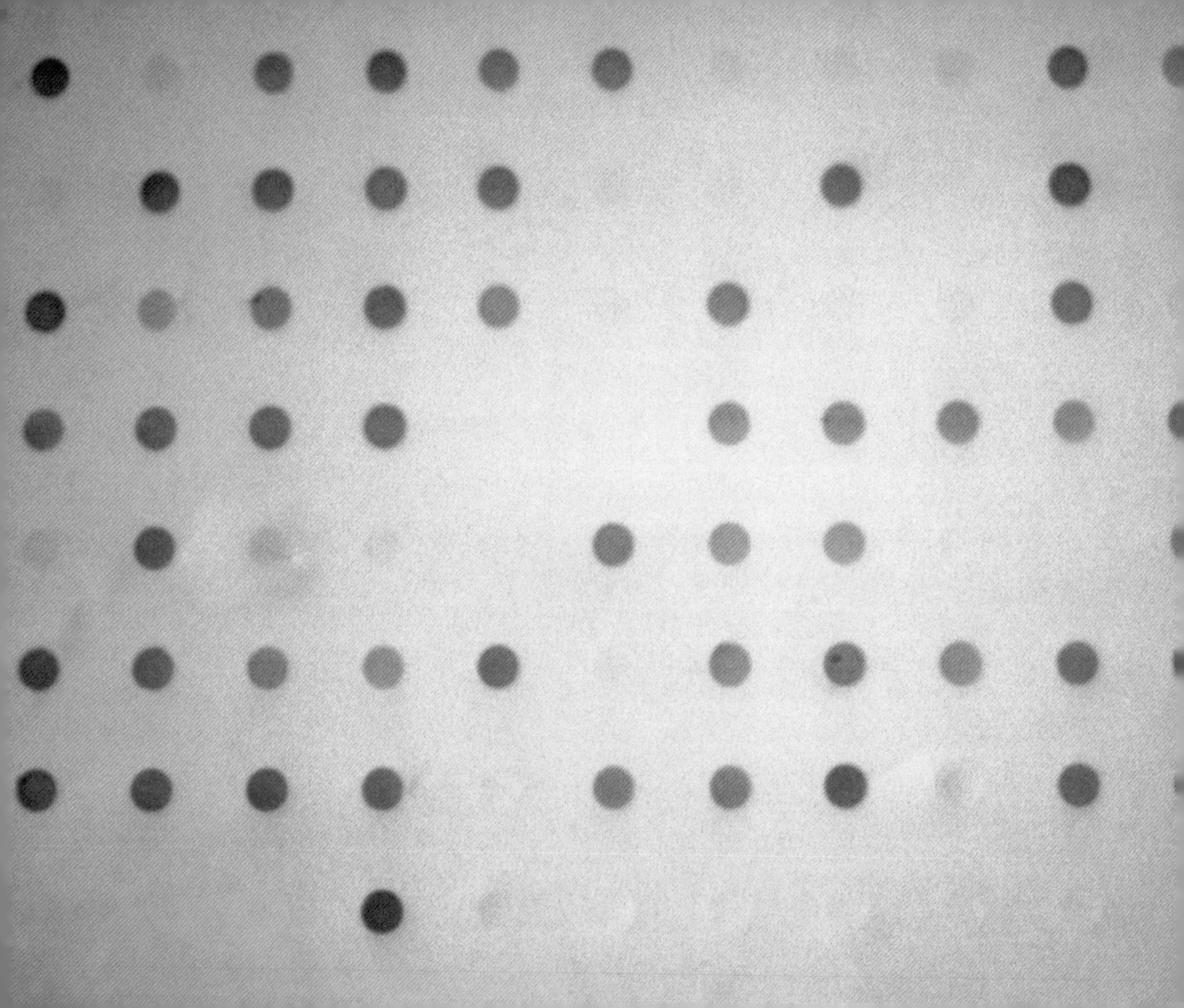

Key Event
Crack cocaine first produced

Almost one hundred years after cocaine was first used as a local anesthetic, crack cocaine entered the world as a recreational drug. It was relatively cheap and gave an intense high making it incredibly addictive.

Cocaine is derived from the coca leaf. In Inca times it was considered to have mystical properties and was reserved for royalty. It stimulates the central nervous system, inducing a huge high in the mid brain due to its ability to block the breakdown of dopamine – a chemical messenger that induces feelings of pleasure.

Crack cocaine production involves dissolving cocaine powder in water and either ammonia or sodium bicarbonate. The mixture is boiled and the solid residue is then dried and cut up into small lumps or "rocks". The name "crack" comes from the noise made when the rocks are heated and then smoked in a crack pipe. The drug itself is described as a freebase – the base form of cocaine as opposed to the salt form. The chemical properties of crack mean that it is insoluble in water and therefore not suitable for either injecting or snorting. Due to a low vaporization point, when crack is smoked, it's actually a combination of combustion and vaporization that gives the user this intense and rapid high.

The high usually only lasts five to ten minutes, after which time the levels of dopamine in the brain rapidly decrease, leading to severe bouts of depression. Longer term health dangers include cardiovascular problems through associated increases in heart rate and blood pressure.

Andrew Impey

Date 1983

Country Bahamas

Why it's key This freebase form of cocaine is immediately absorbed into the blood stream, reaching the brain in a matter of seconds, resulting in a much more rapid and intense high.

Key Invention **The G-force sucks**
Dyson cleans up

James Dyson, an art student and designer, found the air filter in his workshop was constantly clogging with powder particles. This was the same problem that meant he had to change the bag in his vacuum cleaner even when it wasn't full – a loss of suction.

When visiting a saw mill which used an industrial cyclone to draw in dust, trapping it in a vortex of air before funneling it out, he saw this could be the solution to his workshop problems. Not wanting to shell out £75,000 for one, he went about making his own; his work seeded the idea of a vacuum cleaner, which used whirlwinds of air to suck in dust particles.

After 5,127 prototypes, and several failed attempts to find a manufacturer, Dyson launched a vacuum cleaner using cyclone technology in Japan; British investors were worried about disrupting the market for hoover bags. The bright pink Japanese model, the G-Force, was massively popular. The new, super-efficient and slickly designed vacuum cleaner became a status symbol, trading at an inflated price of £2,000. After winning the 1991 International Design Fair, the Dyson Cyclone has gone from strength to strength, gaining a huge market share and becoming a household name. With a knighthood and a billion or so in the bank James Dyson well and truly smashed the stereotype of pecuniarily disadvantaged inventors.

Kate Oliver

Date 1983

Scientist James Dyson

Nationality British

Why It's Key A brave entrepreneur rethinks the idea of the vacuum cleaner, incorporating new technology as appropriate, and designs a fun, efficient and ergonomic product.

Key Discovery
Genetic marker for Huntington's disease found

Huntington's disease is an inherited disease of the central nervous system, usually manifesting itself when a patient is between thirty-five and forty-five years of age. In the early 1980s, there was no predictive test for Huntington's disease. Through the work of American scientists Nancy Wexler and James Gusella, however, a genetic marker was found which allowed the offspring of Huntington's parents to see if they would develop and perhaps pass on the disease.

In 1981, Wexler, of Columbia University and the Hereditary Disease Foundation, traveled to a Venezuelan village with an extremely high rate of Huntington's disease. She constructed a family tree of over 3,000 village members, and obtained blood and skin samples from 570 living descendants. Wexler passed these samples on to Gusella at Massachusetts General Hospital, who was working on finding a marker for Huntington's disease. By using pieces of DNA called probes, Gusella's group tediously examined Wexler's samples and located an area of distinctive DNA on chromosome four. It was thought that if these varying DNA regions frequently corresponded to incidence of disease, then the disease gene was located near this region. This was the first time a disease gene had been mapped to a specific chromosome based on inheritance and genetic markers alone.

This significant scientific achievement allowed families of Huntington's patients to find out if they were also at risk from the disease. With the discovery of the exact mutation sequence in 1993, scientists are now a few steps closer to finding a cure.

Rebecca Hernandez

Date 1983

Scientists Nancy Wexler, James Gusella

Nationality American

Why It's Key A genetic marker allowed families to determine their future potential for the disease, and brought scientists one step closer to a cure.

Key Event
Switching on the Internet

During the 1970s, computer networks were springing up all over the world. In Britain, the CERCnet academic network did a similar job to America's ARPANET, connecting academic and research sites together, while Western Union, the British Post Office, and Tymnet were busy creating the first modern intercontinental network.

If these networks could be connected together, then researchers in California could share results with colleagues in Cambridge. Small businesses could rent time on academic computers. Unfortunately, the different networks used different ways of letting their computers talk to each other – different networking "protocols." A common language, an Esperanto of networking, was needed. Step forward the Internet Protocol suite. Developed by Vint Cerf and Bob Kahn at the Advanced Research Projects Agency, this collection of standards hides the differences between networks, allowing them to talk to each other or "internetwork."

On January 1, 1983, ARPANET officially switched over to the new Internet Protocol, easing the connection of new computers and networks to create one big happy family. This allowed ARPANET and other networks to form an internetwork, which grew into today's Internet. The Protocol proved so flexible and extensible that it's been in use ever since, connecting everything from supercomputers to washing machines. Email, the World Wide Web, instant messaging, online games – if it's moving from one computer to another, the chances are it's moving by the rules of the Internet Protocol.

Matt Gibson

Date 1983

Nationality Global

Why It's Key Laid the foundations for the exponential growth of the Internet, connecting a global community of scientists and researchers, and letting their students poke each other on Facebook during lectures.

Key Person **Barbara McClintock**
Queen of the jumping genes

You may have seen "wild maize" used as a centerpiece for a meal; it stands out from the bland yellow maize by having red, brown, and even blue kernels. The colors come from pigments in the endosperm – the tasty bit of the corn which stores nutrients for the seed. Each ear of maize is formed by multiple flowers, and if the flowers are fertilized by sperm carrying different genes for endosperm pigment, different color kernels will arise.

If you look closely at a colored kernel, you might see even more variation: blue among red; purple against brown. Barbara McClintock (1902–1992) looked very closely at them and saw discordance: multiple colors that could not be explained by Mendel's laws of inheritance. McClintock suspected that the variegation was caused by instability of the genes that coded for the pigment. She eventually identified the series of genes that determine pigmentation, located on chromosome number 9. Changes in the kernels' color occurred when a small piece of chromosome 9 moved from one place to another, close to a gene coding for a pigment, turning it off. McClintock called these sections of genetic material "control elements." These are now more commonly referred to as "jumping genes" or transposons which many other organisms have. They are responsible for much of the antibiotic resistance of bacteria; enable certain parasites to evade their host's immune system; and some even lead to cancers.

Stuart M. Smith

Date 1983

Nationality American

Why She's Key Her discovery of jumping genes led to great advances in genetic engineering, and earned her the Nobel Prize for Physiology or Medicine in 1983.

opposite Barbara McClintock with an ear of spotty corn

Key Experiment
Chimps learn "language"

In 1983, psychologists David and Ann James Premack demonstrated that they had successfully been able to teach language to chimpanzees. The couple had wanted to better understand the fundamentals of language used by animals and humans, and identify the dividing line between animal and more complex human language.

The Premacks worked with a number of chimps, but their most gifted pupil was Sarah. She achieved a vocabulary of around 130 words that she used with 75–80 per cent reliability. She was taught nouns, verbs, adjectives, and concepts, such as "same" and "different." Sarah's language was not vocal; it consisted of plastic tokens representing words. The tokens were different shapes and colors and did not resemble the words they represented. She was also taught to associate objects with different tokens by placing the token near the object. As tokens representing verbs and adjectives were added to her vocabulary, and she was able to build up sentences using the correct word order, the astute ape was even able to indicate matching and non-matching items, answer questions, and understand conditional sentences.

The Premacks' experiments showed not only that chimps could understand the meaning of the words, but also that they could recognize a word without the actual object/action being visible to prompt them. They indicated that great apes are capable of using a simple language that includes some features of more complicated human language. Maybe humans aren't so special after all...

Emma Norman

Date 1983

Scientists David James Premack, Ann James Premack

Nationality American

Why It's Key Ape brains can learn and reproduce human-style language.

Key Experiment
Cross species transplantation

In 1984, the heart of a baboon was transplanted into a child who eventually became known as Baby Fae. Ultimately, the experiment was a failure, the child lived only a few weeks, but it did bring to the public eye one of the more controversial topics in medicine today – cross species transplantation. Known more specifically as xenotransplantation, this process involves the transplantation of tissue from non-human species into humans.

The benefits to such a procedure are obvious – dozens of people on organ transplant lists die every day waiting for compatible donor organs. If a ready supply of potential liver donors could be kept as medical animals, then this problem would immediately go away. Furthermore, it may not even have to be a whole organ – there is some speculation that transplanting even cells or tissues of a non-human species into a human could be of some benefit.

The procedures are fraught with difficulties, however. Technically, transplants are difficult surgeries, but the big issue after transplantation is rejection – the recipient's immune system attacking the new organ. Transplant patients often have take drugs that suppress their bodies' ability to attack the new organ, but also impair their ability to fight off diseases, leaving them vulnerable to infection.

Ethical issues also come in to play. Transplanting a non-human organ into a human may allow certain viruses to mutate and become infectious to humans. Not only is this potentially deadly to the organ recipient, but also to the population as a whole.

B. James McCallum

Date 1984

Scientist Leonard Lee Bailey

Nationality American

Why It's Key Tissues from non-human species may eventually save hundreds of lives – but only if scientists can work out the details.

Key Experiment **Continental drift**
No longer a passing thought

The continental drift theory proposes that all modern-day continents were once one "supercontinent" called Pangaea which broke up some 200 million years ago. Evidence supporting the theory included the discovery of matching fossils and other geological features of the coastlines of different continents, and indications of dramatic climate changes – for example, glacial deposits in Africa. While the evidence for the continents having been connected and then moving apart was strong, there was a lack of hard proof of the phenomenon. That was until 1984.

In that year, NASA scientists from the Crustal Dynamics Project, based at NASA's Goddard Space Flight Center, released the first actual measurements proving that the continents were moving. The data showed the changing distances between continents with great accuracy. Two methods were used.

The first method identified the distances between radio telescopes by measuring the difference in time that it took a signal to reach each telescope. The second used satellites covered in reflectors to bounce laser signals back to Earth. Telescopes next to the lasers picked up the reflected light, the time difference giving a precise measure of distance. The data showed that North America and Europe are moving apart at a rate of 1.5 centimeters each year. Australia, which lies on the Indian plate, was measured to be moving toward the Pacific plate at a rate of 7 centimeters per year, and the Pacific plate – with California on board – was leaving North America at a rate of 4 centimeters per year.

Emma Norman

Date 1984

Scientists NASA

Country Worldwide

Why It's Key The theories of continental drift and plate tectonics are proven.

opposite The Afar triangle in Dijbouti, East Africa, where three of the Earth's rift systems meet and the crust is slowly pulling apart

Key Invention **Flash memory** Takeaway data

In the recent past, students looking for cheap storage for their vinyl record albums would liberate plastic milk crates from behind supermarkets to use as cabinets. Today, it's possible to own hundreds of music recordings, photographs, and other kinds of information which don't require a dedicated wall-full of storage. You can even carry a hefty chunk of it with you without having to push it in a cart. How? Through the use of flash memory.

In 1984, D. Fujio Masuoka invented flash memory while working at Toshiba. Two basic kinds of memory which could be used by computers and other digital devices already existed – the hard disk and RAM. Both of these worked well and are still used in electronic devices today, but they each had disadvantages. RAM (Random Access Memory), for example, allowed data to be stored and accessed quickly, but the information would be lost when the power was turned off.

Masuoka's Flash Memory was inexpensive and had the ability to write, erase, and store data on a small card which would retain the information when the power to the device was turned off.

Today, flash memory is used in mobile phones, cameras, music players, and other devices. The capacity of a single flash memory chip now runs into gigabytes of data – and in the future could potentially replace laptop hard disks.

Andrey Kobilnyk

Date 1984

Scientist Fujio Masuoka

Nationality Japanese

Why It's Key Cheap, portable devices which can read and write digitally stored music, photos, movies, and other data would not be possible without flash memory.

Key Discovery **Man and chimp aren't all that different**

It's an accepted fact that humans evolved from primates, and apes are our closest biological relatives. But Charles Sibley and Jon Ahlquist were the first to show that chimps are nearest to humans on the family tree, coming out ahead of gorillas in what had previously been a three-way tie.

To reach their conclusion, Sibley and Ahlquist used a technique called DNA-DNA hybridization. When heated, the two strands of the DNA double helix can be separated. When cooled, they come back together again. This even works with DNA strands from two very similar, but distinct, species – when they cool, the strands form a hybrid. Scientists can use this technique to work out just how close the match between two species really is. By comparing the results from hybridizing human DNA with samples taken from a range of apes, Sibley and Ahlquist showed that chimps come out top of the evolutionary tree. But their results came under fire from fellow scientists who claimed that the data did not show a significantly strong link between chimps and humans.

Today, further research has shown that we probably did share our last ancestor with chimps – and in fact, early humans and chimps may have interbred for another million years before finally separating. Looking around today, you may sometimes wonder if it's not still going on...

Kat Arney

Date 1984

Scientists Charles Sibley, Jon Ahlquist

Nationality American

Why It's Key Scientists declared they had rewritten the history of human evolution, but their claims were controversial.

opposite Cheetah the chimpanzee, aged 59, tucking in

Key Event **Apple bring the mouse and pull down menus home**

During the advert break after the third quarter of the Super Bowl is a fairly good time to advertise any product. The competition is usually one of the most watched pieces of television in the United States in any given year, and advertisers have to pay through the nose to reach such a vast audience. CNN reported in 2007 that the CBS network was charging up to US$2.6 million for a thirty-second advertisement.

That sort of a financial investment can be a big risk, but big risks often come with big rewards, as Apple Inc. discovered in 1984. During the 1984 Super Bowl, Apple ran a now iconic US$1.5 million advert depicting a gray world based on George Orwell's classic novel *1984*. Directed by Ridley Scott (*Blade Runner*, *Alien,* and *Gladiator*), the advert showed a heroine fighting the oppressive "Big Brother" (IBM) by hurling a sledgehammer into a giant screen, freeing the citizens from his control. The Los Angeles Raiders went on to beat the Washington Redskins 38 – 9.

Apple wanted to show that its new Macintosh computer empowered people by making computers affordable for everyone, rather than just businesses. Although the several-thousand-dollar machine was by no means cheap, it was a hugely successful marketing strategy, introducing the public – albeit a more affluent middle-class public – to the world of pull down menus and mice. Today, Apple is one of the world's most successful computer companies, with a 2007 net income of nearly US$3.5 billion.

Jim Bell

Date 1984

Country USA

Why It's Key The Macintosh computer was the first affordable home computer as we would recognize it today.

Key Publication **Superstring theory** Solving the problems of physics with a few extra dimensions

String theories model all particles, force carriers (bosons), and matter particles (fermions) as different vibrational patterns on a tiny, tiny string. Their major appeal is that one particular string vibration corresponds to a particle of the sort that we expect to transmit the force of gravity. This would provide a way to join the two incompatible cornerstones of modern physics – general relativity, which describes the action of gravity, and quantum physics, which describes everything else. But string theory has a few knotty problems: the biggest of which is that it predicts that matter particles don't exist – a bit upsetting for some and clearly not true.

In their 1984 paper "Covariant description of superstrings," Professors Michael Green and John Schwarz modified string theory to be "superstring theory," by assuming that all particles come in pairs of one fermion and one boson. So electrons (fermions) have a force-carrying partner called the selectron (boson), and the W-boson has a matter particle called the wino. While this is pleasingly symmetric (if a bit crowded), and does describe all the interactions we observe, in order for the theory to work, the Universe has to have ten dimensions. Where they could fit is not explained. To make things even more complicated, there are four different variants even within superstring theory, none of which make differing predictions we can test. Any physical consequences of the theories only become observable at massive energies that haven't existed since the Universe was a tiny fraction of a second old, so they're a bit tricky to recreate.

Kate Oliver

Publication "Covariant Description of Superstrings"

Date 1984

Authors Michael Green, John Schwarz

Nationality British, American

Why It's Key A triumph of mathematics over observable evidence, string theory is gorgeously elegant and beats other theories by having no awkward infinities. Unfortunately, it also fails abjectly by making no observable predictions.

Key Event **Fusing embryos**
Goat + sheep = geep

What do you call a cross between a goat and a sheep? A geep. Well, that punch line may not have them rolling in the aisles, but the geep certainly looks quite funny. The first one was bred in a Cambridge laboratory by Carole Fehilly, along with Elizabeth Tucker and her future husband, Steen Willadsen.

The geep wasn't a mere hybrid, born from the inappropriate liaison of a sheep and a goat. It was created by fusing embryos from the two animals when they were just a handful of cells. Unlike a hybrid, which has two parents, the geep had four. Animals like the geep are called chimeras, after the mythological Greek monster with a lion's head, goat's body, and snake's tail. Their bodies contain a mix of cells from at least two genetically distinct species.

On the surface, it was easy to see which parts came from which cells. Those that grew from sheep cells were covered in wool, while those that came from the goat were covered in hair. All in all, the animal looked like a goat wearing a patchy wool jumper.

Despite its comedic appearance, the geep could have a role to play in saving endangered species from extinction. If the endangered species' cells are mixed with those of a common animal, the chimeric embryo could gestate in a surrogate mother without fear of rejection.

Ed Yong

Date 1984

Country UK

Why It's Key Threw open the door for future work on chimeric animals and sparked much of the controversy that continues to plague the field today.

Key Discovery **Buckyballs**
Kicking off nanochemistry

So-called for their soccer ball shaped structure and perfect symmetry, buckyballs are what could be described as sexy molecules. When the tiny carbon spheres were first identified in the 1980s, they made magazine covers, causing an impact not just within chemistry circles, but right across the scientific world.

Given its correct name, Buckminster fullerene lacks some of the star quality that sent it rocketing to fame, but nonetheless it secured British chemist Harry Kroto a Nobel Prize, along with his American colleagues Robert Curl and Richard Smalley. Originally it was thought that buckyballs – molecules composed of sixty carbon atoms arranged as a hollow cage – were a manmade creation. It later transpired, when scientists started unearthing them in geological formations, that nanoballs had been around for longer than us. They are now known to show up everywhere, from the flame of a candle to interstellar dust.

Despite promising research in the area of medicines, many argue that buckyballs have yet to find any useful practical application, and their glitzy media career has been tainted by questions over toxicity. In 2005, researchers announced they had crushed buckyballs to create a substance harder than diamond. The following year they became wheels for a nano-sized car, and, of late, they have been making headlines in quantum computing. Perhaps their most important contribution to date is having opened chemistry's eyes to a new nanoscale world.

Hayley Birch

Date 1985

Scientists Harry Kroto, Robert Curl, Richard Smalley

Nationality British, American

Why It's Key Brought chemistry into the media spotlight and gave chemists a new nanoscale perspective.

Key Event **Birmingham SkyLink**
600 meters of train levitation

Electromagnetic levitation – or Maglev for short – is what you get when you force magnetic poles together. They repel, keeping a few millimeters apart. You get no friction between them, no wear and tear as one passes over the top of the other, and no noise from them rubbing. These are exactly the features that you'd be keen on if you were a train designer. With the bottom of a train levitating, the maintenance costs are kept down, and locals kept happy at the lack of noise pollution.

The first commercial Maglev train ran from Birmingham International Airport to Birmingham International railway station. Opened in 1984, the train floated fifteen millimeters off the 600-meter-long track; its top speed was measured at 42 kilometers per hour. The Birmingham train was designed as a showcase to demonstrate what Maglev could do; unfortunately few people were interested. As the train became older, companies stopped making spare parts for it, so that in 1995 the train had to stop running, to be replaced with buses and then cable driven carriages.

The most successful Maglev train currently running is the Transrapid, which takes passengers from Pudong International airport, nineteen miles to downtown Shanghai, replacing a forty-minute drive with an eight-minute train ride at speeds of over 400 kilometers per hour.

Douglas Kitson

Date 1984-5

Country UK

Why It's Key Capable of reaching over 250 miles per hour – as fast as jets – while using five times less energy, Maglev trains could be the future of rail.

opposite An unmanned prototype magnetically levitated train in Japan

Key Invention
Genetic fingerprinting

Alec Jeffreys described it as a "Eureka moment." Developing an autoradiograph in a darkroom, he realized that he had stumbled upon a method to "fingerprint" a person's DNA.

Jeffreys, a scientist at the University of Leicester, had been studying heritable DNA variation since the 1970s. His lab studied genetic markers for use in tracing inherited diseases. He used a probe – a short fragment of DNA – to see if it would bind to certain types of DNA, including three human samples from the same family. The first results, though messy, clearly showed that highly variable regions of DNA could be inherited within a family. The discovery was made on September 10, 1984, and Jeffreys claimed that within minutes, his group had come up with a list of possible applications for the technique including paternity tests, criminal investigations, wildlife forensics, and conservation biology.

Jeffreys improved the fingerprinting technique and filed for patents at the end of 1984. In March 1985, *Nature* published the first in a series of four articles on the subject. Press coverage quickly followed and, soon after, Jeffreys' lab was asked to solve an immigration case. They proved the relationship of a Ghanian boy and his mother, and this case established the legal groundwork for using genetic fingerprinting in criminal cases.

Genetic fingerprinting is indeed used today for the applications Jeffreys and his colleagues dreamed of, and has proven to be one of the most important scientific discoveries of the twentieth century.

Rebecca Hernandez

Date 1985

Scientist Alec Jeffreys

Nationality American

Why It's Key Its use in criminal cases, forensics, paternity tests, and numerous other applications puts genetic fingerprinting at the forefront of major scientific achievements.

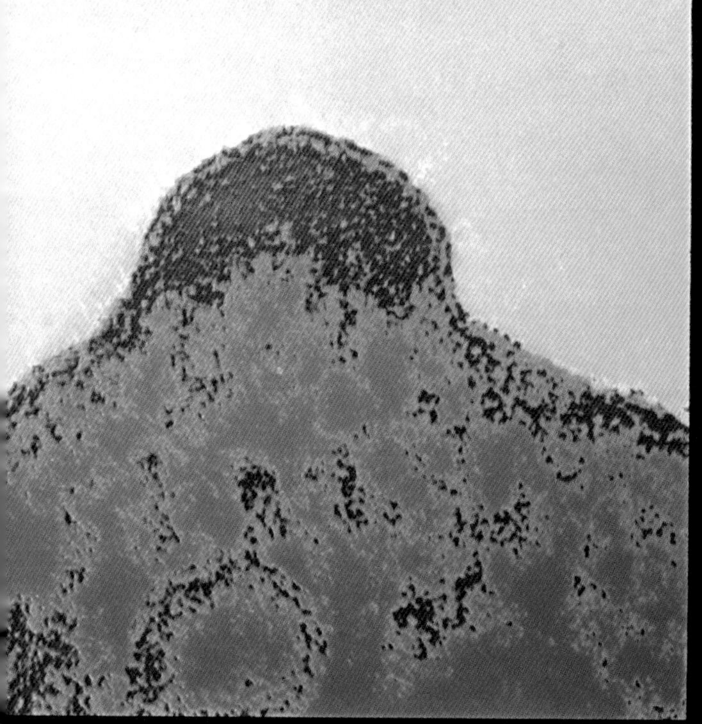
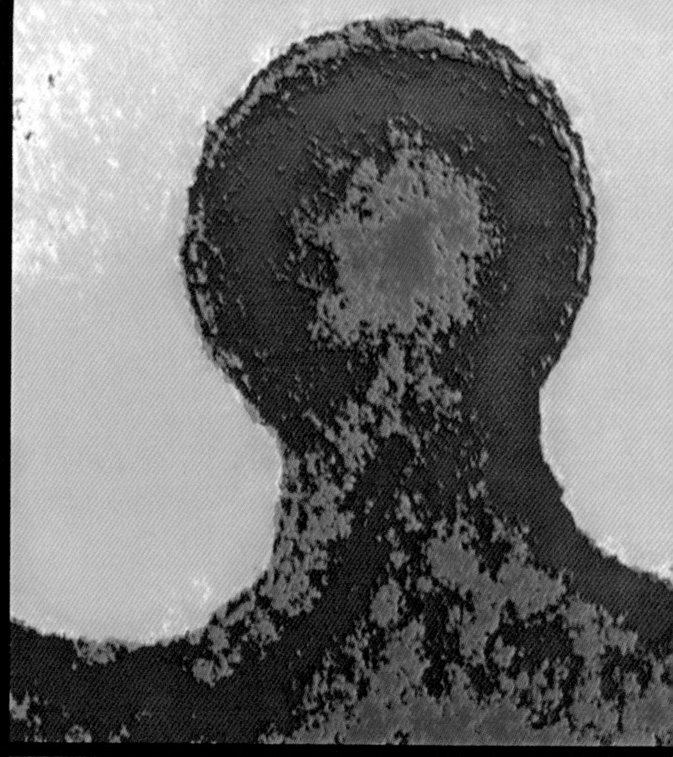
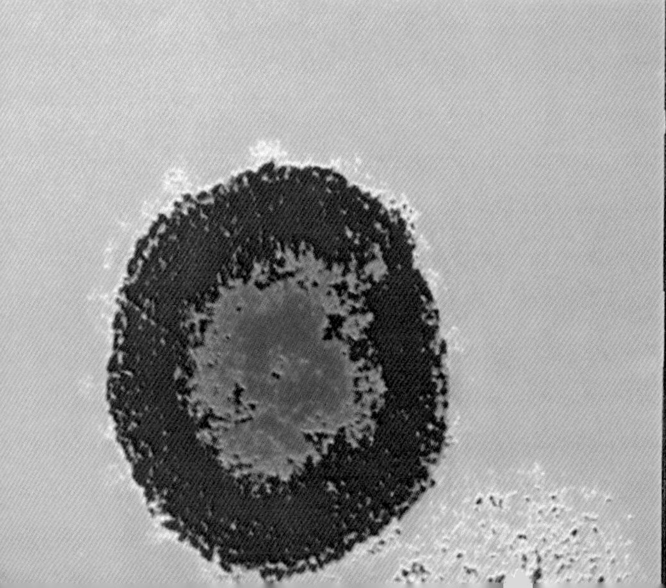
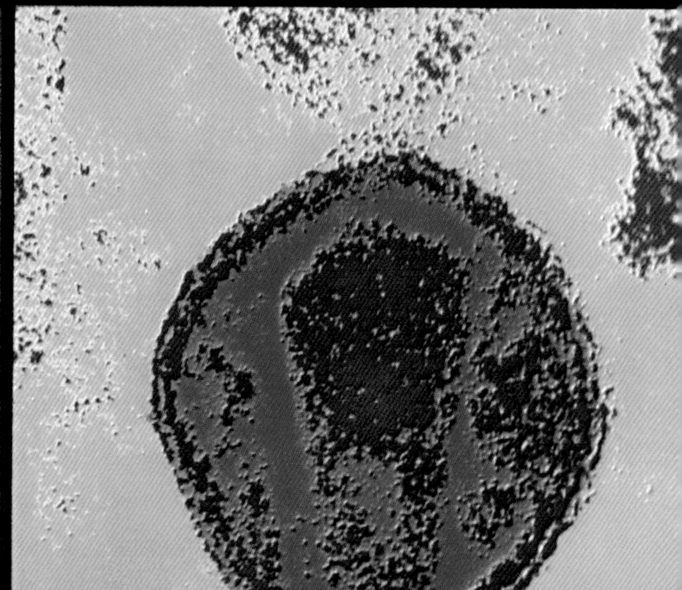

Key Invention **Microsoft Windows is born**
Global dominance beckons

In November 1983, Microsoft announced that Windows was on its way. Expected by April 1984, it would be able to run on systems with two floppy disk drives, 192 kilobytes of RAM, and a mouse. This wasn't what happened, but the eventual Windows did provide a graphical user interface and a multitasking environment, just like they hoped.

At the time, the typical way to use a computer was by typing in your commands, a graphical user interface means you can use a mouse to click on the programs you want to run and also see your files on screen. Apple computers already had this feature, as did VisiCorp's VisiOn, which was the first PC based interface. However, VisiOn lacked developers willing to make programs that ran on it, which prevented the system from being a success.

Multitasking environments allowed users to switch between many programs running at once. Windows had a calendar, a clock, Notepad, Paint, even a game – Reversi – which is less fun than Minesweeper, but still impressive for the time.

In November 1985, almost two years after the proposed release date, Microsoft released Windows 1.0. Instead of the expected requirements, the program needed a minimum of 256 kilobytes of RAM and a hard drive, testing the limits of the current hardware. In spite of this, Windows has gone on to become the most popular operating system available, used on more than 90 per cent of desktops and notebooks worldwide.

Douglas Kitson

Date 1985

Scientists Microsoft

Nationality American

Why It's Key The first of the Windows operating systems – a legacy that would go on to be the most widely used system in computing.

Key Discovery **HIV isolated**
A global killer unveiled

By the end of 2006, 40 million people worldwide – more than the entire population of Canada – were living with human immunodeficiency virus (HIV). The rate of HIV infection continues to rise, despite an international armada of researchers battling against the tide of its pandemic.

Nevertheless, progress has been made, much of which stems from scientists' understanding of the AIDS virus itself. Doctors are now able to administer cocktails of drugs that can keep the virus at bay for years, even decades. Vast bodies of work have been devoted to characterizing the molecules it consists of, in the hope of developing a vaccine that might halt the spread of disease. So far, however, a cure remains elusive. HIV was isolated in the lab two years after scientists first recognized AIDS, although who was responsible for finally pinning down the virus continued to be a hot topic of debate throughout the 1980s. Two teams of researchers – one headed by Luc Montagnier in Paris, the other by Robert Gallo in Washington – published genetic sequences of the viruses they had isolated in January 1985. But it was Gallo who was granted the first patent for a blood test.

The disagreement erupted into a full-blown feud when Montagnier's lab decided to take Gallo to court over the issue. After a long and very public dispute, a compromise was reached: both names would appear on the patents. Income from these would be shared between their funding bodies and a newly established AIDS research foundation.

Hayley Birch

Date 1985

Scientists Robert Gallo, Luc Montagnier

Nationality American, French

Why It's Key Opened the floodgates for research on the mechanics of HIV/AIDS infection, leading to significant progress in prevention, diagnosis, and treatment.

opposite The formation sequence of an HIV particle

Key Discovery
There's a hole in the ozone layer

Ozone is a toxic form of oxygen, consisting of three oxygen atoms, which occurs naturally in the atmosphere. While it's poisonous to us at ground level, a layer of ozone in the upper atmosphere acts as a shield, protecting Earth from potentially harmful rays such as ultraviolet B radiation from the sun.

Measurements of this ozone layer prior to 1975 had shown that it was a complete blanket with no holes. A decade later British geophysicist Joe Farman and his team of scientists were on a research mission in Antarctica to measure the ozone layer, acting on the warnings of two chemists. Their results indicated that it was depleted; holes had started to appear. Subsequent studies proved that certain manufactured gases, such as CFCs (chloroflourocarbons), harm the ozone layer. Most industrial nations have now completely abolished the use of these chemicals and instead use hydrochlorofluorocarbons, which are far less harmful. The Montreal Protocol of 1987 is an agreement of leading countries to take steps to minimize ozone depletion, though some developing countries still use harmful CFCs and there is renewed concern about their impact.

There are several holes which, although the sizes and shapes change, are thought to cover an area of around 22 million square kilometers – around three times the size of the United States. There is currently research underway to assess whether people living under the hole are more at risk of medical problems related to the increased ultra violet light reaching the surface of the Earth in these locations.

Vicky West

Date 1985

Scientist Joe Farman

Country Antarctica

Why It's Key Revealed that the atmospheric layer protecting Earth from harmful UV rays has been compromised due to man-made pollution.

Key Event
French secret service sink *Rainbow Warrior*

On the evening of July 10, 1985, while docked in Auckland Harbour, New Zealand, the eleven-strong crew of Greenpeace's flagship *Rainbow Warrior* were preparing for one of their direct action campaigns. They were planning to disrupt nuclear tests in the French test zone at Moruroa Atoll in the South Pacific.

Despite Greenpeace's opposition to such weapons, it has always strived to conduct its direct action campaigns in accordance with its name; that is, peacefully. But in this case, the crew's peaceful intentions had rubbed somebody up the wrong way. That evening, without warning, two explosions ripped through *Rainbow Warrior*'s hull as it sat in the harbor. Ten of those aboard escaped but the ship rapidly sank with crew member and photographer Fernando Pereira, who died trying to rescue his camera equipment.

The New Zealand authorities soon discovered that French secret service agents were responsible for planting two limpet mines under the ship, making it the first act of state sponsored terrorism in New Zealand.

The event did not deter Greenpeace. In fact, the organization gained publicity and funding out of it. And when France finally admitted their involvement after trying to cover it up, Greenpeace received compensation. This paid for a new *Rainbow Warrior* in 1989 which continues to campaign against nuclear testing, whaling, and other "environmental crimes," as well as providing relief in disaster zones.

James Urquhart

Date 1985

Country New Zealand

Why It's Key Despite the initial setback and loss of life, it promoted the goals of Greenpeace and of New Zealand's already established stance against nuclear technology.

opposite Rainbow Warrior in the docks at Auckland after being ripped apart by the explosion

Key Experiment
Fibre optics revolutionize wired communication

Long-distance telephony has always had a fundamental problem – somehow you have to get the signal along wires from here to there without losing it. The use of copper wires and coaxial cables had been successful, but as more and more data was being transmitted through wires, engineers needed a new system. The answer was fiber optics.

Fiber optics work by sending light through an optical fiber (a long glass or plastic fiber) which forms an electromagnetic carrier wave. This carrier wave can then be modulated to transfer information. Currently telephone calls, Internet, and cable television signals are all transmitted by optical fiber. It outperforms the more traditional copper networks by having a much lower attenuation, meaning less signal loss over distance and interference between signals.

The principle of how optical fibers work has been known since the 1840s, but they were only practically engineered from 1970 onwards. With the rapid advance of telecommunication it was important that the fiber network could keep up – this was conclusively proved by AT&T Bell Laboratories in 1985 when it sent 1.7 billion bits per second of data down a single optical fiber. That's equal to 300,000 simultaneous telephone calls, or 200 television channels.

The high data capacity of optical fibers led directly to computer networking and the development of the Internet – and where would we be today without that?

Helen Potter

Date 1985

Scientists AT&T Bell Laboratories

Country USA

Why It's Key Enabled the fast transfer of large amounts of data, leading to the development of the Internet.

Key Discovery **Markers of disease**
The cystic fibrosis gene is located

Cystic fibrosis is an inherited disease that affects a sufferer's lungs and digestive system. It is caused by a faulty gene producing very thick mucus, which clogs the lungs and can cause fatal bacterial infections. The discovery of the defective gene, in 1985, was one of the most important steps toward understanding the physiology of the disease, and for the future possibility of a cure.

The search for the cystic fibrosis gene was a bit like looking for a needle in a haystack. Unlike some other genetic diseases, scientists didn't have any clues, such as damaged chromosomes, to point them in the right direction. They instead had to use gene "markers" – identifiable segments of DNA common among the families of sufferers – to narrow down the possible locations of the gene.

In 1985, there was a breakthrough when a team led by Chinese geneticist Lap-Chee Tsui, pinpointed the location of the faulty gene to somewhere on chromosome 7. The researchers then analyzed the DNA in this section bit by bit, until eventually they found a mutation (change) in the normal DNA sequence – this was the cystic fibrosis gene.

It later emerged that this gene codes for a protein called CFTR (cystic fibrosis transmembrane conductance regulator) which is important in the production of sweat, digestive juices, and mucus. The full sequence of the gene was finally described in 1989, and proved invaluable for research into new, more effective treatments.

Hannah Isom

Date 1985

Scientist Lap-Chee Tsui

Nationality Chinese

Why It's Key Helped scientists understand the physiology of this devastating and fatal disease and improve hopes of finding a cure.

opposite Willie Frenzel in the tent filled with a mist that helped him breath. At 5 he was the symbol of the Breath of Life campaign of the National Cystic Fibrosis research Foundation.

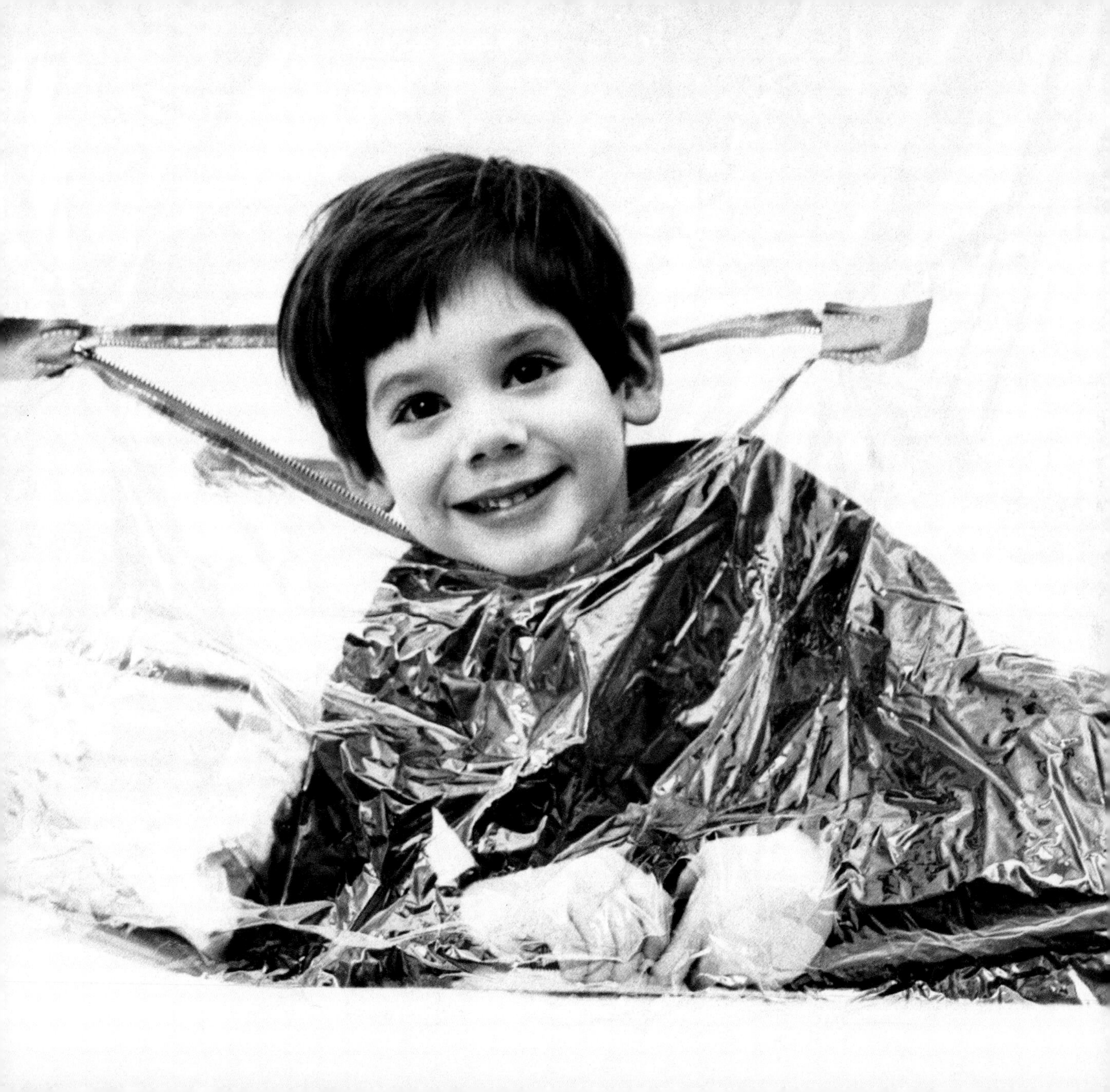

Key Event **Miniature literature** Dickens on the head of a pin

A prolific American scientist once famously demanded to know, "Why cannot we write the entire twenty-four volumes of the *Encyclopaedia Britannica* on the head of a pin?" Making the enquiry was none other than Nobel Prize-winning physicist, Richard Feynman. The audience were the members of the American Physical Society, listening spellbound at Feynman's now legendary after dinner lecture, "There's Plenty of Room at the Bottom," delivered around Christmas time as the 1950s drew to a close.

Fast forward a couple of decades, and science was in a better position to offer Feynman an answer. Nanotechnology was just beginning carve its own, appropriately small place in the scientific world. In 1981, K. Eric Drexler, who went on to receive the first doctorate in the field, published a paper on molecular engineering which is widely regarded as the first journal article on nanotechnology, although the word is never mentioned.

But still no one had risen to "Feynman's challenge." At the end of his talk, he had offered a US$1,000 prize to anyone who could reduce the words of just a single page to pinhead size. He said at the time that he didn't expect to wait long before he awarded it. In the end, the money barely made it out of Feynman's bank account before he died.

Feynman mailed the prize-winning check in 1985. It was made out to Tom Newman, a graduate student who used his electron beam machine to miniaturize the first page of *A Tale of Two Cities* while his supervisor was out of town.

Hayley Birch

Date 1985

Country USA

Why It's Key
Nanotechnology will be one of the most important technologies of the twenty-first century; this was the first time it entered the public eye.

Key Invention **Virtual reality**

The idea of mechanically simulating reality has been around since at least 1962, when Morton Heilig's "Sensorama Simulator" would take its riders on a tour of New York on a virtual motorbike, with all the sights and sounds, and even the wind blowing through their hair – and all without moving an inch.

It took until the 1980s for the words "virtual reality" to appear, though. The phrase was popularized by pioneers like Jaron Lanier, of VPL – Visual Programming Languages, Inc. By then, computers had become powerful enough to create artificial worlds that users could move through and interact with.

Given real depth by "VR" goggles, that projected a different computer-generated image onto each eye for true 3-D vision, these alternate realities let humans move through every kind of world; from lifelike models of real buildings, to fantasy landscapes with alternative rules of physics.

Interfaces like VPL's DataGlove let you interact with the objects in these worlds. The DataGlove was a real glove that could sense the position of your hand and fingers. By being able to tell exactly where your fingers were, the system could draw a moving "echo" of your hand inside its world, letting you pick things up and move them around as if you were really there. The true key to virtual reality lies in the senses, and it was in the 1980s that we started exploring all the different ways we could "be" in virtual worlds, especially through sight and touch.

Matt Gibson

Date 1985

Scientist Jaron Lanier

Nationality American

Why It's Key VR technology has been used to let architects walk through buildings that are still on the drawing board; and in the world of games, motion-sensing controllers can trace their origins back to the VR gloves of the 1980s.

Key Event
World's first GM field trials

1985 saw the first ever field trials of genetically modified (GM) crops, signalling a new era in agriculture and sparking fierce political debate.

When the first GM plant was generated in 1983 the possibilities for crop plants seemed endless, with promises of increased yield, improved shelf life, and better nutrition. However, the technology was met with much opposition due to concerns over potentially negative impacts on agriculture, economics, public health, and the environment. Despite this, many people believed the benefits far outweighed the potential risks, and continued to develop GM crops which, before they could be grown commercially, required testing in the field.

It was not long after the very first field trials in 1985 that the world also saw its first group of GM crop saboteurs: Earth First. This encounter was to set the stage for future protests by the anti-GM lobby, with countless GM field trials delayed, interrupted, and completely destroyed. In the face of this vociferous opposition, increasing numbers of GM crops have been tested and approved for use. By 2002, approximately 18 per cent of the worlds cultivated land was planted to GM crops and it was believed that only those who never ate processed foods were eating a GM-free diet.

More than twenty years on and the anti-GM lobby appear as influential as ever; in November 2007, the Italian National Institute for Research on Food and Nutrition was accused of suppressing results favourable to GM maize. The debate continues.

Becky Poole

Date 1985

Country USA

Why It's Key Field trials of GM crops made their commercialization possible, changing the future of world agriculture and sparking an often militant international debate.

Key Event
The launch of the Discovery Channel

In the 1970s and early 1980s cable television in the United States was beginning to become divided up into channels providing certain kinds of programmes. CNN, for example, showed news, ESPN sports, and MTV music. Journalist and film-maker John Hendricks saw a gap in the market. With funding from the BBC and a number of American investors, the Discovery Channel was launched on June 17 1985, to provide "real-world programming" to viewers across America.

Discovery was the world's first channel to be devoted solely to science and nature documentaries; over two decades later it is one of the most widely distributed cable networks in the world. Discovery networks are shown in 33 different languages in 170 countries and territories, beaming into 431 million homes across the globe.

A number of widely acclaimed series, films, and specials have been produced by the Discovery networks; in 2007 the *Planet Earth* series won four Emmy Awards with three other Discovery programmes winning Annual News and Documentary Emmy Awards. Permeating so many homes, Discovery's numerous channels have provided science and technology education for millions of home. It is arguably science's, and the environment's, most important mouthpiece.

By 2007, Hendricks had donated US$6 million to the Lowell Observatory for the construction of the Discovery Channel Telescope which, once it is completed in 2009, will be one of the USA's largest telescopes.

Shamini Bundell

Date 1985

Country USA

Why It's Key The first time a company was founded exclusively for the provision of scientific and factual television entertainment.

Key Person **Elizabeth Blackburn**
Where the ends justify the means

Every chromosome is capped by special structures called telomeres, like the plastic bits on the end of shoelaces. They prevent chromosomes from fraying or fusing together. Today, telomere research is one of the most exciting branches of molecular biology and much of this work was inspired by an Australian scientist called Elizabeth Blackburn (b. 1948).

In the 1970s, Blackburn started working on telomeres by studying a single-celled microbe called *Tetrahymena*, whose abundance of telomeres were easier to study than the small number found in human cells. Blackburn soon found that telomeres are actually short, repeated bits of DNA, but her work really took off in 1985 when, together with student Carol Greider, she discovered a protein called telomerase that lengthens telomeres.

That discovery formed the basis for future work that linked telomeres to both aging and cancer. Telomerase isn't active in most cells, so their telomeres shorten as they age. This makes the cells' chromosomes more vulnerable, and eventually stops them from dividing (and renewing) altogether. On the other hand, cells with indefinitely active telomerase can rejuvenate their telomeres and are effectively immortal. These include stem cells and, more sinisterly, the vast majority of cancer cells. Several cancer researchers have their sights set on telomerase as a target for future drugs.

Today, Blackburn is contributing heavily to the field she kick-started, not least by inspiring new generations of female scientists. In a typically male-biased field, telomere research is still dominated by women.

Ed Yong

Date 1985

Nationality Australian

Why She's Key Discovered a protein that could explain both how cells age, and how they turn cancerous.

Key Event **Around the world in 9 days, 3 minutes, and 44 seconds**

It may have taken Phileas Fogg only eighty days to travel the world, but it took Dick Rutan and Jeana Yaeger considerably less than that. In 1986, they were able to fly around the world, non-stop, in just over a week.

This astonishing feat took the two pilots to the edge of their physical and mental abilities. They flew for nine days, three minutes, and forty-four seconds, without refueling, over a distance of 40,210 kilometers. With just a few gallons of fuel remaining, they touched down at Edwards Air Force Base on December 23, smashing the previous record of 20,168 kilometers.

Their plane, the Voyager, contained seventeen fuel tanks, but was designed to be as light as possible while remaining strong enough to circumnavigate the globe. It wasn't all plain sailing for Rutan and Yaeger, however. Even before take off, the wings were so heavy with fuel that they scraped along the runway, putting the whole trip in jeopardy. Violent storms and typhoons, as well as a refusal to enter international airspace, made the achievement all the more remarkable.

This ultimate long-haul flight marked the breaking of one of the last remaining aviation records. Both pilots and plane have since been honored; Rutan and Yaeger being awarded the Presidential Citizens Medal by the then president Ronald Regan, and Voyager taking pride of place at the Smithsonian National Air and Space Museum in Washington DC.

Nathan Dennison

Date 1986

Country USA

Why It's Key One of the last aviation records is broken and the Earth no longer seems that big.

opposite The Rutan Voyager on its record-breaking flight

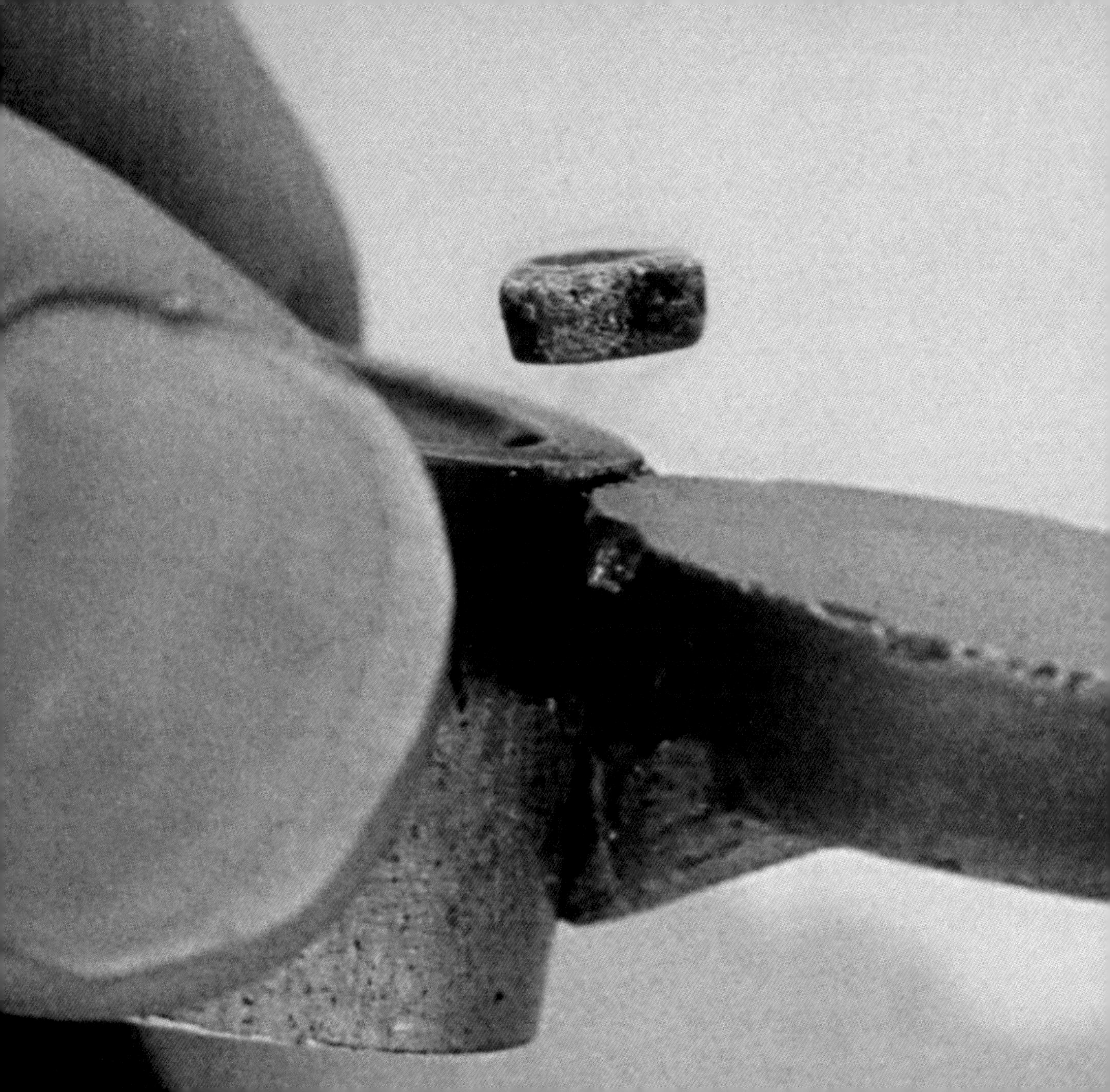

Key Discovery **High-temperature superconductivity**
The path of least resistance

The term high-temperature does not usually bring -238 degrees Celsius to mind. However, in a superconducting context, it represented dizzy new heights. In 1986, two physicists working in a Swiss hilltop laboratory discovered a material that could conduct electricity over large distances at this temperature. In a conducting material, electrons move, resulting in a flow of electricity and a weak magnetic field. There are many things electrons can bump into; this is known as resistance. When conductive material is cooled, the resistance can disappear, resulting in superconductivity. No energy loss occurs and the magnetic field becomes propelled outwards.

Previously superconductivity had required temperatures of -269 degrees Celsius – the boiling point of helium. By the 1970s, the temperature had been increased to -250 degrees Celsius, using a superconducting alloy, but it was still impractical and expensive to cool metals to this temperature. Physicists Georg Bednorz and Karl Müller's breakthrough came from the use of a ceramic oxide, consisting of copper oxide, barium, and lanthanum. After cooling it down to -238 degrees Celsius, they found the resistance disappeared. This is because the strong atomic forces can pull electrons from atom to atom. They kick-started an explosive research effort, resulting in superconductivity temperatures of -183 degrees Celsius. The cheaper coolant liquid nitrogen could now be used, opening up the practical application of superconductivity. The superconducting temperature continues to rise today.

Riaz Bhunnoo

Date 1986

Scientists Georg Bednorz, Karl Müller

Nationality German, Swiss

Why It's Key High temperature superconductivity has practical applications including efficient electrical wires, ultra-fast computers, medical imaging devices, and high-speed trains that float on magnetic rails.

opposite Demonstrating magnetic levitation of superconductors

Key Event **Launch ends in tragedy for Challenger**
"Teacher in Space"

Five years into the U.S. Shuttle Program, and things were beginning to get a little routine. Sure, it took a great deal of time, effort, and money to launch the vehicles, but after twenty-four flights to low-Earth orbit, the public's love affair with astronautics had waned.

A spark of interest returned when, on January 28, 1986, NASA sent a civilian, a high-school teacher no less, on her way to orbit. She was taking experiments designed by children into space, to inspire the next generation of scientists and engineers. But Christine McAuliffe and her six companions never made it.

Just seventy-three seconds into the flight, the vehicle was consumed in a ball of fire. A malfunction in one of the shuttle's solid rocket boosters had triggered a rapid disintegration of the main fuel tank. It later emerged that the crew had probably survived the initial catastrophe, only to perish when their cabin fell to the Atlantic.

Investigators concluded that an "O-ring" seal in the booster had eroded, allowing hot gases to escape and rupture the fuel tank. More damningly, the enquiry pointed a finger at NASA's management practices, which allowed corners to be cut in an effort to meet budgetary pressures.

NASA took two-and-a-half years to recover sufficiently to resume the Shuttle Program, which took another knock in 2003 with the disintegration of sister shuttle Columbia on re-entry. Belatedly, Barbara Morgan became the first teacher in space in 2007.

Matt Brown

Date 1986

Country USA

Why It's Key The Challenger disaster brought home the dangers of space travel to a complacent NASA and an apathetic public.

Key Event **Magic medicines**
FDA bites the bullet

A magic bullet. This was the treatment that Nobel Prize-winning scientist and medic Paul Ehrlich dreamed of towards the end of the nineteenth century; a substance that could target and destroy, with a high degree of specificity, the agent of a disease.

When, in 1986, an entirely new type of therapeutic went on sale, it was touted by some as the first of Ehrlich's "magic bullet" medicines. The U.S. Food and Drug Administration had approved the use of monoclonal antibodies – often dubbed "MAbs" – that could be used to stop a patient rejecting an organ transplant. The first drug to hit the market was called Orthoclone OKT3, manufactured by Ortho Biotech.

Antibodies come in all shapes and sizes, designed for defending the body against millions of possible enemies. One antibody seeks out one enemy. In the case of transplant rejection, the attackers are the body's own T cells; cells for which OKT3 has a particular affinity. It works by latching onto these T cells and preventing them from overreacting to the appearance of the "invading" organ.

Until 1975, scientists hadn't worked out a way of obtaining a pure source of one type of antibody. But thanks to the work of Georges Köhler and César Milstein, it was now possible to produce clones of single antibody-producing cells in mice – hence, monoclonal. "MAbs" haven't turned out to be quite the magic bullets Ehrlich imagined, but they've since been used to treat patients with everything from psoriasis to cancer.

Hayley Birch

Date 1986

Country USA

Why It's Key Monoclonal antibodies were a new type of treatment designed to get right to the heart of the problem.

Key Event
The Chernobyl nuclear disaster

For friends and foes of nuclear power alike it was their worst nightmare come true. At 1:23 am local time on April 26, 1986, a reactor at the Chernobyl nuclear power station in Ukraine exploded, releasing radioactive material into the atmosphere.

The previous day, the reactor crew at Chernobyl-4 had started to prepare for testing of the reactor's turbines. They wanted to find out exactly how long the turbines would spin for if the reactor lost main electrical power. Such tests had occurred before without problems, even though it was known that the reactor could become unstable at low power. During the test a huge power surge occurred due to a design flaw, and the reactor ruptured as a result of the immense pressure of steam that had built up inside. A second explosion allowed a sudden inrush of air to set fire to the graphite in the reactor core. The fire burned for nine days despite the 5,000 tons of lead, sand, and other inhibitors that were dropped onto the exposed reactor by helicopters. Radioactive dust was carried by the wind over Ukraine, Belarus, Russia, Scandinavia, and Europe, and the increased background radiation was detectable even by simple school laboratory Geiger counters.

On May 2, 45,000 residents were evacuated from within a 10-kilometer radius of Chernobyl, and a further 116,000 were moved over the next couple of days as the evacuation zone was extended to 30 kilometers. In the following years another 200,000 people were resettled from the region due to soil contamination.

David Hawksett

Date 1986

Country USSR (now Ukraine)

Why It's Key Fifty-seven workers were killed as a direct result of the infamous accident, which has been central to the arguments of anti-nuclear power campaigners ever since.

opposite Debris in reactor number 4 at the Chernobyl power station

Key Event
Mir launches world peace

On February 20, 1986, just five days before the twenty-seventh Congress of the Communist Party of the Soviet Union, and only weeks after the U.S. Challenger space shuttle disaster, the Mir Space Station was sent into orbit above our heads. Neither its timing nor its political impact could have been better. Mir translates to both "world" and "peace," but despite the almost too-good-to-be-true name, what made Mir so special was its modular design; it could get bigger with the snap of a safety hatch. The Communists were conquering space.

Both the Americans and the Soviets had already put space stations into orbit. But these mini-laboratories had not been able to accommodate a permanent presence in space, let alone facilitate expansions. Mir, on the other hand, had several docking bays, each designed to connect with additional modules, as well as the spacecraft that would shuttle cosmonauts between Earth and space. Indeed, the modular design was clearly utilized; having started as a mere 20 ton tube, by the end of its life, Mir was to weigh a mammoth 137 tons.

Mir's significance is underlined by the events of the Cold War that occurred either side of its 1986 launch. The Soviet invasion of Afghanistan, in 1979, and Reagan's "Star Wars" initiative of 1983, had not left the world feeling peaceful. But the Intermediate Range Nuclear Forces Treaty, signed by Reagan and Gorbachev in 1987, and the fall of the Berlin Wall in 1989, restored a great deal of confidence in our future.
Christopher Booroff

Date 1986

Country USSR

Why It's Key Mir opened up a new era of science in space, and helped to bring to an end an era of icy hostility on Earth.

opposite Mir Space Station

Key Publication **Out of Africa**
The search for Eve

In 1987, anthropologists Rebecca Cann, Mark Stoneking, and Allan Wilson published a profound theory of human evolution called the "Out of Africa" hypothesis. The study used DNA to trace the maternal lineage of a sample group of 147 people from populations of five geographic areas of the world – Africans, Asians, Caucasians, Aboriginal Australians, and Aboriginal New Guineans. Mitochondrial DNA is passed on exclusively from mother to child. This means that a sequence of mitochondrial DNA can remain virtually unchanged, barring the odd mutation, for generations, and can be used to trace historical maternal lineage.

The Out of Africa study assumed that the DNA in different maternal lines mutates at a steady rate over a very long period of time, so the different lines slowly diverge, becoming less alike as time goes on. Based on the level of DNA divergence in different populations, the study concluded that all current-day humans could be related back to Africa to one common ancestor, nicknamed Eve, who lived about 200,000 years ago. The theory proposes that the modern humans then spread out of Africa into Asia and Europe.

The Out of Africa theory is one of many human evolution theories, however Cann's study prompted much further research using mitochondrial DNA to investigate human lineage. As more information about different populations' genetics filters in from around the world, a better picture of human evolution will emerge.
Emma Norman

Date 1987

Scientists Rebecca Cann, Mark Stoneking, Allan Wilson

Nationality American

Why It's Key Produced a methodology that is widely believed by the scientific community to be a valuable technique for tracing the origins of humans.

Key Experiment
The Internet goes broadband

We can be forgiven for not always knowing our ADSL from our DSL or ISDN, but an Asymmetric Digital Subscriber Line (ADSL) is now how many of us access the Internet. The principal for DSL comes directly from the famous paper by mathematician Claude Shannon, "A Mathematical Theory of Communication," published in 1948. But Joseph Lechleider, a researcher at Bellcore, USA, is considered by many to be the father of the technology that allows broadband signals to be sent down a telephone wire.

Inspired by Shannon's work, Lechleider demonstrated mathematically that it would be possible. His research at Bellcore saw him add wideband digital signals on top of the existing analogue voice signals carried by a conventional twisted copper telephone line. He originally saw it as an opportunity to provide video on demand to home users directly over their telephone cable, without the need for a separate cable television system.

Lechleider's next breakthrough was to recognize that customers would benefit greatly from a subscriber line which allowed much higher rates of information download than upload, and the "A" in "ADSL" was born.

Although the potential benefits of DSL were obvious, it was years before it started to take off. This was due to several factors, including the relatively undeveloped state of the Internet, and the limited processing power of computers. And of course, it was much more profitable for provider companies to rent out a second telephone line for Internet access.

David Hawksett

Date 1987

Scientist Joseph Lechleider

Nationality American

Why It's Key ADSL technology is nowadays the prime competition for the cable broadband Internet worldwide.

Key Discovery **World's oldest embryo unearthed**
150 million-years-old fossilized dinosaur egg reveal

Between 65 and 230 million years ago dinosaurs roamed the planet and there are references to the discovery of their fossilized remains dating back over two thousand years. However, the first scientific description of a dinosaur fossil wasn't made until 1824.

Some specific locations, such as the Cleveland-Lloyd quarry in the American state of Utah, have been at the center of an unprecedented amount of fossil records. Since 1928, the quarry has yielded thousands of bones, mainly belonging to the meat eating allosaurs.

In addition to bones, eggs have also been exposed in the sedimentary rocks in Utah and their mineralized remains have led to a greater understanding of the life cycle of dinosaurs. One particular egg, which contained a fossilized embryo, received a great deal of attention in 1987 and after detailed examination, it was concluded that is was the world's oldest known embryo at 150 million years old.

The egg was actually taken to a nearby hospital, for a CAT scan. The resulting three-dimensional images clearly showed the early-stage development of an embryo – head, tail, and body – looking remarkably similar to a tadpole.

Scientists believe that many dinosaur species were very fast-developing and precocial – being able to leave the nest almost immediately and fend for themselves. However, some species may well have remained in the nest like most bird species and relied upon their parents for food and protection.

Andrew Impey

Date 1987

Scientists Wade Miller and colleagues

Nationality American

Why It's Key Paleontologists used advanced medical technology to confirm the oldest known dinosaur embryo ever discovered.

Key Invention **Prozac**

They say that in the 1960s people took acid to make the world weird; now the world is weird and people take Prozac to make it normal. This antidepressant drug, which first hit the market in 1987, is now keeping millions of individuals on the straight and narrow.

According to the World Health Organization: "5–10 per cent of persons in a community at a given time are in need of help for depression." This is regardless of race; however, the prevalence is strongly centered on people aged twenty to forty and is also slightly higher in women.

Prozac is the registered name for fluoxetine hydrochloride, a chemical that was first developed back in the 1970s, which acts on nerve cells in the brain. It works by manipulating the levels of a chemical called serotonin, which is released by the brain. This chemical is associated with sleep, appetite, aggression, and mood and is one of many different neurotransmitters which carry messages between nerve cells. Once the message has been delivered, the neurotransmitters are destroyed in a process called re-uptake. Prozac inhibits the destruction, effectively increasing the levels of serotonin in the brain.

Scientists know that this increase in serotonin results in a lightening of a patient's mood, but much work is still to be done before we truly understand why the brain undergoes this feel-good transformation. In addition, Prozac isn't without its possible side effects which include diarrhoea, insomnia, and lowered sex drive – surely that's enough to make anyone depressed.
Andrew Impey

Date 1987

Scientist Eli Lilly Ltd

Nationality American

Why It's Key Used in the treatment of depression, bulimia, and obsessive compulsive disorder; Prozac is currently the world's most widely prescribed antidepressant.

Key Event **Woodstock of physics** The resurrection of superconductors

When Paul Chu and his colleagues reported a new class of materials that could superconduct at higher than liquid nitrogen temperatures (77K), a jolt shot through the physics community. Though superconductivity was a well-documented phenomenon, historically it required superlow temperatures. These temperatures could only be achieved using expensive compressed helium, rendering superconductivity an interesting, but ultimately impractical, technology. Liquid nitrogen, however, is cheaper by volume than beer so can be utilized for applications outside the laboratory.

With the promise of high temperature superconductivity in the air, research groups all over the world began pouring their efforts into developing new superconducting compounds. This rampant interest and resulting avalanche of findings led the American Physical Society to organize a special session on the topic, allowing parties five minutes to deliver their spiel.The session was almost 4,000 physicists strong; the venue was so overcrowded that the New York hotel had to set up TVs for the overflow. 51 speakers expounded results, often to exalted cheers, until 3 am. The rock festival-like fervour of the gathering elicited the nickname "The Woodstock of Nerds" by journalists; it was quickly renamed "The Woodstock of Physics" by image-conscious scientists.

The aftermath was equally impressive. Government funding skyrocketed in the field, and journals were swamped with papers expressing new superconducting compounds.
Kellye Curtis

Date 1987

Country USA

Why It's Key Rekindled interest in superconductors and gave hope for amazing practical applications, such as ultra-efficient electrical transmission, high speed rail transport, and superfast computing.

Key Event **Patent granted to cover a genetically engineered mouse**

In 1988, the United States Patent and Trademark Office granted a patent to Harvard College, for a strain of mouse that was prone to getting cancer. Called the "Harvard Mouse" or "OncoMouse," it contains a gene, the oncogene, which significantly increases the rodents' susceptibility to cancer. This made the strain very useful for cancer research, so much so that, in addition to its patent, OncoMouse is now a registered trademark.

This was the first time such a patent had been applied for and, understandably, it caused some concern around the world. Although the patents were granted without hesitation or question by the U.S. courts, other countries were not as quick to pass it off. The European Patent Office initially refused the patent in 1989 on the basis of a European Patent Convention stating that animals could not be patented. However, on appeal the ruling was overturned and the patent was granted in 1992.

This was the first time a living creation had been patented, raising questions about whether future genetic modifications, as well as DNA itself, could be owned. Regardless of the debates, companies quickly started filing for patents covering other genetically modified organisms, with the significant potential opportunities for drug development in mind.

David Hall

Date 1988

Country USA

Why It's Key The first time a patent had been filed for a living creature, which led to ethical arguments about future genetic creations around the world.

Key Publication ***A Brief History of Time***

When renowned physicist Stephen Hawking published *A Brief History of Time* in 1988, it contained an introduction by legendary American astronomer Carl Sagan. In it he recalls being in London in 1974 for a meeting on extraterrestrial life, at the prestigious Royal Society. During a break he noticed an even larger meeting in another room, where he saw a young man in a wheelchair slowly signing his name in an ancient book which also included that of Isaac Newton. Sagan was witnessing the induction of a new fellow into the Royal Society – "Stephen Hawking was a legend even then."

Hawking's book on cosmology was aimed at the general public. He had been advised that every equation in a book "reduces the readership by half!" and so Hawking, in a staggering achievement, and despite his disability, covered this intensely mathematical subject in a book that had just one equation – perhaps the most famous of all, $E=mc^2$.

Upon launch, it received critical acclaim for its lucid and clear descriptions of some of the hardest concepts to grasp in cosmology, from the nature of the direction of time flow, to black holes and particle physics. It stayed on the *Sunday Times* bestseller list for a record-breaking 237 weeks. Within five years, there had been forty hardback editions in the USA and thirty-nine in the UK, each selling millions of copies. Many, including Hawking himself, jest that many copies are destined merely for a household coffee table to impress visitors!

David Hawksett

Publication *A Brief History of Time*

Date 1988

Author Stephen Hawking

Nationality British

Why It's Key Hawking's book has done more to popularize cosmology, and physics in general, than possibly any other book in the last hundred years.

opposite Stephen Hawking

Key Event
The first transatlantic fiber optic

In 1988, a consortium of companies led by AT&T, France Telecom, and British Telecom completed a project to lay the first fiber optic transatlantic cable.

The last half of the nineteenth century had seen the installation of short lengths of underwater cable between England and France, across the English Channel. Laying cables across the Atlantic presented more of a problem as they were easily damaged and broken.

In 1956, the first transatlantic telephone cable (TAT-1) was completed. It could carry thirty-six simultaneous voice telephone calls between the UK and North America. Soon, four other cables were laid across the Atlantic, with call capacity ranging from 48 to 845 simultaneous calls. In 1976, TAT-6 was completed, which could carry 4,800 simultaneous calls, and by 1978, TAT-7 with 4,000 calls. The first fiber optic transatlantic cable, TAT-8, was designed to carry ten times this many – an astonishing 40,000 simultaneously made calls.

Fiber optics works by carrying light through an inner glass or plastic core. Around this is another layer of glass which guides the light through the inner core. The outside of the cable is wrapped in protective materials to prevent breakage and damage. Fiber optic cables have great advantages over metal wires in that they have lower signal loss over large distances, meaning that fewer signal amplifiers are needed. They're also thinner, carry more data at higher rates, and are immune to electrical interference.

Andrey Kobilnyk

Date 1988

Country USA, UK, France

Why It's Key Modern global voice and data communication was only made affordable and possible by high speed underwater fiber optic cables.

Key Event **Seikan tunnel vision**
Japan's feat of engineering

Hollowing out a tunnel underneath the sea is a massively difficult and risky business. So when the Seikan Tunnel opened in 1988, linking the Japanese islands of Honshu and Hokkaido, it marked one of the most astounding accomplishments of human engineering. Stretching 53.85 kilometers and dropping to a depth of 240 meters below the Tsugaru Strait, it is the longest and deepest rail tunnel in the world.

Preliminary geological surveys were conducted as early as 1946, but the project never really got off the ground – or under it – after the treacherous ferry crossing, the only link between the islands, claimed 1,430 lives during a typhoon in 1954. By 1971, however, the main excavation was underway, but with hazardous consequences. Huge volumes of water, as much as 70 tons per minute, would breach the walls, inundating the developing tunnel and endangering lives. Technological innovation, however, soon came to the rescue. New grouting techniques stopped the water seeping in, and following any excavation work, the walls were immediately "shotcreted" by spraying concrete at a very high pressure to support the bedrock. Even so, thirty-three lives were lost during the project, and massive pumps still continually work to keep the water at bay.

By the time the tunnel was complete, Japan's social and economic climate had changed, with many people preferring the cheaper and faster option of air travel.. Nonetheless it remains a valuable part of Japan's transport infrastructure and a symbol of the nation's remarkable ingenuity.

James Urquhart

Date 1988

Country Japan

Why It's Key The world's longest and deepest rail tunnel required pioneering engineering techniques, revolutionizing the art of tunnel building in the process.

Key Person **Ray Kurzweil**
The ultimate thinking machine

In 1965 a high school pupil appeared on an American TV show called I've Got a Secret. He played a short piece of music on an upright piano and challenged a panel of celebrities to guess the secret behind his composition. They were stumped. That's because this precocious youngster had built his own computer program to analyze the patterns in music by famous composers and then compose original melodies in a similar style.

Ever since his early exploits, Kurzweil has been at the forefront of enabling computers to recognize abstract patterns. He created the first print-to-speech machines that could read text in any font out loud, to help the blind. This was an innovation that would lead to a lasting friendship with musician Stevie Wonder and the first machines that could accurately reproduce the rich, complex tones of acoustic instruments such as grand pianos.

Now recognized as one of the world's foremost authorities on artificial intelligence, Kurzweil has broadened his horizons and looked to the future. Being a man who correctly predicted the rise and dominance of the Internet, the fall of the Soviet Union and even chess-master Gary Kasparov's defeat to a computer, Kurzweil's predictions are given more credit than most. So when he says that computers will be more intelligent than people by 2050 and that he'll live forever thanks to scientifically-driven immortality, who are we to argue?

William Scribe

Date 1988

Nationality American

Why He's Key Kurzweil has made innumerable advances in the fields of artificial intelligence, text and speech recognition software, and electronic music. He might also live forever.

Key Event **A precedent-setting case of DNA fingerprinting**

On a cold evening in November 1983, fifteen-year-old schoolgirl Lynda Mann was raped and murdered in the small town of Narborough in Leicestershire, England. A subsequent investigation led only to the fact that the semen samples on the body and clothing were of the type A blood group. No other leads or evidence were found and the case eventually went cold.

Three years later Dawn Ashworth, also fifteen, was found murdered just a mile from where Mann's body was discovered. Evidence also pointed to a killer with type A blood, and investigators assumed the same person murdered both girls.

The prime suspect was seventeen-year-old Richard Buckland who worked at the local mental hospital. Buckland was arrested and ended up confessing to Ashworth's murder, but claimed innocence in the case of Mann's.

Around this time, the technique of genetic fingerprinting had been invented by Alec Jeffreys at the University of Leicester, but it had never been used in a criminal case before. Jeffreys took Buckland's DNA and proved that he was actually not the killer. Police then decided to take voluntary blood samples from local men with type A blood, around 5,000 samples. It soon emerged that a local man named Colin Pitchfork had persuaded a friend to go in his place for the sample collection; he was turned into the police and confessed to the crime. Pitchfork's DNA was an exact match to that found on both bodies, and he was convicted of murder in 1988.

Rebecca Hernandez

Date 1988

Country UK

Why It's Key The Pitchfork case was the first criminal case in which genetic fingerprinting was used to exonerate a suspect as well as convict one.

Key Person **Gertrude Elion**
From pickles to leukemia

In retrospect, Gertrude Elion (1918–1999) was an unlikely candidate to have developed some of the most important medicines of the twentieth century. Her parents had immigrated to New York City from Eastern Europe in the early 1900s; her father studying dentistry and eventually practicing in their apartment. The stock market crash left the family in dire financial straits, and it was unlikely that young Gertrude would be able to further her education. Her grades, however, secured her a place at Hunter College, New York, where she chose to study chemistry.

After a series of odd jobs, which included teaching chemistry at high school level, and checking the color of mayonnaise and the acidity of pickles for an industrial food service, she joined George Hitchings' lab in 1944. Elion would later share half of the 1988 Nobel Prize in Medicine and Physiology with her new boss.

In the 1940s, it was known that in order to reproduce, cells had to incorporate certain substrates to synthesize nucleic acid. Elion and Hitchings realized that if medications that resembled these substrates were incorporated into cells instead, they might halt further synthesis, and could therefore be used to target rogue or diseased cells, preventing them from replicating. Together they developed drugs to combat leukemia, malaria, gout, and herpes, as well as the anti-rejection medicine azathioprone, which is used to suppress the immune system after a transplant operation.

Stuart M. Smith

Date 1988

Nationality American

Why She's Key Elion helped create many key medications.

opposite Gertrude Elion in her lab after winning the Nobel Prize

Key Event
Worldwide ban on ivory trade

Two hundred years ago, there might have been as many as 27 million elephants inhabiting sub-Saharan Africa. Historically, killing one was no mean feat; African elephants are the largest of all terrestrial mammals on the planet and have a propensity to trample would-be attackers underfoot. The advent of high-powered firearms, however, rather swung the balance in the hunters' favor.

As elephant ivory became an easier resource to get hold of, demand – initially from Europe and the United States; latterly from Japan and China – soared; elephants had become big business. By 1979, there were just 1.3 million African elephants left and, despite strict quotas, that number halved to just 625,000 individuals over the following decade, due mainly to illegal poaching. By the close of the1980s, the elephant's fate was beginning to look precarious. In 1989, 115 countries took the decision at a CITES (Convention on International Trade in Endangered Species) meeting in Geneva, Switzerland, to ban the worldwide trade in ivory altogether. At once, it became almost impossible for illegal traders to pass their poached ivory off as legal, since none of it was. The US and European markets for ivory were virtually closed down and the Asian markets were seriously reduced.

Since the ban, elephant numbers haven't increased, but their dramatic decline has been halted. A 2007 report by the World Conservation Union (IUCN) estimated that there are still around 600,000 elephants left in Africa. For the time being, their future seems much safer.

Mark Steer

Date 1989

Country Switzerland

Why It's Key Halting the global trade in ivory decreased the pressure of illegal poaching on elephants and halted their slide into extinction.

Key Discovery **The SRY gene** Determining sex

Chromosomes X and Y are considered a double act in human sex determination. But Y took center stage in the 1980s when scientists discovered its key purpose was choosing a human fetus' sex. The world over, scientists were racing to be first to locate the exact gene on the Y chromosome that controlled gender. In 1987, an American and Finnish research team led by David Page claimed to have located the sex gene, dubbing it "ZFY."

Squabbles over the gene started soon after Page's announcement, with reports popping up that said ZFY wasn't truly the sex gene. The race to locate the gender gene continued, with Page's team keen to rectify any mistake. In a scientific showdown played out in the premier journal *Nature*, Page's team and another, led by British researchers Peter Goodfellow and Robin Lovell-Badge, both presented their "we found it" cases. Goodfellow and Lovell-Badge, it turned out, had the stronger case, and are now widely recognized as having tracked down the gender gene. They called their gene SRY (sex-determining region on Y gene); a name that stuck.

The sex-determining gene works like a light switch. It turns combinations of other genes on or off so that testes can form. Without SRY, the formation of ovaries continues unabated – humans, like many species, are "default set" to female. SRY did away with thousands of years of speculation and superstitions regarding what, and who, determines sex.

Raychelle Burks

Date 1989

Scientists Peter Goodfellow, Robin Lovell-Badge

Nationality British

Why It's Key SRY showed maleness is all in the genes.

opposite Computer artwork of an X (left) and a Y (right) chromosome

Key Event **First successful liver transplant from a living donor**

The liver has amazing powers of regeneration and can re-grow from as little as a quarter of its original size. That makes it particularly suitable for transplants. Originally, the only liver donors were dead ones, but this severely limited the supply of fresh livers.

Following a failed attempt in 1986, an Australian team were the first to successfully transplant a liver from one living person to another in 1989. The recipient was Iichirou Tsuruyama, a Japanese boy who was suffering from a fatal liver disease called biliary artesia. With bile unable to flow from his liver to his gallbladder, he had been waiting for a donor for several months and had only a few years to live.

In August, a team of Australian surgeons headed by Russell Strong removed Iichirou's entire liver and replaced it with a piece taken from his mother. The entire procedure took ten hours, and both mother and child survived. Strong rightly predicted that the technique would be commonplace within years. A few months later, American surgeons transplanted a piece of liver from Teresa Smith to her daughter Alyssa, who also suffered from biliary artesia. She too made a full recovery and graduated from high school in 2006.

At first, living donor liver transplants were only used to treat children, but the procedure spread to adults too, and today it's used to treat patients with a range of advanced liver diseases, like cirrhosis and liver cancer.

Ed Yong

Date 1989

Country Australia

Why It's Key Pioneered a technique that saved hundreds of lives, and greatly reduced the number of people waiting for liver transplants.

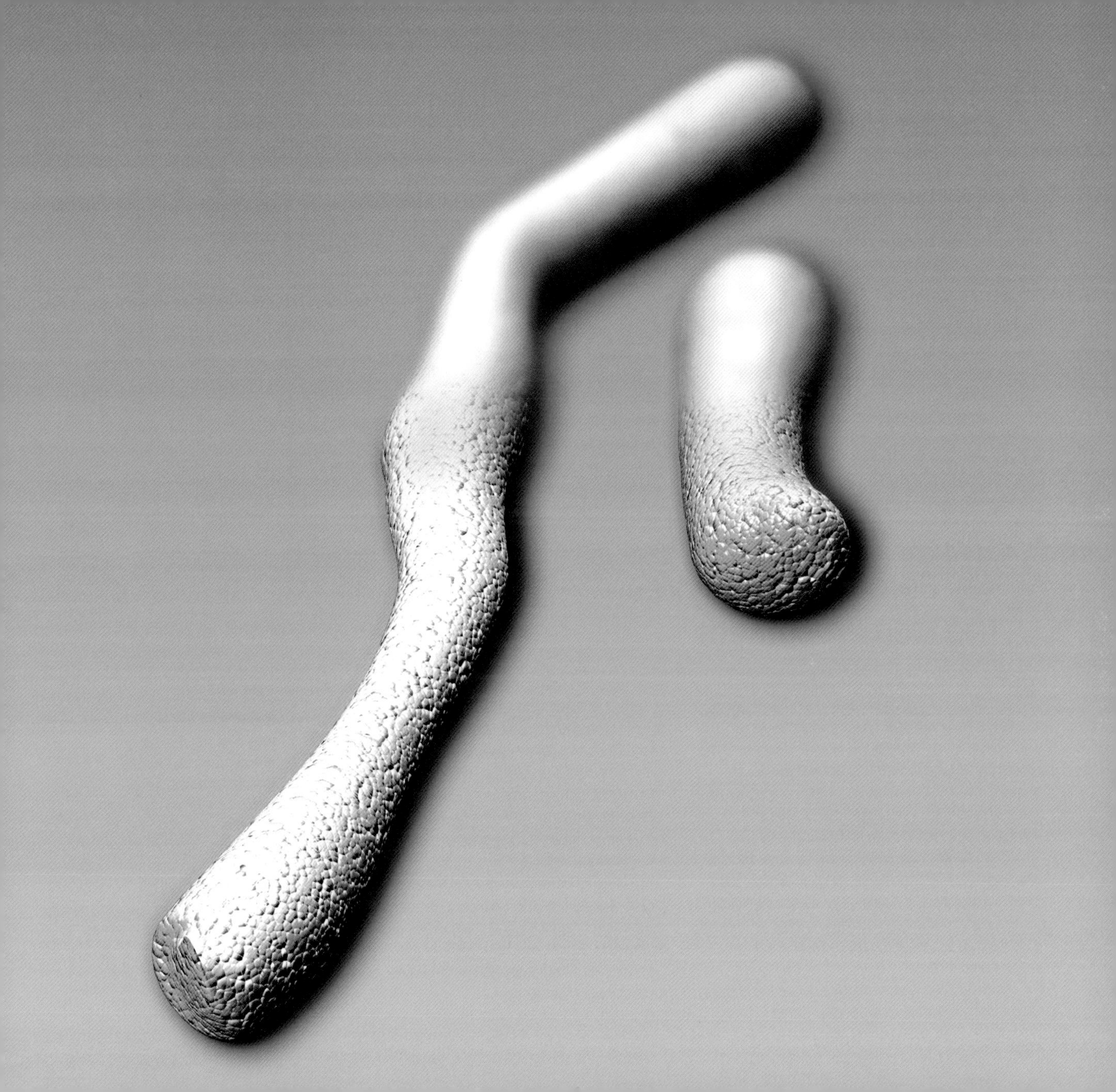

Key Event
Exxon Valdez oil spill

The *Exxon Valdez* oil spill was smaller than the *Amoco Cadiz* spill in 1978 and the spill in the Persian Gulf in 1991, but it remains one of the most famous oil spill disasters in history. En route from Alaska to California, the ill-fated tanker had strayed out of the shipping lanes, running aground in Prince William Sound. With the aid of a pilot, the huge tanker had safely maneuvred through the treacherous Valdez Narrows and, with the hard part seemingly over, the pilot then left the ship's crew to continue their journey.

Having left the shipping lanes to avoid icebergs, a ship's mate was left at the helm with exact instructions of when to rejoin the intended course. Despite this, the tanker was holed on the Bligh Reef, just after midnight on March 24, 1989.

A total of 11 million gallons of crude oil, equivalent to 240,000 barrels, was discharged into the sea, the majority of which escaped within just six hours of the accident. The clean-up operation didn't start in earnest until April and lasted over six months, involving over 11,000 personnel. Exxon was taken to court in 1994 but appeal after appeal means the punitive damages that run into billions are still unpaid.

The oil barons claimed that the damage was never as bad as first feared, but scientists say 26,000 gallons of oil remain in the sediment. Fish and seabird populations continue to show signs of the effect of persistent toxicity – highlighted through reduced reproductive success – and some biologists claim it could take decades for the area to fully recover.

Andrew Impey

Date 1989

Country USA

Why It's Key Thousands of seabirds, otters, and seals died in an accident which raised questions over the speed with which habitats can recover from a man-made disaster.

opposite Oil slick from the Exxon Valdez

Key Person **Brian Sykes**
The iceman cometh

Researching your ancestral history got a whole lot more interesting in 1989, when Bryan Sykes, published his first report on retrieving DNA from ancient bones. As Professor of Human Genetics at the University of Oxford, he has pioneered the use of genetic material to reconstruct human prehistory. He has been part of several high-profile cases, in particular Oetzi the Iceman and the Cheddar Man.

Oetzi came to the public attention when he was discovered stuck in the ice by two German tourists in 1991. He died around 3300 BCE and was preserved almost perfectly in ice. It has been determined that he died from an arrow wound that severed the artery under his collar bone.

Cheddar Man was the name given to the human remains removed from the Cheddar Gorge in south west England in 1903. The bones were dated to about 7150 BCE, making it Britain's oldest complete skeleton. Sykes compared sections of DNA extracted from one of the Cheddar Man's molars to people still living in the same area. Much to everyone's surprise, he found three exact matches. Maybe modern-day Brits aren't related to Middle-Eastern agriculturalists after all, but rather to European hunter-gatherers who later adopted an agricultural lifestyle.

In two other famous cases, Sykes debunked the claims of a woman who claimed to be related to the Russian Imperial family, and claimed that an American accountant, Tom Robinson, was direct descendent of Genghis Khan – although this has since been part of a long process of verification.

David Hall

Date 1989

Nationality British

Why He's Key The pioneer of tracing human history through genetics, Sykes opened a new window onto our past.

Key Event **Toxic treatment**
Botox gets the green light

In 1980, Dr Allen Scott, an American ophthalmologist, decided that injecting patients with one of the world's most poisonous substances was an acceptable thing to do. He was pioneering a new treatment for the medical conditions of strabismus (cross-eyes) and blepharospasm (uncontrollable eye closure).

The substance was botulinum toxin type A, known for its muscle paralyzing properties, and associated with botulism – a deadly disease caused by eating food infected with the toxin-producing bacteria *Clostridium botulinum*. Scott discovered that administering localized injections of the neurotoxin in small, harmless doses could relax the muscles around the eye and provide a temporary remedy. Subsequent trials showed the poison was safe and effective as a treatment for these conditions, which convinced the U.S. Food and Drug Administration (FDA) to officially approve its use, as "Oculinum," in 1989. Alastair and Jean Carruthers, however, noticed an unusual side-effect – it erased the frown lines and wrinkles associated with facial expressions. In 2002, the FDA approved "Botox" as a de-wrinkler, starting the phenomenal cosmetic craze.

Botox has also proven to be extremely effective for other, more pressing, conditions such as excess sweating, neck and facial spasms, and abnormal stomach and bladder function, for which, in each case, its use has also been approved by the FDA. And with studies in 2007 showing great potential for the toxin as a treatment for arthritis, and possibly depression, it seems there may be much more to Botox than meets the eye.

James Urquhart

Date 1989

Country USA

Why It's Key Approval of a treatment for medical purposes that would go on to change the face of celebrity.

Key Experiment
Science gives cold fusion the cold shoulder

In 1989, Stanley Pons and Martin Fleischmann, two chemists from the University of Utah, announced that they had produced nuclear fusion at room temperature. Many scientists fell over themselves in a rush to tell them how wrong they were.

Fusion occurs when two atoms merge together to form a heavier one, releasing a large amount of energy in the process. The energy we feel from the Sun is a result of hydrogen atoms fusing together to form helium in its core. Current theories suggest that fusion requires temperatures of millions of degrees Celsius in order to "squeeze" atoms together. If this occurred at much lower temperatures it would be "cold" fusion – as contrasted to the "hot" fusion in the Sun.

Pons and Fleischmann reported that electrodes made out of palladium immersed in a jar of room temperature deuterium (heavy hydrogen) resulted in a higher rise in the temperature of the deuterium than expected, which they believed was explained by energy released during the fusion of some of the deuterium atoms.

At first, scientists could not duplicate these results. Later, some did, but not consistently. Perhaps due to the perceived challenge to existing scientific knowledge, or threat to reputations and research grants, many scientists reacted emotionally to the possibility of cold fusion, and Pons and Fleischmann were widely discredited. After the dust settled, research into cold fusion continued. Undoubtedly, science as a whole suffers if new challenges are dismissed without investigation.

Andrey Kobilnyk

Date 1989

Scientists Stanley Pons, Martin Fleischmann

Nationality American

Why It's Key If possible, cold fusion may provide a source of cheap and clean power. The search to find a working method continues.

opposite Professors Stanley Pons (left) and Martin Fleischmann (right)

Key Event
First arrest for Internet sabotage

The first Internet virus was, allegedly, a harmless intellectual exercise. As a student at Cornell University, Robert Tappan Morris (Junior – his dad, of the same name, worked for the National Security Agency, coincidentally) wrote a programme that would enter another computer via security holes in commonly used programmes, copy itself to the hard drive if it wasn't already there, run innocuously in the background, and infect the next computer.

At this level it would have been a pest, but Morris realized that a computer could resist infection by falsely reporting it was already infected. So he set the worm to replicate itself on random occasions, even if it appeared that it was already in the computer. This caused an explosion in the number of worms as computers got infected with multiple copies of the programme.

With instances of the worm growing exponentially, it didn't take long for the processing power of many machines to be used up, bringing them, and any system relying on them, to a grinding halt. The Internet crashed. According to the U.S. general accounting office, between US$100,000 and US$10,000,000 were lost due to lack of Internet access.

Robert Morris became the first person to be charged under the Computer Fraud and Abuse act. His sentence, however – 400 hours of community service and US$10,050 – was lenient compared to what the law allowed. And it doesn't seem to have stopped Morris from becoming a professor of computer science at the Massachusetts Institute of Technology, where he initially released the worm.

Kate Oliver

Date 1989

Country USA

Why It's Key The downside of increased networking becomes apparent, and digital crimes become a consideration for the courts.

Key Event
Galileo launched

The Galileo mission was a truly massive undertaking, worthy of the father of modern astronomy's name. On its fourteen-year journey from Earth, via Venus, to Jupiter – taking in a couple of asteroids and a comet en route – the two-ton craft covered an amazing 5.41 billion kilometers. At a cost of an estimated US$1.39 billion, that works out as a fairly reasonable US$2 per mile; not much more than a taxi ride.

On this epic voyage, Galileo observed a comet collision, became the first craft to visit an asteroid or orbit Jupiter, and provided invaluable new data about the giant planet and its moons.

It was also apt that the craft named after the discoverer of the largest moons of Jupiter brought back so much invaluable information about them. Not only did it discover possible liquid water oceans on Europa, but it also found evidence for water on Callisto; a magnetic field around Ganymede; and extreme volcanic activity on Io.

The spacecraft took a rather convoluted route to Jupiter, as it was not possible to put enough fuel on the five-meter craft to go straight there. Instead, Galileo was to slingshot once around Venus and twice around Earth to gain enough momentum to make the journey. Along the way, it was able to pass asteroids 951 Gaspra (October 29, 1991) and 243 Ida (August 28, 1993). Images of Gaspra showed a body with an irregular shape about twenty kilometers across, with a surprisingly low number of craters on its surface.

Jim Bell

Date 1989

Country USA

Why It's Key Galileo was an amazingly ambitious project, and has greatly improved our understanding of our Solar System.

opposite The Galileo spaceprobe's antenna being tested at the Jet Propulsion laboratory

Key Event
Human Genome Project set up

In 1990, biology suddenly went "big." Expected to cost US$3 billion and last for fifteen years, the Human Genome Project (HGP) – coordinated by the US Department of Energy and National Institutes of Health – was the largest international collaboration ever undertaken in the biological sciences, involving thousands of scientists. Their aim: to map the genetic landscape and sequence the 3 billion letters, or "base-pairs", that constitute all the genetic information in our twenty-four chromosomes – the human genome.

This was an enormous task, especially as when lab work began, sequencing technology was slow and required a great degree of manual preparation. The public consortium's project used a method in which DNA was laboriously broken into manageable chunks of about 40,000 base pairs. These chunks were then "shotgunned"; a way of breaking them into smaller, randomly sized pieces of DNA. The smaller pieces were then sequenced to discover the order of base pairs and the resulting data was "stitched" back together to reveal the overall sequence.

In 1998, Craig Venter's Celera Genomics declared they would enter the race. Venter was confident of producing the sequence in three years and at a fraction of the cost of the publicly funded effort. He pioneered a new technique called "whole genome shotgun sequencing," eliminating the laborious first step and instead he shotgunned the whole genome. In 2001, the Celera team published their version of the genome for the bargain price of US$300 million – a tenth of the cost of the public consortium's project.

Andrey Kobilnyk

Date 1990

Country 16 countries worldwide

Why It's Key The start of a major project to crack the code representing our own genetic make-up.

Key Person **Tim Berners-Lee**
Inventing www

Sir Timothy John Berners-Lee (OM KBE FRS FREng FRSA) (b.1955) was rated by America's *Time* magazine as one of hundred most important people of the twentieth century. In the scientists and thinkers section, he occupies the same space as Sigmund Freud and the Wright brothers, but to many, he has been far more influential than even these giants of the modern age. Berners–Lee is the man who, in 1990, invented the World Wide Web – the invention that has changed the way that we live our lives perhaps more than any other invention of the last century.

Born to two mathematicians on June 8, 1955, Berners-Lee studied physics at Oxford, and showed an early zeal for exploiting the uses of computers by hacking into the university's machine. He was banned from using it, but his enthusiasm was obviously undiminished. At CERN, the European particle physics laboratory in Switzerland, he conceived a way to link computers, their users, and packages of code called "hypertext" with the Internet – a worldwide series of computer networks – that would allow users to find, share, and retrieve information as never before. It is said that there are now as many web pages in the world as there are people.

Berners-Lee then ensured that the world would have the web for free, and continues to take a leading role in developing and safeguarding his invention, and its successors, such as the "Semantic Web", which promises major advances in how computers and humans will communicate in the future.

Arran Frood

Date 1990

Nationality British

Why He's Key Berners-Lee's vision and practical skills in bringing the world wide web to fruition where other pioneer's dreams have remained theoretical have made him a legend in his own time.

opposite Tim Berners-Lee

if s ⊨ a.post and u ⊨ ∃i: top(f(a)(i)).post then
∀a' ∈ acts(A): s ⊨ a'.pre implies u ⊨ ∃i: bot(f(a')(i)).pre
u ⊨ ∃i: bot(f(a)(i)).pre
L = {(R1, G1), ... (Rn, Gn)}

Key Event **Gene therapy** Quick fix for rogue cells?

One of the most well studied genetic diseases of our time is Adenosine Deaminase (ADA) deficiency. Sufferers of the disorder have a fault in a gene on chromosome number 20, which codes for an enzyme essential to maintaining a healthy immune system. As a result, patients are left vulnerable to infections that would be easily fought off in individuals without the defective gene.

In 1990, a group of pioneering doctors used this information in an attempt to produce a new kind of treatment for ADA deficiency. Their test patient was a four-year-old girl with a severe form of the disease, and their treatment process was dubbed "gene therapy," now a well-recognized technique. In this first trial, a blood sample containing immune cells was taken from the child. Healthy versions of the affected genes were then inserted in the cell's chromosomes causing them to manufacture the enzyme required to beat the disease. These "fixed" cells were then injected back into the patient.

The therapy was regarded as a revolutionary success, as the patient's immune system function was restored, although it acted only as a temporary solution, requiring ongoing treatment to maintain levels of healthy cells. The study set a benchmark for future research, however, giving way to an increasing number of gene therapy trials. Although the technique has since suffered some major setbacks, including a number of associated deaths, some trials have shown great promise for the treatment of diseases including Parkinson's and various types of cancer.

Hannah Welham

Date 1990

Country USA

Why It's Key A first attempt at a controversial new treatment for genetic disease.

Key Experiment **Tracy the transgenic sheep**

Of all the famous sheep in history, the most famous is undoubtedly Dolly, the first cloned sheep, whose creation generated a media frenzy in the late 1990s. But before Dolly there was another sheep that arguably deserves as much media acclaim. Let's not forget Tracy the transgenic sheep.

Animals that contain DNA from other animals are called "transgenic." These animals are created by applying, for example, human DNA to zygotes (fertilized eggs) and hoping that the human DNA integrates. The need for a more effective way of creating transgenic animals was what led the Roslin Institute – at that time a subsidiary of the Institute of Animal Physiology and Genetics Research – to the research that ultimately created Dolly. The goal of the Roslin team, lead by Ian Wilmut and Keith Campbell, was to produce an animal that would be able to manufacture human proteins for use in medicines. Proteins cannot be made artificially in a laboratory environment, so blood donations are depended on for medicinal proteins. By introducing the "code" for a protein into a genetically engineered animal, it is possible to vastly increase the amount of this protein that can be produced, improving the possibilities for treatment in humans.

Tracy was successfully engineered to produce milk that contained alpha-1-antitrypsin, an enzyme that is used in the treatment of lung conditions such as emphysema and cystic fibrosis.

Barney Grenfell

Date 1990

Scientists Ian Wilmut, Keith Campbell

Nationality British

Why It's Key Tracy was one of the forerunners in the production of transgenic animals for medicinal purposes. Problems with her creation also inspired the research that lead to the development of Dolly, the first cloned sheep.

Key Invention **Dycam Model 1**
The first commercial digital camera

Long gone are the days of biting disappointment that occurred when opening an envelope of your freshly developed holiday photos, only to find that each and every one of them is blurred, overexposed, or grimly unflattering. Yes, thanks to the filmless digital camera, you can snap, snap, snap to your heart's content and always be guaranteed a top quality album. But how did this come to be?

The idea of a filmless camera had been around for quite some time, but the first real digital prototype was completed in 1975 by Kodak engineer Steven Sasson. Its first picture – a still, 0.01 megapixel, black-and-white image – took twenty-three seconds to record and was stored on digital cassette tape. The camera was developed as an experiment, and was never intended for commercial sale; a good job considering that it weighed nearly four kilograms and was about the size of a human head.

It was fifteen years later when Dycam launched the first digital cameras for sale. The Dycam Model 1 and the Fotoman (two cameras whose only difference was the color of their casing) may have looked more like the video entry systems you'd normally see on a tower block of flats, but they were actually incredibly sophisticated. You could use them to take photos, connect the camera unit to a PC, and then download the images onto the computer's hard drive – in much the same way you would with a contemporary camera.

Chris Lochery

Date 1990

Scientists Steven Sasson, Dycam

Nationality American

Why It's Key The Dycam Model 1 and Fotoman were the Adam and Eve of commercial digital photography.

Key Experiment **The discovery of RNA interference**
Flower power

In 1990, an experiment on petunias yielded surprising results that changed our understanding of the mechanisms controlling gene expression within cells.

Plant biologists in America were attempting to increase petal pigmentation in petunias by introducing an additional copy of a gene controlling the color of flowers. Adding the chalcone synthase gene, they expected to produce more pigmented petals, but instead the opposite happened. Some of the genetically-modified flowers became less pigmented, with a number of white petals instead of the usual purple. The levels of chemical messengers (mRNA) coding for pigments were actually reduced. This work was summarized in a paper written by Carolyn Napoli, Richard Jorgensen, and Claude Lemieux, and was an early study of the process now known as RNA interference (RNAi). Further experimentation in other labs revealed the secret. mRNA is usually in the form of a "single-stranded" molecule; if double-stranded mRNA is present in a cell, it must have come from another source, such as a virus. If double-stranded mRNA is present in a cell where there is a matching DNA sequence, it will switch off the gene in a series of steps, hence the pale petals seen in the petunias.

The discovery of RNAi has great potential in both research and medicine. Scientists can use it to study the effects of "knocking-out" a specific gene in an organism – blocking its activity – to study its effect. In medicine, silencing a defective, disease-causing gene may help provide treatment for genetic disorders.

Katie Giles

Date 1990

Scientists Carolyn Napoli, Richard Jorgensen, Claude Lemieux

Nationality American, Canadian

Why It's Key Flower-powered discovery of an intracellular process with great potential in research and medicine.

Key Event **Hubble launches**
New telescope is out of this world

Double, double, toil and trouble; fire burn, and cauldron bubble. The Hubble Space Telescope has proven to be plenty of toil, but, with stunning images of the fiery balls of gas that light up our night sky, and crystal ball-like views into the Universe's past, few people would deny that Hubble was worth the trouble.

On April 26, 1990, the crew of space shuttle Discovery placed Hubble into orbit 600 kilometers above sea level. The telescope, a joint project between NASA and the European Space Agency (ESA), takes advantage of the fact that it floats above the distorting effects of the Earth's atmosphere. However, a rogue speck of paint meant that the first images Hubble recorded were blurred, meaning that astronauts has to be sent up to carry out history's most impressive bit of DIY in order to rectify the problem.

Given that Hubble bears the name of the man that first provided evidence that our Universe is expanding, it is fitting that one of the telescope's greatest achievements should be to determine the rate of that expansion.

By measuring this, scientists can estimate how long ago the Big Bang occurred – and therefore the age of the Universe – and also predict the future of the cosmos; will it continue to expand or will it one day collapse? Double, double, toil and trouble; fire burn, and cauldron bubble.

Christopher Booroff

Date 1990

Country USA

Why It's Key The Hubble Space Telescope has been a huge success, providing inspirational images to the wider community, and reams of data for scientists to analyze.

opposite The Hubble Space Telescope

Key Invention **World Wide Web**
Invented at CERN in Switzerland

The arrival of the World-Wide Web (www) heralded the beginning of the modern era we now live in – the Information Age. Like the nuclear age and the space age, brought on by Hiroshima and Sputnik respectively, the www is the work of physicists, and is an advance that has changed the way that people live more than any invention since the industrial revolution.

The www was invented at CERN, the European particle physics laboratory complex that spans the border of France and Switzerland, by British physicist Tim Berners-Lee and colleagues, principally Belgian Robert Cailliau. Berners-Lee saw CERN as a minature version of the world in a few years time: it took too long to find information, and too much of it was being lost. His original proposal is now the stuff of legend. In it, he details with startling lucidity a new networking interface that would allow people across the world to access, change, share, and store information more easily: "a 'web' of notes with links between them is far more useful than a fixed hierarchical system." His proposal also credits earlier pioneers of the "docuverse," notably Ted Nelson who coined the term "hypertext," which is what the first "ht" of http stands for.

Although it was initially overlooked, the system was working less than two years after the original proposal was made. Berners-Lee's August 6, 1991 posting is considered the official birthday of the www: "The WorldWideWeb (www) project aims to allow links to be made to any information anywhere."

Arran Frood

Date 1990

Scientists Tim Berners-Lee, Robert Cailliau, et al

Nationality UK, Belgium

Why It's Key An invention that has changed the way that people live, meet, discover, consume and interact more than any other of the modern era.

Key Publication **Carbon nanotubes**
Rolling into the nano-age

It sometimes escapes our notice that 95 per cent of the world's population don't speak English as a first language. But when it comes to scientific papers, publishing in English can mean the difference between worldwide acclaim and relative obscurity; which is why Messieurs (if you'll pardon the French) Radushkevich and Lukyanovich probably wish they hadn't published what might have been a landmark paper in Russian, in the middle of the Cold War.

As it stands, their discovery – tiny tubes made entirely of carbon – is often credited to Sumio Iijima, a Japanese physicist working with NEC. Iijima's description of carbon nanotubes in 1991 bore a striking resemblance to images presented by the Russians nearly forty years earlier in 1952. Back in the fifties, nanotechnology was a mere glint in physicist Richard Feynman's eye. And Radushkevich and Lukyanovich simply weren't equipped with the magnifying power needed to analyze the structures, which were 50 nanometers in diameter – 1,600 times thinner than a human hair. Iijima's paper, by contrast, was published in English, at a time when scientists were beginning to see the world from a nano perspective.

Despite being minute, nanotubes are actually giant molecules; sheets of carbon atoms rolled into cylinders and closed at one end. Some have single walls, while others are made up of many layers. They exhibit great strength, have extraordinary electrical properties, and are currently the subject of frenzied study by experts who believe they can use them to develop incredibly powerful quantum computers.

Hayley Birch

Date 1991, *Nature*

Author Sumio Iijima

Nationality Japanese

Why It's Key Although not the first description of carbon nanotubes, Iijima's paper had a massive impact on carbon science. Nanotubes are now finding employment everywhere, from tennis racquets to quantum computers.

opposite Colored transmission electron micrograph showing the capped end and multi-layer wall of a nanotube

Key Invention
Entanglement based secure communication

Sending private messages has always been an issue for diplomats and the military. Today, when so much information is on the Internet, how do we prevent others from eavesdropping on our communications? How can we encrypt our messages?

An early method called for the use of "one time pads" – initially just with paper pads – in which both the sender and receiver of a message had a "key" which would be used only to decode one message and then thrown away. But what if the sender and receiver didn't have a key? Or if the key was intercepted by an eavesdropper?

In 1991, Polish physicist Artur Ekert developed a method by which a key could be sent using a property of quantum mechanics called "entanglement." Quantum entanglement occurs when two particles are created in a manner where each has an opposite property to the other. One of the two particles would be retained by the sender and the other sent to the receiver. Each entangled pair would be completely randomly generated.

Someone could still eavesdrop on the stream of particles and learn part of the key – but one of the strange properties of quantum mechanics is that the act of measuring a particle will change it. By comparing their keys the sender and receiver will see which parts had been intercepted and discard them. They could then send a message using their unique key – essentially a one time pad – which would be perfectly secure.

Andrey Kobilnyk

Date 1991

Scientist Artur Ekert

Nationality Polish

Why It's Key Secrecy and privacy are more at stake than ever in our digital world – quantum key encryption, if made cheap and possible over long distances, could provide a secure and quick solution.

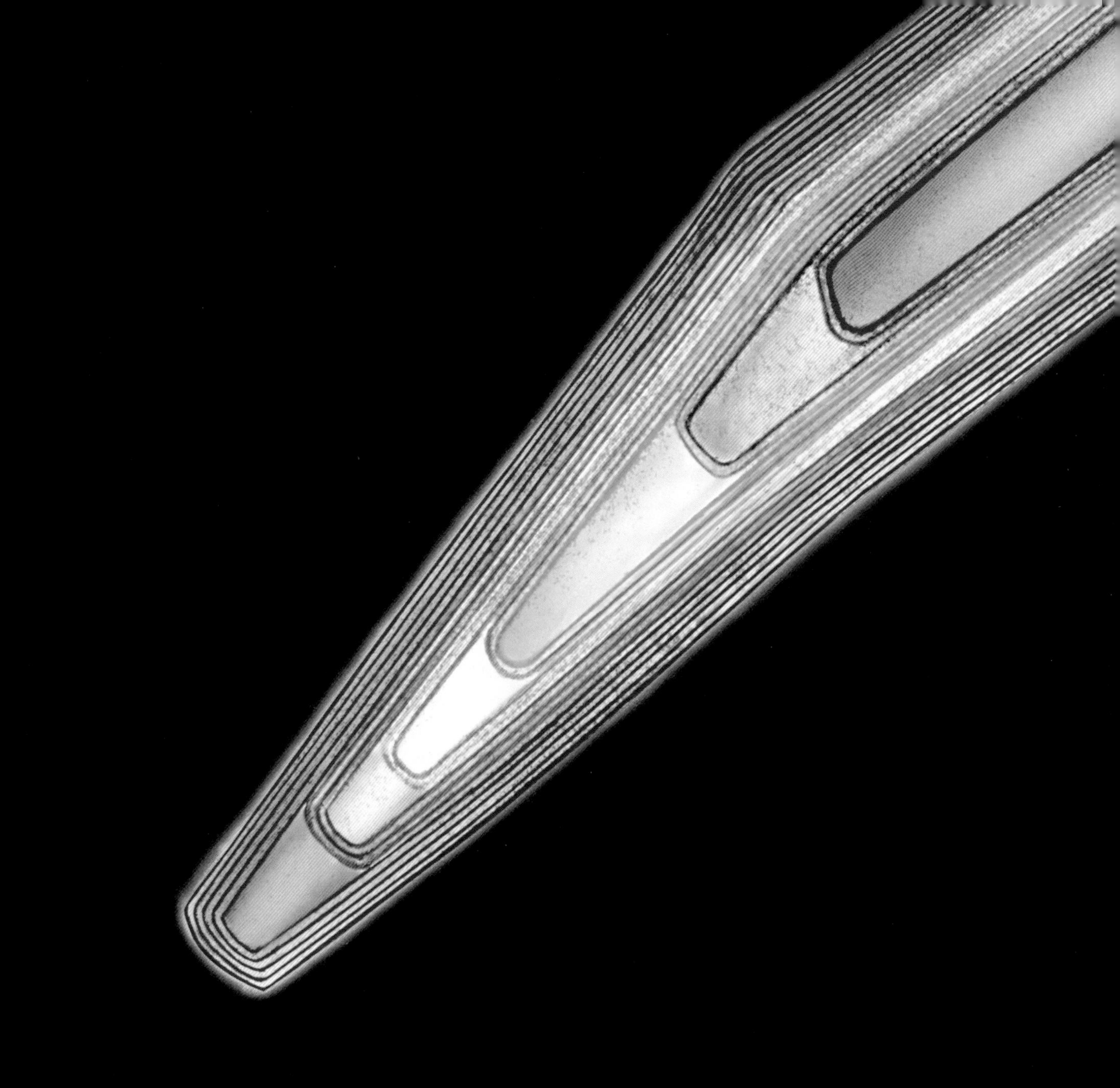

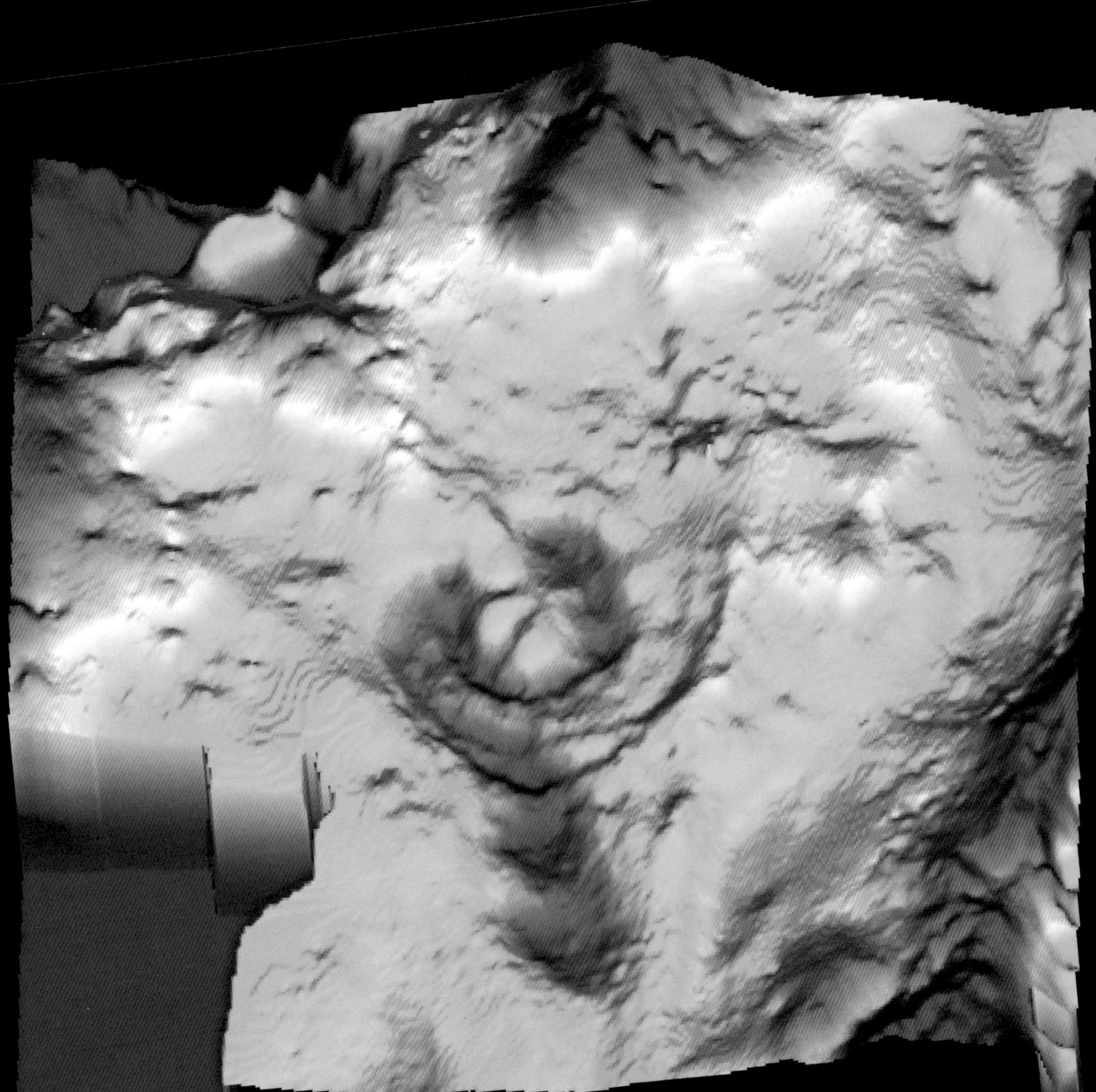

Key Discovery **Chicxulub crater**
Site of dinosaur-killing meteor crash

About 65 million years ago, a huge lump of rock slammed into the Earth's surface from outer space, putting paid to 70 per cent of all species on the planet, including those big, scary dinosaurs.

For many years, scientists had been speculating about what caused the demise of the dinosaurs, but it was a chance discovery in 1978 that would add weight to one theory in particular – an asteroid strike.

Glen Penfield was a geophysicist working for a Mexican oil company which was prospecting for drilling locations along the Yucatan Peninsula. While analyzing the survey data, he noticed an arc with extraordinary symmetry, forming a circle 180 kilometers wide, with the village of Puerto Chicxulub at the center. It would be over ten years, however, before, using a combination of seismic monitoring, geological sampling, and computer modeling, enough evidence was amassed to persuade others that either a comet or an asteroid, measuring roughly 10 kilometers in diameter, impacted on the Earth's surface, causing giant tsunamis, global firestorms, and filling the atmosphere with thick dense ash.

The impact would have been two million times more powerful than any bomb that has ever been exploded and it is now largely accepted as the cause of the mass extinction during the cretaceous period. However, some scientists still aren't convinced and the debate is set to rumble on well into the future.

Andrew Impey

Date 1991

Scientists Glen Penfield, Alan Hildebrand

Country Mexico

Why It's Key Provided evidence that the Earth was indeed struck by an asteroid the size of a small city, which probably led to the extinction of the dinosaurs 65 million years ago.

opposite A gravity anomaly map of the Chicxulub impact crater

Key Invention **Linux for you**
From Linus and GNU

Released in the early 1980s, for a long time DOS was the only affordable operating system for most home computer users. The Unix system provided a superior alternative but, by going for the big money, it priced itself out of reach of small users. By the 1990s, most universities used Unix systems and many computer scientists wanted to run it at home. The only cheap option was Minix, similar to Unix, written for educational purposes. Minix was limited, but it had one big advantage for the aspiring programmers and hackers: Unlike DOS and Unix, the code was "open source" – available for the public to change and adapt.

The hero of the open source movement was Richard Stallman, who in 1983, created the GNU project, providing free, quality software. By 1991, the GNU project had created a number of the tools needed to make an operating system, but was still years from finishing. In September of that year, Linus Torvalds, a student at the University of Helsinki, released a new operating system, similar in many ways to Unix, onto the embryonic Internet. The open source Linux was free to be tweaked and tested by the techno-world. Updated versions were written and its community grew from then on.

It rapidly gained a reputation for providing secure, reliable, and near virus-immune computers which were much less likely to crash than those using other systems. Powered by the programs of GNU, Linux grew beyond the teaching tool that was Minix into a useable system.

Douglas Kitson

Date 1991

Scientist Linus Torvalds

Nationality Finnish

Why It's Key Linus Torvalds isn't a billionaire like Bill Gates, but his operating system, Linux, runs four out of five of the world's top supercomputers.

Key Discovery
Oetzi the iceman

In September 1991, a frozen body was discovered by German hikers Helmut and Erika Simon in the Alps near the Austrian-Italian border. Initially mistaken for the corpse of a modern hiker, he was removed roughly without careful archaeological analysis. Only later did the authorities realize that the corpse was much older than they could have imagined; dating back 5,300 years, it was the oldest frozen mummy ever found.

The iceman was named Oetzi (Ötzi) after the area in which he was found, Oetztal, just inside the Italian border. Oetzi was extraordinarily well preserved by the ice; he was male, just 5 foot 3 inches (160 centimeters) tall, and even his age could be calculated from his bone density – the diminutive mountain man was in his forties when he died. Oetzi was also found with some belongings – a copper axe, cap, woven grass cloak, goatskin leggings, bearskin hat, bow and arrow, and a stone-tipped knife. Analysis of the contents of his intestines revealed his last meals – ibex (wild goat), red deer, and grain.

There has been much speculation about how he died, but the most compelling evidence came in 2001. X-rays of the body discovered a flint arrowhead embedded in Oetzi's shoulder, indicating he died whilst fleeing attackers who had shot him in the back. By 2007, researchers were convinced that the arrowhead had pierced one of Oetzi's arteries; he had rapidly bled to death.

Oetzi can be seen in South Tyrol Museum of Archaeology, Bolzano, Italy.

Emma Norman

Date 1991

Country Austria/Italy

Why It's Key Oetzi is the oldest iceman discovered so far and he provides a fascinating insight into early human life.

opposite The frozen corpse named "Oetzi"

Key Event **A little misinformation goes a long way**
Public hears red wine protects against heart disease

In 1991, newsman Morley Safer told 33.7 million Americans that "alcohol – in particular red wine – reduces the risk of heart disease." Speaking on the radio program, *60 Minutes*, he said that heart disease was far less common in France than in the USA, despite the fatty diets of the French. Red wine was the solution to this "French Paradox" and, according to Safer, its benefits were "all but confirmed." Listeners were convinced and within four weeks, sales of the "wonder"-beverage had rocketed by 44 per cent. The alcohol industry quickly capitalized, with many wineries starting to describe their products as "health foods."

It was a massive oversimplification that had long-lasting effects. To this day, many people view alcohol, and particularly red wine, as a healthy choice, even though later studies found that the benefits of light drinking against heart disease were exaggerated and only apply to older people. In contrast, alcohol definitely causes several other diseases including cancer, cirrhosis, and hypertension, while drinking too much actually causes heart disease.

According to the World Health Organization, the "French Paradox" itself was greatly overestimated and Pierre Ducimetière, one of the scientists who coined the phrase, has disavowed it based on modern data. French rates of heart disease are unremarkable for Europe and it's still the country's biggest cause of death. Safer did get one thing right though – the French do drink more alcohol than Americans. But they also pay a high price for it with high levels of alcohol-related problems, cirrhosis, and liver disease.

Ed Yong

Date 1991

Country USA

Why It's Key Kick-started a campaign of misinformation that health professionals are still dealing with.

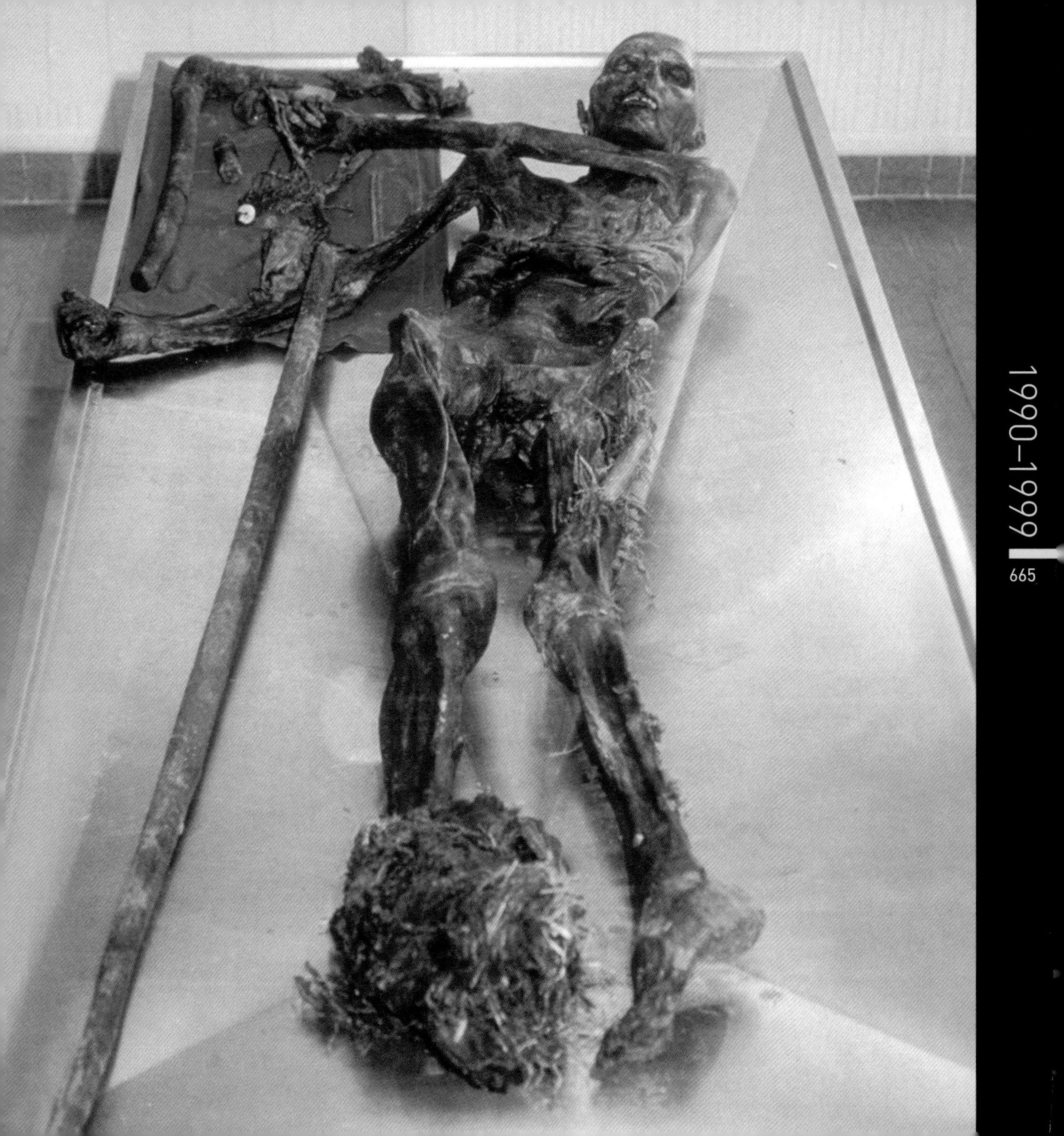

Key Event **South America hit by first cholera epidemic for a century**

Cholera is an ancient disease, often seen after natural and man-made disasters. From time-to-time global epidemics ravage populations across the world before receding. Seven pandemics have been recorded since 1817; the last began in 1961. In 1991 it reached Peru, which had been free from cholera for over a century, killing 700 and sickening 100,000.

The impact of cholera is explosive on every level. The sudden massive diarrhea kills even the fittest person within hours. It is so ferocious and fast that panic is unavoidable, disrupting social structure and paralyzing commerce. In the two years that cholera rampaged through Latin America, US$770 million were lost in food-trade embargoes and tourism. Over a million people became ill, and many thousands died. The precise cause was never established. Certainly the bacterium *Vibrio cholera* was to blame, but the source is uncertain. Seafood may have been contaminated by effluent discharged from ships; or water or sewage treatment procedures may have failed. Robust public health prevention strategies are essential with cholera because, although it is easy to treat (just needing good rehydration fluids) treatment must be immediate – which is not always possible during an outbreak.

Evidence suggests, however, that preventive measures do not always work, so research must continue into the fundamental aspects of the disease, like the trigger that makes the bacterium stop moving and instead stick to the gut lining to secrete its fatal toxin. The new strain that emerged in recent years also warrants very close monitoring.

S. Maria Hampshire

Date 1991

Country Peru

Why It's Key This was the worst cholera epidemic in the Western hemisphere for seventy years, despite non-stop efforts to control the spread of the disease with public health interventions.

opposite Colored transmission electron micrograph of Vibrio cholerae bacteria, the cause of cholera

Key Event **The Sanger Institute Opens** World leader in genetic research

The Sanger Institute is intrinsically linked with the Human Genome Project. Established in 1992, its founding director was Sir John Sulston, a leading figure in the world of genetics thanks to his work on a tiny nematode worm, *Caenorhabditis elegans*, which is made up of only about a thousand cells.

The Sanger Institute was not alone in the genome sequencing project, but rather the UK's part of an international publicly funded consortium, which included the United States, Germany, China, Japan, and France. The final draft of the human genome was finally released in 2003, coinciding with the 50th Anniversary of Watson and Crick's groundbreaking discovery of the double helix structure of DNA. The Sanger Institute had been responsible for around one third of the sequencing, including chromosomes 1, 6, 9, 10, 11, 13, 20, 22, and X. Since the publication of the genome, the Sanger Institute – now officially called the Wellcome Trust Sanger Institute – has continued working on sequencing other organisms' genomes, as well as developing tools that help to analyze the meaning of these genetic maps. The institute now divides its research areas into five distinct areas: Human Genetics, Model Organisms (animals useful in studying human disease), Pathogens, Bioinformatics, and Sequencing. And as well as cutting-edge scientific research, the institute is involved in public engagement with science, hosting seminars and workshops, and providing resources to help further the public's understanding of genetics.

Jim Bell

Date 1992

Country UK

Why It's Key The Sanger Institute was an integral part of the publication of the Human Genome, a vital map that holds the key to future steps forward in human health care, and understanding human nature.

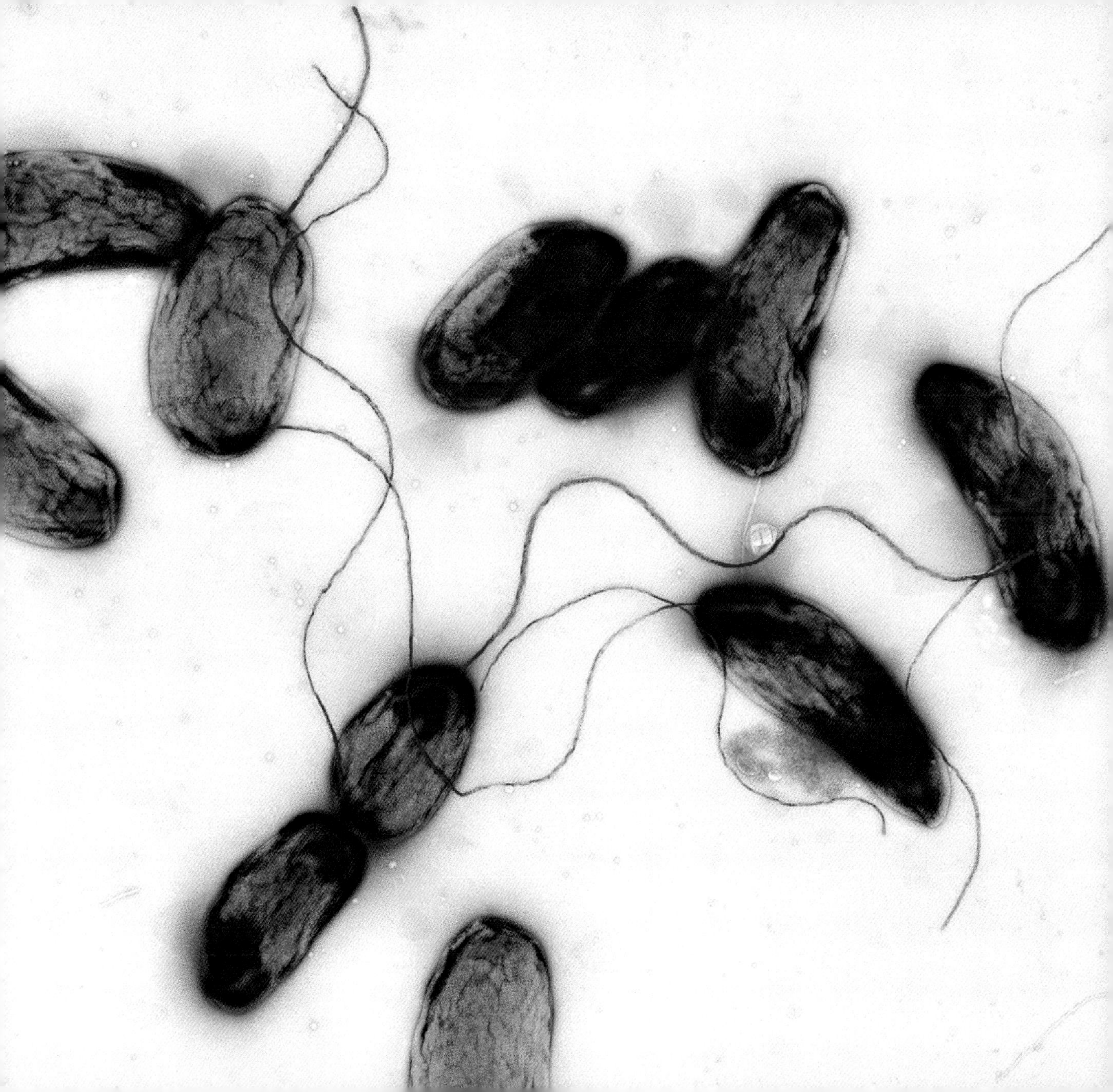

Key Discovery **Finding extrasolar planets**
Our Solar System is not unique

Humans have wondered about planets around other stars for thousands of years. The ancient Greek philosopher Democritus believed there were innumerable worlds which differ in size. He was right, but he wasn't proven so for another 2,400 years, by which time he had almost certainly lost interest.

While innumerable claims were made from the 1850s onwards, the existence of extrasolar planets wasn't confirmed until 1992, when Aleksander Wolszczan and Dale Frail made precise timing measurements of pulses from a rapidly spinning neutron star. They found anomalies that could only be explained if the star was being orbited by two planets, each with roughly three times the mass of the Earth.

In 1995, Michel Mayor and Didier Queloz were the first to prove the existence of an extrasolar planet around a normal star, 51 Pegasi. They worked out that variations in the star's velocity could be explained by the existence of a planet approximately the size of Jupiter orbiting the star every four days.

In recent years, there has been a flurry of discoveries. By 2007, astronomers had discovered 250 planets around normal stars and four around neutron stars, and there is little sign that the rate of discovery is tailing off. Current observations suggest that at least 10 per cent of stars like the Sun have planets.

Eric Schulman

Date 1992

Scientists Aleksander Wolszczan, Dale Frail

Nationality Polish, Canadian

Why It's Key Gives us multiple examples of solar systems, helping us to understand planetary formation and evolution.

Key Event **Rio Earth Summit**
The world's wake up call

The United Nations Conference on Environment and Development, also called the Earth Summit, was held in June 1992 in Rio de Janeiro, Brazil. Its aim was to bring together countries from across the world to recognize the need for future sustainable development, and action to prevent global warming.

The key message the summit aimed to deliver was that our attitudes and behavior toward the environment and development had to change in order to prevent a bleak future. Governments across the world needed to rethink economic development and find ways to stop the destruction of irreplaceable natural resources, and causes of pollution.

The scale of the Rio Earth Summit, and the range of issues tackled there, were much greater than previous UN environment conferences; 172 governments sent representatives, as well as 2,400 non-governmental representatives. One of the key documents adopted at the summit was a framework convention on climate change. It was this that paved the way for the Kyoto Protocol to set mandatory limits on countries' emissions of greenhouse gases. 175 countries have so far ratified the Protocol, however only thirty-six of them, plus the EEC, are actually required to reduce their emissions to the limits set.

The summit brought the issues of sustainability and climate change into the international political limelight, arguably for the first time. Sadly, the poor level of commitment to actually delivering on promises is one indication that a shift in attitude is still needed; climate change is still a very real threat.

Emma Norman

Date 1992

Country Brazil

Why It's Key It brought the environmental impacts of human development – and the need for action – onto the international political agenda, and led to agreements tackling the very real threat of climate change.

Key Experiment **Farming animals for organs**
The first successful "xenotransplant"

Xenotransplantation, the use of animal organs and tissues in human hosts, holds great potential. In 1992, Duke Medical Center, USA, first used pig livers in humans as a bridge (a holding measure) in very ill patients awaiting a transplant with a human liver. One patient whose temporary porcine liver was kept outside of her body – and plumbed into her own blood vessels – survived long enough to receive a human transplant. Another patient had the liver transplanted into her body, but died thirty-two hours later. In these early trials, the use of immune-suppressing drugs required to stop organ rejection often resulted in death of the recipient, due to overwhelming infection. In light of this, the concept of genetically engineered animals, whose organs would be "humanized" and therefore not rejected, was a source of great excitement. An English biotechnology company created the world's first transgenic pig, Astrid, with human genes expressed on her organs. It was a rival company, however, that was the first to be given approval for use of transgenic organs in humans – in 1995 the Food and Drug Administration allowed the use of genetically modified pig livers in humans.

But the great xenotransplantation revolution has not taken place; results are mixed, with many patients faring badly, contracting animal-related infections. The authorities' confidence in this technology has been shaken, and strict regulations brought in. However, the lack of donor organs and growing need for transplants worldwide means xenotransplantation still merits research.

Katie Giles

Date 1992

Scientists Jeffrey Platt

Country USA

Why It's Key A breakthrough which could save thousands of lives – using animals as a source of organs for those in need of transplants.

Key Discovery **The saola**
Vietnam's secret mammal

The saola or Vu Quang ox was first discovered when researchers found three sets of long horns in hunter's houses in 1992. It is one of the world's rarest mammals and lives in the forested Annamite Mountain range on the border of Vietnam and Laos.

Saolas live in small groups of up to five individuals in the mountain forests during the rainy season, and move down to the lowlands in winter. Their food consists of mainly plant matter, including leaves and stems.

They stand around 85 centimeters in height, and weigh about 90 kilograms. Their brown fur has a thick black stripe along the back, and they have white patches on their feet. Their face also has a mixture of black stripes and white blotches, and they have long slightly backward-curved horns which can grow to around 50 centimeters. They are thought to have the largest scent glands of any animal.

Although a few have been killed by hunters, they are very shy creatures and rarely come close to villages or human activity. No one is quite sure whether they are cattle, antelope, or goats, and only eleven have been recorded alive. All attempts at keeping them in captivity have failed so far.

The Annamite Range is home to many other newly discovered, or rediscovered, species, including a striped rabbit, a huge ox, and the Indo-Chinese tiger.

David Hall

Date 1992

Country Vietnam/Laos

Why It's Key This was the first discovery of a large mammal in half a century, and encouraged countless other biological expeditions in the same region.

Key Event **Microbolometer unveiled** Top secret invention makes imaging a snap

When Samuel Langely invented the bolometer in 1880, he could hardly have imagined it would give rise to today's high-tech thermal imaging. In fact, it wasn't until a hundred years after its creation that the technology finally made it into the mainstream.

A bolometer measures a metal's change in electrical resistance due to radiation (light). Essentially a thermometer for radiation, a bolometer is a simple way to measures all types of light – body heat, sunlight, and fire, to name a few – using simple electronics.

In the mid-1980s under the stamp of "Top Secret," U.S. government-contracted scientists at Honeywell created a new and improved microbolometer. This tiny device fitted perfectly into teeny-tiny spaces like miniature cameras and sensors. Crucially, despite often being subject to harsh radiation, it required no large or expensive cooling units. Thermal imaging was born, quickly finding its way into secret missions being carried out by the military, and of course, into not-so-secret missions being exposed by the military.

Microbolometer technology came to the masses in 1992 when it was declassified by the U.S. government. Almost overnight, devices were licensed, patented, and at work just about everywhere. Numerous applications popped up in science and industry including microscopes, mobile phones, cameras, automobiles, airplanes, personal security systems, and home electronics. In little over a decade what was once a top secret technology had infiltrated our every day lives.

Raychelle Burks

Date 1992

Country USA

Why It's Key Snapping quality photos under tough conditions became as easy as saying "cheese!"

opposite Thermogram of a runner, showing variations in temperature – red being the warmest and blue the coldest

Key Invention **Flat screen TV sets demonstrated for the first time**

The plasma display panel (PDP) was invented at the University of Illinois in 1964. Although it was no doubt terribly exciting at the time, the device probably wouldn't stir up much of a reaction in your modern consumer, as the phrase "plasma panel" is now synonymous with grand, sleek cinematic style TV screens which you can hang up on your wall, protruding no further than a framed photograph.

Of course, the word plasma actually refers to the ionized gas found within the individual PDP, but when presented with the glorious and magical full-color image of a plasma screen television, it is easy to forget that science has anything to do with it.

The people we can thank for crafting the elegant and slim display units we associate with the word plasma today are the Japanese technology giants Fujitsu. It was a long hard slog for them, taking over three decades to research, develop, and finally manufacture the technology which allowed for a full color super-size screen, but they finally managed it in 1993.

First put on show to the general public in 1995, these plasma display television units – much like the ones we know and love today – were a far cry from the small, monochromatic panels of the Illinois labs.

Chris Lochery

Date 1993

Scientists Fujitsu

Nationality Japanese

Why It's Key Opened up a whole new world of television design and televisual technology – which has since led to the creation of larger, sleeker television sets and High-Definition TV.

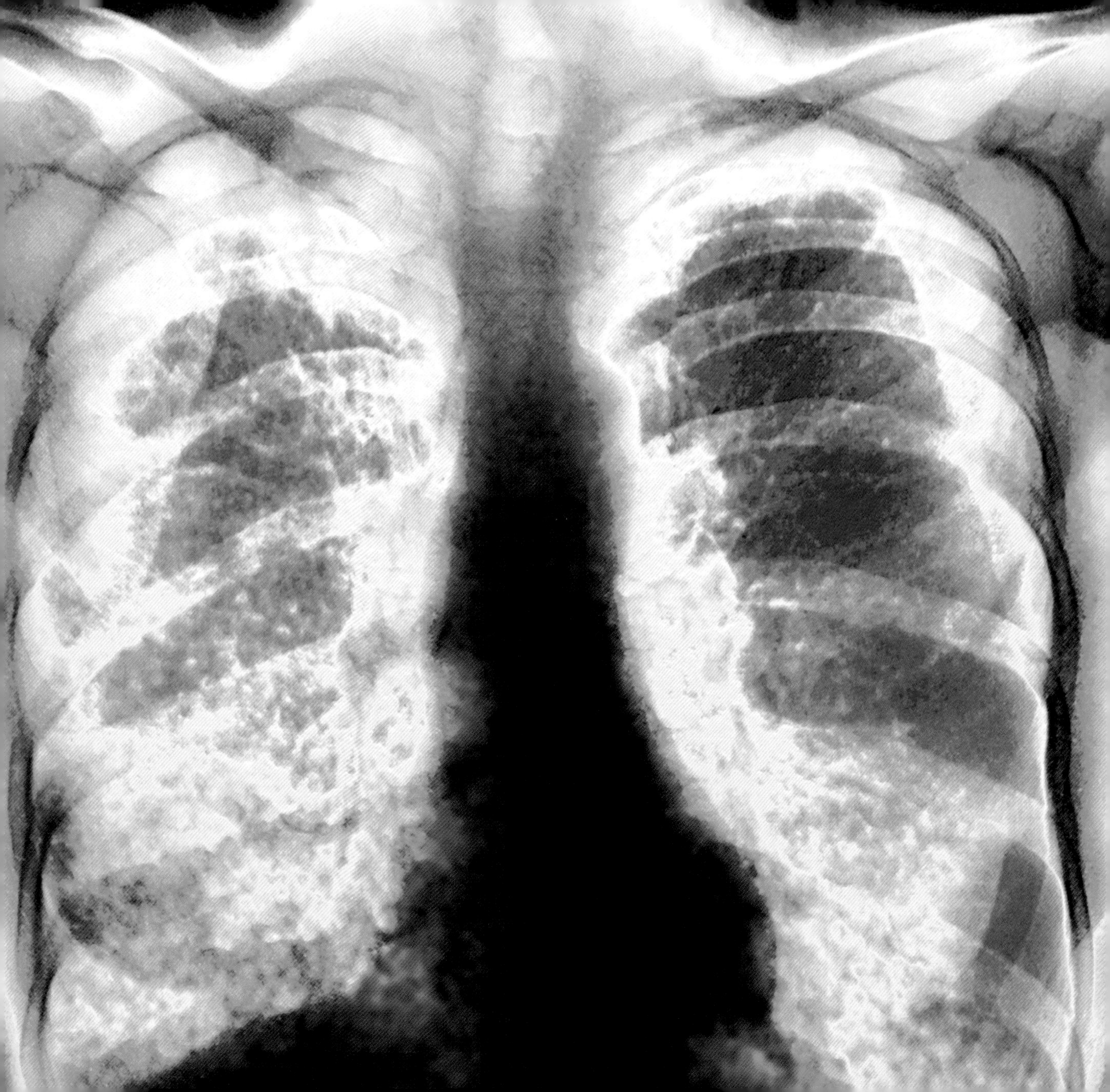

Key Event **TB is back**
Return of the king's evil

Also known as "white plague" and "king's evil," tuberculosis has been around a very long time. Egyptian mummies dating from 2400 BCE have been found to have "Pott's disease" or TB of the spine; Hippocrates wrote about "phthisis" in 460 BCE and even warned physicians about treating patients in late stages, as their inevitable deaths would hurt the doctor's reputation; and in the seventeenth and eighteenth centuries, TB was more commonly known as "consumption," causing as many as one quarter of all the deaths in Europe. Although many treatments were attempted, management generally consisted of isolating infected persons in a sanitarium until they healed, or died.

In 1944, the new antibiotic streptomycin was administered to a patient, who went on to make a full and rapid recovery. Within months, the bacteria had developed resistance to streptomycin, but other drugs were developed, and remained effective if given in combination. The incidence of TB was in steady decline in developed countries. By the mid 1980s, however, the incidence of TB had begun to rise again. Numerous factors contributed to the increase such as immigration, the rise in HIV, and the emergence of multi-drug-resistant tuberculosis. Now certain strains of TB are resistant to every common anti-TB medication.

In 1993, the World Health Organization declared a worldwide TB emergency. Currently, an estimated 2 billon people, one third of humanity, are infected. There are 8 million new infections each year, and 1.6 million deaths from TB: one every fifteen seconds.

Stuart M. Smith

Date 1993

Country Global

Why It's Key TB now infects one in three people worldwide.

opposite Colored X-ray of a patient with TB. The purple and green areas are the lungs and the white patches are the areas affected by the disease.

Key Invention
Intel gives chips a powerful boost

Most computer owners are familiar with the term Pentium, as these processors are the most widely used general computing chips in the world. Pentium chips, introduced by the Intel Corporation in 1993, created quite a stir when they were first released. At the time, they were the fastest and most powerful microprocessors ever made.

The Pentium group was led by an engineer named Vinod Dahm who, along with several other engineers, began work on the project in 1990. As a replacement for Intel's 486 processor, the 64-bit Pentium chip would be five times more powerful than the original i486, and 300 times more powerful than Intel's famous 8088 processor, developed in the late 1970s and used in the original IBM PCs. The Pentium chip was first used in various products from about two dozen companies, mostly for servers, and not personal computers. Initially, prices were high and availability was low, but over the years this relationship quickly reversed. Computer makers were keen to offer the Pentium chip to their customers as a simple replacement to the 486 processor, but analysts warned that this quick fix would only work if the computer itself was capable of handling the larger amount of processing that the new Pentium chip was able to do. Most customers opted to wait for new personal computers which already contained the chip.

Over the years, the Pentium chips have undergone numerous permutations and "Pentium" is now the trademarked name of many different types of microprocessors sold by the company.

Rebecca Hernandez

Date 1993

Scientist Vinod Dahm

Nationality Indian

Why It's Key Pentium chips laid the groundwork for powerful processing, and are still the most popular computer chips in the world.

Key Publication
Preparing for the quantum computer

Cryptography is the art of coding and making information secure. In 1994, Peter Shor of AT&T's Bell Laboratories in New Jersey discovered an algorithm that could break through nearly all current security; the only problem was that he needed a quantum computer to do it.

Most current coding systems, like those that protect online purchases, rely on the long time it takes normal computers to process large numbers. Shor's algorithm could do this in seconds but it is designed to work on a quantum computer of which only prototypes currently exist. Unlike normal computers, quantum computers use the quantum states of atoms to represent 0s, 1s, or a combination of 0 and 1. When quantum computers become a reality, they will perform many calculations in parallel and offer much more computing capacity than even today's best supercomputers. The nature of quantum mechanics means that all quantum algorithms, including Shor's, are probabilistic – they give the answer with a high probability and if the process is repeated then the error reduces.

Nearly all current encryption methods would no longer be secure if a quantum computer was used to try and crack them. Most scientists believe that quantum computers will become practical in the next few decades so new quantum encryption methods will need to be prepared for when computing moves into this new quantum age.

Leila Sattary

Date 1994

Scientist Peter Shor

Nationality American

Why It's Key Deemed that current encryption methods will become useless when the quantum computer eventually becomes a reality.

Key Discovery **Fermat's Last Theorem**
An ancient problem solved

The ancient Greek text *Arithmetica*, written by Diophantus in the third century, posed many mathematical problems. Much of Diophantus' work led to useful equations such as Pythagoras' theorem for triangles: $a^2 + b^2 = c^2$. In the 1600s, Pierre de Fermat, a lawyer and mathematician, scribbled in the side of his copy of the *Arithemetica* that it was impossible to find a similar formula for higher powers than 2, and that he had found a "wonderful proof" of this, which the margin was too narrow to contain. Fermat's proof was never published and the problem became known as "Fermat's Last Theorem" and one of the most famous problems in mathematics.

It was already known that Pythagoras' formula had integer solutions, for example, $3^2 + 4^2 = 5^2$, but for higher powers than 2 (e.g. $a^3 + b^3 = c^3$) it became increasingly difficult to prove. Over the ages many mathematicians including Fermat himself, Euler, Dirichlet, and Legendre, proved that a formula did not exist for particular powers, but no proof was found to describe higher powers in general. Between 1908 and 1911, due to a series of cash prizes offered by mathematical institutions, over a thousand incorrect proofs for Fermat's last theorem were produced.

Fermat's "wonderful proof" was almost certainly wrong, as it was twentieth-century methods that finally formed a definitive proof. In 1994, over 350 years after Fermat's original note, Andrew Wiles used elliptical curves, complex mathematics, and the power of supercomputers to finally find a proof that would have been totally alien to Fermat himself.

Leila Slattary

Date 1994

Scientist Andrew Wiles

Nationality British

Why It's Key Andrew Wiles solved a very famous mathematical problem, building on centuries of work and creating new mathematics to do so.

Key Invention
USB

Remember the mind-boggling maze of wires that were computer connections? You needed a parallel port for your printers and zip drives, serial ports for modems, digital cameras, and everything else that needed a faster connection came with their own device cards, for which there was little space in the computer case. Installing them was complicated to say the least.

Then came the Universal Serial Bus. Backed by Intel, Microsoft, US Robotics and Philips, and brought to prominence by Apple, it has become indispensable to modern IT. With its simple connectors that feed upstream to computers and downstream to the devices, it was now possible to link up your mouse, keyboard, printers, scanners, flash drives, music players, etc much more easily. When linked via USB to different devices, the host computer assigns each a separate address (enumeration) then determines the data process required by each individual device connected. The USB divides the available bandwidth accordingly into frames so every device gets the amount it needs.

Because USB cables can provide their own power, the potential for usage is greater than simply storage or connection. You can get a storage device that is also a camera, or toward the more wacky end of the spectrum, a USB cable-powered desktop aquarium to liven up the working day blues. With USB 3.0 products planned for 2009, potential for performance of computer peripherals is even greater. The simplest ideas are always the best.

Fiona Kellagher

Date 1994-5

Country USA

Why It's Key Revolutionized and simplified data connection between machines.

Key Event **Tomato paves the way for GM foods**
Other vegetables will have to ketch-up

In the early 1990s, humankind's greatest dream became a reality; the quest for a slightly tastier tomato had reached its conclusion with the Flavr Savr.

Ordinarily tomatoes are picked unripe from the vine; this is to avoid bruising caused by rough handling of the tender fruits, but also to prolong their shelf life, as once picked, tomatoes spoil quickly. In their unripened state they are harder and therefore less prone to damage, but they are also inedible. Unripe tomatoes are blasted with ethylene gas, which acts as a ripening agent. The ripened tomatoes are then placed, unbruised, on the shelves for us to eat, but at what cost? The ethylene ripening process, it is argued, causes a significant loss of flavor in the tomato.

Scientists at Calgene, a Californian biotech company, developed a tomato with a delayed ripening gene, which meant it spoiled less when picked, and so it could stay on the vine longer, creating more of that wonderful natural flavor. They sought and received FDA (Food and Drug Administration) approval for the manufacture of the Flavr Savr tomato, and began widespread growth of the new "super-tomato."

Unfortunately a combination of poor business planning – much money had to be invested into transport of the delicate fruit – and growing consumer concerns over the safety of GM crops, as well as worries that a genetic marker in Flavr Savr would lead to antibiotic resistant bacteria, meant Flavr Savr was not a commercial success.

Barney Grenfell

Date 1994

Country USA

Why It's Key Flavr Savr was the first genetically modified plant to get pre-market FDA approval.

Key Discovery
The Wollemi pine, thought extinct

Despite a burgeoning human population, there are still a few pockets of habitat around the globe that have escaped the prying eyes of humankind. Dense rainforests are a prime example, and occasionally deliver a bolt from the blue.

In 1994, an expert wildlife officer, David Noble, was abseiling in a gorge in Australia's Wollemi National Park (just 160 kilometers from Sydney), when he stumbled across a group of trees that took him completely by surprise. What grabbed Noble's attention was that he hadn't got a clue what species they were.

The trees were over forty meters in height and exhibited a strange, bubbly, chocolate-coloured bark. Samples were taken and analyzed at the Royal Botanic Gardens in Sydney, and, to everyone's amazement, it turned out the trees belonged to a group thought to have died out two million years ago. The Wollemi pine has since been given its own genus and, in honor of the location where they were found and the man who discovered them, the tree's scientific name is *Wollemia nobilis.*

It is currently one of the world's oldest and rarest trees. Just three stands of the trees have been located in the wild, but they appear to be multi-stemmed with a connecting root system, so it is hard to be sure of exact numbers – it's possible that each stand is made up of a single individual. One thing is for sure though, the exact location of the remaining trees is a closely guarded secret.

Andrew Impey

Date 1994

Scientist David Noble

Nationality Australian

Why It's Key A major botanical find, proving yet again that there is still much to learn about the planet that we inhabit.

opposite The Wollemi pine

Key Publication ***Extinction Rates***
Compares current with historical extinction rates

Humans have had a unique impact on the ecology of this planet, but just how much damage are we doing? Are we planetary scourges of biodiversity – guilty of specicide most foul – or are shifting ecosystems simply wreaking havoc on plant and animal species alike, as they have done for millennia?

The debate over the nature and scale of human impact on species numbers is fiercely political. Environmentalists insist we are in the midst of a biodiversity crisis, the likes of which have never before been seen on Earth. Industrialists have a tendency to downplay human impact. Science, though, is apolitical... at least in theory. Knowing whether or not there is a biodiversity crisis on this planet is dependent on an understanding of the variables in question; on realistic estimations of populations; and understanding extinction rates. The February 1995 book *Extinction Rates* is a collection of the main conclusions of a symposium organized by the book's authors, ecologists John Lawton and Robert May.

The book evaluates different methods of population estimation, and compares current with historical extinction rates. It shows that a comparison of current levels of population decline may not match the great extinctions of the past, but that comparing current with past rates is not straightforward. The authors argue that, while some groups of species are doing better than we thought, others are in catastrophic decline. The causes and consequences of biodiversity loss are not well enough understood for quick solutions to be found.

Jamie F. Lawson

Publication *Extinction Rates*

Date 1995

Authors John Lawton, Robert May

Nationality British, Australian

Why It's Key Injected a much needed dose of science into an extremely emotive issue, providing a dispassionate assessment of human impact on the ecology of Earth.

Key Event
Hemophilus influenza genome sequenced

Bogged down by the technical difficulties of genome sequencing, scientists had cracked the genetic codes of only a handful of viruses by the early 1990s. All that changed when J. Craig Venter developed "shotgun sequencing." He took aim at living organisms – viruses, by contrast, are considered to somewhere between dead and alive – and scored a direct hit by sequencing the bacterium *Haemophilus influenzae*.

About ten times larger than the average virus, *H. flu* is comprised of over 1.8 million base pairs, and shotgun sequencing identified each one. Venter's shotgun method first dices DNA, creating a collection of randomly cut pellets of genetic information, then looks for spots where pellets were once connected, so that sequenced pellets can be matched to form an intact and fully mapped genome.

With *H. flu*'s genome pieced together and shotgun sequencing proving to be a success, genome sequencing went from a Herculean task to a straightforward job that could be done in a year or less. A few months after *H. flu*'s genome was published, the genome of *Mycoplasma genitalium*, another bacterium, was solved. The next few years brought rapid advancements, with the genomes of yeast species, worms, and the fruit fly decoded. And with *H. flu* under his belt, Venter and his shotgun method were soon to become instrumental in sequencing the most sought after genome of all: ours.

Raychelle Burks

Date 1995

Country USA

Why It's Key Showed genome sequencing didn't have to be a long and labored process, and paved the way for sequencing more complex organisms' genomes.

opposite Colored scanning electron micrograph showing the flu bacteria lying inside a human nose

Key Experiment **Bose-Einstein condensation**
Creating the coldest ever recorded place to study atoms

Eric Cornell and Carl Wieman were the first to achieve Bose-Einstein condensation (BEC), seventy years after its existence was predicted by the two physicists in whose honor it was named. In 1995, Cornelll and Wieman cooled a cloud of atoms by slowing them down from hundreds of kilometers per hour to just one meter per hour, causing their temperature to plummet to the lowest ever recorded and create a new phase of matter – BEC.

At high temperatures, atoms in a gas zoom around and behave like footballs – discrete objects that change direction and velocity when they collide. When atoms are cooled they behave more like waves than discrete particles. As the temperature becomes lower, the wavelengths of the atoms become longer and eventually these "matter waves" overlap, forming BEC.

Cornell and Wieman first "laser cooled" the atoms by firing photons at them; as atoms slow down they get colder and every time an atom tries to escape, the lasers push it back into place. They then used evaporative cooling, which is much like cooling coffee, to allow the hottest particles to escape and reach temperatures just billionths of a degree above absolute zero (-273 degrees Celsius). This is colder even than the deepest parts of space; in fact, the coldest place in the Universe, unless extraterrestrials are doing similar experiments.

Theoreticians have since used BEC results to model black holes, superfluids, and superconductivity.

Leila Sattary

Date 1995

Scientists Eric Cornell, Carl Wieman

Nationality American

Why It's Key The discovery of BEC has proved the existence of a new phase of matter, and forms the basis for many quantum mechanics experiments.

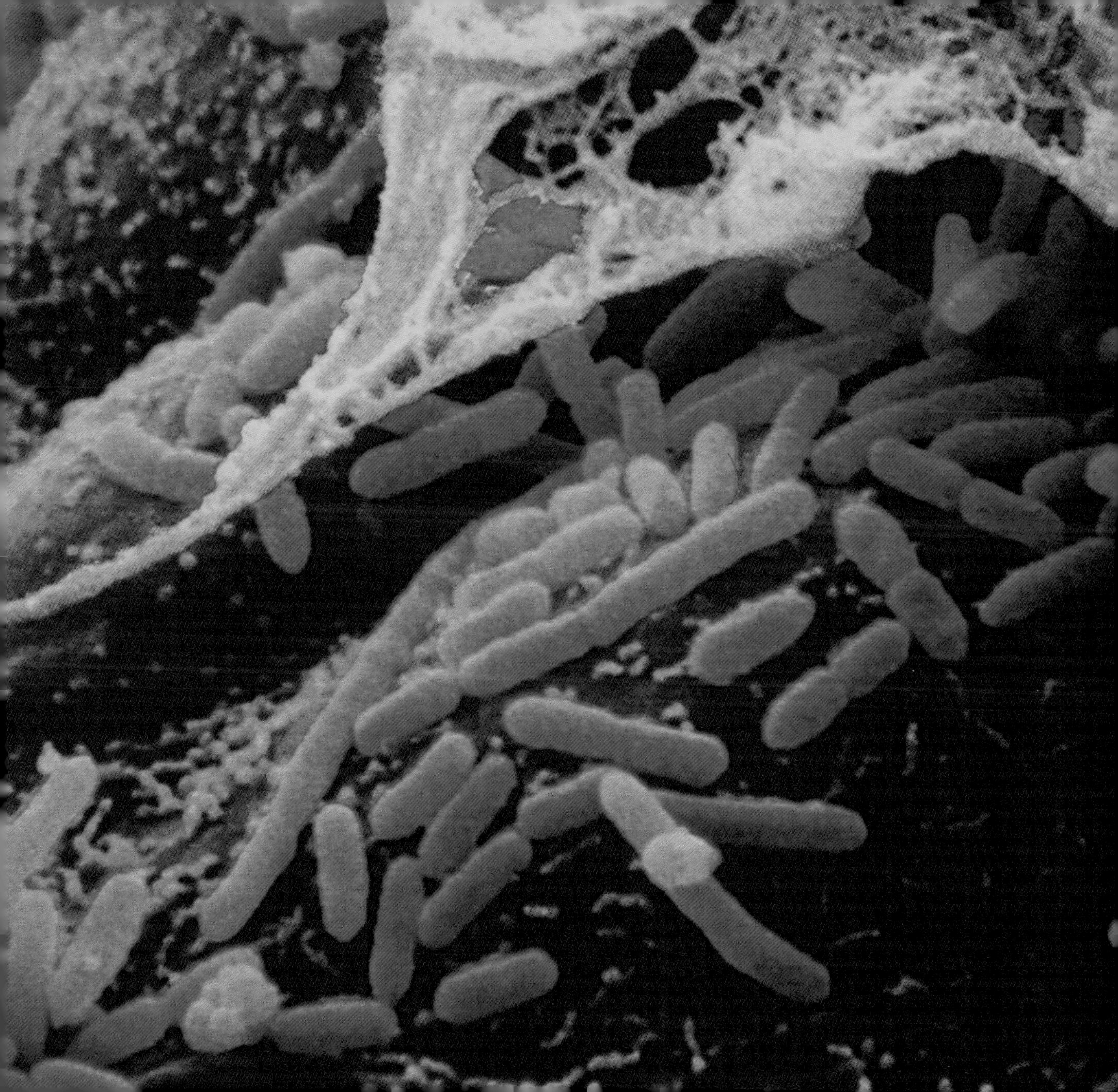

Key Invention

DVDs spin into action

In 1995, some of the world's largest electronics empires were poised and ready to fight it out, eager to see whose innovation would be the next big thing in home video. The united forces of Sony and Philips were vying with another major electronics consortium headed by Toshiba and Time Warner. Each group had developed their own type of data disk storage technology.

Wisely, in December of 1995, the competitors sat down and settled on a technology they would support as an industry, and the DVD – then the Digital Video Disc; now the Digital Versatile Disk – was born.

Compact Discs – or CDs – had already been available for a number of years, and while being suitable for storing many kinds of data, including music, the CD just didn't have the capability to contain a motion picture with sound and picture quality equivalent to the current standard, VHS tape. The new DVD format was to be the same physical size as a CD, 120 x 1.2 milimeters, but capable of storing more than five times as much information – up to 4.7 gigabytes.

Using a form of compression called MPEG-2, the DVD could store movies up to 133 minutes in length. However, there were other features which ensured that everyone wanted DVDs: unlike VHS tapes, recording quality did not fade over time; image and sound were of higher quality; and physically, the discs were far smaller than video tapes. And of course, DVDs never had to be rewound. Ever.

Andrey Kobilnyk

Date 1995

Scientists Sony, Philips, Matsushita, Toshiba, and others

Nationality International

Why It's Key A revolution in affordable high quality data storage, primarily used for the consumer home video market.

opposite DVDs

Key Publication

M-theory sparks second superstring revolution

Edward Witten's M-theory helped to further understanding of superstring theory, essentially advancing our understanding of the Universe at large. But what is superstring theory, and what does the M in M-theory stand for? Even theoretical physicists wouldn't be able to give you a straight answer.

String theory is based around the idea that all fundamental particles consist of tiny vibrating strings of energy, where different vibrations are different fundamental particles. The original theory had five versions and required twenty-six dimensions to make sense, far more than the number we are able to perceive.

Witten proposed in his M-theory, that these five versions were variations of the same theory and that only eleven dimensions were required for it to work - as I'm sure you can work out, that's a loss of fifteen dimensions. As to what the M in M-theory stands for, even Witten hasn't answered this question. It has been speculated that it stands for magic, mystery, or matrix; all good candidates for such an intriguing theory.

He also introduced the theory that the pulsating strings could expand to form membranes, or m-branes. This theory has been used to explain the Big Bang, by imagining that branes the size of a Universe could collide in a super energy connection, causing a huge explosion. The M-theory has been called unscientific as it cannot be experimentally tested by scientists. It is nonetheless a fascinating idea, predicting parallel Universes that are different to ours by the merest vibration of a string.

Nathan Dennison

Date 1995

Author Edward Witten

Nationality American

Why It's Key Another weird and wonderful explanation of what makes up our Universe and how we came to be.

Key Event
GPS systems are go

It all started like this. Wanting to know where the first artificial satellite to orbit the Earth – the Soviet-launched Sputnik – had got to on its travels, two physicists from Baltimore figured out a way of locating its position by measuring the changes in the radio frequency that it emitted.

Years later, another physicist, also from Baltimore, realized that if the exact position of the satellite was known, then another object on the ground could figure out its own location simply by turning this same equation on its head, and using the satellite as a fixed point.

This Global Positioning System (GPS) was used almost exclusively for military purposes until 1983, when a commercial Korean airliner was shot down after accidentally straying into Soviet territory. President Ronald Reagan promptly declared the GPS system should be made freely available for civilian use to avert any similar catastrophes. Whilst access was finally granted, up until the year 2000, civilians were relegated to using a restricted, less accurate system.

A fully operational system comprising a constellation of twenty-four satellites capable of tracking people anywhere in the world was announced by the NAVSTAR program on July 17, 1995. But the story doesn't end there: New plans to modernize the system, making it even more accurate, have been announced with the aim of being complete by 2013.

Chris Lochery

Date 1995

Country USA

Why It's Key Satellite navigation is now used in all walks of life, in such far-ranging situations as military maneuvers, paramedic emergencies, and simple Sunday afternoon drives.

Key Discovery
Top quark finally found

Having eluded physicists for almost two decades, the final quark of the set of six was observed in 1995, at Fermilab near Chicago. Gerardus 't Hooft and Martinus Veltman shared the Nobel Prize for Physics in 1999 for their discovery.

Six types of quarks, and their six antiquark counterparts, are the smallest building blocks in physics. Quarks are never found naturally on their own; protons and neutrons, for example, are each made up of three quarks. The Standard Model, which describes the particles and forces that determine the fundamental nature of matter and energy, predicted the mass of the top quark to be much greater than the other five quarks. In fact, the top quark was found to have a mass as large as a gold atom, which makes it unstable and difficult to observe.

To create top quarks, the groups at Fermilab collided beams of protons and antiprotons at speeds almost as fast as the speed of light. At a high enough energy level, collisions can produce top quark and top antiquark pairs. The Tevatron particle accelerator at Fermilab is the only accelerator with enough energy to produce top quarks, although the opening of the Large Hadron Collider at CERN in 2008 will provide further experiments.

The discovery of the top quark confirmed the Standard Model's predictions and took us one step closer to the discovery of the mysterious Higgs boson or "God particle" – the only one of the Standard Model's particles yet to be observed.

Leila Sattary

Date 1995

Scientists Gerardus 't Hooft, Martinus Veltman

Nationality Dutch

Why It's Key The final quark to be discovered, and an experiment that bolstered scientists' confidence in their pursuit of the Higgs boson.

Key Event **More dots**
The arrival of high definition television

High Definition Television delivers much better viewing quality than traditional television, building up its pictures from a million or more tiny dots. Before HDTV, televisions only used a few hundred thousand dots for the same size of screen.

The first experiments in HDTV were carried out during the 1970s. Despite its better picture quality, however, it took a while to catch on. This is because – all else being equal – to get twice the picture quality, a single HDTV channel would push two normal channels off the air, and broadcasters and the public alike wanted more channels rather than a better picture.

In the 1990s, researchers like MPEG – the Moving Picture Experts Group – worked on taking traditional analogue TV signals and turning them into a digital stream of zeroes and ones. They also compressed them, even studying how our eyes worked to figure out which bits of the picture could be safely left out without us noticing. They compressed the signal so well that several high-definition digital channels could fit into the space of a single analogue station.

That made plenty of room for the introduction of HDTV, and groups like America's "Grand Alliance" started setting the standards that everyone from the broadcasters to the television manufacturers would have to follow to make sure the new systems worked. The first standards arrived in the mid-1990s, with the U.S. ATSC system – the replacement for the old analogue NTSC standard – being approved in 1996.

Matt Gibson

Date 1996

Country USA

Why It's Key The picture quality of analogue television hadn't improved in decades. But work on digital, high-definition television during the 1990s finally let us make massive improvements to picture quality.

Key Event **(R)Evolutionary thoughts**
Did God create the Big Bang?

In October, 1950, Pope Pius XII stated that, "The Teaching Authority of the Church does not forbid that ... research and discussions, on the part of men experienced in both fields, take place with regard to the doctrine of evolution." Almost fifty years later, in October 1996, Pope John Paul II would – albeit cautiously – echo the sentiments of his predecessor. Did this signal a convergence of science and religion; an end to years of dogmatic belief in one or the other?

Many scientists believe that the Universe began with an explosion, 10–15 billion years ago. Background radiation, the fact that all elements are built of protons and neutrons, and the observed expansion of the Universe are offered as proof of this "Big Bang" theory. All life has evolved from original, simpler forms.
In his statement, Pope John Paul spoke of the various "theories" of evolution: materialist; reductionist; as well as purely spiritualist. The final judgment, he stated, "is within the competence of philosophy and, beyond that, of theology." He adhered to Pius XII's original point that if the physical origins of the human body do indeed come from living matter that previously existed, then the spiritual soul is created directly by God.

However, theories of evolution that view the "spirit" as developing or emerging from living matter, or as a simple secondary result of natural forces upon that matter are, he stated, "incompatible with the truth about man. They are therefore unable to serve as the basis for the dignity of the human person."

Mike Davis

Date 1996

Country Italy

Why It's Key This address confirmed the Catholic Church's willingness to find an acceptable balance between religious belief and modern scientific research; but not at the expense of fundamental teachings in the former.

Key Discovery **Did birds descend from dinosaurs?** Getting in a flap

When, in 1996, the fossil of a small dinosaur called *Sinosauropteryx*, which appeared to have been covered in delicate feather-like structures, was found, a wave of related discoveries soon followed. Together, these radically changed the way in which the idea of birds being descended from dinosaurs was viewed.

It was the man known as Darwin's bulldog, Thomas Huxley, who first suggested in 1868 that birds evolved from dinosaurs, but the idea was slow to catch on. *Sinosauropteryx*, however, would change all that. First reported by Chinese paleontologists, Ji Qiang and Ji Shuan, the fossil of this new dinosaur, which clearly couldn't fly, showed in great detail the impressions left by its covering of primitive feathers. Two years later a similar feathered dinosaur, later named *Caudipteryx*, was found, and following this, *Sinornithosaurus*; each discovery strengthening Huxley's original argument.

In 1999, scientists thought they'd hit the jackpot when another fossil was discovered in China. *Archaeoraptor* showed both dinosaur-like and bird-like features, providing strong evidence of the link between dinosaurs and birds. The debate had been resolved – until *Archaeoraptor* was found to be fake. The fake fossil was a combination of the upper body of a bird and the tail of a previously unidentified dinosaur.

While the debate still isn't totally over, the evidence that resulted from the finding of *Sinosauropteryx* has undoubtedly helped bolster the dino-bird theory. Today, descent from dinosaurs is the most widely accepted explanation for the evolution of birds, 140 years after it was first proposed.

Eleanor Hullis

Date 1996

Scientists Ji Qiang, Ji Shuan

Nationality Chinese

Why It's Key The discovery of *Sinosauropteryx* was the first tangible evidence suggesting an evolutionary link between dinosaurs and birds.

opposite Detail of the fossilized skull of *Sinosauropteryx prima*, 120 million years old.

Key Event **Hello Dolly** Birth of the world's first mammalian clone

On July 5, 1996, Ian Wilmut and his team at the Roslin Institute, Edinburgh, made history as they witnessed the birth of the world's first mammalian clone: Dolly the sheep.

The previous year, the team at Roslin had succeeded in cloning two identical sheep from one fertilized egg, but the challenge was to clone an organism from a fully developed adult cell. In order to do so, they had to re-program an adult cell with the DNA of Dolly's mother.

They began by isolating adult cells from the udders of a donor sheep. After extracting the nucleus – where the cell's genetic information is housed – from an udder cell, they injected it into a fertilized sheep egg using a microscopic needle. The fertilized egg was then transplanted into the womb of a surrogate sheep and, a few months later, a miracle was born. Wilmut decided to name the sheep Dolly because the cell had originated from a mammary gland (udder), and he could not think of a finer pair of "glands" than those of country and western singer, Dolly Parton. Sadly, Dolly – the sheep, that is – passed away on Valentine's day in 2003, when she was only six years old. Her untimely death has raised questions about the safety of cloning. There was some evidence that, because Dolly's mother was already six years old when her DNA was extracted, poor Dolly was "old before her time."

Hannah Isom

Date 1996

Country UK

Why It's Key The first mammal to be cloned from an adult animal cell.

Key Discovery **Life on Mars?** Meteorites may have the answer

Meteorites that originate from Mars are known as Nakhlites, after one of the first found in Nakhla, Egypt in 1911. They are believed to come from crystallized lava flows 1.3 billion years old. While these pieces of planetary rock may not contain alien creatures, they do hint that life may have once existed elsewhere in our Solar System.

The first hints that meteors might act as a window to life outside our planet came in 1969. A carbon-rich Martian meteorite was found that contained elements and chemicals similar to those found on Earth. Since carbon is essential to nearly all processes of life, its presence could indicate the existence of life on the red planet. Amino acids, the building blocks of all proteins were also detected, as well as fullerenes, or "bucky-balls," a type of carbon molecule.

In 1996, the meteorite ALH84001, which had been found twelve years earlier in Antarctica, shot to fame when a NASA scientist claimed that it contained traces of fossilized life. Examining the rock under an electron microscope, David McKay had spotted what looked suspiciously like chains of bacterial cells. Exactly what the structures are still remains open to debate, with skeptics claiming they could be the result of ancient contamination following the meteor's arrival on Earth, or even just peculiar rock formations.

The find reignited interest about life on Mars; even President Clinton stated at the time, "If this discovery is confirmed, it will surely be one of the most stunning insights into our universe that science has ever uncovered."

Nathan Dennison

Date 1996

Scientist David McKay

Nationality American

Why It's Key Provided some evidence to suggest that life elsewhere in the Universe exists, or at least existed, even within our own solar system.

opposite Colored scanning electron micrograph of a possible microfossil of a primitive organism from a meteorite that came from Mars, found in 1984

Key Discovery **Lake Vostok**

At over 240 kilometers long and 56 kilometers wide, Lake Vostok covers roughly the same area as Lake Ontario. By volume it is the sixth biggest lake in the world. It is also the world's most secluded body of water; no boats have ever cruised its coastline; no divers charted its depths. That's because Lake Vostok lies under 4 kilometers of ice, beneath the coldest place on Earth.

In the 1950s, Soviet scientists founded the Vostok Station at the "Southern Pole of Inaccessibility" on Antarctica; by the mid-1980s it had endured the coldest temperatures ever recorded on Earth, -89.2 degrees celsius. But its biggest surprise still lay in store. Satellite images from the European ERS-1 satellite revealed the existence, and astonishing size, of Lake Vostok in 1996.

Cut off from the rest of the world for 15 million years, the lake contains freshwater supersaturated with oxygen, which may harbour organisms unlike any we've ever encountered before. It might even give us hints about the possibility of life on Jupiter's ice-bound moon, Europa.

But herein lies the problem. How do you study a completely isolated and uncontaminated environment without contaminating it? Russian plans to drill into the water have been put on hold while scientists struggle to find a way to lower a robotic probe into the waters without taking any oil or foreign bacteria with it. Lake Vostok will one day yield her secrets; the hope is that she'll do so without herself being harmed.

Mark Steer

Date 1996

Scientists Andrei Kapitsa, Gordon Robin, Martin Siegert

Nationality Russian, Australian, British

Why It's Key Lake Vostok may contain clues as to the inhabitants of some of the outer reaches of the Solar System. But we've yet to discover its icy secrets.

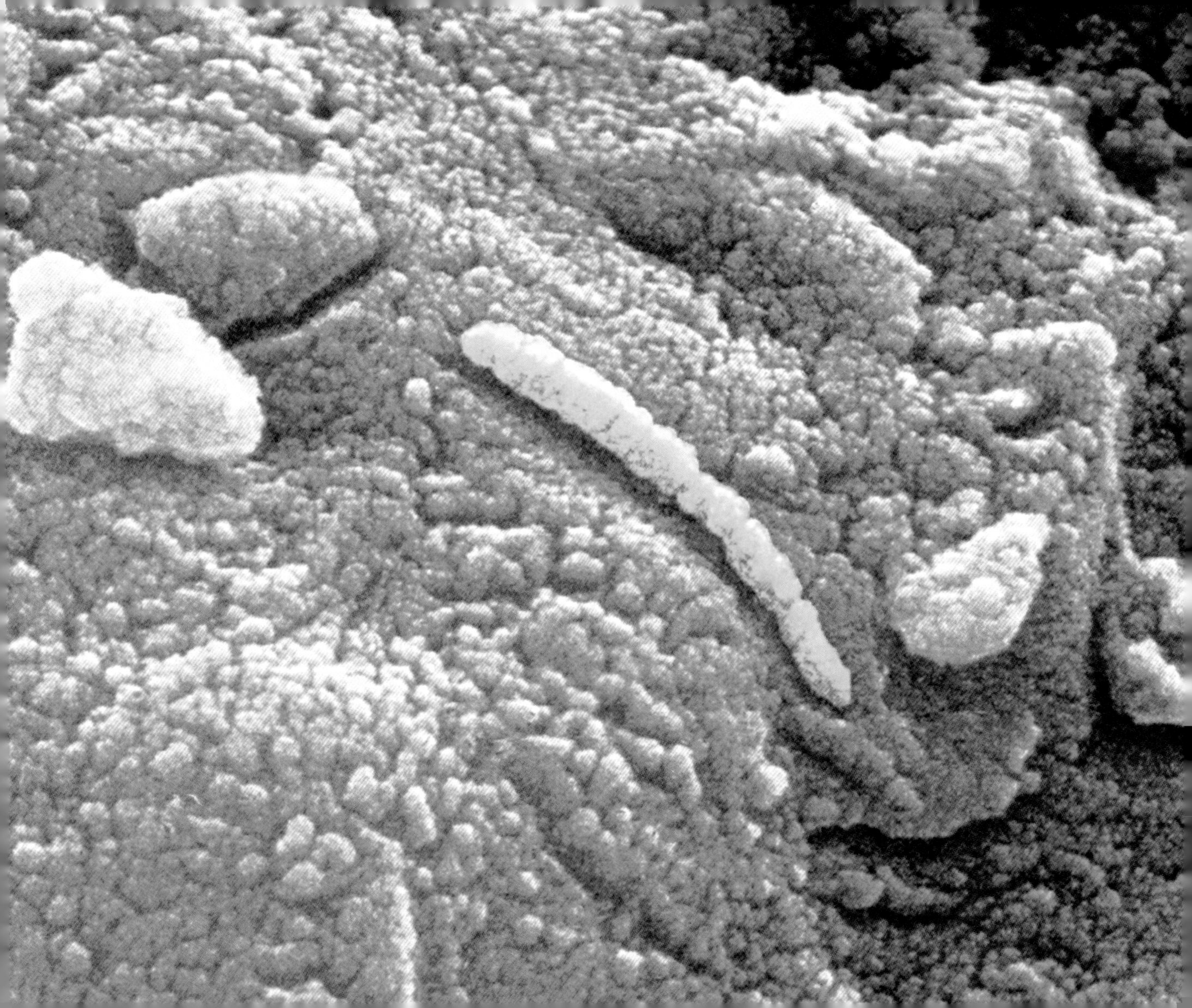

Key Discovery **Those birds can make tools**
Stone the crows!

In 1996, Australian ecologist Gavin Hunt brought the previously innocuous New Caledonian crow to the attention of scientists and the media worldwide. Members of this unremarkable-looking species, found only on the small islands of New Caledonia, were shown not only to use tools, but also to manufacture at least two very sophisticated types.

A small number of animals are known to use tools, but tool manufacture is far rarer and was once thought a defining characteristic of humans. When seen in animals, it tends to be no more complex than breaking a twig from a branch and removing the leaves – even in our closest cousins, chimpanzees. New Caledonian crows, however, show the most advanced tool manufacture of any non-human animal; for example spending many minutes carefully crafting hooked ends to their twig tools to aid prey capture.

In the twelve years that have passed since Hunt's publication, much research has been carried out to determine more about these remarkable birds. Controlled experiments on captive crows have uncovered even more striking abilities – individuals select appropriate tools for particular tasks, one subject spontaneously made her own hooked tools by bending wire, and some individuals will use tools in sequence to obtain a goal.

The New Caledonian crow has found fame indeed: establishing itself a prime position in the upper ranks of "clever creatures" and, in doing so, has challenged us to rethink our ideas about what sets humans apart from other animals.

Jo Wimpenny

Date 1996

Scientist Gavin Hunt

Country New Caledonia

Why It's Key The presence of sophisticated tool use and manufacture in a species of bird forces us to question our primate-centered views on animal cognition.

opposite A New Caledonian crow

Key Event **Deforestation**
The UN's stark warning

Deforestation is defined as the conversion of natural areas of forest to areas of non-forest, such as farmland, pasture, or urban space. In many areas, the removal of large areas of forest has led to significant reductions in biodiversity and has had knock-on effects in terms of climate change, desertification, and water resources.

One of the largest causes of deforestation over the last twenty years is slash-and-burn activity, where large areas of forest are cleared for short-term crops and settlements, then left as the inhabitants move on. The resulting land is usually nutrient poor after crops have been grown, or is quickly eroded by rains, meaning that the original forest is unable to grow back.

Another pressure on forests is the growing demand for wood for construction and paper products. This is particularly the case in the Far East. Brazil has lost a lot of forest to slash-and-burn and to road building schemes, while Madagascar has lost about 94 per cent to mining and other human activities.

The United Nations Commission on Sustainable Development was started in 1992. Five years later, they estimated that 4.6 million hectares of rainforest and 6.1 million hectares of moist deciduous forest were being converted to other land uses every year. This was highest in Asia, followed by Latin America and the Caribbean. While the deforestation rate remains high in Brazil and Indonesia, forest plantation schemes have been trying to redress the balance in recent years.

David Hall

Date 1997

Location Asia, Latin America, the Carribean

Why It's Key The first accurate figures published on the amount of natural forest being lost around the world, which highlighted the need for international action.

Key Experiment
Quantum teleportation

Quantum teleportation is not the "Beam me up Scotty" type of teleportation that you might imagine. It's actually taking something, describing it to the tiniest level – the level of photons and quarks – and using this description to copy it somewhere else.

Descriptions at the quantum level are difficult; Heisenberg's Uncertainty Principle says it's impossible – you cannot know where something is and how fast it's moving at the same time, so you'll never be able to copy it. Getting round this fact requires a concept called entanglement; two particles, joined but not touching, where action on one has effects on the other; an area of physics described by Einstein as "spooky actions at a distance."

Starting here, a research team from the University of Innsbruck entangled a cloud of caesium atoms with light. Next, they mixed the light with what they wanted to teleport, which was a weak light pulse. A weak light pulse has very little in it, so it's easy to describe. The weak pulse and the entangled light were then measured. Across the lab this had an effect on the caesium cloud. Using the description, the quantum states of the pulse could now be recreated by the team, in the cloud, completing the teleportation.

While this isn't beaming up a solid object, it's extremely useful in communication, especially in quantum cryptography. By sending information like this it's impossible to eavesdrop without the entangled receiver, making the message completely secure.

Douglas Kitson

Date 1997

Scientist Dirk Bouwmeester

Nationality Dutch

Why It's Key Completely secure communication is what you need for confidential information, like credit card details, or top secret secrets

Key Invention **Gas-powered fuel cells**
A growing trend

Someone finding this book on the shelves of a second-hand shop many years hence might well exclaim, "Wow, you know there used to be vehicles powered by small explosions of something called "gas," that pumped out toxic fumes as they moved. But it wasn't gas at all, it was this stuff made from fossils!"

Right now, however, the journey from fossil fuel dependency to fuel cell freedom is ongoing. Many significant developments chart the route; Sir William Grove's experiments in the nineteenth century; Francis Bacon's discoveries starting in the 1930s, and NASA's use of related technology in the 1960s, to name just a few. A transitional landmark was reached in 1997 when, on October 21, a team from the U.S. Department of Energy, Los Alamos National Laboratory, the Plug Power Company, and management consultants, Arthur D. Little, successfully demonstrated the first-ever gasoline-powered, fuel cell electric engine for the automobile.

Via an onboard fuel processor, and in keeping with the general principles of other such cells, the unit generated electricity through an electrochemical process converting the chemical energy of hydrogen (from fuel) and oxygen (from the air) into electricity. This hybrid system produced negligible amounts of sulfur and nitrogen oxides, and less than half the amount of "greenhouse" gases that a conventional internal combustion engine produces. The system could use ethanol, from corn and other crops, and thus the potential for vehicles to use less fossil fuel – and, ultimately, none at all – was born.

Mike Davis

Date 1997

Scientists US Department of Energy, Los Alamos National Laboratory, Plug Power Company, Arthur D. Little

Nationality American

Why It's Key The invention paved the way for subsequent fuel-cell engines and vehicles; we can finally being to wean ourselves from the dependency on fossil fuels.

Key Event
The “ear mouse” is genetically engineered

Many of us will know the iconic image of a laboratory mouse proudly sporting a human ear on its back. And many of us will have heard the vociferous complaints of the anti-genetic engineering lobby, decrying it as a travesty of science. But what lay behind all this?

The mouse was created by brothers Dr Charles and Joseph Vacanti, as part of their work to develop transplantable tissues for humans. Why the focus on ears in particular? Well, around one in a thousand people are born with small, misshapen, or even completely missing ears and ears are easily damaged in fights and accidents. It was as good a place to start as any.

The ear was in fact produced by grafting cow cartilage cells within a biodegradable polymer scaffold onto the mouse. No genetic engineering was ever involved, and the resulting ear would not have been suitable for transplanting into humans, as the cow cells would be rejected by the immune system.

The “ear mouse,” as it became known, got tangled up with genetic engineering when an anti-genetics pressure group placed an advert in the *New York Times*. The full-page spread showed the mouse, complete with the caption “This is an actual photo of a genetically engineered mouse with a human ear on its back.” Although it was completely untrue – the mouse was not genetically modified, and the cartilage cells were actually from a cow – it seeded the idea that this kind of manipulation was possible.

Kat Arney

Date 1997

Scientists Charles and Joseph Vacanti

Nationality American

Why It’s Key Spin over substance caused confusion and misinformation, damaging science and frightening the public.

Key Event **Along came Polly**
More advanced than Dolly

A year after the groundbreaking birth of Dolly, the same team of scientists, Ian Wilmut and Keith Campbell of the Roslin Institute in Edinburgh, introduced the world to Polly. Like Dolly, Polly was a cloned sheep, but, with the addition of a human gene, she could produce a useful human protein in her milk. Although animals that could express foreign genes (transgenic) had been around for years, Polly was the first to be produced by cloning.

Wilmut and Campbell used a sheep fetus cell and engineered it to express a human gene that makes a protein called coagulation factor IX. This protein is very important in the clotting of blood; people with hemophilia type B are lacking in it. The fetus containing the modified gene was then implanted into a female sheep and, a few months later, Polly was born.

The birth of Polly led to the development of a technique known as “pharming” where factor IX and other important proteins can be produced in the milk of livestock. This allowed patients access to a cheaper and safer form of clotting factor than was previously available.

Though Dolly will always be remembered as the biggest breakthrough, Polly was arguably the more important sheep. The applications of transgenic animal technology are vast, including the growth of new organs for transplants, that will not be rejected.

Hannah Isom

Date 1997

Country UK

Why It’s Key First transgenic animal to be produced as a result of cloning.

Key Event
Garry Kasparov defeated by a machine

It's understandable that Garry Kasparov, possibly the greatest chess player in history, didn't sportingly shake his opponent's hand after being defeated on May 11, 1997. Not because, as the reigning world champion, he had just lost his first match in twelve years, but rather beacuse his opponent had no hand to shake. It was a machine called Deep Blue.

IBM began developing Deep Blue as a chess-playing computer in 1989, in order to research parallel processing – a kind of computing that carries out many instructions simultaneously. Deep Blue first challenged Kasparov in 1996, making history as the first machine to win a chess game against a reigning world champion under normal time controls. Kasparov beat the machine fair and square over a six-game match.

By the 1997 rematch, IBM had doubled Deep Blue's "brute force" processing power and enhanced its "chess knowledge." It could now analyze 200 million positions per second to find optimal moves on the board. Conversely, Kasparov could explore just three positions per second. Deep Blue could now also adapt to a switch in Kasparov's game strategy, something that Kasparov used to his advantage in 1996. After some surprising blunders, a psyched-out and mentally drained Kasparov lost the rematch, 2.5 games to 3.5.

Afterwards, while Kasparov suggested IBM had cheated, many hailed it as a defining moment in humanity. A machine had outsmarted a human at the peak of their mental ability. But despite the implication, Deep Blue wasn't intelligent or able to think. It was simply programmed to calculate the best move.

James Urquhart

Date 1997

Country USA

Why It's Key The first time a reigning world chess champion had been defeated by a computer under regular competition conditions, scrutinized by millions of people for its fascinating man-versus-machine theme.

opposite Kasparov plays Deep Blue in one of many chess games

Key Event
The Kyoto Protocol

In the annals of international diplomacy, the Kyoto Protocol – named after the Japanese city in which it was agreed in December 1997 – ranks as one of the greatest international treaties ever negotiated. It's certainly top of the environmental charts, at least until a successor takes over in 2012.

The protocol fleshed out the United Nations Framework Convention on Climate Change (UNFCCC), an international treaty which demanded that countries reduce their emission of greenhouse gases responsible for global warming. Under Kyoto, developed countries committed to cutting their emissions of six key greenhouse gases by 5.2 per cent from 1990 levels, as measured by an average across 2008–2012. Each country had its own target, and each could meet it in various ways, such as by planting trees or setting up markets to trade rights to emit greenhouse gases. Developing countries merely had to monitor and report their emissions. By 2007, 174 countries (and the European Union) had ratified the protocol – but significantly, not the United States; the world's largest greenhouse gas emitter pulled out in 2001. China, another leading emitter, doesn't have to cut emissions because it is a developing country in Kyoto terms.

Whether or not the remaining countries meet their protocol targets – and it seems many won't – Kyoto's influence on preventing global warming is still unclear. Beset by political gameplaying and accusations of being both too modest and holding back economic development, Kyoto is still a remarkable feat which has set a framework for future climate negotiations.

Richard Van Noorden

Date 1997

Country Japan

Why It's Key The landmark climate negotiation forced countries to start cutting greenhouse gas emissions.

Key Discovery *Homo sapiens idaltu*
The age of modern man

Our knowledge of human evolution is limited by the number of fossil discoveries. In the 1960s it was believed that modern human beings, *Homo sapiens*, had been around for 130,000 years, after some fossils were found in Kibish, Ethiopia.

However, in 1997, it appeared that even older *Homo sapiens* bones had been found. The skulls of two adults and a child, dated to be around 160,000 years old were pulled out of sediments near the village of Herto, Ethiopia. The skulls were not an exact match to those of people living today; they were slightly larger with more pronounced brow ridges. The skulls were assigned to a new sub-species, *Homo sapiens idaltu*.

There are two theories describing the origin of modern humans – the "out of Africa" theory which says that all modern humans came from Africa, and the multi-regionalism theory which says that modern humans developed from several species in different countries. Genetic data has always shown that modern humans originated in Africa sometime in the last 200,000 years; the skull discovery supports this data by filling in the gap in the fossil record. Experts argue that if modern features in humans already existed in Africa 160,000 years ago, we could not have descended from species in different countries.

In 2005, the Kibish bones were re-analyzed and, after looking into the volcanic ash layers surrounding the bones, researchers discovered that the bones were in fact 195,000 years old. These are now thought to be the oldest *Homo sapiens* bones known.

Riaz Bhunnoo

Date 1997

Scientist Tim White

Country Ethiopia

Why It's Key The bones discovered are the oldest known fossils of modern human beings. They provide strong evidence that *Homo sapiens* existed around 200,000 years ago and originated in Africa.

Key Experiment **The attraction of a vacuum**
Measuring the Casimir effect

Two flat plates face each other; they're in a vacuum, neither one has a charge, and there's nothing up anybody's sleeve, but they still attract each other. This was the odd prediction of Dutch physicist, Hendrik Casimir.

Casimir's idea was based on the quantum mechanical assumption that the energy in a vacuum cannot always be exactly zero – it must have a range of different possible values. These possible energies manifest themselves as packets of energy – called photons – with different wavelengths. In empty space all wavelengths are possible, but between our two plates, some are suppressed because they don't fit into the space between the surfaces. Leaving out the suppressed frequencies means that there are fewer photons in the gap between the plates than there are on the outside. In effect this means that the "photon pressure" is less in the gap than it is elsewhere; the plates are drawn together. Although Casimir proposed this theory in 1948, equipment wasn't good enough to confirm it until 1997, when Steve Lamoreaux at the University of Washington measured the force between a lens and a glass plate 0.001 millimeters apart; his results supporting Casimir's theory.

What was once just a scientific curiosity is becoming increasingly interesting to physicists concerned with the fundamentals of the Universe. As measurements of the effect are still being refined, it's thought they could give us a window into studying the higher dimensions predicted by some Grand Theories of Everything.

Kate Oliver

Date 1997

Scientist Hendrik Casimir

Nationality Dutch

Why It's Key Researchers hope precise measurements of the Casimir effect will shed light on the number of dimensions the Universe has, and how the forces within it act over tiny distances.

Key Event **International Space Station takes shape in orbit**

Easily visible to the naked eye, the International Space Station is the largest man-made structure ever assembled in space. The research facility weighs over 230 tons, and orbits around 290 kilometers above our heads at over 27,360 kilometers per hour.

The first section was lofted to orbit on November 20, 1998. Zarya, meaning "dawn," launched from the Baikonur Cosmodrome in Kazakhstan, carrying the hopes of sixteen nations. Just two weeks later, the USA launched the second piece, a connecting node known as Unity. A Russian life support module followed, allowing habitation from November 2000. Since then, the ISS has been permanently manned and has grown considerably in size. It is both the most sophisticated platform for space sciences ever built, and a testing ground for international space collaboration. As well as sections from the two space superpowers, it also includes laboratories from Europe and Japan, and a construction boom from Canada.

Among other things, the laboratories use the microgravity conditions to look for new drugs and investigate effects of weightlessness on the human body. The 2003 Columbia shuttle disaster brought a hiatus to construction, however, and the whole venture has been called into question for cost overruns and a limited ambition. Nevertheless, the program is now back on track and due for completion in 2010. Russia recently announced that they may continue to develop their section of the ISS after this date.

Matt Brown

Date 1998

Nationality International

Why It's Key Permanently manned since November 2000, the ISS is humanity's only toehold in the final frontier, and a symbol of international cooperation in troubled times.

Key Discovery **Dark forces at work**
Everything is getting faster and faster

Curled up, as you are, with a good book in a comfortable chair, where the Universe came from might be the least of your concerns – but what about where it's going? In 1998, two groups of cosmologists discovered that the expansion of the Universe – which began 13.7 billion years ago in the Big Bang – is getting faster.

Hold on to your chair, because this is a bit wild. Imagine ants on the surface of a balloon. As the balloon expands, the distance between the ants becomes greater. For us, the ants represent galaxies – many stars gravitationally bound to each other. During the billions of years it takes for light to reach us, space expands between the galaxies. We know this because light from distant galaxies appears to be shifted toward the red part of the electromagnetic spectrum; more distant galaxies show a higher red shift, as space has had a chance to expand more during the light's journey to us.

Light from distant exploding stars called supernovae has been shown to be dimmer than is predicted by the red shifts of their parent galaxies. The conclusion that scientists came to in 1998 was that approximately nine billion years ago, space began to expand faster. Why? Cosmologists claim some sort of unknown "dark energy," capable of stretching space-time, is at work.

It's currently thought that dark energy might account for 70 per cent of the entire Universe, which means that there's an awful lot of stuff out there that we know absolutely nothing about.

Andrey Kobilnyk

Date 1998

Scientist Adam G. Riess, Saul Perlmutter, Brian Schmidt

Nationality American, Australian

Why It's Key Dark energy raises new questions for cosmologists and may potentially represent a yet undiscovered force in nature.

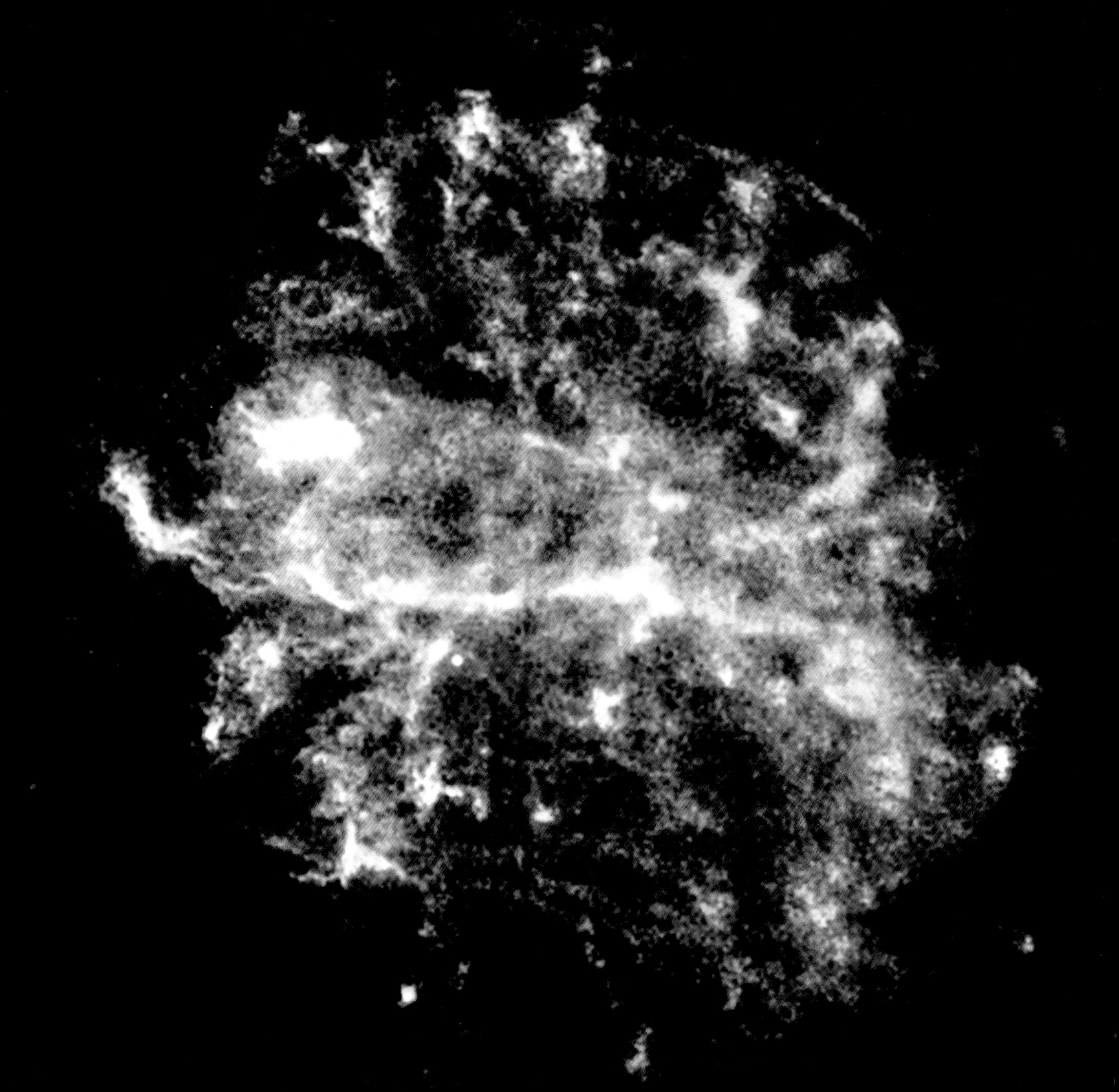

Key Experiment **How much mass must a neutrino need?**

Neutrinos are extremely tiny, neutrally-charged subatomic particles. The only forces they respond to are gravity and the 'weak force', which is the force that causes the nuclei of radioactive elements to split.

There are three different types, or flavours, of neutrino, each related to a different type of particle - electron neutrinos, muon neutrinos and tau neutrinos. In 1957, Italian-born physicist Bruno Pontecorvo suggested the possibility that neutrinos could change from one of the three types into another - he called these neutrino oscillations. These oscillations, however, left physicists with a problem; for them to occur, neutrinos have to have a mass - an idea that didn't fit with existing theories.

Studying neutrinos is really tricky. Since they don't really interact with other types of matter, neutrinos are exceedingly hard to catch. It wasn't until 1998 that an international team of scientists, using a neutrino observatory called Super-Kamiokande (basically a 50,000 ton tank of highly purified water residing 1 km underground in a Japanese mine), finally managed to confirm that neutrino oscillations occurred, and therefore that neutrinos must have a mass.

We still don't know how heavy neutrinos actually are, but simply the fact that they do have mass had profound implications for our knowledge about the Universe. Not only does it hint that all matter in the Universe has mass, but maybe, just maybe, neutrinos could make up the mysterious dark matter.

Douglas Kitson

Date 1998

Scientists Yoshiyuki Fukuda and colleagues

Country Japan

Why It's Key Neutrinos are an important component of particle physics, explaining how the sun and other nuclear reactions work. Understanding their properties is vital to understanding the Universe..

opposite Colour X-ray telescope image of supernova remnant G29.0+1.8. The gas shell is expanding after being blown from an exploding star.

Key Event **Construction begins on the Large Hadron Collider** Smashing!

What would you do if you want to explore the forces of nature, shed some light on "dark matter," and understand why particles have mass? If your answer is "build a 27-kilometer-long, doughnut-shaped particle-smashing machine underneath Geneva" and you can put three billion euros on your VISA card, then you might be on the right path: The Large Hadron Collider (LHC), due to be in operation by May 2008, is exactly such a beast.

To answer the questions we've set for ourselves, theoretical physics tells us that we have to better understand the origins of our Universe, created 14.7 billion years ago during the Big Bang. From earlier machines similar to the LHC, we have been able to recreate on a small scale the high-energy conditions which were present in the first fractions of a second after the Big Bang. The LHC will work by producing a stream of protons, accelerated to something approaching the speed of light and guided by magnets through the 27 kilometer ring, which is a near vacuum. Traveling in the opposite direction will be another stream of protons. When these two streams smack into each other, we will be able observe the way matter and energy behaved within billionths of a second after the Big Bang.

Physicists believe that the LHC will reveal some of the properties of our Universe which were "set" as it cooled to produce the conditions necessary to make possible the Universe we see today.

Andrey Kobilnyk

Date 1998

Country Switzerland

Why It's Key The Large Hadron Collider will be able to recreate, on a small scale, the energies and events which were found in the first billionths of a second after the Big Bang.

Key Discovery **Normal living mice produced from freeze dried mouse sperm**

Artificial insemination is not a new concept. An Arab prince reportedly artificially inseminated his prized mare with sperm obtained surreptitiously from an enemy's equally prized stallion in the fourteenth century. Lazzaro Spallanzani is credited with the first truly documented artificial insemination in 1784 – that of a dog. However, more than two hundred years later, two Japanese biochemists working in Hawaii took it to a new level.

In 1998, Teruhiko Wakayama and Ryuzo Yanagimachi freeze-dried mouse sperm and left it at room temperature for up to three months. Some of the desiccated sperm even made a plane trip back and forth between Hawaii and Tokyo. The pair then reconstituted the sperm with distilled water and injected it into viable eggs. The result was normal healthy offspring who were able to breed and produce subsequent normal healthy offspring. Why is this revolutionary? The sperm used to produce these mice was completely dead, and had been for months.

The sperm of humans and many other large animals can be frozen in liquid nitrogen and remain viable for quite some time, but mouse sperm does not remain viable when subjected to this process. This leads to huge groups of genetically engineered mice being kept as breeding stock. Wakayama and Yanagimachi's technique allows researchers to keep only the freeze-dried sperm of such mice for future use. The sperm could even be placed in the mail and produce genetically engineered mice half way across the globe.

B. James McCallum

Date 1998

Scientists Teruhiko Wakayama, Ryuzo Yanagimachi

Nationality Japanese

Why It's Key Freeze dried mouse sperm that is officially dead is revived and used to fertilize an egg. Who knows what they'll freeze dry next...

opposite A scanning electron micrograph of a cluster of human embryonic stem cells

Key Experiment **An embryonic controversy** Human stem cells grown in the lab

The human body houses 220 different types of cells. Liver cells, nerve cells, skin cells, sperm cells – they're all perfectly adapted for the functions they perform. But each of the 100 million, million cells that make up you and I start out the same. And this is what makes human stem cells such valuable tools for scientists. By taking a cell that effectively represents a blank canvas and molding it into whatever shape is required, researchers hope to cure any number of otherwise incurable diseases.

When two separate U.S. teams grew human embryonic stem cells in the lab in November 1998, there was immediately talk that their work might lead to treatments for diseases such as Parkinson's and leukemia. But ever since a debate has run about the ethics of using cells from human embryos in research.

The equivalent cells in mice had been cultured back in the early 1980s, but this was the first time scientists had shown human embryonic stem cells could be kept alive in the lab for a sustained periods of time. The two rival teams – from the University of Wisconsin-Madison and Johns Hopkins University – used different methods, taking the cells from slightly different points in development of the embryo. But both were able to observe their stem cells differentiate to form cells that in a human would have served different functions in the body.

In 2005, the US government stopped funding research using anything but the existing human embryonic stem cell lines. The controversy continues.

Kate Oliver

Date 1998

Scientists James Thomson, John Gearhart

Nationality American

Why It's Key Stem cells had already grabbed headlines, but without a method for growing them in the lab, scientists couldn't exploit their valuable research potential.

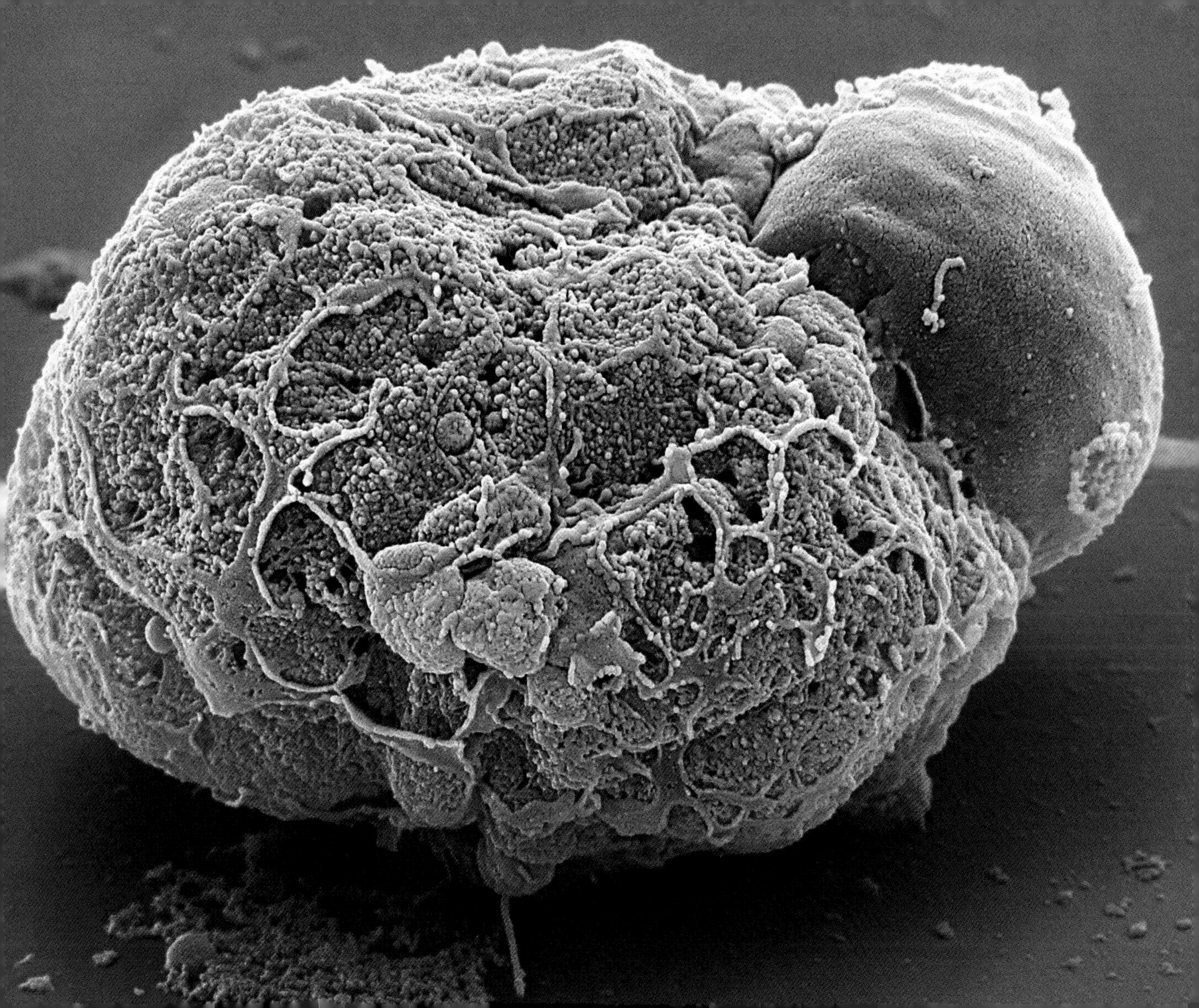

Key Publication **Open Source** Something for nothing

The computer industry has seen massive developments in technology over the past few decades, both in the hardware itself but also, largely through the Internet, in the way that we produce and consume software. This development called for a revision of many of the conventional rules that govern the production of things like software, and enabled individual producers to collaborate with each other across the globe. It is in the spirit of open collaboration and beneficial adaptation of ideas that the Open Source Initiative (OSI) developed, out of the free software movement.

Open source culture crosses many genres; examples can be found in the arts, especially music and visual arts – anything where a relaxed approach to copyright can be applied. Allowing for creativity through collaboration. The Open Source Initiative, founded in 1998, primarily by Bruce Perens and Eric S. Raymond, relates specifically to the development and distribution of software. The premise is that software and its source code should be free and available for all to obtain and use, allowing programmers to make modifications as they see fit. This, in turn, allows for the development of better software that is more relevant to individual companies/individuals.

Copyright laws and protections govern generic software, such as that produced by Microsoft. This, the OSI would argue, hampers the development of software; if you have the monopoly, you have no motivation to develop better software, and you are also pushing up the cost to individual consumers.

Barney Grenfell

Date 1998

Authors Bruce Perens, Eric S. Raymond

Nationality American

Why It's Key Open Source is a completely different approach to software development and distribution, that flies in the face of traditional consumer-driven philosophies.

Key Experiment **RNA: The interfering worm that sparked a revolution**

A small, transparent worm seems an unlikely character in one of the most exciting science stories of the late twentieth century. But without these tiny wrigglers – known to scientists as *Caenorhabditis elegans* – Andrew Fire and Craig Mello would not have earned the 2006 Nobel Prize in Physiology or Medicine.

In 1998, Fire and Mello, along with other colleagues, published a paper in the journal *Nature*, showing that injecting a certain form of RNA (ribonucleic acid) into worm cells could shut down the activity of genes. RNA, closely related to DNA, is the chemical message that is produced when genes are "read." The technique Fire and Mello pioneered has come to be known, appropriately, as RNA interference.

The pair found that adding RNA that corresponded to the DNA of a particular gene could shut down that specific gene, and no others. The discovery that you could turn off genes at will opened up a world of possibilities for biologists. Switching off a gene at a particular point in a worm's development, for example, could reveal much more about its role than if the gene had been faulty all along.

Further experiments carried out by other scientists showed that this technique – and variations of it – could be used to control genes in many different species, from fruit flies to mammals. Researchers are now investigating the potential of RNA interference in future medical treatments for diseases including HIV, Hepatitis C, and cancer.

Kat Arney

Date 1998

Scientists Andrew Fire, Craig Mello

Nationality American

Why It's Key Finding a simple way to turn genes off has hugely advanced our understanding of the roles of different genes, and could be the key to future medical treatments.

Key Discovery **Refreshing rocket fuel found on the Moon**

On March 5, 1998, NASA scientists announced that a source of rocket fuel may have been found on the Moon. Within hours, news channels, internet forums, and pubs were hosting discussions on the economic benefits of returning to our nearest neighbor. This fuel that suddenly had the world talking was something already found in several places on Earth – water.

Water is made up of hydrogen and oxygen, the two main components of rocket fuel. What made this fuel special, however, was that it was already in space; the cost of lifting-off from the Moon to distant planets would be far less than the costs associated with Earth-based launches, due to the significantly weaker gravitational field at the lunar surface.

Although many other materials would need to be shipped to the Moon, having the fuel waiting there would be a huge cost saver when considering the quantities required. Another added benefit is that water is the most refreshing drink known to man, even better than beer, whether we admit it or not. The water that had been spotted was frozen, but defrosting ice is not something rocket scientists find too challenging.

Although subsequent tests have cast doubt over the 1998 findings, hopes still remain high that ice, and therefore valuable commodities, exist on the Moon. Even in small, impractical quantities, any confirmation of water elsewhere in the Solar System would be of great importance to science; where there is water, there is often life.

Christopher Booroff

Date 1998

Scientists NASA, Alan Binder

Nationality American

Why It's Key Finding rocket fuel on the Moon would drastically reduce the cost of interplanetary space flight and turn the Moon itself into a more viable location for a research facility.

Key Event **Fastest wind speeds ever recorded on Earth**

On May 3, 1999, over fifty tornados ripped across the Great Plains of Oklahoma and Kansas in the United States, causing forty-four deaths and US$1.6 billion in damage. But while most people sought refuge, a small team of daring atmospheric scientists were hunting twisters along the open country roads. They made a record-breaking measurement of 512 kilometers per hour – the fastest wind speed ever recorded on Earth.

Led by Joshua Wurman of the University of Oklahoma, the team pursued a developing super-cell thunder storm. In the dark. Armed with two truck-mounted Doppler radars, they were able to remain at a safe distance from danger and found themselves in a prime position to get detailed information on the formation of a tornado spawned from the storm.

Doppler radars work by firing pulses of microwaves that bounce off rain, dust, and other objects in the air, providing an accurate estimate of wind speed. The team did concede, though, that the wind speed could have been 16 kilometers per hour more or less than their 510 kilometers per hour measurement.

Previously, the fastest wind speed documented was 458 kilometers per hour in 1991, measured using a static – and less precise – Doppler radar during a tornado, also in Oklahoma. Since Doppler radars measure winds that are at least 19.8 meters above the ground, the fastest recorded ground-level wind speed remains at 370 kilometers per hour, measured at Mount Washington, New Hampshire in 1934.

James Urquhart

Date 1999

Scientist Joshua Wurman

Nationality American

Why It's Key Being able to make such measurements with Doppler radars is a breakthrough that will help solve the mysteries of tornado formation and decay, enhancing our knowledge of nature and enabling better forecasting.

Key Event **Once around the world**
Piccard and Jones circumnavigate in a balloon

One year after a rooster, a duck, and a sheep flew 3.25 kilometers in eight minutes in a hot air balloon, the Montgolfier brothers sent aloft Pilatre de Rozier and Marquis d'Arlandes for the first ever manned balloon flight. The Frenchmen traveled for twenty-five minutes and covered eight kilometers. Fast forward 206 years from that historic flight, and a Swiss psychiatrist and his British co-pilot flew another balloon 40,806 kilometers further. Bertrand Piccard and Brian Jones left the Swiss Alpine village of Chateau d'Oex on March 1, 1999, and landed near D'khla in the Egyptian desert nearly three weeks later.

During their 19-day, 21-hour and 47-minute flight, the two men circumnavigated the world in the "Breitling Orbiter 3," and shattered all previous distance, altitude, and time-in-the-air records.

The animals, and the Frenchmen, had primitive balloons made of cloth and paper. By contrast, the "Breitling Orbiter 3" was the most advanced balloon of its time. Designed and built by Cameron Balloons of Bristol, England, the "Breitling Orbiter 3" stood fifty-five meters tall when fully inflated. Like most long-distance balloons, it was a Rozier, a hybrid balloon filled with helium and hot air. Pilatre de Rozier, who not only shared the first manned balloon flight, but invented the hybrid balloon, died in 1785 when his hydrogen/hot air hybrid balloon exploded in mid-air during an attempt to cross the English Channel.

Stuart M. Smith

Date 1999

Scientists Brian Jones, Bertrand Piccard

Nationality British, Swiss

Why It's Key One of the last of the great aviation challenges is completed.

opposite Breitling Orbiter 3

Key Publication **Hotspots**
The Earth's most endangered places

It is a truth (almost) universally acknowledged that biodiversity is a Good Thing. It is also almost universally accepted that species are becoming extinct at a much faster rate than normal. Mankind is on the cusp of causing a "mass extinction" event not unlike, in terms of severity, that which wiped out the dinosaurs.

In a world where conservation budgets are insufficient given the number of species threatened with extinction, it is crucial to identify conservation priorities. British ecologist Norman Myers first defined the biodiversity hotspot concept in the late 1980s to try and answer the conservationist's dilemma: which areas are the most important to conserve? A decade of discussion later, in 1999, the theory had come of age.

To count as a biodiversity hotspot, an area must contain over 1,500 species of plant found nowhere else on Earth and have lost more than 70 per cent of its original natural habitat. To date thirty-four hotspots have been identified around the world from the forests of the tropical Andes to the coastal climes of the Mediterranean basin. Covering just 2.3 per cent of the world's land area, more than half of the world's species are found within these areas.

Whilst not the be all and end all of conservation, identifying hotspots enables conservationists to make informed decisions about where best to concentrate their resources.

Mark Steer

Publication *Hotspots. Earth's Biologically Richest and Most Endangered Terrestrial Ecoregions.*

Date 1999

Authors Russell Mittermeier, Norman Myers, and colleagues

Why It's Key Identified the areas of the world most in need of our protection.

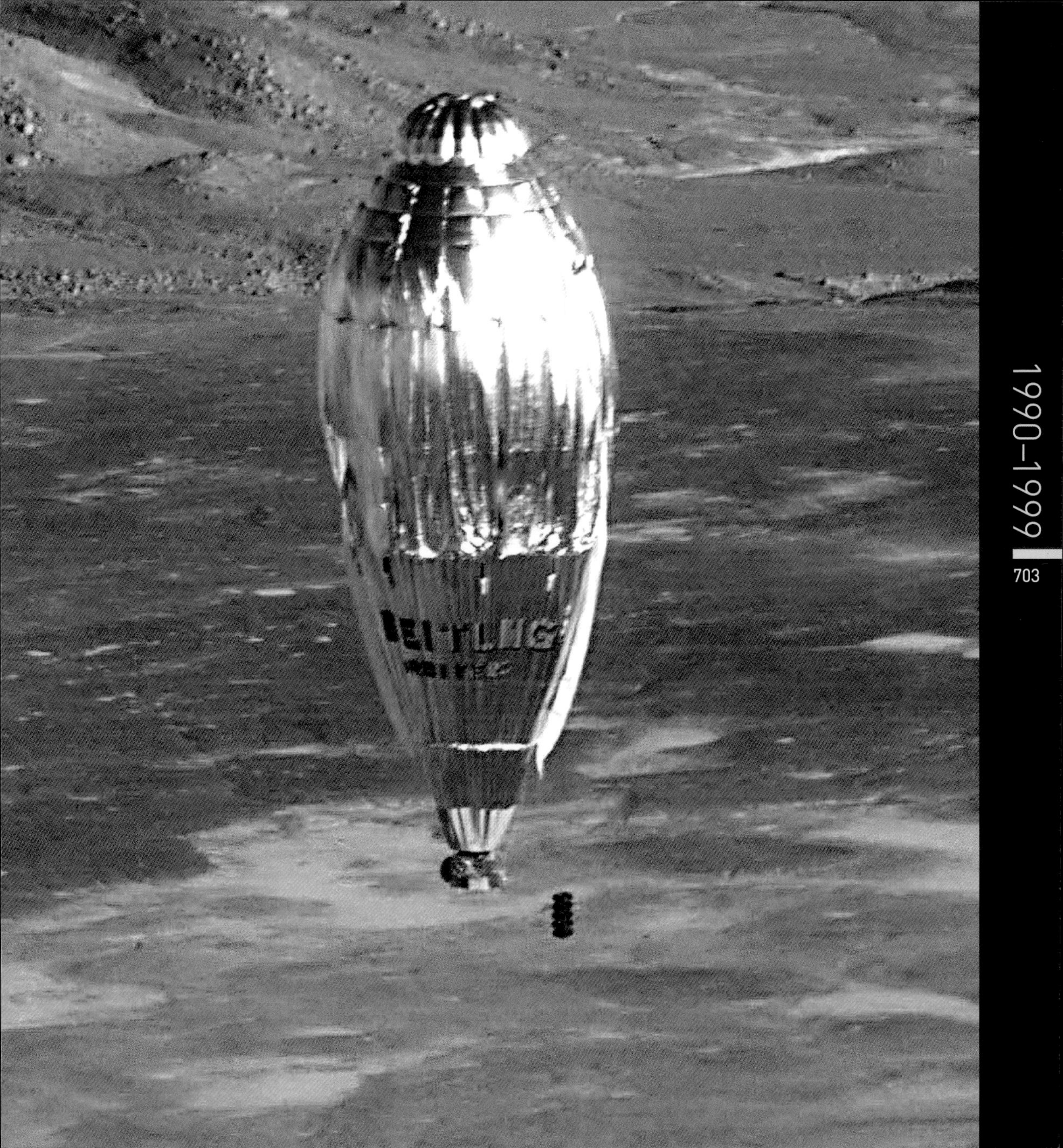

Key Invention **Email on your phone**
Fruit of the future

Cast your mind back to the early 1990's. This is a long time after dinosaurs ruled the Earth, but way before your phone could take pictures, surf the Internet, and play all of your MP3s.

In the early half of the decade email still hadn't taken off on PCs, but phone and pager manufacturers were already working on ways to send and receive email messages without wires. One of these companies was Research In Motion, set up in 1984 in Canada. After spending years working on wireless data transmission, in 1996 they introduced their first wireless hand-held system, called the Interactive Pager.

To find a larger market for their pager, RIM started working with a branding agency called Lexicon. A strategist at Lexicon noticed the gadget's keyboard looked like berry seeds. In 1999, the BlackBerry 850 Wireless Handheld was launched. It brought together a miniature keyboard, wireless networking, and email. But what RIM were providing was a complete package of device, software, the servers, and airtime, making the BlackBerry unique.

Its launch changed the business world forever; work got its teeth a little further into people's home lives, and made it even more important to keep up to date. At the end of 2007, RIM announced they had passed 10 million subscribers. A "killer app" is a computer term for a something so significant to the users that any system seems incomplete without it. Wireless email made business properly mobile, and the BlackBerry was its killer app.

Douglas Kitson

Date 1999

Scientists RIM

Nationality Canadian

Why It's Key Email on the go. One of the first big things that telephones could do that wasn't simply a phone call.

Key Publication **WHO warns of rising disease vectors**
Mosquitoes and ticks to benefit from climate change

It's a sad irony that the animals that harm our health will reap the greatest benefits from the climate change that we ourselves are bringing about. While rising global temperatures may spell the end for endearing species like polar bears and corals, they will be a boon to ticks, mosquitoes, rats, and other carriers of disease.

A report from the World Health Organization (WHO) in 1999 warned that global warming is bad news for both planetary and personal health. Changes in rainfall and humidity accompany the rising temperatures and create more of the wet and damp conditions that many pests and parasites thrive in.

Over the twentieth century, the average temperature in Europe rose by about 0.8 degrees Celsius, giving many disease carriers a foothold. Malaria was once common in Europe, and wetter conditions could see the disease making a comeback. Parts of western Asia and eastern Europe, including Turkey, Azerbaijan, and Tajikistan, are already under the threat of malarial mosquitoes, which are likely to spread westwards. Meanwhile, longer summers have already caused an increase in cases of tick-borne diseases like encephalitis and Lyme disease. Warmer climates in northern Europe would allow the sandflies of the Mediterranean to spread northwards, carrying visceral leishmaniasis with them.

In 1999, the WHO called for European governments to join forces in monitoring the spread of these diseases and find ways of reducing the health costs of climate change.

Ed Yong

Date 1999

Scientist R. Sari Kovats

Nationality Swiss

Why It's Key Raised the alarm over the impact of climate change on our own health, as well as that of the planet.

opposite A mosquito draws blood

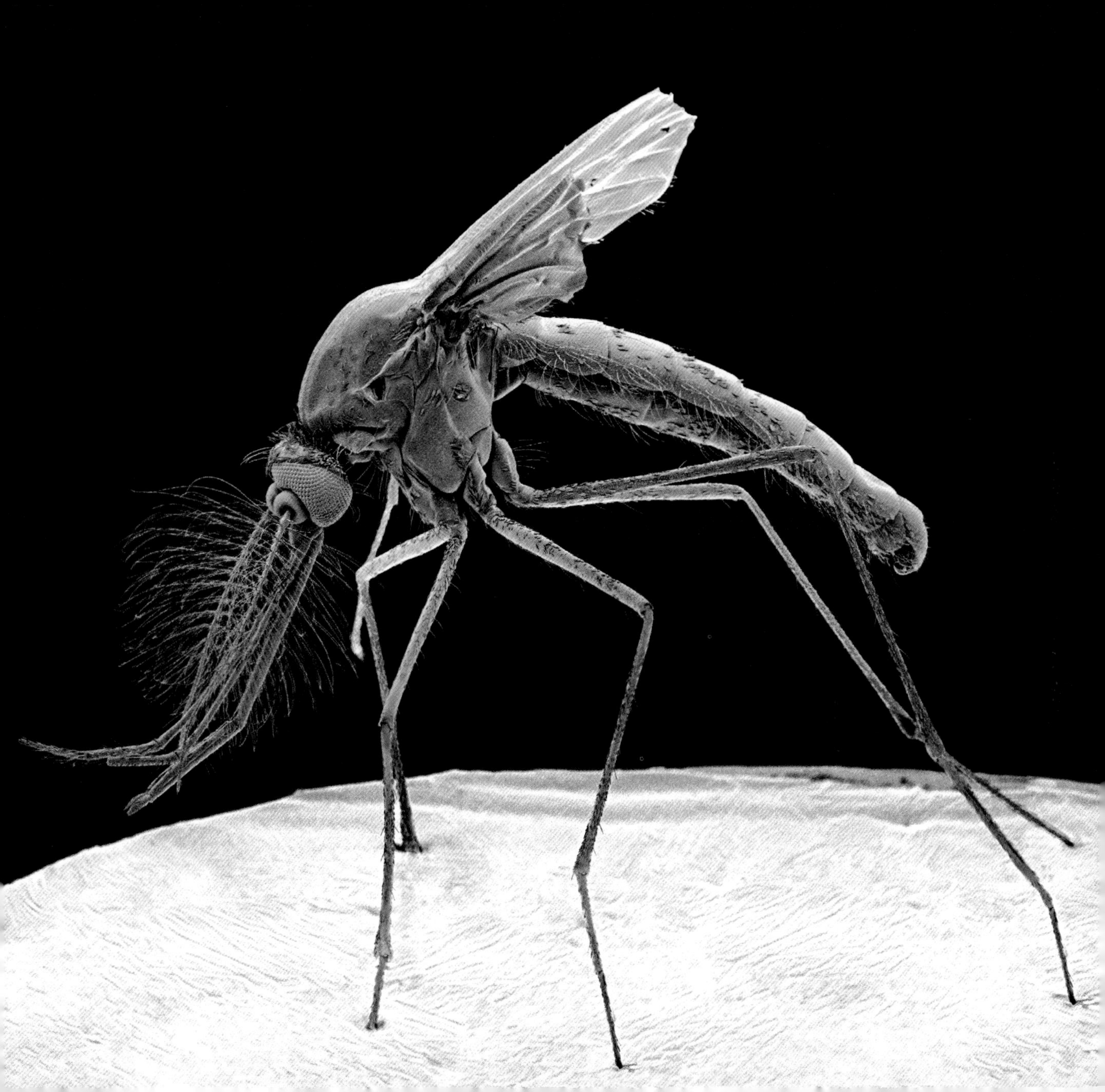

Key Event **Small world**
Global population reaches six billion

At two minutes past midnight on October 12, 1999, in a hospital in Sarajevo, Bosnia and Herzegovina, a very special baby was born. Fatima Nevic's eight-pound baby boy was declared the six billionth living person in the world.

According to information released by the United Nations to celebrate the "Day of Six Billion," the global population did not reach one billion until 1804. Since then, however, the number of people in the world has been expanding at an increasingly rapid rate. The leap from one billion to two billion took 123 years, whereas the leap from five billion to six billion took just twelve.

In 1999, the global population was gaining around 78 million more people each year, which is the same as the populations of France, Greece, and Sweden combined. The rate of growth has been such that the number of people alive on Earth has roughly doubled in the forty years from 1959, and increased threefold in the seventy-two years from 1927.

Current projections from the U.S. Census Bureau show that the number of people in the world will continue to grow in the twenty-first century, albeit at a lower rate than that seen in the last century. According to estimates, we'll only have to wait until 2042 until we're welcoming the nine billionth person into the world.

Christina Giles

Date 1999

Country Global

Why It's Key A landmark that exemplifies the massive increase in the world's population over the last century.

Key Experiment **Alex the Parrot**
The bird-brain-box

An African gray parrot named Alex (1976–2007) was the subject of an intensive, long-running behavioural research programme conducted by psychologist Professor Irene Pepperberg.

Alex's achievements are legendary and described in many research papers and a book, *The Alex Studies* (1999). He could distinguish and categorize objects by texture, color, material and abstract relational features like shape, bigger or smaller, same or different, and number categories up to six. When challenged with a problem, Alex had been trained to give his answers in English, delivered in a distinctive American accent. In the early 1980s, a time when most people thought birds lacked intelligence and could only learn by trial and error, these were astounding findings. The key question is: how did Alex do it? Was there real understanding of concepts and abstract relationships? In the hours of daily training over more than twenty-five years, what methods, strategies, and processing skills had Alex developed in the computational center in his brain? Continuous debate and controversy accompanied his increasingly prolific output through his life, but the investigation into his cognitive powers is now at an end; sadly Alex died in 2007, aged 31.

This particular African gray parrot will undoubtedly be remembered fondly. The scientific world lost a star research subject, and the answer to the key question remains unresolved. The last words might appropriately be attributed to Alex's spontaneous interjection when he had had enough of the day's training session: "wanna go back."

Arthur Goldsmith

Date 1999

Scientist Irene Pepperberg

Nationality American

Why It's Key Showed that animal learning is much more complex than we previously thought and ignited fierce academic debate about the nature of intelligence.

Key Event
Millennium Bug

Since time immemorial, people have been forecasting impending apocalypse – some even offering rather impressive information as to the specific times and dates. Generally, they fail to hold the attention of even the most gullible half-wit, but at the turn of the Millennium, growing unease began to spread through otherwise rational, logical, and sensible sectors of society. Public consensus was that, when the clock struck midnight at New Year, we were in for a crisis of Biblical proportions. In March 1997, a US Department of Defense computer began serving certain contractors – due to deliver goods to the government in 2000 – with overdue notices, claiming that the Department had been waiting in for these deliveries for ninety-seven years.

After the media got its grubby hands on other stories of similarly unremarkable administrative cock-ups, a heady mix of new technology and old-fashioned scaremongering quickly turned the world crazy. Suddenly it became the considered opinion of many people that computers would rather fizz, pop, and set themselves on fire than count past 1999.

With the 20:20 vision of hindsight, it has become transparently clear that we had massively underestimated technology. Almost everything electrical, from tiny digital wristwatches to gargantuan global communication satellite systems managed to flip over to 01/01/2000 without so much as a glitch. Apart from in Australia, that is, where bus ticket validation machines in two states stopped working. That aside, however, things were absolutely fine.

Chris Lochery

Date 1999

Country Global

Why It's Key Armageddon was neatly side-stepped - but the chances are it was never really headed in our direction anyhow.

Key Discovery **Sticky feet**
Getting to grips with geckos

Geckos are very attractive little lizards; there's no denying it. The reason, however, isn't because of their looks, but because of their extraordinary ability to scurry up walls and dangle from ceilings. "Attraction" is how these creatures defy gravity.

No one knew how geckos could climb until 2000 when a team of American biologists and engineers, led by Robert Full and Kellar Autumn, solved the puzzle. It was previously known that geckos' toes have millions of microscopic and densely packed hair-like fibers called setae. And that the tip of each seta has around one thousand smaller pads called spatulae. Full and his colleagues discovered that these pads, when close enough to a surface, induce van der Waals forces – weak attractions between molecules when they are brought very close together. With the combined adhesive force produced by billions of individual pads, their sticking power is immense: about 1000 times more than is required to keep them from falling. Simply by peeling their feet in a certain way, geckos instantly become "unstuck," allowing them to move at speed yet remain attached to a surface like a vertical wall.

The discovery kick-started the current research race to create synthetic dry adhesive tapes using nanotechnology to mimic the gecko's foot. Thes would have extraordinary sticking power, even underwater and in a vacuum, while being easily removable and re-usable. Several prototypes have shown great promise, and with possibilities including wall-climbing robots, adhesive gloves for astronauts, and waterproof bandages, the gecko remains inspirationally attractive.

James Urquhart

Date 2000

Scientist Robert Full, Kellar Autumn

Nationality American

Why It's Key Explained how geckos climb smooth surfaces, prompting a race to create revolutionary adhesives with innumerable potential uses.

HONDA
ASIMO

Key Invention Golden rice

The production of genetically modified "golden rice" was first announced in 2000 by Xudong Ye and his colleagues at the Swiss Federal Institute of Technology. Termed "golden" because it contained high levels of vitamin A which dyes the rice yellow, it was designed to combat dietary deficiencies in the developing world.

A major health problem in the developing world, vitamin A deficiency afflicts at least 180 million women and children in Asia, causing 500,000 children to go blind every year. This is because normal rice, which often makes up two thirds of calorific diet in many poor countries, almost completely lacks the compound.

By inserting two genes into the genome of the rice plants, the scientists, led by Professor Ingo Potrykus, had managed to dramatically increase the production of vitamin A within each grain of rice. Potrykus campaigned vigorously to remove restrictions preventing poor farmers from benefiting from the research carried out by biotech corporations.

Despite its potential advantages, however, golden rice has not yet been widely distributed due to concerns the crop will not tackle the root cause of poor diet: poverty. This biotechnology project, many say, has directed funds away from more appropriate approaches to improve health by reducing poverty. Some fear the humanitarian face of the project will act as a "Trojan horse" for other genetically modified organisms which are controlled by large corporations. Other criticisms include the fear of it cross-breeding with wild plants, and unknown side-effects.

Mel Wilson

Date 2000

Scientist Ingo Potrykus

Nationality German

Why It's Key Raises the possibility that genetically modified foods can be put to humanitarian use.

Key Invention ASIMO Robo-helper

It's a robot called ASIMO (Advanced Step in Innovative Mobility) but it doesn't look like a giant industrial metal arm, or a bulky box of lights on wheels. In fact, it looks a lot like us.

ASIMO can walk like a human – and avoid obstacles, which in itself is an enormous achievement. Designed by researchers at Honda, the idea, eventually, is that it will "create a partner for people, a new kind of robot with a positive function in society." Many researchers investigating how humans interact with machines believe we prefer robots that look and "feel" more like us.

At 120 centimeters tall, ASIMO is about the size of an average seven year-old. Unveiled in 2000, it was designed to be big enough to interact with humans in a home or office environment without getting in the way. It could operate a light switch, reach items on a desk or kitchen counter, and carry objects around.

Advanced recognition technology enabled ASIMO to recognize and track the movements of humans. It could identify postures and gestures such as pointing, and even shake hands with you. And if you waved at it, ASIMO waved back. It even had the ability to recognize up to ten faces registered in its memory – and would address these people by name.

Someday in the near future, you might have a helper robot like ASIMO. And you can throw away the remote control – it knows its own name, so if you call, it will come...

Andrey Kobilnyk

Date 2000

Scientist Honda

Nationality Japanese

Why It's Key Helper robots may be useful in the future to automate routine chores and extend the independence of our ageing populations.

opposite Honda's humanoid robot Asimo

Key Event **It's alive!** 250-million-year-old bacteria revived

In 2000, Russell Vreeland from the University of West Chester, Pennsylvania, courted controversy when he published results suggesting that he had successfully revived a bacterium that had lain dormant in salt crystals for 250 million years.

Vreeland and his team collected salt crystals from the Salado Formation in New Mexico by drilling into an air intake shaft for the Waste Isolation Pilot Plant – the first underground storage facility for waste from the manufacture of nuclear weapons. These crystals contained tiny pockets of salt solution from which the bacteria were isolated. By confirming the age of the salt beds, and that the crystals hadn't dissolved and reformed, Vreeland was able to date his bacterium to at least 250 million years old, making it the oldest living organism, beating the previous contender – bacterial spores from a bee preserved in amber – by an impressive 210 million years.

The bacterium, a strain of *Bacillus* catchily known as strain 2-9-3, caused debate as it was suggested that it was a contaminant from a much later time frame rather than a true prehistoric organism. This was supported by a team from Tel Aviv University in 2001, who analyzed the DNA sequence of strain 2-9-3 and found that it was virtually identical to that of a common "modern" bacterium. Vreeland, however, maintains that his bacterium is not a contaminant, estimating "that the chances of it being a contaminant are less than one in a million."

Helen Potter

Date 2000

Country Mexico

Why It's Key The discovery of the world's oldest living organisms?

Key Person **Bill Gates** Microsoft's PC king

Bill Gates has been described as a lot of things; a geek, a philanthropist, an entrepreneur, a hard-nosed business man, even "part-Einstein, part-John McEnroe." And these are certainly among some of the kindest descriptions you'll hear.

At fifty-one years of age, his accumulated fortune, according to Forbes magazine's list of billionaires, was a staggering US$56 billion. Gates has made himself a comfortable seat atop the pile of the planet's wealthiest individuals, but under it all is a man who dropped out of Harvard at age twenty and only modestly admits to a being a little smarter than average.

Here is a man who in 2000 created one of the world's biggest charities, and who has ploughed billions into scientific research. As co-founder, and previously Chief Executive, of Microsoft, he is responsible for the operating system that the vast majority of personal computers run on today. Yet many doubt his integrity and, unsurprisingly, he has often found himself criticized for his business dealings.

Gates was voted *Time* magazine's Person of the Year in 2005 for his charitable endeavors, along with his wife Melinda and, some might say bizarrely, Bono, lead singer of the Irish rock band, U2. In an interview with the magazine, Bono is quoted as saying, "He's changing the world twice. And the second act for Bill Gates may be the one that history regards more."

Hayley Birch

Date 2000

Nationality American

Why He's Key One of the world's most influential entrepreneurs, whose substantial contributions to personal computing – and charitable causes – are undeniable.

Key Publication **Analysis of the genome of flowering plant *Arabidopsis thaliana***

The field of plant genetics received a real boost in December 2000. It was during this month that the full genomic sequence of the plant *Arabidopsis thaliana* was published in the journal *Nature*. The work was performed by scientists from the United States, Japan, and Europe working for the Arabidopsis Genome Initiative (AGI), who began this huge task in 1996.

While most non-scientists had never heard of the tiny weed *Arabidopsis*, plant scientists had been using it for years as a model in the laboratory. A small plant, it matures quickly and reproduces copiously; a perfect model for studying plant genetics. In addition, it was believed that many of the genes found in *Arabidopsis* would have counterparts in other important plant species such as wheat, rice, and corn.

The *Arabidopsis* genome was found to contain around 120 million base pairs of DNA and around 15,000 different genes. It was of particular note to the scientists studying the genome that there was very little "junk DNA" present – that is, DNA that contains no genes – unlike in the human genome sequence.

The *Arabidopsis* project marked a significant achievement not only in plant genetics, but also in the fields of information technology and bioinformatics. Without the processing capability of modern computers and networks, the project never would have been completed in such a short period of time, and at the time of its publication, it was the most complete sequence of a eukaryote – an organism with complex cell structures – ever published.

Rebecca Hernandez

Publication *Nature*

Date 2000

Author Arabidopsis Genome Initiative

Nationality American, Japanese, European

Why It's Key Marked a significant achievement in the areas of plant biology, genetics, and bioinformatics, and provided more data on a useful model species.

Key Event **A fish out of water**
Robot wired up to fish brain

The phrase "disembodied brain" is one you're more likely to come across in an episode of Doctor Who, or a bad eighties horror movie, than in a laboratory. But Ferdinando Mussa-Ivaldi's laboratory is no ordinary laboratory. And what his team created there in 2000 might easily have sprung – or swum – straight out of a science fiction film.

Mussa-Ivaldi wired a robot and the brain of a fish together in such a way that the resulting "cyborg" was able to sense light and move towards it. This was the first time scientists had been able to establish a two-way connection between brain tissue and a robot.

Now, any science fiction fan who knows his cybermen will tell you this wouldn't be possible with a human brain. Your grey cells simply wouldn't function without the rest of your body. But a lamprey – that's a different matter. These blood-sucking fish have an uncanny ability to keep the cogs of their brains whirring even when nothing else remains. Mussa-Ivaldi and his team were able keep lampreys' brains alive for several weeks by plunging them into oxygenated salt water.

Although the team's fish-bot could only master basic manoeuvres, the experiment marked a fundamental step towards understanding how electronic and biological systems can communicate with each other. Whether or not we live to regret it, their research may help us to one day create true human-machine hybrids.

Ferdinando Mussa-Ivaldi

Date 2000

Scientist Ferdinando Mussa-Ivaldi

Nationality Italian-American,

Why It's Key Understanding how biological systems "talk" to electronic devices is key to developing useful prostheses, such as bionic limbs, but may also allow us to build the cyborgs of science fiction novels.

Key Event **The Clay Millennium Problems** Mathematics for our millennium

For the turn of the year 2000, as the rest of the world celebrated with champagne, fireworks, and all-night parties, the Clay Mathematics Institute (CMI) named a prize of US$1 million for the solving of each of seven unanswered mathematical problems.

The seven problems are: the Birch & Swinnerton-Dyer Conjecture, the Hodge Conjecture, the Navier-Stokes Equation, the Poincaré Conjecture, the Riemann Hypothesis, the Yang-Mills Theory, and the P vs NP Problem. These problems relate to everything from prime numbers to topography, via quantum mechanics and the movements of fluids.

Some of the problems have remained unsolved for a great deal of time. The Riemann Hypothesis, for example, was first conceived in 1859, the same year as Darwin's *On The Origin Of The Species*.

At the time of writing, the Poincaré Conjecture – which says that any enclosed 3D shape can be squashed into a sphere – is in fact thought to have been solved. The eccentric Russian mathematician responsible, Grigory Perelman, has yet to pursue the prize money, despite living in virtual poverty; following some unfavorable experiences in the past, Perelman explained in 2006 that he preferred to distance himself from the wider mathematics community.

At the CMI Millennium Meeting on May 24, 2000, mathematician Timothy Gower lectured about the importance of mathematics; the Institute's intention in setting up the prize fund was to highlight exactly this. But the unclaimed prizes also stand testament to another aspect of mathematics – its difficulty.

Neal Anthwal

Date 2000

Country USA

Why It's Key Cash prizes were offered for the solving of seven key mathematical problems.

Key Event **First crew enter the International Space Station**

To Let: Magnificent newly built three bedroom detached property with beautiful ocean views – US$20 million per week. No previous tenants. Minimum contract twenty weeks.

Whether it was rising fuel prices, the deterioration of pop music, or the spectacular ocean views which drove an American and two Russians to move into the International Space Station (ISS) on November 2, 2000, is unimportant. What is of great significance is that this event marked a new era in the spaceage; a truly international project had been accomplished.

The ISS is a space research facility constructed by the United States, Russia, Japan, Canada, and most of Europe (Germany, France, and Italy being the major contributors). Orbiting the Earth at approximately 390 kilometers above sea level, the station contributes vital research toward understanding, among other things, the long-term implications of space flight for the human body. Our intellectual capabilities as a civilization are huge, however even the wealthiest of nations would struggle to individually implement our most ambitious projects. Therefore the greatest benefit derived from this multi-billion-dollar project will likely be seen in future collaborations, as we begin to pool our resources with increasing success.

In the years since Bill Shepherd, Yuri Gidzenko, and Sergei Krikalev boarded the ISS for their 140-day tenancy, it has remained inhabited at all times, maintaining a permanent human presence in space. It is the fact that our stay on Earth is unlikely to be indefinite that makes this achievement so valuable.

Christopher Booroff

Date 2000

Country Russia, USA

Why It's Key International collaborations will be necessary to fulfill our intellectual potential as a race, and the International Space Station is a vital starting point.

opposite The International Space Station orbits above the Earth

Key Publication

Dating the Universe

Why did the Frenchman stare at uranium? It may sound like the kind of nervous joke told on a first date, but the answer is to determine the age of the Universe. On February 7, 2001, by targeting an elusive spectral line, Roger Cayrel, of Observatoire de Paris-Meudon, and his colleagues stated a vastly improved estimate for the age of the Universe: 12.5 billion years ± 3 billion.

Metals are formed by supernova explosions, so when astronomers pointed their telescopes at a star near the edge of the Milky Way, its deficiency in metals revealed that it was very old, formed during the early stages of the Universe. However, a small amount of uranium-238, which would usually be obscured by other metals, was detected. The half-life of the radioactive element uranium-238 is 4.5 billion years, meaning that, due to radioactive decay, its abundance will halve every 4.5 billion years. Cayrel's team calculated that the star they were observing would have reached the quantity of uranium-238 they were observing after 12.5 billion years of life. In a paper published in *Nature*, the team reasoned that the Universe must be marginally older than the star.

Much like trying to figure out the age of your blind date, a variety of techniques can be used in estimating the age of the Universe however Cayrel's analysis was by far the most precise to date, and laboratory work with uranium-238 is expected to further reduce the uncertainties.

Christopher Booroff

Publication "Measurement of Stellar Age from Uranium Decay"

Date 2001

Author Roger Cayrel

Nationality French

Why It's Key Knowing the age of the Universe has huge theological implications, and is key to several areas of cosmology and astronomy.

Key Event

Four days of surgery for two separate bodies

In April 2001, Nepalese twins Ganga and Jamuna Shrestha found themselves in different rooms for the first time. The eleven-month-old girls were conjoined twins, born with a rare condition called craniopagus where their skulls and parts of their brains were fused together at the back. To give them a better chance of successful and normal lives, a team of Singaporean neurosurgeons, led by Keith Goh, attempted to separate them.

Only one such operation had previously been successful but in that case the twins' brains had not been fused. Ganga and Jamuna posed a much bigger challenge. They shared a common network of blood vessels that had to be meticulously and steadily separated, while their brain regions had to be carefully divided so that each girl received the right complement of higher functions. Goh's team used modern brain-scanning techniques to visualize the twins' joined brains and spent almost half a year rehearsing the surgery in a virtual operating theater. Singaporeans raised over US$350,000 to meet any expenses and the cost of future treatments, and the team finally performed the grueling landmark operation in shifts, starting on a Friday afternoon and finishing, over eighty hours later, on the following Tuesday.

While the surgery was considered a success, it was a bittersweet one. Ganga, formerly the feistier twin, picked up an infection that has left her practically comatose. Jamuna fared better but while she can speak, she was unable to walk as of 2003. She is still very attached to her sister.

Ed Yong

Date 2001

Country Singapore

Why It's Key Showcased both the incredible power of modern surgical techniques and the massive risks involved in separating conjoined twins.

Key Event **How to sequence a genome**
Of mice and men

Since the early 1900s, mice have been used as experimental subjects in labs. Today, with the knowledge that the DNA of a lab mouse is 90 per cent the same as to that of humans, mice are widely used as animal models for human illnesses.

J. Craig Venter, president and chief scientific officer of Celera Genomics, aimed to "read" the full human genetic code – and alongside it, the genetic code of the lab mouse. The company innovated on a method called whole genome shotgun sequencing to churn out the sequences much quicker and cheaper than believed possible, and, in the case of the mouse genome, pipping to the post a public consortium striving towards the same goals.

By identifying DNA that was common to those suffering from a specific disease, scientists would theoretically be able to locate identical or similar gene sequences that produced the same illness in a mouse. Mice with this particular DNA feature could then be bred or cloned as experimental subjects and used to test various treatments for the illness. As the genetic composition of a mouse is so similar to that of a human, what cures a mouse could, potentially, work for us too.

The publication of the draft human sequence on February 12, 2001, was followed by the announcement that the entire mouse genome had been sequenced on April 27 that same year. Venter decided, somewhat controversially, to allow only paying subscribers access to this second sequence, while his public rivals were posting their mouse data on the internet.

Andrey Kobilnyk

Date 2001

Country USA

Why It's Key Sequencing the mouse genome meant researchers could compare it to human DNA – potentially speeding up research into cures for human illnesses.

Key Event **Viagra sales exceed US$1 billion**
Men rejoice; the Internet despairs

Viagra is undoubtedly one of the most famous drugs of all time. Google lists 49 million web pages that carry its name, while only 15 million mention aspirin. The drug is synonymous with dodgy websites and spam emails, but it has also helped millions of men give their sex lives a lift.

Viagra is the trademark name of sildenafil citrate, a drug first developed by pharmaceutical company Pfizer to treat hypertension and angina. It failed in that respect, but early clinical trials showed that it had an unusual side effect – men who took it developed strong erections. Viagra blocks an enzyme that breaks down cGMP, a signaling molecule essential for erections. It relaxes muscles that line the penis's arteries, blood rushes in, and hey presto: It's standing to attention.

In a brilliant marketing turnaround, Pfizer decided to launch Viagra as a treatment for erectile dysfunction; in 1998, the U.S. Food and Drug Administration approved it as the first ever pill for treating impotence. As the only other option was a squirm-inducing injection, demand for the drug was massive. Between 1999 and 2001, sales of Viagra exceeded US$1 billion, and in the UK, it will soon be available on the high street without a prescription. The chemist Boots began trialing over-the-counter sales of the drug in early 2007.

Despite its success, Viagra isn't an aphrodisiac – it can't make you want to have sex, it just helps to get the job done.

Ed Yong

Date 2001

Country USA

Why It's Key Provided the first approved pill for treating impotence

Key Publication
The human genome sequence

The publication of the human genome sequence signalled an uneasy truce between the public and private sector organizations vying to be the first to spell out the 3-billion-letter code. Following a series of bitter feuds, two draft sequences were finally published – just a day apart – in the journals *Science* and *Nature*.

James Watson, famous for his discovery of the structure of DNA, originally led the public sector bid, but was replaced following a row with private sector rival Craig Venter. At one point, a collaboration between Venter's company Celera Genomics and the public consortium looked likely, but this plan was shelved when relations between Venter and Watson's successor, Francis Collins, also broke down.

Despite these controversies, the 2001 publication of sequence marked an astounding achievement by both organizations. Perhaps the most surprising outcome of the projects was that humans have only about 30,000 genes – scientists had previously expected humans to have more than three times that number.

The sequence, however, is far from the end of the story – more than a string of code is needed to understand the complex interactions that occur between our DNA and the environment in which it must function. A vast body of work is now focused on interpreting the code and finding ways in which we might use it to combat disease.

James Urquhart

Publication "A Physical Map of the Human Genome," "The Sequence of the Human Genome"

Date 2001

Nationality American

Why It's Key Lessons were learned not just about the nature of the human genome: it contained far fewer genes than predicted, but about large scale science.

opposite Computer screen display of a human DNA sequence represented as a series of colored bands

Key Event
First tourist in space

Hands up those who wanted to be an astronaut when they were young. Well, Dennis Tito did and, forty years after Yuri Gagarin became the first man in space, his wish came true.

Tito had always had a strong interest in space exploration. His early career involved working in a jet propulsion lab and tracking NASA's Mars probes. He was in love with space. But he loved money more and ended up running his own investment firm. By the age of forty Tito had made his first million and by the age of sixty, his immense personal fortune would give him the opportunity of a lifetime.

He was an American hero of the business world, but it was the Russians who were to offer Tito a chance to realize his dream of space travel. The offer did have a few caveats though, the biggest being the US$20 million price of the return ticket. The Americans were disgusted with the whole affair, saying that space was no place for amateurs. The Russians took little notice, claiming it was their rocket and therefore they could take whomever they wanted.

Tito spent just under eight days in space; six days were spent in the International Space Station and in that time he orbited the Earth 128 times. Apparently he doesn't like the term "tourist" and would rather be called an "independent researcher." NASA was a little more scathing when they claimed he'd be the last "space cowboy" visiting the jointly owned space station for a while.

Andrew Impey

Date 2001

Country Russia

Why It's Key The first paying space tourist whose trip opened the floodgates for companies vying for a slice of the market in deep space package tours.

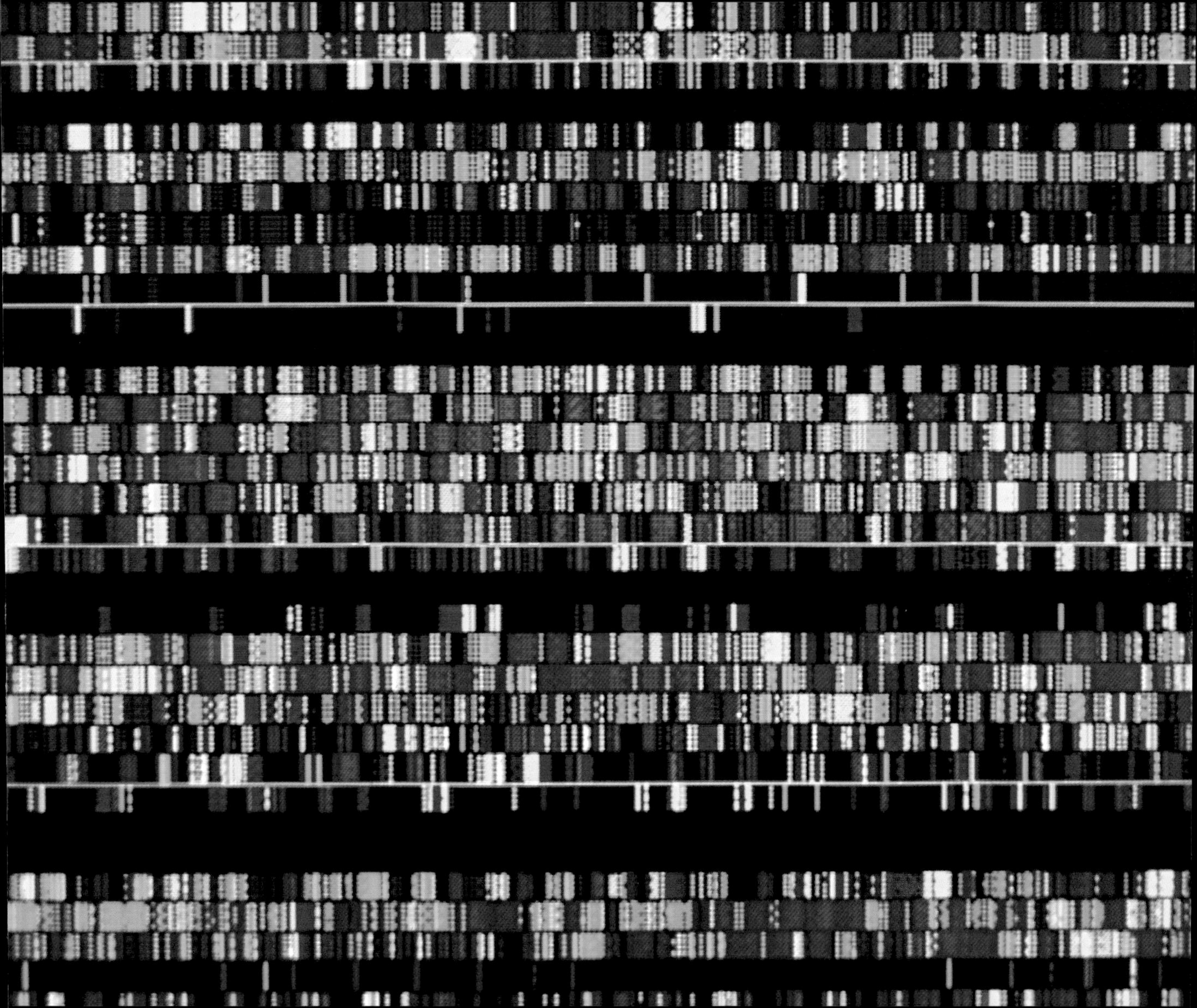

Key Event **Mir is decommissioned**
World peace takes the plunge

Crowds of distraught Russians gathered across Moscow to protest against the decision to decommission Mir. But it was too late. Despite its former glory, the Russian space program was short of cash and its limited resources were being diverted to the International Space Station (ISS). Their own orbiting laboratory, Mir, was about to meet its watery grave in a ball of flames.

Much like a broken down oil tanker floating in one of our oceans, Mir could not be left in orbit. So despite the risks involved, it was decided that this 137-ton bowling ball, this emblem of Russian pride, would be brought back to Earth, largely burning up on re-entry to the Earth's atmosphere. The procedure was a huge success; having traveled along its intended path, the debris that survived the descent hit their ocean target with welcome accuracy. Mir translates to both "peace" and "world" and it was therefore fitting that the decommissioning ceremony passed relatively peacefully as the world gazed on. The resounding success of the project left the media free to report on the deeper issues, such as the pain felt by the Russian people. Mir was seen as a symbol of both Russian scientific and political greatness, as well as a sign of economic riches expired.

The precision and control with which Mir was guided towards the Pacific Ocean, just east of Fiji, was an emotional final triumph for a hugely successful project.

Christopher Booroff

Date 2001

Country Russia

Why It's Key Space station Mir lasted fifteen years and housed thousands of experiments. It was an icon of the 1990s, and its decommissioning turned into a fireworks display for the world.

Key Experiment
Scientists clone human embryos

Human cloning is often thought of as something that occurs in bad science fiction films, but it became a reality in 2001, at least at the embryonic stage, due to the efforts of scientists at Advanced Cell Technology, Inc. The company was able to successfully remove the DNA from a human egg, which contains just one set of chromosomes, and replace it with DNA from a human body cell, which contains two.

The resulting hybrid cell then reverted to a completely embryonic state, forming cellular structures that only occur at this stage of development. Subsequently, the embryo divided into four to six cell structures, as is usual for a developing fertilized egg. At this very rearly stage, the researchers were able to isolate pluripotent stem cells from the embryo – cells with the capacity to become any tissue.

The company's goal is not to clone a human, but to fine tune the stem cells and guide their development into tissues that can then be transplanted back into humans. Though there are many human embryonic stem cell lines available, the ACT technology represents advancement in at least one big area – compatibility. If one can take DNA from an individual and grow it into new tissues, as has been done in this case, the cloned cells are far less likely to be rejected by that person's immune system if they're implanted back into their body. In other words, one could grow a perfectly matched transplant organ. The same cannot be said for existing human embryonic stem cell lines.

B. James McCallum

Date 2001

Scientists Advanced Cell Technology

Country USA

Why It's Key It may solve any immunogenic problems that ultimately might occur in stem cell organ growth.

Key Person **Craig Venter**
Genetic maverick

In 2000, John Craig Venter (b.1946) became the first biotech billionaire. A surfer and high-school dropout, Venter served in Vietnam before returning home to restart his education. He achieved a bachelor's degree in biochemistry in 1972, and a PhD in physiology and pharmacology in 1975.

In the 1990s, his ability as a scientist, project manager, and businessman resulted in major innovations in genomics. Venter founded the Institute for Genomic Research in 1992, which three years later was the first to reveal the entire genome of an organism: the bacterium *Haemophilus influenzae*. Celera, the next company Venter set up, published a first draft of the human genome in 2001 for just US$300 million – far less than the internationally funded US$3 billion Human Genome Project. A new technique, pioneered by Venter, called "whole genome shotgun sequencing" was crucial. The idea that Celera could "own" – and charge for access to – human genetic data caused an outcry. But by the time this was possible, so much human genome data was freely available that it was no longer a viable business model.

Venter went on to form his own research group – the J. Craig Venter Institute – and in 2005, Synthetic Genomics – a company researching the programming of DNA to produce artificial organisms. According to Venter, synthetic biological machines may have the ability to produce environmentally friendly fuels in the near future. In 2007, Venter revealed that the first complete human genome belonging to a single person had been sequenced: his own.

Andrey Kobilnyk

Date 2001

Nationality American

Why He's Key Craig Venter has been instrumental and controversial in the modern science of genomics, as both a scientist and a businessman.

Key Person **Sir Paul Nurse**
Yeast provides the key to cancer

Colleagues were skeptical when Paul Nurse (b. 1949) came to work at the Imperial Cancer Research Fund – now Cancer Research UK. At the time, he was doing research on a microscopic yeast commonly used to brew African beer, but not thought to be very relevant to cancer. He was known as a bright scientist, however, so he was given a chance to prove it. He did just that, going on to win a joint share in the 2001 Nobel Prize in Physiology or Medicine, along with Tim Hunt and Leland Hartwell.

Nurse's research focused on the cell cycle – the molecular "engine" that drives cell division. This process is essential to life, enabling new cells to be made when and where they are needed. But if it goes awry, cancer can result. Through painstaking genetic research, Nurse helped to identify one of the key molecules involved in the yeast cell cycle, known as cdc2. In 1987, he and his team went on to show that a virtually identical molecule also controls the cell cycle in humans. The two are so alike that the human version of cdc2 can be used to replace a faulty version in yeast.

This breakthrough showed that virtually all living cells, from yeast to man, are driven by the same basic engine. It also vindicated Nurse's choice of research subject, proving that research using yeast or other model organisms could have direct relevance to cancer research for human benefit.

Kat Arney

Date 2001

Nationality British

Why He's Key Showed that virtually all organisms are controlled by the same molecular "engine" and opened up new avenues in the fight against cancer.

Key Experiment

Stem cells for heart failure

In late 2001, Dr Hans F. R. Dohmann at the Pro-Cardiaco Hospital in Rio de Janeiro, Brazil was joined by his colleague Dr. Emerson C. Perin, from the Texas Medical Center, USA, to begin a novel therapy for heart failure: using stems cells to re-grow heart tissue damaged during heart attacks. These areas of scar tissue cannot contract, so they prevent the heart from pumping blood efficiently. Patients were given medications to increase the amount of stem cells present in their bone marrow. The bone marrow was then harvested, and the stem cells separated and cultured. These cells were injected into the most damaged areas of the heart through a small catheter threaded through the arteries.

Of the fourteen patients enrolled in the study, thirteen improved and four of five patients awaiting heart transplants were well enough to be taken off the waiting list. Stem cells are present in tissues other than just bone marrow. An earlier study in Paris in 2000 used stem cells from thigh muscle, which also appeared to help heart function. However, it may have contributed to later problems, as skeletal muscle contracts differently to cardiac muscle.

Research is still underway. In February 2007, physicians in Spain found a new way to collect those all important stem cells – they harvested a patient's fat with liposuction. Stem cells were then separated, cultured, and injected into the heart as before. If successful in treating heart disease, this could really help those of us trying to rationalize that second buttery croissant.

Stuart M. Smith

Date 2001

Scientist Hans Dohmann, Emerson Perin

Nationality Brazilian

Why It's Key Could be one of the first practical uses of stem cells.

opposite Bone marrow, scanning electron micrograph

Key Invention

Robotic surgery

While most of us would still consider surgery performed by a robot as science fiction, the reality is that more and more technology shows up in operating rooms every day. Surgeons are turning to technology to assist with technically challenging procedures and also to help prevent fatigue-associated mishaps. The result? Devices that look like they're straight out of *Star Wars*.

Today's robotic surgeons are far from independent practitioners – they require a human physician to guide their every move. They do, however, offer some distinct advantages over the plain old flesh and blood variety. Robots can avoid the tremors of human hands, no matter how small, and remarkably precise movements can be performed by the robotic implement, sparing delicate tissues rough handling. Robotic surgery uses extremely small incisions, which can speed recovery time of the patient. And lastly, the surgeon need not be in the same room, or even on the same continent, as the patient – successful operations have been performed across the Atlantic. In 2001 robotic surgery really came of age when surgeons based in New York successfully performed an operation on a sixty-eight-year-old woman in Stuttgart.

So what kind of surgeries are robots performing these days? Everything from the removal of the prostate to the replacement of heart valves. And while you are unlikely to have your neighborhood sawbones replaced by R2D2 anytime soon, the advantages of robotic-assisted surgery are clear, and only likely to be expanded upon in the future.

James McCallum

Date 2001

Scientists Jacques Marescaux and colleagues

Nationality French

Why It's Key Robotic surgery allows procedures to be minimally invasive, speeding recovery time and decreasing complications and discomfort.

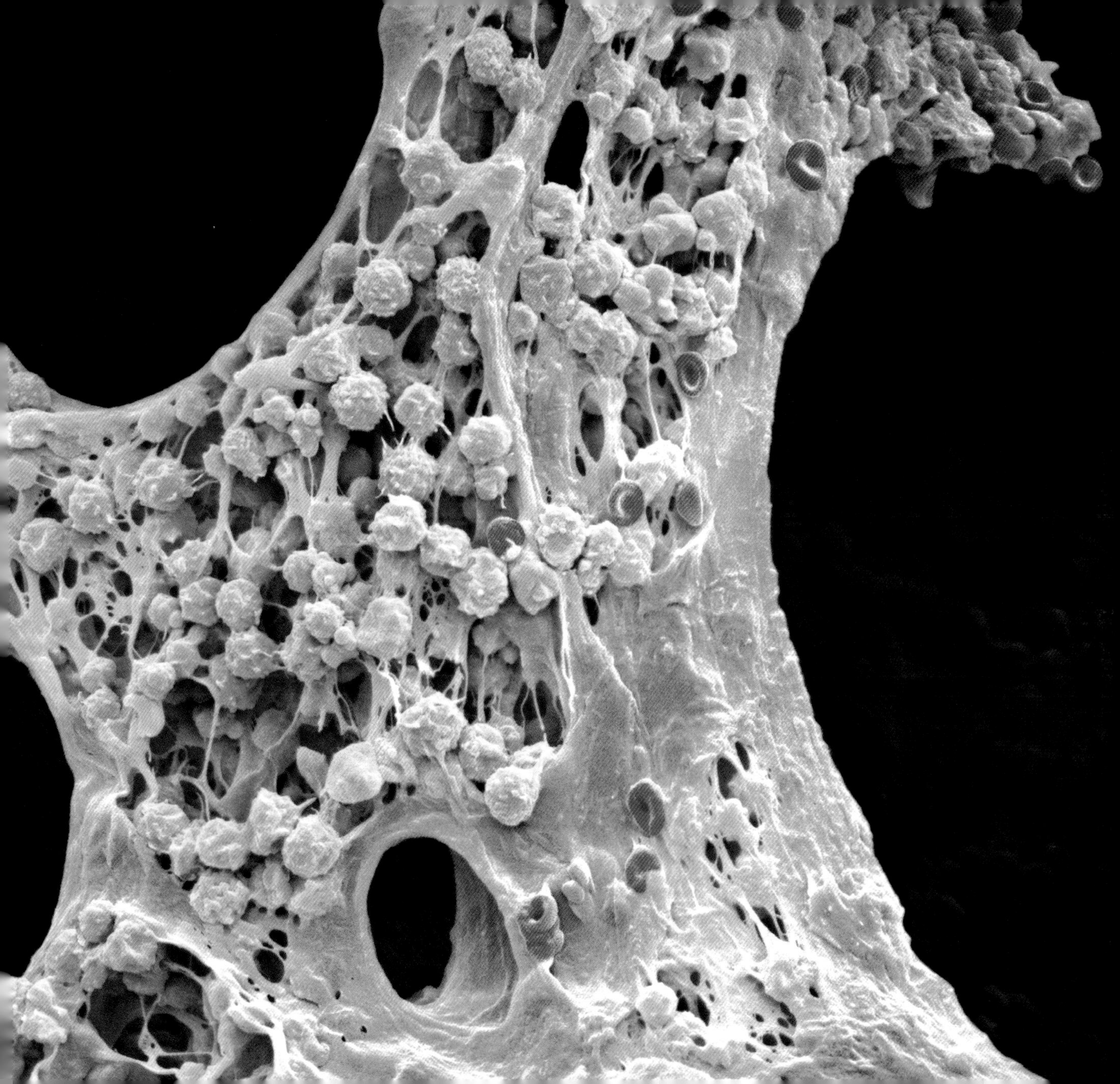

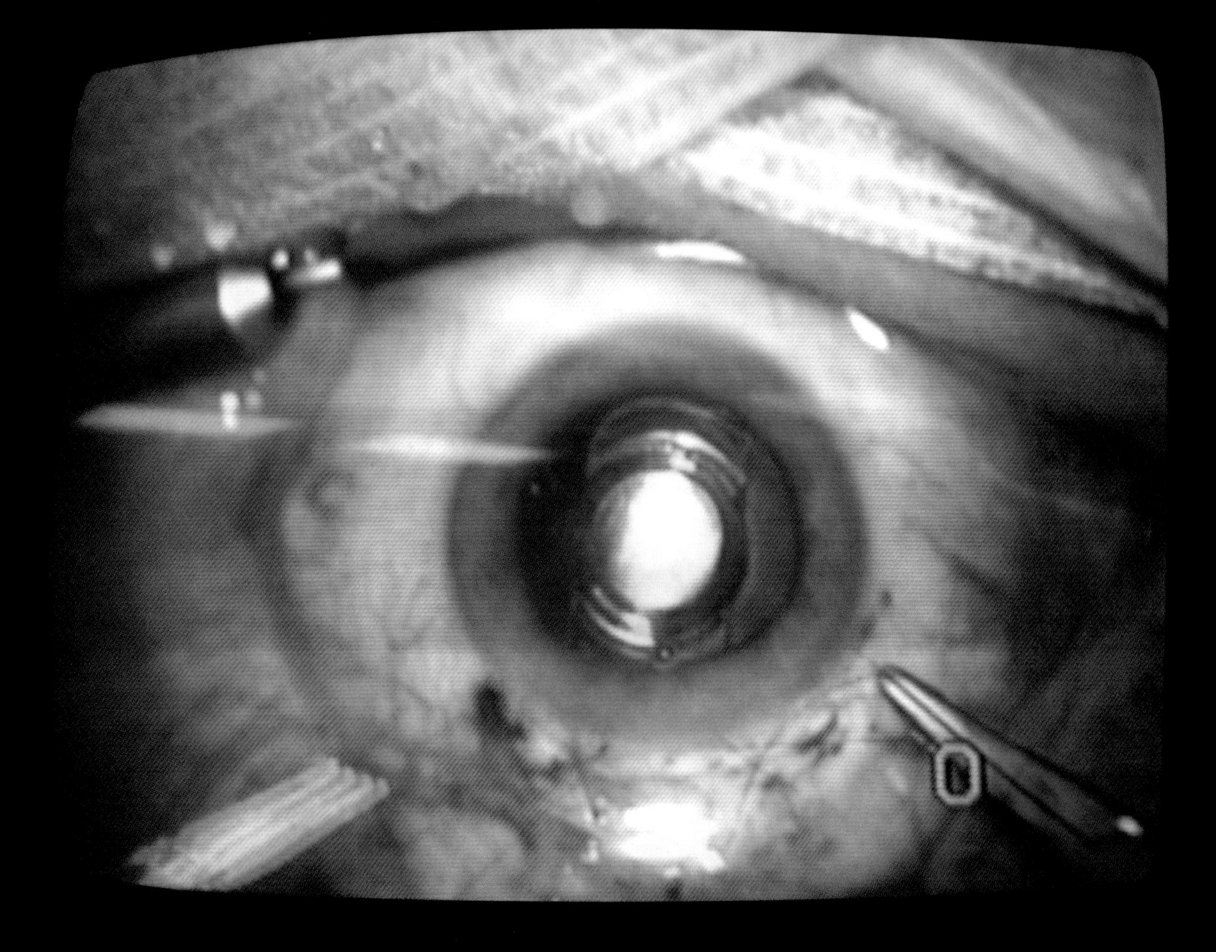

Key Invention
First bionic eyes tested

Prosthetic body parts have been around for many decades. With advancing technology, however, scientists have been able to develop more advanced prosthetics – bionic implants – that can actually mimic the original function of the body part.

In 2002, scientists in America moved away from the development of bionic limbs and began testing a new invention – the bionic eye. Externally, the device resembles a pair of glasses with a miniature camera attached. Images are transmitted from the camera to a tiny four-square-millimeter chip that sits on the retina, via a radio implanted next to the patient's eye. The chip contains a number of electrodes which stimulate the ganglion cells that transmit visual information to the optic nerve. An early version of the bionic eye was fitted to six patients with retinitis pigmentosa (a degenerative eye disease suffered by 1.5 million people worldwide), and results were very promising. The patients reported that not only could they see light and dark, but they could differentiate between simple items, see motion, and avoid walking into large objects.

This early model used sixteen electrodes, but scientists are now ready to test a model with sixty. The new model is hoped to go on sale by 2009, and is expected to cost £15,000. But scientists believe that an even more advanced model, with one thousand electrodes, could restore sight to the extent that blind people will be able to recognize faces.

Faith Smith

Date 2002

Scientist Mark Humayun

Nationality American

Why It's Key At some point in the future, this technology could potentially allow the blind to see.

opposite Implantable Miniature Telescope (IMT), a micro-sized telescope approximately the size of a pea, set inside the eye to magnify images onto the retina

Key Invention
Quantum dots

If there's one technology that is likely to define the twenty-first century, it is nanotechnology, and quantum dots could be at the forefront of change. These tiny, light-emitting chunks of silicon (or other semiconductor) contain as few as a hundred atoms, and could potentially be the key to producing everything from super-fast computers to ultra-efficient solar energy cells. They are so small that, if you could line them up side by side, over 3 million would fit on the tip of your nose.

One potential application of the dots is in the development of optical computers – computers which use light instead of electricity to store and manipulate data. They would allow engineers to build switches and logic gates that were much smaller and faster than current versions, maybe increasing computer speeds a million-fold. Another hope is that the dots will act as biological markers, searching out specific molecules, for instance, those unique to Alzheimer's sufferers, and illuminating when they find them. Many of these molecules only occur in very low concentrations, so it'll be much easier to spot them if they're glowing.

Although the concept of quantum dots was being discussed in the early 1980s, it has taken years of research and development to arrive at a point where quantum dots have become technologically useful and commercially viable. The announcement in December 2002, however, that scientists had created functioning markers for breast cancer, heralded the beginning of a dotty future.

Mark Steer

Date 2002

Scientist Xingyong Wu

Nationality Chinese

Why It's Key Provided practical uses for quantum mechanics: building lasers, tracing the routes and changes in molecules, and developing quantum computers.

Key Discovery **Toumai uncovered** Our oldest relative?

In 2002, a group of scientists, led by Michel Brunet from the University of Poitiers, France, unearthed the bones of the creature they later named *Sahelpithecus tchadensis*. Possibly the oldest hominid (human-like creature) ever discovered, the discovery called into question our previous understanding of human origins.

In the North African deserts of northern Chad, the team found a skull, pieces of jawbone, and some teeth of the ancient ape nicknamed Toumai, meaning "hope of life" in the African language Goran. The skull shows a combination of primitive ape features, such as the shape of the skull and size of the braincase, and more modern human characteristics, such as a pronounced brow ridge, short face, and small teeth.

It was believed at the time that humans evolved from a common ancestor to chimpanzees around five to seven million years ago. But Toumai was already showing distinctly human-like features six to seven million years ago, so anthropologists have been forced to push our chimp-like common ancestors further back into the mist's of time.

It wasn't just Toumai's age that caused great interest. The skull was found over 1,000 kilometers west of the East African sites where early hominid investigations have been focused. And what's more, Toumai's modern features might actually indicate that the fossils found in East Africa aren't our direct ancestors at all.

Emma Norman

Date 2002

Scientist Michel Brunet

Country Chad

Why It's Key An important discovery helping us to trace our ancient ancestors.

Key Discovery **Water found on Mars**

More than a year after the launch of Mars Odyssey, a robotic spacecraft which has been orbiting the planet ever since, scientists were amazed by data indicating the presence of vast quantities of sub-surface ice.

Named as a tribute to the renowned science fiction writer Arthur C. Clarke, Odyssey entered orbit around Mars on October 24, 2001. The orbiter began collecting data the following year, after a period of "aero-braking" – basically dragging the spacecraft through the Martian atmosphere to slow it down – brought it closer to the planet's surface.

Measurements taken by Odyssey have enabled scientists to map the mineral and chemical element composition of Mars, leading to the exciting discovery of buried ice. Although the Martian climate is too cold and its atmosphere too thin to sustain liquid bodies of surface water, scientists had long suspected that it was present in the planet's history.

Using a suite of instruments collectively known as the "gamma-ray spectrometer" the spacecraft can detect elements present in the soil such as carbon, silicon, and hydrogen. A high density of hydrogen, a proxy for water, was found to be concentrated around one meter below the surface, and appears to be most abundant at the South and North Poles. Odyssey also functions as a relay for information from the two land-based rover modules, and has yielded more photographs of Mars than any previous mission.

Mel Wilson

Date 2002

Scientists NASA

Nationality American

Why It's Key Finding water is the first step to answering the great mystery of whether life may ever have existed on the red planet.

opposite Composition of elements in the top meter of the surface of Mars in summer (water is blue). Data comes from the gamma-ray spectrometer on board Mars Odyssey

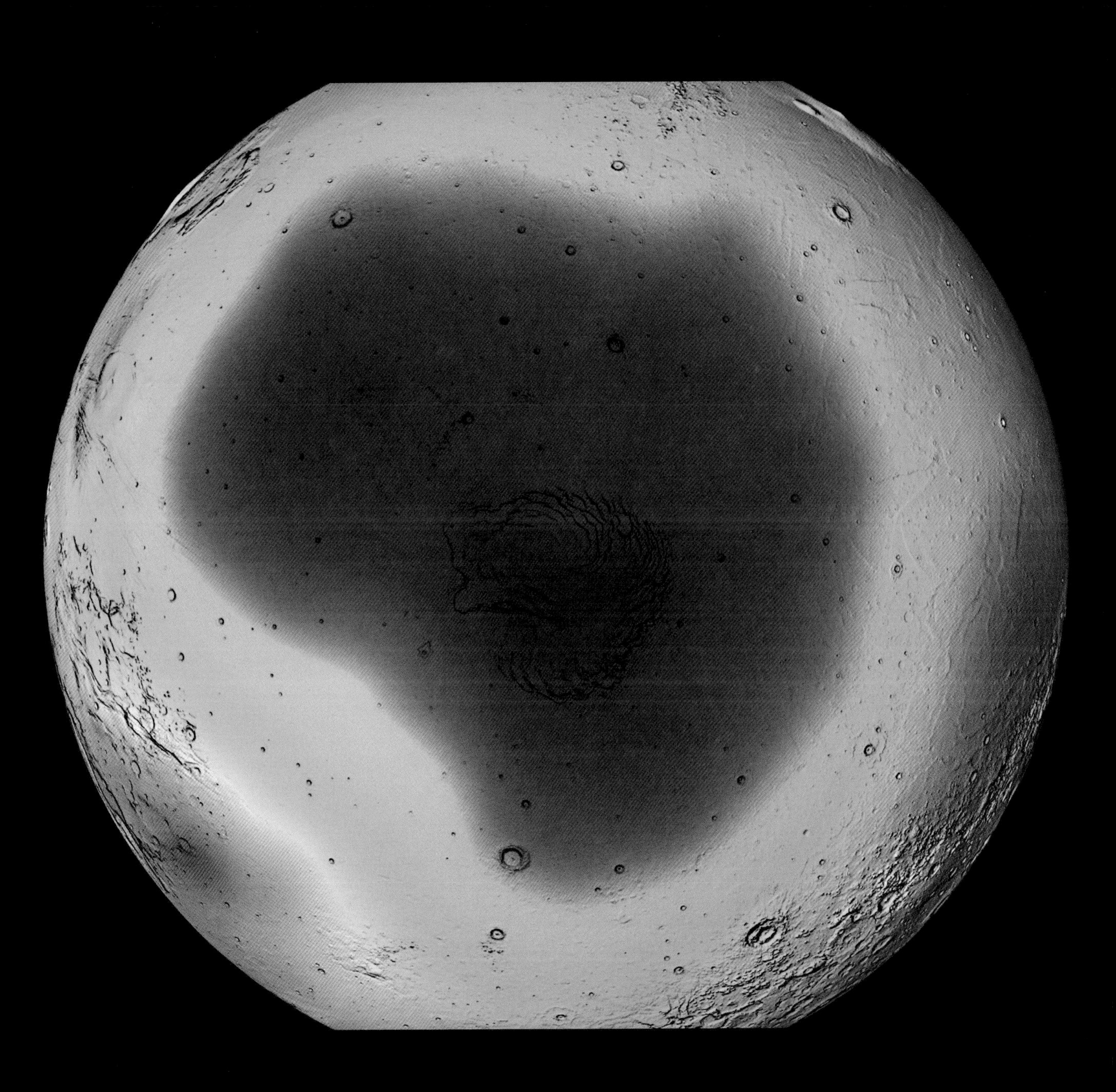

Key Event

Pi calculated to 1.24 trillion digits

Professor Yasumasa Kanada held the Guinness world record for calculating the physical constant pi to 206.158 billion decimal places, in 1999. Not content to rest on his laurels, he broke this record three years later by going all the way to 1.24 trillion digits.

Pi, which is the ratio of the diameter of a circle to its circumference, is normally estimated as 3.14. But pi actually has an infinite number of digits, and more frustratingly, all of them appear to be random. How has such chaos sprung from such a simple ratio? It has been the subject of study and obsession among mathematicians since before 200 BCE, when Archimedes tried to untangle the mystery of the constant using geometry. The closest he got was 3.1419.

Since then, the accuracy and number of digits has continued to increase, and the history of pi is intrinsically linked to the history of mathematics and computation. In fact, computing pi to a specific number of digits is often used to test the speed of a computer.

So, what are the implications of all these new digits? To anyone who's not a follower of pi, probably not too much. The big impact stems from the methods used to find the numbers and the computation of pi serves as a measure of the ability of supercomputers.

Kellye Curtis

Date 2002

Country Japan

Why It's Key Computing pi to more and more digits pushes the boundaries of computation, and serves as a test for supercomputer speed. It's also delicious nerd trivia.

Key Event

First successful scramjet flight

Ever since the conception of the jet engine in the 1930s, there has been a constant quest for speed The U.S. Air Force Blackbird – a reconnaissance aircraft – could travel at more than three times the speed of sound using conventional ramjets. Rockets, of course, could go even faster. But the concept of a supersonic scramjet involves flying without rockets, using oxygen from the atmosphere, air intakes on the vehicle, and a combustion chamber. The main difference is that the scramjet would operate at more than twice the speed of the Blackbird and possibly up to 17,000 miles per hour and maybe even beyond.

Scramjets are difficult things to get right, and one of the main hurdles is control. At such huge speeds, the drag caused by collecting oxygen from the atmosphere is immensely powerful. Additionally, the air collected is very hot, making it hard to control fuel combustion, not forgetting that the air passes through the engine at thousands of miles per hour.

The credit for the first practical demonstration of scramjets belongs to the University of Queensland, Australia. On July 30, 2002 they launched their experimental scramjet engine into space attached to a rocket. As the vehicle plummeted back to Earth it picked up speed and, by the time it reached 35 kilometers above the ground it was traveling at Mach 7.6, fast enough for the scramjet to be fed hydrogen fuel and ignited. The scramjet worked as planned, even if only for a few seconds, and a new frontier in aviation was opened.

David Hawksett

Date 2002

Country Australia

Why It's Key Demonstrated that the scramjet worked and could one day provide an alternative to rockets for entering space.

opposite X-43A aircraft (black) is accelerated by a Pegasus booster rocket (white) on March 27, 2004

Key Publication
Bubble fusion

It's possible that nuclear fusion could be the energy source of the future. The process of squeezing hydrogen atoms until they fuse together causes a lot of energy to be released – just like in the heart of our nearest star, the Sun.

In recent years, a few claims have been made that fusion can occur in different ways to how it does in the Sun. One of these claims was made in 2002 by Rusi Taleyarkhan, a scientist at the Oak Ridge National Laboratory in the United States.

Taleyarkhan had produced an apparatus that bombarded a liquid with ultrasonic vibrations. The liquid itself was made of deuterium and acetone – deuterium, a "heavy" form of hydrogen, is well suited as a candidate for fusion. Taleyarkhan believed that the vibrations were able to collapse gas bubbles in the liquid – in what is called "acoustic cavitation." In the process of collapsing, the temperature in a very small area would reach millions of degrees Celsius, high enough that it would fuse hydrogen atoms from the liquid and release energy.

To become accepted as truth, science demands that experiments should be able to be replicated by others and produce the same results. Controversially, as of 2007, attempts to validate the original experiment have resulted in a string of reported failures and subsequent rebuttals by Taleyarkhan. So far, no scientist outside of Taleyarkhan's lab has stepped forward to validate the original experiment and the claim that fusion is occurring in this manner.

Andrey Kobilnyk

Date 2002

Author Rusi Taleyarkhan

Nationality American

Why It's Key Helped us gain a better understanding of nuclear fusion. Whether it occurs in collapsing bubbles or in other unexpected forms, it may help us to achieve the ultimate goal of producing a fusion reactor – a low carbon, low pollution energy source for the future.

Key Discovery
The cells that cause jet lag

In 1866 the German anatomist Max Schultze, a master of microscopy, announced that there were two types of light-sensitive cell in the eye's retina: rods and cones. Rods detect low levels of light, but won't pick up colors. Cones, on the other hand, only work well in bright light, but do pick up different colors. Humans have three different types of cones which pick up red, blue and green light. All the colors that we see are created by different proportions of these cones sending information to our brain.

For well over a century rods and cones were thought of as the only photosensitive cells in our eyes, but, at the turn of the twenty-first century, scientists realized they had been overlooking a third type of light-detecting cells.

First noted in 1991, the cells lie deeper within the retina than the rods and cones and look like the underside of a canopy of twisted tree branches. However, it took another decade of careful experimentation before anyone would truly believe that they actually were light-sensitive. Incontrovertible proof came from an experiment using mice which had been genetically engineered to have no rods or cones but could still sense changes in light levels.

These new cells, which labor under the less-than-snappy name of intrinsically photosensitive ganglion cells, link to the parts of the brain which are involved in setting our body clock. They are also important in causing your pupils to expand and contract when light levels change.

William Scribe

Date 2002

Scientists Russell Foster, Ignacio Provencio, Robert Lucas and others

Nationality British, American

Why It's Key Despite over two hundred years of constant research, the eye can still spring surprises. Photosensitive ganglion cells keep our body clocks ticking in time… until we go on a long plane ride.

Key Invention **Tactical lasers**
Prototype shoots down incoming shell

On November 4, 2002, the face of the modern-day battlefield changed forever. As science fiction crossed over into the realms of science fact, a tactical high-energy laser was used to shoot an incoming artillery shell straight out of the sky.

The stunt was in fact a demonstration, which took place at the White Sands test range in New Mexico. It was successful in showing that lasers could be used to remove an enemy threat before it had time to reach its target.

Sci-fi stories aside, lasers aren't quite the brightly colored beams of light we may have come to imagine. A deuterium-fluoride mix produces infrared radiation that is actually invisible to the human eye, but has enough energy to destroy an incoming artillery shell.

Both solid state and chemical lasers have been proposed as weaponry to add to the arsenals of armies. Currently, they are only one per cent efficient, with most of the energy being lost as heat, but future developments could see them on the back of helicopters or ships when entering battle. An Airborne Laser has been developed for a Boeing 747 freighter aircraft on a nose mounted rotating turret. Using a chemical (oxygen-iodine) laser, the flying laser pen could shoot down enemy aircraft at a range of hundreds of miles. At US$3,000 a zap, this kind of technology doesn't come cheap, but the benefits to national security could be priceless.

Nathan Dennison

Date 2002

Scientists US military

Country USA

Why It's Key High energy lasers could save the lives of countless soldiers and civilians in years to come.

Key Event
SARS hits the headlines

Italian physician Carlo Urbani was the first person to identify the respiratory disease SARS (severe acute respiratory syndrome) as a dangerous and highly contagious disease. Urbani worked for the medical organization Médecins Sans Frontières and had received the Nobel Peace Prize on their behalf in 1999.

In 2003, he was asked to look at Johnny Chen, an American businessman who had been brought into a hospital in Hanoi suffering from what the doctors there thought was influenza. Several of the medical staff who treated Chen soon developed the same symptoms despite basic hospital procedures. Urbani quickly saw that it was not the flu, called for isolation of the patient, and also notified the World Health Organization (WHO), It was this move that probably saved the lives of millions of people around the world.

SARS is thought to have originated from China in 2002, but only reached the public spotlight after the death of Mr Chen in February 2003. During the 2002–2003 pandemic, SARS infected some 8,096 known people and caused 774 deaths, a mortality rate of 9.6 per cent.

Urbani was based in Hanoi with the World Health Organization, where he worked on infectious diseases. On March 11, 2003, while flying from Hanoi to a conference in Bangkok, Urbani fell ill and was taken to hospital. In a cruel twist of fate, he had himself contracted SARS, and died eighteen days later.

David Hall

Date 2003

Country Vietnam

Why It's Key A newly discovered infectious respiratory disease, which united the world in an effort to study, fight, and defeat a potential epidemic.

Key Person **Nora Volkow**
Director of the National Institute on Drug Abuse

Not only famous for being the great-granddaughter of the renowned Russian revolutionary Leon Trotsky, Nora Volkow (b. 1956) has been instrumental in demonstrating that, far from being a lifestyle choice, drug addiction is actually a disease of the human brain.

Her first contentious claim, made early in her career, was that cocaine was a dangerous drug – shocking in the 1980s, because it was widely believed to be safe. Her second claim, which raises controversy even today, is that addiction is a disease, whether it involves taking drugs, smoking, drinking, sex, gambling, or eating. Backing up her extraordinary claim is an entire career spent scanning the brains of people affected by different conditions, including different types of addiction. She has shown abnormalities in parts of the brain cortex and in its wiring, and has linked these signs to specific addictive behaviors and long-term reactions to addictive substances. It is clear that certain people are more prone to becoming addicted.

The likelihood of their becoming addicted all depends on a brain chemical called dopamine, which is associated with motivation and pleasure. Dopamine undergoes huge surges – a powerful reward – when certain drugs are taken. All the common drug addictions, from marijuana to heroin, plus many sedative and stimulant drugs used as therapy, are at least partly a result of this chemical. What also matters is the sensitivity of individual people's dopamine response, because it can be modified by their specific environmental, genetic, age, and social factors.

S. Maria Hampshire

Date 2003

Nationality Mexican

Why She's Key Volkow's idea that drug addiction is actually a complex disease means thousands of people will get more help and support, both socially and medically.

Key Event
The Columbia disaster

NASA's Space Shuttle Columbia became the second tragic space shuttle disaster when, on February 1, 2003, it disintegrated upon re-entry into the Earth's atmosphere, killing all seven crew members. Due to conclude its twenty-eighth mission at 9.16 am EST, the shuttle was thirty-eight miles above the ground on approach to the Kennedy Space Center, Florida, traveling at eight times the speed of sound (20,100 kilometers per hour). Ground control received their last broken message from the mission commander Rick Husband "Roger, uh, bu-" at 8.59 am when Columbia broke apart above north central Texas. The sound of the explosion was reported as observers on the ground recorded footage of the debris hurtling ahead of smoke trails near to Dallas.

Columbia was the oldest space shuttle in NASA's fleet of four and had made its 113th launch on January 16, on a sixteen-day scientific research mission orbiting Earth. Footage of the launch revealed a piece of foam falling from the main propellant tank had struck the leading edge of the left wing. "Foam shedding," as it was termed, had occurred on many previous launches, but was not considered a significant safety concern. On this occasion, however, the shuttle's thermal protection system sustained damage, which caused hot gas to penetrate the wing, leading to the break-up.

The investigation board subsequently found that engineers' requests for information to investigate the damage were not met by management, who, the report said, were led by the belief that repairs could not be made even if investigation found it necessary.

Mel Wilson

Date 2003

Country USA

Why It's Key Although it has become commonplace in the public consciousness, the dangerously experimental nature of space travel was highlighted by the Columbia disaster.

opposite The crew of the shuttle mission STS-107 – who died aboard Columbia

BROWN
CLARK
CHAWLA
ANDERSON
RAMON
HUSBAND
MCCOOL
STS 107

-200
T (μK)
+200

Key Event
USA and Russia welcome China into space

On October 16, 2003, China became the third nation on Earth to have a manned space program. Alongside the Russian Federation and the United States, it is one of only three countries able to independently launch manned spacecraft.

Just like in the Russian and American programs, the Chinese launch vehicles began life as inter-continental ballistic missiles. A single use spacecraft, Shenzhou V, carried taikonaut – the Chinese equivalent of an astronaut or cosmonaut – Yang Liwei on his 21.5-hour flight. Shenzhou V was loosely based on the Russian Soyuz craft – the mainstay of the Russian space program for the last forty years.

Despite a warm welcome to the exclusive space club, the Chinese government were quick to point out the "failings" of their Russian and American counterparts. According to an official Chinese government statement: "Unlike that of Soviet cosmonaut Yuri Gagarin, the first man in space, and Alan Shepard, the first American astronaut, Yang's family background was by no means humble or poor by Chinese standards"; a peculiar sentiment for a Communist country to express.

As for Yang Liwei, he has since had an asteroid named after him and received the accolade of "Space Hero." And the official line? "Good health, intelligence, amazing willpower, and eagerness to succeed, as well as a happy family life, have made Yang Liwei China's first astronaut."

Jim Bell

Date 2003

Country China

Why It's Key Less a major scientific breakthrough, more a massive political statement; this is an important reminder that space is not just for the Russians and Americans.

Key Experiment **The Wilkinson Microwave Anisotropy Probe** Meet the oldest photons in the Universe

Scientists 150 years ago didn't know whether the Universe was millions of years old, billions of years old, or infinitely old. Astronomers twenty years ago didn't know whether the Universe was 10 billion years old or 20 billion years old. Now, thanks to the Wilkinson Microwave Anisotropy Probe (WMAP) astronomers have measured the age of the Universe to a precision of one per cent: it is between 13.58 and 13.89 billion years old.

The WMAP satellite has been observing the cosmic microwave background radiation at five different frequencies since 2001. This background radiation is composed of ultraviolet photons of light, from about 380,000 years after the Big Bang, that have been stretched into microwaves by the thousand-fold expansion of the Universe since that time. The variation in the amount of radiation coming from different parts of the sky constrains models of the Universe, which enables astronomers to estimate the age, composition, and expansion rate of the Universe.

WMAP observations imply that the Universe is composed of about four per cent normal matter, about 20 per cent dark matter, and about 75 per cent dark energy. Since dark matter and dark energy are not yet understood, the nature of the vast majority of the Universe is a mystery that remains to be solved.

Eric Schulman

Date 2003

Scientists NASA

Country USA

Why It's Key Provided precise estimates of the age and composition of the Universe.

opposite Whole sky projection from the WMAP shows the Milky Way (red band) – and cosmic microwave background. Colors show temperature variations (plus or minus 200 millionths of a Kelvin) in the background radiation. Dark blue is cool, progressing through blue, green, and yellow to red (hot)

Key Event **Get a *Second Life***

Since virtual realities appeared in the 1980s, computer games have been the most popular artificial worlds. Three-dimensional immersive games have evolved; from 1992's landmark shoot-em-up *Wolfenstein 3D*, through to the current craze for MMORPGs – "massively multiplayer online role-playing games" – like *World of Warcraft*.

So what is *Second Life*? Well according to its creator, Philip Linden, "I'm not a gamer, and *Second Life* isn't a game." Launched by Linden Labs in 2003, *Second Life* is different. You can't win or lose, and nobody's keeping score. Instead of delivering a fully-formed game, Linden decided to create an empty world and let its residents fill it up.

Linden put the landscape in place, and defined the laws of physics, but almost everything else in *Second Life* is created by the residents themselves. On arrival, you can customize everything, from changing the angle of your nose to buying some land and then building a house from scratch. Some people use bodies faithful to their real appearance, whereas some fly around as dragons – wings, fiery breath, and all. If you want to play a game in *Second Life*, you make up your own rules.

So we've established that it's not a game, but *Second Life* itself seems to be winning; its population is now in the millions, spread over a virtual landscape the size of four New Yorks.

Matt Gibson

Date 2003

Country USA

Why It's Key The first massively popular virtual reality pastime outside the gaming world, *Second Life* provides a glimpse of the future of entertainment.

Key Discovery **Methuselah**
The old man of the Universe

Methuselah is a figure from the Old Testament who is said to have lived for 969 years. According to the Bible, this makes him the oldest person ever to have lived. It was rather apt then that a planet 2.5 times the mass of Jupiter, and thought to be around 12.7 billion years old, should be named after the Hebrew patriarch.

12.7 billion years is very, very old, even in cosmic terms. Current estimates age the Universe at around 13.7 billion years old, which means that, as long as Methuslah's birth date is accurate to within a few million years, planets started to form much earlier than we previously thought. This is important because it suggests that planets might be more abundant in the Universe than we realize; and more planets means more possibilities for life elsewhere.

Before the discovery of Methuselah by the Hubble space telescope – in the M4 globular star cluster some 12,350 light years away – it had been thought that planets required heavy elements like iron to form. Such elements are made by nuclear fusion reactions in dying stars. As Methuselah is so old, it is likely to be made mainly of hydrogen and helium since there would not have been any opportunity for heavier elements to have come into existence at that point.

Locked in a graceful dance with a pair of stars, Methuselah was ejected to the outer reaches of the M4 cluster, where its companion – a "flashing" star called a pulsar – first caught the attention of the Hubble.

Jim Bell

Date 2003

Scientists Steinn Sigurdsson and colleagues

Nationality Icelandic

Why It's Key Completely altered our view of the distant history of the Universe, and increased the likelihood of finding planets in orbit around other stars.

Key Event
Biofuel debate

We are becoming painfully aware of the impact that human activity is having on our planet. Regardless of the climate change debate, it is undeniable that as human populations grow we are using up ever more of the planet's limited resources.

Biofuels are one of the possible renewable alternatives that could replace fossil fuels as main sources of vehicle fuel. They can be derived from crops such as sugar cane, corn, and even cotton. Although bio-ethanol and bio-diesel do release carbon dioxide when burned, the carbon they release is absorbed by the fuel crop during its growth, making the whole cycle essentially "carbon neutral."

Unfortunately it seems that biofuels may not deliver all that they promise; the process of growing and refining the fuels currently produces large amounts of pollution and is appallingly inefficient. In terms of land, the space required to grow enough biofuel to power the world would be almost unimaginable. And it is estimated that the grain required to fill the tank of a 4x4 vehicle just once could feed a person for a year. Nevertheless a lot of money has been ploughed into biofuels projects, such as the multi-fuel car, the Volkswagen Gol Total Flex. Introduced in 2003, it is extremely popular in Brazil, and can be powered by either gasoline or bio-ethanol.

It remains to be seen whether biofuels can provide a solution to the energy crisis, or whether the current biofuels boom should be regarded as a white elephant painted green.

Jim Bell

Date 2003

Country Brazil

Why It's Key One way or another, biofuels are an integral part of the current environmental crisis, but the fact that they are being seriously considered and invested in shows at least that there is a real desire to consider renewable energy sources.

Key Experiment Chinese scientists create human embryos with rabbit eggs

In 2003, a team of scientists at Shanghai's Second Medical University succeeded in extracting human cells from embryos of human-animal origin. The embryos were part human, part rabbit, made up of rabbit cells containing human DNA, and although they only existed for a few days – after a week the embryos were destroyed – they yielded both human stem cells, and a new method for producing them. Huizhen Sheng and his team had effectively reset the cells, turning back their internal clocks to a stage when they could become any type of cell in the body, from a blood cell to a brain cell.

The research team took human skin cells and transferred the nuclei – where the DNA is housed – into rabbit egg cells that had had their own nuclei removed. The human DNA was reprogrammed to form embryos and the "memory" of the differentiated skin cells was removed.

The resulting embryos were hailed by some as the first human-animal "chimeras," after the Greek mythological creature, which had a lion's head, a goat's body, and the tail of a serpent. Scientifically speaking, chimeras are in fact animals with two different populations of cells, such as the "geep" created in 1984, which was a mixture of goat and sheep cells. Sheng's experiments did, however, bring scientists one step closer to that end. Over the following years, researchers succeeded in producing human-mouse chimeras and finally, in 2007, a controversial sheep, 15 per cent of whose cells were human.

Hannah Isom

Date 2003

Scientist Huizhen Sheng

Nationality Chinese

Why It's Key Presented the possibility for a cheap source of human stem cells and new models of disease, and raised many ethical questions about the creation of human-animal hybrids.

Key Invention **Organic LEDs**
The future's bright

In 2004, Sony announced mass production of organic LEDs. They promise that screens made with them will be brighter, respond quicker, and give a better range of colours. Magic!

Organic LEDs (if chemistry isn't your first language, organic compounds are those that contain the same elements as living things, namely carbon, hydrogen, and oxygen) work in just the same way as traditional LEDs, by exciting electrons and waiting for them to drop down to lower energy states, releasing a photon. The difference is that in organic LEDs, the electrons travel through organic molecules instead of layers of metal.

Given how much control we have over molecular structure, it is now perfectly possible to take an organic molecule that conducts electrons and alter it to be water soluble, bond with nitrogen, shaped like a smiley face, or do whatever we want. With this amount of control over the microscopic properties of the LED, the finished component can have a wide variety of optional extras: for example, organic LEDs can be made flexible – they could make a roll-up TV, or electronic newspaper. And because they're already made of the same elements you find in the human body, they are also candidates for bio-compatible prosthetics, such as artificial eyes. But mainly, they are a whole lot cheaper than inorganic LEDs, which is why, when Sony decided an LED screen was their next development, their natural choice was organics.

Kate Oliver

Date 2004

Scientist Sony

Nationality Japanese

Why It's Key A cheaper, brighter, ultimately customizable light source, many times more efficient than traditional bulbs, and just waiting to be built into new, exciting technology.

opposite A panel of organic light emitting diodes (OLEDs)

Key Invention
Self-replicating robot

Since the mathematician John von Neumann formally proposed the idea of self-replicating robots in 1955, scientists have dreamt of creating a machine capable of replicating or even repairing itself. In 2004, a team from Cornell University brought this dream closer to reality by creating the first robot that could make a copy of itself.

The team unveiled their creation in the science journal *Nature* in May 2005. The machine was based around a series of modular blocks they called "molecubes." Each was packed with a motor allowing the ten-centimeter cube to swivel over a diagonal shear line. An individual block was programmed with the information required to build a new version of the four-block machine that was held together by magnets, a task which the blocks were able to do within three minutes. Though the machine was relatively simple – it was just four blocks after all, with the sole ability to make a copy of itself and nothing else – it hints at the potential for self-repairing robots in the future, especially with regards to space travel. For instance, if a normal robot sent out into space breaks, we cannot easily travel to fix it. However, if the robot had the capacity to recreate one or more of its components, it may be able to replace the broken part and continue. The ultimate vision of this technology is to send self-replicating probes to Earth's near-neighbors, as von Neumann proposed more than half a century ago.

Steve Robinson

Date 2004

Scientists Vistor Zykov, Hop Lipson and colleagues

Nationality Russian, American

Why It's Key In the future, self-replicating machines will revolutionize robotics, and could herald the key to exploring the galaxy.

Key Event **MMR and autism**
There is no link

The 1998 MMR controversy was a disaster for doctors, parents, and scientists. That February, the UK media picked up on research – led by Dr Andrew Wakefield and published in the medical journal *The Lancet* – linking incidences of autism to the widely-used childhood vaccine for measles, mumps, and rubella. Following a media frenzy, in which Wakefield was reported to have advocated using single jabs for each disease – hitherto untested – the number of children receiving the MMR vaccine dropped by more than 10 per cent in some areas of the UK.

What wasn't well communicated was the fact that Wakefield's study had focused on just twelve children, nine of whom were suffering from autism when they were referred to him. In the paper, Wakefield himself had stressed the need for further investigations to examine the "possible relations" of their symptoms to the vaccine. A fourteen-year-long Finnish study of the safety of the MMR vaccine, which concluded there was no data to support a link with autism, went almost unnoticed. By 2004, the triple vaccine had been used to protect 500 million people in ninety countries from the dangers of measles, mumps, and rubella. A succession of larger scientific studies interrogating the supposed link had trashed the notion that the vaccine might cause autism, but still there remained an air of nervousness surrounding the jab.

Finally, in March 2004, *The Lancet* published a retraction by ten of the thirteen authors of the original study. Wakefield now faces charges brought against him by the General Medical Council.

Hayley Birch

Date 2004

Country UK

Why It's Key Proof that everything you read in the papers – and in highly respected medical journals – should be taken with a pinch of salt.

Key Invention **First podcast indicates growing dominance of iPods**

After being temporarily dismissed as techno-speak, podcasting sailed into the mainstream thanks to the efforts of entrepreneur and broadcaster Adam Curry. Today Curry is known as the presenter of the Daily Source Code podcast, in which he mixes music with his own personal stories and thoughts.

Podcasts, in their basic form, are audio files that can be downloaded from the Internet and transferred onto a personal media player such as an iPod. What Curry did was to create a computer application that could find and deliver these files to a media player automatically.

Within months of his application hitting the Web, people who had never even considered broadcasting were hosting their own shows, and, in the space of a year, Google search results for the word "podcast" had soared to 61 million from a measly 6,000. The technology was freely available; anyone could broadcast, so they did, triggering an explosion in amateur, radio-style broadcasts. There was even talk that podcasts could pose a threat to live radio.

The movement gained extra impetus when technology giants Apple made a place for podcasts within their downloadable music player, iTunes. By the end of 2005, the editors of the New Oxford American Dictionary had voted "podcast" Word of the Year. Apple then attempted, unsuccessfully, to apply trademark status to the terms "pod" and "podcast."

Hayley Birch

Date 2004

Scientist Adam Curry

Nationality American

Why It's Key The public are given the chance to become broadcasters; Apple continues its quest for world domination.

opposite i-pod

MENU

Key Discovery
Binary pulsars

In 2003, astronomers using the Parkes radio telescope in New South Wales, Australia, discovered a doomed star system: two neutron stars orbiting each other every 147 minutes, destined to destroy each other in 85 million years. The system is called PSR J0737-3039 and is the sixth known binary pulsar.

Einstein's theory of general relativity predicts that closely-orbiting objects as massive as neutron stars – the collapsed remnants of huge stars – emit large amounts of gravitational radiation. These gravitational waves remove energy from the binary system, causing the orbit to get smaller and smaller until the two neutron stars collide.

Rapidly-rotating neutron stars emit beams of radio waves from their magnetic poles. Like the Earth, whose magnetic and geographic poles are separated by about ten degrees, neutron stars often have misaligned magnetic and rotational axes. This causes the radio beams to sweep across space. When the Earth is within the area swept by the beam, radio telescopes detect a pulse of intense radiation as the beam sweeps by.

In 2004, the PSR J0737-3039 team discovered that both neutron stars in the system are pulsars. It is the first double pulsar found, and will allow astronomers to measure the properties of both objects in the binary system. Over the next decade, observations will allow astronomers to test gravitational physics with unprecedented precision.

Eric Schulman

Date 2004

Scientist Andrew Lyne and colleagues

Nationality British

Why It's Key Provides a so-far unique laboratory for performing tests of general relativity, neutron star structure and evolution, and other astrophysical theories.

Key Event
Geologists unveil a wet Martian history

Two geologists, each with one arm, a magnifying glass, and eyes a mere 1.5 meters from the ground, plod along at a snail's pace across the Martian surface. Their names are Spirit and Opportunity, and they arrived on Mars in 2004 on January 4 and 25 respectively, as part of NASA's Mars Exploration Program. If you haven't guessed already: They're robots.

Spirit was sent to a possible former lake, while Opportunity was tasked with investigating the mineral deposits of Meridiani Planum. Both robots were primarily designed to carry out tests on rocks and soils in a bid to determine whether or not water had once flowed on Mars. The selection of the equipment at their disposal included panoramic cameras, thermal emission spectrometers, magnets, microscopes, and rock abrasion tools. Their first major breakthrough came when Opportunity detected hematite while analyzing its first few samples. Hematite usually only forms in the presence of water, but not always. Conclusive proof came soon after in the form of the detection of jarosite, a mineral which can only form in water. Further evidence of aquatic activity has followed, and late in 2007 both robots were still busy at work, having already hugely exceeded their planned ninety-day mission remits.

On Earth, as a rule, where you find water, you find life. Given the evidence already presented by the Mars Exploration Rovers, as they are jointly known, the red planet may well present us with some of the most significant and exciting discoveries of the next century.

Christopher Booroff

Date 2004

Country USA

Why It's Key Evidence of water on Mars tells us where to look for signs of life, either past or present, while furthering our understanding of the Solar System.

opposite Mars Exploration Rover Opportunity examines bedrock on the Martian surface

Key Discovery **Graphene**
Beyond pencil lead

The "lead" in your pencil is made of carbon – in a form called graphite. Now unless you have a background in nanochemistry, you could be forgiven for thinking graphite and graphene are one and the same thing. It's an easy mistake to make; essentially they're both made from sheets of carbon atoms. So why does one make an excellent – if rather dull – writing material, while the other excites the imaginations of aircraft creators and electronics enthusiasts?

The difference is actually quite simple. Graphite is composed of lots of layers of carbon atoms, loosely bonded together between the layers, so that when you drag it across a piece of paper some of the bonds are broken, leaving bits of graphite – as words or drawings – on your page. Graphene, on the other hand, is made from vast, one-atom-thick sheets of carbon atoms. And it's far from flaky. Since its discovery in 2004, scientists have been busy designing ultra-strong materials and researching ways to make new, ultra-fast components for computers and other electronic devices.

In modern computer chips, copper is used to connect the different layers because it is a good electrical conductor. But as chips continue to shrink and copper connections have to be made smaller, they will become less and less efficient. This is just one area where graphene has already shown promise – electrons have been shown to flow through it at near light speed. It could even one day replace the silicon in silicon chips.

Hayley Birch

Date 2004

Scientist Andre Geim

Nationality Russian

Why It's Key Graphite is made of the same stuff as diamonds, and could one day be more valuable than them, or at least more useful.

Key Event **Small steps to the Moon in preparation for a giant leap to Mars**

In 1961, President John F. Kennedy proposed landing a man on the Moon and returning him safely to Earth. On January 14, 2004 President George W. Bush announced plans to take astronauts back to the Moon... and leave them there.

During his speech, Bush identified key targets aimed at enabling future trips to Mars and beyond. Key to achieving these goals will be conducting research on the long-term implications of working in reduced gravity. This will be provided by the completion of the International Space Station (ISS) by 2010, and the establishment of a permanent lunar base by 2020. It was also announced that the space shuttle will be replaced with spacecraft capable of carrying astronauts to "other worlds" by 2014.

It is widely recognized that such bold projects may require international collaborations; the ISS and potentially a lunar research facility would provide opportunities to strengthen the relationships and techniques required to successfully pool our resources. A further reduction in economic burdens could be gained through mining raw materials. If fuel can be extracted from the Moon, then the cost of future missions to Mars could be greatly reduced due to the gravitational field being weaker at the lunar surface than on Earth.

The potential rewards of lucrative minerals, or even the discovery of extraterrestrial life, has finally proven too tempting for the U.S. government. In the meantime, the Moon appears set to become the largest stepping stone ever used.

Christopher Booroff

Date 2004

Country USA

Why It's Key Could lead to the discovery of life elsewhere in our Solar System, while developing some of the technology required to one day leave Earth for a new home.

Key Discovery **"Hobbit" fossils found**
One up for the little guy

Few discoveries of modern times have caused more uproar than when, in October 2004, a joint Australian-Indonesian team of paleoanthropologists working on the Indonesian island of Flores reported they had discovered the bones of a "hobbit." They named the human-like creature, who stood just one meter tall and used stone tools, as a new species – *Homo floresiensis* – estimated the bones to be somewhere between 38,000 and 13,000 years old. Further excavations of the same caves revealed another seven individuals.

The hobbits sparked worldwide interest; if true, these mini-people would have co-existed with their cousins, *Homo sapiens*. What's more, the inhabitants of Flores tell legends of the existence of Ebu Gogo – tiny people matching the description of the hobbits – on the island. That they could still be living somewhere deep in the forests of Indonesia suddenly seemed a possibility. But not everyone was convinced that the bones were those of a new species. Critics, including Indonesia's "King of Paleoanthropology" Teuku Jacob, insisted that the bones belonged to humans suffering microcephaly – a condition which causes people to develop unusually small heads. The debate turned acrimonious when Jacob was accused of taking the bones without permission and failing to return them.

Claim and counter-claim about the hobbits flew back and forth between researchers. In 2007, two studies put the answer beyond doubt. Independent analyses of the skulls and wrists showed that they were separate species – the little man is here to stay.

Mark Steer

Date 2004

Scientists Peter Brown, Mike Morewood

Country Indonesia

Why It's Key Arguably the most extraordinary anthropological discovery since the *Australopithecus afarensis* fossil, Lucy, was discovered.

Key Event
Giant squid caught on camera

Tales of giant sea creatures have been told since the time of the *Odyssey*, in which Homer describes a twelve-tentacled serpent with six heads. Writing in 1755, the bishop of Bergen in Norway described a monster a mile long, which he called a "Kraken" (a word made famous by John Wyndham in his science fiction novel *The Kraken Wakes*) and, in 1770, the Royal Society heard Charles Douglas' account of "an animal of 25 fathoms long." Amazingly, among the fables and fabrications, there may just lie an inkling of truth.

The biggest squid ever caught was hauled in at Island Bay, New Zealand in 1880. It measured 18.5 meters – not quite 25 fathoms (around 46 meters), but longer than two buses parked end to end. It wasn't until 2004, however, that a live giant squid was caught on camera for the first time. Researchers baited the eight meter long beast off the coast of Japan, but were left with only a film full of stills and a severed tentacle.

In 2006, the same team, led by Japanese biologist Tsunemi Kubodera, shot the first film of a giant squid as they struggled to haul aboard an animal of a mere seven meters. They tracked down the huge invertebrate by following in the wake of larger predators – sperm whales, some of the only animals in the oceans capable of making a meal of a giant squid. The huge invertebrates are in fact an essential part of the sperm whale's diet, making them an interesting subject for study as another link in the food chains of marine ecosystems.

Hayley Birch

Date 2004

Country New Zealand

Why It's Key The first views of a living giant squid, and an intriguing addition to our picture of underwater ecosystems.

Key Event
Boxing Day tsunami

On Boxing Day 2004, an earthquake, recorded between 9.1 and 9.3 on the Richter scale, hit the Indian Ocean, 240 kilometers off the coast of Sumatra. Over 1,600 kilometers of the India tectonic plate slipped under the Burma plate, releasing an immense amount of energy and displacing an estimated 30 cubic kilometers of water. Although it was a massive earthquake, it was the water that proved most devastating.

The waves that spread across the Indian Ocean are thought to have measured up to thirty meters high by the time they reached the coastlines of countries such as Sumatra, Indonesian and Thailand. Tsunamis travel quickly across open water at a height of only a foot or so, but upon reaching shallower water, slow down due to friction with the rising seabed, and grow in height until arriving at the shore. We will probably never know exactly how many people were killed in the disaster, but current estimates put the figure at 230,000. Entire towns and even cities were totally destroyed, such as the Sri Lankan city of Galle. Many areas are still struggling to recover from the damage caused to industry and infrastructure, but the psychological trauma will undoubtedly last even longer.

There was a massive reaction to the disaster from the international community, and billions of dollars in aid were pledged. Aid has been directed not only at looking after victims of the disaster but also at creating a system of early warning buoys that will hopefully prevent such devastating loss of life in years to come.
Jim Bell

Date 2004

Location Southeast Asia

Why It's Key A poignant reminder of our vulnerability to the power and might of nature – in such an event, we are as fragile as any other living thing on the planet.

opposite Water recedes as the first of six tsunamis rolls towards Hat Rai Lay Beach, near Krabi in southern Thailand

Key Event
Voyager 2 enters the heliosheath

The Sun is continually ejecting a stream of charged particles – plasma – from its upper atmosphere into space. This is the solar wind. For roughly 10 billion kilometers these particles career outwards at over a million kilometers per hour. However, after a while they start to collide with the ions, atoms, and molecules that inhabit the interstellar medium, and slow down to more reasonable, subsonic, speeds before eventually stopping altogether.

This "bubble" of solar wind which encompasses the Solar System is called the heliosphere and beyond it the Sun has no influence. For the first time in our history, a manmade object is nearing the very edge of the bubble and is preparing to leave our home star behind. The space probe Voyager 1 crossed into a zone known as the heliosheath in December 2004. In this zone the solar winds have slowed down to subsonic speeds and the interstellar medium is beginning to penetrate. Voyager 1 is now over 15 billion kilometers from the Sun – more than twice the distance from the Sun to Pluto; scientists believe that it will cross the final boundary in 2015, and are hopeful that it will still have the power to send back information about the greatest beyond there is.

But long after its power has died, Voyager 1's trip of discovery will continue. Along with its twin, Voyager 2, which crossed into the heliosheath in 2007, the probe carries a golden record that contains pictures and sounds of Earth. It's an interstellar calling-card for anyone willing to listen.
Mark Steer

Date 2004

Country USA

Why It's Key Voyager 1 really is learning about life on the edge, and if interstellar travel is ever possible, it'll be partly thanks to the information sent back by this pioneering probe.

Key Discovery
Junk DNA: or is it?

Science has still to answer many difficult questions. How did the Universe begin? What is dark matter? Why haven't extraterrestrials made contact yet? But among these great mysteries, there remains one slightly less epic, but nonetheless fascinating question: Why does most of our DNA seem to do diddly squat?

It might seem a small detail when there are other pressing matters at stake, but only about one per cent of our DNA has any obvious function. Scientists have been arguing for years about what the rest of the three billion base pairs are up to.

In 2004, researchers at the University of California, Santa Cruz, thought they'd stumbled on a clue. They had found long stretches of genetic code in rodents, which were almost identical to sections of human DNA. This must mean, they reasoned, that these pieces of code were important somehow – evolution must have wanted to keep them for a purpose. But just a few months later, scientists at a meeting in New York sat agog as researchers from the Lawrence Berkeley National Laboratory told how they had deleted massive chunks of these "conserved" regions in mice with no discernible effect. A year later, the mice were still showing no outward signs of their inherent defects. It looked liked the Santa Cruz findings could have been a red herring.

So how much of our DNA really is worth keeping? The answer is: We still don't know. Even with the entire genome sequenced, our lack of understanding is still quite astounding.

Hayley Birch

Date 2004

Scientist Gill Bejerano

Nationality Israeli

Why It's Key Only one hundredth of our DNA codes for proteins, which begs the question: What the hell is the rest of it doing?

Key Experiment **Molecular DNA computer Doctors** The future of medicine?

Despite being made of totally different materials than the usual electronics, mouse, and keyboard combinations, biomolecular computing is possible. Living organisms, like us, have systems that carry out processes controlled by some kind of input.

The medicine of the future is a "Doctor in the form of a cell"; little organisms made of only a few strands of DNA and chemicals that float throughout your body. These mini "doctors" would be designed sin such a way that, if they find a problem, they'd produce the drugs our bodies need to fix it.

Perfecting this nanotechnology is still a long way off, but the first steps have been taken. Ehud Shapiro and his team at the Weizmann Institute of Science in Israel first produced a DNA computer in 2002, and in 2004 built one which releases a cancer-fighting drug if it finds cancerous conditions. The biomolecular computer has short DNA strands that search the body's genes and bind with the parts involved in specific cancers. The computer then observes what it's attached to, to see if there are any cancerous symptoms. If the gene has turned bad, then the computer releases a therapeutic piece of DNA that binds to the cancer and suppresses it.

While this was all carried out in a test tube, it serves as a proof of a principle that could change the face of medical treatment, as well as having other potential uses in biochemical sensing and genetic engineering.

Douglas Kitson

Date 2004

Scientist Ehud Shapiro

Nationality Israeli

Why It's Key The future of medical diagnosis and treatment.

Key Discovery **Supersolids?**

In 2004, at Pennsylvania State University, the first solid Bose-Einstein condensate (BEC) was created. Researchers found that when they cooled a form of helium, helium-4, past its transition to a solid, it changed phase into a BEC. Coined a "supersolid," a solid BEC acts like a superfluid; it can move without undergoing frictional forces – the ultimate quantum lubricant.

When helium is very cold it is a very odd substance; to turn into a solid it needs to be cooled and then put under great pressure and even then, in its solid state, it can be squeezed like rubber. Physicists Eun-Seong Kim and Moses Chan observed that the solid helium BEC could move through another solid which had holes. Usually when we cool materials, the atoms within them move closer together, which is why liquids turn to solids when they get colder. Even if you get to absolute zero – the temperature at which there is no energy left in the system – there would still be small gaps in a crystal because of a phenomenon called zero-point energy.

There is still some controversy as to whether what Kim and Chan observed was really a supersolid, and experiments are ongoing to finally settle the question of the existence of supersolidity. Their experiment confirms an earlier prediction that even solids can become BECs under the right conditions, and the observations are causing scientists to rethink how to distinguish between solids and liquids in the quantum world.

Leila Sattary

Date 2004

Scientists Eun-Seong Kim, Moses Chan

Nationality American

Why It's Key A new phase of matter for quantum mechanics, causing a rethink of fundamental distinctions between states.

Key Event **Web 2.0**
Nobody really knows what it means

In October 2004, the first Web 2.0 conference was held in San Francisco for business leaders to discuss the changing ways in which users interacted with commercial enterprises online. The term "Web 2.0" was coined to describe the radical changes in the way people were using websites; rather than finding and reading static pages, users were now able to modify and create content. Examples are sites created almost entirely by user content such as wikis, blogs, and sites hosting online auctions, image sharing, and social networking.

Tim O'Reilly, founder of O'Reilly Media and host for the 2004 conference, considers Web 2.0 as a way in which businesses can make the best use of the strengths of the World Wide Web, including using the social networking inherent in many of the sites mentioned above. British polymath Stephen Fry, on the other hand, describes it as an example of genuine interactivity in "the reciprocity between the user and the provider." Moving away from the economic focus, it can be thought of as a way of facilitating creativity, collaboration, and sharing among Internet users.

Although the name implies a radically new kind of Internet, the changes in use have developed slowly over time (beginning well before the 2004 conference) and do not rely on any single piece of new technology. Inventor of the World Wide Web, Tim Berners-Lee does not think the development merits the new name, claiming that "Web 2.0" is a "piece of jargon... nobody really knows what it means."

Shamini Bundell

Date 2004

Country USA

Why It's Key A revolution in the way people use and interact with the Internet.

Key Event **First privately funded spaceflight SpaceShipOne**

Out of this world explorations have always been out of most people's grasp. Unless you're an astronaut with an advanced degree in physics, chances are you won't be heading spacewards any time soon. However, Scaled Composite's SpaceShipOne opened up the possibilities for change in all of this.

Burt Rutan, one of the most influential aircraft designers of the twentieth century pioneered the first privately-funded manned spaceflight using innovative technology. It employed a hybrid rocket motor combining the stability of a solid rocket motor with the controllability of a liquid one. It featured a unique feathered re-entry system whereby the rear half of the wing and the twin tail booms folded upward along a hinge running the length of the wing, which increased drag whilst maintaining stability and minimal heat build-up. Although it generated huge interest, SpaceShipOne only completed three spaceflights, two within 14 days of each other, winning it the Ansari X prize. A fourth was scheduled for October 13, but Rutan, concerned about keeping the craft's systems intact, cancelled it and all other flights.

This hasn't stopped the idea of commercial space journeys progressing. Richard Branson, business mogul and adventure junkie, has long set his sights on spaceflight for the (albeit very rich) masses. Along with Rutan and Scaled Composites, he is developing SpaceShipTwo, based on technology used on the first craft. Virgin Galactic is aiming to transport people on private space flights by the end of 2009. So forget the PhD, all you'll need is a spare $US200,000. Easy.

Fiona Kellagher

Date 2004

Scientist Bert Rutan

Nationality American

Why It's Key Innovative technology made for a safer flight and has brought the idea of space travel well and truly into the realms of possibility rather than a science fiction fantasy.

Key Event **Huygens probe lands on Titan**

The probe sails through space, its dormant systems waiting. As it approaches its target, it prepares for atmospheric entry. Blasting through the upper layers of the thick atmosphere, the heat shield slows the craft from about 26,000 kilometers per hour to just 1.5 times the speed of sound. The parachutes soon open and the probe begins to drift serenely, through clouds, over the landscape, toward its final landing site. This is no Earth re-entry mission, but the journey of the Huygens probe. On January 14, 2005, it landed on Titan, Saturn's largest moon, and become the most distant human object to land on an alien world.

The Cassini-Huygens Probe – named after Giovanni Domenico Cassini, discoverer of four Saturn moons, and physicist Christiaan Huygens – left Earth in October 1997 on a seven-year voyage to study Saturn and Titan. The surface of Titan remained a mystery: traditional telescopes could not pierce its thick layer of orange cloud. Its atmosphere is a complex, heaving mix of hydrocarbons, and some scientists speculated that there might even be "lakes" composed not of water, but of liquid methane.

Seven years, and billions of miles, after the launch, the Huygens probe detached from the main Cassini craft, traveling the remaining 2.5 million miles itself. After landing with a "splat" on Titan "mud," it began to return images of the surface. We have since learned that huge methane lakes exist as predicted: some measure more than 100 kilometres wide. Titan remains the furthest world to have been touched by humanity.

Steve Robinson

Date 2005

Country Europe/USA

Why It's Key The Huygens probe has allowed us to study a distant world, whose clouds, mountains, and lakes seem eerily similar to Earth's.

opposite Technicians with the heat shield of the Huygens probe

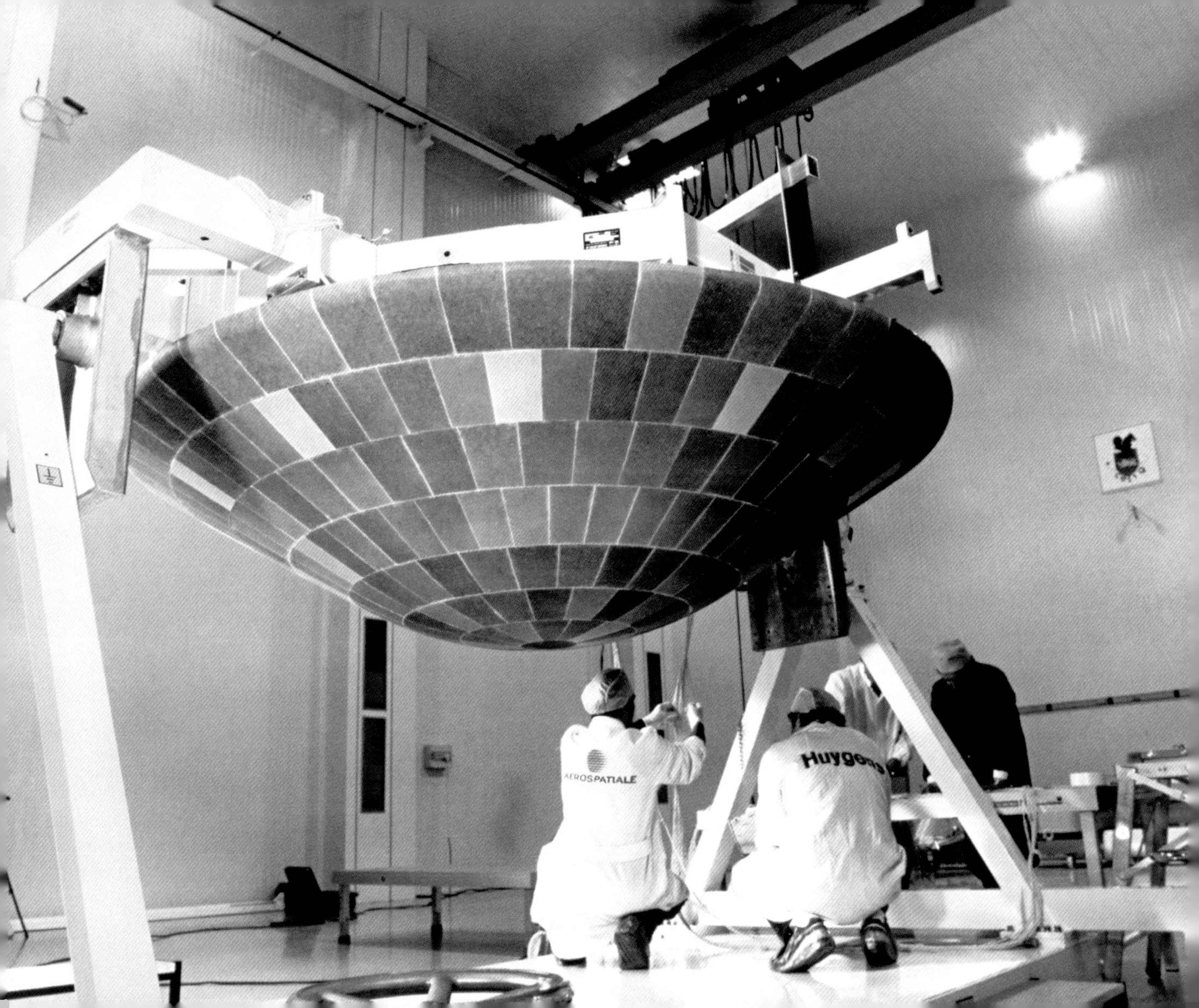
AEROSPATIALE

Key Discovery **Life in the driest place on Earth**
Bacteria survive in Mars-like soil

Environments on Earth don't get much more hostile than the Atacama desert; the driest place on the planet. The region is sandwiched in a "rain shadow" by the Andes Mountains to the east and coastal mountains bordering the Pacific Ocean to the west. Most parts get just a few millimeters of rain every few years, and some haven't seen water for centuries.

It's the closest environment that this planet has to Mars; so close that NASA uses the Atacama as a model for the red planet. And yet, even here, there is life. Deep in the Atacama's soils sleep bacteria, sitting out the drought and awaiting the eventual smattering of rain.

In 2004, a team of scientists led by Raina Maier managed to culture bacteria recovered from the otherwise lifeless soil. They were originally searching for chemicals that would hint at the past presence of plants, and collected soil samples from a depth of 20-30 centimeters along a 200 kilometer stretch of desert. Even though they took every precaution to make sure the samples weren't contaminated, they still managed to grow bacteria from these samples back in the lab, using sterile water.

Maier's work contradicted a study published just a year earlier, suggesting that the Atacama's arid heart was sterile. In the years that separate the Atacama's infrequent bursts of rain, Maier speculates that the bacteria go into suspended animation. They're there, but as she says, "They're not partying."

Ed Yong

Date 2004

Scientist Raina Maier

Nationality American

Why It's Key Showed that even the most barren environment on Earth can support life, and hinted that Mars explorers could find life there by looking more closely.

Key Invention **Right up close**
The superlens

So much of what we know about the world around us depends on what we can see of it. Or rather, what we can see through an ultra-high-resolution microscope. Without their trusty lenses, modern nano-scientists would be akin to chefs without stoves; musketeers without rapiers.

But as we charge, swords raised, into the nano-era, caught up in the chase for the quantum computer chip, we find that something is holding us back – we can't see. Optics, it would appear, has been dragging its feet somewhat.

Thanks to Xiang Zhang at the University of California, Berkeley, however, it looks like we'll be going in eyes wide open after all. In 2005, Zhang's team made the first "superlens," a lens that provided far greater resolution than a conventional microscope. Using superlenses, the researchers have been able to distinguish wires only seventy nanometers apart – that's a hundred times less than the distance across the smallest human cell, the red blood cell.

Zhang achieved this none-too-small feat by looking for "evanescent" light waves, which disappear so quickly after bouncing off an object – within just a few nanometres – that they're almost impossible to catch. He developed a lens made of silver that could collect these "near-field" waves and use them to produce an image. Further improvements have allowed scientists to turn tricksy evanescent light into bog standard light, meaning images can now be processed further away from an object.

Hayley Birch

Date 2005

Scientist Xiang Zhang

Nationality Chinese

Why It's Key Superlenses allow scientists to zoom in on a hitherto unseen world; the field of superoptics may have applications in communications, electronics, and computing.

Key Invention **Wireless electricity**
Banishing the low battery beep

Marin Soljačić doesn't like being woken up by his mobile phone. Especially when it's not even a call, but a reminder that he's forgotten to plug it in to charge it. Soljačić, a researcher at the Massachusetts Institute of Technology, has set out to rid us of this modern-day plague by making sure we never have to plug anything in again. He's researching WiTricity: wireless electricity.

Following in Nikola Tesla's footsteps, Soljačić 's team are looking into ways of transmitting electricity without all those irritating wires. Soljačić 's dream is to walk into his house and have his phone simply begin charging, hooking its electricity out of the air without wires, and, crucially, without having to rely on an absent-minded professor to plug it in. Relying on magnetic fields WiTricity uses resonance for efficiency – the same phenomenon by which an opera singer can shatter a wineglass. In WiTricity's case, however, it's magnetic rather than sonic resonance that is the key. The transmitter resonates at a particular frequency. If there's no receiver nearby, then the transmitter efficiently re-absorbs the energy. If a receiver is nearby, then it absorbs the power instead.

So far, the team have managed to light up a 60-watt bulb from a distance of two meters. They hope that distance can be increased and that the antennas can be made small enough to fit into a mobile phone – and that should be the last we hear of the "LOW BATTERY" beep.

Matt Gibson

Date 2005

Scientist Marin Soljačić

Nationality Croatian

Why It's Key In the wireless era, the last remaining tangle of wires under our desks are the power cords. Without them, we may never trip over a cable again.

Key Discovery **Neutron star bursts**
Explosions of cosmic proportions

In late 2004, astrophysicists observed a once-in-a-lifetime cosmological event. In a fraction of a second, a neutron star on the other side of the galaxy released a gamma-ray burst containing more energy than the Sun releases in a quarter of a million years. Had it been closer to the Earth, some scientists believe it could have resulted in a mass extinction.

The explosion, revealed by scientists at a NASA press conference the following year, was more than a hundred times more powerful than the two short gamma-ray bursts observed previously in 1979 and 1998.

Gamma-ray bursts are one of the great remaining mysteries of the Universe. These flashes of energy appear across the Universe randomly and are thought to be telltale signs that a high-mass star has collapsed to form a black hole, sometimes billions of light years away.

The Earth's atmosphere absorbs most incoming gamma rays and, although they can be detected from the ground, satellites and space-based instruments such as NASA's SWIFT observatory have been launched specifically to study them. In 2004, SWIFT telescopes were pointing at the neutron star SGR-1806-20, some 50,000 light years away. A type of neutron star called a "magnetar," it has the most powerful magnetic field of any object known. The collision of two neutron stars in a binary system or between a neutron star and a black hole is thought to have released the gamma-ray burst – the most energetic event that humankind had ever recorded.

Arran Frood

Date 2005

Scientists Bryan Gaensler and colleagues

Nationality Australian

Why It's Key Provided scientists with crucial data to help understand the nature of cosmic explosions.

Key Invention **The nanocar**
Single molecule transport

Among other bizarre and amazing mini-things to come out of the nanotechnology boom, is the nanocar. Legendary physicist Richard Feynman promoted the idea of mechanically manipulating matter on an atomic scale back in the 1950s, and the nanocar, developed by a team at Rice University, Texas, is a remarkable example of his vision. At only four nanometers across, it is little wider than a strand of DNA. To put this in context; a human hair measures around 80,000 nanometers.

Within a single molecule, the nanocar has a chassis with axles and pivoting suspension; its wheels are comprised of buckyballs – spheres made of sixty carbon atoms. It is the first car-shaped, single-molecule, nano-object to roll perpendicular to its axles. Using scanning tunneling microscopy the team observed the nanocar rolling, not sliding, on a surface of gold when heated to two hundred degrees Celsius.

Envisaging a billion miniscule factories, Feynman influenced the later popular work of K. Eric Drexler in his book *Engines of Creation: The Coming Era of Nanotechnology* (1986). Although it does not have a molecular motor, the nanocar, according to its creators, is part of a concept to create tiny trucks for tiny factories after all. While this might all sound like the science fiction of Hollywood, the same team has already succeeded in producing a truck able to carry a "payload."

Mel Wilson

Date 2005

Scientists James Tour, Kevin Kelly

Nationality American

Why It's Key The first nanocar to roll on wheels, a possible precursor to the nano-truck, perhaps to function in nano-factories of the future? As a single-molecule, the nanocar takes "low-scale industry" to new levels of low.

opposite Buckminsterfullerene molecule, (computer artwork)

Key Experiment **Slow light**
A Quantum Leap

In 2005, physicist Matthew Sellars and his group succeeded in slowing light from its usual 300,000 kilometers per second, down to just a few hundred meters per second. This isn't just a neat trick; it could provide new ways of storing and processing information, and pave the way for considerably faster and more powerful computers.

Sellars used a silicate crystal, doped with a rare earth metal called praseodymium, which stores the light. They used one laser to provide the light, and a second laser to cause an interaction between the crystal and the light, which allows it to be "stored" for a few seconds as it passes through. The second laser can be turned on and off to store the light then read it out again. The researchers stored the light for a few seconds, but saw the potential to store it for much longer. The next step is to try and store a single photon, which could be used as a "qubit."

Sellar's work has taken a crucial step toward the next generation of computers. In a normal computer the information is stored in a "bit" which can have the value of either 0 or 1. A quantum computer would use "qubits" which can be 0, 1, or a bit of both, and would increase computational power dramatically. The advance from conventional computers to a fully functioning quantum computer would be comparable to the difference between the abacus and today's PCs.

Leila Sattary

Date 2005

Scientist Matthew Sellars

Nationality Australian

Why It's Key Slow light is an important step toward the realization of a quantum computer, which would be considerably faster and more powerful than today's machines.

Key Event **Deep Impact** Hollywood in outer space

The 1998 movie *Deep Impact* depicted the collision of a large comet with the Earth. Life imitated art seven years later, when a space probe with (coincidentally) the same name slammed a projectile into the side of comet 9P/Tempel.

The Deep Impact spacecraft was launched by NASA on January 12, 2005. Its mission was to get the first look at the interior of a comet, by blowing a hole in its surface. The probe reached its target just six months later. In another unintentional nod to Hollywood, the craft fired its impactor at Tempel on July 4; Independence Day.

The copper projectile, which weighed 350 kilograms, hit the icy rock at a relative speed of 10.3 kilometers per second. The images were spectacular, both from the Deep Impact mother craft and from Earth-based telescopes. The impact kicked up a plume of dust and ice, whose composition gave astronomers an unprecedented insight into the comet's structure. Later images showed an impact crater 100 meters wide and 30 meters deep.

The results not only gave us new insights into the origins and composition of these icy bodies. One day, a comet or other large object may have Earth's name on it. Remembering the Michael Bay film *Armageddon*, the experience and results gained from the Deep Impact mission may prove valuable if life ever does imitate the movies.

Matt Brown

Date 2005

Country USA

Why It's Key A unique glimpse into a comet that reached the parts other missions couldn't.

opposite NASA's Deep Impact space probe successfully impacted the comet Tempel 1

Key Event **Fusion reactor brings nations together**

The industrialization that has drained the planet of fossil fuels over the last few centuries has turned the search for an alternative power source into a frantic scrabble. In 2005, the announcement that a new experimental nuclear fusion reactor was to be built came not a moment too soon.

In one of the most expensive scientific collaborations ever, the EU, United States, Russia, Japan, South Korea, and China decided on a home for a potentially huge source of electricity – the US$12.5 billion International Thermonuclear Experimental Reactor (ITER) was to reside in Cadarache, France.

Nuclear fusion is the process that occurs in the Sun; essentially the opposite of nuclear fission, which is used in current reactors to pull atoms apart. Building the ITER is equivalent to building a star on Earth. At incredibly high temperatures, it is hoped, atoms can be forced to fuse together, releasing the binding energy within the individual atoms.

Using readily available deuterium and tritium as fuels currently requires more energy than it actually releases. But one kilogram of fusion fuel would be equivalent to using 10 million kilograms of fossil fuels, so the success of the experimental reactor is of vast importance. Natural reserves of fossil fuels are finite, meaning a new sustainable energy source is required and that ITER may well provide the solution. If it works, it could show that fusion is a viable energy technology that has fewer safety implications than fission reactors. It is, however, predicted that electricity-producing fusion plants will not occur before 2050.

Nathan Dennison

Date 2005

Country France

Why It's Key The ITER could provide a way to meet all our energy requirements, more efficiently than traditional nuclear fission reactors, and at little cost to the environment.

Key Person
Dr. Randall and her amazing sandwich

Lisa Randall (b.1962) is a particle physicist and cosmologist – she studies the most fundamental particles that make up everything around us, the forces that act between them, and how these fundamentals have changed (if at all) since the birth of the Universe. Although she is wellknown within the field, with her work frequently cited and her name regularly making "ones to watch" lists, she is probably best known to the public for her 2005 book *Warped Passages: Unraveling the Universe's Hidden Dimensions*, an explanation of recent developments in theoretical physics. The book explores the previous century in physics, up to and including the mind-bending concepts of string theory and tiny extra dimensions curled up upon themselves.

Randall is not satisfied with these theories, arguing that they make no testable predictions. Her technique is more experimental, producing models of different concepts for the structure of our world and then poking them until they make predictions that can be confirmed or denied. One theory hypothesizes that our four-dimensional Universe, and another like it, sandwich a five-dimensional space between them. The sandwich filling would distort forces between the Universes, potentially making them act extremely weakly in one Universe – just like gravity does in ours.

Dr. Randall's willingness to engage with the public on matters of science, her pragmatic approach to theory testing, and her reputation in the field have earned her several scientific awards, a bestseller, and tenure at Princeton, Caltech, and MIT.

Kate Oliver

Date 2005

Nationality American

Why She's Key A highly unusual physicist, both for her determination to share her ideas and wonder at the Universe with the layman, and for being a string theorist who demands evidence.

Key Publication Qubytes and quantum computing
The future of code breaking

On December 1, 2005, a team of computer scientists from Austria claimed to have produced the world's first qubyte, having a huge impact on the world of computing.

Since the introduction of the integrated circuit in the 1970s, computers have stored their information as bits, collected into bytes. To a computer, a bit is seen as either a 0 or a 1 (binary code) and nothing else. A qubit, however, differs from a regular bit in that it can either be a 0 or a 1, or a "superposition" of both a 0 and a 1. This would mean that a "quantum computer" could reduce the time it takes to process a task, since it already has every possible answer stored in its superposition of qubits; it would simply have to locate the correct answer.

The vast improvement in speed achieved by quantum computers will afford ultra-fast password crackers; whereas a conventional computer must systematically guess and try a password, a quantum computer can achieve the same result in a fraction of the time due to the interaction of its qubytes. Although this is only one potential use of large-scale quantum computers, it is widely regarded to be one of the most important. Military organizations and government institutions are investing vast amounts of time and money into researching the possibility of using quantum computers as decryption tools. Gone are the days of Colossus-type decipherers stationed at Bletchley Park.

Gavin Hammond

Publication "Scalable Multi-Particle Entanglement of Trapped Ions"

Date 2005

Author Hartmut Häffner and colleagues

Nationality Austrian

Why It's Key Qubytes allow for the manufacture of quantum computers, which will be substantially faster than even the supercomputers of today.

Key Event
Bird flu spreads mass panic

In 2005, avian influenza, or bird flu, was carried out of Southeast Asia to Europe and Africa. Concern spread that a flu pandemic could occur on the scale of the 1918 Spanish Flu, which killed an estimated 40–50 million people. Incidents of the "H5N1" deadly bird flu strain rose sharply in 2006, with thousands of new cases reported across Europe and Africa, leading to panic stockpiling of antiviral drugs in many western countries, and a rush to find a vaccine for the virus.

The first detected case was in Hong Kong in 1997, but it did not resurface again until 2003. The disease soon became endemic in wild bird populations in many Asian countries. Birds' migratory routes brought the virus from Asia to Russia. By October 2005, reports confirmed the H5N1 strain in Turkey, Romania, and Croatia. The disease was rampant in wild bird populations: ten countries across two continents confirmed fresh cases of the disease in a single week. Many countries were on the alert to look for signs of the virus in bird populations by the end of the year.

By November 2007, there had been 335 reported cases of the disease in humans resulting in 206 deaths in ten countries. Most human cases were the result of close contact with infected birds, and so far no major outbreaks amongst human populations have occurred. The stockpiling of anti-viral drugs continues, and the World Health Organization has formed an Influenza Pandemic Task Force to deal with any future human outbreak.

Steve Robinson

Date 2005

Country Global

Why It's Key Bird flu is a serious concern; estimates suggest that between 5 and 150 million people could die from a global bird flu pandemic.

Key Event **Cosmos 1**
Sailing on the solar wind

It's such a pretty idea that many would love to see it work – and someday, it just may.

Cosmos 1 was a US$4 million privately funded project led by the Planetary Society. The plan was to send a 100-kilogram spacecraft into low earth orbit and then to unfurl a 30-meter-diameter mylar (reflective plastic) sail and allow energy from the Sun to gradually push it into a higher orbit. Unfortunately the method chosen of sending Cosmos 1 into space – a refitted ballistic missile launched from a Russian nuclear submarine – failed to achieve the necessary initial orbit, and the spacecraft was lost.

Similar to boats being pushed by the wind on the oceans of the Earth, solar sails rely on the "solar wind" – a constant stream of photons and other particles radiating from the sun in all directions. Solar sails aren't intended to lift spacecraft from the surface of the Earth, but they may work very well for journeys between stars. In theory, "sailing" to another star has a big advantage over rocket propulsion – no fuel is required. During departure, while still relatively close to a sun, a low-mass, no-fuel spacecraft with a vast solar sail could gain quite a bit of acceleration, potentially achieving velocities in excess of 160,900 kilometers per hour. Upon arrival, the spacecraft can flip over and use the "push" of the photons being emitted from the star at its destination to slow itself down.

So for the longest of journeys, it looks like traveling "light" may be the way to go.

Andrey Kobilnyk

Date 2005

Country USA

Why It's Key Solar sails might make interstellar travel more feasible.

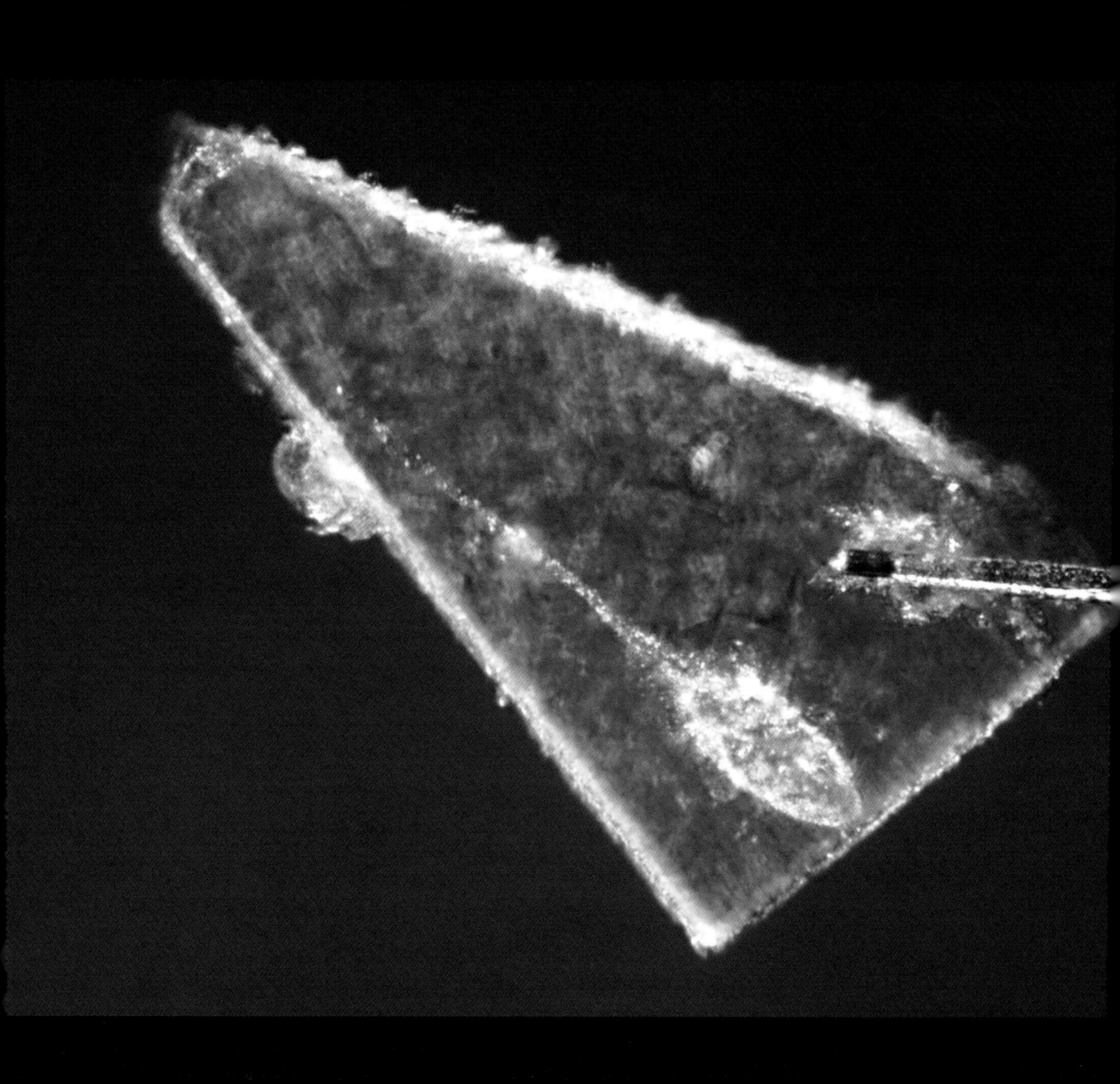

Key Event
First facial transplant surgery

The face is generally the first thing we look at when we meet someone; indeed, large areas of our brains are dedicated to recognizing faces, and they are integral to our non-verbal communication and social interaction.

Little wonder then, that people who incur damage to the face, through injury or disease, suffer tremendous social stigma. In 2005, however, a medical first offered a ray of hope to those traumatized by serious facial disfigurements.

The first facial transplant took place on November 27. Maxillofacial surgeon Professor Bernard Devauchelle undertook a five-hour operation, replacing the nose and mouth of Isabelle Dinoire – which had been torn off by a dog – with a similarly shaped triangle of flesh from a living but brain-dead donor. This operation was the first facial transplant, although face replant surgery had taken place almost a decade earlier, reattaching the face of an Indian girl who had her face ripped off by a threshing machine.

The problems with face transplants are numerous, and the process is not without its detractors. As with any transplant, there is a risk of infection afterwards, and patients will have to take a cocktail of immunosuppressive drugs for the rest of their lives to prevent their bodies from rejecting the foreign tissue. With facial transplants there are additional ethical factors to consider; donors must be physically alive at the time of facial removal, and the idea of "wearing" someone else's face is in itself a troubling concept for many people.

Barney Grenfell

Date 2005

Country France

Why It's Key Since Denoire's first facial transplant, more have been performed. This medical procedure remains the best hope for many unfortunate people suffering from serious facial disfigurement.

Key Event **Spacecraft chases comet**
I've been tailing you for months

NASA's Stardust spacecraft was launched into space in 1999 with a unique mission: to intercept a comet and return to Earth with samples of its tail. The journey was to take seven years and cover over three billion miles.

Roughly the size of your average fridge, Stardust's flight path crossed the tail of Comet P/Wild-2 in 2004. The spacecraft's primary objectives were to collect interstellar dust and samples from the comet. Collecting particles the size of sand grains may not seem that difficult, or indeed dangerous, but when they are traveling at almost ten times the speed of a bullet, that makes it a little more tricky. At such velocities, their shape or chemical composition may be changed, or they may even be vaporized. To overcome this problem a silicon-based aerogel collector – essentially a large frying pan full of gel – was deployed; the gel acting as both an airbag and a collection vessel.

In January 2006, a capsule containing the samples re-entered the Earth's atmosphere and landed in Utah's Salt Lake desert.

The contents are still being eagerly examined by astronomers around the world, but have already thrown up some surprises. Some particles were a million times larger than typical space dust grains. One hypothesis for why these were found is that comets might pick up a lot of rocky matter as they swing around the Sun and through the inner Solar System. As research continues, it may transpire that comets are the vacuum cleaners of space.

Andrew Impey

Date 2006

Country USA

Why It's Key The first ever spacecraft to return to Earth with a sample from a comet. These samples are already revealing clues about the early history of the Solar System.

opposite Light micrograph of a section of aerogel from the Stardust spacecraft showing the track of a particle of captured comet dust (running bottom right to upper left)

Key Discovery *Tiktaalik roseae*
A fish out of water

When scientists in 2006 published their fossil discoveries of *Tiktaalik roseae* – a part fish, part amphibian creature that lived around 375 million years ago – they added an extremely important piece to the evolutionary puzzle, by explaining how and why tetrapods evolved from fish.

Previously, fossil records showed a gap between *Panderichthys*, a 385-million-year-old fish with primitive signs of evolving land dwelling features, and *Acanthostega*, the earliest known four-limbed land vertebrate – or tetrapod – which lived about 365 million years ago.

In 1998, American paleontologists Neil Shubin and Edward Daeschler led a team out to Ellesmere Island in the Canadian Arctic, 600 miles from the North Pole, and began searching for the "missing link" that would fill the gap. And after years of fossil hunting, they were justly rewarded: In 2004, they discovered several well preserved fossils of a strange creature that would offer a glimpse into our evolutionary ancestry and help explain the transition from water to land. Two years later, after a lot of delicate chiseling and careful analyses, they published their findings.

Tikaalik blurred the boundary between two very different forms of life. Like fish, it had fins, scales, gills, and a primitive jaw. But it also had characteristics of a land-dweller such as shoulder, elbow, and wrist joints in the fins; the beginnings of a neck; and a flat crocodile-like head with eyes positioned on top. These features suggest that it lived in shallow water and could support its own body weight.

James Urquhart

Date 2006

Scientists Neil Shubin, Edward Daeschler

Nationality American

Why It's Key Marked a significant step in evolutionary understanding of how tetrapods – including all amphibians, reptiles, mammals, birds, and even dinosaurs – evolved from fish.

opposite Tiktaalik roseae, better known as the "fishapod"

Key Event Merck's anti-cervical cancer vaccine approved by FDA

The age-old saying, "prevention is better than cure" certainly rings true when considering the possible impact of a cervical cancer vaccine. Developed by German pharmaceutical company Merck, to work against Human Papilloma Virus (HPV), which causes the disease, its approval by the Food and Drug Administration (FDA) in 2006 raises the possibility of complete eradication.

HPV is the most common sexually transmitted infection in the United States and, globally, cervical cancer is the second most common cancer among women. The vaccine, Gardasil, is most effective at protecting women who have never previously been exposed to any strain of the HPV virus. It is therefore of major significance that the FDA approved the drug for use in nine to twenty-six year olds; vaccinating girls before they become sexually active could mean cervical cancer is eliminated within one generation.

Gardasil's success depends considerably on its cost. Regular screening in the UK and United States often means the cancer is detected early and treatment at this stage has a high success rate, but it is in the developing world where most deaths from cervical cancer occur, making it vital that the vaccine becomes affordable and globally available.

The potential impact this vaccine could have on women's reproductive health is phenomenal. Vaccinations could put an end to routine smear tests and screening., and lead, eventually, to cervical cancer joining smallpox as a disease found only in medical history books.

Ceri Harrop

Date 2006

Country USA

Why It's Key The very first anti-cancer vaccine that has the potential to eliminate the second most common malignancy in women worldwide.

Key Experiment **Ununoctium**
The heaviest element created

Ununoctium is a super-heavy element that was first synthesized in 2006 by a joint team of Russian and American researchers working at the Joint Institute for Nuclear Research at Dubna, Russia. They had been colliding two other elements together, californium-249 atoms and calcium-48 ions, when they produced the new element. The name literally means "one-one-eight" and is pronounced oon-oon-OCT-i-em.

Only three atoms of ununoctium have ever been detected. It has an atomic number of 118; the highest number currently assigned to a discovered element, and has been assigned the symbol Uuo. It is tricky to make, and has a very short half-life – less than one millisecond – making it even harder to study. However, even though the half-life is extremely short, it is longer than theoreticians had predicted.

Seven years previously, the element had had a false dawn. In 1999, researchers at the Lawrence Berkeley National Laboratory discovered elements 116 and 118, but because other scientists were not able to come to the same conclusion, they retracted their paper. In 2002, the lab announced that the original data was made up.

The work in Russia was more robust, although the element remains very hard to synthesize because of the complex processes involved, coupled with a very small fusion reaction probability – billions of calcium ions need to be shot at the californium to get a single reaction fusing the two atoms together.

David Hall

Date 2006

Scientist Yuri Tsolakovich Oganessian

Nationality Russian

Why It's Key New elements are still being discovered, and the higher-than-expected half-life of Uuo hints that some massive molecules might have isotopes which are relatively stable.

Key Event **Pluto is designated a dwarf planet**
A defining moment

What is a planet? It sounds simple enough but, surprisingly, scientists didn't have a definite answer until August 24, 2006. On this date, the International Astronomical Union (IAU) held a vote to define the term "planet." Consequently, Pluto was stripped of its title as the ninth planet of the Solar System.

Ever since American Clyde Tombaugh discovered Pluto in the outer reaches of the Solar System in 1930, it had been considered a planet. But doubts over Pluto's planetary status flared up in the 1990s when several objects of comparable size were discovered in what is now known as the Kuiper belt – an outer region of the Solar System beyond the current eight planets. Then, in 2005, a crushing blow came when Eris was discovered, a celestial body of similar appearance but slightly larger and more distant than Pluto. It meant that if Pluto was a planet then Eris must be too.

But the new IAU definition prevented this. It proposed that a celestial object is a planet only if it orbits the Sun, forms a spherical shape, and has cleared its orbit from other objects. Since Pluto's elliptical orbit overlaps with Neptune's it was placed in a new category of "dwarf planet"(along with Eris and Ceres), which resides in the asteroid belt between Mars and Jupiter.

Controversy ensued, however. Many astronomers clung onto the idea of Pluto as a planet, suggesting that the vote, in which only 424 scientists took part, and the definition, were highly flawed.

James Urquhart

Date 2006

Country Czech Republic

Why It's Key Introduced new scientific nomenclature that re-designated Pluto as a "dwarf planet," despite it being culturally and historically recognized as the ninth planet of the Solar System.

opposite Computer artwork of the dwarf planet Pluto

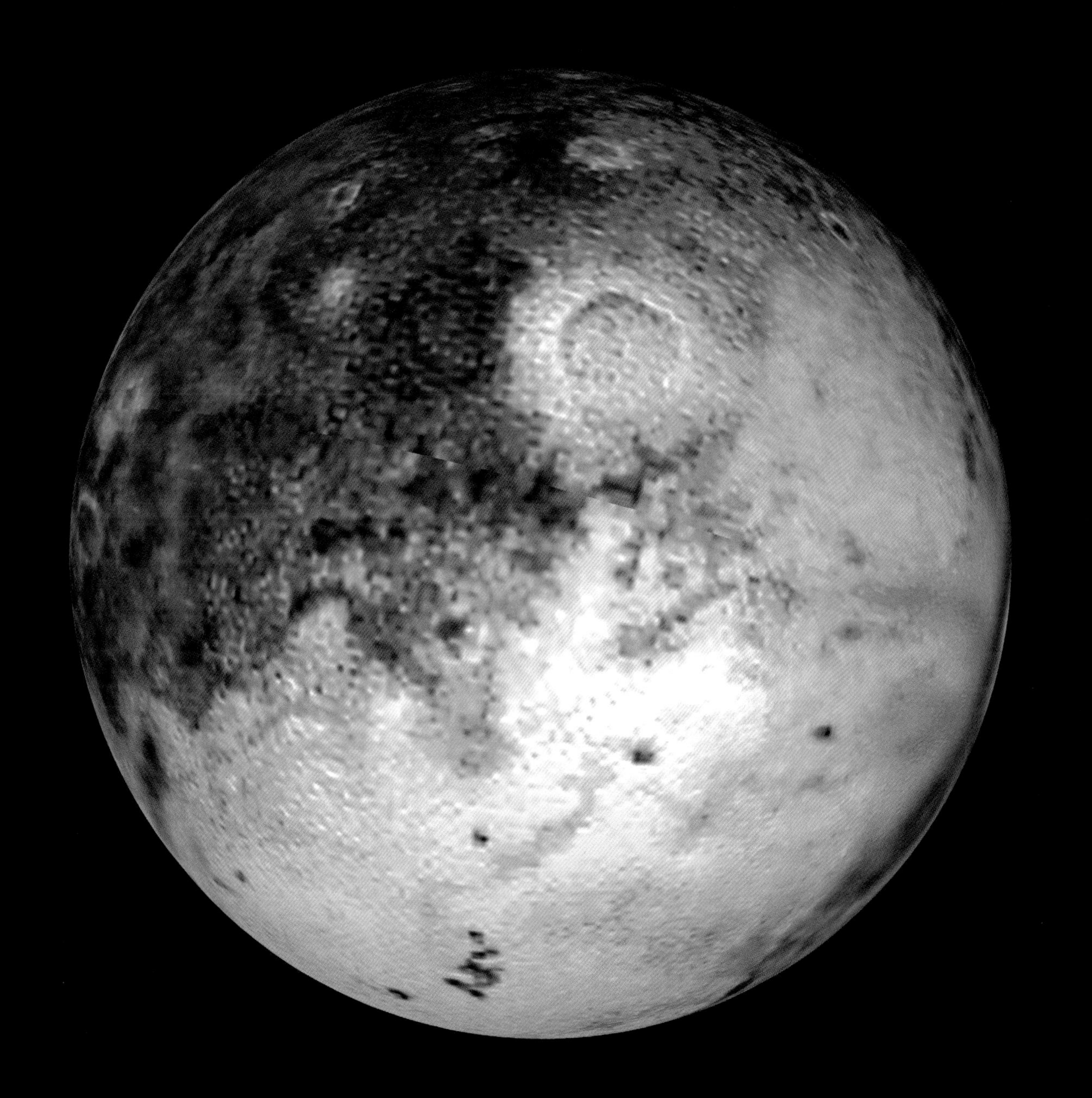

Key Discovery **Self sustaining bacteria** The world beneath our feet

When a self-sustaining community of bacteria was found in the rocky depths of a South African goldmine in 2006, it had profound implications for the existence of extraterrestrial life. No, scientists didn't think that these organisms came from outer-space. Rather, the discovery hinted at the kind of life that might exist on other worlds, such as Mars, and where it might be found – namely, deep underground.

Life in the deep subsurface was nothing new. Scientists had previously established that micro-organisms live deep underground in many parts of the world, surviving extreme temperatures and pressures on somewhat restricted and unique diets. But what wasn't known was whether subterranean bacteria were new arrivals that had hitchhiked from the surface and were destined for extinction, or whether they were part of an ancient underground biosphere, able to exist indefinitely and completely disconnected from any surface-derived resources. The goldmine offered an excellent opportunity to answer this question. Tullis Onstott and Lisa Pratt, along with their research team, learned of a water-filled fracture about three kilometers down that was uncontaminated by human activities. Knowing that bacteria already existed deep within the mine they set forth to discover the water's secrets.

A vast number of bacterial species were found and further analysis indicated that they had been there for millions of years, permanent and self-sustaining, relying on radiation from uranium ores to survive, rather than sunlight. So if extraterrestrials do exist, the chances are that they live life to the extreme.

James Urquhart

Date 2006

Scientists Tullis Onstott, Lisa Pratt

Nationality South African

Why It's Key The discovery that life can exist deep inside Earth fueled optimism that life forms could survive in the extreme, and seemingly lifeless, environments of other planets, such as Mars.

Key Discovery **Twin satellites give new information on global warming**

Thanks to the mass media, we're all familiar with the concept of climate change. But how do scientists measure it? Do they simply stick an elbow in the oceanic bath water and note down the temperature? Not quite. One way of determining the extent of it is by monitoring the speed of glacial melting, particularly in the ice sheets of Antarctica and Greenland. These ice sheets are the largest bodies of ice on the planet, covering over 15 million square kilometers, and were once thought to be untouchable by global warming. More recent evidence shows this is not the case.

Measurement of the mass of the ice sheets has previously been carried out by laser altimeter and radar techniques; from these, glaciologists could extrapolate how quickly the ice sheets were melting. But in 2002, a new project began that enabled much more accurate and direct measurement of their mass, leading to an alarming discovery.

The project in question is entitled GRACE (Gravity Recovery and Climate Experiment), a NASA project which measures changes in gravity via two satellites that orbit the earth every 94 minutes. The satellites can detect localized differences in the Earth's gravitational pull caused by movements of water, including that of melting ice.

The results of GRACE's monitoring show that, contrary to previous assertions, the ice sheets might be affected by global warming after all: the Greenland sheet is actually melting at a rate of 239 cubic kilometers a year, equating to a rise in sea levels of 0.5 millimeters a year.

Barney Grenfell

Date 2006

Scientists Jianli Chen, Clark Wilson, Byron Tapley

Location Greenland

Why It's Key Developments in satellite technology, specifically the ability to measure changes in Earth's localized gravitational pull, allow more accurate measurement of changes in the ice sheets.

opposite Arctic sea ice maximum extent. White area is 2006 maximum, yellow shows the cumulative maximum 1979-2006

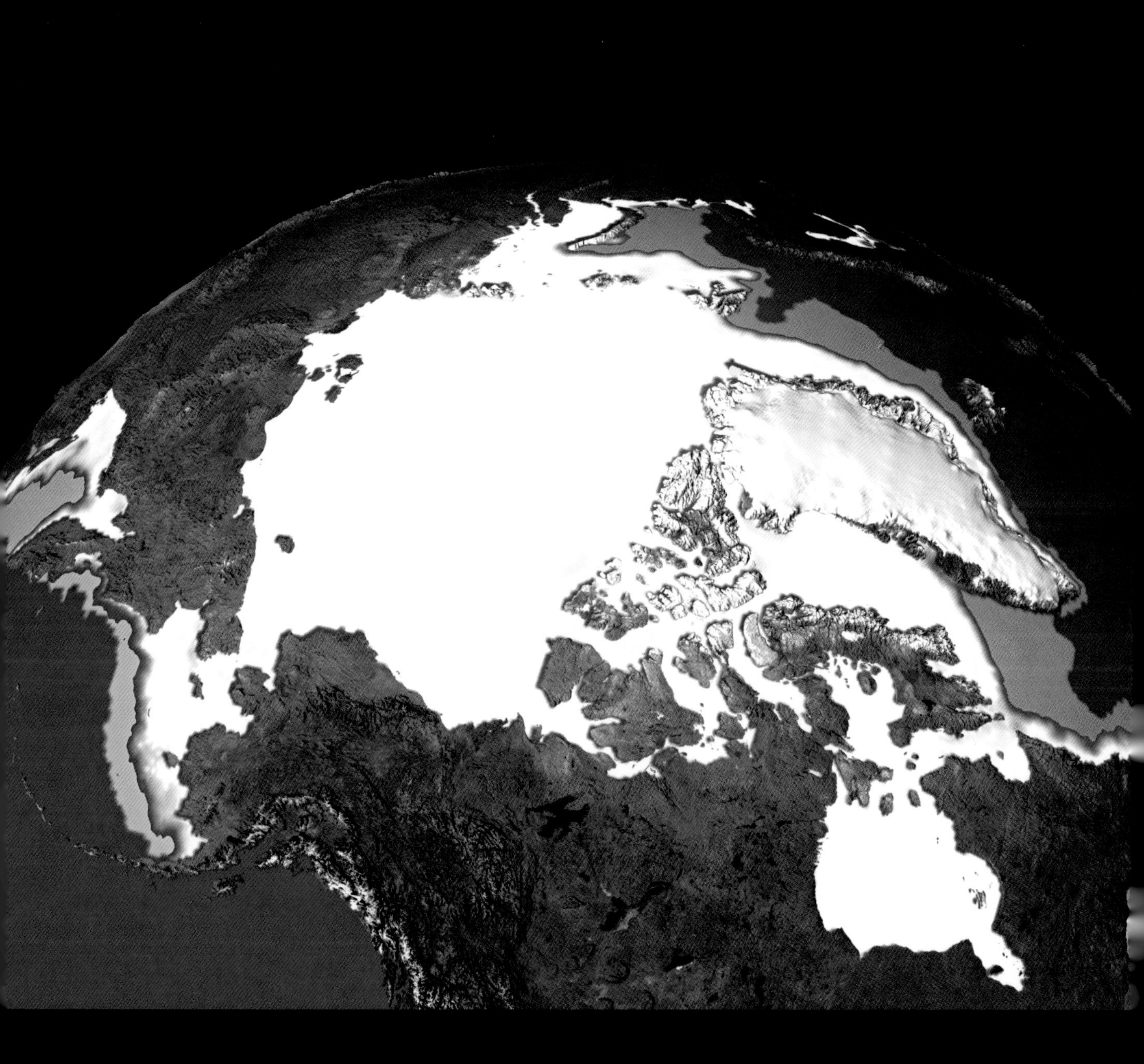

Hurricane Katrina
August 29, 2005
Photo: NOAA

Key Discovery
DNA reveals some long-lost relatives

It is a common perception that we humans were responsible for the demise of the Neanderthals thousands of years ago. The Neanderthals, often viewed as a wild, brutish species, are believed to have been out-bred by the more intelligent *Homo sapiens*. But a study by scientist Svante Pääbo showed that we are in fact cousins of the Neanderthals, last sharing a common ancestor some 450,000 years ago. As late as 30,000 years ago, it is believed modern humans and Neanderthals coexisted in Europe.

Pääbo took DNA samples from a 38,000-year-old Croatian male's femur – a Y chromosome had already been identified – and used a technique called pyrosequencing to sequence the man's DNA. In pyrosequencing, different light signals are generated dependent upon which DNA base pair or "letter" of the genetic code is present. Pääbo's work revealed that 99.5 per cent of the Neanderthal DNA was the same as human DNA, suggesting that we haven't evolved as much as we may have thought.

Being able to obtain Neanderthal genome information means that anthropologists need no longer rely solely on the discovery of fossilized remains. The recovered sequences could be used to provide biological information about attributes that aren't present in fossils, such as skin color and even, perhaps, speech. This discovery has opened up the possibility of genetic time travel, using DNA to discover more about our prehistoric relations.

Nathan Dennison

Date 2006

Scientist Svante Pääbo

Nationality Swedish

Why It's Key Sequencing of Neanderthal DNA reveals more about our evolutionary history.

Key Event *An Inconvenient Truth*
The power of film

A science documentary acclaimed as one of the must-see films of the year? Unheard of! That is, until *An Inconvenient Truth* blitzed triumphantly through cinemas in 2006. The film – and its accompanying best-selling book – achieved what thousands of scientists hadn't: it explained vividly how urgent it was for humanity to tackle global warming.

Its impact was due to its unlikely star, the former U.S. vice-president Albert Arnold (Al) Gore. Bouncing back from his defeat by George W. Bush in the 2000 U.S. presidential election, Gore decided to become an environmental activist, delivering inspirational lectures on climate change across the world. Gore and director Davis Guggenheim blended a soft-focus narrative of the politician's personal journey, with the facts and dire predictions of his lectures. Global warming would cause rising sea-levels, droughts, storms, epidemics, conflict, and species extinction. Humans were creating this potential catastrophe by burning fossil fuels and releasing ever more carbon dioxide into the atmosphere. We had to stop it.

The film received critical acclaim, grossed US$49 million worldwide, and won two Oscars. Later, scientists questioned the accuracy of some of Gore's claims, but agreed he'd got the fundamental message right. His efforts were granted equal recognition as scientists of the Intergovernmental Panel on Climate Change, with whom he shared the 2007 Nobel Peace Prize. The scientists had painstakingly gathered all the evidence on global warming: an inconvenient truth, perhaps, for the scientists to swallow.

Richard Van Noorden

Date 2006

Country USA

Why It's Key Politician's film publicizes the imminent catastrophe of climate change

opposite *An Inconvenient Truth*

Key Experiment **Invisibility Cloak created**

We see objects because light bounces off them and hits our eyes. Light normally travels in straight lines, but if we bent light to take it around an object, and then released it on the other side, heading out in the same direction as though nothing had happened, the light wouldn't bounce into our eyes, and the object would therefore be invisible.

That doesn't sound too hard does it? Unfortunately, however, getting light to act in this way is pretty tricky. Tricky, but not impossible. In 2006, engineers David Schurig and David Smith of Duke University designed a cloak that allowed microwaves – which are electromagnetic waves, just like light – to slide around a copper cylinder as though they were water. The two engineers had built the cloak based on designs they had produced in collaboration with the English physicist Sir John Pendry, who founded the field of metamaterials. These are bizarre substances that do things you wouldn't expect in nature, like bend light.

Whilst Schurig and Smith's cloak only worked for microwaves of a single frequency and only masked a copper disk the size of a couple of chocolate buttons, it was one of the most extraordinary advances of modern times. Metamaterials, like those used for the cloak, are the building blocks of an amazing future.

Douglas Kitson

Date 2006

Scientists David Schurig, David Smith, John Pendry

Nationality American, British

Why It's Key Be it an invisible jet like Wonder Woman, or sneaking around like Harry Potter, who wouldn't want invisibility?

Key Discovery **Counting electrons**
Single file

Electrical current is measured by ammeters, which track how much charge flows past a certain point. However, if you could see closely enough, you'd see that it's actually millions of electrons passing through the ammeter that give a reading. This isn't really a problem if your electronics experiences consist of blobbing solder on by hand and occasionally burning a hole in your workbench, but in the world of precise electronics – that of microchips and nanometers – knowing what all your electrons are up to could be a pretty big deal.

It's therefore just as well that Toshima Fujisawa and colleagues at NTT Basic Research Laboratories in Atsugi, Japan, found a way to do it.

Using quantum dots – crystals that are only a few nanometers across – the current can be slowed to only let electrons through one at a time. A quantum point contact measures the charge in each quantum dot, from which the Japanese scientists were able to tell if there was an electron in there or not. By checking every 20 milliseconds, the researcher was able to count the flow of each electron as they pass through the dots, as well as finding out which direction they're going.

As well as potentially becoming a very sensitive ammeter, probing the paths of electrons in the tiniest of microchips, this technique also could mean big things for quantum computing. The behavior of individual electrons could be used to create unbelievably powerful, yet sensibly sized, computers of the future.

Douglas Kitson

Date 2006

Scientist Toshima Fujisawa

Nationality Japanese

Why It's Key Watching the behavior of individual electrons is a whole new level of precision in electronics, and is good news for quantum computers.

Key Experiment
A computer based on quantum bang-bang

Imagine a computer billions of times faster than your laptop, able to crack the toughest codes, whose processors are built at an atomic scale. This is the realm of the quantum computer. Scientists are making inroads to harness the power of the atom for computation, and small-scale examples already exist.

In a traditional computer, a "bit" of information is either in one state or another – call it "on" and "off," or 1 and 0. But if you do your computing at an atomic level, you can make use of quantum uncertainty. The bit of information can be in multiple states – both on and off – at the same time. These quantum bits, or qubits, would allow billion-fold increases in speed over current processors.

All good in theory, but a quantum computer is difficult to build. As soon as a qubit is observed, or interacts with its environment, the uncertain state "collapses" into one state or another, thus losing its advantage over a conventional system. So a way is needed to isolate the qubit from its surroundings.

In January 2006, a team at Oxford University caged a nitrogen atom qubit inside a football-like carbon shell, known as a buckyball. This inert casing partially limited the qubit's interaction with the wider world, but not quite enough. Their eventual solution was to hit the qubit with what they termed a "quantum bang-bang," a repeated pulse of microwave radiation. By giving the qubit a regular "kick" in this way, they effectively decoupled it from environmental effects.

Matt Brown

Date 2006

Scientist John Morton, Simon Benjamin

Nationality British

Why It's Key It's still early days, but the bang-bang method might just be the kick needed to enable powerful quantum computers.

Key Event Hwang Woo Suk found guilty of falsifying data about stem cells

In 2004, Korean scientist Hwang Woo-suk claimed to have cloned the first human embryo. But he was soon to become embroiled in one of the largest investigations into scientific fraud in living memory.

When Hwang published his two seminal papers in the journal *Science*, he became an overnight hero in the field of genetic research. Not only did he claim to have produced the first cloned human embryo, he also claimed to have used it to develop human stem cells. This was a major breakthrough in the race to cure diseases such as Parkinson's, cancer, and spinal cord injuries.

At least it would have been, had his own research institution not published a damning report in January 2006 that exposed Hwang as a fake. It was revealed that the key data in his two landmark papers had in fact been fabricated. Hwang tried desperately to displace the blame, claiming that his junior researchers had deceived him, but the damage had already been done and his reputation as a leading scientist was left in tatters.

The revelations came as a huge blow to his admirers, in particular the Korean scientific community who had been so proud of his achievements. Hwang was eventually fired from his position as professor at Seoul National University, where he had conducted his ill-fated research.

Hannah Isom

Date 2006

Country Korea

Why It's Key One of the biggest scientific scandals in living memory was uncovered.

Key Invention
A breakthrough in bionic limbs

Prosthetic limbs have been around for centuries – an ancient Roman false leg has even been found in a tomb dating from 300 BCE. Since then of course, artificial body parts have developed in leaps and bounds, aiding amputees in both leaping and bounding. The problem with most modern artificial limbs is that their attachment to the stump means they rub against the skin during movement and cause pain and discomfort to the wearer.

In 2006, a team at University College London announced the development of a method for attaching prosthetics directly onto the skeleton of the patient. Direct contact between the artificial limb and the remaining bone would provide a firm attachment that wouldn't irritate the skin. A vital step toward providing amputees with replacement body parts as close to the original as possible, this advance could pave the way for new limbs to be directly connected to the central nervous system.

Previously, no one had been able to find a way of allowing skin to heal naturally around an implant without a high risk of infection. But the UCL scientists had a bright idea. They examined the way in which deer's antlers are able to protrude through the skin and found that the tissue formed a tight mesh around the porous bone. They used this as a basis for designing their own implants. A titanium rod attaching directly onto the bone was used in trials with people who had lost thumbs and fingers, with promising results.

Shamini Bundell

Date 2006

Country UK

Why It's Key The biggest development in artificial limbs since the invention of the wooden leg.

opposite Claudia Mitchell (left) demonstrates the functionality of her "bionic arm" during a news conference with the arm's developer Dr. Todd Kuiken

Key Experiment
The two billion degree mystery

In 2006, the U.S. Department of Energy National Nuclear Security Administration's Sandia laboratory heated a gas to two billion degrees Celsius; a temperature more than one hundred times hotter than the Sun. Although the results were unexpected, they are seen as a major stepping stone in the development of nuclear fusion.

Carefully splitting atoms – nuclear fission – may already be one of many answers to fossil fuel shortages. Nuclear fusion, on the other hand, may be the most significant solution to the energy demands of solving world poverty, interstellar space flight, and plasma screen televisions. Fusion naturally takes place in stars and involves combining atoms to release huge amounts of energy. Reliably replicating this on Earth would change civilization forever, by removing one of its main barriers – energy production. One of Sandia's machines, "Z," researches nuclear fusion by using a magnetic field to condense plasma (ionized gas) to a thickness equal to that of the lead of a pencil, releasing energy in the form of X-rays. Temperatures of several million degrees Celsius are usually achieved; however in early 2006 following a reconfiguration of the wires used to produce the magnetic field, temperatures of two billion degrees Celsius were recorded.

Exactly how this event was triggered is unknown, but Malcolm Haines of Imperial College London has theorized that microturbulences formed by the magnetic field could be the cause. Should this mystery be solved, and other obstacles overcome, we will have fueled a new era in civilization.

Christopher Booroff

Date 2006

Scientist Chris Deeney

Nationality British

Why It's Key Understanding this experiment may eventually lead to cheaper energy, but could initially assist in maintaining a safe and secure nuclear weapons stockpile.

www.ric.org
Rehabilitation Institute of Chicago

Key Publication **Piezoelectric materials used to generate energy**

Increasingly, efforts are being made to find new ways of generating electricity that do not depend on fossil fuels, such as using sun, wind, and water to generate large proportions of the energy we consume.

Recently, another possible renewable source has been suggested for generating environmentally friendly, sustainable electricity: people. Or, more specifically, the vibrations created through their movement. In 2006, materials scientists at the Georgia Institute of Technology, led by Doctor Zhong Lin Wang, researched the use of piezoelectric technology to generate small amounts of electricity by inserting devices into people's shoes. This electricity could then be used to power small nano-devices.

Piezoelectric technology harnesses the power of movement by converting kinetic energy (movement) into electric energy, making use of the fact that some materials, mostly crystals and some ceramics, produce an electric potential when stress is applied to them. More recent ideas for generating electricity in this way include building this piezoelectric technology into the floor. Even if only the smallest amount of energy was generated by each footstep, thousands of people stepping thousands of times through a busy urban center – a shopping mall or train station – would generate significant power, enough to provide energy for lighting or heating.

The project "Pacesetters," by architectural firm Facility, is experimenting with means of generating electricity from the kinetic energy created during every day life, such as piezoelectric staircases and floors.

Barney Grenfell

Date 2006

Scientist Zhong Lin Wang

Nationality Chinese

Why It's Key New renewable sources of electricity that make use of naturally occurring resources are becoming increasingly important. Piezoelectric technology offers one way of harnessing these sources of energy.

opposite Oscillating quartz crystal

Key Invention **Nanotube paper battery**

In 2007, researchers at Rensselaer Polytechnic Institute in New York State developed a prototype battery made from paper and nanotubes. The devices are flexible and could eventually be used to power nano-devices and even scaled up to power cars. The prototype is the size of a postage stamp, and can provide enough juice to illuminate a small light bulb.

Carbon nanotubes are a member of the fullerene structural family, which includes spherical buckyballs – molecules composed entirely of carbon. Unlike buckyballs, nanotubes are cylindrical and their unique structure gives them bonding stronger than that found in diamonds. As the nanotubes are immensely robust, flexible, and electrically conductive, they can be used in nanoscale electronics and made so that they are just one atom thick.

The paper battery uses the properties of nanotubes to create a battery that can be bent, folded, or cut and still function. Nanotubes were attached to tiny pieces of paper only a few tens of microns thick that were then coated with aluminum foil. When two of these were arranged with the foil sides together they were found to act like a capacitor – an electrical component which stores charge: the foil was the conducting plates and the nanotubes the electrodes.

As the paper batteries are mainly made from paper and carbon, they could be used to power pacemakers which, instead of toxic liquids, use blood as the electrolyte. Carbon nanotubes are expensive to make, but if costs were reduced, large scale newspaper sized batteries could make electric cars commonplace.

Leila Sattary

Date 2007

Scientists Victor L. Pushparaj, Lijie Ci, Robert J. Linhardt

Nationality Indian, Chinese, American

Why It's Key A method of producing electricity for nano-devices, with the possibility of eventually being scaled up to power larger devices.

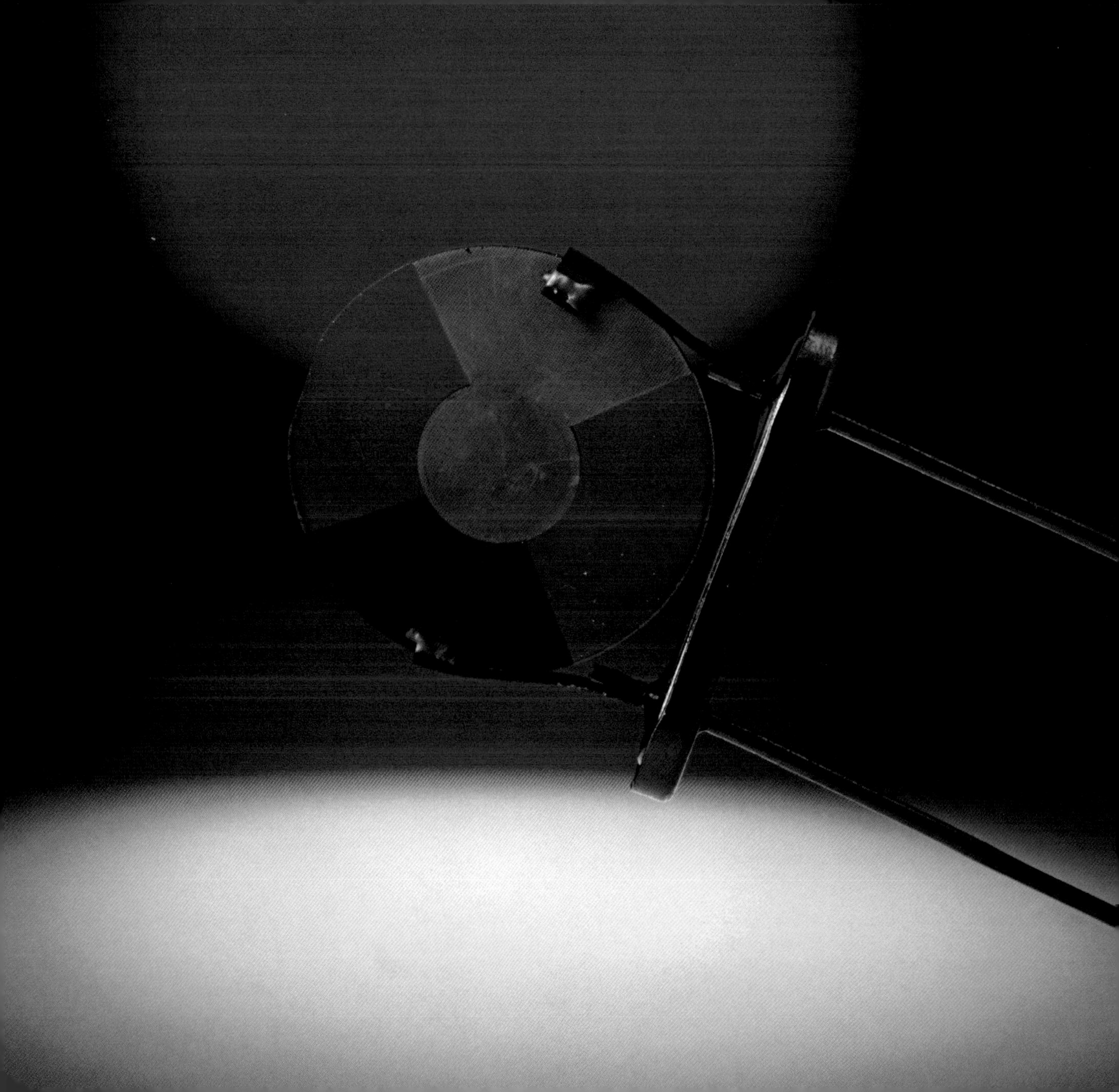

Key Invention

Hitachi breaks one-terabyte storage barrier

Since the first consumer PC was created, computer storage capacity has grown steadily. By August 2007, the latest innovation of hard drive memory hit the market in the form of Hitachi's Deskstar 7K1000 drive, offering a massive one terabyte of memory – enough to store a quarter of a million MP3 files.

The first hard drive was introduced by IBM in 1956, weighed a ton, and held just five megabytes of data. Now drives can hold up to two hundred thousand times as much data. They work by magnetizing a circular disk in a setup not dissimilar from a record player. The data is encoded into binary characters using the "head," which changes the direction of magnetization of tracks on the spinning disc. The head is also able to read data already embedded in the magnetic material coating the disc. The disk spins at very high speeds and the technology involved in reading and writing data is incredibly precise. Such is the proximity and speed of the head relative to the disc, that one technology expert made the analogy that if the head were an aircraft it would fly at 900 kilometers an hour, just 0.2 micrometers (or one fifth of a millionth of a meter) from the ground!

Hitachi's hard drive marks the start of the "terabyte era." It took the industry thirty-five years to reach one gigabyte storage, yet only sixteen more to reach one terabyte, which is equal to a thousand gigabytes. Though there is a physical limit to how much data you can squeeze onto a disc, industry experts believe that this trend will continue for a long time yet.

Steve Robinson

Date 2007

Scientists Hitachi

Nationality Japanese

Why It's Key Hitachi have ushered in a new era in digital storage capacity with their one terabyte hard drive.

Key Discovery

Solid light and superfast computers

Every year, as winter approaches and the weather becomes colder, liquid water slowly turns into solid ice. An analogous "phase transition" does not occur with light; photons from the Sun reach our planet with the same speed, irrespective of the outside temperature.

However, researchers at the universities of Melbourne and Cambridge have developed a theory that allows light to undergo such a phase transition, resulting in solid light. The theory describes how to create a crystal of light using, for example, very thin sheets of diamond. Usually, photons of light do not interact with each other; however, photons in a crystal do interact with each other, repelling each other just like electrons do in an electrical circuit.

This repulsion between photons means that scientists have a degree of control over the photons which could, for example, be beneficial in computing. Currently, computer processors operate by driving electrons round a circuit, so the speed of the processor is (partly) determined by the speed at which the electrons propagate through the wires. A computer made out of diamond would not only be pretty to look at – and pretty expensive – but would also allow a photonic current to "flow" through its components, resulting in an extremely fast machine.

A system comprising controllable photons would also give scientists a "laboratory" in which to study quantum mechanics, something which today's comparatively slow computers are unable to provide.

Gavin Hammond

Date 2007

Scientists Andrew Greentree, Jared Cole, Lloyd Hollenberg, Charles Tahan

Nationality Australian, British

Why It's Key Solid light gives scientists control over photons, which do not normally interact with each other. This means extremely fast computers can be built out of thin lattices of diamond.

Key Discovery
Hydrogen fuel cells

Diesel and gasoline, as well as being polluting, will become economically impractical fuels later this century. One of the most widely touted alternatives is hydrogen, a green and readily obtainable fuel. Hydrogen, however, is difficult to store and transport safely – think of the Hindenburg airship crash – and large infrastructure changes would be needed to make its widespread use economical.

Nevertheless, scientists are chipping away at the barriers. Various solids and liquids, including metal hydrides and carbon nanotubes, are candidates for the safe and compact storage of hydrogen. Perhaps the most promising method – deriving hydrogen on demand from water – was announced in 2007 by Professor Jerry Woodall of Perdue University.

In 1967, Woodall serendipitously found that adding water to an aluminum-gallium alloy liberates large amounts of hydrogen gas. His team have now honed the process to something that could be economically viable. Small pellets of the alloy are mixed with water, releasing hydrogen for use in a fuel cell or combustion engine. The gallium is untouched by the reaction, acting solely as a catalyst. The aluminum, on the other hand, is oxidized by the process and would need periodic recycling via carbon-neutral means.

The fuel and kit could fit into a normal-sized car, and costs could be comparable to gasoline, given a few improvements to fuel cell technology and aluminum recycling. It's promising stuff, but the petroleum industry may take some convincing.

Matt Brown

Date 2007

Scientist Jerry Woodall

Nationality American

Why It's Key It's still early days, but if hydrogen fuel cells work, one of the biggest challenges facing our way of life could be solved.

Key Discovery
New stem cell found in amniotic fluid

Until 2007, stem cell research had mainly centered on two types of cells – the controversial but fast-growing embryonic stem cell, and the slower-growing mature stem cells, but Anthony Atala and others working at Wake Forest University and Harvard, changed all that. Around the turn of the twenty-first century, Dr Atala and his group found what they thought were stem cells floating around in amniotic fluid – the liquid that surrounds fetuses as they are developing in the womb – and also in placental tissue. It took them another seven years to prove that these were stem cells, but it may have been well worth the wait.

Amniotic fluid stem cells, or AFS cells for short, are potentially a windfall to researchers in that they reproduce at nearly the same rate as the embryonic stem cells, but do not carry the same political stigma, as no human embryo would need to be harmed in order to harvest them. Also working in favor of the AFS cells is the fact that they don't seem to mature into tumors – something that cannot be said of embryonic cell lines. Unlike the more mature cells, however, they still appear to be truly able to differentiate into any of the body's tissues, and they do it rapidly. The result is perhaps the best of both worlds – a juvenile that still grows fast, but doesn't yet know what it wants to be when it grows up.

B. James McCallum

Date 2007

Scientist Anthony Atala

Nationality Peruvian

Why It's Key Amniotic fluid stem cells may just be the most practical stem cell yet: pluripotent, apolitical, and fast-growing.

LIVE EARTH
THE CONCERTS FOR A CLIMATE IN CRISIS
TM

Key Publication **Global warming**
The debate is finally over

"Friday, 2nd February 2007 may go down in history as the day when the question mark was removed from the question of whether climate change has anything to do with human activities," said Achim Steiner, executive director of the United Nations Environment Program. It was on that day that the Intergovernmental Panel on Climate Change (IPCC) published a summary of its fourth report since being set up in 1988 by the World Meteorological Organization and the UN. The report stated, in no uncertain terms, that it is "very likely" global warming is being caused by humans.

Such strong words from this independent scientific body sent ripples throughout the international community, and earned the panel the 2007 Nobel Peace Prize. Since the late 1960s, an ever-increasing number of scientists and activists had argued that rising global temperatures were a direct and dangerous consequence of human's activities. But high-profile critics had kept the matter under discussion for forty years, giving politicians and decision-makers ample opportunity to drag their heels over taking any meaningful action. The IPCC's report, however, was a significant nail in the skeptics' coffin. Within four months, even U.S. president George W. Bush had been forced to back down from his position of denial.

Whether governments now find the courage to make real progress toward lessening the effects of a soaring human population on the Earth's atmosphere is not known, but the scientific consensus is now clear. If the politicians fail to act, the future will be bleak.

Mark Steer

Date 2007

Authors IPCC

Nationality International

Why It's Key All reasonable doubt is cast aside: humans are causing climate change.

Key Event **Live Earth**
Rock stars go green

On July 7, 2007, former U.S. Vice President Al Gore, the chair of the Alliance for Climate Protection, organized a monumental music event that brought together an estimated two billion people worldwide to combat the climate crisis.

Live Earth was a twenty-four-hour music event broadcast to all seven continents in a bid to encourage governments, corporations, and individuals to take action to reduce their "carbon footprints" and help prevent climate change. The concerts, which took place in major cities including London, New York, Sydney, Tokyo, Rio de Janeiro, and Johannesburg, were host to some of the biggest names in music. Acts such as Madonna, The Police, and the Red Hot Chili Peppers turned out to call for a mass global effort to reduce carbon emissions.

But the event caused controversy when the media called for an expert to calculate the carbon footprint of the event. Due mainly to the excessive emissions from private jets used to fly the artists to concerts, the footprint was calculated at a massive 37,500 tons. To put this into perspective, the average Briton has a footprint of 10 tons per year.

Gore hit back at complaints, ensuring the event was carbon neutral by spending approximately US$2 million in offsetting, and funding a reforestation and reagricultural project in Mozambique. The message of the event was clear, however – everyone can do their bit to reduce carbon emissions, even by a simple action such as turning off a light... or taking a train instead of a private jet.

Faith Smith

Date 2007

Country International

Why It's Key Brought the climate crisis to the attention of billions of people worldwide.

opposite Live Earth

Key Invention
X-48B blended wing aircraft gets off the ground

The drive toward more fuel-efficient aircraft has begun in earnest, with both economic and climatic motivations behind the research. In 2007, developments took a key turn as Boeing performed a successful test flight of their prototype blended wing aircraft, the X-48B.

Traditional aircraft have a standard "tube and wing" design, whereby lift is generated over the two wings. "Flying wing" designs consist of a single wing with no separate central body, unlike the tube of standard craft. The blended wing design uses features from both these by literally blending the wings with the main body of the aircraft, so that the whole of the craft contributes to lift and reduces drag. This means less fuel needs to be used to keep the craft in the air, increasing fuel efficiency. Ground level noise pollution caused by the aircraft is also reduced if the engines are placed on top of the wings, as sound waves are directed up rather than down.

Years of research by Boeing's Phantom Works research and development wing, cooperating with NASA and the U.S. Air Force Research Laboratory, culminated in the test flight in 2007. Following the successful wind-tunnel testing of the X48B model in 2006, the prototype craft flew for the first time, for thirty-one minutes, over California the following July. Fears over instability inherent to the blended wing design were allayed by the successful flight, which now promises much for the next generation of fuel-efficient, noiseless aircraft.

Steve Robinson

Date 2007

Scientists Boeing, NASA

Nationality American

Why It's Key The successful test flight proves the blended wing concept works, and the possibility of a fuel-efficient, quiet aircraft is now greater than ever.

opposite Sub-sized X-48B blended wing concept demonstrator in a wind tunnel at NASA's Langley Research Center, Virginia, USA

Key Invention
Human exoskeleton

Can you imagine what life would be like if your skeleton was outside of your body? While we use our skeletal frameworks as a sort of coat hanger for flesh, there are a plethora of creatures that support and, perhaps more importantly, protect the soft bits of their bodies with an "exoskeleton," for example crustaceans, spiders, and insects.

For decades, humans have been fascinated with the idea of creating their own exoskeletons, but not necessarily for providing shell-like protection – a powered exoskeleton would lend extra strength and mobility to its wearer. Japanese researchers at Matsushita Electric have recently designed and produced prototypes of a partial exoskeleton suit that could be worn by a human. While it is obvious that such an invention would afford huge benefits to paralyzed people and the infirm, it is exciting to realize the potential of such a suit elsewhere.

Nurses could make use of powered upper arms to lift patients in and out of hospital beds; soldiers could run into battle wielding huge weaponry; and movie enthusiasts could relive some of their favourite moments from films, maybe imitating the superhuman strength acquired by The Hulk. We'll have to wait some time, however, before a full-body suit is available, and affordable. At the time of writing, Matsushita Electric were selling their "Power Pedal" legs for US$167,000, so start saving!

Gavin Hammond

Date 2007

Scientists Matsushita Electric

Nationality Japanese

Why It's Key Extra strength and mobility can be readily applied to a number of practical applications, particularly in the medical and military professions.

X-48B
Boeing Phantom Works

Key Event The coldest swim in the world
Taking a dip at the North Pole

Next time the boiler breaks down and you have to take a cold shower, steel yourself and think of Lewis Gordon Pugh, who on July 15, 2007 swam a kilometer at the North Pole wearing only goggles and trunks.

It took Pugh 18 minutes and 50 seconds to cover the distance, in water of temperatures between minus 1.7 and zero degrees Celsius. These sub-freezing temperatures would kill most swimmers, but British-born Pugh – nicknamed "The Polar Bear" – has the remarkable ability of being able to raise his core body temperature, reportedly by up to 1.4 degrees Celsius, before he jumps into the water.

Pugh is the only man ever to have completed long-distance swims in every ocean on Earth, but he described the North Pole challenge – another world first – as his most challenging swim to date. "It was like jumping into a dark black hole – it was frightening. I was in excruciating pain from beginning to end," he said.

Not a bundle of fun then, but Pugh was swimming for a greater cause: to highlight the effects of global warming on polar ice caps. Only in recent years have cracks in the sea ice extended all the way to 90 degrees North in the Arctic Ocean. Pugh called the swim a triumph and a tragedy; a triumph that he could swim in such ferocious conditions, and a tragedy that it was possible to swim at the North Pole.

Richard Van Noorden

Date 2007

Nationality British

Why It's Key An endurance feat uniquely highlights the reality of melting polar ice caps.

opposite Lewis Gordon Pugh becomes the first man to swim in the waters of the North Pole

Key Discovery The darkest material ever produced
The future's black

In some respects, science is a lot like competitive sport. Teams of researchers are always striving to be first, or fastest, or to build the toughest materials and solve the trickiest problems. But while Formula One racing drivers pit their wits against each other to make it around a three-mile track in the shortest time, some scientists vie with each other to create the darkest substance known to man.

For most people, black is black is black. But not for researchers at the Rensselaer Polytechnic Institute in New York. Not content with ordinary, lighter, shades of black, in 2008, they decided to have a shot at breaking the world record for the darkest material by creating a blacker-than-black nanocarpet from tiny carbon nanotubes. The result was a triumph – their material was darker by four times than the next darkest.

But surely black is black is black? Not so. It all depends on the amount of light that a material is absorbing. The hollow tubes of the Rensselaer scientists' nanocarpet absorb more than 99.9 per cent of the light that hits them; your average black paint, by contrast, only manages to suck in around 95 per cent.

More importantly though, what's the use? Well, if the same principle works for other wavelengths, as well as visible light, this material might prove useful for creating more efficient solar panels; and in defense, for fashioning ultra-sneaky stealth coatings.

Hayley Birch

Date 2008

Scientist Pulickel Ajayan

Nationality Indian

Why It's Key Besides achieving a place in the record books, scientists may have created something that will help solve the energy crisis.

INDEX

Page numbers in *italics* refer to illustrations

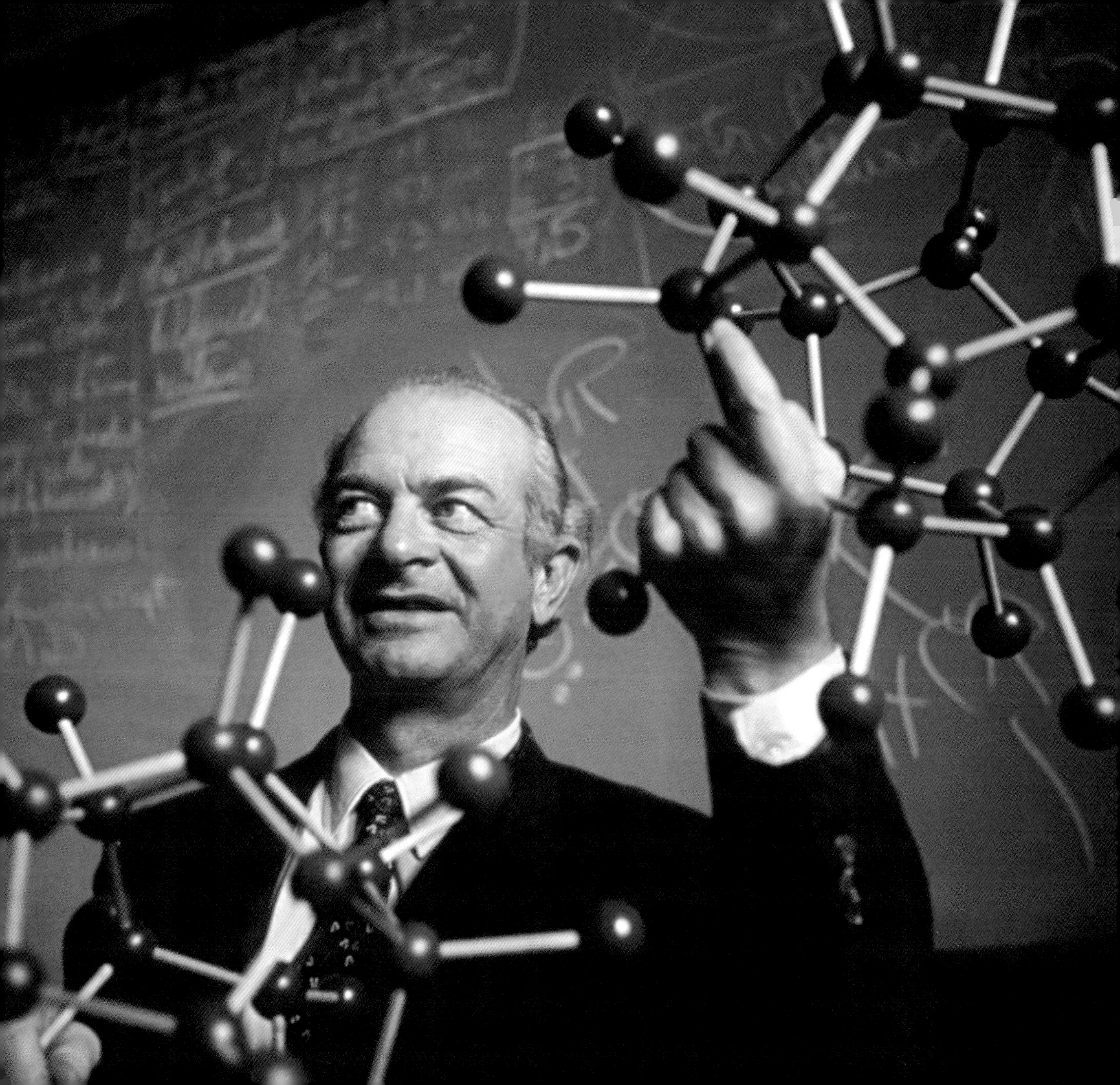

F

G

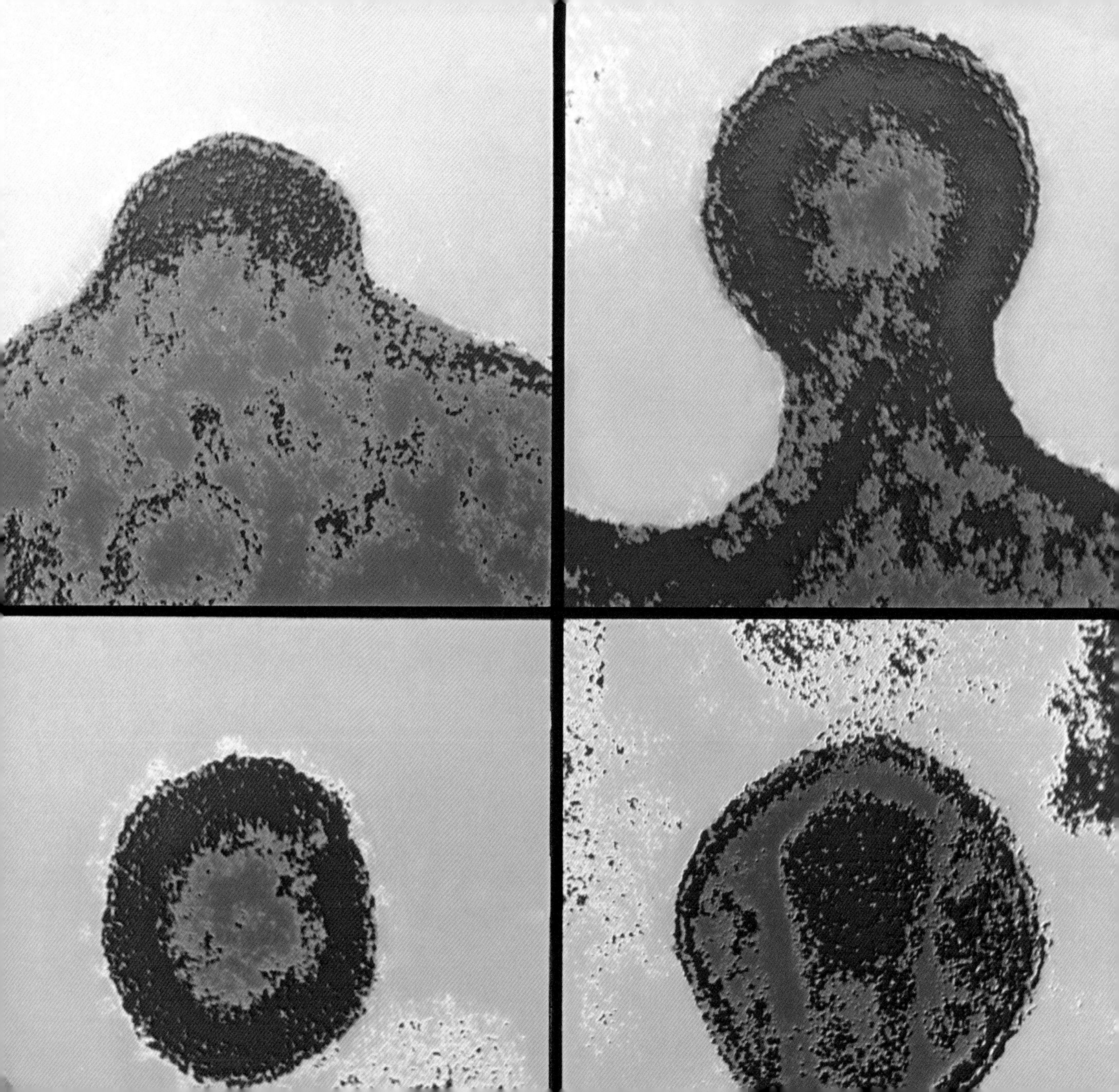

M

N

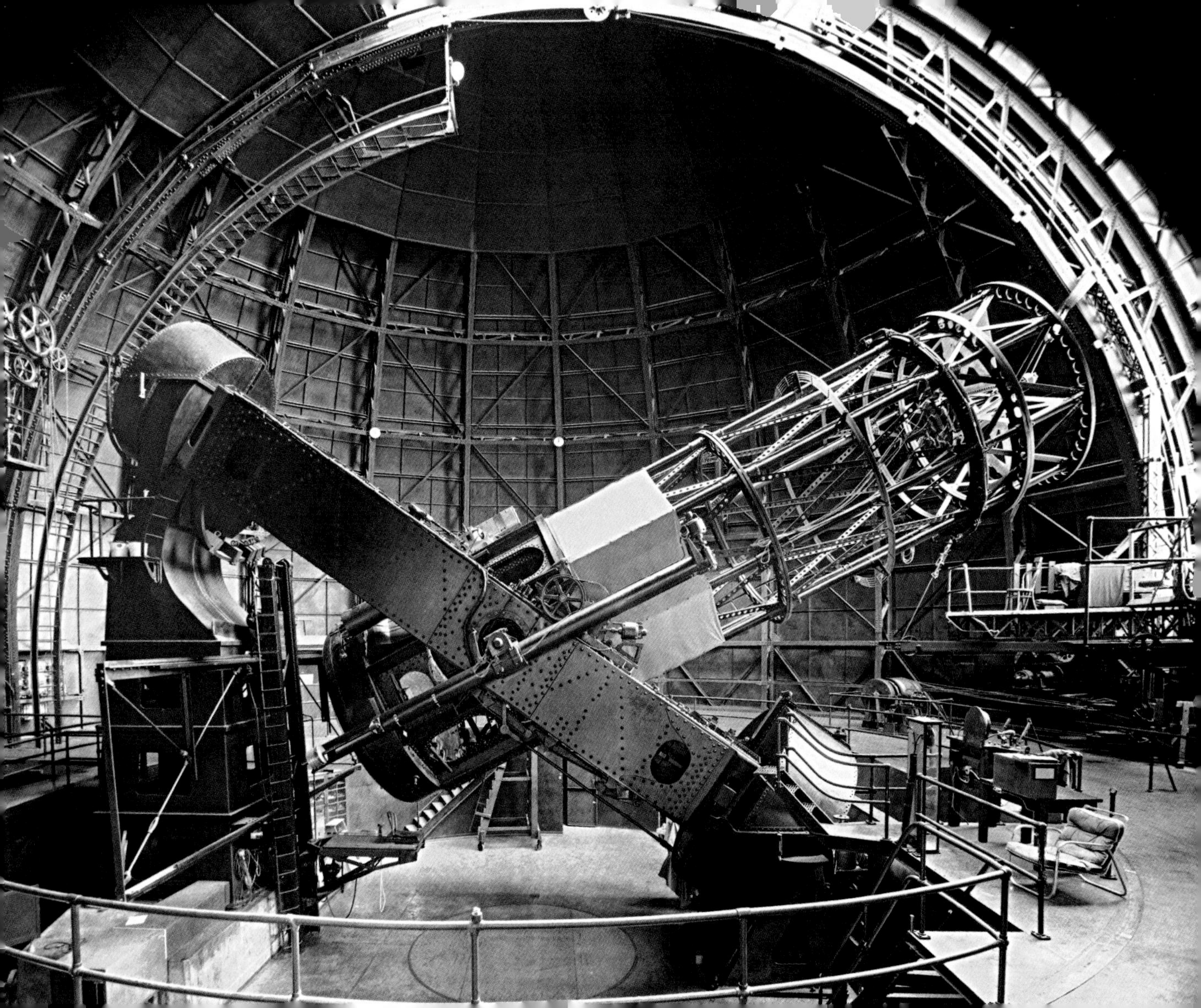

X

Y

X